Math Fin Law 4

Mathematical Financial Law

Public Listed Firm Rule No.12576-16333

by

Steve Asikin

Steve Asikin ISBN 14: 978-1514685136, ISBN 10: 1514685132

MATH FIN LAW 4

Mathematical Financial Law
Public Listed Firm Rule No. 12576-16333

By Steve Asikin

Creative Team:Steve Asikin

A Polimedia Paperback
First Published in Indonesia 2015
By Polimedia Publishing
This 1st edition published in 2015
Jalan Srengseng Sawah, Jagakarsa
Jakarta 12640, INDONESIA

ISBN-13: 978-1514685136
ISBN-10: 1514685132

PREFACE

This essays is about 111111 Most Useful and Common Rules in Mathematical Financial Laws. In investment decisions, synthesizing the financial measure requirements (quantitative construction), should be the most important element in the fundamental financial healthiness. The massive amount of numbers in a company's financial statements must not be bewildered and should never intimidating sincere investors. However, through financial ratio synthesis, we must be able to work with these numbers as a "healthy financial blue print".

Fundamental financial ratio analysis is fundamental (important) but just be the basement of the financial skyscraper. To build a strong financial structure, we need to do the financial synthesis to construct our far more advanced financial high structures.

The nicest thing about it, is that only requires simple Junior Highschool Algebra. We just need to be consistent with its less number treatment of constellations. Its saddest part is that many professors and doctors are no longer able to understand Junior Highschool Algebra, and hiding in Accounting jargons.

Please go back to its Introduction for glossary, at anywhere inside the book, so then we can still understand what are those mean with simple Algebraic Rule there. We try to avoid some sentences multiplied, added, eliminated divided, rooted or powered by another sentences, that make many traditional finance book successful (delivering few simple things, seen as difficult and scary), but less useful and seldom be accurate.

Steve Asikin ISBN 14: 978-1514685136, ISBN 10: 1514685132

TABLE OF CONTENTS

Steve Asikin ISBN 14: 978-1514685136, ISBN 10: 1514685132

INTRODUCTION:

The Math Fin Law Terminologies

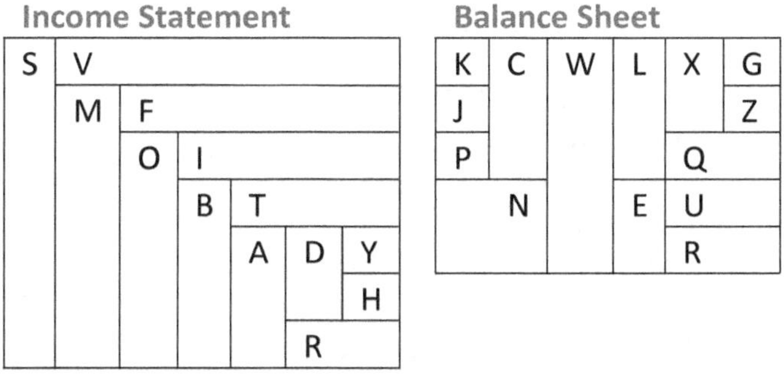

Please carefully look that all of 26 characters (**A** to **Z**) used once, except the **R**, that in Accountancy must be appeared in both diagrams at same identical value, with glossary (of well known abbreviations) in the next page. Only 2 (two) of them are historical as **U**=**E'** and **S'** means carried forward legal numbers from previous year.

The 13 (thirteen) common measures (**c**, **d**, **f**, **g**, **i**, **j**, **l**, **p**, **q**, **s**, **t**, **v**, and **y**), are put up front as desired criteria along the way, so then later analyst could be surprised.

In Balance Sheet, we found that **W**= **C**+**N**= **K**+**J**+**P**+**N**= **L**+**E**= **X**+**Q**+**U**+**R**= **G**+**Z**+**Q**+**U**+**R** as well, so then we put **W** in the middle of its T-Account balance.

In Income Statement, same applied as **S**= **V**+**F**+**I** +**T**+**Y**+**H**+**R**= **V**+**M**= **V**+**F**+ **O**= **V**+**F**+**I**+**B**= **V**+**F**+**I**+**T**+**A**= **V**+**F**+**I**+**B**= **V**+**F**+**I**+**T**+**D**+**R**.

Steve Asikin ISBN 14: 978-1514685136, ISBN 10: 1514685132

A= After Tax Income;
B= Before Tax Income,
C= Current Assets, **c**= C/X= Current Ratio,
D= Dividend Paid, **d**= D/A=Dividend Portion or Payout,
E= Equity or Capital, **E'**= Equity of Past Year,
F= Fixed Cost, **f**= F/S=Fixed Portion,
G= Goods or Trade Payables,
    g= 360G/V= Goods Payable Days,
H= Home Taken Dividend,
I= Interest Expense, **i**= I/S= Interest Portion,
J= Jobs or Account Receivable,
    j= 360J/S= Jobs or Trade Account ReceivableDays,
K= Kind of Cash,
L= Liabilities or Debt, **l**= L/E= Leverage or Gearing Ratio,
M= Margin of Contribution,
N= Non Current Assets,
O= Operational Surplus,
P= Procured Inventories,
    p= 360P/V= Procured Inventory Days,
Q= Quoted Longterm Liabilities,
    q= [C-P]/X= Quick or Acid Test Ratio,
R= Retained Earnings,
S= Sales or Revenues, **S'**= Sales of Past Year,
    s= [S/S']-1= Sales Growth,
T= Tax Paid, **t**= T/B= Tax Rate,
U= Utilized or Starting Capital,
V= Variable Cost, **v**= V/S= Variable Portion,
W= Wealth or Total Assets,
X= Xpress or Current Debt,
Y= Yielding Tax of Dividend,
    y= Y/D= Yielding Tax Rate of Dividend,
Z= Zero Trade Current Debt.

Steve Asikin ISBN 14: 978-1514685136, ISBN 10: 1514685132

CHAPTER-19:

Q= Quoted Longterm Debt Optimization

Continued from Book 1,2 and 3:

Rule-12576:

If both (**/**), (**U**), (**S**), (**f**), (**V**), (**I**), (**T**), (**A**), (**d**) and (**X**) are known, then its Quoted Longterm Debt Planned is:

Q= **/**{**U**+**S**[1-**f**]-**V**-**I**-**T**-**Ad**}-**X**

Rule-12577:

If both (**Q**), (**U**), (**S**), (**f**), (**V**), (**I**), (**T**), (**A**), (**d**) and (**X**) are known, then its Leverage or Gearing Ratio Planned is:

/= [**Q**+**X**]/{**U**+**S**[1-**f**]-**V**-**I**-**T**-**Ad**}

Rule-12578:

If both (**/**), (**Q**), (**S**), (**f**), (**V**), (**I**), (**T**), (**A**), (**d**) and (**X**) are known, then its Utilized or Starting Capital must be:

U= [**Q**+**X**]/**/**-{**S**[1-**f**]-**V**-**I**-**T**-**Ad**}

Rule-12579:

If both (**/**), (**U**), (**Q**), (**f**), (**V**), (**I**), (**T**), (**A**), (**d**) and (**X**) are known, then its Sales or Revenue Planned is:

S= {**V**+**I**+**T**+**Ad**+[**Q**+**X**]/**/**-**U**}/[1-**f**]

Rule-12580:

If both (**/**), (**U**), (**S**), (**Q**), (**V**), (**I**), (**T**), (**A**), (**d**) and (**X**) are known, then its Fixed Portion Planned is:

f= 1-{**V**+**I**+**T**+**Ad**+[**Q**+**X**]/**/**-**U**}/**S**

Steve Asikin ISBN 14: 978-1514685136, ISBN 10: 1514685132

Rule-12581:

If both (**/**), (**U**), (**$**), (**Q**), (**Q**), (**I**), (**T**), (**A**), (**d**) and (**X**) are known, then its Variable Portion Planned is:

V= {**$**[1-**f**]-**I**-**T**-**Ad**}-{[**Q**+**X**]/**/U**}

Rule-12582:

If both (**/**), (**U**), (**$**), (**Q**), (**V**), (**Q**), (**T**), (**A**), (**d**) and (**X**) are known, then its Interest Expense Planned is:

I= {**$**[1-**f**]-**V**-**T**-**Ad**}-{[**Q**+**X**]/**/U**}

Rule-12583:

If both (**/**), (**U**), (**$**), (**Q**), (**V**), (**I**), (**Q**), (**A**), (**d**) and (**X**) are known, then its Tax Planned is:

T= {**$**[1-**f**]-**V**-**I**-**Ad**}-{[**Q**+**X**]/**/U**}

Rule-12584:

If both (**/**), (**U**), (**$**), (**Q**), (**V**), (**I**), (**T**), (**Q**), (**d**) and (**X**) are known, then its After Tax Income Planned is:

A= {**$**[1-**f**]-**V**-**I**-**T**-[**Q**+**X**]/**/U**}/**d**

Rule-12585:

If both (**/**), (**U**), (**$**), (**Q**), (**V**), (**I**), (**T**), (**A**), (**Q**) and (**X**) are known, then its Dividend Payout Planned is:

d= {**$**[1-**f**]-**V**-**I**-**T**-[**Q**+**X**]/**/U**}/**A**

Rule-12586:

If both (**/**), (**U**), (**$**), (**Q**), (**V**), (**I**), (**T**), (**A**), (**d**) and (**Q**) are known, then its Xpress or Current Debt Planned is:

X= **/**{**U**+**$**[1-**f**]-**V**-**I**-**T**-**Ad**}-**Q**

Steve Asikin ISBN 14: 978-1514685136, ISBN 10: 1514685132

Rule-12587:

If both (**/**), (**U**), (**$**), (**f**), (**V**), (**I**), (**T**), (**A**), (**d**), (**P**), (**c**) and (**q**) are known, then its Quoted Longterm Debt Planned is:

Q= **/**{**U**+**$**[1-**f**]-**V**-**I**-**T**-**Ad**}-**P**/[**c**-**q**]

Rule-12588:

If both (**Q**), (**U**), (**$**), (**f**), (**V**), (**I**), (**T**), (**A**), (**d**), (**P**), (**c**) and (**q**) are known, then its Leverage or Gearing Ratio Planned is:

/= {**Q**+**P**/[**c**-**q**]}/{**U**+**$**[1-**f**]-**V**-**I**-**T**-**Ad**}

Rule-12589:

If both (**/**), (**Q**), (**$**), (**f**), (**V**), (**I**), (**T**), (**A**), (**d**), (**P**), (**c**) and (**q**) are known, then its Utilized or Starting Capital must be:

U= {**Q**+**P**/[**c**-**q**]}/**/**-{**$**[1-**f**]-**V**-**I**-**T**-**Ad**}

Rule-12590:

If both (**/**), (**U**), (**Q**), (**f**), (**V**), (**I**), (**T**), (**A**), (**d**), (**P**), (**c**) and (**q**) are known, then its Sales or Revenue Planned is:

$= (**V**+**I**+**T**+**Ad**+{**Q**+**P**/[**c**-**q**]}/**/**-**U**)/[1-**f**]

Rule-12591:

If both (**/**), (**U**), (**$**), (**Q**), (**V**), (**I**), (**T**), (**A**), (**d**), (**P**), (**c**) and (**q**) are known, then its Fixed Portion Planned is:

f= 1-(**V**+**I**+**T**+**Ad**+({**Q**+**P**/[**c**-**q**]}/**/**-**U**)/**$**

Steve Asikin ISBN 14: 978-1514685136, ISBN 10: 1514685132

Rule-12592:

If both (**/**), (**U**), (**$**), (**f**), (**Q**), (**I**), (**T**), (**A**), (**d**), (**P**), (**c**) and (**q**) are known, then its Variable Cost Planned is:

V= {**$**[1-**f**]-**I**-**T**-**Ad**}-({**Q**+**P**/[**c**-**q**]}/**/**-**U**)

Rule-12593:

If both (**/**), (**U**), (**$**), (**f**), (**V**), (**Q**), (**T**), (**A**), (**d**), (**P**), (**c**) and (**q**) are known, then its Interest Expense Planned is:

I= {**$**[1-**f**]-**V**-**T**-**Ad**}-({**Q**+**P**/[**c**-**q**]}/**/**-**U**)

Rule-12594:

If both (**/**), (**U**), (**$**), (**f**), (**V**), (**I**), (**Q**), (**A**), (**d**), (**P**), (**c**) and (**q**) are known, then its Tax Planned is:

T= {**$**[1-**f**]-**V**-**I**-**Ad**}-({**Q**+**P**/[**c**-**q**]}/**/**-**U**)

Rule-12595:

If both (**/**), (**U**), (**$**), (**f**), (**V**), (**I**), (**T**), (**Q**), (**d**), (**P**), (**c**) and (**q**) are known, then its After Tax Income Planned is:

A= (**$**[1-**f**]-**V**-**I**-**T**-{**Q**+**P**/[**c**-**q**]}/**/**-**U**)/**d**

Rule-12596:

If both (**/**), (**U**), (**$**), (**f**), (**V**), (**I**), (**T**), (**A**), (**Q**), (**P**), (**c**) and (**q**) are known, then its Dividend Payout Planned is:

d= (**$**[1-**f**]-**V**-**I**-**T**-{**Q**+**P**/[**c**-**q**]}/**/**-**U**)/**A**

Steve Asikin ISBN 14: 978-1514685136, ISBN 10: 1514685132

Rule-12597:

If both (**/**), (**U**), (**$**), (**f**), (**V**), (**I**), (**T**), (**A**), (**d**), (**Q**), (**c**) and (**q**) are known, then its Procured Inventory Planned is:

P= [**c**-**q**](**/**{**U**+**$**[1-**f**]-**V**-**I**-**T**-**Ad**}-**Q**)

Rule-12598:

If both (**/**), (**U**), (**$**), (**f**), (**V**), (**I**), (**T**), (**A**), (**d**), (**P**), (**Q**) and (**q**) are known, then its Current Ratio Planned is:

c= **q**+**P**/(**/**{**U**+**$**[1-**f**]-**V**-**I**-**T**-**Ad**}-**Q**)

Rule-12599:

If both (**/**), (**U**), (**$**), (**f**), (**V**), (**I**), (**T**), (**A**), (**d**), (**P**), (**c**) and (**Q**) are known, then its Quick or Acid Test Ratio Planned is:

q= **c**-**P**/(**/**{**U**+**$**[1-**f**]-**V**-**I**-**T**-**Ad**}-**Q**)

Rule-12587:

If both (**/**), (**U**), (**$**), (**f**), (**V**), (**I**), (**T**), (**A**), (**d**), (**p**), (**c**) and (**q**) are known, then its Quoted Longterm Debt Planned is:

Q= **/**{**U**+**$**[1-**f**]-**V**-**I**-**T**-**Ad**}-**Vp**/{360[**c**-**q**]}

Rule-12588:

If both (**Q**), (**U**), (**$**), (**f**), (**V**), (**I**), (**T**), (**A**), (**d**), (**p**), (**c**) and (**q**) are known, then its Leverage or Gearing Ratio Planned is:

/= (**Q**+**Vp**/{360[**c**-**q**]})/{**U**+**$**[1-**f**]-**V**-**I**-**T**-**Ad**}

Steve Asikin ISBN 14: 978-1514685136, ISBN 10: 1514685132

Rule-12589:

If both (**/**), (**Q**), (**$**), (**f**), (**V**), (**I**), (**T**), (**A**), (**d**), (**p**), (**c**) and (**q**) are known, then its Utilized or Starting Capital must be:

U= (**Q**+**Vp**/{360[**c**-**q**]})/**/**-{**$**[1-**f**]-**V**-**I**-**T**-**Ad**}

Rule-12590:

If both (**/**), (**U**), (**Q**), (**f**), (**V**), (**I**), (**T**), (**A**), (**d**), (**p**), (**c**) and (**q**) are known, then its Sales or Revenue Planned is:

$= [**V**+**I**+**T**+**Ad**+(**Q**+**Vp**/{360[**c**-**q**]})/**/**-**U**]/[1-**f**]

Rule-12591:

If both (**/**), (**U**), (**$**), (**Q**), (**V**), (**I**), (**T**), (**A**), (**d**), (**p**), (**c**) and (**q**) are known, then its Fixed Portion Planned is:

f= 1-[**V**+**I**+**T**+**Ad**+(**Q**+**Vp**/{360[**c**-**q**]})/**/**-**U**]/**$**

Rule-12592:

If both (**/**), (**U**), (**$**), (**f**), (**Q**), (**I**), (**T**), (**A**), (**d**), (**p**), (**c**) and (**q**) are known, then its Variable Cost Planned is:

V= (**/**{**U**+**$**[1-**f**]-**V**-**I**-**T**-**Ad**}-**Q**)/(**p**/{360[**c**-**q**]}+**/**)

Rule-12593:

If both (**/**), (**U**), (**$**), (**f**), (**V**), (**Q**), (**T**), (**A**), (**d**), (**p**), (**c**) and (**q**) are known, then its Interest Expense Planned is:

I= {**$**[1-**f**]-**V**-**T**-**Ad**}-[(**Q**+**Vp**/{360[**c**-**q**]})/**/**-**U**]

Steve Asikin ISBN 14: 978-1514685136, ISBN 10: 1514685132

Rule-12594:

If both (**/**), (**U**), (**$**), (**f**), (**V**), (**I**), (**Q**), (**A**), (**d**), (**p**), (**c**) and (**q**) are known, then its Tax Planned is:

T= {**$**[1-**f**]-**V**-**I**-**Ad**}-[(**Q**+**Vp**/{360[**c**-**q**]})/**/**-**U**]

Rule-12595:

If both (**/**), (**U**), (**$**), (**f**), (**V**), (**I**), (**T**), (**Q**), (**d**), (**p**), (**c**) and (**q**) are known, then its After Tax Income Planned is:

A= [**$**[1-**f**]-**V**-**I**-**T**-(**Q**+**Vp**/{360[**c**-**q**]})/**/**-**U**]/**d**

Rule-12596:

If both (**/**), (**U**), (**$**), (**f**), (**V**), (**I**), (**T**), (**A**), (**Q**), (**p**), (**c**) and (**q**) are known, then its Dividend Payout Planned is:

d= [**$**[1-**f**]-**V**-**I**-**T**-(**Q**+**Vp**/{360[**c**-**q**]})/**/**-**U**]/**A**

Rule-12597:

If both (**/**), (**U**), (**$**), (**f**), (**V**), (**I**), (**T**), (**A**), (**d**), (**Q**), (**c**) and (**q**) are known, then its Procured Inventory Days Planned:

p= 360[**c**-**q**](**/**{**U**+**$**[1-**f**]-**V**-**I**-**T**-**Ad**}-**Q**)/**V**

Rule-12598:

If both (**/**), (**U**), (**$**), (**f**), (**V**), (**I**), (**T**), (**A**), (**d**), (**p**), (**Q**) and (**q**) are known, then its Current Ratio Planned is:

c= **q**+**Vp**/[360(**/**{**U**+**$**[1-**f**]-**V**-**I**-**T**-**Ad**}-**Q**)]

Steve Asikin ISBN 14: 978-1514685136, ISBN 10: 1514685132

Rule-12599:

If both (**/**), (**U**), (**$**), (**f**), (**V**), (**I**), (**T**), (**A**), (**d**), (**p**), (**c**) and (**Q**) are known, then its Quick or Acid Test Ratio Planned is:

q= **c-Vp/[**360(**/**{**U**+**$**[1-**f**]-**V-I-T-Ad**}-**Q**)**]**

Rule-12600:

If both (**/**), (**U**), (**$**), (**f**), (**V**), (**I**), (**T**), (**A**), (**d**), (**v**), (**p**), (**c**) and (**q**) are known, then its Quoted Longterm Debt Planned is:

Q= **/**{**U**+**$**[1-**f**]-**V-I-T-Ad**}-**$vp**/{360[**c-q**]}

Rule-12601:

If both (**Q**), (**U**), (**$**), (**f**), (**V**), (**I**), (**T**), (**A**), (**d**), (**v**), (**p**), (**c**) and (**q**) are known, then its Leverage or Gearing Ratio Planned is:

/= (**Q**+**$vp**/{360[**c-q**]})/{**U**+**$**[1-**f**]-**V-I-T-Ad**}

Rule-12602:

If both (**/**), (**Q**), (**$**), (**f**), (**V**), (**I**), (**T**), (**A**), (**d**), (**v**), (**p**), (**c**) and (**q**) are known, then its Utilized or Starting Capital must be:

U= (**Q**+**$vp**/{360[**c-q**]})/**/**-{**$**[1-**f**]-**V-I-T-Ad**}

Rule-12603:

If both (**/**), (**U**), (**Q**), (**f**), (**V**), (**I**), (**T**), (**A**), (**d**), (**v**), (**p**), (**c**) and (**q**) are known, then its Sales or Revenue Planned is:

$= (**/**{**U-V-I-T-Ad**}-**Q**)/(**vp**/{360[**c-q**]}-**/**[1-**f**])

Steve Asikin ISBN 14: 978-1514685136, ISBN 10: 1514685132

Rule-12604:

If both (**/**), (**U**), (**$**), (**Q**), (**V**), (**I**), (**T**), (**A**), (**d**), (**v**), (**p**), (**c**) and (**q**) are known, then its Fixed Portion Planned is:

f= 1-[V+I+T+Ad+(Q+$vp/{360[c-q]})/⁄U]/$

Rule-12605:

If both (**/**), (**U**), (**$**), (**f**), (**Q**), (**I**), (**T**), (**A**), (**d**), (**v**), (**p**), (**c**) and (**q**) are known, then its Variable Portion Planned is:

V= {$[1-f]-I-T-Ad}-[(Q+$vp/{360[c-q]})/⁄U]

Rule-12606:

If both (**/**), (**U**), (**$**), (**f**), (**V**), (**Q**), (**T**), (**A**), (**d**), (**v**), (**p**), (**c**) and (**q**) are known, then its Interest Expense Planned is:

I= {$[1-f]-V-T-Ad}-[(Q+$vp/{360[c-q]})/⁄U]

Rule-12607:

If both (**/**), (**U**), (**$**), (**f**), (**V**), (**I**), (**Q**), (**A**), (**d**), (**v**), (**p**), (**c**) and (**q**) are known, then its Tax Planned is:

T= {$[1-f]-V-I-Ad}-[(Q+$vp/{360[c-q]})/⁄U]

Rule-12608:

If both (**/**), (**U**), (**$**), (**f**), (**V**), (**I**), (**T**), (**Q**), (**d**), (**v**), (**p**), (**c**) and (**q**) are known, then its After Tax Income Planned is:

A= [$[1-f]-V-I-T-(Q+$vp/{360[c-q]})/⁄U]/d

Steve Asikin ISBN 14: 978-1514685136, ISBN 10: 1514685132

Rule-12609:

If both (**/**), (**U**), (**$**), (**f**), (**V**), (**I**), (**T**), (**A**), (**Q**), (**v**), (**p**), (**c**) and (**q**) are known, then its Dividend Payout Planned is:

d= [**$**[1-**f**]-**V**-**I**-**T**-(**Q**+**$vp**/{360[**c**-**q**]})/**/**-**U**]/**A**

Rule-12610:

If both (**/**), (**U**), (**$**), (**f**), (**V**), (**I**), (**T**), (**A**), (**d**), (**Q**), (**p**), (**c**) and (**q**) are known, then its Variable Portion Planned is:

v= **V**/**$**

or it can also be found as

v= 360[**c**-**q**](**/**{**U**+**$**[1-**f**]-**V**-**I**-**T**-**Ad**}-**Q**)/[**$p**]

Rule-12611:

If both (**/**), (**U**), (**$**), (**f**), (**V**), (**I**), (**T**), (**A**), (**d**), (**v**), (**Q**), (**c**) and (**q**) are known, then its Procured Inventory Days Planned:

p= 360[**c**-**q**](**/**{**U**+**$**[1-**f**]-**V**-**I**-**T**-**Ad**}-**Q**)/[**$v**]

Rule-12612:

If both (**/**), (**U**), (**$**), (**f**), (**V**), (**I**), (**T**), (**A**), (**d**), (**v**), (**p**), (**Q**) and (**q**) are known, then its Current Ratio Planned is:

c= **q**+**$vp**/[360(**/**{**U**+**$**[1-**f**]-**V**-**I**-**T**-**Ad**}-**Q**)]

Rule-12613:

If both (**/**), (**U**), (**$**), (**f**), (**V**), (**I**), (**T**), (**A**), (**d**), (**v**), (**p**), (**c**) and (**Q**) are known, then its Quick or Acid Test Ratio Planned is:

q= **c**-**$vp**/[360(**/**{**U**+**$**[1-**f**]-**V**-**I**-**T**-**Ad**}-**Q**)]

Steve Asikin ISBN 14: 978-1514685136, ISBN 10: 1514685132

Rule-12614:

If both (**/**), (**U**), (**$**), (**f**), (**V**), (**I**), (**T**), (**A**), (**d**), (**$'**), (**v**), (**p**), (**s**), (**c**) and (**q**) are known, then its Quoted Longterm Debt Planned is:

Q= **/**{**U**+**$**[1-**f**]-**V**-**I**-**T**-**Ad**}-**$'vp**[1+**s**]/{360[**c**-**q**]}

Rule-12615:

If both (**Q**), (**U**), (**$**), (**f**), (**V**), (**I**), (**T**), (**A**), (**d**), (**$'**), (**v**), (**p**), (**s**), (**c**) and (**q**) are known, then its Leverage or Gearing Ratio Planned is:

/= (**Q**+**$'vp**[1+**s**]/{360[**c**-**q**]})/{**U**+**$**[1-**f**]-**V**-**I**-**T**-**Ad**}

Rule-12616:

If both (**/**), (**Q**), (**$**), (**f**), (**V**), (**I**), (**T**), (**A**), (**d**), (**$'**), (**v**), (**p**), (**s**), (**c**) and (**q**) are known, then its Utilized or Starting capital must be:

U= (**Q**+**$'vp**[1+**s**]/{360[**c**-**q**]})/**/**-{**$**[1-**f**]-**V**-**I**-**T**-**Ad**}

Rule-12617:

If both (**/**), (**U**), (**Q**), (**f**), (**V**), (**I**), (**T**), (**A**), (**d**), (**$'**), (**v**), (**p**), (**s**), (**c**) and (**q**) are known, then its Sales or Revenue Planned is:

$**= (**Q**-**/**{**U**-**V**-**I**-**T**-**Ad**}+**$'vp[1+**s**]/{360[**c**-**q**]})
/{**/**[1-**f**]}

Rule-12618:

If both (**/**), (**U**), (**$**), (**Q**), (**V**), (**I**), (**T**), (**A**), (**d**), (**$'**), (**v**), (**p**), (**s**), (**c**) and (**q**) are known, then its Fixed Cost Planned is:

f= 1-[**V**+**I**+**T**+**Ad**
+(**Q**+**$'vp**[1+**s**]/{360[**c**-**q**]})/**/**-**U**]/**$**

Steve Asikin ISBN 14: 978-1514685136, ISBN 10: 1514685132

Rule-12619:

If both (**/**), (**U**), (**$**), (**f**), (**Q**), (**I**), (**T**), (**A**), (**d**), (**$'**), (**v**), (**p**), (**s**), (**c**) and (**q**) are known, then its Variable Cost Planned is:

V= {**$**[1-**f**]-**I**-**T**-**Ad**}
-[(**Q**+**$'vp**[1+**s**]/{360[**c**-**q**]})/**/U**]

Rule-12620:

If both (**/**), (**U**), (**$**), (**f**), (**V**), (**Q**), (**T**), (**A**), (**d**), (**$'**), (**v**), (**p**), (**s**), (**c**) and (**q**) are known, then its Interest Expense Planned is:

I= {**$**[1-**f**]-**V**-**T**-**Ad**}
-[(**Q**+**$'vp**[1+**s**]/{360[**c**-**q**]})/**/U**]

Rule-12621:

If both (**/**), (**U**), (**$**), (**f**), (**V**), (**I**), (**Q**), (**A**), (**d**), (**$'**), (**v**), (**p**), (**s**), (**c**) and (**q**) are known, then its Tax Planned is:

T= {**$**[1-**f**]-**V**-**I**-**Ad**}
-[(**Q**+**$'vp**[1+**s**]/{360[**c**-**q**]})/**/U**]

Rule-12622:

If both (**/**), (**U**), (**$**), (**f**), (**V**), (**I**), (**T**), (**Q**), (**d**), (**$'**), (**v**), (**p**), (**s**), (**c**) and (**q**) are known, then its After Tax Income Planned is:

A= [**$**[1-**f**]-**V**-**I**-**T**-(**Q**+**$'vp**[1+**s**]/{360[**c**-**q**]})/**/U**]/**d**

Rule-12623:

If both (**/**), (**U**), (**$**), (**f**), (**V**), (**I**), (**T**), (**A**), (**Q**), (**$'**), (**v**), (**p**), (**s**), (**c**) and (**q**) are known, then its Dividend Payout Planned is:

d= [**$**[1-**f**]-**V**-**I**-**T**-(**Q**+**$'vp**[1+**s**]/{360[**c**-**q**]})/**/U**]/**A**

Steve Asikin ISBN 14: 978-1514685136, ISBN 10: 1514685132

Rule-12624:

If both (**/**), (**U**), (**$**), (**f**), (**V**), (**I**), (**T**), (**A**), (**d**), (**Q**), (**v**), (**p**), (**s**), (**c**) and (**q**) are known, then its Sales Past must be:

$'= 360[c-q](/{U+$[1-f]-V-I-T-Ad}-Q)/{vp[1+s]}

Rule-12625:

If both (**/**), (**U**), (**$**), (**f**), (**V**), (**I**), (**T**), (**A**), (**d**), (**$'**), (**Q**), (**p**), (**s**), (**c**) and (**q**) are known, then its Variable Portion Planned is:

v= V/$

or it can also be found as

v= 360[c-q](/{U+$[1-f]-V-I-T-Ad}-Q)/{$'p[1+s]}

Rule-12626:

If both (**/**), (**U**), (**$**), (**f**), (**V**), (**I**), (**T**), (**A**), (**d**), (**$'**), (**v**), (**Q**), (**s**), (**c**) and (**q**) are known, then its Procured Inventory Days Planned:

p= 360[c-q](/{U+$[1-f]-V-I-T-Ad}-Q)/{$'v[1+s]}

Rule-12627:

If both (**/**), (**U**), (**$**), (**f**), (**V**), (**I**), (**T**), (**A**), (**d**), (**$'**), (**v**), (**Q**), (**s**), (**c**) and (**q**) are known, then its Sales Growth Planned is:

s= $/$'-1

on it can also be found as

s= 360[c-q](/{U+$[1-f]-V-I-T-Ad}-Q)/[$'vp]-1

Steve Asikin ISBN 14: 978-1514685136, ISBN 10: 1514685132

Rule-12628:

If both (**/**), (**U**), (**$**), (**f**), (**V**), (**I**), (**T**), (**A**), (**d**), (**$'**), (**v**), (**p**), (**s**), (**Q**) and (**q**) are known, then its Current Ratio Planned is:

c= **q**+**$'vp**[1+**s**]/[360(**/**{**U**+**$**[1-**f**]-**V**-**I**-**T**-**Ad**}-**Q**)]

Rule-12629:

If both (**/**), (**U**), (**$**), (**f**), (**V**), (**I**), (**T**), (**A**), (**d**), (**$'**), (**v**), (**p**), (**s**), (**Q**) and (**q**) are known, then its Current Ratio Planned is:

q= **c**-**$'vp**[1+**s**]/[360(**/**{**U**+**$**[1-**f**]-**V**-**I**-**T**-**Ad**}-**Q**)]

Rule-12630:

If both (**/**), (**U**), (**$**), (**f**), (**V**), (**I**), (**t**), (**D**) and (**X**) are known, then its Quoted Longterm Debt Planned is:

Q= **/**(**U**+{**$**[1-**f**]-**V**-**I**}[1-**t**]-**D**)-**X**

Rule-12631:

If both (**Q**), (**U**), (**$**), (**f**), (**V**), (**I**), (**t**), (**D**) and (**X**) are known, then its Leverage or Gearing Ratio Planned is:

/= [**Q**+**X**]/(**U**+{**$**[1-**f**]-**V**-**I**}[1-**t**]-**D**)

Rule-12632:

If both (**/**), (**Q**), (**$**), (**f**), (**V**), (**I**), (**t**), (**D**) and (**X**) are known, then its Utilized or Starting Capital must be:

U= [**Q**+**X**]/**/**-({**$**[1-**f**]-**V**-**I**}[1-**t**]-**D**)

Rule-12633:

If both (**/**), (**U**), (**Q**), (**f**), (**V**), (**I**), (**t**), (**D**) and (**X**) are known, then its Sales or Revenue Planned is:

$= (**V**+**I**+{**D**+[**Q**+**X**]/**/**-**U**}/[1-**t**])/[1-**f**]

Steve Asikin ISBN 14: 978-1514685136, ISBN 10: 1514685132

Rule-12634:

If both (**/**), (**U**), (**S**), (**Q**), (**V**), (**I**), (**t**), (**D**) and (**X**) are known, then its Fixed Portion Planned is:

f= 1-(**V**+**I**+{**D**+[**Q**+**X**]/**/U**}/[1-**t**])/**S**

Rule-12635:

If both (**/**), (**U**), (**S**), (**f**), (**Q**), (**I**), (**t**), (**D**) and (**X**) are known, then its Variable Cost Planned is:

V= {**S**[1-**f**]-**I**}-{**D**+[**Q**+**X**]/**/U**}/[1-**t**]

Rule-12636:

If both (**/**), (**U**), (**S**), (**f**), (**V**), (**Q**), (**t**), (**D**) and (**X**) are known, then its Interest Expense Planned is:

I= {**S**[1-**f**]-**V**}-{**D**+[**Q**+**X**]/**/U**}/[1-**t**]

Rule-12637:

If both (**/**), (**U**), (**S**), (**f**), (**V**), (**I**), (**Q**), (**D**) and (**X**) are known, then its Tax Rate Planned is:

t= 1-{**D**+[**Q**+**X**]/**/U**}/{**S**[1-**f**]-**V**-**I**}

Rule-12638:

If both (**/**), (**U**), (**S**), (**f**), (**V**), (**I**), (**t**), (**Q**) and (**X**) are known, then its Dividend Payout Planned is:

D= **U**+{**S**[1-**f**]-**V**-**I**}[1-**t**]-[**Q**+**X**]/**/**

Rule-12639:

If both (**/**), (**U**), (**S**), (**f**), (**V**), (**I**), (**t**), (**D**) and (**Q**) are known, then its Xpress or Gearing Ratio Planned is:

X= **/**(**U**+{**S**[1-**f**]-**V**-**I**}[1-**t**]-**D**)-**Q**

Steve Asikin ISBN 14: 978-1514685136, ISBN 10: 1514685132

Rule-12640:

If both (**/**), (**U**), (**$**), (**f**), (**V**), (**I**), (**t**), (**D**), (**P**), (**c**) and (**q**) are known, then its Quoted Longterm Debt Planned is:

Q= **/**(**U**+{**$**[1-**f**]-**V**-**I**}[1-**t**]-**D**)-**P**/[**c**-**q**]

Rule-12641:

If both (**Q**), (**U**), (**$**), (**f**), (**V**), (**I**), (**t**), (**D**), (**P**), (**c**) and (**q**) are known, then its Leverage or Gearing Ratio Planned is:

/= {**Q**+**P**/[**c**-**q**]}/(**U**+{**$**[1-**f**]-**V**-**I**}[1-**t**]-**D**)

Rule-12642:

If both (**/**), (**Q**), (**$**), (**f**), (**V**), (**I**), (**t**), (**D**), (**P**), (**c**) and (**q**) are known, then its Utilized or Starting Capital must be:

U= {**Q**+**P**/[**c**-**q**]}/**/**-({**$**[1-**f**]-**V**-**I**}[1-**t**]-**D**)

Rule-12643:

If both (**/**), (**U**), (**Q**), (**f**), (**V**), (**I**), (**t**), (**D**), (**P**), (**c**) and (**q**) are known, then its Sales or Revenue Planned is:

$= [**V**+**I**+(**D**+{**Q**+**P**/[**c**-**q**]}/**/**-**U**)/[1-**t**]]/[1-**f**]

Rule-12644:

If both (**/**), (**U**), (**$**), (**Q**), (**V**), (**I**), (**t**), (**D**), (**P**), (**c**) and (**q**) are known, then its Fixed Portion Planned is:

f= 1-[**V**+**I**+(**D**+{**Q**+**P**/[**c**-**q**]}/**/**-**U**)/[1-**t**]]/**$**

Steve Asikin ISBN 14: 978-1514685136, ISBN 10: 1514685132

Rule-12645:

If both (/), (U), ($), (f), (Q), (I), (t), (D), (P), (c) and (q) are known, then its Variable Cost Planned is:

V= {$[1-f]-I}-(D+{Q+P/[c-q]}//-U)/[1-t]

Rule-12646:

If both (/), (U), ($), (f), (V), (Q), (t), (D), (P), (c) and (q) are known, then its Interest Expense Planned is:

I= {$[1-f]-V}-(D+{Q+P/[c-q]}//-U)/[1-t]

Rule-12647:

If both (/), (U), ($), (f), (V), (I), (Q), (D), (P), (c) and (q) are known, then its Tax Planned is:

t= 1-(D+{Q+P/[c-q]}//-U)/{$[1-f]-V-I}

Rule-12648:

If both (/), (U), ($), (f), (V), (I), (t), (Q), (P), (c) and (q) are known, then its Dividend Planned is:

D= U+{$[1-f]-V-I}[1-t]-{Q+P/[c-q]}//

Rule-12649:

If both (/), (U), ($), (f), (V), (I), (t), (D), (Q), (c) and (q) are known, then its Procured Inventory Planned is:

P= [c-q][/(U+{$[1-f]-V-I}[1-t]-D)-Q]

Rule-12650:

If both (/), (U), ($), (f), (V), (I), (t), (D), (P), (Q) and (q) are known, then its Current Ratio Planned is:

c= q+P/[/(U+{$[1-f]-V-I}[1-t]-D)-Q]

Steve Asikin ISBN 14: 978-1514685136, ISBN 10: 1514685132

Rule-12651:

If both (**/**), (**U**), (**$**), (**f**), (**V**), (**I**), (**t**), (**D**), (**P**), (**c**) and (**Q**) are known, then its Quick or Acid Test Ratio Planned is:

q= **c**-**P**/[**/**(**U**+{**$**[1-**f**]-**V**-**I**}[1-**t**]-**D**)-**Q**]

Rule-12652:

If both (**/**), (**U**), (**$**), (**f**), (**V**), (**I**), (**t**), (**D**), (**p**), (**c**) and (**q**) are known, then its Quoted Longterm Debt Planned is:

Q= **/**(**U**+{**$**[1-**f**]-**V**-**I**}[1-**t**]-**D**)-**Vp**/{360[**c**-**q**]}

Rule-12653:

If both (**Q**), (**U**), (**$**), (**f**), (**V**), (**I**), (**t**), (**D**), (**p**), (**c**) and (**q**) are known, then its Leverage or Gearing Ratio Planned is:

/= (**Q**+**Vp**/{360[**c**-**q**]})/(**U**+{**$**[1-**f**]-**V**-**I**}[1-**t**]-**D**)

Rule-12654:

If both (**/**), (**Q**), (**$**), (**f**), (**V**), (**I**), (**t**), (**D**), (**p**), (**c**) and (**q**) are known, then its Utilized or Starting Capital must be:

U= (**Q**+**Vp**/{360[**c**-**q**]})/**/**-({**$**[1-**f**]-**V**-**I**}[1-**t**]-**D**)

Rule-12655:

If both (**/**), (**U**), (**Q**), (**f**), (**V**), (**I**), (**t**), (**D**), (**p**), (**c**) and (**q**) are known, then its Sales or Revenue Planned is:

$= {**V**+**I**+[**D**+(**Q**+**Vp**/{360[**c**-**q**]})/**/**-**U**]/[1-**t**]}/[1-**f**]

Steve Asikin ISBN 14: 978-1514685136, ISBN 10: 1514685132

Rule-12656:

If both (**/**), (**U**), (**$**), (**Q**), (**V**), (**I**), (**t**), (**D**), (**p**), (**c**) and (**q**) are known, then its Fixed Portion Planned is:

f= 1-{**V**+**I**+[**D**+(**Q**+**Vp**/{360[**c**-**q**]})/**/**-**U**]/[1-**t**]}/**$**

Rule-12657:

If both (**/**), (**U**), (**$**), (**f**), (**Q**), (**I**), (**t**), (**D**), (**p**), (**c**) and (**q**) are known, then its Variable Cost Planned is:

V= [**/**(**U**+{**$**[1-**f**]-**I**}[1-**t**]-**D**)-**Q**]
/(**p**/{360[**c**-**q**]}+**/**[1-**t**])

Rule-12658:

If both (**/**), (**U**), (**$**), (**f**), (**V**), (**Q**), (**t**), (**D**), (**p**), (**c**) and (**q**) are known, then its Interest Expense Planned is:

I= {**$**[1-**f**]-**V**}-[**D**+(**Q**+**Vp**/{360[**c**-**q**]})/**/**-**U**]/[1-**t**]

Rule-12659:

If both (**/**), (**U**), (**$**), (**f**), (**V**), (**I**), (**Q**), (**D**), (**p**), (**c**) and (**q**) are known, then its Tax Rate Planned is:

t= 1-[**D**+(**Q**+**Vp**/{360[**c**-**q**]})/**/**-**U**]/{**$**[1-**f**]-**V**-**I**}

Rule-12660:

If both (**/**), (**U**), (**$**), (**f**), (**V**), (**I**), (**t**), (**Q**), (**p**), (**c**) and (**q**) are known, then its Dividend Payout Planned is:

D= **U**+{**$**[1-**f**]-**V**-**I**}[1-**t**]-(**Q**+**Vp**/{360[**c**-**q**]})/**/**

Rule-12661:

If both (**/**), (**U**), (**$**), (**f**), (**V**), (**I**), (**t**), (**D**), (**Q**), (**c**) and (**q**) are known, then its Procured Inventory Days Planned:

p= 360[**c**-**q**][**/**(**U**+{**$**[1-**f**]-**V**-**I**}[1-**t**]-**D**)-**Q**]/**V**

Steve Asikin ISBN 14: 978-1514685136, ISBN 10: 1514685132

Rule-12662:

If both (/), (**U**), (**$**), (**f**), (**V**), (**I**), (**t**), (**D**), (**p**), (**Q**) and (**q**) are known, then its Current Ratio Planned is:

c= **q**+**Vp**/{360[/(**U**+{**$**[1-**f**]-**V**-**I**}[1-**t**]-**D**)-**Q**]}

Rule-12663:

If both (/), (**U**), (**$**), (**f**), (**V**), (**I**), (**t**), (**D**), (**p**), (**c**) and (**Q**) are known, then its Quick or Acid Test Ratio Planned is:

q= **c**-**Vp**/{360[/(**U**+{**$**[1-**f**]-**V**-**I**}[1-**t**]-**D**)-**Q**]}

Rule-12664:

If both (/), (**U**), (**$**), (**f**), (**V**), (**I**), (**t**), (**D**), (**v**), (**p**), (**c**) and (**q**) are known, then its Quoted Longterm Debt Planned is:

Q= /(**U**+{**$**[1-**f**]-**V**-**I**}[1-**t**]-**D**)-**$vp**/{360[**c**-**q**]}

Rule-12665:

If both (**Q**), (**U**), (**$**), (**f**), (**V**), (**I**), (**t**), (**D**), (**v**), (**p**), (**c**) and (**q**) are known, then its Leverage or Gearing Ratio Planned is:

/= (**Q**+**$vp**/{360[**c**-**q**]})/(**U**+{**$**[1-**f**]-**V**-**I**}[1-**t**]-**D**)

Rule-12666:

If both (/), (**Q**), (**$**), (**f**), (**V**), (**I**), (**t**), (**D**), (**v**), (**p**), (**c**) and (**q**) are known, then its Utilized or Strating Capital must be:

U= (**Q**+**$vp**/{360[**c**-**q**]})//({**$**[1-**f**]-**V**-**I**}[1-**t**]-**D**)

Steve Asikin ISBN 14: 978-1514685136, ISBN 10: 1514685132

Rule-12667:

If both (**/**), (**U**), (**Q**), (**f**), (**V**), (**I**), (**t**), (**D**), (**v**), (**p**), (**c**) and (**q**) are known, then its Sales or Revenue Planned is:

$= [**/**(**U**-[**V**+**I**][1-**t**]-**D**)-**Q**]
/(**vp**/{360[**c**-**q**]}-**/**[1-**f**][1-**t**])

Rule-12668:

If both (**/**), (**U**), (**$**), (**Q**), (**V**), (**I**), (**t**), (**D**), (**v**), (**p**), (**c**) and (**q**) are known, then its Fixed Portion Planned is:

f= 1-{**V**+**I**+[**D**+(**Q**+**$vp**/{360[**c**-**q**]})/**/-U**]/[1-**t**]}/**$**

Rule-12669:

If both (**/**), (**U**), (**$**), (**f**), (**Q**), (**I**), (**t**), (**D**), (**v**), (**p**), (**c**) and (**q**) are known, then its Variable Cost Planned is:

V= {**$**[1-**f**]-**I**}-[**D**+(**Q**+**$vp**/{360[**c**-**q**]})/**/-U**]/[1-**t**]

Rule-12670:

If both (**/**), (**U**), (**$**), (**f**), (**V**), (**Q**), (**t**), (**D**), (**v**), (**p**), (**c**) and (**q**) are known, then its Interest Expense Planned is:

I= {**$**[1-**f**]-**V**}-[**D**+(**Q**+**$vp**/{360[**c**-**q**]})/**/-U**]/[1-**t**]

Rule-12671:

If both (**/**), (**U**), (**$**), (**f**), (**V**), (**I**), (**Q**), (**D**), (**v**), (**p**), (**c**) and (**q**) are known, then its Tax Rate Planned is:

t= 1-[**D**+(**Q**+**$vp**/{360[**c**-**q**]})/**/-U**]/{**$**[1-**f**]-**V**-**I**}

Steve Asikin ISBN 14: 978-1514685136, ISBN 10: 1514685132

Rule-12672:

If both (/), (U), ($), (f), (V), (I), (t), (Q), (v), (p), (c) and (q) are known, then its Dividend Planned is:

D= U+{$[1-f]-V-I}[1-t]-(Q+$vp/{360[c-q]})//

Rule-12673:

If both (/), (U), ($), (f), (V), (I), (t), (D), (Q), (p), (c) and (q) are known, then its Variable Portion Planned is:

v= V/$

or it can also be found as

v= 360[c-q][/(U+{$[1-f]-V-I}[1-t]-D)-Q]/[$p]

Rule-12674:

If both (/), (U), ($), (f), (V), (I), (t), (D), (v), (Q), (c) and (q) are known, then its Procured Inventory Days Planned:

p= 360[c-q][/(U+{$[1-f]-V-I}[1-t]-D)-Q]/[$v]

Rule-12675:

If both (/), (U), ($), (f), (V), (I), (t), (D), (v), (p), (Q) and (q) are known, then its Current Ratio Planned is:

c= q+$vp/{360[/(U+{$[1-f]-V-I}[1-t]-D)-Q]}

Rule-12676:

If both (/), (U), ($), (f), (V), (I), (t), (D), (v), (p), (c) and (Q) are known, then its Quick or Acid Test Ratio Planned is:

q= c-$vp/{360[/(U+{$[1-f]-V-I}[1-t]-D)-Q]}

Steve Asikin ISBN 14: 978-1514685136, ISBN 10: 1514685132

Rule-12677:

If both (**/**), (**U**), (**$**), (**f**), (**V**), (**I**), (**t**), (**D**), (**$'**), (**v**), (**p**), (**s**), (**c**) and (**q**) are known, then its Quoted Longterm Debt Planned is:

Q= **/**(**U**+{**$**[1-**f**]-**V**-**I**}[1-**t**]-**D**)-**$'vp**[1+**s**]/{360[**c**-**q**]}

Rule-12678:

If both (**Q**), (**U**), (**$**), (**f**), (**V**), (**I**), (**t**), (**D**), (**$'**), (**v**), (**p**), (**s**), (**c**) and (**q**) are known, then its Leverage or Gearing Ratio Planned is:

/= (**Q**+**$'vp**[1+**s**]/{360[**c**-**q**]})
/(**U**+{**$**[1-**f**]-**V**-**I**}[1-**t**]-**D**)

Rule-12679:

If both (**/**), (**Q**), (**$**), (**f**), (**V**), (**I**), (**t**), (**D**), (**$'**), (**v**), (**p**), (**s**), (**c**) and (**q**) are known, then its Utilized or Starting Capital must be:

U= (**Q**+**$'vp**[1+**s**]/{360[**c**-**q**]})/**/**
-({**$**[1-**f**]-**V**-**I**}[1-**t**]-**D**)

Rule-12680:

If both (**/**), (**U**), (**Q**), (**f**), (**V**), (**I**), (**t**), (**D**), (**$'**), (**v**), (**p**), (**s**), (**c**) and (**q**) are known, then its Sales or Revenue Planned is:

$= [**/**(**U**-[**V**+**I**][1-**t**]-**D**)-**Q**]
/(**vp**/{360[**c**-**q**]}-**/**[1-**f**][1-**t**])

Steve Asikin ISBN 14: 978-1514685136, ISBN 10: 1514685132

Rule-12681:

If both (**/**), (**U**), (**S**), (**Q**), (**V**), (**I**), (**t**), (**D**), (**S'**), (**v**), (**p**), (**s**), (**c**) and (**q**) are known, then its Fixed Portion Planned is:

f= 1-{**V**+**I**+[**D**+(**Q**+**S'vp**[1+**s**]/{360[**c**-**q**]})/**/-U**]/[1-**t**]}/**S**

Rule-12682:

If both (**/**), (**U**), (**S**), (**f**), (**Q**), (**I**), (**t**), (**D**), (**S'**), (**v**), (**p**), (**s**), (**c**) and (**q**) are known, then its Variable Cost Planned is:

V= {**S**[1-**f**]-**I**}-[**D**+(**Q**+**S'vp**[1+**s**]/{360[**c**-**q**]})/**/-U**]/[1-**t**]

Rule-12683:

If both (**/**), (**U**), (**S**), (**f**), (**V**), (**Q**), (**t**), (**D**), (**S'**), (**v**), (**p**), (**s**), (**c**) and (**q**) are known, then its Interest Expense Planned is:

I= {**S**[1-**f**]-**V**}-[**D**+(**Q**+**S'vp**[1+**s**]/{360[**c**-**q**]})/**/-U**]/[1-**t**]

Rule-12684:

If both (**/**), (**U**), (**S**), (**f**), (**V**), (**I**), (**Q**), (**D**), (**S'**), (**v**), (**p**), (**s**), (**c**) and (**q**) are known, then its Tax Rate Planned is:

t= 1-[**D**+(**Q**+**S'vp**[1+**s**]/{360[**c**-**q**]})/**/-U**]/{**S**[1-**f**]-**V**-**I**}

Steve Asikin ISBN 14: 978-1514685136, ISBN 10: 1514685132

Rule-12685:

If both (**/**), (**U**), (**S**), (**f**), (**V**), (**I**), (**t**), (**Q**), (**S'**), (**v**), (**p**), (**s**), (**c**) and (**q**) are known, then its Dividend Planned is:

D= U+{S[1-f]-V-I}[1-t]
-(Q+S'vp[1+s]/{360[c-q]})//

Rule-12686:

If both (**/**), (**U**), (**S**), (**f**), (**V**), (**I**), (**t**), (**D**), (**Q**), (**v**), (**p**), (**s**), (**c**) and (**q**) are known, then its Sales Past must be:

S'= 360[c-q][/(U+{S[1-f]-V-I}[1-t]-D)-Q]
/{vp[1+s]}

Rule-12687:

If both (**/**), (**U**), (**S**), (**f**), (**V**), (**I**), (**t**), (**D**), (**S'**), (**Q**), (**p**), (**s**), (**c**) and (**q**) are known, then its Variable Portion Planned is:

v= 360[c-q][/(U+{S[1-f]-V-I}[1-t]-D)-Q]
/{S'p[1+s]}

Rule-12688:

If both (**/**), (**U**), (**S**), (**f**), (**V**), (**I**), (**t**), (**D**), (**S'**), (**v**), (**Q**), (**s**), (**c**) and (**q**) are known, then its Procured Inventory Days Planned:

p= 360[c-q][/(U+{S[1-f]-V-I}[1-t]-D)-Q]
/{S'v[1+s]}

Steve Asikin ISBN 14: 978-1514685136, ISBN 10: 1514685132

Rule-12689:

If both (**/**), (**U**), (**$**), (**f**), (**V**), (**I**), (**t**), (**D**), (**$'**), (**v**), (**p**), (**Q**), (**c**) and (**q**) are known, then its Sales Growth Planned is:

s= **$**/**$'**-1

or it can also be found as

s= 360[**c**-**q**][**/**(**U**+{**$**[1-**f**]-**V**-**I**}[1-**t**]-**D**)-**Q**]/{**$'vp**]-1

Rule-12690:

If both (**/**), (**U**), (**$**), (**f**), (**V**), (**I**), (**t**), (**D**), (**$'**), (**v**), (**p**), (**s**), (**Q**) and (**q**) are known, then its Current Ratio Planned is:

c= **q**+**$'vp**[1+**s**]

/{360[**/**(**U**+{**$**[1-**f**]-**V**-**I**}[1-**t**]-**D**)-**Q**]}

Rule-12691:

If both (**/**), (**U**), (**$**), (**f**), (**V**), (**I**), (**t**), (**D**), (**$'**), (**v**), (**p**), (**s**), (**c**) and (**Q**) are known, then its Quick or Acid Test Ratio Planned is:

q= **c**-**$'vp**[1+**s**]/{360[**/**(**U**+{**$**[1-**f**]-**V**-**I**}[1-**t**]-**D**)-**Q**]}

Rule-12692:

If both (**/**), (**U**), (**$**), (**f**), (**V**), (**I**), (**t**), (**A**), (**d**) and (**X**) are known, then its Quoted Longterm Debt Planned is:

Q= **/**(**U**+{**$**[1-**f**]-**V**-**I**}[1-**t**]-**Ad**)-**X**

Rule-12693:

If both (**Q**), (**U**), (**$**), (**f**), (**V**), (**I**), (**t**), (**A**), (**d**) and (**X**) are known, then its Leverage or Gearing Ratio Planned is:

/= [**Q**+**X**]/(**U**+{**$**[1-**f**]-**V**-**I**}[1-**t**]-**Ad**)

Steve Asikin ISBN 14: 978-1514685136, ISBN 10: 1514685132

Rule-12694:

If both (/), (**Q**), (**$**), (**f**), (**V**), (**I**), (**t**), (**A**), (**d**) and (**X**) are known, then its Utilized or Starting Capital must be:

U= [**Q**+**X**]/**/**({**$**[1-**f**]-**V**-**I**}[1-**t**]-**Ad**)

Rule-12695:

If both (/), (**U**), (**Q**), (**f**), (**V**), (**I**), (**t**), (**A**), (**d**) and (**X**) are known, then its Sales or Revenue Planned is:

$= (**V**+**I**+{**Ad**+[**Q**+**X**]/**/U**}/[1-**t**])/[1-**f**]

Rule-12696:

If both (/), (**U**), (**$**), (**Q**), (**V**), (**I**), (**t**), (**A**), (**d**) and (**X**) are known, then its Fixed Portion Planned is:

f= 1-(**V**+**I**+{**Ad**+[**Q**+**X**]/**/U**}/[1-**t**])/**$**

Rule-12697:

If both (/), (**U**), (**$**), (**f**), (**Q**), (**I**), (**t**), (**A**), (**d**) and (**X**) are known, then its Variable Cost Planned is:

V= {**$**[1-**f**]-**I**}-{**Ad**+[**Q**+**X**]/**/U**}/[1-**t**]

Rule-12698:

If both (/), (**U**), (**$**), (**f**), (**V**), (**Q**), (**t**), (**A**), (**d**) and (**X**) are known, then its Interest Expense Planned is:

I= {**$**[1-**f**]-**V**}-{**Ad**+[**Q**+**X**]/**/U**}/[1-**t**]

Rule-12699:

If both (/), (**U**), (**$**), (**f**), (**V**), (**I**), (**Q**), (**A**), (**d**) and (**X**) are known, then its Tax Rate Planned is:

t= 1-{**Ad**+[**Q**+**X**]/**/U**}/{**$**[1-**f**]-**V**-**I**}

Steve Asikin ISBN 14: 978-1514685136, ISBN 10: 1514685132

Rule-12700:

If both (/), (U), ($), (f), (V), (I), (t), (Q), (d) and (X) are known, then its After Tax Income Planned is:

A= ({$[1-f]-V-I}[1-t]-[Q+X]//-U)/d

Rule-12701:

If both (/), (U), ($), (f), (V), (I), (t), (A), (Q) and (X) are known, then its Dividend Portion Planned is:

d= ({$[1-f]-V-I}[1-t]-[Q+X]//-U)/A

Rule-12702:

If both (/), (U), ($), (f), (V), (I), (t), (A), (d) and (Q) are known, then its Xpress or Current Debt Planned is:

X= /(U+{$[1-f]-V-I}[1-t]-Ad)-Q

Rule-12703:

If both (/), (U), ($), (f), (V), (I), (t), (A), (d), (P), (c) and (q) are known, then its Quoted Longterm Debt Planned is:

Q= /(U+{$[1-f]-V-I}[1-t]-Ad)-P/[c-q]

Rule-12704:

If both (/), (U), ($), (f), (V), (I), (t), (A), (d), (P), (c) and (q) are known, then its Quoted Longterm Debt Planned is:

/= {Q+P/[c-q]}/(U+{$[1-f]-V-I}[1-t]-Ad)

Steve Asikin ISBN 14: 978-1514685136, ISBN 10: 1514685132

Rule-12705:

If both (**/**), (**Q**), (**$**), (**f**), (**V**), (**I**), (**t**), (**A**), (**d**), (**P**), (**c**) and (**q**) are known, then its Utilized or Starting capital must be:

U= {Q+P/[c-q]}//({$[1-f]-V-I}[1-t]-Ad)

Rule-12706:

If both (**/**), (**U**), (**Q**), (**f**), (**V**), (**I**), (**t**), (**A**), (**d**), (**P**), (**c**) and (**q**) are known, then its Sales or Revenue Planned is:

$= [V+I+(Ad+{Q+P/[c-q]}//U)/[1-t]]/[1-f]

Rule-12707:

If both (**/**), (**U**), (**$**), (**Q**), (**V**), (**I**), (**t**), (**A**), (**d**), (**P**), (**c**) and (**q**) are known, then its Fixed Portion Planned is:

f= 1-[V+I+(Ad+{Q+P/[c-q]}//U)/[1-t]]/$

Rule-12708:

If both (**/**), (**U**), (**$**), (**f**), (**Q**), (**I**), (**t**), (**A**), (**d**), (**P**), (**c**) and (**q**) are known, then its Variable Cost Planned is:

V= {$[1-f]-I}-(Ad+{Q+P/[c-q]}//U)/[1-t]

Rule-12709:

If both (**/**), (**U**), (**$**), (**f**), (**V**), (**Q**), (**t**), (**A**), (**d**), (**P**), (**c**) and (**q**) are known, then its Interest Expense Planned is:

I= {$[1-f]-V}-(Ad+{Q+P/[c-q]}//U)/[1-t]

Steve Asikin ISBN 14: 978-1514685136, ISBN 10: 1514685132

Rule-12710:

If both (/), (U), ($), (f), (V), (I), (Q), (A), (d), (P), (c) and (q) are known, then its Tax Rate Planned is:

t= 1-(Ad+{Q+P/[c-q]}/ƚU)/{$[1-f]-V-I}

Rule-12711:

If both (/), (U), ($), (f), (V), (I), (t), (Q), (d), (P), (c) and (q) are known, then its After Tax Income Planned is:

A= ({$[1-f]-V-I}[1-t]-{Q+P/[c-q]}/ƚU)/d

Rule-12712:

If both (/), (U), ($), (f), (V), (I), (t), (A), (Q), (P), (c) and (q) are known, then its Dividend Payout Planned is:

d= ({$[1-f]-V-I}[1-t]-{Q+P/[c-q]}/ƚU)/A

Rule-12713:

If both (/), (U), ($), (f), (V), (I), (t), (A), (d), (Q), (c) and (q) are known, then its Procured Inventory Planned is:

P= [c-q][/U+{$[1-f]-V-I}[1-t]-Ad)-Q]

Rule-12714:

If both (/), (U), ($), (f), (V), (I), (t), (A), (d), (P), (Q) and (q) are known, then its Current Ratio Planned is:

c= q+P/[/U+{$[1-f]-V-I}[1-t]-Ad)-Q]

Steve Asikin ISBN 14: 978-1514685136, ISBN 10: 1514685132

Rule-12715:

If both (**ʎ**), (**U**), (**$**), (**f**), (**V**), (**I**), (**t**), (**A**), (**d**), (**P**), (**c**) and (**Q**) are known, then its Quick or Acid Test Ratio Planned is:

q= **c**-**P**/[**ʎ**(**U**+{**$**[1-**f**]-**V**-**I**}[1-**t**]-**Ad**)-**Q**]

Rule-12716:

If both (**ʎ**), (**U**), (**$**), (**f**), (**V**), (**I**), (**t**), (**A**), (**d**), (**p**), (**c**) and (**q**) are known, then its Quoted Longterm Debt Planned is:

Q= **ʎ**(**U**+{**$**[1-**f**]-**V**-**I**}[1-**t**]-**Ad**)-**Vp**/{360[**c**-**q**]}

Rule-12717:

If both (**Q**), (**U**), (**$**), (**f**), (**V**), (**I**), (**t**), (**A**), (**d**), (**p**), (**c**) and (**q**) are known, then its Leverage or Gearing Ratio Planned is:

ʎ= (**Q**+**Vp**/{360[**c**-**q**]})/(**U**+{**$**[1-**f**]-**V**-**I**}[1-**t**]-**Ad**)

Rule-12718:

If both (**ʎ**), (**Q**), (**$**), (**f**), (**V**), (**I**), (**t**), (**A**), (**d**), (**p**), (**c**) and (**q**) are known, then its Utilized or Starting Capital must be:

U= (**Q**+**Vp**/{360[**c**-**q**]})/**ʎ**-({**$**[1-**f**]-**V**-**I**}[1-**t**]-**Ad**)

Rule-12719:

If both (**ʎ**), (**U**), (**Q**), (**f**), (**V**), (**I**), (**t**), (**A**), (**d**), (**p**), (**c**) and (**q**) are known, then its Sales or Revenue Planned is:

$= {**V**+**I**+[**Ad**+(**Q**+**Vp**/{360[**c**-**q**]})/**ʎ**-**U**]/[1-**t**]}/[1-**f**]

Steve Asikin ISBN 14: 978-1514685136, ISBN 10: 1514685132

Rule-12720:

If both (**/**), (**U**), (**$**), (**Q**), (**V**), (**I**), (**t**), (**A**), (**d**), (**p**), (**c**) and (**q**) are known, then its Fixed Portion Planned is:

f= 1-{V+I+[Ad+(Q+Vp/{360[c-q]})/-U]/[1-t]}/$

Rule-12721:

If both (**/**), (**U**), (**$**), (**f**), (**Q**), (**I**), (**t**), (**A**), (**d**), (**p**), (**c**) and (**q**) are known, then its Variable Cost Planned is:

V= [/U+{$[1-f]-I}[1-t]-Ad)-Q]
/(p/{360[c-q]}+/[1-t])

Rule-12722:

If both (**/**), (**U**), (**$**), (**f**), (**V**), (**Q**), (**t**), (**A**), (**d**), (**p**), (**c**) and (**q**) are known, then its Interest Expense Planned is:

I= {$[1-f]-V}-[Ad+(Q+Vp/{360[c-q]})/-U]/[1-t]

Rule-12723:

If both (**/**), (**U**), (**$**), (**f**), (**V**), (**I**), (**Q**), (**A**), (**d**), (**p**), (**c**) and (**q**) are known, then its Tax Rate Planned is:

t= 1-[Ad+(Q+Vp/{360[c-q]})/-U]/{$[1-f]-V-I}

Rule-12724:

If both (**/**), (**U**), (**$**), (**f**), (**V**), (**I**), (**t**), (**Q**), (**d**), (**p**), (**c**) and (**q**) are known, then its After Tax Income Planned is:

A= [{$[1-f]-V-I}[1-t]-(Q+Vp/{360[c-q]})/-U]/d

Steve Asikin ISBN 14: 978-1514685136, ISBN 10: 1514685132

Rule-12725:

If both (**/**), (**U**), (**$**), (**f**), (**V**), (**I**), (**t**), (**A**), (**Q**), (**p**), (**c**) and (**q**) are known, then its Dividend Payout Planned is:

d= [{**$**[1-**f**]-**V**-**I**}[1-**t**]-(**Q**+**Vp**/{360[**c**-**q**]})/**/**-**U**]/**A**

Rule-12726:

If both (**/**), (**U**), (**$**), (**f**), (**V**), (**I**), (**t**), (**A**), (**d**), (**Q**), (**c**) and (**q**) are known, then its Procured Inventory Days Planned:

p= 360[**c**-**q**][**/**(**U**+{**$**[1-**f**]-**V**-**I**}[1-**t**]-**Ad**)-**Q**]/**V**

Rule-12727:

If both (**/**), (**U**), (**$**), (**f**), (**V**), (**I**), (**t**), (**A**), (**d**), (**p**), (**Q**) and (**q**) are known, then its Current Ratio Planned is:

c= **q**+**Vp**/{360[**/**(**U**+{**$**[1-**f**]-**V**-**I**}[1-**t**]-**Ad**)-**Q**]}

Rule-12728:

If both (**/**), (**U**), (**$**), (**f**), (**V**), (**I**), (**t**), (**A**), (**d**), (**p**), (**c**) and (**Q**) are known, then its Quick or Acid Test Ratio Planned is:

q= **c**-**Vp**/{360[**/**(**U**+{**$**[1-**f**]-**V**-**I**}[1-**t**]-**Ad**)-**Q**]}

Rule-12729:

If both (**/**), (**U**), (**$**), (**f**), (**V**), (**I**), (**t**), (**A**), (**d**), (**v**), (**p**), (**c**) and (**q**) are known, then its Quoted Longterm Debt Planned is:

Q= **/**(**U**+{**$**[1-**f**]-**V**-**I**}[1-**t**]-**Ad**)-**$vp**/{360[**c**-**q**]}

Steve Asikin ISBN 14: 978-1514685136, ISBN 10: 1514685132

Rule-12730:

If both (**Q**), (**U**), (**$**), (**f**), (**V**), (**I**), (**t**), (**A**), (**d**), (**v**), (**p**), (**c**) and (**q**) are known, then its Leverage or Gearing Ratio Planned is:

/= (Q+$vp/{360[c-q]})/(U+{$[1-f]-V-I}[1-t]-Ad)

Rule-12731:

If both (**/**), (**Q**), (**$**), (**f**), (**V**), (**I**), (**t**), (**A**), (**d**), (**v**), (**p**), (**c**) and (**q**) are known, then its Utilized or Starting Capital must be:

U= (Q+$vp/{360[c-q]})//-({$[1-f]-V-I}[1-t]-Ad)

Rule-12732:

If both (**/**), (**U**), (**Q**), (**f**), (**V**), (**I**), (**t**), (**A**), (**d**), (**v**), (**p**), (**c**) and (**q**) are known, then its Sales Revenue Planned is:

$= (/{U-[V+I][1-t]-Ad}-Q)/(vp/{360[c-q]}-/[1-f])

Rule-12733:

If both (**/**), (**U**), (**$**), (**Q**), (**V**), (**I**), (**t**), (**A**), (**d**), (**v**), (**p**), (**c**) and (**q**) are known, then its Fixed Portion Planned is:

f= 1-{V+I+[Ad+(Q+$vp/{360[c-q]})//-U]/[1-t]}/$

Rule-12734:

If both (**/**), (**U**), (**$**), (**f**), (**Q**), (**I**), (**t**), (**A**), (**d**), (**v**), (**p**), (**c**) and (**q**) are known, then its Variable Cost Planned is:

V= {$[1-f]-I}-[Ad+(Q+$vp/{360[c-q]})//-U]/[1-t]

Steve Asikin ISBN 14: 978-1514685136, ISBN 10: 1514685132

Rule-12735:

If both (**/**), (**U**), (**S**), (**f**), (**V**), (**Q**), (**t**), (**A**), (**d**), (**v**), (**p**), (**c**) and (**q**) are known, then its Interest Expense Planned is:

I= {**S**[1-**f**]-**V**}-**[Ad**+(**Q**+**Svp**/{360[**c-q**]})/**/-U]**/[1-**t**]

Rule-12736:

If both (**/**), (**U**), (**S**), (**f**), (**V**), (**I**), (**Q**), (**A**), (**d**), (**v**), (**p**), (**c**) and (**q**) are known, then its Tax Rate Planned is:

t= 1-**[Ad**+(**Q**+**Svp**/{360[**c-q**]})/**/-U]**/{**S**[1-**f**]-**V-I**}

Rule-12737:

If both (**/**), (**U**), (**S**), (**f**), (**V**), (**I**), (**t**), (**Q**), (**d**), (**v**), (**p**), (**c**) and (**q**) are known, then its After Tax Income Planned is:

A= **[**{**S**[1-**f**]-**V-I**}[1-**t**]-(**Q**+**Svp**/{360[**c-q**]})/**/-U]/d**

Rule-12738:

If both (**/**), (**U**), (**S**), (**f**), (**V**), (**I**), (**t**), (**A**), (**Q**), (**v**), (**p**), (**c**) and (**q**) are known, then its Dividend Payout Planned is:

d= **[**{**S**[1-**f**]-**V-I**}[1-**t**]-(**Q**+**Svp**/{360[**c-q**]})/**/-U]/A**

Rule-12739:

If both (**/**), (**U**), (**S**), (**f**), (**V**), (**I**), (**t**), (**A**), (**d**), (**Q**), (**p**), (**c**) and (**q**) are known, then its Variable Portion Planned is:

s= **V/S**

or it can also be found as

v= 360[**c-q**]**[/(U**+{**S**[1-**f**]-**V-I**}[1-**t**]-**Ad**)-**Q]**/[**Sp**]

Steve Asikin ISBN 14: 978-1514685136, ISBN 10: 1514685132

Rule-12740:

If both (**/**), (**U**), (**S**), (**f**), (**V**), (**I**), (**t**), (**A**), (**d**), (**v**), (**Q**), (**c**) and (**q**) are known, then its Procured Inventory Days Planned:

p= 360[**c-q**][**/**(**U**+{**S**[1-**f**]-**V-I**}[1-**t**]-**Ad**)-**Q**]/[**Sv**]

Rule-12741:

If both (**/**), (**U**), (**S**), (**f**), (**V**), (**I**), (**t**), (**A**), (**d**), (**v**), (**p**), (**Q**) and (**q**) are known, then its Current Ratio Planned is:

c= **q**+**Svp**/{360[**/**(**U**+{**S**[1-**f**]-**V-I**}[1-**t**]-**Ad**)-**Q**]}

Rule-12742:

If both (**/**), (**U**), (**S**), (**f**), (**V**), (**I**), (**t**), (**A**), (**d**), (**v**), (**p**), (**c**) and (**Q**) are known, then its Quick or Acid Test Ratio Planned is:

q= **c**-**Svp**/{360[**/**(**U**+{**S**[1-**f**]-**V-I**}[1-**t**]-**Ad**)-**Q**]}

Rule-12743:

If both (**/**), (**U**), (**S**), (**f**), (**V**), (**I**), (**t**), (**A**), (**d**), (**S'**), (**v**), (**p**), (**s**), (**c**) and (**q**) are known, then its Quoted Longterm Debt Planned is:

Q= **/**(**U**+{**S**[1-**f**]-**V-I**}[1-**t**]-**Ad**)
-**S'vp**[1+**s**]/{360[**c-q**]}

Rule-12744:

If both (**Q**), (**U**), (**S**), (**f**), (**V**), (**I**), (**t**), (**A**), (**d**), (**S'**), (**v**), (**p**), (**s**), (**c**) and (**q**) are known, then its Leverage or Gearing Ratio Planned is:

/= (**Q**+**S'vp**[1+**s**]/{360[**c-q**]})
/(**U**+{**S**[1-**f**]-**V-I**}[1-**t**]-**Ad**)

Steve Asikin ISBN 14: 978-1514685136, ISBN 10: 1514685132

Rule-12745:

If both (/), (Q), ($), (f), (V), (I), (t), (A), (d), ($'), (v), (p), (s), (c) and (q) are known, then its Utilized or Starting Capital must be:

U= (Q+$'vp[1+s]/{360[c-q]})//
-({$[1-f]-V-I}[1-t]-Ad)

Rule-12746:

If both (/), (U), ($), (f), (V), (I), (t), (A), (d), (Q), (v), (p), (s), (c) and (q) are known, then its Sales or Revenue Planned is:

$= [Q-/(U-[V+I][1-t]-Ad)
+$'vp[1+s]/{360[c-q]}]/{[1-f] [1-t]}

Rule-12747:

If both (/), (U), ($), (Q), (V), (I), (t), (A), (d), ($'), (v), (p), (s), (c) and (q) are known, then its Fixed Portion Planned is:

f= 1-{V+I+[Ad+(Q+$'vp[1+s]/{360[c-q]})//-U]
/[1-t]}/$

Rule-12748:

If both (/), (U), ($), (f), (Q), (I), (t), (A), (d), ($'), (v), (p), (s), (c) and (q) are known, then its Variable Cost Planned is:

V= {$[1-f]-I}
-[Ad+(Q+$'vp[1+s]/{360[c-q]})//-U]/[1-t]

Steve Asikin ISBN 14: 978-1514685136, ISBN 10: 1514685132

Rule-12749:

If both (**/**), (**U**), (**S**), (**f**), (**V**), (**Q**), (**t**), (**A**), (**d**), (**S'**), (**v**), (**p**), (**s**), (**c**) and (**q**) are known, then its Interest Expense Planned is:

I= {**S**[1-**f**]-**V**}
-[**Ad**+(**Q**+**S'vp**[1+**s**]/{360[**c-q**]})/**/U**]/[1-**t**]

Rule-12750:

If both (**/**), (**U**), (**S**), (**f**), (**V**), (**I**), (**Q**), (**A**), (**d**), (**S'**), (**v**), (**p**), (**s**), (**c**) and (**q**) are known, then its Tax Rate Planned is:

t= 1-[**Ad**+(**Q**+**S'vp**[1+**s**]/{360[**c-q**]})/**/U**]
/{**S**[1-**f**]-**V**-**I**}

Rule-12751:

If both (**/**), (**U**), (**S**), (**f**), (**V**), (**I**), (**t**), (**Q**), (**d**), (**S'**), (**v**), (**p**), (**s**), (**c**) and (**q**) are known, then its After Tax Income Planned is:

A= [{**S**[1-**f**]-**V**-**I**}[1-**t**]
-(**Q**+**S'vp**[1+**s**]/{360[**c-q**]})/**/U**]/**d**

Rule-12752:

If both (**/**), (**U**), (**S**), (**f**), (**V**), (**I**), (**t**), (**A**), (**Q**), (**S'**), (**v**), (**p**), (**s**), (**c**) and (**q**) are known, then its Dividend Planned is:

d= [{**S**[1-**f**]-**V**-**I**}[1-**t**]
-(**Q**+**S'vp**[1+**s**]/{360[**c-q**]})/**/U**]/**A**

Steve Asikin ISBN 14: 978-1514685136, ISBN 10: 1514685132

Rule-12753:

If both (**/**), (**U**), (**$**), (**f**), (**V**), (**I**), (**t**), (**A**), (**d**), (**Q**), (**v**), (**p**), (**s**), (**c**) and (**q**) are known, then its Sales Past must be:

$'**= 360[**c-q**][**/**(**U**+{**$[1-**f**]-**V-I**}[1-**t**]-**Ad**)-**Q**]
/{**vp**[1+**s**]}

Rule-12754:

If both (**/**), (**U**), (**$**), (**f**), (**V**), (**I**), (**t**), (**A**), (**d**), (**$'**), (**Q**), (**p**), (**s**), (**c**) and (**q**) are known, then its Variable Portion Planned is:

v= **V/$**

or it can also be found as

v= 360[**c-q**][**/**(**U**+{**$**[1-**f**]-**V-I**}[1-**t**]-**Ad**)-**Q**]
/{**$'p**[1+**s**]}

Rule-12755:

If both (**/**), (**U**), (**$**), (**f**), (**V**), (**I**), (**t**), (**A**), (**d**), (**$'**), (**v**), (**Q**), (**s**), (**c**) and (**q**) are known, then its Procured Inventory Days Planned:

p= 360[**c-q**][**/**(**U**+{**$**[1-**f**]-**V-I**}[1-**t**]-**Ad**)-**Q**]
/{**$'v**[1+**s**]}

Rule-12756:

If both (**/**), (**U**), (**$**), (**f**), (**V**), (**I**), (**t**), (**A**), (**d**), (**$'**), (**v**), (**p**), (**Q**), (**c**) and (**q**) are known, then its Sales Growth Planned is:

s= 360[**c-q**][**/**(**U**+{**$**[1-**f**]-**V-I**}[1-**t**]-**Ad**)-**Q**]
/[**$'vp**]-1

Steve Asikin ISBN 14: 978-1514685136, ISBN 10: 1514685132

Rule-12757:

If both (**/**), (**U**), (**$**), (**f**), (**V**), (**I**), (**t**), (**A**), (**d**), (**$'**), (**v**), (**p**), (**s**), (**Q**) and (**q**) are known, then its Current Ratio Planned is:

c= **q**+**$'vp**[1+**s**]
/{360[**/**(**U**+{**$**[1-**f**]-**V**-**I**}[1-**t**]-**Ad**)-**Q**]}

Rule-12758:

If both (**/**), (**U**), (**$**), (**f**), (**V**), (**I**), (**t**), (**A**), (**d**), (**$'**), (**v**), (**p**), (**s**), (**c**) and (**Q**) are known, then its Qquick or Acid Test Ratio Planned is:

q= **c**-**$'vp**[1+**s**]
/{360[**/**(**U**+{**$**[1-**f**]-**V**-**I**}[1-**t**]-**Ad**)-**Q**]}

Rule-12759:

If both (**/**), (**U**), (**$**), (**f**), (**V**), (**I**), (**t**), (**d**) and (**X**) are known, then its Quoted Longterm Debt Planned is:

Q= **/**(**U**+{**$**[1-**f**]-**V**-**I**}[1-**t**][1-**d**])-**X**

Rule-12760:

If both (**Q**), (**U**), (**$**), (**f**), (**V**), (**I**), (**t**), (**d**) and (**X**) are known, then its Leverage or Gearing Ratio Planned is:

/= [**Q**+**X**]/(**U**+{**$**[1-**f**]-**V**-**I**}[1-**t**][1-**d**])

Rule-12761:

If both (**/**), (**Q**), (**$**), (**f**), (**V**), (**I**), (**t**), (**d**) and (**X**) are known, then its Utilized or Starting Capital must be:

U= [**Q**+**X**]/**/**-{**$**[1-**f**]-**V**-**I**}[1-**t**][1-**d**]

Steve Asikin ISBN 14: 978-1514685136, ISBN 10: 1514685132

Rule-12762:

If both (**/**), (**U**), (**Q**), (**f**), (**V**), (**I**), (**t**), (**d**) and (**X**) are known, then its Sales or Revenue Planned is:

$= (**V**+**I**+{[**Q**+**X**]/**/U**}/{[1-**t**][1-**d**]})/[1-**f**]

Rule-12763:

If both (**/**), (**U**), (**$**), (**Q**), (**V**), (**I**), (**t**), (**d**) and (**X**) are known, then its Fixed Rate Planned is:

f= 1-(**V**+**I**+{[**Q**+**X**]/**/U**}/{[1-**t**][1-**d**]})/**$**

Rule-12764:

If both (**/**), (**U**), (**$**), (**f**), (**Q**), (**I**), (**t**), (**d**) and (**X**) are known, then its Variable Portion Planned is:

V= {**$**[1-**f**]-**I**}-{[**Q**+**X**]/**/U**}/{[1-**t**][1-**d**]}

Rule-12765:

If both (**/**), (**U**), (**$**), (**f**), (**V**), (**Q**), (**t**), (**d**) and (**X**) are known, then its Interest Expense Planned is:

I= {**$**[1-**f**]-**V**}-{[**Q**+**X**]/**/U**}/{[1-**t**][1-**d**]}

Rule-12766:

If both (**/**), (**U**), (**$**), (**f**), (**V**), (**I**), (**Q**), (**d**) and (**X**) are known, then its Leverage or Tax Rate Planned is:

t = 1-{**Q**+**X**]/**/U**}/([1-**d**]{**$**[1-**f**]-**V**-**I**})

Rule-12767:

If both (**/**), (**U**), (**$**), (**f**), (**V**), (**I**), (**t**), (**Q**) and (**X**) are known, then its Dividend Payout Planned is:

d = 1-{**Q**+**X**]/**/U**}/([1-**t**]{**$**[1-**f**]-**V**-**I**})

Steve Asikin ISBN 14: 978-1514685136, ISBN 10: 1514685132

Rule-12768:

If both (**Ł**), (**U**), (**$**), (**f**), (**V**), (**I**), (**t**), (**d**) and (**Q**) are known, then its Xpress or Current Debt Planned is:

X= **Ł**(**U**+{**$**[1-**f**]-**V**-**I**}[1-**t**][1-**d**])-**Q**

Rule-12769:

If both (**Ł**), (**U**), (**$**), (**f**), (**V**), (**I**), (**t**), (**d**), (**P**), (**c**) and (**q**) are known, then its Leverage or Gearing Ratio Planned is:

Q= **Ł**(**U**+{**$**[1-**f**]-**V**-**I**}[1-**t**][1-**d**])-**P**/[**c**-**q**]

Rule-12770:

If both (**Q**), (**U**), (**$**), (**f**), (**V**), (**I**), (**t**), (**d**), (**P**), (**c**) and (**q**) are known, then its Leverage or Gearing Ratio Planned is:

Ł= {**Q**+**P**/[**c**-**q**]}/(**U**+{**$**[1-**f**]-**V**-**I**}[1-**t**][1-**d**])

Rule-12771:

If both (**Ł**), (**Q**), (**$**), (**f**), (**V**), (**I**), (**t**), (**d**), (**P**), (**c**) and (**q**) are known, then its Utilized or Starting Capital must be:

U={**Q**+**P**/[**c**-**q**]}/**Ł**-{**$**[1-**f**]-**V**-**I**}[1-**t**][1-**d**]

Rule-12772:

If both (**Ł**), (**U**), (**Q**), (**f**), (**V**), (**I**), (**t**), (**d**), (**P**), (**c**) and (**q**) are known, then its Sales or Revenue Planned is:

$= [**V**+**I**+({**Q**+**P**/[**c**-**q**]}/**Ł**-**U**)/{[1-**t**][1-**d**]}]/[1-**f**]

Steve Asikin ISBN 14: 978-1514685136, ISBN 10: 1514685132

Rule-12773:

If both (**/**), (**U**), (**$**), (**Q**), (**V**), (**I**), (**t**), (**d**), (**P**), (**c**) and (**q**) are known, then its Fixed Portion Planned is:

f= 1-**[V**+**I**+({**Q**+**P**/[**c**-**q**]}/**/U**)/{[1-**t**][1-**d**]}**]**/**$**

Rule-12774:

If both (**/**), (**U**), (**$**), (**f**), (**Q**), (**I**), (**t**), (**d**), (**P**), (**c**) and (**q**) are known, then its Variable Cost Planned is:

V= {**$**[1-**f**]-**I**}-({**Q**+**P**/[**c**-**q**]}/**/U**)/{[1-**t**][1-**d**]}

Rule-12775:

If both (**/**), (**U**), (**$**), (**f**), (**V**), (**Q**), (**t**), (**d**), (**P**), (**c**) and (**q**) are known, then its Interest Expense Planned is:

I= {**$**[1-**f**]-**V**}-({**Q**+**P**/[**c**-**q**]}/**/U**)/{[1-**t**][1-**d**]}

Rule-12776:

If both (**/**), (**U**), (**$**), (**f**), (**V**), (**I**), (**Q**), (**d**), (**P**), (**c**) and (**q**) are known, then its Tax Rate Planned is:

t = 1-({**Q**+**P**/[**c**-**q**]}/**/U**)/ ([1-**d**]{**$**[1-**f**]-**V**-**I**})

Rule-12777:

If both (**/**), (**U**), (**$**), (**f**), (**V**), (**I**), (**t**), (**Q**), (**P**), (**c**) and (**q**) are known, then its Dividend Payout Planned is:

d = 1-({**Q**+**P**/[**c**-**q**]}/**/U**)/ ([1-**t**]{**$**[1-**f**]-**V**-**I**})

Rule-12778:

If both (**/**), (**U**), (**$**), (**f**), (**V**), (**I**), (**t**), (**d**), (**Q**), (**c**) and (**q**) are known, then its Procured Inventory Planned is:

P= [**c**-**q**]**[/U**+{**$**[1-**f**]-**V**-**I**}[1-**t**][1-**d**])-**Q]**

Steve Asikin ISBN 14: 978-1514685136, ISBN 10: 1514685132

Rule-12779:

If both (**/**), (**U**), (**$**), (**f**), (**V**), (**I**), (**t**), (**d**), (**P**), (**Q**) and (**q**) are known, then its Current Ratio Planned is:

c= **q**+**P**/[**/**(**U**+{**$**[1-**f**]-**V**-**I**}[1-**t**][1-**d**])-**Q**]

Rule-12780:

If both (**/**), (**U**), (**$**), (**f**), (**V**), (**I**), (**t**), (**d**), (**P**), (**c**) and (**Q**) are known, then its Quick or Acid Test Ratio Planned is:

q= **c**-**P**/[**/**(**U**+{**$**[1-**f**]-**V**-**I**}[1-**t**][1-**d**])-**Q**]

Rule-12781:

If both (**/**), (**U**), (**$**), (**f**), (**V**), (**I**), (**t**), (**d**), (**p**), (**c**) and (**q**) are known, then its Quoted Longterm Debt Planned is:

Q= **/**(**U**+{**$**[1-**f**]-**V**-**I**}[1-**t**][1-**d**])-**Vp**/{360[**c**-**q**]}

Rule-12782:

If both (**Q**), (**U**), (**$**), (**f**), (**V**), (**I**), (**t**), (**d**), (**p**), (**c**) and (**q**) are known, then its Leverage or Gearing Ratio Planned is:

/= (**Q**+**Vp**/{360[**c**-**q**]})/(**U**+{**$**[1-**f**]-**V**-**I**}[1-**t**][1-**d**])

Rule-12783:

If both (**/**), (**Q**), (**$**), (**f**), (**V**), (**I**), (**t**), (**d**), (**p**), (**c**) and (**q**) are known, then its Utilized or Starting Capital must be:

U=(**Q**+**Vp**/{360[**c**-**q**]})/**/**-{**$**[1-**f**]-**V**-**I**}[1-**t**][1-**d**]

Steve Asikin ISBN 14: 978-1514685136, ISBN 10: 1514685132

Rule-12784:

If both (**/**), (**U**), (**Q**), (**f**), (**V**), (**I**), (**t**), (**d**), (**p**), (**c**) and (**q**) are known, then its Sales or Revenue Planned is:

S= {**V**+**I**+[(**Q**+**Vp**/{360[**c-q**]})/**/-U**] /{[1-**t**][1-**d**]}}/[1-**f**]

Rule-12785:

If both (**/**), (**U**), (**S**), (**Q**), (**V**), (**I**), (**t**), (**d**), (**p**), (**c**) and (**q**) are known, then its Fixed Portion Planned is:

f= 1-{**V**+**I**+[(**Q**+**Vp**/{360[**c-q**]})/**/-U**] /{[1-**t**][1-**d**]}}/**S**

Rule-12786:

If both (**/**), (**U**), (**S**), (**f**), (**Q**), (**I**), (**t**), (**d**), (**p**), (**c**) and (**q**) are known, then its Variable Cost Planned is:

V= [**/**(**U**+{**S**[1-**f**]-**I**}[1-**t**][1-**d**])-**Q**] /(**p**/{360[**c-q**]}+**/**[1-**t**][1-**d**])

Rule-12787:

If both (**/**), (**U**), (**S**), (**f**), (**V**), (**Q**), (**t**), (**d**), (**p**), (**c**) and (**q**) are known, then its Interest Expense Planned is:

I= {**S**[1-**f**]-**V**}-[(**Q**+**Vp**/{360[**c-q**]})/**/-U**] /{[1-**t**][1-**d**]}

Rule-12788:

If both (**/**), (**U**), (**S**), (**f**), (**V**), (**I**), (**Q**), (**d**), (**p**), (**c**) and (**q**) are known, then its Tax Rate Planned is:

t = 1-[(**Q**+**Vp**/{360[**c-q**]})/**/-U**]/([1-**d**]{**S**[1-**f**]-**V**-**I**})

Steve Asikin ISBN 14: 978-1514685136, ISBN 10: 1514685132

Rule-12789:

If both (/), (U), ($), (f), (V), (I), (t), (Q), (p), (c) and (q) are known, then its Dividend Payout Planned is:

d = 1-[(Q+Vp/{360[c-q]})//U]/([1-t]{$[1-f]-V-I})

Rule-12790:

If both (/), (U), ($), (f), (V), (I), (t), (d), (Q), (c) and (q) are known, then its Procured Inventory Days planned:

p= 360[c-q][/(U+{$[1-f]-V-I}[1-t][1-d])-Q]/V

Rule-12791:

If both (/), (U), ($), (f), (V), (I), (t), (d), (p), (Q) and (q) are known, then its Current Ratio Planned is:

c= q+Vp/{360[/(U+{$[1-f]-V-I}[1-t][1-d])-Q]}

Rule-12792:

If both (/), (U), ($), (f), (V), (I), (t), (d), (p), (c) and (Q) are known, then its Quick or Acid Test Ratio Planned is:

q= c-Vp/{360[/(U+{$[1-f]-V-I}[1-t][1-d])-Q]}

Rule-12793:

If both (/), (U), ($), (f), (V), (I), (t), (d), (v), (p), (c) and (q) are known, then its Quoted Longterm Debt Planned is:

Q= /(U+{$[1-f]-V-I}[1-t][1-d])-$vp/{360[c-q]}

Steve Asikin ISBN 14: 978-1514685136, ISBN 10: 1514685132

Rule-12794:

If both (**Q**), (**U**), (**$**), (**f**), (**V**), (**I**), (**t**), (**d**), (**v**), (**p**), (**c**) and (**q**) are known, then its Leverage or Gearing Ratio Planned is:

/= (**Q**+**$vp**/{360[**c**-**q**]})/(**U**+{**$**[1-**f**]-**V**-**I**}[1-**t**][1-**d**])

Rule-12795:

If both (**/**), (**Q**), (**$**), (**f**), (**V**), (**I**), (**t**), (**d**), (**v**), (**p**), (**c**) and (**q**) are known, then its Utilized or Starting Capital must be:

U=(**Q**+**$vp**/{360[**c**-**q**]})/**/**-{**$**[1-**f**]-**V**-**I**}[1-**t**][1-**d**]

Rule-12796:

If both (**/**), (**U**), (**Q**), (**f**), (**V**), (**I**), (**t**), (**d**), (**v**), (**p**), (**c**) and (**q**) are known, then its Sales or Revenue Planned is:

$= **[/U**-[**V**+**I**][1-**t**][1-**d**])-**Q]**
/(**vp**/{360[**c**-**q**]}-**/**[1-**f**][1-**t**][1-**d**])

Rule-12797:

If both (**/**), (**U**), (**$**), (**Q**), (**V**), (**I**), (**t**), (**d**), (**v**), (**p**), (**c**) and (**q**) are known, then its Fixed Portion Planned is:

f= 1-**{V**+**I**+**[**(**Q**+**$vp**/{360[**c**-**q**]})/**/**-**U]**
/{[1-**t**][1-**d**]}**}**/**$**

Rule-12798:

If both (**/**), (**U**), (**$**), (**f**), (**Q**), (**I**), (**t**), (**d**), (**v**), (**p**), (**c**) and (**q**) are known, then its Variable Cost Planned is:

V= {**$**[1-**f**]-**I**}-**[**(**Q**+**$vp**/{360[**c**-**q**]})/**/**-**U]**
/{[1-**t**][1-**d**]}

Steve Asikin ISBN 14: 978-1514685136, ISBN 10: 1514685132

Rule-12799:

If both (**/**), (**U**), (**$**), (**f**), (**V**), (**Q**), (**t**), (**d**), (**v**), (**p**), (**c**) and (**q**) are known, then its Interest Expense Planned is:

I= {**$**[1-**f**]-**V**}-[(**Q**+**$vp**/{360[**c**-**q**]})/**/**-**U**]
/{[1-**t**][1-**d**]}

Rule-12800:

If both (**/**), (**U**), (**$**), (**f**), (**V**), (**I**), (**Q**), (**d**), (**v**), (**p**), (**c**) and (**q**) are known, then its Tax Rate Planned is:

t = 1-[(**Q**+**$vp**/{360[**c**-**q**]})/**/**-**U**]
/([1-**d**]{**$**[1-**f**]-**V**-**I**})

Rule-12801:

If both (**/**), (**U**), (**$**), (**f**), (**V**), (**I**), (**t**), (**Q**), (**v**), (**p**), (**c**) and (**q**) are known, then its Dividend Payout Planned is:

d = 1-[(**Q**+**$vp**/{360[**c**-**q**]})/**/**-**U**]
/([1-**t**]{**$**[1-**f**]-**V**-**I**})

Rule-12802:

If both (**/**), (**U**), (**$**), (**f**), (**V**), (**I**), (**t**), (**d**), (**Q**), (**p**), (**c**) and (**q**) are known, then its Variable Portion Planned is:

v= **V**/**$**

or it can also be found as

v= 360[**c**-**q**][**/**(**U**+{**$**[1-**f**]-**V**-**I**}[1-**t**][1-**d**])-**Q**]/[**$p**]

Steve Asikin ISBN 14: 978-1514685136, ISBN 10: 1514685132

Rule-12803:

If both (**/**), (**U**), (**S**), (**f**), (**V**), (**I**), (**t**), (**d**), (**v**), (**Q**), (**c**) and (**q**) are known, then its Procured Inventory Planned is:

p= 360[**c**-**q**][**/**(**U**+{**S**[1-**f**]-**V**-**I**}[1-**t**][1-**d**])-**Q**]/[**Sv**]

Rule-12804:

If both (**/**), (**U**), (**S**), (**f**), (**V**), (**I**), (**t**), (**d**), (**v**), (**p**), (**Q**) and (**q**) are known, then its Current Ratio Planned is:

c= **q**+**Svp**/{360[**/**(**U**+{**S**[1-**f**]-**V**-**I**}[1-**t**][1-**d**])-**Q**]}

Rule-12805:

If both (**/**), (**U**), (**S**), (**f**), (**V**), (**I**), (**t**), (**d**), (**v**), (**p**), (**c**) and (**Q**) are known, then its Quick or Acid Test Ratio Planned is:

q= **c**-**Svp**/{360[**/**(**U**+{**S**[1-**f**]-**V**-**I**}[1-**t**][1-**d**])-**Q**]}

Rule-12806:

If both (**/**), (**U**), (**S**), (**f**), (**V**), (**I**), (**t**), (**d**), (**S'**) (**v**), (**p**), (**s**), (**c**) and (**q**) are known, then its Quoted Longterm Debt Planned is:

Q= **/**(**U**+{**S**[1-**f**]-**V**-**I**}[1-**t**][1-**d**])
-**S'vp**[1+**s**]/{360[**c**-**q**]}

Rule-12807:

If both (**Q**), (**U**), (**S**), (**f**), (**V**), (**I**), (**t**), (**d**), (**S'**) (**v**), (**p**), (**s**), (**c**) and (**q**) are known, then its Leverage or Gearing Ratio Planned is:

/= (**Q**+**S'vp**[1+**s**]/{360[**c**-**q**]})
/(**U**+{**S**[1-**f**]-**V**-**I**}[1-**t**][1-**d**])

Steve Asikin ISBN 14: 978-1514685136, ISBN 10: 1514685132

Rule-12808:

If both (**/**), (**Q**), (**$**), (**f**), (**V**), (**I**), (**t**), (**d**), (**$'**) (**v**), (**p**), (**s**), (**c**) and (**q**) are known, then its Utilized or Starting Capital must be:

U=(**Q**+**$'vp**[1+**s**]/{360[**c**-**q**]})/**/**
-{**$**[1-**f**]-**V**-**I**}[1-**t**][1-**d**]

Rule-12809:

If both (**/**), (**U**), (**Q**), (**f**), (**V**), (**I**), (**t**), (**d**), (**$'**) (**v**), (**p**), (**s**), (**c**) and (**q**) are known, then its Sales or Revenue Planned is:

$= **[Q**-**/**(**U**-[**V**+**I**][1-**t**][1-**d**])
+**$'vp**[1+**s**]/{360[**c**-**q**]}**]**/{**/**[1-**f**] [1-**t**][1-**d**]}

Rule-12810:

If both (**/**), (**U**), (**$**), (**Q**), (**V**), (**I**), (**t**), (**d**), (**$'**) (**v**), (**p**), (**s**), (**c**) and (**q**) are known, then its Fixed Portion Planned is:

f= 1-**{V**+**I**+**[**(**Q**+**$'vp**[1+**s**]/{360[**c**-**q**]})/**/**-**U]**
/{[1-**t**][1-**d**]}**}**/**$**

Rule-12811:

If both (**/**), (**U**), (**$**), (**f**), (**Q**), (**I**), (**t**), (**d**), (**$'**) (**v**), (**p**), (**s**), (**c**) and (**q**) are known, then its Variable Cost Planned is:

V= {**$**[1-**f**]-**I**}-**[**(**Q**+**$'vp**[1+**s**]/{360[**c**-**q**]})/**/**-**U]**
/{[1-**t**][1-**d**]}

Steve Asikin ISBN 14: 978-1514685136, ISBN 10: 1514685132

Rule-12812:

If both (**/**), (**U**), (**$**), (**f**), (**V**), (**Q**), (**t**), (**d**), (**$'**) (**v**), (**p**), (**s**), (**c**) and (**q**) are known, then its Interest Expense Planned is:

I= {**$**[1-**f**]-**V**}-[(**Q**+**$'vp**[1+**s**]/{360[**c-q**]})/**/-U**]
/{[1-**t**][1-**d**]}

Rule-12813:

If both (**/**), (**U**), (**$**), (**f**), (**V**), (**I**), (**Q**), (**d**), (**$'**) (**v**), (**p**), (**s**), (**c**) and (**q**) are known, then its Tax Rate Planned is:

t = 1-[(**Q**+**$'vp**[1+**s**]/{360[**c-q**]})/**/-U**]
/([1-**d**]{**$**[1-**f**]-**V-I**})

Rule-12814:

If both (**/**), (**U**), (**$**), (**f**), (**V**), (**I**), (**t**), (**Q**), (**$'**) (**v**), (**p**), (**s**), (**c**) and (**q**) are known, then its Dividend Payout Planned is:

d = 1-[(**Q**+**$'vp**[1+**s**]/{360[**c-q**]})/**/-U**]
/([1-**t**]{**$**[1-**f**]-**V-I**})

Rule-12815:

If both (**/**), (**U**), (**$**), (**f**), (**V**), (**I**), (**t**), (**d**), (**Q**) (**v**), (**p**), (**s**), (**c**) and (**q**) are known, then its Sales Past must be:

$'**= 360[**c-q**][**/U**+{**$[1-**f**]-**V-I**}[1-**t**][1-**d**])-**Q**]
/{**vp**[1+**s**]}

Steve Asikin ISBN 14: 978-1514685136, ISBN 10: 1514685132

Rule-12816:

If both (**/**), (**U**), (**S**), (**f**), (**V**), (**I**), (**t**), (**d**), (**S'**) (**Q**), (**p**), (**s**), (**c**) and (**q**) are known, then its Variable Portion Planned is:

s= **V**/**S**

or it can also be found as

v= 360[**c-q**][**/**(**U**+{**S**[1-**f**]-**V-I**}[1-**t**][1-**d**])-**Q**]
/{**S'p**[1+**s**]}

Rule-12817:

If both (**/**), (**U**), (**S**), (**f**), (**V**), (**I**), (**t**), (**d**), (**S'**) (**v**), (**Q**), (**s**), (**c**) and (**q**) are known, then its Procured Inventory Days planned:

p= 360[**c-q**][**/**(**U**+{**S**[1-**f**]-**V-I**}[1-**t**][1-**d**])-**Q**]
/{**S'v**[1+**s**]}

Rule-12818:

If both (**/**), (**U**), (**S**), (**f**), (**V**), (**I**), (**t**), (**d**), (**S'**) (**v**), (**p**), (**Q**), (**c**) and (**q**) are known, then its Sales Growth Planned is:

s= **S**/**S'**-1

or it can also be found as

s= 360[**c-q**][**/**(**U**+{**S**[1-**f**]-**V-I**}[1-**t**][1-**d**])-**Q**]
/[**S'vp**]-1

Rule-12819:

If both (**/**), (**U**), (**S**), (**f**), (**V**), (**I**), (**t**), (**d**), (**S'**) (**v**), (**p**), (**s**), (**Q**) and (**q**) are known, then its Current Ratio Planned is:

c= **q**+**S'vp**[1+**s**]
/{360[**/**(**U**+{**S**[1-**f**]-**V-I**}[1-**t**][1-**d**])-**Q**]}

Steve Asikin ISBN 14: 978-1514685136, ISBN 10: 1514685132

Rule-12820:

If both (**/**), (**U**), (**$**), (**f**), (**V**), (**I**), (**t**), (**d**), (**$'**) (**v**), (**p**), (**s**), (**c**) and (**Q**) are known, then its Quick or Acid Test Ratio Planned is:

q= **c**-**$'vp**[1+**s**]
/{360[**/**(**U**+{**$**[1-**f**]-**V**-**I**}[1-**t**][1-**d**])-**Q**]}

Rule-12821:

If both (**/**), (**U**), (**$**), (**f**), (**i**), (**V**), (**T**), (**D**) and (**X**) are known, then its Quoted Longterm Debt Planned is:

Q= **/**{**U**+**$**[1-**f**-**i**]-**V**-**T**-**D**}-**X**

Rule-12822:

If both (**Q**), (**U**), (**$**), (**f**), (**i**), (**V**), (**T**), (**D**) and (**X**) are known, then its Leverage or Gearing Ratio Planned is:

/= [**Q**+**X**]/{**U**+**$**[1-**f**-**i**]-**V**-**T**-**D**}

Rule-12823:

If both (**/**), (**Q**), (**$**), (**f**), (**i**), (**V**), (**T**), (**D**) and (**X**) are known, then its Utilized or Starting Capital must be:

U= [**Q**+**X**]/**/**-{**$**[1-**f**-**i**]-**V**-**T**-**D**}

Rule-12824:

If both (**/**), (**U**), (**Q**), (**f**), (**i**), (**V**), (**T**), (**D**) and (**X**) are known, then its Sales or Revenue Planned is:

$= {**V**+**T**+**D**+[**Q**+**X**]/**/**-**U**}/[1-**f**-**i**]

Rule-12825:

If both (**/**), (**U**), (**$**), (**Q**), (**i**), (**V**), (**T**), (**D**) and (**X**) are known, then its Fixed Portion Planned is:

f = 1-**i**-{**V**+**T**+**D**+[**Q**+**X**]/**/**-**U**}/**$**

Steve Asikin ISBN 14: 978-1514685136, ISBN 10: 1514685132

Rule-12826:

If both (**/**), (**U**), (**$**), (**f**), (**Q**), (**V**), (**T**), (**D**) and (**X**) are known, then its Interest Portion Planned is:

i = 1-**f**-{**V**+**T**+**D**+[**Q**+**X**]/**/**-**U**}/**$**

Rule-12827:

If both (**/**), (**U**), (**$**), (**f**), (**i**), (**Q**), (**T**), (**D**) and (**X**) are known, then its Variable Cost Planned is:

V= {**$**[1-**f**-**i**]-**T**-**D**}-[**Q**+**X**]/**/**-**U**}

Rule-12828:

If both (**/**), (**U**), (**$**), (**f**), (**i**), (**V**), (**Q**), (**D**) and (**X**) are known, then its Tax Planned is:

T= {**$**[1-**f**-**i**]-**V**-**D**}-[**Q**+**X**]/**/**-**U**}

Rule-12829:

If both (**/**), (**U**), (**$**), (**f**), (**i**), (**V**), (**T**), (**Q**) and (**X**) are known, then its Dividend Planned is:

D= {**$**[1-**f**-**i**]-**V**-**T**}-[**Q**+**X**]/**/**-**U**}

Rule-12830:

If both (**/**), (**U**), (**$**), (**f**), (**i**), (**V**), (**T**), (**D**) and (**Q**) are known, then its Xpress or Current Debt Planned is:

X= **/**{**U**+**$**[1-**f**-**i**]-**V**-**T**-**D**}-**Q**

Rule-12831:

If both (**/**), (**U**), (**$**), (**f**), (**i**), (**V**), (**T**), (**D**), (**P**), (**c**) and (**q**) are known, then its Quoted Longterm Debt Planned is:

Q= **/**{**U**+**$**[1-**f**-**i**]-**V**-**T**-**D**}-**P**/[**c**-**q**]

Steve Asikin ISBN 14: 978-1514685136, ISBN 10: 1514685132

Rule-12832:

If both (**Q**), (**U**), (**$**), (**f**), (**i**), (**V**), (**T**), (**D**), (**P**), (**c**) and (**q**) are known, then its Leverage or Gearing Ratio Planned is:

/= {**Q**+**P**/[**c**-**q**]}/{**U**+**$**[1-**f**-**i**]-**V**-**T**-**D**}

Rule-12833:

If both (**/**), (**Q**), (**$**), (**f**), (**i**), (**V**), (**T**), (**D**), (**P**), (**c**) and (**q**) are known, then its Utilized or Starting Capital must be:

U= {**Q**+**P**/[**c**-**q**]}/**/**-{**$**[1-**f**-**i**]-**V**-**T**-**D**}

Rule-12834:

If both (**/**), (**U**), (**Q**), (**f**), (**i**), (**V**), (**T**), (**D**), (**P**), (**c**) and (**q**) are known, then its Sales or Revenue Planned is:

$**= (**V**+**T**+**D**+{**Q**+**P**/[**c**-**q**]}/**/**-**U**)/**$ [1-**f**-**i**]

Rule-12835:

If both (**/**), (**U**), (**$**), (**Q**), (**i**), (**V**), (**T**), (**D**), (**P**), (**c**) and (**q**) are known, then its Fixed Portion Planned is:

f = 1-**i**-(**V**+**T**+**D**+{**Q**+**P**/[**c**-**q**]}/**/**-**U**)/**$**

Rule-12836:

If both (**/**), (**U**), (**$**), (**f**), (**Q**), (**V**), (**T**), (**D**), (**P**), (**c**) and (**q**) are known, then its Interest Portion Planned is:

i = 1-**f**-(**V**+**T**+**D**+{**Q**+**P**/[**c**-**q**]}/**/**-**U**)/**$**

Rule-12837:

If both (**/**), (**U**), (**$**), (**f**), (**i**), (**Q**), (**T**), (**D**), (**P**), (**c**) and (**q**) are known, then its Variable Cost Planned is:

V= {**$**[1-**f**-**i**]-**T**-**D**}-({**Q**+**P**/[**c**-**q**]}/**/**-**U**)

Steve Asikin ISBN 14: 978-1514685136, ISBN 10: 1514685132

Rule-12838:

If both (**/**), (**U**), (**$**), (**f**), (**i**), (**V**), (**Q**), (**D**), (**P**), (**c**) and (**q**) are known, then its Tax Planned is:

T= {**$**[1-**f**-**i**]-**V**-**D**}-({**Q**+**P**/[**c**-**q**]}/**/**-**U**)

Rule-12839:

If both (**/**), (**U**), (**$**), (**f**), (**i**), (**V**), (**T**), (**Q**), (**P**), (**c**) and (**q**) are known, then its Dividend Planned is:

D= {**$**[1-**f**-**i**]-**V**-**T**}-({**Q**+**P**/[**c**-**q**]}/**/**-**U**)

Rule-12840:

If both (**/**), (**U**), (**$**), (**f**), (**i**), (**V**), (**T**), (**D**), (**Q**), (**c**) and (**q**) are known, then its Procured Inventory Planned is:

P= [**c**-**q**](**/**{**U**+**$**[1-**f**-**i**]-**V**-**T**-**D**}-**Q**)

Rule-12841:

If both (**/**), (**U**), (**$**), (**f**), (**i**), (**V**), (**T**), (**D**), (**P**), (**Q**) and (**q**) are known, then its Current Ratio Planned is:

c= **q**+**P**/(**/**{**U**+**$**[1-**f**-**i**]-**V**-**T**-**D**}-**Q**)

Rule-12842:

If both (**/**), (**U**), (**$**), (**f**), (**i**), (**V**), (**T**), (**D**), (**P**), (**c**) and (**Q**) are known, then its Quick or Acid Test Ratio Planned is:

q= **c**-**P**/(**/**{**U**+**$**[1-**f**-**i**]-**V**-**T**-**D**}-**Q**)

Rule-12843:

If both (**/**), (**U**), (**$**), (**f**), (**i**), (**V**), (**T**), (**D**), (**p**), (**c**) and (**q**) are known, then its Quoted Longterm Debt Planned is:

Q= **/**{**U**+**$**[1-**f**-**i**]-**V**-**T**-**D**}-**Vp**/{360[**c**-**q**]}

Steve Asikin ISBN 14: 978-1514685136, ISBN 10: 1514685132

Rule-12844:

If both (**Q**), (**U**), (**$**), (**f**), (**i**), (**V**), (**T**), (**D**), (**p**), (**c**) and (**q**) are known, then its Leverage or Gearing Ratio Planned is:

/= (**Q**+**Vp**/{360[**c**-**q**]})/{**U**+**$**[1-**f**-**i**]-**V**-**T**-**D**}

Rule-12845:

If both (**/**), (**Q**), (**$**), (**f**), (**i**), (**V**), (**T**), (**D**), (**p**), (**c**) and (**q**) are known, then its Utilized or Starting Capital must be:

U= (**Q**+**Vp**/{360[**c**-**q**]})/**/**-{**$**[1-**f**-**i**]-**V**-**T**-**D**}

Rule-12846:

If both (**/**), (**U**), (**Q**), (**f**), (**i**), (**V**), (**T**), (**D**), (**p**), (**c**) and (**q**) are known, then its Sales or Revenue Planned is:

$**= [**V**+**T**+**D**+(**Q**+**Vp**/{360[**c**-**q**]})/**/**-**U**]/**$ [1-**f**-**i**]

Rule-12847:

If both (**/**), (**U**), (**$**), (**Q**), (**i**), (**V**), (**T**), (**D**), (**p**), (**c**) and (**q**) are known, then its Fixed Portion Planned is:

f = 1-**i**-[**V**+**T**+**D**+{**Q**+**P**/[**c**-**q**]}/**/**-**U**]/**$**

Rule-12848:

If both (**/**), (**U**), (**$**), (**f**), (**Q**), (**V**), (**T**), (**D**), (**p**), (**c**) and (**q**) are known, then its Interest Portion Planned is:

i = 1-**f**-[**V**+**T**+**D**+(**Q**+**Vp**/{360[**c**-**q**]})/**/**-**U**]/**$**

Rule-12849:

If both (**/**), (**U**), (**$**), (**f**), (**i**), (**Q**), (**T**), (**D**), (**p**), (**c**) and (**q**) are known, then its Variable Cost Planned is:

V= (**/**{**U**+**$**[1-**f**-**i**]-**T**-**D**}-**Q**)/(**p**/{360[**c**-**q**]}+**/**)

Steve Asikin ISBN 14: 978-1514685136, ISBN 10: 1514685132

Rule-12850:

If both (/), (U), ($), (f), (i), (V), (Q), (D), (p), (c) and (q) are known, then its Tax Planned is:

T= [$[1-f-i]-V-D}-(Q+Vp/{360[c-q]})/ /-U]

Rule-12851:

If both (/), (U), ($), (f), (i), (V), (T), (Q), (p), (c) and (q) are known, then its Dividend Planned is:

D= [$[1-f-i]-V-T}-(Q+Vp/{360[c-q]})/ /-U]

Rule-12852:

If both (/), (U), ($), (f), (i), (V), (T), (D), (Q), (c) and (q) are known, then its Procured Inventory Days planned:

p= 360[c-q](/{U+$[1-f-i]-V-T-D}-Q)/V

Rule-12853:

If both (/), (U), ($), (f), (i), (V), (T), (D), (p), (Q) and (q) are known, then its Current Ratio Planned is:

c= q+Vp/[360(/{U+$[1-f-i]-V-T-D}-Q)]

Rule-12854:

If both (/), (U), ($), (f), (i), (V), (T), (D), (p), (c) and (Q) are known, then its Quick or Acid Test Ratio Planned is:

q= c-Vp/[360(/{U+$[1-f-i]-V-T-D}-Q)]

Steve Asikin ISBN 14: 978-1514685136, ISBN 10: 1514685132

Rule-12855:

If both (**/**), (**U**), (**$**), (**f**), (**i**), (**V**), (**T**), (**D**), (**v**), (**p**), (**c**) and (**q**) are known, then its Quoted Longterm Debt Planned is:

Q= **/**{**U**+**$**[1-**f**-**i**]-**V**-**T**-**D**}-**$vp**/{360[**c**-**q**]}

Rule-12856:

If both (**Q**), (**U**), (**$**), (**f**), (**i**), (**V**), (**T**), (**D**), (**v**), (**p**), (**c**) and (**q**) are known, then its Leverage or Gearing Ratio Planned is:

/= (**Q**+**$vp**/{360[**c**-**q**]})/{**U**+**$**[1-**f**-**i**]-**V**-**T**-**D**}

Rule-12857:

If both (**/**), (**Q**), (**$**), (**f**), (**i**), (**V**), (**T**), (**D**), (**v**), (**p**), (**c**) and (**q**) are known, then its Utilized or Starting Capital must be:

U= (**Q**+**$vp**/{360[**c**-**q**]})/**/**-{**$**[1-**f**-**i**]-**V**-**T**-**D**}

Rule-12858:

If both (**/**), (**U**), (**Q**), (**f**), (**i**), (**V**), (**T**), (**D**), (**v**), (**p**), (**c**) and (**q**) are known, then its Sales or Revenue Planned is:

$= (**/**{**U**-**V**-**T**-**D**}-**Q**)/(**vp**/{360[**c**-**q**]}-**/**[1-**f**-**i**])

Rule-12859:

If both (**/**), (**U**), (**$**), (**Q**), (**i**), (**V**), (**T**), (**D**), (**v**), (**p**), (**c**) and (**q**) are known, then its Fixed Portion Planned is:

f = 1-**i**-[**V**+**T**+**D**+(**Q**+**$vp**/{360[**c**-**q**]})/**/**-**U**]/**$**

Steve Asikin ISBN 14: 978-1514685136, ISBN 10: 1514685132

Rule-12860:

If both (**/**), (**U**), (**$**), (**f**), (**Q**), (**V**), (**T**), (**D**), (**v**), (**p**), (**c**) and (**q**) are known, then its Interest Portion Planned is:

i = 1-**f**-[**V**+**T**+**D**+(**Q**+**$vp**/{360[**c**-**q**]})/**/-U**]/**$**

Rule-12861:

If both (**/**), (**U**), (**$**), (**f**), (**i**), (**Q**), (**T**), (**D**), (**v**), (**p**), (**c**) and (**q**) are known, then its Variable Cost Planned is:

V= [**$**[1-**f**-**i**]-**T**-**D**}-(**Q**+**$vp**/{360[**c**-**q**]})/**/-U**]

Rule-12862:

If both (**/**), (**U**), (**$**), (**f**), (**i**), (**V**), (**Q**), (**D**), (**v**), (**p**), (**c**) and (**q**) are known, then its Tax Planned is:

T= [**$**[1-**f**-**i**]-**V**-**D**}-(**Q**+**$vp**/{360[**c**-**q**]})/**/-U**]

Rule-12863:

If both (**/**), (**U**), (**$**), (**f**), (**i**), (**V**), (**T**), (**Q**), (**v**), (**p**), (**c**) and (**q**) are known, then its Dividend Planned is:

D= [**$**[1-**f**-**i**]-**V**-**T**}-(**Q**+**$vp**/{360[**c**-**q**]})/**/-U**]

Rule-12864:

If both (**/**), (**U**), (**$**), (**f**), (**i**), (**V**), (**T**), (**D**), (**Q**), (**p**), (**c**) and (**q**) are known, then its Variable Portion Planned is:

v= **V**/**$**

or it can also be found as

v= 360[**c**-**q**](**/**{**U**+**$**[1-**f**-**i**]-**V**-**T**-**D**}-**Q**)/[**$p**]

Steve Asikin ISBN 14: 978-1514685136, ISBN 10: 1514685132

Rule-12865:

If both (**/**), (**U**), (**$**), (**f**), (**i**), (**V**), (**T**), (**D**), (**v**), (**Q**), (**c**) and (**q**) are known, then its Procured Inventory Days Planned:

p= 360[**c-q**](**/**{**U**+**$**[1-**f-i**]-**V-T-D**}-**Q**)/[**$v**]

Rule-12866:

If both (**/**), (**U**), (**$**), (**f**), (**i**), (**V**), (**T**), (**D**), (**v**), (**p**), (**Q**) and (**q**) are known, then its Current Ratio Planned is:

c= **q**+**$vp**/[360(**/**{**U**+**$**[1-**f-i**]-**V-T-D**}-**Q**)]

Rule-12867:

If both (**/**), (**U**), (**$**), (**f**), (**i**), (**V**), (**T**), (**D**), (**v**), (**p**), (**c**) and (**Q**) are known, then its Quick or Acid Test Ratio Planned is:

q= **c-$vp**/[360(**/**{**U**+**$**[1-**f-i**]-**V-T-D**}-**Q**)]

Rule-12868:

If both (**/**), (**U**), (**$**), (**f**), (**i**), (**V**), (**T**), (**D**), (**$'**), (**v**), (**p**), (**s**), (**c**) and (**q**) are known, then its Quoted Longterm Debt Planned is:

Q= **/**{**U**+**$**[1-**f-i**]-**V-T-D**}-**$'vp**[1+**s**]/{360[**c-q**]}

Rule-12869:

If both (**Q**), (**U**), (**$**), (**f**), (**i**), (**V**), (**T**), (**D**), (**$'**), (**v**), (**p**), (**s**), (**c**) and (**q**) are known, then its Leverage or Gearing Ratio Planned is:

/= (**Q**+**$'vp**[1+**s**]/{360[**c-q**]})/{**U**+**$**[1-**f-i**]-**V-T-D**}

Steve Asikin ISBN 14: 978-1514685136, ISBN 10: 1514685132

Rule-12870:

If both (/), (Q), ($), (f), (i), (V), (T), (D), ($'), (v), (p), (s), (c) and (q) are known, then its Utilized or Starting capital must be:

U= (Q+$'vp[1+s]/{360[c-q]})//{$[1-f-i]-V-T-D}

Rule-12871:

If both (/), (U), (Q), (f), (i), (V), (T), (D), ($'), (v), (p), (s), (c) and (q) are known, then its Sales or Revenue Planned is:

$= (Q-/{U-V-T-D}+$'vp[1+s]/{360[c-q]}/(/[1-f-i])

Rule-12872:

If both (/), (U), ($), (f), (i), (V), (T), (D), ($'), (v), (p), (s), (c) and (q) are known, then its Fixed Portion Planned is:

f = 1-i-[V+T+D+(Q+$'vp[1+s]/{360[c-q]})//U]/$

Rule-12873:

If both (/), (U), ($), (f), (Q), (V), (T), (D), ($'), (v), (p), (s), (c) and (q) are known, then its Interest Portion Planned is:

i = 1-f-[V+T+D+(Q+$'vp[1+s]/{360[c-q]})//U]/$

Rule-12874:

If both (/), (U), ($), (f), (i), (Q), (T), (D), ($'), (v), (p), (s), (c) and (q) are known, then its Variable Cost Planned is:

V= [$[1-f-i]-T-D}-(Q+$'vp[1+s]/{360[c-q]})//U]

Steve Asikin ISBN 14: 978-1514685136, ISBN 10: 1514685132

Rule-12875:

If both (/), (U), ($), (f), (i), (V), (Q), (D), ($'), (v), (p), (s), (c) and (q) are known, then its Tax Planned is:

T= [$[1-f-i]-V-D}-(Q+$'vp[1+s]/{360[c-q]})/ /U]

Rule-12876:

If both (/), (U), ($), (f), (i), (V), (T), (Q), ($'), (v), (p), (s), (c) and (q) are known, then its Dividend Planned is:

D= [$[1-f-i]-V-T}-(Q+$'vp[1+s]/{360[c-q]})/ /U]

Rule-12877:

If both (/), (U), ($), (f), (i), (V), (T), (D), (Q), (v), (p), (s), (c) and (q) are known, then its Sales Past must be:

$'= 360[c-q](/{U+$[1-f-i]-V-T-D}-Q)/{vp[1+s]}

Rule-12878:

If both (/), (U), ($), (f), (i), (V), (T), (D), ($'), (Q), (p), (s), (c) and (q) are known, then its Variable Portion Planned is:

v= V/$

or it can also be found as

v= 360[c-q](/{U+$[1-f-i]-V-T-D}-Q)/{$'p[1+s]}

Rule-12879:

If both (/), (U), ($), (f), (i), (V), (T), (D), ($'), (v), (Q), (s), (c) and (q) are known, then its Procured Inventory Days planned:

p= 360[c-q](/{U+$[1-f-i]-V-T-D}-Q)/{$'v[1+s]}

Steve Asikin ISBN 14: 978-1514685136, ISBN 10: 1514685132

Rule-12880:

If both (**/**), (**U**), (**$**), (**f**), (**i**), (**V**), (**T**), (**D**), (**$'**), (**v**), (**p**), (**Q**), (**c**) and (**q**) are known, then its Sales Growth Planned is:

v= **$**/**$'**-1

or it can also be found as

s= 360[**c-q**](**/**{**U**+**$**[1-**f-i**]-**V-T-D**}-**Q**)/[**$'vp**]-1

Rule-12881:

If both (**/**), (**U**), (**$**), (**f**), (**i**), (**V**), (**T**), (**D**), (**$'**), (**v**), (**p**), (**s**), (**Q**) and (**q**) are known, then its Current Ratio Planned is:

c= **q**+**$'vp**[1+**s**]/[360(**/**{**U**+**$**[1-**f-i**]-**V-T-D**}-**Q**)]

Rule-12882:

If both (**/**), (**U**), (**$**), (**f**), (**i**), (**V**), (**T**), (**D**), (**$'**), (**v**), (**p**), (**s**), (**c**) and (**Q**) are known, then its Quick or Acid Test Ratio Planned is:

q= **c-$'vp**[1+**s**]/[360(**/**{**U**+**$**[1-**f-i**]-**V-T-D**}-**Q**)]

Rule-12883:

If both (**/**), (**U**), (**$**), (**f**), (**i**), (**V**), (**T**), (**d**) and (**X**) are known, then its Quoted Longterm Debt Planned is:

Q= **/**{**U**+**$**[1-**f-i**]-**V-T**}[1-**d**]-**X**

Rule-12884:

If both (**Q**), (**U**), (**$**), (**f**), (**i**), (**V**), (**T**), (**d**) and (**X**) are known, then its Leverage or Gearing Ratio Planned is:

/= [**Q**+**X**]/{**U**+**$**[1-**f-i**]-**V-T**}[1-**d**]

Steve Asikin ISBN 14: 978-1514685136, ISBN 10: 1514685132

Rule-12885:

If both (**/**), (**Q**), (**$**), (**f**), (**i**), (**V**), (**T**), (**d**) and (**X**) are known, then its Utilized or Starting Capital must be:

U= [**Q**+**X**]/**/**-{**$**[1-**f**-**i**]-**V**-**T**}[1-**d**]

Rule-12886:

If both (**/**), (**U**), (**Q**), (**f**), (**i**), (**V**), (**T**), (**d**) and (**X**) are known, then its Sales or Revenue Planned is:

$= (**V**+**T**+{[**Q**+**X**]/**/**-**U**}/[1-**d**])/[1-**f**-**i**]

Rule-12887:

If both (**/**), (**U**), (**$**), (**Q**), (**i**), (**V**), (**T**), (**d**) and (**X**) are known, then its Fixed Portion Planned is:

f= 1-**i**-(**V**+**T**+{[**Q**+**X**]/**/**-**U**}/[1-**d**])/**$**

Rule-12888:

If both (**/**), (**U**), (**$**), (**f**), (**Q**), (**V**), (**T**), (**d**) and (**X**) are known, then its Interest Portion Planned is:

i= 1-**f**-(**V**+**T**+{[**Q**+**X**]/**/**-**U**}/[1-**d**])/**$**

Rule-12889:

If both (**/**), (**U**), (**$**), (**f**), (**i**), (**Q**), (**T**), (**d**) and (**X**) are known, then its Variable Cost Planned is:

V= **$**[1-**f**-**i**]-**T**-{[**Q**+**X**]/**/**-**U**}/[1-**d**]

Rule-12890:

If both (**/**), (**U**), (**$**), (**f**), (**i**), (**V**), (**Q**), (**d**) and (**X**) are known, then its Tax Planned is:

T= **$**[1-**f**-**i**]-**V**-{[**Q**+**X**]/**/**-**U**}/[1-**d**]

Steve Asikin ISBN 14: 978-1514685136, ISBN 10: 1514685132

Rule-12891:

If both (**/**), (**U**), (**S**), (**f**), (**i**), (**V**), (**T**), (**Q**) and (**X**) are known, then its Dividend Payout Planned is:

d= 1-{[**Q**+**X**]/**/**-**U**}/{**S**[1-**f**-**i**]-**V**-**T**}

Rule-12892:

If both (**/**), (**U**), (**S**), (**f**), (**i**), (**V**), (**T**), (**d**) and (**Q**) are known, then its Xpress or Current Debt Planned is:

X= **/**{**U**+**S**[1-**f**-**i**]-**V**-**T**}[1-**d**]-**Q**

Rule-12893:

If both (**/**), (**U**), (**S**), (**f**), (**i**), (**V**), (**T**), (**d**), (**P**), (**c**) and (**q**) are known, then its Quoted Longterm Debt Planned is:

Q= **/**{**U**+**S**[1-**f**-**i**]-**V**-**T**}[1-**d**]-**P**/[**c**-**q**]

Rule-12894:

If both (**Q**), (**U**), (**S**), (**f**), (**i**), (**V**), (**T**), (**d**), (**P**), (**c**) and (**q**) are known, then its Leverage or Gearing Ratio Planned is:

/= {**Q**+**P**/[**c**-**q**]}/{**U**+**S**[1-**f**-**i**]-**V**-**T**}[1-**d**]

Rule-12895:

If both (**/**), (**Q**), (**S**), (**f**), (**i**), (**V**), (**T**), (**d**), (**P**), (**c**) and (**q**) are known, then its Utilized or Starting Capital must be:

U= {**Q**+**P**/[**c**-**q**]}/**/**-{**S**[1-**f**-**i**]-**V**-**T**}[1-**d**]

Steve Asikin ISBN 14: 978-1514685136, ISBN 10: 1514685132

Rule-12896:

If both (**/**), (**U**), (**Q**), (**f**), (**i**), (**V**), (**T**), (**d**), (**P**), (**c**) and (**q**) are known, then its Sales or Revenue Planned is:

$= [V+T+({Q+P/[c-q]}/**/**U)/[1-d]]/[1-f-i]

Rule-12897:

If both (**/**), (**U**), (**$**), (**Q**), (**i**), (**V**), (**T**), (**d**), (**P**), (**c**) and (**q**) are known, then its Fixed Portion Planned is:

f= 1-i-[V+T+({Q+P/[c-q]}/**/**U)/[1-d]]/$

Rule-12898:

If both (**/**), (**U**), (**$**), (**f**), (**Q**), (**V**), (**T**), (**d**), (**P**), (**c**) and (**q**) are known, then its Interest Portion Planned is:

i= 1-f-[V+T+({Q+P/[c-q]}/**/**U)/[1-d]]/$

Rule-12899:

If both (**/**), (**U**), (**$**), (**f**), (**i**), (**Q**), (**T**), (**d**), (**P**), (**c**) and (**q**) are known, then its Variable Cost Planned is:

V= $[1-f-i]-T-({Q+P/[c-q]}/**/**U)/[1-d]

Rule-12900:

If both (**/**), (**U**), (**$**), (**f**), (**i**), (**V**), (**Q**), (**d**), (**P**), (**c**) and (**q**) are known, then its Tax Planned is:

T= $[1-f-i]-V-({Q+P/[c-q]}/**/**U)/[1-d]

Rule-12901:

If both (**/**), (**U**), (**$**), (**f**), (**i**), (**V**), (**T**), (**Q**), (**P**), (**c**) and (**q**) are known, then its Dividend Payout Planned is:

d= 1-({Q+P/[c-q]}/**/**U)/{$[1-f-i]-V-T}

Steve Asikin ISBN 14: 978-1514685136, ISBN 10: 1514685132

Rule-12902:

If both (**/**), (**U**), (**$**), (**f**), (**i**), (**V**), (**T**), (**d**), (**Q**), (**c**) and (**q**) are known, then its Procured Inventory Planned is:

P= [**c**-**q**](**/**{**U**+**$**[1-**f**-**i**]-**V**-**T**}[1-**d**]-**Q**)

Rule-12903:

If both (**/**), (**U**), (**$**), (**f**), (**i**), (**V**), (**T**), (**d**), (**P**), (**Q**) and (**q**) are known, then its Current Ratio Planned is:

c= **q**+**P**/(**/**{**U**+**$**[1-**f**-**i**]-**V**-**T**}[1-**d**]-**Q**)

Rule-12904:

If both (**/**), (**U**), (**$**), (**f**), (**i**), (**V**), (**T**), (**d**), (**P**), (**c**) and (**Q**) are known, then its Quick or Acid Test Ratio Planned is:

q= **c**-**P**/(**/**{**U**+**$**[1-**f**-**i**]-**V**-**T**}[1-**d**]-**Q**)

Rule-12905:

If both (**/**), (**U**), (**$**), (**f**), (**i**), (**V**), (**T**), (**d**), (**p**), (**c**) and (**q**) are known, then its Quoted Longterm Debt Planned is:

Q= **/**{**U**+**$**[1-**f**-**i**]-**V**-**T**}[1-**d**]-**Vp**/{360[**c**-**q**]}

Rule-12906:

If both (**Q**), (**U**), (**$**), (**f**), (**i**), (**V**), (**T**), (**d**), (**p**), (**c**) and (**q**) are known, then its Leverage or Gearing Ratio Planned is:

/= (**Q**+**Vp**/{360[**c**-**q**]})/{**U**+**$**[1-**f**-**i**]-**V**-**T**}[1-**d**]

Steve Asikin ISBN 14: 978-1514685136, ISBN 10: 1514685132

Rule-12907:

If both (**/**), (**Q**), (**$**), (**f**), (**i**), (**V**), (**T**), (**d**), (**p**), (**c**) and (**q**) are known, then its Utilized or Starting Capital must be:

U= (Q+Vp/{360[c-q]})/ / {$[1-f-i]-V-T}[1-d]

Rule-12908:

If both (**/**), (**U**), (**Q**), (**f**), (**i**), (**V**), (**T**), (**d**), (**p**), (**c**) and (**q**) are known, then its Sales or Revenue Planned is:

$= {V+T+[(Q+Vp/{360[c-q]})/ / U]/[1-d]}/[1-f-i]

Rule-12909:

If both (**/**), (**U**), (**$**), (**Q**), (**i**), (**V**), (**T**), (**d**), (**p**), (**c**) and (**q**) are known, then its Fixed Portion Planned is:

f= 1-i-{V+T+[(Q+Vp/{360[c-q]})/ / U]/[1-d]}/$

Rule-12910:

If both (**/**), (**U**), (**$**), (**f**), (**Q**), (**V**), (**T**), (**d**), (**p**), (**c**) and (**q**) are known, then its Interest Portion Planned is:

i= 1-f-{V+T+[(Q+Vp/{360[c-q]})/ / U]/[1-d]}/$

Rule-12911:

If both (**/**), (**U**), (**$**), (**f**), (**i**), (**Q**), (**T**), (**d**), (**p**), (**c**) and (**q**) are known, then its Variable Cost Planned is:

V= (/{U+$[1-f-i]-T}[1-d]-Q)
/(p/{360[c-q]}+/[1-d])

Rule-12912:

If both (**/**), (**U**), (**$**), (**f**), (**i**), (**V**), (**Q**), (**d**), (**p**), (**c**) and (**q**) are known, then its Tax Planned is:

T= $[1-f-i]-V-[(Q+Vp/{360[c-q]})/ / U]/[1-d]

Steve Asikin ISBN 14: 978-1514685136, ISBN 10: 1514685132

Rule-12913:

If both (**/**), (**U**), (**$**), (**f**), (**i**), (**V**), (**T**), (**Q**), (**p**), (**c**) and (**q**) are known, then its Dividend Payout Planned is:

d= 1-[(**Q**+**Vp**/{360[**c**-**q**]})/**/**-**U**]/{**$**[1-**f**-**i**]-**V**-**T**}

Rule-12914:

If both (**/**), (**U**), (**$**), (**f**), (**i**), (**V**), (**T**), (**d**), (**Q**), (**c**) and (**q**) are known, then its Procured Inventory Days Planned:

p= 360[**c**-**q**](**/**{**U**+**$**[1-**f**-**i**]-**V**-**T**}[1-**d**]-**Q**)/**V**

Rule-12915:

If both (**/**), (**U**), (**$**), (**f**), (**i**), (**V**), (**T**), (**d**), (**p**), (**Q**) and (**q**) are known, then its Current Ratio Planned is:

c= **q**+**Vp**/[360(**/**{**U**+**$**[1-**f**-**i**]-**V**-**T**}[1-**d**]-**Q**)]

Rule-12916:

If both (**/**), (**U**), (**$**), (**f**), (**i**), (**V**), (**T**), (**d**), (**p**), (**c**) and (**Q**) are known, then its Quick or Acid Test Ratio Planned is:

q= **c**-**Vp**/[360(**/**{**U**+**$**[1-**f**-**i**]-**V**-**T**}[1-**d**]-**Q**)]

Rule-12917:

If both (**/**), (**U**), (**$**), (**f**), (**i**), (**V**), (**T**), (**d**), (**v**), (**p**), (**c**) and (**q**) are known, then its Quoted Longterm Debt Planned is:

Q= **/**{**U**+**$**[1-**f**-**i**]-**V**-**T**}[1-**d**]-**$vp**/{360[**c**-**q**]}

Steve Asikin ISBN 14: 978-1514685136, ISBN 10: 1514685132

Rule-12918:

If both (**Q**), (**U**), (**$**), (**f**), (**i**), (**V**), (**T**), (**d**), (**v**), (**p**), (**c**) and (**q**) are known, then its Leverage or Gearing Ratio Planned is:

/= (**Q**+**$vp**/{360[**c**-**q**]})/{**U**+**$**[1-**f**-**i**]-**V**-**T**}[1-**d**]

Rule-12919:

If both (**/**), (**Q**), (**$**), (**f**), (**i**), (**V**), (**T**), (**d**), (**v**), (**p**), (**c**) and (**q**) are known, then its Utilized or Starting Capital must be:

U= (**Q**+**$vp**/{360[**c**-**q**]})/**/**-{**$**[1-**f**-**i**]-**V**-**T**}[1-**d**]

Rule-12920:

If both (**/**), (**U**), (**q**), (**f**), (**i**), (**V**), (**T**), (**d**), (**v**), (**p**), (**c**) and (**q**) are known, then its Sales or Revenue Planned is:

$= (**/**{**U**-**V**-**T**}[1-**d**]-**Q**)
/(**vp**/{360[**c**-**q**]}-**/**[1-**f**-**i**][1-**d**])

Rule-12921:

If both (**/**), (**U**), (**$**), (**Q**), (**i**), (**V**), (**T**), (**d**), (**v**), (**p**), (**c**) and (**q**) are known, then its Fixed Portion Planned is:

f= 1-**i**-{**V**+**T**+[(**Q**+**$vp**/{360[**c**-**q**]})/**/**-**U**]/[1-**d**]}/**$**

Rule-12922:

If both (**/**), (**U**), (**$**), (**f**), (**Q**), (**V**), (**T**), (**d**), (**v**), (**p**), (**c**) and (**q**) are known, then its Interest Portion Planned is:

i= 1-**f**-{**V**+**T**+[(**Q**+**$vp**/{360[**c**-**q**]})/**/**-**U**]/[1-**d**]}/**$**

Steve Asikin ISBN 14: 978-1514685136, ISBN 10: 1514685132

Rule-12923:

If both (**/**), (**U**), (**$**), (**f**), (**i**), (**Q**), (**T**), (**d**), (**v**), (**p**), (**c**) and (**q**) are known, then its Variable Cost Planned is:

V= $[1-f-i]-Y-[(Q+$vp/{360[c-q]})/⁄U]/[1-d]

Rule-12924:

If both (**/**), (**U**), (**$**), (**f**), (**i**), (**V**), (**Q**), (**d**), (**v**), (**p**), (**c**) and (**q**) are known, then its Tax Planned is:

T= $[1-f-i]-V-[(Q+$vp/{360[c-q]})/⁄U]/[1-d]

Rule-12925:

If both (**/**), (**U**), (**$**), (**f**), (**i**), (**V**), (**T**), (**Q**), (**v**), (**p**), (**c**) and (**q**) are known, then its Dividend Payout Planned is:

d= 1-[(Q+$vp/{360[c-q]})/⁄U]/{$[1-f-i]-V-T}

Rule-12926:

If both (**/**), (**U**), (**$**), (**f**), (**i**), (**V**), (**T**), (**d**), (**Q**), (**p**), (**c**) and (**q**) are known, then its Variable Portion Planned is:

v= V/$

or it can also be found as

v= 360[c-q](/{U+$[1-f-i]-V-T}[1-d]-Q)/[$p]

Rule-12927:

If both (**/**), (**U**), (**$**), (**f**), (**i**), (**V**), (**T**), (**d**), (**v**), (**Q**), (**c**) and (**q**) are known, then its Procured Inventory Days planned:

p= 360[c-q](/{U+$[1-f-i]-V-T}[1-d]-Q)/[$v]

Steve Asikin ISBN 14: 978-1514685136, ISBN 10: 1514685132

Rule-12928:

If both (**/**), (**U**), (**$**), (**f**), (**i**), (**V**), (**T**), (**d**), (**v**), (**p**), (**Q**) and (**q**) are known, then its Current Ratio Planned is:

c= **q**+**$vp**/[360(**/**{**U**+**$**[1-**f**-**i**]-**V**-**T**}[1-**d**]-**Q**)]

Rule-12929:

If both (**/**), (**U**), (**$**), (**f**), (**i**), (**V**), (**T**), (**d**), (**v**), (**p**), (**c**) and (**Q**) are known, then its Quick or Acid Test Ratio Planned is:

q= **c**-**$vp**/[360(**/**{**U**+**$**[1-**f**-**i**]-**V**-**T**}[1-**d**]-**Q**)]

Rule-12930:

If both (**/**), (**U**), (**$**), (**f**), (**i**), (**V**), (**T**), (**d**), (**$'**), (**v**), (**p**), (**s**), (**c**) and (**q**) are known, then its Quoted Longterm Debt Planned is:

Q= **/**{**U**+**$**[1-**f**-**i**]-**V**-**T**}[1-**d**]-**$'vp**[1+**s**]/{360[**c**-**q**]}

Rule-12931:

If both (**Q**), (**U**), (**$**), (**f**), (**i**), (**V**), (**T**), (**d**), (**$'**), (**v**), (**p**), (**s**), (**c**) and (**q**) are known, then its Leverage or Gearing Ratio Planned is:

/= (**Q**+**$'vp**[1+**s**]/{360[**c**-**q**]})
/{**U**+**$**[1-**f**-**i**]-**V**-**T**}[1-**d**]

Rule-12932:

If both (**/**), (**Q**), (**$**), (**f**), (**i**), (**V**), (**T**), (**d**), (**$'**), (**v**), (**p**), (**s**), (**c**) and (**q**) are known, then its Utilized or Starting Capital must be:

U= (**Q**+**$'vp**[1+**s**]/{360[**c**-**q**]})/**/**
-{**$**[1-**f**-**i**]-**V**-**T**}[1-**d**]

Steve Asikin ISBN 14: 978-1514685136, ISBN 10: 1514685132

Rule-12933:

If both (/), (U), (S), (f), (i), (V), (T), (d), (S'), (v), (p), (s), (c) and (q) are known, then its Sales or Revenue Planned is:

S= (Q-/{U-V-T}[1-d]+S'vp[1+s]/{360[c-q]})
/{/[1-f-i][1-d]}

Rule-12934:

If both (/), (U), (S), (Q), (i), (V), (T), (d), (S'), (v), (p), (s), (c) and (q) are known, then its Fixed Portion Planned is:

f= 1-i-{V+T+[(Q+S'vp[1+s]/{360[c-q]})/-U]
/[1-d]}/S

Rule-12935:

If both (/), (U), (S), (f), (Q), (V), (T), (d), (S'), (v), (p), (s), (c) and (q) are known, then its Interest Portion Planned is:

i= 1-f-{V+T+[(Q+S'vp[1+s]/{360[c-q]})/-U]
/[1-d]}/S

Rule-12936:

If both (/), (U), (S), (f), (i), (Q), (T), (d), (S'), (v), (p), (s), (c) and (q) are known, then its Variable Cost Planned is:

V= S[1-f-i]-Y-[(Q+S'vp[1+s]/{360[c-q]})/-U]
/[1-d]

Steve Asikin ISBN 14: 978-1514685136, ISBN 10: 1514685132

Rule-12937:

If both (/), (U), (S), (f), (i), (V), (Q), (d), (S'), (v), (p), (s), (c) and (q) are known, then its Tax Planned is:

T= S[1-f-i]-V-[(Q+S'vp[1+s]/{360[c-q]})/ƒ-U]
/[1-d]

Rule-12938:

If both (/), (U), (S), (f), (i), (V), (T), (Q), (S'), (v), (p), (s), (c) and (q) are known, then its Dividend Payout Planned is:

d= 1-[(Q+S'vp[1+s]/{360[c-q]})/ƒ-U]
/{S[1-f-i]-V-T}

Rule-12939:

If both (/), (U), (S), (f), (i), (V), (T), (d), (Q), (v), (p), (s), (c) and (q) are known, then its Sales Past must be:

S'= 360[c-q](/{U+S[1-f-i]-V-T}[1-d]-Q)/{vp[1+s]}

Rule-12940:

If both (/), (U), (S), (f), (i), (V), (T), (d), (S'), (Q), (p), (s), (c) and (q) are known, then its Variable Portion Planned is:

s= V/S

or it can also be found as

v= 360[c-q](/{U+S[1-f-i]-V-T}[1-d]-Q)/{S'p[1+s]}

Rule-12941:

If both (/), (U), (S), (f), (i), (V), (T), (d), (S'), (v), (Q), (s), (c) and (q) are known, then its Procured Inventory Days planned:

p= 360[c-q](/{U+S[1-f-i]-V-T}[1-d]-Q)/{S'v[1+s]}

Steve Asikin ISBN 14: 978-1514685136, ISBN 10: 1514685132

Rule-12942:

If both (**/**), (**U**), (**$**), (**f**), (**i**), (**V**), (**T**), (**d**), (**$'**), (**v**), (**p**), (**Q**), (**c**) and (**q**) are known, then its Sales Growth Planned is:

s= **$**/**$'**-1

or it can also be found as

s= 360[**c-q**](**/**{**U**+**$**[1-**f-i**]-**V-T**}[1-**d**]-**Q**)/[**$'vp**]-1

Rule-12943:

If both (**/**), (**U**), (**$**), (**f**), (**i**), (**V**), (**T**), (**d**), (**$'**), (**v**), (**p**), (**s**), (**Q**) and (**q**) are known, then its Current Ratio Planned is:

c= **q**+**$'vp**[1+**s**]/[360(**/**{**U**+**$**[1-**f-i**]-**V-T**}[1-**d**]-**Q**)]

Rule-12944:

If both (**/**), (**U**), (**$**), (**f**), (**i**), (**V**), (**T**), (**d**), (**$'**), (**v**), (**p**), (**s**), (**c**) and (**Q**) are known, then its Quick or Acid Test Ratio Planned is:

q= **c-$'vp**[1+**s**]/[360(**/**{**U**+**$**[1-**f-i**]-**V-T**}[1-**d**]-**Q**)]

Rule-12945:

If both (**/**), (**U**), (**$**), (**f**), (**i**), (**V**), (**T**), (**A**), (**d**) and (**X**) are known, then its Quoted Longterm Debt Planned is:

Q= **/**{**U**+**$**[1-**f-i**]-**V-T-Ad**}-**X**

Rule-12946:

If both (**Q**), (**U**), (**$**), (**f**), (**i**), (**V**), (**T**), (**A**), (**d**) and (**X**) are known, then its Leverage or Gearing Ratio Planned is:

/= [**Q**+**X**]/{**U**+**$**[1-**f-i**]-**V-T-Ad**}

Steve Asikin ISBN 14: 978-1514685136, ISBN 10: 1514685132

Rule-12947:

If both (**/**), (**Q**), (**$**), (**f**), (**i**), (**V**), (**T**), (**A**), (**d**) and (**X**) are known, then its Utilized or Starting Capital must be:

U= [**Q**+**X**]/**/**{**$**[1-**f**-**i**]-**V**-**T**-**Ad**}

Rule-12948:

If both (**/**), (**U**), (**Q**), (**f**), (**i**), (**V**), (**T**), (**A**), (**d**) and (**X**) are known, then its Sales or Revenue Planned is:

$= {**V**+**T**+**Ad**+[**Q**+**X**]/**/U**}/[1-**f**-**i**]

Rule-12949:

If both (**/**), (**U**), (**$**), (**Q**), (**i**), (**V**), (**T**), (**A**), (**d**) and (**X**) are known, then its Fixed Portion Planned is:

f= 1-**i**-{**V**+**T**+**Ad**+[**Q**+**X**]/**/U**}/**$**

Rule-12950:

If both (**/**), (**U**), (**$**), (**f**), (**Q**), (**V**), (**T**), (**A**), (**d**) and (**X**) are known, then its Interest Portion Planned is:

i= 1-**f**-{**V**+**T**+**Ad**+[**Q**+**X**]/**/U**}/**$**

Rule-12951:

If both (**/**), (**U**), (**$**), (**f**), (**i**), (**Q**), (**T**), (**A**), (**d**) and (**X**) are known, then its Variable Cost Planned is:

V= {**$**[1-**f**-**i**]-**T**-**Ad**}-[**Q**+**X**]/**/U**

Rule-12952:

If both (**/**), (**U**), (**$**), (**f**), (**i**), (**V**), (**Q**), (**A**), (**d**) and (**X**) are known, then its Tax Planned is:

T= {**$**[1-**f**-**i**]-**V**-**Ad**}-[**Q**+**X**]/**/U**

Steve Asikin ISBN 14: 978-1514685136, ISBN 10: 1514685132

Rule-12953:

If both (/), (U), ($), (f), (i), (V), (T), (Q), (d) and (X) are known, then its After Tax Income Planned is:

A= {$[1-f-i]-V-T-[Q+X]/ /-U}/d

Rule-12954:

If both (/), (U), ($), (f), (i), (V), (T), (A), (Q) and (X) are known, then its Dividend Payout Planned is:

d= {$[1-f-i]-V-T-[Q+X]/ /-U}/A

Rule-12955:

If both (/), (U), ($), (f), (i), (V), (T), (A), (d) and (Q) are known, then its Xpress or Current Debt Planned is:

X= /{U+$[1-f-i]-V-T-Ad}-Q

Rule-12956:

If both (/), (U), ($), (f), (i), (V), (T), (A), (d), (P), (c) and (q) are known, then its Quoted Longterm Debt Planned is:

Q= /{U+$[1-f-i]-V-T-Ad}-P/[c-q]

Rule-12957:

If both (Q), (U), ($), (f), (i), (V), (T), (A), (d), (P), (c) and (q) are known, then its Leverage or Gearing Ratio Planned is:

/= {Q+P/[c-q]}/{U+$[1-f-i]-V-T-Ad}

Steve Asikin ISBN 14: 978-1514685136, ISBN 10: 1514685132

Rule-12958:

If both (/), (**Q**), (**$**), (**f**), (**i**), (**V**), (**T**), (**A**), (**d**), (**P**), (**c**) and (**q**) are known, then its Utilized or Starting Capital must be:

U= {Q+P/[c-q]}//-{$[1-f-i]-V-T-Ad}

Rule-12959:

If both (/), (**U**), (**Q**), (**f**), (**i**), (**V**), (**T**), (**A**), (**d**), (**P**), (**c**) and (**q**) are known, then its Sales or Revenue Planned is:

$= [V+T+Ad+{Q+P/[c-q]}//-U]/[1-f-i]

Rule-12960:

If both (/), (**U**), (**$**), (**Q**), (**i**), (**V**), (**T**), (**A**), (**d**), (**P**), (**c**) and (**q**) are known, then its Fixed Portion Planned is:

f= 1-i-[V+T+Ad+{Q+P/[c-q]}//-U]/$

Rule-12961:

If both (/), (**U**), (**$**), (**f**), (**Q**), (**V**), (**T**), (**A**), (**d**), (**P**), (**c**) and (**q**) are known, then its Interest Portion Planned is:

i= 1-f-[V+T+Ad+{Q+P/[c-q]}//-U]/$

Rule-12962:

If both (/), (**U**), (**$**), (**f**), (**i**), (**Q**), (**T**), (**A**), (**d**), (**P**), (**c**) and (**q**) are known, then its Variable Cost Planned is:

V= {$[1-f-i]-T-Ad}-({Q+P/[c-q]}//-U)

Steve Asikin ISBN 14: 978-1514685136, ISBN 10: 1514685132

Rule-12963:

If both (**/**), (**U**), (**$**), (**f**), (**i**), (**V**), (**Q**), (**A**), (**d**), (**P**), (**c**) and (**q**) are known, then its Tax Planned is:

T= {**$**[1-**f**-**i**]-**V**-**Ad**}-({**Q**+**P**/[**c**-**q**]}/**/**-**U**)

Rule-12964:

If both (**/**), (**U**), (**$**), (**f**), (**i**), (**V**), (**T**), (**Q**), (**d**), (**P**), (**c**) and (**q**) are known, then its After Tax Income Planned is:

A= (**$**[1-**f**-**i**]-**V**-**T**-{**Q**+**P**/[**c**-**q**]}/**/**-**U**)/**d**

Rule-12965:

If both (**/**), (**U**), (**$**), (**f**), (**i**), (**V**), (**T**), (**A**), (**Q**), (**P**), (**c**) and (**q**) are known, then its Dividend Payout Planned is:

d= (**$**[1-**f**-**i**]-**V**-**T**-{**Q**+**P**/[**c**-**q**]}/**/**-**U**)/**A**

Rule-12966:

If both (**/**), (**U**), (**$**), (**f**), (**i**), (**V**), (**T**), (**A**), (**d**), (**Q**), (**c**) and (**q**) are known, then its Procured Inventory Planned is:

P= [**c**-**q**](**/**{**U**+**$**[1-**f**-**i**]-**V**-**T**-**Ad**}-**Q**)

Rule-12967:

If both (**/**), (**U**), (**$**), (**f**), (**i**), (**V**), (**T**), (**A**), (**d**), (**P**), (**Q**) and (**q**) are known, then its Current Ratio Planned is:

c= **q**+**P**/(**/**{**U**+**$**[1-**f**-**i**]-**V**-**T**-**Ad**}-**Q**)

Steve Asikin ISBN 14: 978-1514685136, ISBN 10: 1514685132

Rule-12968:

If both (**/**), (**U**), (**$**), (**f**), (**i**), (**V**), (**T**), (**A**), (**d**), (**P**), (**c**) and (**Q**) are known, then its Quick or Acid Test Ratio Planned is:

q= **c**-**P**/(**/**{**U**+**$**[1-**f**-**i**]-**V**-**T**-**Ad**}-**Q**)

Rule-12969:

If both (**/**), (**U**), (**$**), (**f**), (**i**), (**V**), (**T**), (**A**), (**d**), (**p**), (**c**) and (**q**) are known, then its Quoted Longterm Debt Planned is:

Q= **/**{**U**+**$**[1-**f**-**i**]-**V**-**T**-**Ad**}-**Vp**/{360[**c**-**q**]}

Rule-12970:

If both (**Q**), (**U**), (**$**), (**f**), (**i**), (**V**), (**T**), (**A**), (**d**), (**p**), (**c**) and (**q**) are known, then its Leverage or Gearing Ratio Planned is:

/= (**Q**+**Vp**/{360[**c**-**q**]})/{**U**+**$**[1-**f**-**i**]-**V**-**T**-**Ad**}

Rule-12971:

If both (**/**), (**Q**), (**$**), (**f**), (**i**), (**V**), (**T**), (**A**), (**d**), (**p**), (**c**) and (**q**) are known, then its Utilized or Starting Capital must be:

U= (**Q**+**Vp**/{360[**c**-**q**]})/**/**-{**$**[1-**f**-**i**]-**V**-**T**-**Ad**}

Rule-12972:

If both (**/**), (**U**), (**Q**), (**f**), (**i**), (**V**), (**T**), (**A**), (**d**), (**p**), (**c**) and (**q**) are known, then its Sales or Revenue Planned is:

$= [**V**+**T**+**Ad**+(**Q**+**Vp**/{360[**c**-**q**]})/**/**-**U**]/[1-**f**-**i**]

Steve Asikin ISBN 14: 978-1514685136, ISBN 10: 1514685132

Rule-12973:

If both (**/**), (**U**), (**$**), (**Q**), (**i**), (**V**), (**T**), (**A**), (**d**), (**p**), (**c**) and (**q**) are known, then its Fixed Portion Planned is:

f= 1-**i**-[**V**+**T**+**Ad**+(**Q**+**Vp**/{360[**c**-**q**]})/**/**-**U**]/**$**

Rule-12974:

If both (**/**), (**U**), (**$**), (**f**), (**Q**), (**V**), (**T**), (**A**), (**d**), (**p**), (**c**) and (**q**) are known, then its Interest Portion Planned is:

i= 1-**f**-[**V**+**T**+**Ad**+(**Q**+**Vp**/{360[**c**-**q**]})/**/**-**U**]/**$**

Rule-12975:

If both (**/**), (**U**), (**$**), (**f**), (**i**), (**Q**), (**T**), (**A**), (**d**), (**p**), (**c**) and (**q**) are known, then its Variable Cost Planned is:

V= (**/**{**U**+**$**[1-**f**-**i**]-**T**-**Ad**}-**Q**)/(**p**/{360[**c**-**q**]}+**/**)

Rule-12976:

If both (**/**), (**U**), (**$**), (**f**), (**i**), (**V**), (**Q**), (**A**), (**d**), (**p**), (**c**) and (**q**) are known, then its Tax Planned is:

T= {**$**[1-**f**-**i**]-**V**-**Ad**}-[(**Q**+**Vp**/{360[**c**-**q**]})/**/**-**U**]

Rule-12977:

If both (**/**), (**U**), (**$**), (**f**), (**i**), (**V**), (**T**), (**Q**), (**d**), (**p**), (**c**) and (**q**) are known, then its After Tax Income Planned is:

A= [**$**[1-**f**-**i**]-**V**-**T**-(**Q**+**Vp**/{360[**c**-**q**]})/**/**-**U**]/**d**

Steve Asikin ISBN 14: 978-1514685136, ISBN 10: 1514685132

Rule-12978:

If both (**/**), (**U**), (**$**), (**f**), (**i**), (**V**), (**T**), (**A**), (**Q**), (**p**), (**c**) and (**q**) are known, then its Dividend Payout Planned is:

d= [**$**[1-**f**-**i**]-**V**-**T**-(**Q**+**Vp**/{360[**c**-**q**]})/**/**-**U**]/**A**

Rule-12979:

If both (**/**), (**U**), (**$**), (**f**), (**i**), (**V**), (**T**), (**A**), (**d**), (**Q**), (**c**) and (**q**) are known, then its Procured Inventory Days planned:

p= 360[**c**-**q**](**/**{**U**+**$**[1-**f**-**i**]-**V**-**T**-**Ad**}-**Q**)/**V**

Rule-12980:

If both (**/**), (**U**), (**$**), (**f**), (**i**), (**V**), (**T**), (**A**), (**d**), (**p**), (**Q**) and (**q**) are known, then its Current Ratio Planned is:

c= **q**+**Vp**/[360(**/**{**U**+**$**[1-**f**-**i**]-**V**-**T**-**Ad**}-**Q**)]

Rule-12981:

If both (**/**), (**U**), (**$**), (**f**), (**i**), (**V**), (**T**), (**A**), (**d**), (**p**), (**c**) and (**Q**) are known, then its Quick or Acid Test Ratio Planned is:

q= **c**-**Vp**/[360(**/**{**U**+**$**[1-**f**-**i**]-**V**-**T**-**Ad**}-**Q**)]

Rule-12982:

If both (**/**), (**U**), (**$**), (**f**), (**i**), (**V**), (**T**), (**A**), (**d**), (**v**), (**p**), (**c**) and (**q**) are known, then its Quoted Longterm Debt Planned is:

Q= **/**{**U**+**$**[1-**f**-**i**]-**V**-**T**-**Ad**}-**$vp**/{360[**c**-**q**]}

Steve Asikin ISBN 14: 978-1514685136, ISBN 10: 1514685132

Rule-12983:

If both (**Q**), (**U**), (**$**), (**f**), (**i**), (**V**), (**T**), (**A**), (**d**), (**v**), (**p**), (**c**) and (**q**) are known, then its Leverage or Gearing Ratio Planned is:

/= (Q+$vp/{360[c-q]})/{U+$[1-f-i]-V-T-Ad}

Rule-12984:

If both (/), (**Q**), (**$**), (**f**), (**i**), (**V**), (**T**), (**A**), (**d**), (**v**), (**p**), (**c**) and (**q**) are known, then its Utilized or Starting Capital must be:

U= (Q+$vp/{360[c-q]})//-{$[1-f-i]-V-T-Ad}

Rule-12985:

If both (/), (**U**), (**Q**), (**f**), (**i**), (**V**), (**T**), (**A**), (**d**), (**v**), (**p**), (**c**) and (**q**) are known, then its Sales or Revenue Planned is:

$= {/[U-V-T-Ad]-Q}/(vp/{360[c-q]}-/[1-f-i])

Rule-12986:

If both (/), (**U**), (**$**), (**Q**), (**i**), (**V**), (**T**), (**A**), (**d**), (**v**), (**p**), (**c**) and (**q**) are known, then its Fixed Portion Planned is:

f= 1-i-[V+T+Ad+(Q+$vp/{360[c-q]})//-U]/$

Rule-12987:

If both (/), (**U**), (**$**), (**f**), (**Q**), (**V**), (**T**), (**A**), (**d**), (**v**), (**p**), (**c**) and (**q**) are known, then its Interest Portion Planned is:

i= 1-f-[V+T+Ad+(Q+$vp/{360[c-q]})//-U]/$

Steve Asikin ISBN 14: 978-1514685136, ISBN 10: 1514685132

Rule-12988:

If both (**/**), (**U**), (**$**), (**f**), (**i**), (**Q**), (**T**), (**A**), (**d**), (**v**), (**p**), (**c**) and (**q**) are known, then its Variable Cost Planned is:

V= {**$**[1-**f**-**i**]-**T**-**Ad**}-[(**Q**+**$vp**/{360[**c**-**q**]})/**/**-**U**]

Rule-12989:

If both (**/**), (**U**), (**$**), (**f**), (**i**), (**V**), (**Q**), (**A**), (**d**), (**v**), (**p**), (**c**) and (**q**) are known, then its Tax Planned is:

T= {**$**[1-**f**-**i**]-**V**-**Ad**}-[(**Q**+**$vp**/{360[**c**-**q**]})/**/**-**U**]

Rule-12990:

If both (**/**), (**U**), (**$**), (**f**), (**i**), (**V**), (**T**), (**Q**), (**d**), (**v**), (**p**), (**c**) and (**q**) are known, then its After Tax Income Planned is:

A= [**$**[1-**f**-**i**]-**V**-**T**-(**Q**+**$vp**/{360[**c**-**q**]})/**/**-**U**]/**d**

Rule-12991:

If both (**/**), (**U**), (**$**), (**f**), (**i**), (**V**), (**T**), (**A**), (**Q**), (**v**), (**p**), (**c**) and (**q**) are known, then its Dividend Payout Planned is:

d= [**$**[1-**f**-**i**]-**V**-**T**-(**Q**+**$vp**/{360[**c**-**q**]})/**/**-**U**]/**A**

Rule-12992:

If both (**/**), (**U**), (**$**), (**f**), (**i**), (**V**), (**T**), (**A**), (**d**), (**Q**), (**p**), (**c**) and (**q**) are known, then its Variable Portion Planned is:

v= **V**/**$**

or it can also be found as

v= 360[**c**-**q**](**/**{**U**+**$**[1-**f**-**i**]-**V**-**T**-**Ad**}-**Q**)/[**$p**]

Steve Asikin ISBN 14: 978-1514685136, ISBN 10: 1514685132

Rule-12993:

If both (**/**), (**U**), (**$**), (**f**), (**i**), (**V**), (**T**), (**A**), (**d**), (**v**), (**Q**), (**c**) and (**q**) are known, then its Procured Inventory Days Planned:

p= 360[**c-q**](**/**{**U**+**$**[1-**f-i**]-**V-T-Ad**}-**Q**)/[**$v**]

Rule-12994:

If both (**/**), (**U**), (**$**), (**f**), (**i**), (**V**), (**T**), (**A**), (**d**), (**v**), (**p**), (**Q**) and (**q**) are known, then its Current Ratio Planned is:

c= **q**+**$vp**/[360(**/**{**U**+**$**[1-**f-i**]-**V-T-Ad**}-**Q**)]

Rule-12995:

If both (**/**), (**U**), (**$**), (**f**), (**i**), (**V**), (**T**), (**A**), (**d**), (**v**), (**p**), (**c**) and (**Q**) are known, then its Quick or Acid Test Ratio Planned is:

q= **c**-**$vp**/[360(**/**{**U**+**$**[1-**f-i**]-**V-T-Ad**}-**Q**)]

Rule-12996:

If both (**/**), (**U**), (**$**), (**f**), (**i**), (**V**), (**T**), (**A**), (**d**), (**$'**), (**v**), (**s**), (**p**), (**c**) and (**q**) are known, then its Quoted Longterm Debt Planned is:

Q= **/**{**U**+**$**[1-**f-i**]-**V-T-Ad**}-**$'vp**[1+**s**]/{360[**c-q**]}

Rule-12997:

If both (**Q**), (**U**), (**$**), (**f**), (**i**), (**V**), (**T**), (**A**), (**d**), (**$'**), (**v**), (**s**), (**p**), (**c**) and (**q**) are known, then its Leverage or Gearing Ratio Planned is:

/= (**Q**+**$'vp**[1+**s**]/{360[**c-q**]})/{**U**+**$**[1-**f-i**]-**V-T-Ad**}

Steve Asikin ISBN 14: 978-1514685136, ISBN 10: 1514685132

Rule-12998:

If both (**/**), (**Q**), (**$**), (**f**), (**i**), (**V**), (**T**), (**A**), (**d**), (**$'**), (**v**), (**s**), (**p**), (**c**) and (**q**) are known, then its Utilized or Starting Capital must be:

U= (**Q**+**$'vp**[1+**s**]/{360[**c**-**q**]})/**/**{**$**[1-**f**-**i**]-**V**-**T**-**Ad**}

Rule-12999:

If both (**/**), (**U**), (**Q**), (**f**), (**i**), (**V**), (**T**), (**A**), (**d**), (**$'**), (**v**), (**s**), (**p**), (**c**) and (**q**) are known, then its Sales or Revenue Planned is:

$**= (**Q**-**/**{**U**-**V**-**T**-**Ad**}+**$'vp[1+**s**]/{360[**c**-**q**]}
/{**/**[1-**f**-**i**]}

Rule-13000:

If both (**/**), (**U**), (**$**), (**Q**), (**i**), (**V**), (**T**), (**A**), (**d**), (**$'**), (**v**), (**s**), (**p**), (**c**) and (**q**) are known, then its Fixed Portion Planned is:

f= 1-**i**-**[V**+**T**+**Ad**+(**Q**+**$'vp**[1+**s**]/{360[**c**-**q**]})/**/U]**/**$**

Rule-13001:

If both (**/**), (**U**), (**$**), (**f**), (**Q**), (**V**), (**T**), (**A**), (**d**), (**$'**), (**v**), (**s**), (**p**), (**c**) and (**q**) are known, then its Interest Portion Planned is:

i= 1-**f**-**[V**+**T**+**Ad**+(**Q**+**$'vp**[1+**s**]/{360[**c**-**q**]})/**/U]**/**$**

Rule-13002:

If both (**/**), (**U**), (**$**), (**f**), (**i**), (**Q**), (**T**), (**A**), (**d**), (**$'**), (**v**), (**s**), (**p**), (**c**) and (**q**) are known, then its Variable Cost Planned is:

V= {**$**[1-**f**-**i**]-**T**-**Ad**}
-**[**(**Q**+**$'vp**[1+**s**]/{360[**c**-**q**]})/**/U]**

Steve Asikin ISBN 14: 978-1514685136, ISBN 10: 1514685132

Rule-13003:

If both (/), (U), ($), (f), (i), (V), (Q), (A), (d), ($'), (v), (s), (p), (c) and (q) are known, then its Tax Planned is:

T= {$[1-f-i]-V-Ad}
-[(Q+$'vp[1+s]/{360[c-q]})/ /U]

Rule-13004:

If both (/), (U), ($), (f), (i), (V), (T), (Q), (d), ($'), (v), (s), (p), (c) and (q) are known, then its After Tax Income Planned is:

A= [$[1-f-i]-V-T-(Q+$'vp[1+s]/{360[c-q]})/ /U]/d

Rule-13005:

If both (/), (U), ($), (f), (i), (V), (T), (A), (Q), ($'), (v), (s), (p), (c) and (q) are known, then its Dividend Payout Planned is:

d= [$[1-f-i]-V-T-(Q+$'vp[1+s]/{360[c-q]})/ /U]/A

Rule-13006:

If both (/), (U), ($), (f), (i), (V), (T), (A), (d), (Q), (v), (s), (p), (c) and (q) are known, then its Sales Past must be:

$'= 360[c-q](/{U+$[1-f-i]-V-T-Ad}-Q)/{vp[1+s]}

Steve Asikin ISBN 14: 978-1514685136, ISBN 10: 1514685132

Rule-13007:

If both (/), (U), ($), (f), (i), (V), (T), (A), (d), ($'), (Q), (s), (p), (c) and (q) are known, then its Variable Portion Planned is:

v= V/$

or it can also be found as

v= 360[c-q](/{U+$[1-f-i]-V-T-Ad}-Q)/{$'p[1+s]}

Rule-13008:

If both (/), (U), ($), (f), (i), (V), (T), (A), (d), ($'), (v), (s), (Q), (c) and (q) are known, then its Procured Inventory Days planned:

p= 360[c-q](/{U+$[1-f-i]-V-T-Ad}-Q)/{$'v[1+s]}

Rule-13009:

If both (/), (U), ($), (f), (i), (V), (T), (A), (d), ($'), (v), (Q), (p), (c) and (q) are known, then its Sales Growth Planned is:

s= $/$'-1

or it can also be found as

s= 360[c-q](/{U+$[1-f-i]-V-T-Ad}-Q)/[$'vp]-1

Rule-13010:

If both (/), (U), ($), (f), (i), (V), (T), (A), (d), ($'), (v), (s), (p), (Q) and (q) are known, then its Current Ratio Planned is:

c= q+$'vp[1+s]/[360(/{U+$[1-f-i]-V-T-Ad}-Q)]

Steve Asikin ISBN 14: 978-1514685136, ISBN 10: 1514685132

Rule-13011:

If both (/), (U), ($), (f), (i), (V), (T), (A), (d), ($'), (v), (s), (p), (c) and (Q) are known, then its Quick or Acid Test Ratio Planned is:

q= c-$'vp[1+s]/[360(/{U+$[1-f-i]-V-T-Ad}-Q)]

Rule-13012:

If both (/), (U), ($), (f), (i), (V), (t), (D) and (X) are known, then its Quoted Longterm Debt Planned is:

Q= /(U+{$[1-f-i]-V}[1-t]-D)-X

Rule-13013:

If both (Q), (U), ($), (f), (i), (V), (t), (D) and (X) are known, then its Leverage or Gearing Ratio Planned is:

/= [Q+X]/(U+{$[1-f-i]-V}[1-t]-D)

Rule-13014:

If both (/), (Q), ($), (f), (i), (V), (t), (D) and (X) are known, then its Utilized or Starting Capital must be:

U= [Q+X]/-({$[1-f-i]-V}[1-t]-D)

Rule-13015:

If both (/), (U), (Q), (f), (i), (V), (t), (D) and (X) are known, then its Sales or Revenue Planned is:

$= (V+{D+[Q+X]/-U}/[1-t])/[1-f-i]

Rule-13016:

If both (/), (U), ($), (Q), (i), (V), (t), (D) and (X) are known, then its Fixed Portion Planned is:

f= 1-i-(V+{D+[Q+X]/-U}/[1-t])/$

Steve Asikin ISBN 14: 978-1514685136, ISBN 10: 1514685132

Rule-13017:

If both (/), (**U**), (**$**), (**f**), (**Q**), (**V**), (**t**), (**D**) and (**X**) are known, then its Interest Portion Planned is:

i= 1-**f**-(**V**+{**D**+[**Q**+**X**]/**/-U**}/[1-**t**])/**$**

Rule-13018:

If both (/), (**U**), (**$**), (**f**), (**i**), (**Q**), (**t**), (**D**) and (**X**) are known, then its Variable Cost Planned is:

V= **$**[1-**f**-**i**]-{**D**+[**Q**+**X**]/**/-U**}/[1-**t**]

Rule-13019:

If both (/), (**U**), (**$**), (**f**), (**i**), (**V**), (**Q**), (**D**) and (**X**) are known, then its Tax Rate Planned is:

t= 1-{**D**+[**Q**+**X**]/**/-U**}/{**$**[1-**f**-**i**]-**V**}

Rule-13020:

If both (/), (**U**), (**$**), (**f**), (**i**), (**V**), (**t**), (**Q**) and (**X**) are known, then its Dividend Planned is:

D= {**$**[1-**f**-**i**]-**V**}[1-**t**]-[**Q**+**X**]/**/-U**}

Rule-13021:

If both (/), (**U**), (**$**), (**f**), (**i**), (**V**), (**t**), (**D**) and (**Q**) are known, then its Xpress or Current Debt Planned is:

X= /{**U**+**$**[1-**f**-**i**]-**V**}[1-**t**]-**D**}-**Q**

Rule-13022:

If both (/), (**U**), (**$**), (**f**), (**i**), (**V**), (**t**), (**D**), (**P**), (**c**) and (**q**) are known, then its Quoted Longterm Debt Planned is:

Q= /{**U**+**$**[1-**f**-**i**]-**V**}[1-**t**]-**D**}-**P**/[**c**-**q**]

Steve Asikin ISBN 14: 978-1514685136, ISBN 10: 1514685132

Rule-13023:

If both (**Q**), (**U**), (**$**), (**f**), (**i**), (**V**), (**t**), (**D**), (**P**), (**c**) and (**q**) are known, then its Leverage or Gearing Ratio Planned is:

Ł= {**Q**+**P**/[**c**-**q**]}/(**U**+{**$**[1-**f**-**i**]-**V**}[1-**t**]-**D**)

Rule-13024:

If both (**Ł**), (**Q**), (**$**), (**f**), (**i**), (**V**), (**t**), (**D**), (**P**), (**c**) and (**q**) are known, then its Utilized or Starting Capital must be:

U= {**Q**+**P**/[**c**-**q**]}/**Ł**({**$**[1-**f**-**i**]-**V**}[1-**t**]-**D**)

Rule-13025:

If both (**Ł**), (**U**), (**Q**), (**f**), (**i**), (**V**), (**t**), (**D**), (**P**), (**c**) and (**q**) are known, then its Sales or Revenue Planned is:

$= [**V**+(**D**+{**Q**+**P**/[**c**-**q**]}/**ŁU**)/[1-**t**]]/[1-**f**-**i**]

Rule-13026:

If both (**Ł**), (**U**), (**$**), (**Q**), (**i**), (**V**), (**t**), (**D**), (**P**), (**c**) and (**q**) are known, then its Fixed Portion Planned is:

f= 1-**i**-[**V**+(**D**+{**Q**+**P**/[**c**-**q**]}/**ŁU**)/[1-**t**]]/**$**

Rule-13027:

If both (**Ł**), (**U**), (**$**), (**f**), (**Q**), (**V**), (**t**), (**D**), (**P**), (**c**) and (**q**) are known, then its Interest Portion Planned is:

i= 1-**f**-[**V**+(**D**+{**Q**+**P**/[**c**-**q**]}/**ŁU**)/[1-**t**]]/**$**

Rule-13028:

If both (**Ł**), (**U**), (**$**), (**f**), (**i**), (**Q**), (**t**), (**D**), (**P**), (**c**) and (**q**) are known, then its Variable Cost Planned is:

V= **$**[1-**f**-**i**]-(**D**+{**Q**+**P**/[**c**-**q**]}/**ŁU**)/ [1-**t**]

Steve Asikin ISBN 14: 978-1514685136, ISBN 10: 1514685132

Rule-13029:

If both (**/**), (**U**), (**$**), (**f**), (**i**), (**V**), (**Q**), (**D**), (**P**), (**c**) and (**q**) are known, then its Tax Rate Planned is:

t= 1-(**D**+{**Q**+**P**/[**c**-**q**]}/**/**-**U**)/{**$**[1-**f**-**i**]-**V**}

Rule-13030:

If both (**/**), (**U**), (**$**), (**f**), (**i**), (**V**), (**t**), (**Q**), (**P**), (**c**) and (**q**) are known, then its Dividend Planned is:

D= (**$**[1-**f**-**i**]-**V**}[1-**t**]-{**Q**+**P**/[**c**-**q**]}/**/**-**U**)

Rule-13031:

If both (**/**), (**U**), (**$**), (**f**), (**i**), (**V**), (**t**), (**D**), (**Q**), (**c**) and (**q**) are known, then its Procured Inventory Planned is:

P= [**c**-**q**](**/**{**U**+**$**[1-**f**-**i**]-**V**}[1-**t**]-**D**}-**Q**)

Rule-13032:

If both (**/**), (**U**), (**$**), (**f**), (**i**), (**V**), (**t**), (**D**), (**P**), (**Q**) and (**q**) are known, then its Current Ratio Planned is:

c= **q**+**P**/(**/**{**U**+**$**[1-**f**-**i**]-**V**}[1-**t**]-**D**}-**Q**)

Rule-13033:

If both (**/**), (**U**), (**$**), (**f**), (**i**), (**V**), (**t**), (**D**), (**P**), (**c**) and (**Q**) are known, then its Quick or Acid Test Ratio Planned is:

q= **c**-**P**/(**/**{**U**+**$**[1-**f**-**i**]-**V**}[1-**t**]-**D**}-**Q**)

Rule-13034:

If both (**/**), (**U**), (**$**), (**f**), (**i**), (**V**), (**t**), (**D**), (**p**), (**c**) and (**q**) are known, then its Quoted Longterm Debt Planned is:

Q= **/**{**U**+**$**[1-**f**-**i**]-**V**}[1-**t**]-**D**}-**Vp**/{360[**c**-**q**]}

Steve Asikin ISBN 14: 978-1514685136, ISBN 10: 1514685132

Rule-13035:

If both (**Q**), (**U**), (**$**), (**f**), (**i**), (**V**), (**t**), (**D**), (**p**), (**c**) and (**q**) are known, then its Leverage or Gearing Ratio Planned is:

/= (Q+Vp/{360[c-q]})/(U+{$[1-f-i]-V}[1-t]-D)

Rule-13036:

If both (**/**), (**Q**), (**$**), (**f**), (**i**), (**V**), (**t**), (**D**), (**p**), (**c**) and (**q**) are known, then its Utilized or Starting Capital must be:

U= (Q+Vp/{360[c-q]})//-({$[1-f-i]-V}[1-t]-D)

Rule-13037:

If both (**/**), (**U**), (**Q**), (**f**), (**i**), (**V**), (**t**), (**D**), (**p**), (**c**) and (**q**) are known, then its Sales or Revenue Planned is:

$= {V+[D+(Q+Vp/{360[c-q]})//-U]/[1-t]}/[1-f-i]

Rule-13038:

If both (**/**), (**U**), (**$**), (**Q**), (**i**), (**V**), (**t**), (**D**), (**p**), (**c**) and (**q**) are known, then its Fixed Portion Planned is:

f= 1-i-{V+[D+(Q+Vp/{360[c-q]})//-U]/[1-t]}/$

Rule-13039:

If both (**/**), (**U**), (**$**), (**f**), (**Q**), (**V**), (**t**), (**D**), (**p**), (**c**) and (**q**) are known, then its Interest Portion Planned is:

i= 1-f-{V+[D+(Q+Vp/{360[c-q]})//-U]/[1-t]}/$

Steve Asikin ISBN 14: 978-1514685136, ISBN 10: 1514685132

Rule-13040:

If both (/), (U), ($), (f), (i), (Q), (t), (D), (p), (c) and (q) are known, then its Variable Cost Planned is:

V= (/{U+$[1-f-i]}[1-t]-D}-Q)
/(p/{360[c-q]}+/[1-t])

Rule-13041:

If both (/), (U), ($), (f), (i), (V), (Q), (D), (p), (c) and (q) are known, then its Tax Rate Planned is:

t= 1-[D+(Q+Vp/{360[c-q]})//-U]/{$[1-f-i]-V}

Rule-13042:

If both (/), (U), ($), (f), (i), (V), (t), (Q), (p), (c) and (q) are known, then its Dividend Planned is:

D= [$[1-f-i]-V}[1-t]-(Q+Vp/{360[c-q]})//-U]

Rule-13043:

If both (/), (U), ($), (f), (i), (V), (t), (D), (Q), (c) and (q) are known, then its Procured Inventory Days planned:

p= 360[c-q](/{U+$[1-f-i]-V}[1-t]-D}-Q)/V

Rule-13044:

If both (/), (U), ($), (f), (i), (V), (t), (D), (p), (Q) and (q) are known, then its Current Ratio Planned is:

c= q+Vp/[360(/{U+$[1-f-i]-V}[1-t]-D}-Q)]

Steve Asikin ISBN 14: 978-1514685136, ISBN 10: 1514685132

Rule-13045:

If both (**/**), (**U**), (**$**), (**f**), (**i**), (**V**), (**t**), (**D**), (**p**), (**c**) and (**Q**) are known, then its Quick or Acid Test Ratio Planned is:

q= **c**-**Vp**/[360(**/**{**U**+**$**[1-**f**-**i**]-**V**}[1-**t**]-**D**}-**Q**)]

Rule-13046:

If both (**/**), (**U**), (**$**), (**f**), (**i**), (**V**), (**t**), (**D**), (**v**), (**p**), (**c**) and (**q**) are known, then its Quoted Longterm Debt Planned is:

Q= **/**{**U**+**$**[1-**f**-**i**]-**V**}[1-**t**]-**D**}-**$vp**/{360[**c**-**q**]}

Rule-13047:

If both (**Q**), (**U**), (**$**), (**f**), (**i**), (**V**), (**t**), (**D**), (**v**), (**p**), (**c**) and (**q**) are known, then its Leverage or Gearing Ratio Planned is:

/= (**Q**+**$vp**/{360[**c**-**q**]})/(**U**+{**$**[1-**f**-**i**]-**V**}[1-**t**]-**D**)

Rule-13048:

If both (**/**), (**Q**), (**$**), (**f**), (**i**), (**V**), (**t**), (**D**), (**v**), (**p**), (**c**) and (**q**) are known, then its Utilized or Starting Capital must be:

U= (**Q**+**$vp**/{360[**c**-**q**]})/**/**-({**$**[1-**f**-**i**]-**V**}[1-**t**]-**D**)

Rule-13049:

If both (**/**), (**U**), (**Q**), (**f**), (**i**), (**V**), (**t**), (**D**), (**v**), (**p**), (**c**) and (**q**) are known, then its Sales or Revenue Planned is:

$= (**/**{[**U**-**V**][1-**t**]-**D**}-**Q**)
/(**vp**/{360[**c**-**q**]}-**/**[1-**f**-**i**][1-**t**])

Steve Asikin ISBN 14: 978-1514685136, ISBN 10: 1514685132

Rule-13050:

If both (**/**), (**U**), (**$**), (**Q**), (**i**), (**V**), (**t**), (**D**), (**v**), (**p**), (**c**) and (**q**) are known, then its Fixed Portion Planned is:

f= 1-**i**-{**V**+[**D**+(**Q**+**$vp**/{360[**c**-**q**]})/**/-U**]/[1-**t**]}/**$**

Rule-13051:

If both (**/**), (**U**), (**$**), (**f**), (**Q**), (**V**), (**t**), (**D**), (**v**), (**p**), (**c**) and (**q**) are known, then its Interest Portion Planned is:

i= 1-**f**-{**V**+[**D**+(**Q**+**$vp**/{360[**c**-**q**]})/**/-U**]/[1-**t**]}/**$**

Rule-13052:

If both (**/**), (**U**), (**$**), (**f**), (**i**), (**Q**), (**t**), (**D**), (**v**), (**p**), (**c**) and (**q**) are known, then its Variable Cost Planned is:

V= (**/**{**U**+**$**[1-**f**-**i**] }[1-**t**]-**D**}
-**$vp**/{360[**c**-**q**]}-**Q**)/{**/**[1-**t**]}

Rule-13053:

If both (**/**), (**U**), (**$**), (**f**), (**i**), (**V**), (**Q**), (**D**), (**v**), (**p**), (**c**) and (**q**) are known, then its Tax Rate Planned is:

t= 1-[**D**+(**Q**+**$vp**/{360[**c**-**q**]})/**/-U**]/{**$**[1-**f**-**i**]-**V**}

Rule-13054:

If both (**/**), (**U**), (**$**), (**f**), (**i**), (**V**), (**t**), (**Q**), (**v**), (**p**), (**c**) and (**q**) are known, then its Dividend Planned is:

D= [**$**[1-**f**-**i**]-**V**}[1-**t**]-(**Q**+**$vp**/{360[**c**-**q**]})/**/-U**]

Steve Asikin ISBN 14: 978-1514685136, ISBN 10: 1514685132

Rule-13055:

If both (**/**), (**U**), (**$**), (**f**), (**i**), (**V**), (**t**), (**D**), (**Q**), (**p**), (**c**) and (**q**) are known, then its Variable Portion Planned is:

v= **V**/**$**

or it can also be found as

v= 360[**c**-**q**](**/**{**U**+**$**[1-**f**-**i**]-**V**}[1-**t**]-**D**}-**Q**)/[**$p**]

Rule-13056:

If both (**/**), (**U**), (**$**), (**f**), (**i**), (**V**), (**t**), (**D**), (**v**), (**Q**), (**c**) and (**q**) are known, then its Procured Inventory Days planned:

p= 360[**c**-**q**](**/**{**U**+**$**[1-**f**-**i**]-**V**}[1-**t**]-**D**}-**Q**)/[**$v**]

Rule-13057:

If both (**/**), (**U**), (**$**), (**f**), (**i**), (**V**), (**t**), (**D**), (**v**), (**p**), (**Q**) and (**q**) are known, then its Current Ratio Planned is:

c= **q**+**$vp**/**[**360(**/**{**U**+**$**[1-**f**-**i**]-**V**}[1-**t**]-**D**}-**Q**)**]**

Rule-13058:

If both (**/**), (**U**), (**$**), (**f**), (**i**), (**V**), (**t**), (**D**), (**v**), (**p**), (**c**) and (**Q**) are known, then its Quick or Acid Test Ratio Planned is:

q= **c**-**$vp**/**[**360(**/**{**U**+**$**[1-**f**-**i**]-**V**}[1-**t**]-**D**}-**Q**)**]**

Rule-13059:

If both (**/**), (**U**), (**$**), (**f**), (**i**), (**V**), (**t**), (**D**), (**$'**), (**v**), (**p**), (**s**), (**c**) and (**q**) are known, then its Quoted Longterm Debt Planned is:

Q= **/**{**U**+**$**[1-**f**-**i**]-**V**}[1-**t**]-**D**}-**$'vp**[1+**s**]/{360[**c**-**q**]}

Steve Asikin ISBN 14: 978-1514685136, ISBN 10: 1514685132

Rule-13060:

If both (**Q**), (**U**), (**$**), (**f**), (**i**), (**V**), (**t**), (**D**), (**$'**), (**v**), (**p**), (**s**), (**c**) and (**q**) are known, then its Leverage or Gearing Ratio Planned is:

/= (Q+$'vp[1+s]/{360[c-q]})
/(U+{$[1-f-i]-V}[1-t]-D)

Rule-13061:

If both (**/**), (**Q**), (**$**), (**f**), (**i**), (**V**), (**t**), (**D**), (**$'**), (**v**), (**p**), (**s**), (**c**) and (**q**) are known, then its Utilized or Starting Capital must be:

U= (Q+$'vp[1+s]/{360[c-q]})//
-({$[1-f-i]-V}[1-t]-D)

Rule-13062:

If both (**/**), (**U**), (**Q**), (**f**), (**i**), (**V**), (**t**), (**D**), (**$'**), (**v**), (**p**), (**s**), (**c**) and (**q**) are known, then its Sales or Revenue Planned is:

$= (Q-/{U-V}[1-t]-D}+$'vp[1+s]/{360[c-q]})
/{/[1-f-i][1-t]}

Rule-13063:

If both (**/**), (**U**), (**$**), (**Q**), (**i**), (**V**), (**t**), (**D**), (**$'**), (**v**), (**p**), (**s**), (**c**) and (**q**) are known, then its Fixed Portion Planned is:

f= 1-i-{V+[D+(Q+$'vp[1+s]/{360[c-q]})//-U]
/[1-t]}/$

Steve Asikin ISBN 14: 978-1514685136, ISBN 10: 1514685132

Rule-13064:

If both (**/**), (**U**), (**$**), (**f**), (**Q**), (**V**), (**t**), (**D**), (**$'**), (**v**), (**p**), (**s**), (**c**) and (**q**) are known, then its Interest Portion Planned is:

i= 1-**f**-{**V**+[**D**+(**Q**+**$'vp**[1+**s**]/{360[**c**-**q**]})/**/**-**U**]/[1-**t**]}/**$**

Rule-13065:

If both (**/**), (**U**), (**$**), (**f**), (**i**), (**Q**), (**t**), (**D**), (**$'**), (**v**), (**p**), (**s**), (**c**) and (**q**) are known, then its Variable Cost Planned is:

V= (**/**{**U**+**$**[1-**f**-**i**] }[1-**t**]-**D**}-**$'vp**[1+**s**]/{360[**c**-**q**]}-**Q**)/{**/**[1-**t**]}

Rule-13066:

If both (**/**), (**U**), (**$**), (**f**), (**i**), (**V**), (**Q**), (**D**), (**$'**), (**v**), (**p**), (**s**), (**c**) and (**q**) are known, then its Tax Rate Planned is:

t= 1-[**D**+(**Q**+**$'vp**[1+**s**]/{360[**c**-**q**]})/**/**-**U**]/{**$**[1-**f**-**i**]-**V**}

Rule-13067:

If both (**/**), (**U**), (**$**), (**f**), (**i**), (**V**), (**t**), (**Q**), (**$'**), (**v**), (**p**), (**s**), (**c**) and (**q**) are known, then its Dividend Payout Planned is:

D= [**$**[1-**f**-**i**]-**V**}[1-**t**]-(**Q**+**$'vp**[1+**s**]/{360[**c**-**q**]})/**/**-**U**]

Steve Asikin ISBN 14: 978-1514685136, ISBN 10: 1514685132

Rule-13068:

If both (**/**), (**U**), (**$**), (**f**), (**i**), (**V**), (**t**), (**D**), (**Q**), (**v**), (**p**), (**s**), (**c**) and (**q**) are known, then its Sales Past must be:

$'**= 360[**c**-**q**](**/**{**U**+**$[1-**f**-**i**]-**V**}[1-**t**]-**D**}-**Q**)
/{**vp**[1+**s**]}

Rule-13069:

If both (**/**), (**U**), (**$**), (**f**), (**i**), (**V**), (**t**), (**D**), (**$'**), (**Q**), (**p**), (**s**), (**c**) and (**q**) are known, then its Qvariable Portion Planned is:

v= **V/$**

or it can also be found as

v= 360[**c**-**q**](**/**{**U**+**$**[1-**f**-**i**]-**V**}[1-**t**]-**D**}-**Q**)
/{**$'p**[1+**s**]}

Rule-13070:

If both (**/**), (**U**), (**$**), (**f**), (**i**), (**V**), (**t**), (**D**), (**$'**), (**v**), (**Q**), (**s**), (**c**) and (**q**) are known, then its Procured Inventory Days planned:

p= 360[**c**-**q**](**/**{**U**+**$**[1-**f**-**i**]-**V**}[1-**t**]-**D**}-**Q**)
/{**$'v**[1+**s**]}

Rule-13071:

If both (**/**), (**U**), (**$**), (**f**), (**i**), (**V**), (**t**), (**D**), (**$'**), (**v**), (**p**), (**Q**), (**c**) and (**q**) are known, then its Sales Growth Planned is:

s= 360[**c**-**q**](**/**{**U**+**$**[1-**f**-**i**]-**V**}[1-**t**]-**D**}-**Q**)/[**$'vp**]-1

Steve Asikin ISBN 14: 978-1514685136, ISBN 10: 1514685132

Rule-13072:

If both (**/**), (**U**), (**$**), (**f**), (**i**), (**V**), (**t**), (**D**), (**$'**), (**v**), (**p**), (**s**), (**Q**) and (**q**) are known, then its Current Ratio Planned is:

c= **q**+**$'vp**[1+**s**]/[360(**/**{**U**+**$**[1-**f**-**i**]-**V**}[1-**t**]-**D**}-**Q**)]

Rule-13073:

If both (**/**), (**U**), (**$**), (**f**), (**i**), (**V**), (**t**), (**D**), (**$'**), (**v**), (**p**), (**s**), (**c**) and (**Q**) are known, then its Quick or Acid Test Ratio Planned is:

q= **c**-**$'vp**[1+**s**]/[360(**/**{**U**+**$**[1-**f**-**i**]-**V**}[1-**t**]-**D**}-**Q**)]

Rule-13074:

If both (**/**), (**U**), (**$**), (**f**), (**i**), (**V**), (**t**), (**A**), (**d**) and (**X**) are known, then its Xpress or Current Debt Planned is:

Q= **/**(**U**+{**$**[1-**f**-**i**]-**V**}[1-**t**]-**Ad**)-**X**

Rule-13075:

If both (**Q**), (**U**), (**$**), (**f**), (**i**), (**V**), (**t**), (**A**), (**d**) and (**X**) are known, then its Leverage or Gearing Ratio Planned is:

/= [**Q**+**X**]/(**U**+{**$**[1-**f**-**i**]-**V**}[1-**t**]-**Ad**)

Rule-13076:

If both (**/**), (**Q**), (**$**), (**f**), (**i**), (**V**), (**t**), (**A**), (**d**) and (**X**) are known, then its Utilized or Starting Capital must be:

U= [**Q**+**X**]/**/**-({**$**[1-**f**-**i**]-**V**}[1-**t**]-**Ad**)

Steve Asikin ISBN 14: 978-1514685136, ISBN 10: 1514685132

Rule-13077:

If both (**/**), (**U**), (**Q**), (**f**), (**i**), (**V**), (**t**), (**A**), (**d**) and (**X**) are known, then its Sales or Revenue Planned is:

$= (**V**+{**Ad**+[**Q**+**X**]/**/U**}/[1-**t**])/[1-**f**-**i**]

Rule-13078:

If both (**/**), (**U**), (**$**), (**Q**), (**i**), (**V**), (**t**), (**A**), (**d**) and (**X**) are known, then its Fixed Portion Planned is:

f= 1-**i**-(**V**+{**Ad**+[**Q**+**X**]/**/U**}/[1-**t**])/**$**

Rule-13079:

If both (**/**), (**U**), (**$**), (**f**), (**Q**), (**V**), (**t**), (**A**), (**d**) and (**X**) are known, then its Interest Portion Planned is:

i= 1-**f**-(**V**+{**Ad**+[**Q**+**X**]/**/U**}/[1-**t**])/**$**

Rule-13080:

If both (**/**), (**U**), (**$**), (**f**), (**i**), (**Q**), (**t**), (**A**), (**d**) and (**X**) are known, then its Variable Cost Planned is:

V= **$**[1-**f**-**i**]-{**Ad**+[**Q**+**X**]/**/U**}/[1-**t**]

Rule-13081:

If both (**/**), (**U**), (**$**), (**f**), (**i**), (**Q**), (**t**), (**A**), (**d**) and (**X**) are known, then its Variable Cost Planned is:

t = 1-{**Ad**+[**Q**+**X**]/**/U**}/{**$**[1-**f**-**i**]-**V**}

Rule-13082:

If both (**/**), (**U**), (**$**), (**f**), (**i**), (**V**), (**t**), (**Q**), (**d**) and (**X**) are known, then its After Tax Income Planned is:

A= ({**$**[1-**f**-**i**]-**V**}[1-**t**]-[**Q**+**X**]/**/U**)/**d**

Steve Asikin ISBN 14: 978-1514685136, ISBN 10: 1514685132

Rule-13083:

If both (**/**), (**U**), (**$**), (**f**), (**i**), (**V**), (**t**), (**A**), (**Q**) and (**X**) are known, then its Dividend Payout Planned is:

d= ({**$**[1-**f**-**i**]-**V**}[1-**t**]-[**Q**+**X**]/**/U**)/**A**

Rule-13084:

If both (**/**), (**U**), (**$**), (**f**), (**i**), (**V**), (**t**), (**A**), (**d**) and (**Q**) are known, then its Xpress or Current Debt Planned is:

X= **/**(**U**+{**$**[1-**f**-**i**]-**V**}[1-**t**]-**Ad**)-**Q**

Rule-13085:

If both (**/**), (**U**), (**$**), (**f**), (**i**), (**V**), (**t**), (**A**), (**d**), (**P**), (**c**) and (**q**) are known, then its Quoted Longterm Debt Planned is:

Q= **/**{**U**+**$**[1-**f**-**i**]-**V**}[1-**t**]-**Ad**}-**P**/[**c**-**q**]

Rule-13086:

If both (**Q**), (**U**), (**$**), (**f**), (**i**), (**V**), (**t**), (**A**), (**d**), (**P**), (**c**) and (**q**) are known, then its Leverage or Gearing Ratio Planned is:

/= {**Q**+**P**/[**c**-**q**]}/(**U**+{**$**[1-**f**-**i**]-**V**}[1-**t**]-**Ad**)

Rule-13087:

If both (**/**), (**Q**), (**$**), (**f**), (**i**), (**V**), (**t**), (**A**), (**d**), (**P**), (**c**) and (**q**) are known, then its Uti8lized or Starting Capital must be:

U= {**Q**+**P**/[**c**-**q**]}/**/**({**$**[1-**f**-**i**]-**V**}[1-**t**]-**Ad**)

Steve Asikin ISBN 14: 978-1514685136, ISBN 10: 1514685132

Rule-13088:

If both (**/**), (**U**), (**Q**), (**f**), (**i**), (**V**), (**t**), (**A**), (**d**), (**P**), (**c**) and (**q**) are known, then its Sales or Revenue Planned is:

$= [**V**+(**Ad**+{**Q**+**P**/[**c**-**q**]}/**/**-**U**)/[1-**t**]]/[1-**f**-**i**]

Rule-13089:

If both (**/**), (**U**), (**$**), (**Q**), (**i**), (**V**), (**t**), (**A**), (**d**), (**P**), (**c**) and (**q**) are known, then its Fixed Portion Planned is:

f= 1-**i**-[**V**+(**Ad**+{**Q**+**P**/[**c**-**q**]}/**/**-**U**)/[1-**t**]]/**$**

Rule-13090:

If both (**/**), (**U**), (**$**), (**f**), (**Q**), (**V**), (**t**), (**A**), (**d**), (**P**), (**c**) and (**q**) are known, then its Interest Expense Planned is:

i= 1-**f**-[**V**+(**Ad**+{**Q**+**P**/[**c**-**q**]}/**/**-**U**)/[1-**t**]]/**$**

Rule-13091:

If both (**/**), (**U**), (**$**), (**f**), (**i**), (**V**), (**t**), (**A**), (**d**), (**P**), (**c**) and (**q**) are known, then its Quoted Longterm Debt Planned is:

V= **$**[1-**f**-**i**]-(**Ad**+{**Q**+**P**/[**c**-**q**]}/**/**-**U**)/[1-**t**]

Rule-13092:

If both (**/**), (**U**), (**$**), (**f**), (**i**), (**V**), (**Q**), (**A**), (**d**), (**P**), (**c**) and (**q**) are known, then its Tax Rate Planned is:

t = 1-(**Ad**+{**Q**+**P**/[**c**-**q**]}/**/**-**U**)/{**$**[1-**f**-**i**]-**V**}

Steve Asikin ISBN 14: 978-1514685136, ISBN 10: 1514685132

Rule-13093:

If both (**/**), (**U**), (**$**), (**f**), (**i**), (**V**), (**t**), (**Q**), (**d**), (**P**), (**c**) and (**q**) are known, then its After Tax Income Planned is:

A= ({**$**[1-**f**-**i**]-**V**}[1-**t**]-{**Q**+**P**/[**c**-**q**]}/**/**-**U**)/**d**

Rule-13094:

If both (**/**), (**U**), (**$**), (**f**), (**i**), (**V**), (**t**), (**A**), (**Q**), (**P**), (**c**) and (**q**) are known, then its Dividend Planned is:

d= ({**$**[1-**f**-**i**]-**V**}[1-**t**]-{**Q**+**P**/[**c**-**q**]}/**/**-**U**)/**A**

Rule-13095:

If both (**/**), (**U**), (**$**), (**f**), (**i**), (**V**), (**t**), (**A**), (**d**), (**Q**), (**c**) and (**q**) are known, then its Procured Inventory Planned is:

P= [**c**-**q**][**/**(**U**+{**$**[1-**f**-**i**]-**V**}[1-**t**]-**Ad**)-**Q**]

Rule-13096:

If both (**/**), (**U**), (**$**), (**f**), (**i**), (**V**), (**t**), (**A**), (**d**), (**P**), (**Q**) and (**q**) are known, then its Current Ratio Planned is:

c= **q**+**P**/[**/**(**U**+{**$**[1-**f**-**i**]-**V**}[1-**t**]-**Ad**)-**Q**]

Rule-13097:

If both (**/**), (**U**), (**$**), (**f**), (**i**), (**V**), (**t**), (**A**), (**d**), (**P**), (**c**) and (**Q**) are known, then its Quick or Acid test Ratio Planned is:

q= **c**-**P**/[**/**(**U**+{**$**[1-**f**-**i**]-**V**}[1-**t**]-**Ad**)-**Q**]

Steve Asikin ISBN 14: 978-1514685136, ISBN 10: 1514685132

Rule-13098:

If both (**/**), (**U**), (**$**), (**f**), (**i**), (**V**), (**t**), (**A**), (**d**), (**p**), (**c**) and (**q**) are known, then its Quoted Longterm Debt Planned is:

Q= **/**{**U**+**$**[1-**f**-**i**]-**V**}[1-**t**]-**Ad**}-**Vp**/{360[**c**-**q**]}

Rule-13099:

If both (**Q**), (**U**), (**$**), (**f**), (**i**), (**V**), (**t**), (**A**), (**d**), (**p**), (**c**) and (**q**) are known, then its Leverage or Gearing Ratio Planned is:

/= (**Q**+**Vp**/{360[**c**-**q**]})/(**U**+{**$**[1-**f**-**i**]-**V**}[1-**t**]-**Ad**)

Rule-13100:

If both (**/**), (**Q**), (**$**), (**f**), (**i**), (**V**), (**t**), (**A**), (**d**), (**p**), (**c**) and (**q**) are known, then its Utilized or Starting Capital Planned is:

U= (**Q**+**Vp**/{360[**c**-**q**]})/**/**-({**$**[1-**f**-**i**]-**V**}[1-**t**]-**Ad**)

Rule-13101:

If both (**/**), (**U**), (**Q**), (**f**), (**i**), (**V**), (**t**), (**A**), (**d**), (**p**), (**c**) and (**q**) are known, then its Sales or Revenue Planned is:

$= {**V**+[**Ad**+(**Q**+**Vp**/{360[**c**-**q**]})/**/**-**U**]/[1-**t**]}/[1-**f**-**i**]

Steve Asikin ISBN 14: 978-1514685136, ISBN 10: 1514685132

Rule-13102:

If both (**/**), (**U**), (**$**), (**f**), (**i**), (**V**), (**t**), (**A**), (**d**), (**p**), (**c**) and (**q**) are known, then its Fixed Portion Planned is:

f= 1-**i**-{**V**+[**Ad**+(**Q**+**Vp**/{360[**c**-**q**]})/**/**-**U**]/[1-**t**]}/**$**

Rule-13103:

If both (**/**), (**U**), (**$**), (**f**), (**Q**), (**V**), (**t**), (**A**), (**d**), (**p**), (**c**) and (**q**) are known, then its Interest Portion Planned is:

i= 1-**f**-{**V**+[**Ad**+(**Q**+**Vp**/{360[**c**-**q**]})/**/**-**U**]/[1-**t**]}/**$**

Rule-13104:

If both (**/**), (**U**), (**$**), (**f**), (**i**), (**Q**), (**t**), (**A**), (**d**), (**p**), (**c**) and (**q**) are known, then its Variable Cost Planned is:

V= (**/**{**U**+**$**[1-**f**-**i**]}[1-**t**]-**Ad**}-**Q**)
/(**p**/{360[**c**-**q**]}+**/**[1-**t**])

Rule-13105:

If both (**/**), (**U**), (**$**), (**f**), (**i**), (**V**), (**Q**), (**A**), (**d**), (**p**), (**c**) and (**q**) are known, then its Tax Rate Planned is:

t = 1-[**Ad**+(**Q**+**Vp**/{360[**c**-**q**]})/**/**-**U**]/{**$**[1-**f**-**i**]-**V**}

Rule-13106:

If both (**/**), (**U**), (**$**), (**f**), (**i**), (**V**), (**t**), (**Q**), (**d**), (**p**), (**c**) and (**q**) are known, then its After Tax Income Planned is:

A= [{**$**[1-**f**-**i**]-**V**}[1-**t**]-(**Q**+**Vp**/{360[**c**-**q**]})/**/**-**U**]/**d**

Steve Asikin ISBN 14: 978-1514685136, ISBN 10: 1514685132

Rule-13107:

If both (/), (U), ($), (f), (i), (V), (t), (A), (Q), (p), (c) and (q) are known, then its Dividend Payout Planned is:

d= [{$[1-f-i]-V}[1-t]-(Q+Vp/{360[c-q]})/-U]/A

Rule-13108:

If both (/), (U), ($), (f), (i), (V), (t), (A), (d), (Q), (c) and (q) are known, then its Procured Inventory Days planned:

p= 360[c-q][/(U+{$[1-f-i]-V}[1-t]-Ad)-Q]/V

Rule-13109:

If both (/), (U), ($), (f), (i), (V), (t), (A), (d), (p), (Q) and (q) are known, then its Current Ratio Planned is:

c= q+Vp/{360[/(U+{$[1-f-i]-V}[1-t]-Ad)-Q]}

Rule-13110:

If both (/), (U), ($), (f), (i), (V), (t), (A), (d), (p), (c) and (Q) are known, then its Quick or Acid Test Ratio Planned is:

q= c-Vp/{360[/(U+{$[1-f-i]-V}[1-t]-Ad)-Q]}

Rule-13111:

If both (/), (U), ($), (f), (i), (V), (t), (A), (d), (v), (p), (c) and (q) are known, then its Quoted Longterm Debt Planned is:

Q= /(U+{$[1-f-i]-V}[1-t]-Ad)-$vp/{360[c-q]}

Steve Asikin ISBN 14: 978-1514685136, ISBN 10: 1514685132

Rule-13112:

If both (**Q**), (**U**), (**$**), (**f**), (**i**), (**V**), (**t**), (**A**), (**d**), (**v**), (**p**), (**c**) and (**q**) are known, then its Leverage or Gearing Ratio Planned is:

/= (Q+Vp/{360[c-q]})/(U+{$[1-f-i]-V}[1-t]-Ad)

Rule-13113:

If both (**/**), (**Q**), (**$**), (**f**), (**i**), (**V**), (**t**), (**A**), (**d**), (**v**), (**p**), (**c**) and (**q**) are known, then its Utilized or Starting Capital must be:

U= (Q+Vp/{360[c-q]})//-({$[1-f-i]-V}[1-t]-Ad)

Rule-13114:

If both (**/**), (**U**), (**Q**), (**f**), (**i**), (**V**), (**t**), (**A**), (**d**), (**v**), (**p**), (**c**) and (**q**) are known, then its Sales or Revenue Planned is:

$= (/{U-V[1-t]-Ad}-Q)/(vp/{360[c-q]}-/[1-f-i])

Rule-13115:

If both (**/**), (**U**), (**$**), (**Q**), (**i**), (**V**), (**t**), (**A**), (**d**), (**v**), (**p**), (**c**) and (**q**) are known, then its Fixed Portion Planned is:

f= 1-i-{V+[Ad+(Q+$vp/{360[c-q]})//-U]/[1-t]}/$

Rule-13116:

If both (**/**), (**U**), (**$**), (**f**), (**Q**), (**V**), (**t**), (**A**), (**d**), (**v**), (**p**), (**c**) and (**q**) are known, then its Interest Portion Planned is:

i= 1-f-{V+[Ad+(Q+$vp/{360[c-q]})//-U]/[1-t]}/$

Steve Asikin ISBN 14: 978-1514685136, ISBN 10: 1514685132

Rule-13117:

If both (/), (U), (S), (f), (i), (Q), (t), (A), (d), (v), (p), (c) and (q) are known, then its Variable Cost Planned is:

V= (/{U+S[1-f-i]}[1-t]-Ad}
-Svp/{360[c-q]}-Q)/{/[1-t]}

Rule-13118:

If both (/), (U), (S), (f), (i), (V), (Q), (A), (d), (v), (p), (c) and (q) are known, then its Tax Rate Planned is:

t = 1-[Ad+(Q+Svp/{360[c-q]})//-U]/{S[1-f-i]-V}

Rule-13119:

If both (/), (U), (S), (f), (i), (V), (t), (Q), (d), (v), (p), (c) and (q) are known, then its After Tax Income Planned is:

A= [{S[1-f-i]-V}[1-t]-(Q+Svp/{360[c-q]})//-U]/d

Rule-13120:

If both (/), (U), (S), (f), (i), (V), (t), (A), (Q), (v), (p), (c) and (q) are known, then its Dividend Payout Planned is:

d= [{S[1-f-i]-V}[1-t]-(Q+Svp/{360[c-q]})//-U]/A

Rule-13121:

If both (/), (U), (S), (f), (i), (V), (t), (A), (d), (Q), (p), (c) and (q) are known, then its Variable Portion Planned is:

s= V/S

or it can also be found as

v= 360[c-q][/(U+{S[1-f-i]-V}[1-t]-Ad)-Q]/[Sp]

Steve Asikin ISBN 14: 978-1514685136, ISBN 10: 1514685132

Rule-13122:

If both (**/**), (**U**), (**$**), (**f**), (**i**), (**V**), (**t**), (**A**), (**d**), (**v**), (**Q**), (**c**) and (**q**) are known, then its Procured Inventory Days planned:

p= 360[**c-q**][**/**(**U**+{**$**[1-**f-i**]-**V**}[1-**t**]-**Ad**)-**Q**]/[**$v**]

Rule-13123:

If both (**/**), (**U**), (**$**), (**f**), (**i**), (**V**), (**t**), (**A**), (**d**), (**v**), (**p**), (**Q**) and (**q**) are known, then its Current Ratio Planned is:

c= **q**+**$vp**/{360[**/**(**U**+{**$**[1-**f-i**]-**V**}[1-**t**]-**Ad**)-**Q**]}

Rule-13124:

If both (**/**), (**U**), (**$**), (**f**), (**i**), (**V**), (**t**), (**A**), (**d**), (**v**), (**p**), (**c**) and (**Q**) are known, then its Quick or Acid Test Ratio Planned is:

q= **c**-**$vp**/{360[**/**(**U**+{**$**[1-**f-i**]-**V**}[1-**t**]-**Ad**)-**Q**]}

Rule-13125:

If both (**/**), (**U**), (**$**), (**f**), (**i**), (**V**), (**t**), (**A**), (**d**), (**$'**), (**v**), (**p**), (**s**), (**c**) and (**q**) are known, then its Quoted Longterm Debt Planned is:

Q= **/**(**U**+{**$**[1-**f-i**]-**V**}[1-**t**]-**Ad**)
-**$'vp**[1+**s**]/{360[**c-q**]}

Rule-13126:

If both (**Q**), (**U**), (**$**), (**f**), (**i**), (**V**), (**t**), (**A**), (**d**), (**$'**), (**v**), (**p**), (**s**), (**c**) and (**q**) are known, then its Leverage or Gearing Ratio Planned is:

/= (**Q**+**$'vp**[1+**s**]/{360[**c-q**]})
/(**U**+{**$**[1-**f-i**]-**V**}[1-**t**]-**Ad**)

Steve Asikin ISBN 14: 978-1514685136, ISBN 10: 1514685132

Rule-13127:

If both (**/**), (**Q**), (**S**), (**f**), (**i**), (**V**), (**t**), (**A**), (**d**), (**S'**), (**v**), (**p**), (**s**), (**c**) and (**q**) are known, then its Utilized or Starting Capital must be:

U= (**Q**+**S'vp**[1+**s**]/{360[**c**-**q**]})/**/**
-({**S**[1-**f**-**i**]-**V**}[1-**t**]-**Ad**)

Rule-13128:

If both (**/**), (**U**), (**Q**), (**f**), (**i**), (**V**), (**t**), (**A**), (**d**), (**S'**), (**v**), (**p**), (**s**), (**c**) and (**q**) are known, then its Sales or Revenue Planned is:

S= (**Q**-**/**{**U**-**V**[1-**t**]-**Ad**}+**S'vp**[1+**s**]/{360[**c**-**q**]})
/{[1-**f**-**i**][1-**t**]}

Rule-13129:

If both (**/**), (**U**), (**S**), (**Q**), (**i**), (**V**), (**t**), (**A**), (**d**), (**S'**), (**v**), (**p**), (**s**), (**c**) and (**q**) are known, then its Fixed Portion Planned is:

f= 1-**i**-{**V**+[**Ad**+(**Q**+**S'vp**[1+**s**]/{360[**c**-**q**]})/**/**-**U**]
/[1-**t**]}/**S**

Rule-13130:

If both (**/**), (**U**), (**S**), (**f**), (**Q**), (**V**), (**t**), (**A**), (**d**), (**S'**), (**v**), (**p**), (**s**), (**c**) and (**q**) are known, then its Interest Portion Planned is:

i= 1-**f**-{**V**+[**Ad**+(**Q**+**S'vp**[1+**s**]/{360[**c**-**q**]})/**/**-**U**]
/[1-**t**]}/**S**

Steve Asikin ISBN 14: 978-1514685136, ISBN 10: 1514685132

Rule-13131:

If both (**/**), (**U**), (**$**), (**f**), (**i**), (**Q**), (**t**), (**A**), (**d**), (**$'**), (**v**), (**p**), (**s**), (**c**) and (**q**) are known, then its Variable Cost Planned is:

V= (**/**{**U**+**$**[1-**f**-**i**]}[1-**t**]-**Ad**}
-**$'vp**[1+**s**]/{360[**c**-**q**]}-**Q**)/{**/**[1-**t**]}

Rule-13132:

If both (**/**), (**U**), (**$**), (**f**), (**i**), (**V**), (**Q**), (**A**), (**d**), (**$'**), (**v**), (**p**), (**s**), (**c**) and (**q**) are known, then its Tax Rate Planned is:

t = 1-[**Ad**+(**Q**+**$vp**[1+**s**]/{360[**c**-**q**]})/**/**-**U**]
/{**$**[1-**f**-**i**]-**V**}

Rule-13133:

If both (**/**), (**U**), (**$**), (**f**), (**i**), (**V**), (**t**), (**Q**), (**d**), (**$'**), (**v**), (**p**), (**s**), (**c**) and (**q**) are known, then its After Tax Income Planned is:

A= [{**$**[1-**f**-**i**]-**V**}[1-**t**]
-(**Q**+**$'vp**[1+**s**]/{360[**c**-**q**]})/**/**-**U**]/**d**

Rule-13134:

If both (**/**), (**U**), (**$**), (**f**), (**i**), (**V**), (**t**), (**A**), (**Q**), (**$'**), (**v**), (**p**), (**s**), (**c**) and (**q**) are known, then its Dividend Payout Planned is:

d= [{**$**[1-**f**-**i**]-**V**}[1-**t**]
-(**Q**+**$'vp**[1+**s**]/{360[**c**-**q**]})/**/**-**U**]/**A**

Steve Asikin ISBN 14: 978-1514685136, ISBN 10: 1514685132

Rule-13135:

If both (**/**), (**U**), (**$**), (**f**), (**i**), (**V**), (**t**), (**A**), (**d**), (**Q**), (**v**), (**p**), (**s**), (**c**) and (**q**) are known, then its Sales Past must be:

$'**= 360[**c-q**][**/**(**U**+{**$[1-**f-i**]-**V**}[1-**t**]-**Ad**)-**Q**]
/{**vp**[1+**s**]}

Rule-13136:

If both (**/**), (**U**), (**$**), (**f**), (**i**), (**V**), (**t**), (**A**), (**d**), (**$'**), (**Q**), (**p**), (**s**), (**c**) and (**q**) are known, then its Variable Cost Planned is:

v= **V**/**$**

or it can also be found as

v= 360[**c-q**][**/**(**U**+{**$**[1-**f-i**]-**V**}[1-**t**]-**Ad**)-**Q**]
/{**$'p**[1+**s**]}

Rule-13137:

If both (**/**), (**U**), (**$**), (**f**), (**i**), (**V**), (**t**), (**A**), (**d**), (**$'**), (**v**), (**Q**), (**s**), (**c**) and (**q**) are known, then its Procured Inventory Days planned:

p= 360[**c-q**][**/**(**U**+{**$**[1-**f-i**]-**V**}[1-**t**]-**Ad**)-**Q**]
/{**$'v**[1+**s**]}

Rule-13138:

If both (**/**), (**U**), (**$**), (**f**), (**i**), (**V**), (**t**), (**A**), (**d**), (**$'**), (**v**), (**p**), (**Q**), (**c**) and (**q**) are known, then its Sales Growth Planned is:

s= 360[**c-q**][**/**(**U**+{**$**[1-**f-i**]-**V**}[1-**t**]-**Ad**)-**Q**]
/[**$'vp**]-1

Steve Asikin ISBN 14: 978-1514685136, ISBN 10: 1514685132

Rule-13139:

If both (**/**), (**U**), (**S**), (**f**), (**i**), (**V**), (**t**), (**A**), (**d**), (**S'**), (**v**), (**p**), (**s**), (**Q**) and (**q**) are known, then its Current Ratio Planned is:

c= **q**+**S'vp**[1+**s**]
/{360[**/**(**U**+{**S**[1-**f**-**i**]-**V**}[1-**t**]-**Ad**)-**Q**]}

Rule-13140:

If both (**/**), (**U**), (**S**), (**f**), (**i**), (**V**), (**t**), (**A**), (**d**), (**S'**), (**v**), (**p**), (**s**), (**c**) and (**Q**) are known, then its Quick or Acid Test Ratio Planned is:

q= **c**-**S'vp**[1+**s**]
/{360[**/**(**U**+{**S**[1-**f**-**i**]-**V**}[1-**t**]-**Ad**)-**Q**]}

Rule-13141:

If both (**/**), (**U**), (**S**), (**f**), (**i**), (**V**), (**t**), (**d**) and (**X**) are known, then its Quoted Longterm Debt Planned is:

Q= **/**(**U**+{**S**[1-**f**-**i**]-**V**}[1-**t**][1-**d**]}-**X**

Rule-13142:

If both (**Q**), (**U**), (**S**), (**f**), (**i**), (**V**), (**t**), (**d**) and (**X**) are known, then its Leverage or Gearing Ratio Planned is:

/= [**Q**+**X**]/{**U**+{**S**[1-**f**-**i**]-**V**}[1-**t**][1-**d**]}

Rule-13143:

If both (**/**), (**Q**), (**S**), (**f**), (**i**), (**V**), (**t**), (**d**) and (**X**) are known, then its Utilized or Starting Capital must be:

U= [**Q**+**X**]/**/**-{**S**[1-**f**-**i**]-**V**}[1-**t**][1-**d**]

Steve Asikin ISBN 14: 978-1514685136, ISBN 10: 1514685132

Rule-13144:

If both (/), (**U**), (**Q**), (**f**), (**i**), (**V**), (**t**), (**d**) and (**X**) are known, then its Sales or Revenue Planned is:

$= (**V**+{[**Q**+**X**]//**U**}/{[1-**t**][1-**d**]})/[1-**f**-**i**]

Rule-13145:

If both (/), (**U**), (**$**), (**Q**), (**i**), (**V**), (**t**), (**d**) and (**X**) are known, then its Fixed Portion Planned is:

f= 1-**i**-(**V**+{[**Q**+**X**]//**U**}/{[1-**t**][1-**d**]})/**$**

Rule-13146:

If both (/), (**U**), (**$**), (**f**), (**Q**), (**V**), (**t**), (**d**) and (**X**) are known, then its Quoted Longterm Debt Planned is:

i= 1-**f**-(**V**+{[**Q**+**X**]//**U**}/{[1-**t**][1-**d**]})/**$**

Rule-13147:

If both (/), (**U**), (**$**), (**f**), (**i**), (**Q**), (**t**), (**d**) and (**X**) are known, then its Variable Cost Planned is:

V= **$**[1-**f**-**i**]-{[**Q**+**X**]//**U**}/{[1-**t**][1-**d**]}

Rule-13148:

If both (/), (**U**), (**$**), (**f**), (**i**), (**V**), (**Q**), (**d**) and (**X**) are known, then its Tax Rate Planned is:

t= 1-{[**Q**+**X**]//**U**}/({**$**[1-**f**-**i**]-**V**}[1-**d**])

Rule-13149:

If both (/), (**U**), (**$**), (**f**), (**i**), (**V**), (**t**), (**Q**) and (**X**) are known, then its Dividend Payout Planned is:

d= 1-{[**Q**+**X**]//**U**}/({**$**[1-**f**-**i**]-**V**}[1-**t**])

Steve Asikin ISBN 14: 978-1514685136, ISBN 10: 1514685132

Rule-13150:

If both (**/**), (**U**), (**$**), (**f**), (**i**), (**V**), (**t**), (**d**) and (**Q**) are known, then its Xpress or Current Debt Planned is:

X= **/**{**U**+**$**[1-**f**-**i**]-**V**}[1-**t**][1-**d**]}-**Q**

Rule-13151:

If both (**/**), (**U**), (**$**), (**f**), (**i**), (**V**), (**t**), (**d**), (**P**), (**c**) and (**q**) are known, then its Quoted Longterm Debt Planned is:

Q= **/**{**U**+**$**[1-**f**-**i**]-**V**}[1-**t**][1-**d**]}-**P**/[**c**-**q**]

Rule-13152:

If both (**Q**), (**U**), (**$**), (**f**), (**i**), (**V**), (**t**), (**d**), (**P**), (**c**) and (**q**) are known, then its Leverage or Gearing Ratio Planned is:

/= {**Q**+**P**/[**c**-**q**]}/{**U**+{**$**[1-**f**-**i**]-**V**}[1-**t**][1-**d**]}

Rule-13153:

If both (**/**), (**Q**), (**$**), (**f**), (**i**), (**V**), (**t**), (**d**), (**P**), (**c**) and (**q**) are known, then its Utilized or Starting Capital must be:

U= {**Q**+**P**/[**c**-**q**]}/**/**-{**$**[1-**f**-**i**]-**V**}[1-**t**][1-**d**]

Rule-13154:

If both (**/**), (**U**), (**Q**), (**f**), (**i**), (**V**), (**t**), (**d**), (**P**), (**c**) and (**q**) are known, then its Sales or Revenue Planned is:

$= [**V**+({**Q**+**P**/[**c**-**q**]}/**/**-**U**]/{[1-**t**][1-**d**]})/[1-**f**-**i**]

Steve Asikin ISBN 14: 978-1514685136, ISBN 10: 1514685132

Rule-13155:

If both (**/**), (**U**), (**$**), (**Q**), (**i**), (**V**), (**t**), (**d**), (**P**), (**c**) and (**q**) are known, then its Fixed Portion Planned is:

f= 1-**i**-[**V**+({**Q**+**P**/[**c**-**q**]}/**/**-**U**)/{[1-**t**][1-**d**]}]/**$**

Rule-13156:

If both (**/**), (**U**), (**$**), (**f**), (**Q**), (**V**), (**t**), (**d**), (**P**), (**c**) and (**q**) are known, then its Interest Portion Planned is:

i= 1-**f**-[**V**+({**Q**+**P**/[**c**-**q**]}/**/**-**U**)/{[1-**t**][1-**d**]}]/**$**

Rule-13157:

If both (**/**), (**U**), (**$**), (**f**), (**i**), (**Q**), (**t**), (**d**), (**P**), (**c**) and (**q**) are known, then its Variable Cost Planned is:

V= **$**[1-**f**-**i**]-({**Q**+**P**/[**c**-**q**]}/**/**-**U**)/{[1-**t**][1-**d**]}

Rule-13158:

If both (**/**), (**U**), (**$**), (**f**), (**i**), (**V**), (**Q**), (**d**), (**P**), (**c**) and (**q**) are known, then its Tax Rate Planned is:

t= 1-({**Q**+**P**/[**c**-**q**]}/**/**-**U**)/({**$**[1-**f**-**i**]-**V**}[1-**d**])

Rule-13159:

If both (**/**), (**U**), (**$**), (**f**), (**i**), (**V**), (**t**), (**Q**), (**P**), (**c**) and (**q**) are known, then its Dividend Payout Planned is:

d= 1-({**Q**+**P**/[**c**-**q**]}/**/**-**U**)/({**$**[1-**f**-**i**]-**V**}[1-**t**])

Rule-13160:

If both (**/**), (**U**), (**$**), (**f**), (**i**), (**V**), (**t**), (**d**), (**Q**), (**c**) and (**q**) are known, then its Procured Inventory Planned is:

P= [**c**-**q**](**/**{**U**+**$**[1-**f**-**i**]-**V**}[1-**t**][1-**d**]}-**Q**)

Steve Asikin ISBN 14: 978-1514685136, ISBN 10: 1514685132

Rule-13161:

If both (**/**), (**U**), (**$**), (**f**), (**i**), (**V**), (**t**), (**d**), (**P**), (**Q**) and (**q**) are known, then its Current Ratio Planned is:

c= **q**+**P**/(**/**{**U**+**$**[1-**f**-**i**]-**V**}[1-**t**][1-**d**]}-**Q**)

Rule-13162:

If both (**/**), (**U**), (**$**), (**f**), (**i**), (**V**), (**t**), (**d**), (**P**), (**c**) and (**Q**) are known, then its Quick or Acid Test Ratio Planned is:

q= **c**-**P**/(**/**{**U**+**$**[1-**f**-**i**]-**V**}[1-**t**][1-**d**]}-**Q**)

Rule-13163:

If both (**/**), (**U**), (**$**), (**f**), (**i**), (**V**), (**t**), (**d**), (**p**), (**c**) and (**q**) are known, then its Quoted Longterm Debt Planned is:

Q= **/**{**U**+**$**[1-**f**-**i**]-**V**}[1-**t**][1-**d**]}-**Vp**/{360[**c**-**q**]}

Rule-13164:

If both (**Q**), (**U**), (**$**), (**f**), (**i**), (**V**), (**t**), (**d**), (**p**), (**c**) and (**q**) are known, then its Leverage or Gearing Ratio Planned is:

/= (**Q**+**Vp**/{360[**c**-**q**])/{**U**+{**$**[1-**f**-**i**]-**V**}[1-**t**][1-**d**]}

Rule-13165:

If both (**/**), (**Q**), (**$**), (**f**), (**i**), (**V**), (**t**), (**d**), (**p**), (**c**) and (**q**) are known, then its Utilized or Starting Capital must be:

U= (**Q**+**Vp**/{360[**c**-**q**])/**/**-{**$**[1-**f**-**i**]-**V**}[1-**t**][1-**d**]

Steve Asikin ISBN 14: 978-1514685136, ISBN 10: 1514685132

Rule-13166:

If both (/), (**U**), (**Q**), (**f**), (**i**), (**V**), (**t**), (**d**), (**p**), (**c**) and (**q**) are known, then its Sales or Revenue Planned is:

$= {**V**+[(**Q**+**Vp**/{360[**c**-**q**])/**/-U**] /{[1-**t**][1-**d**]}}/[1-**f**-**i**]

Rule-13167:

If both (/), (**U**), (**$**), (**Q**), (**i**), (**V**), (**t**), (**d**), (**p**), (**c**) and (**q**) are known, then its Fixed Portion Planned is:

f= 1-**i**-{**V**+[(**Q**+**Vp**/{360[**c**-**q**])/**/-U**]/{[1-**t**][1-**d**]}}/**$**

Rule-13168:

If both (/), (**U**), (**$**), (**f**), (**Q**), (**V**), (**t**), (**d**), (**p**), (**c**) and (**q**) are known, then its Interest Portion Planned is:

i= 1-**f**-{**V**+[(**Q**+**Vp**/{360[**c**-**q**])/**/-U**]/{[1-**t**][1-**d**]}}/**$**

Rule-13169:

If both (/), (**U**), (**$**), (**f**), (**i**), (**Q**), (**t**), (**d**), (**p**), (**c**) and (**q**) are known, then its Variable Cost Planned is:

V= (/{**U**+**$**[1-**f**-**i**]}[1-**t**][1-**d**]}-**Q**)/(**p**/{360[**c**-**q**]}+/[1-**t**][1-**d**])

Rule-13170:

If both (/), (**U**), (**$**), (**f**), (**i**), (**V**), (**t**), (**d**), (**p**), (**c**) and (**q**) are known, then its Tax Rate Planned is:

t= 1-[(**Q**+**Vp**/{360[**c**-**q**])/**/-U**]/({**$**[1-**f**-**i**]-**V**}[1-**d**])

Rule-13171:

If both (/), (**U**), (**$**), (**f**), (**i**), (**V**), (**t**), (**Q**), (**p**), (**c**) and (**q**) are known, then its Dividend Payout Planned is:

d= 1-[(**Q**+**Vp**/{360[**c**-**q**])/**/-U**]/({**$**[1-**f**-**i**]-**V**}[1-**t**])

Steve Asikin ISBN 14: 978-1514685136, ISBN 10: 1514685132

Rule-13172:

If both (**/**), (**U**), (**$**), (**f**), (**i**), (**V**), (**t**), (**d**), (**Q**), (**c**) and (**q**) are known, then its Procured Inventory Days Planned:

p= 360[**c**-**q**](**/**{**U**+**$**[1-**f**-**i**]-**V**}[1-**t**][1-**d**]}-**Q**)/**V**

Rule-13173:

If both (**/**), (**U**), (**$**), (**f**), (**i**), (**V**), (**t**), (**d**), (**p**), (**Q**) and (**q**) are known, then its Current Ratio Planned is:

c= **q**+**Vp**/[360(**/**{**U**+**$**[1-**f**-**i**]-**V**}[1-**t**][1-**d**]}-**Q**)]

Rule-13174:

If both (**/**), (**U**), (**$**), (**f**), (**i**), (**V**), (**t**), (**d**), (**p**), (**c**) and (**Q**) are known, then its Quick or Acid Test Ratio Planned is:

q= **c**-**Vp**/[360(**/**{**U**+**$**[1-**f**-**i**]-**V**}[1-**t**][1-**d**]}-**Q**)]

Rule-13175:

If both (**/**), (**U**), (**$**), (**f**), (**i**), (**V**), (**t**), (**d**), (**v**), (**p**), (**c**) and (**q**) are known, then its Quoted Longterm Debt Planned is:

Q= **/**{**U**+**$**[1-**f**-**i**]-**V**}[1-**t**][1-**d**]}-**$vp**/{360[**c**-**q**]}

Rule-13176:

If both (**Q**), (**U**), (**$**), (**f**), (**i**), (**V**), (**t**), (**d**), (**v**), (**p**), (**c**) and (**q**) are known, then its Quoted Longterm Debt Planned is:

/= (**Q**+**$vp**/{360[**c**-**q**])/{**U**+{**$**[1-**f**-**i**]-**V**}[1-**t**][1-**d**]}

Steve Asikin ISBN 14: 978-1514685136, ISBN 10: 1514685132

Rule-13177:

If both (**Q**), (**Q**), (**$**), (**f**), (**i**), (**V**), (**t**), (**d**), (**v**), (**p**), (**c**) and (**q**) are known, then its Utilized or Starting Capital must be:

U= (**Q**+**$vp**/{360[**c**-**q**])/**/**-{**$**[1-**f**-**i**]-**V**}[1-**t**][1-**d**]

Rule-13178:

If both (**Q**), (**U**), (**Q**), (**f**), (**i**), (**V**), (**t**), (**d**), (**v**), (**p**), (**c**) and (**q**) are known, then its Sales or Revenue Planned is:

$**= (**/**{**U**+**$[1-**f**-**i**]-**V**}[1-**t**][1-**d**]}-**Q**)
/(**vp**/{360[**c**-**q**]}-**/**[1-**t**][1-**d**])

Rule-13179:

If both (**Q**), (**U**), (**$**), (**Q**), (**i**), (**V**), (**t**), (**d**), (**v**), (**p**), (**c**) and (**q**) are known, then its Fixed Portion Planned is:

f= 1-**i**-{**V**+[(**Q**+**$vp**/{360[**c**-**q**])/**/**-**U**]
/{[1-**t**][1-**d**]}}/**$**

Rule-13180:

If both (**Q**), (**U**), (**$**), (**f**), (**Q**), (**V**), (**t**), (**d**), (**v**), (**p**), (**c**) and (**q**) are known, then itsInterest Portion Planned is:

i= 1-**f**-{**V**+[(**Q**+**$vp**/{360[**c**-**q**])/**/**-**U**]
/{[1-**t**][1-**d**]}}/**$**

Rule-13181:

If both (**Q**), (**U**), (**$**), (**f**), (**i**), (**Q**), (**t**), (**d**), (**v**), (**p**), (**c**) and (**q**) are known, then its Variable Cost Planned is:

V= (**/**{**U**+**$**[1-**f**-**i**]}[1-**t**][1-**d**]}+**$vp**/{360[**c**-**q**]}-**Q**)
/{**/**[1-**t**][1-**d**]}

Steve Asikin ISBN 14: 978-1514685136, ISBN 10: 1514685132

Rule-13182:

If both (**Q**), (**U**), (**$**), (**f**), (**i**), (**V**), (**Q**), (**d**), (**v**), (**p**), (**c**) and (**q**) are known, then its Tax Rate Planned is:

t= 1-**[**(**Q**+**$vp**/{360[**c-q**])/**/-U]**/({**$**[1-**f-i**]-**V**}[1-**d**])

Rule-13183:

If both (**Q**), (**U**), (**$**), (**f**), (**i**), (**V**), (**t**), (**Q**), (**v**), (**p**), (**c**) and (**q**) are known, then its Dividend Payout Planned is:

d= 1-**[**(**Q**+**$vp**/{360[**c-q**])/**/-U]**/({**$**[1-**f-i**]-**V**}[1-**t**])

Rule-13184:

If both (**Q**), (**U**), (**$**), (**f**), (**i**), (**V**), (**t**), (**d**), (**Q**), (**p**), (**c**) and (**q**) are known, then its Variable Portion Planned is:

s= **V**/**$**

or it can also be found as

v= 360[**c-q**](**/**{**U**+**$**[1-**f-i**]-**V**}[1-**t**][1-**d**]}-**Q**)/[**$p**]

Rule-13185:

If both (**Q**), (**U**), (**$**), (**f**), (**i**), (**V**), (**t**), (**d**), (**v**), (**Q**), (**c**) and (**q**) are known, then its Procured Inventory Days Planned:

p= 360[**c-q**](**/**{**U**+**$**[1-**f-i**]-**V**}[1-**t**][1-**d**]}-**Q**)/[**$v**]

Rule-13186:

If both (**Q**), (**U**), (**$**), (**f**), (**i**), (**V**), (**t**), (**d**), (**v**), (**p**), (**Q**) and (**q**) are known, then its Current Ratio Planned is:

c= **q**+**$vp**/**[**360(**/**{**U**+**$**[1-**f-i**]-**V**}[1-**t**][1-**d**]}-**Q**)**]**

Steve Asikin ISBN 14: 978-1514685136, ISBN 10: 1514685132

Rule-13187:

If both (**Q**), (**U**), (**S**), (**f**), (**i**), (**V**), (**t**), (**d**), (**v**), (**p**), (**c**) and (**Q**) are known, then its Quick or Acid Test Ratio Planned is:

q= **c**-**Svp**/[360(**/**{**U**+**S**[1-**f**-**i**]-**V**}[1-**t**][1-**d**]}-**Q**)]

Rule-13188:

If both (**Q**), (**U**), (**S**), (**f**), (**i**), (**V**), (**t**), (**d**), (**S'**), (**v**), (**p**), (**s**), (**c**) and (**q**) are known, then its Quoted Longterm Debt Planned is:

Q= **/**{**U**+**S**[1-**f**-**i**]-**V**}[1-**t**][1-**d**]}
-**S'vp**[1+**s**]/{360[**c**-**q**]}

Rule-13189:

If both (**Q**), (**U**), (**S**), (**f**), (**i**), (**V**), (**t**), (**d**), (**S'**), (**v**), (**p**), (**s**), (**c**) and (**q**) are known, then its Leverage or Gearing Ratio Planned is:

/= (**Q**+**S'vp**[1+**s**]/{360[**c**-**q**])
/{**U**+{**S**[1-**f**-**i**]-**V**}[1-**t**][1-**d**]}

Rule-13190:

If both (**/**), (**Q**), (**S**), (**f**), (**i**), (**V**), (**t**), (**d**), (**S'**), (**v**), (**p**), (**s**), (**c**) and (**q**) are known, then its Utilized or Starting Capital must be:

U= (**Q**+**S'vp**[1+**s**]/{360[**c**-**q**])/**/**
-{**S**[1-**f**-**i**]-**V**}[1-**t**][1-**d**]

Steve Asikin ISBN 14: 978-1514685136, ISBN 10: 1514685132

Rule-13191:

If both (/), (U), (Q), (f), (i), (V), (t), (d), (S'), (v), (p), (s), (c) and (q) are known, then its Sales or Revenue Planned is:

S= (Q-/{U-V}[1-t][1-d]}+S'vp[1+s]/{360[c-q]})
/{/[1-f-i][1-t][1-d]}

Rule-13192:

If both (/), (U), (S), (Q), (i), (V), (t), (d), (S'), (v), (p), (s), (c) and (q) are known, then its Fixed Rate Planned is:

f= 1-i-{V+[(Q+S'vp[1+s]/{360[c-q])/-U]
/{[1-t][1-d]}}/S

Rule-13193:

If both (/), (U), (S), (f), (Q), (V), (t), (d), (S'), (v), (p), (s), (c) and (q) are known, then its Interest Portion Planned is:

i= 1-f-{V+[(Q+S'vp[1+s]/{360[c-q])/-U]
/{[1-t][1-d]}}/S

Rule-13194:

If both (/), (U), (S), (f), (i), (Q), (t), (d), (S'), (v), (p), (s), (c) and (q) are known, then its Variable Cost Planned is:

V= (Q-/{U+S[1-f-i]}[1-t][1-d]}
+S'vp[1+s]/{360[c-q]})/{/[1-t][1-d]}

Steve Asikin ISBN 14: 978-1514685136, ISBN 10: 1514685132

Rule-13195:

If both $(/)$, (U), $(\$)$, (f), (i), (V), (Q), (d), $(\$')$, (v), (p), (s), (c) and (q) are known, then its Tax Rate Planned is:

$$t= 1-[(Q+\$'vp[1+s]/\{360[c-q]\})/\,/-U]\,/(\{\$[1-f-i]-V\}[1-d])$$

Rule-13196:

If both $(/)$, (U), $(\$)$, (f), (i), (V), (t), (Q), $(\$')$, (v), (p), (s), (c) and (q) are known, then its Dividend Payout Planned is:

$$d= 1-[(Q+\$'vp[1+s]/\{360[c-q]\})/\,/-U]\,/(\{\$[1-f-i]-V\}[1-t])$$

Rule-13197:

If both $(/)$, (U), $(\$)$, (f), (i), (V), (t), (d), (Q), (v), (p), (s), (c) and (q) are known, then its Sales Past must be:

$$\$'= 360[c-q](/\{U+\$[1-f-i]-V\}[1-t][1-d]\}-Q)\,/\{vp[1+s]\}$$

Rule-13198:

If both $(/)$, (U), $(\$)$, (f), (i), (V), (t), (d), $(\$')$, (Q), (p), (s), (c) and (q) are known, then its Variable Portion Planned is:

$$v= V/\$$$

or it can also be found as

$$v= 360[c-q](/\{U+\$[1-f-i]-V\}[1-t][1-d]\}-Q)\,/\{\$'p[1+s]\}$$

Steve Asikin ISBN 14: 978-1514685136, ISBN 10: 1514685132

Rule-13199:

If both (**/**), (**U**), (**S**), (**f**), (**i**), (**V**), (**t**), (**d**), (**S'**), (**v**), (**Q**), (**s**), (**c**) and (**q**) are known, then its Procured Inventory Days Planned:

p= 360[**c**-**q**](**/**{**U**+**S**[1-**f**-**i**]-**V**}[1-**t**][1-**d**]}-**Q**)
/{**S'v**[1+**s**]}

Rule-13200:

If both (**/**), (**U**), (**S**), (**f**), (**i**), (**V**), (**t**), (**d**), (**S'**), (**v**), (**p**), (**Q**), (**c**) and (**q**) are known, then its Sales Growth Planned is:

s= **S**/**S'**-1

or it can also be found as

s= 360[**c**-**q**](**/**{**U**+**S**[1-**f**-**i**]-**V**}[1-**t**][1-**d**]}-**Q**)
/[**S'vp**]-1

Rule-13201:

If both (**/**), (**U**), (**S**), (**f**), (**i**), (**V**), (**t**), (**d**), (**S'**), (**v**), (**p**), (**s**), (**Q**) and (**q**) are known, then its Current Ratio Planned is:

c= **q**+**S'vp**[1+**s**]
/[360(**/**{**U**+**S**[1-**f**-**i**]-**V**}[1-**t**][1-**d**]}-**Q**)]

Rule-13202:

If both (**/**), (**U**), (**S**), (**f**), (**i**), (**V**), (**t**), (**d**), (**S'**), (**v**), (**p**), (**s**), (**c**) and (**Q**) are known, then its Quick or Acid Test Ratio Planned is:

q= **c**-**S'vp**[1+**s**]
/[360(**/**{**U**+**S**[1-**f**-**i**]-**V**}[1-**t**][1-**d**]}-**Q**)]

Steve Asikin ISBN 14: 978-1514685136, ISBN 10: 1514685132

Rule-13203:

If both (**/**), (**U**), (**S**), (**f**), (**V**), (**S'**), (**i**), (**s**), (**T**), (**D**) and (**X**) are known, then its Quoted Longterm Debt Planned is:

Q= **/**{**U**+**S**[1-**f**]-**V**-**S'i**[1+**s**]-**T**-**D**}-**X**

Rule-13204:

If both (**Q**), (**U**), (**S**), (**f**), (**V**), (**S'**), (**i**), (**s**), (**T**), (**D**) and (**X**) are known, then its Leverage or Gearing Ratio Planned is:

/= [**Q**+**X**]/{**U**+**S**[1-**f**]-**V**-**S'i**[1+**s**]-**T**-**D**}

Rule-13205:

If both (**/**), (**Q**), (**S**), (**f**), (**V**), (**S'**), (**i**), (**s**), (**T**), (**D**) and (**X**) are known, then its Utilized or Starting Capital must be:

U= [**Q**+**X**]/**/**-{**S**[1-**f**]-**V**-**S'i**[1+**s**]+**T**+**D**

Rule-13206:

If both (**/**), (**U**), (**Q**), (**f**), (**V**), (**S'**), (**i**), (**s**), (**T**), (**D**) and (**X**) are known, then its Sales or Revenue Planned is:

S= {**V**+**S'i**[1+**s**]+**T**+**D**+[**Q**+**X**]/**/**-**U**}/[1-**f**]

Rule-13207:

If both (**/**), (**U**), (**S**), (**Q**), (**V**), (**S'**), (**i**), (**s**), (**T**), (**D**) and (**X**) are known, then its Fix Rate Planned is:

f= 1-{**V**+**S'i**[1+**s**]+**T**+**D**+[**Q**+**X**]/**/**-**U**}/**S**

Steve Asikin ISBN 14: 978-1514685136, ISBN 10: 1514685132

Rule-13208:

If both (/), (U), ($), (f), (Q), ($'), (i), (s), (T), (D) and (X) are known, then its Variable Cost Planned is:

V= $[1-f]-$'i[1+s]-T-D-[Q+X]/ŁU

Rule-13209:

If both (/), (U), ($), (f), (V), (Q), (i), (s), (T), (D) and (X) are known, then its Sales Past must ba:

$'= {$[1-f]-V- T-D-[Q+X]/ŁU}/{i[1+s]}

Rule-13210:

If both (/), (U), ($), (f), (V), ($'), (Q), (s), (T), (D) and (X) are known, then its Interest Portion Planned is:

i= {$[1-f]-V- T-D-[Q+X]/ŁU}/{$'[1+s]}

Rule-13211:

If both (/), (U), ($), (f), (V), ($'), (i), (Q), (T), (D) and (X) are known, then its Sales Growth Planned is:

s= $/$'-1

or it can also be found as

s= {$[1-f]-V- T-D-[Q+X]/ŁU}/[$'i]-1

Rule-13212:

If both (/), (U), ($), (f), (V), ($'), (i), (s), (Q), (D) and (X) are known, then its Tax Planned is:

T= $[1-f]-V-$'i[1+s]-D-[Q+X]/ŁU

Rule-13213:

If both (/), (U), ($), (f), (V), ($'), (i), (s), (T), (Q) and (X) are known, then its Dividend Planned is:

D= $[1-f]-V-$'i[1+s]-T-[Q+X]/ŁU

Steve Asikin ISBN 14: 978-1514685136, ISBN 10: 1514685132

Rule-13214:

If both (**/**), (**U**), (**$**), (**f**), (**V**), (**$'**), (**i**), (**s**), (**T**), (**D**) and (**Q**) are known, then its Quick or Gearing Ratio Planned is:

X= **/**{**U**+**$**[1-**f**]-**V**-**$'i**[1+**s**]-**T**-**D**}-**Q**

Rule-13215:

If both (**/**), (**U**), (**$**), (**f**), (**V**), (**$'**), (**i**), (**s**), (**T**), (**D**), (**P**), (**c**) and (**q**) are known, then its Quoted Longterm Debt Planned is:

Q= **/**{**U**+**$**[1-**f**]-**V**-**$'i**[1+**s**]-**T**-**D**}-**P**/[**c**-**q**]

Rule-13216:

If both (**Q**), (**U**), (**$**), (**f**), (**V**), (**$'**), (**i**), (**s**), (**T**), (**D**), (**P**), (**c**) and (**q**) are known, then its Leverage or Gearing Ratio Planned is:

/= {**Q**+**P**/[**c**-**q**]}/{**U**+**$**[1-**f**]-**V**-**$'i**[1+**s**]-**T**-**D**}

Rule-13217:

If both (**/**), (**Q**), (**$**), (**f**), (**V**), (**$'**), (**i**), (**s**), (**T**), (**D**), (**P**), (**c**) and (**q**) are known, then its Utilized or Starting Capital must be:

U= {**Q**+**P**/[**c**-**q**]}/**/**-{**$**[1-**f**]-**V**-**$'i**[1+**s**]+**T**+**D**

Rule-13218:

If both (**/**), (**U**), (**Q**), (**f**), (**V**), (**$'**), (**i**), (**s**), (**T**), (**D**), (**P**), (**c**) and (**q**) are known, then its Sales or Revenue Planned is:

$**= (**V**+**$'i[1+**s**]+**T**+**D**+{**Q**+**P**/[**c**-**q**]}/**/**-**U**)/[1-**f**]

Steve Asikin ISBN 14: 978-1514685136, ISBN 10: 1514685132

Rule-13219:

If both (**/**), (**U**), (**$**), (**Q**), (**V**), (**$'**), (**i**), (**s**), (**T**), (**D**), (**P**), (**c**) and (**q**) are known, then its Fixed Portion Planned is:

f= 1-(**V**+**$'i**[1+**s**]+**T**+**D**+{**Q**+**P**/[**c**-**q**]}/**/-U**)/**$**

Rule-13220:

If both (**/**), (**U**), (**$**), (**f**), (**Q**), (**$'**), (**i**), (**s**), (**T**), (**D**), (**P**), (**c**) and (**q**) are known, then its Variable Cost Planned is:

V= **$**[1-**f**]-**$'i**[1+**s**]-**T**-**D**-{**Q**+**P**/[**c**-**q**]}/**/-U**

Rule-13221:

If both (**/**), (**U**), (**$**), (**f**), (**V**), (**Q**), (**i**), (**s**), (**T**), (**D**), (**P**), (**c**) and (**q**) are known, then its Sales Past must be:

$'**= (**$[1-**f**]-**V**- **T**-**D**-{**Q**+**P**/[**c**-**q**]}/**/-U**)/{**i**[1+**s**]}

Rule-13222:

If both (**/**), (**U**), (**$**), (**f**), (**V**), (**$'**), (**Q**), (**s**), (**T**), (**D**), (**P**), (**c**) and (**q**) are known, then its Interest Portion Planned is:

i= (**$**[1-**f**]-**V**- **T**-**D**-{**Q**+**P**/[**c**-**q**]}/**/-U**)/{**$'**[1+**s**]}

Rule-13223:

If both (**/**), (**U**), (**$**), (**f**), (**V**), (**$'**), (**i**), (**Q**), (**T**), (**D**), (**P**), (**c**) and (**q**) are known, then its Sales Growth Planned is:

s= **$**/**$'**-1

or it can also be found as

s= (**$**[1-**f**]-**V**- **T**-**D**-{**Q**+**P**/[**c**-**q**]}/**/-U**)/[**$'i**]-1

Steve Asikin ISBN 14: 978-1514685136, ISBN 10: 1514685132

Rule-13224:

If both (**/**), (**U**), (**$**), (**f**), (**V**), (**$'**), (**i**), (**s**), (**Q**), (**D**), (**P**), (**c**) and (**q**) are known, then its Tax Planned is:

T= **$**[1-**f**]-**V**-**$'i**[1+**s**]-**D**-{**Q**+**P**/[**c**-**q**]}/**/**-**U**

Rule-13225:

If both (**/**), (**U**), (**$**), (**f**), (**V**), (**$'**), (**i**), (**s**), (**T**), (**Q**), (**P**), (**c**) and (**q**) are known, then its Dividend Planned is:

D= **$**[1-**f**]-**V**-**$'i**[1+**s**]-**T**-{**Q**+**P**/[**c**-**q**]}/**/**-**U**

Rule-13226:

If both (**/**), (**U**), (**$**), (**f**), (**V**), (**$'**), (**i**), (**s**), (**T**), (**D**), (**Q**), (**c**) and (**q**) are known, then its Procured Inventory Planned is:

P= [**c**-**q**](**/**{**U**+**$**[1-**f**]-**V**-**$'i**[1+**s**]-**T**-**D**}-**Q**)

Rule-13227:

If both (**/**), (**U**), (**$**), (**f**), (**V**), (**$'**), (**i**), (**s**), (**T**), (**D**), (**P**), (**Q**) and (**q**) are known, then its Current Ratio Planned is:

c= **q**+**P**/(**/**{**U**+**$**[1-**f**]-**V**-**$'i**[1+**s**]-**T**-**D**}-**Q**)

Rule-13228:

If both (**/**), (**U**), (**$**), (**f**), (**V**), (**$'**), (**i**), (**s**), (**T**), (**D**), (**P**), (**c**) and (**Q**) are known, then its Qucik or Acid Test Ratio Planned is:

q= **c**-**P**/(**/**{**U**+**$**[1-**f**]-**V**-**$'i**[1+**s**]-**T**-**D**}-**Q**)

Steve Asikin ISBN 14: 978-1514685136, ISBN 10: 1514685132

Rule-13229:

If both (**/**), (**U**), (**$**), (**f**), (**V**), (**$'**), (**i**), (**s**), (**T**), (**D**), (**p**), (**c**) and (**q**) are known, then its Quoted Longterm Debt Planned is:

Q= **/**{**U**+**$**[1-**f**]-**V**-**$'i**[1+**s**]-**T**-**D**}-**Vp**/{360[**c**-**q**]}

Rule-13230:

If both (**Q**), (**U**), (**$**), (**f**), (**V**), (**$'**), (**i**), (**s**), (**T**), (**D**), (**p**), (**c**) and (**q**) are known, then its Leverage or Gearing Ratio Planned is:

/= (**Q**+**Vp**/{360[**c**-**q**]})/{**U**+**$**[1-**f**]-**V**-**$'i**[1+**s**]-**T**-**D**}

Rule-13231:

If both (**/**), (**Q**), (**$**), (**f**), (**V**), (**$'**), (**i**), (**s**), (**T**), (**D**), (**p**), (**c**) and (**q**) are known, then its Utilized or Starting Capital must be:

U= (**Q**+**Vp**/{360[**c**-**q**]})/**/**-{**$**[1-**f**]-**V**-**$'i**[1+**s**]+**T**+**D**

Rule-13232:

If both (**/**), (**U**), (**Q**), (**f**), (**V**), (**$'**), (**i**), (**s**), (**T**), (**D**), (**p**), (**c**) and (**q**) are known, then its Sales or Revenue Planned is:

$**= [**V**+**$'i[1+**s**]+**T**+**D**+(**Q**+**Vp**/{360[**c**-**q**]})/**/**-**U**]
/[1-**f**]

Rule-13233:

If both (**/**), (**U**), (**$**), (**Q**), (**V**), (**$'**), (**i**), (**s**), (**T**), (**D**), (**p**), (**c**) and (**q**) are known, then its Fixed Portion Planned is:

f= 1-[**V**+**$'i**[1+**s**]+**T**+**D**+(**Q**+**Vp**/{360[**c**-**q**]})/**/**-**U**]/**$**

Steve Asikin ISBN 14: 978-1514685136, ISBN 10: 1514685132

Rule-13234:

If both (**/**), (**U**), (**$**), (**f**), (**Q**), (**$'**), (**i**), (**s**), (**T**), (**D**), (**p**), (**c**) and (**q**) are known, then its Variable Cost Planned is:

V= (**/**{**U**+**$**[1-**f**]-**$'i**[1+**s**]-**T**-**D**}-**Q**)/(**p**/{360[**c**-**q**]}+**/**)

Rule-13235:

If both (**/**), (**U**), (**$**), (**f**), (**V**), (**Q**), (**i**), (**s**), (**T**), (**D**), (**p**), (**c**) and (**q**) are known, then its Sales Past must be:

$'**= [**$[1-**f**]-**V**- **T**-**D**-(**Q**+**Vp**/{360[**c**-**q**]})/**/**-**U**]
/{**i**[1+**s**]}

Rule-13236:

If both (**/**), (**U**), (**$**), (**f**), (**V**), (**$'**), (**Q**), (**s**), (**T**), (**D**), (**p**), (**c**) and (**q**) are known, then its Interest Portion Planned is:

i= [**$**[1-**f**]-**V**- **T**-**D**-(**Q**+**Vp**/{360[**c**-**q**]})/**/**-**U**]
/{**$'**[1+**s**]}

Rule-13237:

If both (**/**), (**U**), (**$**), (**f**), (**V**), (**$'**), (**i**), (**Q**), (**T**), (**D**), (**p**), (**c**) and (**q**) are known, then its Sales Growth Planned is:

s= **$**/**$'**-1

or it can also be found as

s= [**$**[1-**f**]-**V**- **T**-**D**-(**Q**+**Vp**/{360[**c**-**q**]})/**/**-**U**]/[**$'i**]-1

Rule-13238:

If both (**/**), (**U**), (**$**), (**f**), (**V**), (**$'**), (**i**), (**s**), (**Q**), (**D**), (**p**), (**c**) and (**q**) are known, then its Tax Planned is:

T= **$**[1-**f**]-**V**-**$'i**[1+**s**]-**D**-(**Q**+**Vp**/{360[**c**-**q**]})/**/**-**U**

Steve Asikin ISBN 14: 978-1514685136, ISBN 10: 1514685132

Rule-13239:

If both (**/**), (**U**), (**$**), (**f**), (**V**), (**$'**), (**i**), (**s**), (**T**), (**Q**), (**p**), (**c**) and (**q**) are known, then its Dividend Planned is:

D= **$**[1-**f**]-**V**-**$'i**[1+**s**]-**T**-(**Q**+**Vp**/{360[**c**-**q**]})/**/**-**U**

Rule-13240:

If both (**/**), (**U**), (**$**), (**f**), (**V**), (**$'**), (**i**), (**s**), (**T**), (**D**), (**Q**), (**c**) and (**q**) are known, then its Procured Inventory Days Planned:

p= 360[**c**-**q**](**/**{**U**+**$**[1-**f**]-**V**-**$'i**[1+**s**]-**T**-**D**}-**Q**)/**V**

Rule-13241:

If both (**/**), (**U**), (**$**), (**f**), (**V**), (**$'**), (**i**), (**s**), (**T**), (**D**), (**p**), (**Q**) and (**q**) are known, then its Current Ratio Planned is:

c= **q**+**Vp**/[360(**/**{**U**+**$**[1-**f**]-**V**-**$'i**[1+**s**]-**T**-**D**}-**Q**)]

Rule-13242:

If both (**/**), (**U**), (**$**), (**f**), (**V**), (**$'**), (**i**), (**s**), (**T**), (**D**), (**p**), (**c**) and (**Q**) are known, then its Quick or Acid Test Ratio Planned is:

q= **c**-**Vp**/[360(**/**{**U**+**$**[1-**f**]-**V**-**$'i**[1+**s**]-**T**-**D**}-**Q**)]

Rule-13243:

If both (**/**), (**U**), (**$**), (**f**), (**V**), (**$'**), (**i**), (**s**), (**T**), (**D**), (**v**), (**p**), (**c**) and (**q**) are known, then its Quoted Longterm Debt Planned is:

Q= **/**{**U**+**$**[1-**f**]-**V**-**$'i**[1+**s**]-**T**-**D**}-**$vp**/{360[**c**-**q**]}

Steve Asikin ISBN 14: 978-1514685136, ISBN 10: 1514685132

Rule-13244:

If both (**Q**), (**U**), (**$**), (**f**), (**V**), (**$'**), (**i**), (**s**), (**T**), (**D**), (**v**), (**p**), (**c**) and (**q**) are known, then its Leverage or Gearing Ratio Planned is:

/= (**Q**+**$vp**/{360[**c**-**q**]})/{**U**+**$**[1-**f**]-**V**-**$'i**[1+**s**]-**T**-**D**}

Rule-13245:

If both (**/**), (**Q**), (**$**), (**f**), (**V**), (**$'**), (**i**), (**s**), (**T**), (**D**), (**v**), (**p**), (**c**) and (**q**) are known, then its Utilized or Starting Capital must be:

U= (**Q**+**$vp**/{360[**c**-**q**]})/**/**-{**$**[1-**f**]-**V**-**$'i**[1+**s**]+**T**+**D**

Rule-13246:

If both (**/**), (**U**), (**Q**), (**f**), (**V**), (**$'**), (**i**), (**s**), (**T**), (**D**), (**v**), (**p**), (**c**) and (**q**) are known, then its Sales or Revenue Planned is:

$**= (**/**{**U**-**V**-**$'i[1+**s**]-**T**-**D**}-**Q**)/(**vp**/{360[**c**-**q**]}-**/**[1-**f**])

Rule-13247:

If both (**/**), (**U**), (**$**), (**Q**), (**V**), (**$'**), (**i**), (**s**), (**T**), (**D**), (**v**), (**p**), (**c**) and (**q**) are known, then its Fixed Portion Planned is:

f= 1-[**V**+**$'i**[1+**s**]+**T**+**D**+(**Q**+**$vp**/{360[**c**-**q**]})/**/**-**U**]/**$**

Rule-13248:

If both (**/**), (**U**), (**$**), (**f**), (**Q**), (**$'**), (**i**), (**s**), (**T**), (**D**), (**v**), (**p**), (**c**) and (**q**) are known, then its Variable Cost Planned is:

V= (**/**{**U**+**$**[1-**f**]-**$'i**[1+**s**]-**T**-**D**}-**$vp**/{360[**c**-**q**]}-**Q**)/**/**

Steve Asikin ISBN 14: 978-1514685136, ISBN 10: 1514685132

Rule-13249:

If both (/), (U), ($), (f), (V), (Q), (i), (s), (T), (D), (v), (p), (c) and (q) are known, then its Sales Past must be:

$'= [$[1-f]-V- T-D-(Q+$vp/{360[c-q]})/ /-U]
/{i[1+s]}

Rule-13250:

If both (/), (U), ($), (f), (V), ($'), (Q), (s), (T), (D), (v), (p), (c) and (q) are known, then its Interest Portion Planned is:

i= [$[1-f]-V- T-D-(Q+$vp/{360[c-q]})/ /-U]
/{$'[1+s]}

Rule-13251:

If both (/), (U), ($), (f), (V), ($'), (i), (Q), (T), (D), (v), (p), (c) and (q) are known, then its Sales Growth Planned is:

s= $/$'-1

or it can also be found

s= [$[1-f]-V- T-D-(Q+$vp/{360[c-q]})/ /-U]/[$'i]-1

Rule-13252:

If both (/), (U), ($), (f), (V), ($'), (i), (s), (Q), (D), (v), (p), (c) and (q) are known, then its Tax Planned is:

T= $[1-f]-V-$'i[1+s]-D-(Q+$vp/{360[c-q]})/ /-U

Rule-13253:

If both (/), (U), ($), (f), (V), ($'), (i), (s), (T), (Q), (v), (p), (c) and (q) are known, then its Dividend Planned is:

D= $[1-f]-V-$'i[1+s]-T-(Q+$vp/{360[c-q]})/ /-U

Steve Asikin ISBN 14: 978-1514685136, ISBN 10: 1514685132

Rule-13254:

If both (/), (U), (S), (f), (V), (S'), (i), (s), (T), (D), (Q), (p), (c) and (q) are known, then its Variable Portion Planned is:

v= V/S

or it can also be found

v= 360[c-q](/{U+S[1-f]-V-S'i[1+s]-T-D}-Q)/[Sp]

Rule-13255:

If both (/), (U), (S), (f), (V), (S'), (i), (s), (T), (D), (v), (Q), (c) and (q) are known, then its Procured Inventory Days Planned:

p= 360[c-q](/{U+S[1-f]-V-S'i[1+s]-T-D}-Q)/[Sv]

Rule-13256:

If both (/), (U), (S), (f), (V), (S'), (i), (s), (T), (D), (v), (p), (Q) and (q) are known, then its Current Ratio Planned is:

c= q+Svp/[360(/{U+S[1-f]-V-S'i[1+s]-T-D}-Q)]

Rule-13257:

If both (/), (U), (S), (f), (V), (S'), (i), (s), (T), (D), (v), (p), (c) and (Q) are known, then its Quick or Acid Test Ratio Planned is:

q= c-Svp/[360(/{U+S[1-f]-V-S'i[1+s]-T-D}-Q)]

Rule-13258:

If both (/), (U), (S), (f), (V), (S'), (i), (s), (T), (D), (S'), (v), (p), (s), (c) and (q) are known, then its Quoted Longterm Debt Planned is:

Q= /{U+S[1-f]-V-S'i[1+s]-T-D}
-S'vp[1+s]/{360[c-q]}

Steve Asikin ISBN 14: 978-1514685136, ISBN 10: 1514685132

Rule-13259:

If both (**Q**), (**U**), (**S**), (**f**), (**V**), (**S'**), (**i**), (**s**), (**T**), (**D**), (**v**), (**p**), (**c**) and (**q**) are known, then its Leverage or Gearing Ratio Planned is:

/= (**Q**+**S'vp**[1+**s**]/{360[**c**-**q**]})
/{**U**+**S**[1-**f**]-**V**-**S'i**[1+**s**]-**T**-**D**}

Rule-13260:

If both (**/**), (**Q**), (**S**), (**f**), (**V**), (**S'**), (**i**), (**s**), (**T**), (**D**), (**v**), (**p**), (**c**) and (**q**) are known, then its Utilized or Starting Capital must be:

U= (**Q**+**S'vp**[1+**s**]/{360[**c**-**q**]})/**/**
-{**S**[1-**f**]-**V**-**S'i**[1+**s**]+**T**+**D**

Rule-13261:

If both (**/**), (**U**), (**Q**), (**f**), (**V**), (**S'**), (**i**), (**s**), (**T**), (**D**), (**v**), (**p**), (**c**) and (**q**) are known, then its Sales or Revenue Planned is:

S= (**Q**-**/**{**U**-**V**-**S'i**[1+**s**]-**T**-**D**}
+**S'vp**[1+**s**]/{360[**c**-**q**]})/{**/**[1-**f**]}

Rule-13262:

If both (**/**), (**U**), (**S**), (**Q**), (**V**), (**S'**), (**i**), (**s**), (**T**), (**D**), (**v**), (**p**), (**c**) and (**q**) are known, then its Fixed Portion Planned is:

f= 1-[**V**+**S'i**[1+**s**]+**T**+**D**
+(**Q**+**S'vp**[1+**s**]/{360[**c**-**q**]})/**/**-**U**]/**S**

Steve Asikin ISBN 14: 978-1514685136, ISBN 10: 1514685132

Rule-13263:

If both (**/**), (**U**), (**$**), (**f**), (**Q**), (**$'**), (**i**), (**s**), (**T**), (**D**), (**v**), (**p**), (**c**) and (**q**) are known, then its Variable Cost Planned is:

V= (**/**{**U**+**$**[1-**f**]-**$'i**[1+**s**]-**T**-**D**}
-**$'vp**[1+**s**]/{360[**c**-**q**]}-**Q**)/**/**

Rule-13264:

If both (**/**), (**U**), (**$**), (**f**), (**V**), (**Q**), (**i**), (**s**), (**T**), (**D**), (**v**), (**p**), (**c**) and (**q**) are known, then its Sales Past must:

$'**= [**$[1-**f**]-**V**- **T**-**D**
-(**Q**+**$'vp**[1+**s**]/{360[**c**-**q**]})/**/**-**U**]/{**i**[1+**s**]}

Rule-13265:

If both (**/**), (**U**), (**$**), (**f**), (**V**), (**$'**), (**Q**), (**s**), (**T**), (**D**), (**v**), (**p**), (**c**) and (**q**) are known, then its Interest Portion Planned is:

i= [**$**[1-**f**]-**V**- **T**-**D**
-(**Q**+**$'vp**[1+**s**]/{360[**c**-**q**]})/**/**-**U**]/{**$'**[1+**s**]}

Rule-13266:

If both (**/**), (**U**), (**$**), (**f**), (**V**), (**$'**), (**i**), (**Q**), (**T**), (**D**), (**v**), (**p**), (**c**) and (**q**) are known, then its Sales Growth Planned is:

s= **$**/**$'**-1

or it can also be found

s= [**$**[1-**f**]-**V**- **T**-**D**
-(**Q**+**$'vp**[1+**s**]/{360[**c**-**q**]})/**/**-**U**]/[**$'i**]-1

Steve Asikin ISBN 14: 978-1514685136, ISBN 10: 1514685132

Rule-13267:

If both (**/**), (**U**), (**$**), (**f**), (**V**), (**$'**), (**i**), (**s**), (**Q**), (**D**), (**v**), (**p**), (**c**) and (**q**) are known, then its Tax Planned is:

T= **$**[1-**f**]-**V**-**$'i**[1+**s**]-**D**
-(**Q**+**$'vp**[1+**s**]/{360[**c**-**q**]})/**/**-**U**

Rule-13268:

If both (**/**), (**U**), (**$**), (**f**), (**V**), (**$'**), (**i**), (**s**), (**T**), (**Q**), (**v**), (**p**), (**c**) and (**q**) are known, then its Dividend Planned is:

D= **$**[1-**f**]-**V**-**$'i**[1+**s**]-**T**
-(**Q**+**$'vp**[1+**s**]/{360[**c**-**q**]})/**/**-**U**

Rule-13269:

If both (**/**), (**U**), (**$**), (**f**), (**V**), (**Q**), (**i**), (**s**), (**T**), (**D**), (**v**), (**p**), (**c**) and (**q**) are known, then its Sales Past must:

$'**= 360[**c**-**q**](**/**{**U**+**$[1-**f**]-**V**-**$'i**[1+**s**]-**T**-**D**}-**Q**)
/{**vp**[1+**s**]}

Rule-13270:

If both (**/**), (**U**), (**$**), (**f**), (**V**), (**$'**), (**i**), (**s**), (**T**), (**D**), (**Q**), (**p**), (**c**) and (**q**) are known, then its Variable Cost Planned is:

v= **V**/**$**

or it can also be found as

v= 360[**c**-**q**](**/**{**U**+**$**[1-**f**]-**V**-**$'i**[1+**s**]-**T**-**D**}-**Q**)
/{**$'p**[1+**s**]}

Steve Asikin ISBN 14: 978-1514685136, ISBN 10: 1514685132

Rule-13271:

If both (/), (U), ($), (f), (V), ($'), (i), (s), (T), (D), (v), (Q), (c) and (q) are known, then its Procured Inventory Days Planned:

p= 360[c-q](/{U+$[1-f]-V-$'i[1+s]-T-D}-Q)
/{$'v[1+s]}

Rule-13272:

If both (/), (U), ($), (f), (V), ($'), (i), (Q), (T), (D), (v), (p), (c) and (q) are known, then its Sales Growth Planned is:

s= 360[c-q](/{U+$[1-f]-V-$'i[1+s]-T-D}-Q)/[$'vp]

Rule-13273:

If both (/), (U), ($), (f), (V), ($'), (i), (s), (T), (D), (v), (p), (Q) and (q) are known, then its Current Ratio Planned is:

c= q+$'vp[1+s]
/[360(/{U+$[1-f]-V-$'i[1+s]-T-D}-Q)]

Rule-13274:

If both (/), (U), ($), (f), (V), ($'), (i), (s), (T), (D), (v), (p), (c) and (Q) are known, then its Quick or Acid Test Ratio Planned is:

q= c-$'vp[1+s]
/[360(/{U+$[1-f]-V-$'i[1+s]-T-D}-Q)]

Steve Asikin ISBN 14: 978-1514685136, ISBN 10: 1514685132

Rule-13275:

If both (**/**), (**U**), (**$**), (**f**), (**V**), (**$'**), (**i**), (**s**), (**T**), (**d**) and (**X**) are known, then its Quoted Longterm Debt Planned is:

Q= **/**(**U**+{**$**[1-**f**]-**V**-**$'i**[1+**s**]-**T**}[1-**d**])-**X**

Rule-13276:

If both (**Q**), (**U**), (**$**), (**f**), (**V**), (**$'**), (**i**), (**s**), (**T**), (**d**) and (**X**) are known, then its Leverage or Gearing Planned is:

/= [**Q**+**X**]/(**U**+{**$**[1-**f**]-**V**-**$'i**[1+**s**]-**T**}[1-**d**])

Rule-13277:

If both (**/**), (**Q**), (**$**), (**f**), (**V**), (**$'**), (**i**), (**s**), (**T**), (**d**) and (**X**) are known, then its Utilized or Starting Capital:

U= [**Q**+**X**]/**/**-{**$**[1-**f**]-**V**-**$'i**[1+**s**]-**T**}[1-**d**]

Rule-13278:

If both (**/**), (**U**), (**Q**), (**f**), (**V**), (**$'**), (**i**), (**s**), (**T**), (**d**) and (**X**) are known, then its Sales or Revenue Planned is:

$**= (**V**+**$'i[1+**s**]+**T**+{[**Q**+**X**]/**/**–**U**}/[1-**d**])/[1-**f**]

Rule-13279:

If both (**/**), (**U**), (**$**), (**Q**), (**V**), (**$'**), (**i**), (**s**), (**T**), (**d**) and (**X**) are known, then its Fixed Portion Planned is:

f= 1-(**V**+**$'i**[1+**s**]+**T**+{[**Q**+**X**]/**/**–**U**}/[1-**d**])/**$**

Rule-13280:

If both (**/**), (**U**), (**$**), (**f**), (**V**), (**$'**), (**i**), (**s**), (**T**), (**d**) and (**X**) are known, then its Variable Cost Planned is:

V= **$**[1-**f**]-**$'i**[1+**s**]-**T**-{[**Q**+**X**]/**/**–**U**}/[1-**d**]

Steve Asikin ISBN 14: 978-1514685136, ISBN 10: 1514685132

Rule-13281:

If both (/), (U), (S), (f), (V), (Q), (i), (s), (T), (d) and (X) are known, then its Sales Past must be:

$S' = (S[1-f]-V- T-\{[Q+X]//-U\}/[1-d])/\{i[1+s]\}$

Rule-13282:

If both (/), (U), (S), (f), (V), (S'), (Q), (s), (T), (d) and (X) are known, then its Interest Portion Planned is:

$i = (S[1-f]-V- T-\{[Q+X]//-U\}/[1-d])/\{S'[1+s]\}$

Rule-13283:

If both (/), (U), (S), (f), (V), (S'), (i), (Q), (T), (d) and (X) are known, then its Sales Growth Planned is:

$s = S/S'-1$

or it can also be found as

$s = (S[1-f]-V- T-\{[Q+X]//-U\}/[1-d])/[S'i]-1$

Rule-13284:

If both (/), (U), (S), (f), (V), (S'), (i), (s), (Q), (d) and (X) are known, then its Tax Planned is:

$T = S[1-f]-V-S'i[1+s]-\{[Q+X]//-U\}/[1-d]$

Rule-13285:

If both (/), (U), (S), (f), (V), (S'), (i), (s), (T), (Q) and (X) are known, then its Dividend Payout Planned is:

$d = 1-\{[Q+X]//-U\}/\{S[1-f]-V-S'i[1+s]-T\}$

Steve Asikin ISBN 14: 978-1514685136, ISBN 10: 1514685132

Rule-13286:

If both $(/), (U), (\$), (f), (V), (\$'), (i), (s), (T), (d)$ and (Q) are known, then its Xpress or Current Debt Planned is:

$$X= /(U+\{\$[1-f]-V-\$'i[1+s]-T\}[1-d])-Q$$

Rule-13287:

If both $(/), (U), (\$), (f), (V), (\$'), (i), (s), (T), (d), (P)$, (c) and (q) are known, then its Quoted Longterm Debt Planned is:

$$Q= /(U+\{\$[1-f]-V-\$'i[1+s]-T\}[1-d])-P/[c-q]$$

Rule-13288:

If both $(Q), (U), (\$), (f), (V), (\$'), (i), (s), (T), (d), (P)$, (c) and (q) are known, then its Leverage or Gearing Ratio Planned is:

$$/= \{Q+P/[c-q]\}/(U+\{\$[1-f]-V-\$'i[1+s]-T\}[1-d])$$

Rule-13289:

If both $(/), (Q), (\$), (f), (V), (\$'), (i), (s), (T), (d), (P)$, (c) and (q) are known, then its Utilized or Starting Capital must be:

$$U= \{Q+P/[c-q]\}/ / -\{\$[1-f]-V-\$'i[1+s]-T\}[1-d]$$

Rule-13290:

If both $(/), (U), (Q), (f), (V), (\$'), (i), (s), (T), (d), (P)$, (c) and (q) are known, then its Sales or Revenue Planned is:

$$\$= [V+\$'i[1+s]+T+(\{Q+P/[c-q]\}/ / -U)/[1-d]]/[1-f]$$

Steve Asikin ISBN 14: 978-1514685136, ISBN 10: 1514685132

Rule-13291:

If both (/), (U), ($), (Q), (V), ($'), (i), (s), (T), (d), (P), (c) and (q) are known, then its Fixed Portion Planned is:

f= 1-[V+$'i[1+s]+T+({Q+P/[c-q]}//–U)/[1-d]]/$

Rule-13292:

If both (/), (U), ($), (f), (Q), ($'), (i), (s), (T), (d), (P), (c) and (q) are known, then its Variable Cost Planned is:

V= $[1-f]-$'i[1+s]-T-({Q+P/[c-q]}//–U)/[1-d]

Rule-13293:

If both (/), (U), ($), (f), (V), (Q), (i), (s), (T), (d), (P), (c) and (q) are known, then its Sales Past must be:

$'= [$[1-f]-V- T-({Q+P/[c-q]}//–U)/[1-d]]
/{i[1+s]}

Rule-13294:

If both (/), (U), ($), (f), (V), ($'), (Q), (s), (T), (d), (P), (c) and (q) are known, then its Interest Portion is:

i= [$[1-f]-V- T-({Q+P/[c-q]}//–U)/[1-d]]
/{$'[1+s]}

Rule-13295:

If both (/), (U), ($), (f), (V), ($'), (i), (Q), (T), (d), (P), (c) and (q) are known, then its Sales Growth Planned is:

s= $/$'-1

or it can also be found as

s= [$[1-f]-V- T-({Q+P/[c-q]}//–U)/[1-d]]/[$'i]-1

Steve Asikin ISBN 14: 978-1514685136, ISBN 10: 1514685132

Rule-13296:

If both (**/**), (**U**), (**$**), (**f**), (**V**), (**$'**), (**i**), (**s**), (**Q**), (**d**), (**P**), (**c**) and (**q**) are known, then its Tax Planned is:

T= **$**[1-**f**]-**V**-**$'i**[1+**s**]-({**Q**+**P**/[**c**-**q**]}/**/**–**U**)/[1-**d**]

Rule-13297:

If both (**/**), (**U**), (**$**), (**f**), (**V**), (**$'**), (**i**), (**s**), (**T**), (**Q**), (**P**), (**c**) and (**q**) are known, then its Dividend Payout Planned is:

d= 1-({**Q**+**P**/[**c**-**q**]}/**/**–**U**)/{**$**[1-**f**]-**V**-**$'i**[1+**s**]-**T**}

Rule-13298:

If both (**/**), (**U**), (**$**), (**f**), (**V**), (**$'**), (**i**), (**s**), (**T**), (**d**), (**Q**), (**c**) and (**q**) are known, then its Procured Inventory Planned is:

P= [**c**-**q**](**/U**+{**$**[1-**f**]-**V**-**$'i**[1+**s**]-**T**}[1-**d**])-**Q**)

Rule-13299:

If both (**/**), (**U**), (**$**), (**f**), (**V**), (**$'**), (**i**), (**s**), (**T**), (**d**), (**P**), (**Q**) and (**q**) are known, then its Current Ratio Planned is:

c= **q**+**P**/(**/U**+{**$**[1-**f**]-**V**-**$'i**[1+**s**]-**T**}[1-**d**])-**Q**)

Rule-13300:

If both (**/**), (**U**), (**$**), (**f**), (**V**), (**$'**), (**i**), (**s**), (**T**), (**d**), (**P**), (**c**) and (**Q**) are known, then its Quick or Acid Test Ratio Planned is:

q= **c**-**P**/(**/U**+{**$**[1-**f**]-**V**-**$'i**[1+**s**]-**T**}[1-**d**])-**Q**)

Steve Asikin ISBN 14: 978-1514685136, ISBN 10: 1514685132

Rule-13301:

If both (**/**), (**U**), (**S**), (**f**), (**V**), (**S'**), (**i**), (**s**), (**T**), (**d**), (**p**), (**c**) and (**q**) are known, then its Variable Cost Planned is:

Q= **/**(**U**+{**S**[1-**f**]-**V**-**S'i**[1+**s**]-**T**}[1-**d**])
-**Vp**/{360[**c**-**q**]}

Rule-13302:

If both (**Q**), (**U**), (**S**), (**f**), (**V**), (**S'**), (**i**), (**s**), (**T**), (**d**), (**p**), (**c**) and (**q**) are known, then its Leverage or Gearing Ratio Planned is:

/= (**Q**+**Vp**/{360[**c**-**q**]})
/(**U**+{**S**[1-**f**]-**V**-**S'i**[1+**s**]-**T**}[1-**d**])

Rule-13303:

If both (**/**), (**Q**), (**S**), (**f**), (**V**), (**S'**), (**i**), (**s**), (**T**), (**d**), (**p**), (**c**) and (**q**) are known, then its Utilized or Starting Capital must be:

U= (**Q**+**Vp**/{360[**c**-**q**]})/**/**
-{**S**[1-**f**]-**V**-**S'i**[1+**s**]-**T**}[1-**d**]

Rule-13304:

If both (**/**), (**U**), (**Q**), (**f**), (**V**), (**S'**), (**i**), (**s**), (**T**), (**d**), (**p**), (**c**) and (**q**) are known, then its Sales or Revenue Planned is:

S= **{V**+**S'i**[1+**s**]+**T**+**[**(**Q**+**Vp**/{360[**c**-**q**]})/**/**–**U]**
/[1-**d**]**}**/[1-**f**]

Steve Asikin ISBN 14: 978-1514685136, ISBN 10: 1514685132

Rule-13305:

If both (**/**), (**U**), (**S**), (**Q**), (**V**), (**S'**), (**i**), (**s**), (**T**), (**d**), (**p**), (**c**) and (**q**) are known, then its Fixed Portion Planned is:

f= 1-{**V**+**S'i**[1+**s**]+**T**+[(**Q**+**Vp**/{360[**c**-**q**]})/**/**–**U**]
/[1-**d**]}/**S**

Rule-13306:

If both (**/**), (**U**), (**S**), (**f**), (**Q**), (**S'**), (**i**), (**s**), (**T**), (**d**), (**p**), (**c**) and (**q**) are known, then its Variable Cost Planned is:

V= (**/**(**U**+{**S**[1-**f**]-**S'i**[1+**s**]-**T**}[1-**d**])-**Q**)
/(**p**/{360[**c**-**q**]}+**/**[1-**d**])

Rule-13307:

If both (**/**), (**U**), (**S**), (**f**), (**V**), (**Q**), (**i**), (**s**), (**T**), (**d**), (**p**), (**c**) and (**q**) are known, then its Sales Past must be:

S'= {**S**[1-**f**]-**V**- **T**-[(**Q**+**Vp**/{360[**c**-**q**]})/**/**–**U**]
/[1-**d**]}/{**i**[1+**s**]}

Rule-13308:

If both (**/**), (**U**), (**S**), (**f**), (**V**), (**S'**), (**Q**), (**s**), (**T**), (**d**), (**p**), (**c**) and (**q**) are known, then its Interest Portion Planned is:

i= {**S**[1-**f**]-**V**- **T**-[(**Q**+**Vp**/{360[**c**-**q**]})/**/**–**U**]
/[1-**d**]}/{**S'**[1+**s**]}

Steve Asikin ISBN 14: 978-1514685136, ISBN 10: 1514685132

Rule-13309:

If both (**/**), (**U**), (**$**), (**f**), (**V**), (**$'**), (**i**), (**Q**), (**T**), (**d**), (**p**), (**c**) and (**q**) are known, then its Variable Cost Planned is:

v= **$**/**V**

or it could also be found

s= {**$**[1-**f**]-**V**- **T**-[(**Q**+**Vp**/{360[**c**-**q**]})/**/**–**U**]
/[1-**d**]}/[**$'i**]-1

Rule-13310:

If both (**/**), (**U**), (**$**), (**f**), (**V**), (**$'**), (**i**), (**s**), (**Q**), (**d**), (**p**), (**c**) and (**q**) are known, then its Tax Planned is:

T= **$**[1-**f**]-**V**-**$'i**[1+**s**]-[(**Q**+**Vp**/{360[**c**-**q**]})/**/**–**U**]
/[1-**d**]

Rule-13311:

If both (**/**), (**U**), (**$**), (**f**), (**V**), (**$'**), (**i**), (**s**), (**T**), (**Q**), (**p**), (**c**) and (**q**) are known, then its Dividend Payout Planned is:

d= 1-[(**Q**+**Vp**/{360[**c**-**q**]})/**/**–**U**]
/{**$**[1-**f**]-**V**-**$'i**[1+**s**]-**T**}

Rule-13312:

If both (**/**), (**U**), (**$**), (**f**), (**V**), (**$'**), (**i**), (**s**), (**T**), (**d**), (**Q**), (**c**) and (**q**) are known, then its Procured Inventory Days Planned:

p= 360[**c**-**q**](**/**(**U**+{**$**[1-**f**]-**V**-**$'i**[1+**s**]-**T**}[1-**d**])-**Q**)/**V**

Steve Asikin ISBN 14: 978-1514685136, ISBN 10: 1514685132

Rule-13313:

If both (**/**), (**U**), (**S**), (**f**), (**V**), (**S'**), (**i**), (**s**), (**T**), (**d**), (**p**), (**Q**) and (**q**) are known, then its Current Ratio Planned is:

c= **q**+**Vp**
/[360(**/U**+{**S**[1-**f**]-**V**-**S'i**[1+**s**]-**T**}[1-**d**])-**Q**)]

Rule-13314:

If both (**/**), (**U**), (**S**), (**f**), (**V**), (**S'**), (**i**), (**s**), (**T**), (**d**), (**p**), (**c**) and (**Q**) are known, then its Quick or Acid Test Ratio Planned is:

q= **c**-**Vp**/[360(**/U**+{**S**[1-**f**]-**V**-**S'i**[1+**s**]-**T**}[1-**d**])-**Q**)]

Rule-13315:

If both (**/**), (**U**), (**S**), (**f**), (**V**), (**S'**), (**i**), (**s**), (**T**), (**d**), (**v**), (**p**), (**c**) and (**q**) are known, then its Quoted Lonmgterm Debt Planned is:

Q= **/U**+{**S**[1-**f**]-**V**-**S'i**[1+**s**]-**T**}[1-**d**])
-**Svp**/{360[**c**-**q**]}

Rule-13316:

If both (**Q**), (**U**), (**S**), (**f**), (**V**), (**S'**), (**i**), (**s**), (**T**), (**d**), (**v**), (**p**), (**c**) and (**q**) are known, then its Leverage or Gearing Ratio Planned is:

/= (**Q**+**Svp**/{360[**c**-**q**]})
/(**U**+{**S**[1-**f**]-**V**-**S'i**[1+**s**]-**T**}[1-**d**])

Steve Asikin ISBN 14: 978-1514685136, ISBN 10: 1514685132

Rule-13317:

If both (**/**), (**Q**), (**$**), (**f**), (**V**), (**$'**), (**i**), (**s**), (**T**), (**d**), (**v**), (**p**), (**c**) and (**q**) are known, then its Utilized or Starting Capital must be:

U= (**Q**+**$vp**/{360[**c**-**q**]})/**/**
-{**$**[1-**f**]-**V**-**$'i**[1+**s**]-**T**}[1-**d**]

Rule-13318:

If both (**/**), (**U**), (**Q**), (**f**), (**V**), (**$'**), (**i**), (**s**), (**T**), (**d**), (**v**), (**p**), (**c**) and (**q**) are known, then its Sales or Revenue Planned is:

$**= **[/**(**U**-{**V**+**$'i[1+**s**]+**T**}[1-**d**])-**Q]**
/(**vp**/{360[**c**-**q**]}-**/**[1-**f**][1-**d**])

Rule-13319:

If both (**/**), (**U**), (**$**), (**Q**), (**V**), (**$'**), (**i**), (**s**), (**T**), (**d**), (**v**), (**p**), (**c**) and (**q**) are known, then its Fixed Portion Planned is:

f= 1-**{V**+**$'i**[1+**s**]+**T**+**[**(**Q**+**$vp**/{360[**c**-**q**]})/**/**–**U]**
/[1-**d**]**}**/**$**

Rule-13320:

If both (**/**), (**U**), (**$**), (**f**), (**Q**), (**$'**), (**i**), (**s**), (**T**), (**d**), (**v**), (**p**), (**c**) and (**q**) are known, then its Variable Cost Planned is:

V= **[/**(**U**+{**$**[1-**f**]-**$'i**[1+**s**]-**T**}[1-**d**])
-**$vp**/{360[**c**-**q**]}-**Q]**/{**/**[1-**d**]}

Steve Asikin ISBN 14: 978-1514685136, ISBN 10: 1514685132

Rule-13321:

If both (**/**), (**U**), (**\$**), (**f**), (**V**), (**Q**), (**i**), (**s**), (**T**), (**d**), (**v**), (**p**), (**c**) and (**q**) are known, then its Sales Past must be:

\$'= {**\$**[1-**f**]-**V**- **T** -[(**Q**+**\$vp**/{360[**c**-**q**]})/**/**–**U**]/[1-**d**]}/{**i**[1+**s**]}

Rule-13322:

If both (**/**), (**U**), (**\$**), (**f**), (**V**), (**\$'**), (**Q**), (**s**), (**T**), (**d**), (**v**), (**p**), (**c**) and (**q**) are known, then its Interest Portion Planned is:

i= {**\$**[1-**f**]-**V**- **T**-[(**Q**+**\$vp**/{360[**c**-**q**]})/**/**–**U**] /[1-**d**]}/{**\$'**[1+**s**]}

Rule-13323:

If both (**/**), (**U**), (**\$**), (**f**), (**V**), (**\$'**), (**i**), (**Q**), (**T**), (**d**), (**v**), (**p**), (**c**) and (**q**) are known, then its Sales Growth Planned is:

s= **\$**/**\$'**-1

or it can also be found as

s= {**\$**[1-**f**]-**V**- **T**-[(**Q**+**\$vp**/{360[**c**-**q**]})/**/**–**U**] /[1-**d**]}/[**\$'i**]-1

Rule-13324:

If both (**/**), (**U**), (**\$**), (**f**), (**V**), (**\$'**), (**i**), (**s**), (**Q**), (**d**), (**v**), (**p**), (**c**) and (**q**) are known, then its Tax Planned is:

T= **\$**[1-**f**]-**V**-**\$'i**[1+**s**] -[(**Q**+**\$vp**/{360[**c**-**q**]})/**/**–**U**]/[1-**d**]

Steve Asikin ISBN 14: 978-1514685136, ISBN 10: 1514685132

Rule-13325:

If both (/), (U), ($), (f), (V), ($'), (i), (s), (T), (Q), (v), (p), (c) and (q) are known, then its Dividend Payout Planned is:

d= 1-[(Q+$vp/{360[c-q]})//–U]
/{$[1-f]-V-$'i[1+s]-T}

Rule-13326:

If both (/), (U), ($), (f), (V), ($'), (i), (s), (T), (d), (v), (p), (c) and (q) are known, then its Variable Portion Planned is:

v= V/$

or it can also be found as

v= 360[c-q](/U+{$[1-f]-V-$'i[1+s]-T}[1-d])-Q)
/[$p]

Rule-13327:

If both (/), (U), ($), (f), (V), ($'), (i), (s), (T), (d), (v), (Q), (c) and (q) are known, then its Procured Inventory Days Planned:

p= 360[c-q](/U+{$[1-f]-V-$'i[1+s]-T}[1-d])-Q)
/[$v]

Rule-13328:

If both (/), (U), ($), (f), (V), ($'), (i), (s), (T), (d), (v), (p), (Q) and (q) are known, then its Current Ratio Planned is:

c= q+$vp
/[360(/U+{$[1-f]-V-$'i[1+s]-T}[1-d])-Q)]

Steve Asikin ISBN 14: 978-1514685136, ISBN 10: 1514685132

Rule-13329:

If both (**/**), (**U**), (**$**), (**f**), (**V**), (**$'**), (**i**), (**s**), (**T**), (**d**), (**v**), (**p**), (**c**) and (**Q**) are known, then its Quick or Acid Test Ratio Planned is:

q= **c**-**$vp**
/[360(**/U**+{**$**[1-**f**]-**V**-**$'i**[1+**s**]-**T**}[1-**d**])-**Q**)]

Rule-13330:

If both (**/**), (**U**), (**$**), (**f**), (**V**), (**$'**), (**i**), (**s**), (**T**), (**d**), (**v**), (**p**), (**c**) and (**q**) are known, then its Variable Cost Planned is:

Q= **/**(**U**+{**$**[1-**f**]-**V**-**$'i**[1+**s**]-**T**}[1-**d**])
-**$'vp**[1+**s**]/{360[**c**-**q**]}

Rule-13331:

If both (**Q**), (**U**), (**$**), (**f**), (**V**), (**$'**), (**i**), (**s**), (**T**), (**d**), (**v**), (**p**), (**c**) and (**q**) are known, then its Leverage or Gearing Ratio Planned is:

/= (**Q**+**$'vp**[1+**s**]/{360[**c**-**q**]})
/(**U**+{**$**[1-**f**]-**V**-**$'i**[1+**s**]-**T**}[1-**d**])

Rule-13332:

If both (**/**), (**Q**), (**$**), (**f**), (**V**), (**$'**), (**i**), (**s**), (**T**), (**d**), (**v**), (**p**), (**c**) and (**q**) are known, then its Utilized or Starting Capital must be:

U= (**Q**+**$'vp**[1+**s**]/{360[**c**-**q**]})/**/**
-{**$**[1-**f**]-**V**-**$'i**[1+**s**]-**T**}[1-**d**]

Steve Asikin ISBN 14: 978-1514685136, ISBN 10: 1514685132

Rule-13333:

If both (**/**), (**U**), (**Q**), (**f**), (**V**), (**$'**), (**i**), (**s**), (**T**), (**d**), (**v**), (**p**), (**c**) and (**q**) are known, then its Sales or Revenue Planned is:

$**= [**Q**-**/**(**U**-{**V**+**$'i[1+**s**]+**T**}[1-**d**])
+**$'vp**[1+**s**]/{360[**c**-**q**]}]/{**/**[1-**f**][1-**d**]}

Rule-13334:

If both (**/**), (**U**), (**$**), (**Q**), (**V**), (**$'**), (**i**), (**s**), (**T**), (**d**), (**v**), (**p**), (**c**) and (**q**) are known, then its Fixed Portion Planned is:

f= 1-{**V**+**$'i**[1+**s**]+**T**
+[(**Q**+**$'vp**[1+**s**]/{360[**c**-**q**]})/**/**–**U**]/[1-**d**]}/**$**

Rule-13335:

If both (**/**), (**U**), (**$**), (**f**), (**V**), (**$'**), (**i**), (**s**), (**T**), (**d**), (**v**), (**p**), (**c**) and (**q**) are known, then its Variable Cost Planned is:

V= [**/**(**U**+{**$**[1-**f**]-**$'i**[1+**s**]-**T**}[1-**d**])
-**$'vp**[1+**s**]/{360[**c**-**q**]}-**Q**]/{**/**[1-**d**]}

Rule-13336:

If both (**/**), (**U**), (**$**), (**f**), (**V**), (**Q**), (**i**), (**s**), (**T**), (**d**), (**v**), (**p**), (**c**) and (**q**) are known, then its Sales Past must be:

$'**= [**/**(**U**+{**$[1-**f**]-**V**-**T**}[1-**d**])-**Q**]
/[[1+**s**](**vp**/{360[**c**-**q**]}-**$'i/**[1+**s**][1-**d**])]

Steve Asikin ISBN 14: 978-1514685136, ISBN 10: 1514685132

Rule-13337:

If both (**/**), (**U**), (**$**), (**f**), (**V**), (**$'**), (**Q**), (**s**), (**T**), (**d**), (**v**), (**p**), (**c**) and (**q**) are known, then its Interest Portion Planned is:

i= {$[1-f]-V- T-[(Q+$'vp[1+s]/{360[c-q]})/**/**–U] /[1-d]}/{$'[1+s]}

Rule-13338:

If both (**/**), (**U**), (**$**), (**f**), (**V**), (**$'**), (**i**), (**Q**), (**T**), (**d**), (**v**), (**p**), (**c**) and (**q**) are known, then its Sales Growth Planned is:

s= [/(U+{$[1-f]-V-T}[1-d])-Q] /[$'(vp/{360[c-q]}-$'i/[1+s][1-d])]-1

Rule-13339:

If both (**/**), (**U**), (**$**), (**f**), (**V**), (**$'**), (**i**), (**s**), (**Q**), (**d**), (**v**), (**p**), (**c**) and (**q**) are known, then its Tax Rate Planned is:

T= $[1-f]-V-$'i[1+s] -[(Q+$'vp[1+s]/{360[c-q]})/**/**–U]/[1-d]

Rule-13340:

If both (**/**), (**U**), (**$**), (**f**), (**V**), (**$'**), (**i**), (**s**), (**T**), (**Q**), (**v**), (**p**), (**c**) and (**q**) are known, then its Dividend Payout Planned is:

d= 1-[(Q+$'vp[1+s]/{360[c-q]})/**/**–U] /{$[1-f]-V-$'i[1+s]-T}

Steve Asikin ISBN 14: 978-1514685136, ISBN 10: 1514685132

Rule-13341:

If both (/), (U), ($), (f), (V), ($'), (i), (s), (T), (d), (Q), (p), (c) and (q) are known, then its Variable Portion Planned is:

v= V/$

or it can also be found as

v= 360[c-q](/U+{$[1-f]-V-$'i[1+s]-T}[1-d])-Q)
/{$'p[1+s]}

Rule-13342:

If both (/), (U), ($), (f), (V), ($'), (i), (s), (T), (d), (v), (Q), (c) and (q) are known, then its Procured Inventory Days planned:

p= 360[c-q](/U+{$[1-f]-V-$'i[1+s]-T}[1-d])-Q)
/{$'v[1+s]}

Rule-13343:

If both (/), (U), ($), (f), (V), ($'), (i), (s), (T), (d), (v), (p), (Q) and (q) are known, then its Current Ratio Planned is:

c= q+$'vp[1+s]
/[360(/U+{$[1-f]-V-$'i[1+s]-T}[1-d])-Q)]

Rule-13344:

If both (/), (U), ($), (f), (V), ($'), (i), (s), (T), (d), (v), (p), (c) and (Q) are known, then its Quick or Acid Test Ratio Planned is:

q= c-$'vp[1+s]
/[360(/U+{$[1-f]-V-$'i[1+s]-T}[1-d])-Q)]

Steve Asikin ISBN 14: 978-1514685136, ISBN 10: 1514685132

Rule-13345:

If both (/), (U), (S), (f), (V), (S'), (i), (s), (T), (A), (d) and (X) are known, then its Variable Cost Planned is:

Q= /{U+S[1-f]-V-S'i[1+s]-T-Ad}-X

Rule-13346:

If both (Q), (U), (S), (f), (V), (S'), (i), (s), (T), (A), (d) and (X) are known, then its Leverage or Gearing Ratio Planned is:

/= [Q+X]/{U+S[1-f]-V-S'i[1+s]-T-Ad}

Rule-13347:

If both (/), (Q), (S), (f), (V), (S'), (i), (s), (T), (A), (d) and (X) are known, then its Utilized or Starting Capital must be:

U= V+S'i[1+s]+T+Ad+[Q+X]/-S[1-f]

Rule-13348:

If both (/), (U), (Q), (f), (V), (S'), (i), (s), (T), (A), (d) and (X) are known, then its Sales or Revenue Planned is:

S= {V+S'i[1+s]+T+Ad+[Q+X]/-U}/[1-f]

Rule-13349:

If both (/), (U), (S), (Q), (V), (S'), (i), (s), (T), (A), (d) and (X) are known, then its Fixed Portion Planned is:

f= 1-{V+S'i[1+s]+T+Ad+[Q+X]/-U}/S

Steve Asikin ISBN 14: 978-1514685136, ISBN 10: 1514685132

Rule-13350:

If both (/), (**U**), (**$**), (**f**), (**Q**), (**$'**), (**i**), (**s**), (**T**), (**A**), (**d**) and (**X**) are known, then its Variable Cost Planned is:

V= **$**[1-**f**]- **$'i**[1+**s**]-**T**-**Ad**-[**Q**+**X**]/**/-U**

Rule-13351:

If both (/), (**U**), (**$**), (**f**), (**V**), (**Q**), (**i**), (**s**), (**T**), (**A**), (**d**) and (**X**) are known, then its Sales Past must be:

$'**={**$[1-**f**]-**V**-**T**-**Ad**-[**Q**+**X**]/**/-U**}/{**i**[1+**s**]}

Rule-13352:

If both (/), (**U**), (**$**), (**f**), (**V**), (**$'**), (**Q**), (**s**), (**T**), (**A**), (**d**) and (**X**) are known, then its Interest Portion Planned is:

i={**$**[1-**f**]-**V**-**T**-**Ad**-[**Q**+**X**]/**/-U**}/{**$'**[1+**s**]}

Rule-13353:

If both (/), (**U**), (**$**), (**f**), (**V**), (**$'**), (**i**), (**Q**), (**T**), (**A**), (**d**) and (**X**) are known, then its Sales Growth Planned is:

s= **$**/**$'**-1

or it can also be found as

s={**$**[1-**f**]-**V**-**T**-**Ad**-[**Q**+**X**]/**/-U**}/[**$'i**]-1

Rule-13354:

If both (/), (**U**), (**$**), (**f**), (**V**), (**$'**), (**i**), (**s**), (**Q**), (**A**), (**d**) and (**X**) are known, then its Tax Planned is:

T= **$**[1-**f**]-**V**-**$'i**[1+**s**]-**Ad**-[**Q**+**X**]/**/-U**

Steve Asikin ISBN 14: 978-1514685136, ISBN 10: 1514685132

Rule-13355:

If both (**/**), (**U**), (**S**), (**f**), (**V**), (**S'**), (**i**), (**s**), (**T**), (**Q**), (**d**) and (**X**) are known, then its After Tax Income Planned is:

A= {**S**[1-**f**]-**V**-**S'i**[1+**s**]-**T**-[**Q**+**X**]/**/-U**}/**d**

Rule-13356:

If both (**/**), (**U**), (**S**), (**f**), (**V**), (**S'**), (**i**), (**s**), (**T**), (**A**), (**Q**) and (**X**) are known, then its Dividend Payout Planned is:

d= {**S**[1-**f**]-**V**-**S'i**[1+**s**]-**T**-[**Q**+**X**]/**/-U**}/**A**

Rule-13357:

If both (**/**), (**U**), (**S**), (**f**), (**V**), (**S'**), (**i**), (**s**), (**T**), (**A**), (**d**) and (**Q**) are known, then its Xpress or Current Debt Planned is:

X= **/**{**U**+**S**[1-**f**]-**V**-**S'i**[1+**s**]-**T**-**Ad**}-**Q**

Rule-13358:

If both (**/**), (**U**), (**S**), (**f**), (**V**), (**S'**), (**i**), (**s**), (**T**), (**A**), (**d**), (**P**), (**c**) and (**q**) are known, then its Variable Cost Planned is:

Q= **/**{**U**+**S**[1-**f**]-**V**-**S'i**[1+**s**]-**T**-**Ad**}-**P**/[**c**-**q**]

Rule-13359:

If both (**/**), (**U**), (**S**), (**f**), (**V**), (**S'**), (**i**), (**s**), (**T**), (**A**), (**d**), (**P**), (**c**) and (**q**) are known, then its Variable Cost Planned is:

/= {**Q**+**P**/[**c**-**q**]}/{**U**+**S**[1-**f**]-**V**-**S'i**[1+**s**]-**T**-**Ad**}

Steve Asikin ISBN 14: 978-1514685136, ISBN 10: 1514685132

Rule-13360:

If both (**/**), (**Q**), (**$**), (**f**), (**V**), (**$'**), (**i**), (**s**), (**T**), (**A**), (**d**), (**P**), (**c**) and (**q**) are known, then its Utilized or Starting Capital must be:

U= **V**+**$'i**[1+**s**]+**T**+**Ad**+{**Q**+**P**/[**c**-**q**]}/**/**-**$**[1-**f**]

Rule-13361:

If both (**/**), (**U**), (**Q**), (**f**), (**V**), (**$'**), (**i**), (**s**), (**T**), (**A**), (**d**), (**P**), (**c**) and (**q**) are known, then its Sales or Revenue Planned is:

$**= (**V**+**$'i[1+**s**]+**T**+**Ad**+{**Q**+**P**/[**c**-**q**]}/**/**-**U**)/[1-**f**]

Rule-13362:

If both (**/**), (**U**), (**$**), (**Q**), (**V**), (**$'**), (**i**), (**s**), (**T**), (**A**), (**d**), (**P**), (**c**) and (**q**) are known, then its Fixed Portion Planned is:

f= 1-(**V**+**$'i**[1+**s**]+**T**+**Ad**+{**Q**+**P**/[**c**-**q**]}/**/**-**U**)/**$**

Rule-13363:

If both (**/**), (**U**), (**$**), (**f**), (**Q**), (**$'**), (**i**), (**s**), (**T**), (**A**), (**d**), (**P**), (**c**) and (**q**) are known, then its Variable Cost Planned is:

V= **$**[1-**f**]- **$'i**[1+**s**]-**T**-**Ad**-{**Q**+**P**/[**c**-**q**]}/**/**-**U**

Rule-13364:

If both (**/**), (**U**), (**$**), (**f**), (**V**), (**Q**), (**i**), (**s**), (**T**), (**A**), (**d**), (**P**), (**c**) and (**q**) are known, then its Sales Past must be:

$'**= (**$[1-**f**]-**V**-**T**-**Ad**-{**Q**+**P**/[**c**-**q**]}/**/**-**U**)/{**i**[1+**s**]}

Steve Asikin ISBN 14: 978-1514685136, ISBN 10: 1514685132

Rule-13365:

If both (**/**), (**U**), (**$**), (**f**), (**V**), (**$'**), (**Q**), (**s**), (**T**), (**A**), (**d**), (**P**), (**c**) and (**q**) are known, then its Interest Portion Planned is:

i= (**$**[1-**f**]-**V**-**T**-**Ad**-{**Q**+**P**/[**c**-**q**]}/**/-U**)/{**$'**[1+**s**]}

Rule-13366:

If both (**/**), (**U**), (**$**), (**f**), (**V**), (**$'**), (**i**), (**Q**), (**T**), (**A**), (**d**), (**P**), (**c**) and (**q**) are known, then its Sales Growth Planned is:

s= **$**/**$'**-1

or it can also be found as

s= (**$**[1-**f**]-**V**-**T**-**Ad**-{**Q**+**P**/[**c**-**q**]}/**/-U**)/[**$'i**]-1

Rule-13367:

If both (**/**), (**U**), (**$**), (**f**), (**V**), (**$'**), (**i**), (**s**), (**Q**), (**A**), (**d**), (**P**), (**c**) and (**q**) are known, then its Tax Planned is:

T= **$**[1-**f**]-**V**-**$'i**[1+**s**]-**Ad**-{**Q**+**P**/[**c**-**q**]}/**/-U**

Rule-13368:

If both (**/**), (**U**), (**$**), (**f**), (**V**), (**$'**), (**i**), (**s**), (**T**), (**Q**), (**d**), (**P**), (**c**) and (**q**) are known, then its After Tax Income Planned is:

A= (**$**[1-**f**]-**V**-**$'i**[1+**s**]-**T**-{**Q**+**P**/[**c**-**q**]}/**/-U**)/**d**

Rule-13369:

If both (**/**), (**U**), (**$**), (**f**), (**V**), (**$'**), (**i**), (**s**), (**T**), (**A**), (**Q**), (**P**), (**c**) and (**q**) are known, then its Dividend Portion Planned is:

d= (**$**[1-**f**]-**V**-**$'i**[1+**s**]-**T**-{**Q**+**P**/[**c**-**q**]}/**/-U**)/**A**

Steve Asikin ISBN 14: 978-1514685136, ISBN 10: 1514685132

Rule-13370:

If both (**/**), (**U**), (**$**), (**f**), (**V**), (**$'**), (**i**), (**s**), (**T**), (**A**), (**d**), (**Q**), (**c**) and (**q**) are known, then its Procured Inventory Planned is:

P= [**c**-**q**](**/**{**U**+**$**[1-**f**]-**V**-**$'i**[1+**s**]-**T**-**Ad**}-**Q**)

Rule-13371:

If both (**/**), (**U**), (**$**), (**f**), (**V**), (**$'**), (**i**), (**s**), (**T**), (**A**), (**d**), (**P**), (**Q**) and (**q**) are known, then its Current Ratio Planned is:

c= **q**+**P**/(**/**{**U**+**$**[1-**f**]-**V**-**$'i**[1+**s**]-**T**-**Ad**}-**Q**)

Rule-13372:

If both (**/**), (**U**), (**$**), (**f**), (**V**), (**$'**), (**i**), (**s**), (**T**), (**A**), (**d**), (**P**), (**c**) and (**Q**) are known, then its Quick or Acid Test Ratio Planned is:

q= **c**-**P**/(**/**{**U**+**$**[1-**f**]-**V**-**$'i**[1+**s**]-**T**-**Ad**}-**Q**)

Rule-13373:

If both (**/**), (**U**), (**$**), (**f**), (**V**), (**$'**), (**i**), (**s**), (**T**), (**A**), (**d**), (**p**), (**c**) and (**q**) are known, then its Quoted Longterm Debt Planned is:

Q= **/**{**U**+**$**[1-**f**]-**V**-**$'i**[1+**s**]-**T**-**Ad**}-**Vp**/{360[**c**-**q**]}

Rule-13374:

If both (**Q**), (**U**), (**$**), (**f**), (**V**), (**$'**), (**i**), (**s**), (**T**), (**A**), (**d**), (**p**), (**c**) and (**q**) are known, then its Leverage or Gearing Ratio Planned is:

/= (**Q**+ **Vp**/{360[**c**-**q**]})
/{**U**+**$**[1-**f**]-**V**-**$'i**[1+**s**]-**T**-**Ad**}

Steve Asikin ISBN 14: 978-1514685136, ISBN 10: 1514685132

Rule-13375:

If both (**/**), (**Q**), (**$**), (**f**), (**V**), (**$'**), (**i**), (**s**), (**T**), (**A**), (**d**), (**p**), (**c**) and (**q**) are known, then its Utilized or Starting Capital must be:

U= **V**+**$'i**[1+**s**]+**T**+**Ad**+(**Q**+**Vp**/{360[**c**-**q**]})/**/**-**$**[1-**f**]

Rule-13376:

If both (**/**), (**U**), (**Q**), (**f**), (**V**), (**$'**), (**i**), (**s**), (**T**), (**A**), (**d**), (**p**), (**c**) and (**q**) are known, then its Sales or Revenue Planned is:

$**= **[V**+**$'i[1+**s**]+**T**+**Ad**+(**Q**+**Vp**/{360[**c**-**q**]})/**/**-**U]**
/[1-**f**]

Rule-13377:

If both (**/**), (**U**), (**$**), (**Q**), (**V**), (**$'**), (**i**), (**s**), (**T**), (**A**), (**d**), (**p**), (**c**) and (**q**) are known, then its Fixed Portion Planned is:

f= 1-**[V**+**$'i**[1+**s**]+**T**+**Ad**
+(**Q**+**Vp**/{360[**c**-**q**]})/**/**-**U]**/**$**

Rule-13378:

If both (**/**), (**U**), (**$**), (**f**), (**Q**), (**$'**), (**i**), (**s**), (**T**), (**A**), (**d**), (**p**), (**c**) and (**q**) are known, then its Variable Cost Planned is:

V= (**/**{**U**+**$**[1-**f**]-**$'i**[1+**s**]-**T**-**Ad**}-**Q**)
/(**p**/{360[**c**-**q**]}+**/**)

Steve Asikin ISBN 14: 978-1514685136, ISBN 10: 1514685132

Rule-13379:

If both (**/**), (**U**), (**$**), (**f**), (**V**), (**Q**), (**i**), (**s**), (**T**), (**A**), (**d**), (**p**), (**c**) and (**q**) are known, then its Sales Past must be:

$'**= [**$[1-**f**]-**V**-**T**-**Ad**-(**Q**+**Vp**/{360[**c**-**q**]})/**/U**]
/{**i**[1+**s**]}

Rule-13380:

If both (**/**), (**U**), (**$**), (**f**), (**V**), (**$'**), (**Q**), (**s**), (**T**), (**A**), (**d**), (**p**), (**c**) and (**q**) are known, then its Interest Portion Planned is:

i= [**$**[1-**f**]-**V**-**T**-**Ad**-(**Q**+**Vp**/{360[**c**-**q**]})/**/U**]
/{**$'**[1+**s**]}

Rule-13381:

If both (**/**), (**U**), (**$**), (**f**), (**V**), (**$'**), (**i**), (**Q**), (**T**), (**A**), (**d**), (**p**), (**c**) and (**q**) are known, then its Sales Growth Planned is:

s= **$**/**$'**-1

or it can also be found as

s= [**$**[1-**f**]-**V**-**T**-**Ad**-(**Q**+**Vp**/{360[**c**-**q**]})/**/U**]/ [**$'i**]-1

Rule-13382:

If both (**/**), (**U**), (**$**), (**f**), (**V**), (**$'**), (**i**), (**s**), (**Q**), (**A**), (**d**), (**p**), (**c**) and (**q**) are known, then its Tax Planned is:

T= **$**[1-**f**]-**V**-**$'i**[1+**s**]-**Ad**-(**Q**+**Vp**/{360[**c**-**q**]})/**/U**

Rule-13383:

If both (**/**), (**U**), (**$**), (**f**), (**V**), (**$'**), (**i**), (**s**), (**T**), (**Q**), (**d**), (**p**), (**c**) and (**q**) are known, then its After Tax Income Planned is:

A= [**$**[1-**f**]-**V**-**$'i**[1+**s**]-**T**-(**Q**+**Vp**/{360[**c**-**q**]})]/**d**

Steve Asikin ISBN 14: 978-1514685136, ISBN 10: 1514685132

Rule-13384:

If both (**/**), (**U**), (**$**), (**f**), (**V**), (**$'**), (**i**), (**s**), (**T**), (**A**), (**Q**), (**p**), (**c**) and (**q**) are known, then its Dividend Portion Planned is:

d= [**$**[1-**f**]-**V**-**$'i**[1+**s**]-**T**-(**Q**+**Vp**/{360[**c**-**q**]})]/**A**

Rule-13385:

If both (**/**), (**U**), (**$**), (**f**), (**V**), (**$'**), (**i**), (**s**), (**T**), (**A**), (**d**), (**Q**), (**c**) and (**q**) are known, then its Procured Inventory Days planned:

p= 360[**c**-**q**](**/**{**U**+**$**[1-**f**]-**V**-**$'i**[1+**s**]-**T**-**Ad**}-**Q**)/**V**

Rule-13386:

If both (**/**), (**U**), (**$**), (**f**), (**V**), (**$'**), (**i**), (**s**), (**T**), (**A**), (**d**), (**p**), (**Q**) and (**q**) are known, then its Current Ratio Planned is:

c= **q**+**Vp**/{360(**/**{**U**+**$**[1-**f**]-**V**-**$'i**[1+**s**]-**T**-**Ad**}-**Q**)}

Rule-13387:

If both (**/**), (**U**), (**$**), (**f**), (**V**), (**$'**), (**i**), (**s**), (**T**), (**A**), (**d**), (**p**), (**c**) and (**Q**) are known, then its Quick or Acid Test Ratio Planned is:

q= **c**-**Vp**/{360(**/**{**U**+**$**[1-**f**]-**V**-**$'i**[1+**s**]-**T**-**Ad**}-**Q**)}

Rule-13388:

If both (**/**), (**U**), (**$**), (**f**), (**V**), (**$'**), (**i**), (**s**), (**T**), (**A**), (**d**), (**v**), (**p**), (**c**) and (**q**) are known, then its Quoted Longterm Debt Planned is:

Q= **/**{**U**+**$**[1-**f**]-**V**-**$'i**[1+**s**]-**T**-**Ad**}-**$vp**/{360[**c**-**q**]}

Steve Asikin ISBN 14: 978-1514685136, ISBN 10: 1514685132

Rule-13389:

If both (**Q**), (**U**), (**$**), (**f**), (**V**), (**$'**), (**i**), (**s**), (**T**), (**A**), (**d**), (**v**), (**p**), (**c**) and (**q**) are known, then its Leverage or Gearing Ratio Planned is:

/= (**Q**+**$vp**/{360[**c**-**q**]})
/{**U**+**$**[1-**f**]-**V**-**$'i**[1+**s**]-**T**-**Ad**}

Rule-13390:

If both (**/**), (**Q**), (**$**), (**f**), (**V**), (**$'**), (**i**), (**s**), (**T**), (**A**), (**d**), (**v**), (**p**), (**c**) and (**q**) are known, then its Utilized or Starting Capital must be:

U= **V**+**$'i**[1+**s**]+**T**+**Ad**
+(**Q**+**$vp**/{360[**c**-**q**]})/**/**-**$**[1-**f**]

Rule-13391:

If both (**/**), (**U**), (**Q**), (**f**), (**V**), (**$'**), (**i**), (**s**), (**T**), (**A**), (**d**), (**v**), (**p**), (**c**) and (**q**) are known, then its Sales or Revenue Planned is:

$**= (**/**{**U**-**V**-**$'i[1+**s**]-**T**-**Ad**}-**Q**)
/(**vp**/{360[**c**-**q**]}-**$/**[1-**f**])

Rule-13392:

If both (**/**), (**U**), (**$**), (**Q**), (**V**), (**$'**), (**i**), (**s**), (**T**), (**A**), (**d**), (**v**), (**p**), (**c**) and (**q**) are known, then its Fixed Portion Planned is:

f= 1-[**V**+**$'i**[1+**s**]+**T**+**Ad**
+(**Q**+**$vp**/{360[**c**-**q**]})/**/**-**U**]/**$**

Steve Asikin ISBN 14: 978-1514685136, ISBN 10: 1514685132

Rule-13393:

If both (/), (U), (S), (f), (Q), (S'), (i), (s), (T), (A), (d), (v), (p), (c) and (q) are known, then its Variable Cost Planned is:

V= = S[1-f]-S'i[1+s]-T-Ad-(Q+Svp/{360[c-q]})/ /U

Rule-13394:

If both (/), (U), (S), (f), (V), (Q), (i), (s), (T), (A), (d), (v), (p), (c) and (q) are known, then its Sales Past must be:

S'= [S[1-f]-V-T-Ad-(Q+Svp/{360[c-q]})/ /U]
/{i[1+s]}

Rule-13395:

If both (/), (U), (S), (f), (V), (S'), (Q), (s), (T), (A), (d), (v), (p), (c) and (q) are known, then its Interest Portion Planned is:

i= [S[1-f]-V-T-Ad-(Q+Svp/{360[c-q]})/ /U]
/{S'[1+s]}

Rule-13396:

If both (/), (U), (S), (f), (V), (S'), (i), (Q), (T), (A), (d), (v), (p), (c) and (q) are known, then its Sales Growth Planned is:

s= S/S'-1

or it can also be found as

s= [S[1-f]-V-T-Ad-(Q+Svp/{360[c-q]})/ /U]
/ [S'i]-1

Steve Asikin ISBN 14: 978-1514685136, ISBN 10: 1514685132

Rule-13397:

If both (**/**), (**U**), (**$**), (**f**), (**V**), (**$'**), (**i**), (**s**), (**Q**), (**A**), (**d**), (**v**), (**p**), (**c**) and (**q**) are known, then its Tax Planned is:

T= **$**[1-**f**]-**V**-**$'i**[1+**s**]-**Ad**-(**Q**+**$vp**/{360[**c**-**q**]})/**/**-**U**

Rule-13398:

If both (**/**), (**U**), (**$**), (**f**), (**V**), (**$'**), (**i**), (**s**), (**T**), (**Q**), (**d**), (**v**), (**p**), (**c**) and (**q**) are known, then its After Tax Income Planned is:

A= **[$**[1-**f**]-**V**-**$'i**[1+**s**]-**T**-(**Q**+**$vp**/{360[**c**-**q**]})**]**/**d**

Rule-13399:

If both (**/**), (**U**), (**$**), (**f**), (**V**), (**$'**), (**i**), (**s**), (**T**), (**A**), (**Q**), (**v**), (**p**), (**c**) and (**q**) are known, then its Dividend Planned is:

d= **[$**[1-**f**]-**V**-**$'i**[1+**s**]-**T**-(**Q**+**$vp**/{360[**c**-**q**]})**]**/**A**

Rule-13400:

If both (**/**), (**U**), (**$**), (**f**), (**V**), (**$'**), (**i**), (**s**), (**T**), (**A**), (**d**), (**Q**), (**p**), (**c**) and (**q**) are known, then its Variable Portion Planned is:

v= 360[**c**-**q**](**/**{**U**+**$**[1-**f**]-**V**-**$'i**[1+**s**]-**T**-**Ad**}-**Q**)/[**$p**]

Rule-13401:

If both (**/**), (**U**), (**$**), (**f**), (**V**), (**$'**), (**i**), (**s**), (**T**), (**A**), (**d**), (**v**), (**Q**), (**c**) and (**q**) are known, then its Procured Inventory Days Planned:

p= 360[**c**-**q**](**/**{**U**+**$**[1-**f**]-**V**-**$'i**[1+**s**]-**T**-**Ad**}-**Q**)/[**$v**]

Steve Asikin ISBN 14: 978-1514685136, ISBN 10: 1514685132

Rule-13402:

If both (**/**), (**U**), (**$**), (**f**), (**V**), (**$'**), (**i**), (**s**), (**T**), (**A**), (**d**), (**v**), (**p**), (**Q**) and (**q**) are known, then its Current Ratio Planned is:

c= **q**+**$vp**/{360(**/**{**U**+**$**[1-**f**]-**V**-**$'i**[1+**s**]-**T**-**Ad**}-**Q**)}

Rule-13403:

If both (**/**), (**U**), (**$**), (**f**), (**V**), (**$'**), (**i**), (**s**), (**T**), (**A**), (**d**), (**v**), (**p**), (**c**) and (**Q**) are known, then its Quick or Acid Test Ratio Planned is:

q= **c**-**$vp**/{360(**/**{**U**+**$**[1-**f**]-**V**-**$'i**[1+**s**]-**T**-**Ad**}-**Q**)}

Rule-13404:

If both (**/**), (**U**), (**$**), (**f**), (**V**), (**$'**), (**i**), (**s**), (**T**), (**A**), (**d**), (**v**), (**p**), (**c**) and (**q**) are known, then its Quoted Longterm Debt Planned is:

Q= **/**{**U**+**$**[1-**f**]-**V**-**$'i**[1+**s**]-**T**-**Ad**}
-**$'vp**[1+**s**]/{360[**c**-**q**]}

Rule-13405:

If both (**Q**), (**U**), (**$**), (**f**), (**V**), (**$'**), (**i**), (**s**), (**T**), (**A**), (**d**), (**v**), (**p**), (**c**) and (**q**) are known, then its Leverage or Gearing Ratio Planned is:

/= (**Q**+**$'vp**[1+**s**]/{360[**c**-**q**]})
/{**U**+**$**[1-**f**]-**V**-**$'i**[1+**s**]-**T**-**Ad**}

Steve Asikin ISBN 14: 978-1514685136, ISBN 10: 1514685132

Rule-13406:

If both (**/**), (**Q**), (**$**), (**f**), (**V**), (**$'**), (**i**), (**s**), (**T**), (**A**), (**d**), (**v**), (**p**), (**c**) and (**q**) are known, then its Utilized or Starting Capital must be:

U= V+$'i[1+s]+T+Ad
+(Q+$'vp[1+s]/{360[c-q]})//$[1-f]

Rule-13407:

If both (**/**), (**U**), (**Q**), (**f**), (**V**), (**$'**), (**i**), (**s**), (**T**), (**A**), (**d**), (**v**), (**p**), (**c**) and (**q**) are known, then its Sales or Revenue Planned is:

$= (/{U-V-$'i[1+s]-T-Ad}-Q)
/(vp/{360[c-q]}-$/[1-f])

Rule-13408:

If both (**/**), (**U**), (**$**), (**Q**), (**V**), (**$'**), (**i**), (**s**), (**T**), (**A**), (**d**), (**v**), (**p**), (**c**) and (**q**) are known, then its Fixed Cost Planned is:

f= 1-[V+$'i[1+s]+T+Ad
+(Q+$'vp[1+s]/{360[c-q]})//U]/$

Rule-13409:

If both (**/**), (**U**), (**$**), (**f**), (**Q**), (**$'**), (**i**), (**s**), (**T**), (**A**), (**d**), (**v**), (**p**), (**c**) and (**q**) are known, then its Variable Cost Planned is:

V= = $[1-f]-$'i[1+s]-T-Ad
-(Q+$'vp[1+s]/{360[c-q]})//U

Steve Asikin ISBN 14: 978-1514685136, ISBN 10: 1514685132

Rule-13410:

If both (**/**), (**U**), (**S**), (**f**), (**V**), (**Q**), (**i**), (**s**), (**T**), (**A**), (**d**), (**v**), (**p**), (**c**) and (**q**) are known, then its Sales Past must be:

S'= (**/**{**U**+**S**[1-**f**]-**V**-**T**-**Ad**}-**Q**)
/[[1+**s**](**vp**/{360[**c**-**q**]}+**i/**)]

Rule-13411:

If both (**/**), (**U**), (**S**), (**f**), (**V**), (**S'**), (**Q**), (**s**), (**T**), (**A**), (**d**), (**v**), (**p**), (**c**) and (**q**) are known, then its Interest Portion Planned is:

i= [**S**[1-**f**]-**V**-**T**-**Ad**
-(**Q**+**S'vp**[1+**s**]/{360[**c**-**q**]})/**/**-**U**]/{**S'**[1+**s**]}

Rule-13412:

If both (**/**), (**U**), (**S**), (**f**), (**V**), (**S'**), (**i**), (**Q**), (**T**), (**A**), (**d**), (**v**), (**p**), (**c**) and (**q**) are known, then its Sales Growth Planned is:

s= **S**/**S'**-1

or it can also be found as

s= (**/**{**U**+**S**[1-**f**]-**V**-**T**-**Ad**}-**Q**)
/[**S'**(**vp**/{360[**c**-**q**]}+**i/**)]-1

Rule-13413:

If both (**/**), (**U**), (**S**), (**f**), (**V**), (**S'**), (**i**), (**s**), (**Q**), (**A**), (**d**), (**v**), (**p**), (**c**) and (**q**) are known, then its Tax Planned is:

T= **S**[1-**f**]-**V**-**S'i**[1+**s**]-**Ad**
-(**Q**+**S'vp**[1+**s**]/{360[**c**-**q**]})/**/**-**U**

Steve Asikin ISBN 14: 978-1514685136, ISBN 10: 1514685132

Rule-13414:

If both (**/**), (**U**), (**$**), (**f**), (**V**), (**$'**), (**i**), (**s**), (**T**), (**Q**), (**d**), (**v**), (**p**), (**c**) and (**q**) are known, then its After Tax Income Planned is:

A= [**$**[1-**f**]-**V**-**$'i**[1+**s**]-**T**
-(**Q**+**$'vp**[1+**s**]/{360[**c**-**q**]})]/**d**

Rule-13415:

If both (**/**), (**U**), (**$**), (**f**), (**V**), (**$'**), (**i**), (**s**), (**T**), (**A**), (**Q**), (**v**), (**p**), (**c**) and (**q**) are known, then its Dividend Payout Planned is:

d= [**$**[1-**f**]-**V**-**$'i**[1+**s**]-**T**
-(**Q**+**$'vp**[1+**s**]/{360[**c**-**q**]})]/**A**

Rule-13416:

If both (**/**), (**U**), (**$**), (**f**), (**V**), (**$'**), (**i**), (**s**), (**T**), (**A**), (**d**), (**Q**), (**p**), (**c**) and (**q**) are known, then its Quoted Longterm Debt Planned is:

v= 360[**c**-**q**](**/**{**U**+**$**[1-**f**]-**V**-**$'i**[1+**s**]-**T**-**Ad**}-**Q**)
/{**$'p**[1+**s**]}

Rule-13417:

If both (**/**), (**U**), (**$**), (**f**), (**V**), (**$'**), (**i**), (**s**), (**T**), (**A**), (**d**), (**v**), (**Q**), (**c**) and (**q**) are known, then itsProcured Inventory Days Planned:

p= 360[**c**-**q**](**/**{**U**+**$**[1-**f**]-**V**-**$'i**[1+**s**]-**T**-**Ad**}-**Q**)
/{**$'v**[1+**s**]}

Steve Asikin ISBN 14: 978-1514685136, ISBN 10: 1514685132

Rule-13418:

If both (**/**), (**U**), (**S**), (**f**), (**V**), (**S'**), (**i**), (**s**), (**T**), (**A**), (**d**), (**v**), (**p**), (**Q**) and (**q**) are known, then its Current Ratio Planned is:

c= **q**+**S'vp**[1+**s**]
/{360(**/**{**U**+**S**[1-**f**]-**V**-**S'i**[1+**s**]-**T**-**Ad**}-**Q**)}

Rule-13419:

If both (**/**), (**U**), (**S**), (**f**), (**V**), (**S'**), (**i**), (**s**), (**T**), (**A**), (**d**), (**v**), (**p**), (**c**) and (**Q**) are known, then its Quick or Acid Test Ratio Planned is:

q= **c**-**S'vp**[1+**s**]
/{360(**/**{**U**+**S**[1-**f**]-**V**-**S'i**[1+**s**]-**T**-**Ad**}-**Q**)}

Rule-13420:

If both (**/**), (**U**), (**S**), (**f**), (**V**), (**S'**), (**i**), (**s**), (**t**), (**D**) and (**X**) are known, then its Quoted Longterm Debt Planned is:

Q= **/**(**U**+{**S**[1-**f**]-**V**-**S'i**[1+**s**]}[1-**t**]-**D**)-**X**

Rule-13421:

If both (**Q**), (**U**), (**S**), (**f**), (**V**), (**S'**), (**i**), (**s**), (**t**), (**D**) and (**X**) are known, then its Leverage or Gearing Ratio Planned is:

/= [**Q**+**X**]/(**U**+{**S**[1-**f**]-**V**-**S'i**[1+**s**]}[1-**t**]-**D**)

Rule-13422:

If both (**/**), (**Q**), (**S**), (**f**), (**V**), (**S'**), (**i**), (**s**), (**t**), (**D**) and (**X**) are known, then its Utilized or Starting Capital must be:

U= **D**+[**Q**+**X**]/**/**-{**S**[1-**f**]-**V**-**S'i**[1+**s**]}[1-**t**]

Steve Asikin ISBN 14: 978-1514685136, ISBN 10: 1514685132

Rule-13423:

If both (**/**), (**U**), (**Q**), (**f**), (**V**), (**$'**), (**i**), (**s**), (**t**), (**D**) and (**X**) are known, then its Sales or Revenue Planned is:

$= (V+$'i[1+s]+{D+[Q+X]/LU}/[1-t])/[1-f]

Rule-13424:

If both (**/**), (**U**), (**$**), (**Q**), (**V**), (**$'**), (**i**), (**s**), (**t**), (**D**) and (**X**) are known, then its Fixed Portion Planned is:

f= 1-(V+$'i[1+s]+{D+[Q+X]/LU}/[1-t])/$

Rule-13425:

If both (**/**), (**U**), (**$**), (**f**), (**Q**), (**$'**), (**i**), (**s**), (**t**), (**D**) and (**X**) are known, then its Variable Cost Planned is:

V= $[1-f]-$'i[1+s]-{D+[Q+X]/LU}/[1-t]

Rule-13426:

If both (**/**), (**U**), (**$**), (**f**), (**V**), (**Q**), (**i**), (**s**), (**t**), (**D**) and (**X**) are known, then its Sales Past must be:

$'= ($[1-f]-V-{D+[Q+X]/LU}/[1-t])/{i[1+s]}

Rule-13427:

If both (**/**), (**U**), (**$**), (**f**), (**V**), (**$'**), (**Q**), (**s**), (**t**), (**D**) and (**X**) are known, then its Interest Portion Planned is:

i= ($[1-f]-V-{D+[Q+X]/LU}/[1-t])/{$'[1+s]}

Rule-13428:

If both (**/**), (**U**), (**$**), (**f**), (**V**), (**$'**), (**i**), (**Q**), (**t**), (**D**) and (**X**) are known, then its Sales Growth Planned is:

s= $/$'-1

or it can also be found as

s= ($[1-f]-V-{D+[Q+X]/LU}/[1-t])/[$'i]-1

Steve Asikin ISBN 14: 978-1514685136, ISBN 10: 1514685132

Rule-13429:

If both (**/**), (**U**), (**$**), (**f**), (**V**), (**$'**), (**i**), (**s**), (**Q**), (**D**) and (**X**) are known, then its Tax Rate Planned is:

t= 1-{**D**+[**Q**+**X**]/**/**-**U**}/{**$**[1-**f**]-**V**-**$'i**[1+**s**]}

Rule-13430:

If both (**/**), (**U**), (**$**), (**f**), (**V**), (**$'**), (**i**), (**s**), (**t**), (**Q**) and (**X**) are known, then its Dividend Payout Planned is:

D= **U**+{**$**[1-**f**]-**V**-**$'i**[1+**s**]}[1-**t**]-[**Q**+**X**]/**/**

Rule-13431:

If both (**/**), (**U**), (**$**), (**f**), (**V**), (**$'**), (**i**), (**s**), (**t**), (**D**) and (**Q**) are known, then its Xpress or Current Debt Planned is:

X= **/**(**U**+{**$**[1-**f**]-**V**-**$'i**[1+**s**]}[1-**t**]-**D**)-**Q**

Rule-13432:

If both (**/**), (**U**), (**$**), (**f**), (**V**), (**$'**), (**i**), (**s**), (**t**), (**D**), (**P**), (**c**) and (**q**) are known, then its Quoted Longterm Debt Planned is:

Q= **/**(**U**+{**$**[1-**f**]-**V**-**$'i**[1+**s**]}[1-**t**]-**D**)-**P**/[**c**-**q**]

Rule-13433:

If both (**Q**), (**U**), (**$**), (**f**), (**V**), (**$'**), (**i**), (**s**), (**t**), (**D**), (**P**), (**c**) and (**q**) are known, then its Leverage or Gearing Ratio Planned is:

/= {**Q**+**P**/[**c**-**q**]}/(**U**+{**$**[1-**f**]-**V**-**$'i**[1+**s**]}[1-**t**]-**D**)

Steve Asikin ISBN 14: 978-1514685136, ISBN 10: 1514685132

Rule-13434:

If both (Ƚ), (Q), ($), (f), (V), ($'), (i), (s), (t), (D), (P), (c) and (q) are known, then its Utilized or Starting Capital must be:

U= D+{Q+P/[c-q]}/Ƚ{$[1-f]-V-$'i[1+s]}[1-t]

Rule-13435:

If both (Ƚ), (U), (Q), (f), (V), ($'), (i), (s), (t), (D), (P), (c) and (q) are known, then its Sales or Revenue Planned is:

$= [V+$'i[1+s]+(D+{Q+P/[c-q]}/ȽU)/[1-t]]/[1-f]

Rule-13436:

If both (Ƚ), (U), ($), (Q), (V), ($'), (i), (s), (t), (D), (P), (c) and (q) are known, then its Fixed Portion Planned is:

f= 1-[V+$'i[1+s]+(D+{Q+P/[c-q]}/ȽU)/[1-t]]/$

Rule-13437:

If both (Ƚ), (U), ($), (f), (Q), ($'), (i), (s), (t), (D), (P), (c) and (q) are known, then its Variable Cost Planned is:

V= $[1-f]-$'i[1+s]- (D+{Q+P/[c-q]}/ȽU)/[1-t]

Rule-13438:

If both (Ƚ), (U), ($), (f), (V), (Q), (i), (s), (t), (D), (P), (c) and (q) are known, then its Sales Past must be:

$'= [$[1-f]-V-(D+{Q+P/[c-q]}/ȽU)/[1-t]]/{i[1+s]}

Steve Asikin ISBN 14: 978-1514685136, ISBN 10: 1514685132

Rule-13439:

If both (/), (U), ($), (f), (V), ($'), (Q), (s), (t), (D), (P), (c) and (q) are known, then its Interest Portion Planned is:

i= [$[1-f]-V-(D+{Q+P/[c-q]}/ /U)/[1-t]]/{$'[1+s]}

Rule-13440:

If both (/), (U), ($), (f), (V), ($'), (i), (Q), (t), (D), (P), (c) and (q) are known, then its Sales Past must be:

s= $/$'-1

or it can also be found as

s=[$[1-f]-V-(D+{Q+P/[c-q]}/ /U)/[1-t]]/[$'i]-1

Rule-13441:

If both (/), (U), ($), (f), (V), ($'), (i), (s), (Q), (D), (P), (c) and (q) are known, then its Tax Rate Planned is:

t= 1-(D+{Q+P/[c-q]}/ /U)/{$[1-f]-V-$'i[1+s]}

Rule-13442:

If both (/), (U), ($), (f), (V), ($'), (i), (s), (t), (Q), (P), (c) and (q) are known, then its Dividend Planned is:

D= U+{$[1-f]-V-$'i[1+s]}[1-t]-{Q+P/[c-q]}//

Rule-13443:

If both (/), (U), ($), (f), (V), ($'), (i), (s), (t), (D), (Q), (c) and (q) are known, then its Procured Inventory Days planned:

P= [c-q][/(U+{$[1-f]-V-$'i[1+s]}[1-t]-D)-Q]

Steve Asikin ISBN 14: 978-1514685136, ISBN 10: 1514685132

Rule-13444:

If both (**/**), (**U**), (**S**), (**f**), (**V**), (**S'**), (**i**), (**s**), (**t**), (**D**), (**P**), (**Q**) and (**q**) are known, then its Current Ratio Planned is:

c= **q**+**P**/[**/**(**U**+{**S**[1-**f**]-**V**-**S'i**[1+**s**]}[1-**t**]-**D**)-**Q**]

Rule-13445:

If both (**/**), (**U**), (**S**), (**f**), (**V**), (**S'**), (**i**), (**s**), (**t**), (**D**), (**P**), (**c**) and (**Q**) are known, then its Quick or Acid Test Ratio Planned is:

q= **c**-**P**/[**/**(**U**+{**S**[1-**f**]-**V**-**S'i**[1+**s**]}[1-**t**]-**D**)-**Q**]

Rule-13446:

If both (**/**), (**U**), (**S**), (**f**), (**V**), (**S'**), (**i**), (**s**), (**t**), (**D**), (**p**), (**c**) and (**q**) are known, then its Quoted Longterm Debt Planned is:

Q= **/**(**U**+{**S**[1-**f**]-**V**-**S'i**[1+**s**]}[1-**t**]-**D**)-**Vp**/{360[**c**-**q**]}

Rule-13447:

If both (**Q**), (**U**), (**S**), (**f**), (**V**), (**S'**), (**i**), (**s**), (**t**), (**D**), (**p**), (**c**) and (**q**) are known, then its Leverage or Gearing Ratio Planned is:

/= (**Q**+**Vp**/{360[**c**-**q**]})
/(**U**+{**S**[1-**f**]-**V**-**S'i**[1+**s**]}[1-**t**]-**D**)

Rule-13448:

If both (**/**), (**Q**), (**S**), (**f**), (**V**), (**S'**), (**i**), (**s**), (**t**), (**D**), (**p**), (**c**) and (**q**) are known, then its Utilized or Starting Capital must be:

U= **D**+(**Q**+**Vp**/{360[**c**-**q**]})/**/**
-{**S**[1-**f**]-**V**-**S'i**[1+**s**]}[1-**t**]

Steve Asikin ISBN 14: 978-1514685136, ISBN 10: 1514685132

Rule-13449:

If both (**/**), (**U**), (**Q**), (**f**), (**V**), (**S'**), (**i**), (**s**), (**t**), (**D**), (**p**), (**c**) and (**q**) are known, then its Sales or Revenue Planned is:

S= {**V**+**S'i**[1+**s**]
+[**D**+(**Q**+**Vp**/{360[**c**-**q**]})/**/U**]/[1-**t**]}/[1-**f**]

Rule-13450:

If both (**/**), (**U**), (**S**), (**Q**), (**V**), (**S'**), (**i**), (**s**), (**t**), (**D**), (**p**), (**c**) and (**q**) are known, then its Fixed Portion Planned is:

f= 1-{**V**+**S'i**[1+**s**]
+[**D**+(**Q**+**Vp**/{360[**c**-**q**]})/**/U**]/[1-**t**]}/**S**

Rule-13451:

If both (**/**), (**U**), (**S**), (**f**), (**Q**), (**S'**), (**i**), (**s**), (**t**), (**D**), (**p**), (**c**) and (**q**) are known, then its Variable Cost Planned is:

V= [**/**(**U**+{**S**[1-**f**]-**S'i**[1+**s**]}[1-**t**]-**D**) –**Q**]
/(**p**/{360[**c**-**q**]}+**/**[1-**t**])

Rule-13452:

If both (**/**), (**U**), (**S**), (**f**), (**V**), (**Q**), (**i**), (**s**), (**t**), (**D**), (**p**), (**c**) and (**q**) are known, then its Sales Past must be:

S'= {**S**[1-**f**]-**V**-[**D**+(**Q**+**Vp**/{360[**c**-**q**]})/**/U**]
/[1-**t**]}/{**i**[1+**s**]}

Steve Asikin ISBN 14: 978-1514685136, ISBN 10: 1514685132

Rule-13453:

If both (**/**), (**U**), (**$**), (**f**), (**V**), (**$'**), (**Q**), (**s**), (**t**), (**D**), (**p**), (**c**) and (**q**) are known, then its Interest Portion Planned is:

i={**$**[1-**f**]-**V**-[**D**+(**Q**+**Vp**/{360[**c**-**q**]})/**/**-**U**] /[1-**t**]}/{**$'**[1+**s**]}

Rule-13454:

If both (**/**), (**U**), (**$**), (**f**), (**V**), (**$'**), (**i**), (**Q**), (**t**), (**D**), (**p**), (**c**) and (**q**) are known, then its Sales Growth Planned is:

s= **$**/**$'**-1

or it can also be found as

s={**$**[1-**f**]-**V**-[**D**+(**Q**+**Vp**/{360[**c**-**q**]})/**/**-**U**] /[1-**t**]}/[**$'i**]-1

Rule-13455:

If both (**/**), (**U**), (**$**), (**f**), (**V**), (**$'**), (**i**), (**s**), (**Q**), (**D**), (**p**), (**c**) and (**q**) are known, then its Tax Rate Planned is:

t= 1-[**D**+(**Q**+**Vp**/{360[**c**-**q**]})/**/**-**U**] /{**$**[1-**f**]-**V**-**$'i**[1+**s**]}

Rule-13456:

If both (**/**), (**U**), (**$**), (**f**), (**V**), (**$'**), (**i**), (**s**), (**t**), (**Q**), (**p**), (**c**) and (**q**) are known, then its Dividend Payout Planned is:

D= **U**+{**$**[1-**f**]-**V**-**$'i**[1+**s**]}[1-**t**] -(**Q**+**Vp**/{360[**c**-**q**]})/**/**

Steve Asikin ISBN 14: 978-1514685136, ISBN 10: 1514685132

Rule-13457:

If both (**/**), (**U**), (**$**), (**f**), (**V**), (**$'**), (**i**), (**s**), (**t**), (**D**), (**Q**), (**c**) and (**q**) are known, then its Procured Inventory Days Planned:

p= 360[**c-q**][**/**(**U**+{**$**[1-**f**]-**V**-**$'i**[1+**s**]}[1-**t**]-**D**)-**Q**]/**V**

Rule-13458:

If both (**/**), (**U**), (**$**), (**f**), (**V**), (**$'**), (**i**), (**s**), (**t**), (**D**), (**p**), (**Q**) and (**q**) are known, then its Current Ratio Planned is:

c= **q**+**Vp**
/{360[**/**(**U**+{**$**[1-**f**]-**V**-**$'i**[1+**s**]}[1-**t**]-**D**)-**Q**]}

Rule-13459:

If both (**/**), (**U**), (**$**), (**f**), (**V**), (**$'**), (**i**), (**s**), (**t**), (**D**), (**p**), (**c**) and (**Q**) are known, then its Quick or Acid Test Ratio Planned is:

q= **c**-**Vp**/{360[**/**(**U**+{**$**[1-**f**]-**V**-**$'i**[1+**s**]}[1-**t**]-**D**)-**Q**]}

Rule-13460:

If both (**/**), (**U**), (**$**), (**f**), (**V**), (**$'**), (**i**), (**s**), (**t**), (**D**), (**v**), (**p**), (**c**) and (**q**) are known, then its Quoted Longterm Debt Planned is:

Q= **/**(**U**+{**$**[1-**f**]-**V**-**$'i**[1+**s**]}[1-**t**]-**D**)
-**$vp**/{360[**c-q**]}

Steve Asikin ISBN 14: 978-1514685136, ISBN 10: 1514685132

Rule-13461:

If both (**Q**), (**U**), (**S**), (**f**), (**V**), (**S'**), (**i**), (**s**), (**t**), (**D**), (**v**), (**p**), (**c**) and (**q**) are known, then its Leverage or Gearing Ratio Planned is:

/= (Q+Svp/{360[c-q]})
/(U+{S[1-f]-V-S'i[1+s]}[1-t]-D)

Rule-13462:

If both (**/**), (**Q**), (**S**), (**f**), (**V**), (**S'**), (**i**), (**s**), (**t**), (**D**), (**v**), (**p**), (**c**) and (**q**) are known, then its Utilized or Starting Capital must be:

U= D+(Q+Svp/{360[c-q]})//
-{S[1-f]-V-S'i[1+s]}[1-t]

Rule-13463:

If both (**/**), (**U**), (**Q**), (**f**), (**V**), (**S'**), (**i**), (**s**), (**t**), (**D**), (**v**), (**p**), (**c**) and (**q**) are known, then its Sales or Revenue Planned is:

S= [/(U-{V+S'i[1+s]}[1-t]-D)-Q]
/(vp/{360[c-q]}-/[1-f] [1-t])

Rule-13464:

If both (**/**), (**U**), (**S**), (**Q**), (**V**), (**S'**), (**i**), (**s**), (**t**), (**D**), (**v**), (**p**), (**c**) and (**q**) are known, then its Fixed Portion Planned is:

f= 1-{V+S'i[1+s]
+[D+(Q+Svp/{360[c-q]})//-U]/[1-t]}/S

Steve Asikin ISBN 14: 978-1514685136, ISBN 10: 1514685132

Rule-13465:

If both (**/**), (**U**), (**$**), (**f**), (**Q**), (**$'**), (**i**), (**s**), (**t**), (**D**), (**v**), (**p**), (**c**) and (**q**) are known, then its Variable Cost Planned is:

V= [/(U+{$[1-f]-$'i[1+s]}[1-t]-D)
-$vp/{360[c-q]}-Q]/{/[1-t]}

Rule-13466:

If both (**/**), (**U**), (**$**), (**f**), (**V**), (**Q**), (**i**), (**s**), (**t**), (**D**), (**v**), (**p**), (**c**) and (**q**) are known, then its Sales Past must be:

$'= {$[1-f]-V-[D+(Q+$vp/{360[c-q]})/-U]
/[1-t]}/{i[1+s]}

Rule-13467:

If both (**/**), (**U**), (**$**), (**f**), (**V**), (**$'**), (**Q**), (**s**), (**t**), (**D**), (**v**), (**p**), (**c**) and (**q**) are known, then its Interest Portion Planned is:

i={$[1-f]-V-[D+(Q+$vp/{360[c-q]})/-U]
/[1-t]}/{$'[1+s]}

Rule-13468:

If both (**/**), (**U**), (**$**), (**f**), (**V**), (**$'**), (**i**), (**Q**), (**t**), (**D**), (**v**), (**p**), (**c**) and (**q**) are known, then its Sales Growth Planned is:

s= $/$'-1

or it can also be found as

s={$[1-f]-V-[D+(Q+$vp/{360[c-q]})/-U]
/[1-t]}/[$'i]-1

Steve Asikin ISBN 14: 978-1514685136, ISBN 10: 1514685132

Rule-13469:

If both (**/**), (**U**), (**S**), (**f**), (**V**), (**S'**), (**i**), (**s**), (**Q**), (**D**), (**v**), (**p**), (**c**) and (**q**) are known, then its Tax Rate Planned is:

t= 1-[**D**+(**Q**+**Svp**/{360[**c**-**q**]})/**/**-**U**]
/{**S**[1-**f**]-**V**-**S'i**[1+**s**]}

Rule-13470:

If both (**/**), (**U**), (**S**), (**f**), (**V**), (**S'**), (**i**), (**s**), (**t**), (**Q**), (**v**), (**p**), (**c**) and (**q**) are known, then its Dividend Planned is:

D= **U**+{**S**[1-**f**]-**V**-**S'i**[1+**s**]}[1-**t**]
-(**Q**+**Svp**/{360[**c**-**q**]})/**/**

Rule-13471:

If both (**/**), (**U**), (**S**), (**f**), (**V**), (**S'**), (**i**), (**s**), (**t**), (**D**), (**Q**), (**p**), (**c**) and (**q**) are known, then its Variable Portion Planned is:

v= **V**/**S**

or it can also be found as

v= 360[**c**-**q**]
[**/**(**U**+{**S**[1-**f**]-**V**-**S'i**[1+**s**]}[1-**t**]-**D**)-**Q**]/[**Sp**]

Rule-13472:

If both (**/**), (**U**), (**S**), (**f**), (**V**), (**S'**), (**i**), (**s**), (**t**), (**D**), (**v**), (**Q**), (**c**) and (**q**) are known, then its Procured Inventory DaysPlanned:

p= 360[**c**-**q**]
[**/**(**U**+{**S**[1-**f**]-**V**-**S'i**[1+**s**]}[1-**t**]-**D**)-**Q**]/[**Sv**]

Steve Asikin ISBN 14: 978-1514685136, ISBN 10: 1514685132

Rule-13473:

If both (**/**), (**U**), (**S**), (**f**), (**V**), (**S'**), (**i**), (**s**), (**t**), (**D**), (**v**), (**p**), (**Q**) and (**q**) are known, then its Current Ratio Planned is:

c= **q**+**Svp**
/{360[**/**(**U**+{**S**[1-**f**]-**V**-**S'i**[1+**s**]}[1-**t**]-**D**)-**Q**]}

Rule-13474:

If both (**/**), (**U**), (**S**), (**f**), (**V**), (**S'**), (**i**), (**s**), (**t**), (**D**), (**v**), (**p**), (**c**) and (**Q**) are known, then its Quick or Acid Test Ratio Planned is:

q= **c**-**Svp**
/{360[**/**(**U**+{**S**[1-**f**]-**V**-**S'i**[1+**s**]}[1-**t**]-**D**)-**Q**]}

Rule-13475:

If both (**/**), (**U**), (**S**), (**f**), (**V**), (**S'**), (**i**), (**s**), (**t**), (**D**), (**v**), (**p**), (**c**) and (**q**) are known, then its Quoted Longterm Debt Planned is:

Q= **/**(**U**+{**S**[1-**f**]-**V**-**S'i**[1+**s**]}[1-**t**]-**D**)
-**S'vp**[1+**s**]/{360[**c**-**q**]}

Rule-13476:

If both (**Q**), (**U**), (**S**), (**f**), (**V**), (**S'**), (**i**), (**s**), (**t**), (**D**), (**v**), (**p**), (**c**) and (**q**) are known, then its Leverage or Gearing Ratio Planned is:

/= (**Q**+**S'vp**[1+**s**]/{360[**c**-**q**]})
/(**U**+{**S**[1-**f**]-**V**-**S'i**[1+**s**]}[1-**t**]-**D**)

Steve Asikin ISBN 14: 978-1514685136, ISBN 10: 1514685132

Rule-13477:

If both (**/**), (**Q**), (**$**), (**f**), (**V**), (**$'**), (**i**), (**s**), (**t**), (**D**), (**v**), (**p**), (**c**) and (**q**) are known, then its Utilized or Starting Capital must be:

U= **D**+(**Q**+**$'vp**[1+**s**]/{360[**c**-**q**]})/**/**
-{**$**[1-**f**]-**V**-**$'i**[1+**s**]}[1-**t**]

Rule-13478:

If both (**/**), (**U**), (**$**), (**f**), (**Q**), (**$'**), (**i**), (**s**), (**t**), (**D**), (**v**), (**p**), (**c**) and (**q**) are known, then its Variable Portion Planned is:

V= **[/**(**U**+{**$**[1-**f**]-**$'i**[1+**s**]}[1-**t**]-**D**)
-**$'vp**[1+**s**]/{360[**c**-**q**]}-**Q]**/{**/**[1-**t**]}

Rule-13479:

If both (**/**), (**U**), (**$**), (**Q**), (**V**), (**$'**), (**i**), (**s**), (**t**), (**D**), (**v**), (**p**), (**c**) and (**q**) are known, then its Fixed Portion Planned is:

f= 1-**{V**+**$'i**[1+**s**]
+**[D**+(**Q**+**$'vp**[1+**s**]/{360[**c**-**q**]})/**/**-**U]**
/[1-**t**]**}**/**$**

Rule-13480:

If both (**/**), (**U**), (**$**), (**f**), (**Q**), (**$'**), (**i**), (**s**), (**t**), (**D**), (**v**), (**p**), (**c**) and (**q**) are known, then its Variable Cost Planned is:

V= **[/**(**U**+{**$**[1-**f**]-**$'i**[1+**s**]}[1-**t**]-**D**)
-**$'vp**[1+**s**]/{360[**c**-**q**]}-**Q]**/{**/**[1-**t**]}

Steve Asikin ISBN 14: 978-1514685136, ISBN 10: 1514685132

Rule-13481:

If both (**/**), (**U**), (**$**), (**f**), (**V**), (**Q**), (**i**), (**s**), (**t**), (**D**), (**v**), (**p**), (**c**) and (**q**) are known, then its Sales Past must be:

$'**= [**/**(**U**+{**$[1-**f**]-**V**}[1-**t**]-**D**)-**Q**]
/([1+**s**](**vp**/{360[**c**-**q**]}+**i/**[1-**t**])]

Rule-13482:

If both (**/**), (**U**), (**$**), (**f**), (**V**), (**$'**), (**Q**), (**s**), (**t**), (**D**), (**v**), (**p**), (**c**) and (**q**) are known, then its Interest Portion Planned is:

i={**$**[1-**f**]-**V**-[**D**+(**Q**+**$'vp**[1+**s**]/{360[**c**-**q**]})/**/**-**U**]
/[1-**t**]}/{**$'**[1+**s**]}

Rule-13483:

If both (**/**), (**U**), (**$**), (**f**), (**V**), (**$'**), (**i**), (**Q**), (**t**), (**D**), (**v**), (**p**), (**c**) and (**q**) are known, then its Sales Growth Planned is:

s= **$**/**$'**-1

or it can also be found as

s= [**/**(**U**+{**$**[1-**f**]-**V**}[1-**t**]-**D**)-**Q**]
/(**$'**(**vp**/{360[**c**-**q**]}+**i/**[1-**t**])]-1

Rule-13484:

If both (**/**), (**U**), (**$**), (**f**), (**V**), (**$'**), (**i**), (**s**), (**Q**), (**D**), (**v**), (**p**), (**c**) and (**q**) are known, then its Tax Rate Planned is:

t= 1-[**D**+(**Q**+**$'vp**[1+**s**]/{360[**c**-**q**]})/**/**-**U**]
/{**$**[1-**f**]-**V**-**$'i**[1+**s**]}

Steve Asikin ISBN 14: 978-1514685136, ISBN 10: 1514685132

Rule-13485:

If both (/), (U), ($), (f), (V), ($'), (i), (s), (t), (Q), (v), (p), (c) and (q) are known, then its Dividend Planned is:

D= U+{$[1-f]-V-$'i[1+s]}[1-t]
-(Q+$'vp[1+s]/{360[c-q]})//

Rule-13486:

If both (/), (U), ($), (f), (V), ($'), (i), (s), (t), (D), (Q), (p), (c) and (q) are known, then its Variable Portion Planned is:

v= V/$

or it can also be found as

v= 360[c-q][/(U+{$[1-f]-V-$'i[1+s]}[1-t]-D)
-Q]/{$'p[1+s]}

Rule-13487:

If both (/), (U), ($), (f), (V), ($'), (i), (s), (t), (D), (v), (Q), (c) and (q) are known, then its Procured Inventory Days Planned:

p= 360[c-q][/(U+{$[1-f]-V-$'i[1+s]}[1-t]-D)
-Q]/{$'v[1+s]}

Rule-13488:

If both (/), (U), ($), (f), (V), ($'), (i), (s), (t), (D), (v), (p), (Q) and (q) are known, then its Current Ratio Planned is:

c= q+$'vp[1+s]
/{360[/(U+{$[1-f]-V-$'i[1+s]}[1-t]-D)-Q]}

Steve Asikin ISBN 14: 978-1514685136, ISBN 10: 1514685132

Rule-13489:

If both (/), (U), ($), (f), (V), ($'), (i), (s), (t), (D), (v), (p), (c) and (Q) are known, then its Quick or Acid Test Ratio Planned is:

q= c-$'vp[1+s]
/{360[/(U+{$[1-f]-V-$'i[1+s]}[1-t]-D)-Q]}

Rule-13490:

If both (/), (U), ($), (f), (V), ($'), (i), (s), (t), (A), (d) and (X) are known, then its Quoted Longterm Debt Planned is:

Q= /(U+{$[1-f]-V-$'i[1+s]}[1-t]-Ad)-X

Rule-13491:

If both (Q), (U), ($), (f), (V), ($'), (i), (s), (t), (A), (d) and (X) are known, then its Leverage or Gearing Ratio Planned is:

/= [Q+X]/(U+{$[1-f]-V-$'i[1+s]}[1-t]-Ad)

Rule-13492:

If both (/), (Q), ($), (f), (V), ($'), (i), (s), (t), (A), (d) and (X) are known, then its Utilized or Starting Capital must be:

U= Ad+[Q+X]/ /-{$[1-f]-V-$'i[1+s]}[1-t]

Rule-13493:

If both (/), (U), (Q), (f), (V), ($'), (i), (s), (t), (A), (d) and (X) are known, then its Sales or Revenue Planned is:

$= (V+$'i[1+s]+{Ad+[Q+X]/ /-U}/[1-t])/[1-f]

Steve Asikin ISBN 14: 978-1514685136, ISBN 10: 1514685132

Rule-13494:

If both (/), (U), ($), (Q), (V), ($'), (i), (s), (t), (A), (d) and (X) are known, then its Fixed Portion Planned is:

f= 1-(V+$'i[1+s]+{Ad+[Q+X]/AU}/[1-t])/$

Rule-13495:

If both (/), (U), ($), (f), (Q), ($'), (i), (s), (t), (A), (d) and (X) are known, then its Variable Cost Planned is:

V= $[1-f]-$'i[1+s]-{Ad+[Q+X]/AU}/[1-t]

Rule-13496:

If both (/), (U), ($), (f), (V), (Q), (i), (s), (t), (A), (d) and (X) are known, then its Sales Past must be:

$'= ($[1-f]-V-{Ad+[Q+X]/AU}/[1-t])/{i[1+s]}

Rule-13497:

If both (/), (U), ($), (f), (V), ($'), (Q), (s), (t), (A), (d) and (X) are known, then its Interest Portion Planned is:

i= ($[1-f]-V-{Ad+[Q+X]/AU}/[1-t])/{$'[1+s]}

Rule-13498:

If both (/), (U), ($), (f), (V), ($'), (i), (Q), (t), (A), (d) and (X) are known, then its Sales Growth Planned is:

s= $/$'-1

or it can also be found as

s= ($[1-f]-V-{Ad+[Q+X]/AU}/[1-t])/[$'i]-1

Steve Asikin ISBN 14: 978-1514685136, ISBN 10: 1514685132

Rule-13499:

If both (**/**), (**U**), (**$**), (**f**), (**V**), (**$'**), (**i**), (**s**), (**Q**), (**A**), (**d**) and (**X**) are known, then its Tax Rate Planned is:

t= 1-{**Ad**+[**Q**+**X**]/**/**-**U**}/{**$**[1-**f**]-**V**-**$'i**[1+**s**]}

Rule-13500:

If both (**/**), (**U**), (**$**), (**f**), (**V**), (**$'**), (**i**), (**s**), (**t**), (**Q**), (**d**) and (**X**) are known, then its After Tax Income Planned is:

A= (**U**+{**$**[1-**f**]-**V**-**$'i**[1+**s**]}[1-**t**]-[**Q**+**X**]/**/**)/**d**

Rule-13501:

If both (**/**), (**U**), (**$**), (**f**), (**V**), (**$'**), (**i**), (**s**), (**t**), (**A**), (**Q**) and (**X**) are known, then its Dividend Payout Planned is:

d= (**U**+{**$**[1-**f**]-**V**-**$'i**[1+**s**]}[1-**t**]-[**Q**+**X**]/**/**)/**A**

Rule-13502:

If both (**/**), (**U**), (**$**), (**f**), (**V**), (**$'**), (**i**), (**s**), (**t**), (**A**), (**d**) and (**Q**) are known, then its Xpress or Current Debt Planned is:

X= **/**(**U**+{**$**[1-**f**]-**V**-**$'i**[1+**s**]}[1-**t**]-**Ad**)-**Q**

Rule-13503:

If both (**/**), (**U**), (**$**), (**f**), (**V**), (**$'**), (**i**), (**s**), (**t**), (**A**), (**d**), (**P**), (**c**), and (**q**) are known, then its Quoted Longterm Debt Planned is:

Q= **/**(**U**+{**$**[1-**f**]-**V**-**$'i**[1+**s**]}[1-**t**]-**Ad**)-**P**/[**c**-**q**]

Steve Asikin ISBN 14: 978-1514685136, ISBN 10: 1514685132

Rule-13504:

If both (**Q**), (**U**), (**S**), (**f**), (**V**), (**S'**), (**i**), (**s**), (**t**), (**A**), (**d**), (**P**), (**c**), and (**q**) are known, then its Leverage or Gearing Ratio Planned is:

ł= {Q+P/[c-q]}/(U+{S[1-f]-V-S'i[1+s]}[1-t]-Ad)

Rule-13505:

If both (**ł**), (**Q**), (**S**), (**f**), (**V**), (**S'**), (**i**), (**s**), (**t**), (**A**), (**d**), (**P**), (**c**), and (**q**) are known, then its Utilized or Starting Capital must be:

U= Ad+{Q+P/[c-q]}/ł-{S[1-f]-V-S'i[1+s]}[1-t]

Rule-13506:

If both (**ł**), (**U**), (**Q**), (**f**), (**V**), (**S'**), (**i**), (**s**), (**t**), (**A**), (**d**), (**P**), (**c**), and (**q**) are known, then its Sales or Revenue Planned is:

S= [V+S'i[1+s]+(Ad+{Q+P/[c-q]}/ł-U)/[1-t]]/[1-f]

Rule-13507:

If both (**ł**), (**U**), (**S**), (**Q**), (**V**), (**S'**), (**i**), (**s**), (**t**), (**A**), (**d**), (**P**), (**c**), and (**q**) are known, then its Fixed Portion Planned is:

f= 1-[V+S'i[1+s]+(Ad+{Q+P/[c-q]}/ł-U)/[1-t]]/S

Rule-13508:

If both (**ł**), (**U**), (**S**), (**f**), (**Q**), (**S'**), (**i**), (**s**), (**t**), (**A**), (**d**), (**P**), (**c**), and (**q**) are known, then its Variable Cost Planned is:

V= S[1-f]-S'i[1+s]-(Ad+{Q+P/[c-q]}/ł-U)/[1-t]

Steve Asikin ISBN 14: 978-1514685136, ISBN 10: 1514685132

Rule-13509:

If both (**/**), (**U**), (**$**), (**f**), (**V**), (**Q**), (**i**), (**s**), (**t**), (**A**), (**d**), (**P**), (**c**), and (**q**) are known, then its Sales Past must be:

$'**= [**$[1-**f**]-**V**-(**Ad**+{**Q**+**P**/[**c**-**q**]}/**/**-**U**)/[1-**t**]]
/{**i**[1+**s**]}

Rule-13510:

If both (**/**), (**U**), (**$**), (**f**), (**V**), (**$'**), (**Q**), (**s**), (**t**), (**A**), (**d**), (**P**), (**c**), and (**q**) are known, then its Interest Portion Planned is:

i= [**$**[1-**f**]-**V**-(**Ad**+{**Q**+**P**/[**c**-**q**]}/**/**-**U**)/[1-**t**]]
/{**$'**[1+**s**]}

Rule-13511:

If both (**/**), (**U**), (**$**), (**f**), (**V**), (**$'**), (**i**), (**Q**), (**t**), (**A**), (**d**), (**P**), (**c**), and (**q**) are known, then its Sales Growth Planned is:

s= [**$**[1-**f**]-**V**-(**Ad**+{**Q**+**P**/[**c**-**q**]}/**/**-**U**)/[1-**t**]]/[**$'i**]-1

Rule-13512:

If both (**/**), (**U**), (**$**), (**f**), (**V**), (**$'**), (**i**), (**s**), (**Q**), (**A**), (**d**), (**P**), (**c**), and (**q**) are known, then its Tax Rate Planned is:

t= 1-(**Ad**+{**Q**+**P**/[**c**-**q**]}/**/**-**U**)/{**$**[1-**f**]-**V**-**$'i**[1+**s**]}

Rule-13513:

If both (**/**), (**U**), (**$**), (**f**), (**V**), (**$'**), (**i**), (**s**), (**t**), (**Q**), (**d**), (**P**), (**c**), and (**q**) are known, then its After Tax Income Planned is:

A= (**U**+{**$**[1-**f**]-**V**-**$'i**[1+**s**]}[1-**t**]-{**Q**+**P**/[**c**-**q**]}/**/**)/**d**

Steve Asikin ISBN 14: 978-1514685136, ISBN 10: 1514685132

Rule-13514:

If both (/), (U), ($), (f), (V), ($'), (i), (s), (t), (A), (Q), (P), (c), and (q) are known, then its Dividend Payout Planned is:

d= (**U**+{**$**[1-**f**]-**V**-**$'i**[1+**s**]}[1-**t**]-{**Q**+**P**/[**c**-**q**]}/**/**)/**A**

Rule-13515:

If both (/), (U), ($), (f), (V), ($'), (i), (s), (t), (A), (d), (Q), (c), and (q) are known, then its Procured Inventory Days planned:

P= [**c**-**q**][**/**(**U**+{**$**[1-**f**]-**V**-**$'i**[1+**s**]}[1-**t**]-**Ad**)-**Q**]

Rule-13516:

If both (/), (U), ($), (f), (V), ($'), (i), (s), (t), (A), (d), (P), (Q), and (q) are known, then its Current Ratio Planned is:

c= **q**+**P**/[**/**(**U**+{**$**[1-**f**]-**V**-**$'i**[1+**s**]}[1-**t**]-**Ad**)-**Q**]

Rule-13517:

If both (/), (U), ($), (f), (V), ($'), (i), (s), (t), (A), (d), (P), (c), and (Q) are known, then its Quick or Acid Test Ratio Planned is:

q= **c**-**P**/[**/**(**U**+{**$**[1-**f**]-**V**-**$'i**[1+**s**]}[1-**t**]-**Ad**)-**Q**]

Rule-13518:

If both (/), (U), ($), (f), (V), ($'), (i), (s), (t), (A), (d), (p), (c), and (q) are known, then its Quoted Longterm Debt Planned is:

Q= **/**(**U**+{**$**[1-**f**]-**V**-**$'i**[1+**s**]}[1-**t**]-**Ad**)
-**Vp**/{360[**c**-**q**]}

Steve Asikin ISBN 14: 978-1514685136, ISBN 10: 1514685132

Rule-13519:

If both (**Q**), (**U**), (**S**), (**f**), (**V**), (**S'**), (**i**), (**s**), (**t**), (**A**), (**d**), (**p**), (**c**), and (**q**) are known, then its Leverage or Gearing Ratio Planned is:

/= (Q+Vp/{360[c-q]})
/(U+{S[1-f]-V-S'i[1+s]}[1-t]-Ad)

Rule-13520:

If both (**/**), (**Q**), (**S**), (**f**), (**V**), (**S'**), (**i**), (**s**), (**t**), (**A**), (**d**), (**p**), (**c**), and (**q**) are known, then its Utilized or Starting Capital must be:

U= Ad+(Q+Vp/{360[c-q]})//
-{S[1-f]-V-S'i[1+s]}[1-t]

Rule-13521:

If both (**/**), (**U**), (**Q**), (**f**), (**V**), (**S'**), (**i**), (**s**), (**t**), (**A**), (**d**), (**p**), (**c**), and (**q**) are known, then its Sales or Revenue Planned is:

S={V+S'i[1+s]+[Ad+(Q+Vp/{360[c-q]})//-U]
/[1-t]}/[1-f]

Rule-13522:

If both (**/**), (**U**), (**S**), (**Q**), (**V**), (**S'**), (**i**), (**s**), (**t**), (**A**), (**d**), (**p**), (**c**), and (**q**) are known, then its Fixed Portion Planned is:

f= 1-{V+S'i[1+s]
+[Ad+(Q+Vp/{360[c-q]})//-U]/[1-t]}/S

Steve Asikin ISBN 14: 978-1514685136, ISBN 10: 1514685132

Rule-13523:

If both (/), (U), ($), (f), (Q), ($'), (i), (s), (t), (A), (d), (p), (c), and (q) are known, then its Variable Cost Planned is:

V= [/U+{$[1-f]-$'i[1+s]}[1-t]-Ad)-Q]
/(p/{360[c-q]}+/[1-t])

Rule-13524:

If both (/), (U), ($), (f), (V), (Q), (i), (s), (t), (A), (d), (p), (c), and (q) are known, then its Sales Past must be:

$'= {$[1-f]-V-[Ad+(Q+Vp/{360[c-q]})/-U]
/[1-t]}/{i[1+s]}

Rule-13525:

If both (/), (U), ($), (f), (V), ($'), (Q), (s), (t), (A), (d), (p), (c), and (q) are known, then its Interest Portion Planned is:

i= {$[1-f]-V-[Ad+(Q+Vp/{360[c-q]})/-U]
/[1-t]}/{$'[1+s]}

Rule-13526:

If both (/), (U), ($), (f), (V), ($'), (i), (Q), (t), (A), (d), (p), (c), and (q) are known, then its Sales Growth Planned is:

s= $/$'-1

or it can also be found as

s= {$[1-f]-V-[Ad+(Q+Vp/{360[c-q]})/-U]
/[1-t]}/[$'i]-1

Steve Asikin ISBN 14: 978-1514685136, ISBN 10: 1514685132

Rule-13527:

If both (**/**), (**U**), (**$**), (**f**), (**V**), (**$'**), (**i**), (**s**), (**Q**), (**A**), (**d**), (**p**), (**c**), and (**q**) are known, then its Tax Rate Planned is:

t= 1-[**Ad**+(**Q**+**Vp**/{360[**c-q**]})/**/**-**U**]
/{**$**[1-**f**]-**V**-**$'i**[1+**s**]}

Rule-13528:

If both (**/**), (**U**), (**$**), (**f**), (**V**), (**$'**), (**i**), (**s**), (**t**), (**Q**), (**d**), (**p**), (**c**), and (**q**) are known, then its After Tax Income Planned is:

A= [**U**+{**$**[1-**f**]-**V**-**$'i**[1+**s**]}[1-**t**]
-(**Q**+**Vp**/{360[**c-q**]})/**/**]/**d**

Rule-13529:

If both (**/**), (**U**), (**$**), (**f**), (**V**), (**$'**), (**i**), (**s**), (**t**), (**A**), (**Q**), (**p**), (**c**), and (**q**) are known, then its Dividend Payout Planned is:

d= [**U**+{**$**[1-**f**]-**V**-**$'i**[1+**s**]}[1-**t**]
-(**Q**+**Vp**/{360[**c-q**]})/**/**]/**A**

Rule-13530:

If both (**/**), (**U**), (**$**), (**f**), (**V**), (**$'**), (**i**), (**s**), (**t**), (**A**), (**d**), (**Q**), (**c**), and (**q**) are known, then its Procured Inventory Days Planned:

p= 360[**c-q**]
[**/**(**U**+{**$**[1-**f**]-**V**-**$'i**[1+**s**]}[1-**t**]-**Ad**)-**Q**]/**V**

Steve Asikin ISBN 14: 978-1514685136, ISBN 10: 1514685132

Rule-13531:

If both (**/**), (**U**), (**S**), (**f**), (**V**), (**S'**), (**i**), (**s**), (**t**), (**A**), (**d**), (**p**), (**Q**), and (**q**) are known, then its Current Ratio Planned is:

c= q+Vp
/{360[/(U+{S[1-f]-V-S'i[1+s]}[1-t]-Ad)-Q]}

Rule-13532:

If both (**/**), (**U**), (**S**), (**f**), (**V**), (**S'**), (**i**), (**s**), (**t**), (**A**), (**d**), (**p**), (**c**), and (**Q**) are known, then its Quick or Acid Test Ratio Planned is:

q= c-Vp
/{360[/(U+{S[1-f]-V-S'i[1+s]}[1-t]-Ad)-Q]}

Rule-13533:

If both (**/**), (**U**), (**S**), (**f**), (**V**), (**S'**), (**i**), (**s**), (**t**), (**A**), (**d**), (**p**), (**v**), (**c**), and (**q**) are known, then its Quoted Longterm Debt Planned is:

Q= /(U+{S[1-f]-V-S'i[1+s]}[1-t]-Ad)
-Svp/{360[c-q]}

Rule-13534:

If both (**Q**), (**U**), (**S**), (**f**), (**V**), (**S'**), (**i**), (**s**), (**t**), (**A**), (**d**), (**p**), (**v**), (**c**), and (**q**) are known, then its Leverage or Gearing Ratio Planned is:

/= (Q+Svp/{360[c-q]})
/(U+{S[1-f]-V-S'i[1+s]}[1-t]-Ad)

Steve Asikin ISBN 14: 978-1514685136, ISBN 10: 1514685132

Rule-13535:

If both (**/**), (**Q**), (**$**), (**f**), (**V**), (**$'**), (**i**), (**s**), (**t**), (**A**), (**d**), (**p**), (**v**), (**c**), and (**q**) are known, then its Utilized or Starting Capital must be:

U= **Ad**+(**Q**+**$vp**/{360[**c**-**q**]})/**/**
-{**$**[1-**f**]-**V**-**$'i**[1+**s**]}[1-**t**]

Rule-13536:

If both (**/**), (**U**), (**Q**), (**f**), (**V**), (**$'**), (**i**), (**s**), (**t**), (**A**), (**d**), (**p**), (**v**), (**c**), and (**q**) are known, then its Sales or Revenue Planned is:

$**= (**/U**-{**V**+**$'i[1+**s**]}[1-**t**]-**Ad**)-**Q**)
/(**vp**/{360[**c**-**q**]}-**/**[1-**f**] [1-**t**])

Rule-13537:

If both (**/**), (**U**), (**$**), (**f**), (**V**), (**$'**), (**i**), (**s**), (**t**), (**A**), (**d**), (**p**), (**v**), (**c**), and (**q**) are known, then its Quoted Longterm Debt Planned is:

f= 1-{**V**+**$'i**[1+**s**]
+[**Ad**+(**Q**+**$vp**/{360[**c**-**q**]})/**/**-**U**]/[1-**t**]}/**$**

Rule-13538:

If both (**/**), (**U**), (**$**), (**f**), (**Q**), (**$'**), (**i**), (**s**), (**t**), (**A**), (**d**), (**p**), (**v**), (**c**), and (**q**) are known, then its Variable Cost Planned is:

V= [**/U**+{**$**[1-**f**]-**$'i**[1+**s**]}[1-**t**]-**Ad**)
-**$vp**/{360[**c**-**q**]}-**$**]/{**/**[1-**t**]}

Steve Asikin ISBN 14: 978-1514685136, ISBN 10: 1514685132

Rule-13539:

If both (**/**), (**U**), (**$**), (**f**), (**V**), (**Q**), (**i**), (**s**), (**t**), (**A**), (**d**), (**p**), (**v**), (**c**), and (**q**) are known, then its Sales Past must be:

$$\$' = \{\$[1-f]-V-[Ad+(Q+\$vp/\{360[c-q]\})/IU]/[1-t]\}/\{i[1+s]\}$$

Rule-13540:

If both (**/**), (**U**), (**$**), (**f**), (**V**), (**$'**), (**Q**), (**s**), (**t**), (**A**), (**d**), (**p**), (**v**), (**c**), and (**q**) are known, then its Interest Portion Planned is:

$$i = \{\$[1-f]-V-[Ad+(Q+\$vp/\{360[c-q]\})/IU]/[1-t]\}/\{\$'[1+s]\}$$

Rule-13541:

If both (**/**), (**U**), (**$**), (**f**), (**V**), (**$'**), (**i**), (**Q**), (**t**), (**A**), (**d**), (**p**), (**v**), (**c**), and (**q**) are known, then its Sales Growth Planned is:

$$s = \$/\$'-1$$

or it can also be found as

$$s = \{\$[1-f]-V-[Ad+(Q+\$vp/\{360[c-q]\})/IU]/[1-t]\}/[\$'i]-1$$

Rule-13542:

If both (**/**), (**U**), (**$**), (**f**), (**V**), (**$'**), (**i**), (**s**), (**Q**), (**A**), (**d**), (**p**), (**v**), (**c**), and (**q**) are known, then its Tax Rate Planned is:

$$t = 1-[Ad+(Q+\$vp/\{360[c-q]\})/IU]/\{\$[1-f]-V-\$'i[1+s]\}$$

Steve Asikin ISBN 14: 978-1514685136, ISBN 10: 1514685132

Rule-13543:

If both (**/**), (**U**), (**$**), (**f**), (**V**), (**$'**), (**i**), (**s**), (**t**), (**Q**), (**d**), (**p**), (**v**), (**c**), and (**q**) are known, then its After Tax Income Planned is:

A= [**U**+{**$**[1-**f**]-**V**-**$'i**[1+**s**]}[1-**t**]
-(**Q**+**$vp**/{360[**c**-**q**]})/**/**]/**d**

Rule-13544:

If both (**/**), (**U**), (**$**), (**f**), (**V**), (**$'**), (**i**), (**s**), (**t**), (**A**), (**Q**), (**p**), (**v**), (**c**), and (**q**) are known, then its Dividend Payout Planned is:

d= [**U**+{**$**[1-**f**]-**V**-**$'i**[1+**s**]}[1-**t**]
-(**Q**+**$vp**/{360[**c**-**q**]})/**/**]/**A**

Rule-13545:

If both (**/**), (**U**), (**$**), (**f**), (**V**), (**$'**), (**i**), (**s**), (**t**), (**A**), (**d**), (**p**), (**Q**), (**c**), and (**q**) are known, then its Variable Portion Planned is:

v= **V**/**$**

or it can also be found as

v= 360[**c**-**q**]
[**/**(**U**+{**$**[1-**f**]-**V**-**$'i**[1+**s**]}[1-**t**]-**Ad**)-**Q**]/[**$p**]

Rule-13546:

If both (**/**), (**U**), (**$**), (**f**), (**V**), (**$'**), (**i**), (**s**), (**t**), (**A**), (**d**), (**Q**), (**v**), (**c**), and (**q**) are known, then its Procured Inventory Days planned:

p= 360[**c**-**q**]
[**/**(**U**+{**$**[1-**f**]-**V**-**$'i**[1+**s**]}[1-**t**]-**Ad**)-**Q**]/[**$v**]

Steve Asikin ISBN 14: 978-1514685136, ISBN 10: 1514685132

Rule-13547:

If both (**/**), (**U**), (**$**), (**f**), (**V**), (**$'**), (**i**), (**s**), (**t**), (**A**), (**d**), (**p**), (**v**), (**Q**), and (**q**) are known, then its Current Ratio Planned is:

c= **q**+**$vp**
/{360[**/**(**U**+{**$**[1-**f**]-**V**-**$'i**[1+**s**]}[1-**t**]-**Ad**)-**Q**]}

Rule-13548:

If both (**/**), (**U**), (**$**), (**f**), (**V**), (**$'**), (**i**), (**s**), (**t**), (**A**), (**d**), (**p**), (**v**), (**c**), and (**Q**) are known, then its Quick or Acid Test Ratio Planned is:

q= **c**-**$vp**
/{360[**/**(**U**+{**$**[1-**f**]-**V**-**$'i**[1+**s**]}[1-**t**]-**Ad**)-**Q**]}

Rule-13549:

If both (**/**), (**U**), (**$**), (**f**), (**V**), (**$'**), (**i**), (**s**), (**t**), (**A**), (**d**), (**p**), (**v**), (**c**), and (**q**) are known, then its Quoted Longterm Debt Planned is:

Q= **/**(**U**+{**$**[1-**f**]-**V**-**$'i**[1+**s**]}[1-**t**]-**Ad**)
-**$'vp**[1+**s**]/{360[**c**-**q**]}

Rule-13550:

If both (**Q**), (**U**), (**$**), (**f**), (**V**), (**$'**), (**i**), (**s**), (**t**), (**A**), (**d**), (**p**), (**v**), (**c**), and (**q**) are known, then its Leverage or Gearing Ratio Planned is:

/= (**Q**+**$'vp**[1+**s**]/{360[**c**-**q**]})
/(**U**+{**$**[1-**f**]-**V**-**$'i**[1+**s**]}[1-**t**]-**Ad**)

Steve Asikin ISBN 14: 978-1514685136, ISBN 10: 1514685132

Rule-13551:

If both (**/**), (**Q**), (**S**), (**f**), (**V**), (**S'**), (**i**), (**s**), (**t**), (**A**), (**d**), (**p**), (**v**), (**c**), and (**q**) are known, then its Utilized or Starting Capital must be:

U= **Ad**+(**Q**+**S'vp**[1+**s**]/{360[**c**-**q**]})/**/**
-{**S**[1-**f**]-**V**-**S'i**[1+**s**]}[1-**t**]

Rule-13552:

If both (**/**), (**U**), (**S**), (**f**), (**Q**), (**S'**), (**i**), (**s**), (**t**), (**A**), (**d**), (**p**), (**v**), (**c**), and (**q**) are known, then its Variable Cost Planned is:

V= [**/U**+{**S**[1-**f**]-**S'i**[1+**s**]}[1-**t**]-**Ad**)
-**S'vp**[1+**s**]/{360[**c**-**q**]}-**Q**]/{**/**[1-**t**]}

Rule-13553:

If both (**/**), (**U**), (**S**), (**Q**), (**V**), (**S'**), (**i**), (**s**), (**t**), (**A**), (**d**), (**p**), (**v**), (**c**), and (**q**) are known, then its Fixed Portion Planned is:

f= 1-{**V**+**S'i**[1+**s**]+[**Ad**
+(**Q**+**S'vp**[1+**s**]/{360[**c**-**q**]})/**/**-**U**]/[1-**t**]}/**S**

Rule-13554:

If both (**/**), (**U**), (**S**), (**f**), (**Q**), (**S'**), (**i**), (**s**), (**t**), (**A**), (**d**), (**p**), (**v**), (**c**), and (**q**) are known, then its Variable Cost Planned is:

V= [**/U**+{**S**[1-**f**]-**S'i**[1+**s**]}[1-**t**]-**Ad**)
-**S'vp**[1+**s**]/{360[**c**-**q**]}-**S**]/{**/**[1-**t**]}

Steve Asikin ISBN 14: 978-1514685136, ISBN 10: 1514685132

Rule-13555:

If both (/), (U), ($), (f), (V), (Q), (i), (s), (t), (A), (d), (p), (v), (c), and (q) are known, then its Sales Past must be:

$' = [/(U+{$[1-f]-V}[1-t]-Ad)-Q]
/[[1+s](vp /{360[c-q]}+i/[1-t])]

Rule-13556:

If both (/), (U), ($), (f), (V), ($'), (Q), (s), (t), (A), (d), (p), (v), (c), and (q) are known, then its Interest Portion Planned is:

i= {$[1-f]-V-[Ad+(Q+$'vp[1+s]/{360[c-q]})/-U]
/[1-t]}/{$'[1+s]}

Rule-13557:

If both (/), (U), ($), (f), (V), ($'), (i), (Q), (t), (A), (d), (p), (v), (c), and (q) are known, then its Sales Growth Planned is:

s= $/$'-1

or it can also be found as

s= [/(U+{$[1-f]-V}[1-t]-Ad)-Q]
/[$'(vp /{360[c-q]}+i/[1-t])]-1

Rule-13558:

If both (/), (U), ($), (f), (V), ($'), (i), (s), (Q), (A), (d), (p), (v), (c), and (q) are known, then its Tax Rate Planned is:

t= 1-[Ad+(Q+$'vp[1+s]/{360[c-q]})/-U]
/{$[1-f]-V-$'i[1+s]}

Steve Asikin ISBN 14: 978-1514685136, ISBN 10: 1514685132

Rule-13559:

If both (/), (U), ($), (f), (V), ($'), (i), (s), (t), (Q), (d), (p), (v), (c), and (q) are known, then its After Tax Income Planned is:

A= [U+{$[1-f]-V-$'i[1+s]}[1-t]
-(Q+$'vp[1+s]/{360[c-q]})/Λ]/d

Rule-13560:

If both (/), (U), ($), (f), (V), ($'), (i), (s), (t), (A), (Q), (p), (v), (c), and (q) are known, then its Dividend Payout Planned is:

d= [U+{$[1-f]-V-$'i[1+s]}[1-t]
-(Q+$'vp[1+s]/{360[c-q]})/Λ]/A

Rule-13561:

If both (/), (U), ($), (f), (V), ($'), (i), (s), (t), (A), (d), (p), (Q), (c), and (q) are known, then its Variable Portion Planned is:

v= V/$

or it can also be found as

v= 360[c-q][Λ(U+{$[1-f]-V-$'i[1+s]}[1-t]-Ad)-Q]
/{$'p[1+s]}

Rule-13562:

If both (/), (U), ($), (f), (V), ($'), (i), (s), (t), (A), (d), (Q), (v), (c), and (q) are known, then its Procured Inventory Days planned:

p= 360[c-q][Λ(U+{$[1-f]-V-$'i[1+s]}[1-t]-Ad)-Q]
/{$'v[1+s]}

Steve Asikin ISBN 14: 978-1514685136, ISBN 10: 1514685132

Rule-13563:

If both (/), (U), ($), (f), (V), ($'), (i), (s), (t), (A), (d), (p), (v), (Q), and (q) are known, then its Current Ratio Planned is:

c= q+$'vp[1+s]
/{360[/(U+{$[1-f]-V-$'i[1+s]}[1-t]-Ad)-Q]}

Rule-13564:

If both (/), (U), ($), (f), (V), ($'), (i), (s), (t), (A), (d), (p), (v), (c), and (Q) are known, then its Quick or Acid Test Ratio Planned is:

q= c-$'vp[1+s]
/{360[/(U+{$[1-f]-V-$'i[1+s]}[1-t]-Ad)-Q]}

Rule-13565:

If both (/), (U), ($), (f), (V), ($'), (i), (s), (t), (d), and (X) are known, then its Quoted Longterm Debt Planned is:

Q= /(U+{$[1-f]-V-$'i[1+s]}[1-t][1-d])-X

Rule-13566:

If both (Q), (U), ($), (f), (V), ($'), (i), (s), (t), (d), and (X) are known, then its Leverage or Gearing Ratio Planned is:

/= [Q+X]/(U+{$[1-f]-V-$'i[1+s]}[1-t][1-d])

Rule-13567:

If both (/), (Q), ($), (f), (V), ($'), (i), (s), (t), (d), and (X) are known, then its Utilized or Starting Capital must be:

U= [Q+X]/-{$[1-f]-V-$'i[1+s]}[1-t][1-d]

Steve Asikin ISBN 14: 978-1514685136, ISBN 10: 1514685132

Rule-13568:

If both (**/**), (**U**), (**Q**), (**f**), (**V**), (**$'**), (**i**), (**s**), (**t**), (**d**), and (**X**) are known, then its Sales or Revenue Planned is:

$**= (**V**+**$'i[1+**s**]+{[**Q**+**X**]/**/**–**U**}/{[1-**t**][1-**d**]})/[1-**f**]

Rule-13569:

If both (**/**), (**U**), (**$**), (**Q**), (**V**), (**$'**), (**i**), (**s**), (**t**), (**d**), and (**X**) are known, then its Fixed Portion Planned is:

f= 1-(**V**+**$'i**[1+**s**]+{[**Q**+**X**]/**/**–**U**}/{[1-**t**][1-**d**]})/**$**

Rule-13570:

If both (**/**), (**U**), (**$**), (**f**), (**Q**), (**$'**), (**i**), (**s**), (**t**), (**d**), and (**X**) are known, then its Variable Cost Planned is:

V= **$**[1-**f**]-**$'i**[1+**s**]-{[**Q**+**X**]/**/**–**U**}/{[1-**t**][1-**d**]}

Rule-13571:

If both (**/**), (**U**), (**$**), (**f**), (**V**), (**Q**), (**i**), (**s**), (**t**), (**d**), and (**X**) are known, then its Sales Past must be:

$'**= (**$[1-**f**]-**V**-{[**Q**+**X**]/**/**–**U**}/{[1-**t**][1-**d**]})/{**i**[1+**s**]}

Rule-13572:

If both (**/**), (**U**), (**$**), (**f**), (**V**), (**$'**), (**Q**), (**s**), (**t**), (**d**), and (**X**) are known, then its Interest Portion Planned is:

i= (**$**[1-**f**]-**V**-{[**Q**+**X**]/**/**–**U**}/{[1-**t**][1-**d**]})/{**$'**[1+**s**]}

Rule-13573:

If both (**/**), (**U**), (**$**), (**f**), (**V**), (**$'**), (**i**), (**Q**), (**t**), (**d**), and (**X**) are known, then its Sales Growth Planned is:

s= **$**/**$'**-1

or it can also be found as

s= (**$**[1-**f**]-**V**-{[**Q**+**X**]/**/**–**U**}/{[1-**t**][1-**d**]})/[**$'i**]-1

Steve Asikin ISBN 14: 978-1514685136, ISBN 10: 1514685132

Rule-13574:

If both (**/**), (**U**), (**$**), (**f**), (**V**), (**$'**), (**i**), (**s**), (**Q**), (**d**), and (**X**) are known, then its Tax Rate Planned is:

t= 1-{[**Q**+**X**]/**/**–**U**}/({**$**[1-**f**]-**V**-**$'i**[1+**s**]}[1-**d**])

Rule-13575:

If both (**/**), (**U**), (**$**), (**f**), (**V**), (**$'**), (**i**), (**s**), (**t**), (**Q**), and (**X**) are known, then its Dividend Payout Planned is:

d= 1-{[**Q**+**X**]/**/**–**U**}/({**$**[1-**f**]-**V**-**$'i**[1+**s**]}[1-**t**])

Rule-13576:

If both (**/**), (**U**), (**$**), (**f**), (**V**), (**$'**), (**i**), (**s**), (**t**), (**d**), and (**Q**) are known, then its Xpress or Current Debt Planned is:

X= **/**(**U**+{**$**[1-**f**]-**V**-**$'i**[1+**s**]}[1-**t**][1-**d**])-**Q**

Rule-13577:

If both (**/**), (**U**), (**$**), (**f**), (**V**), (**$'**), (**i**), (**s**), (**t**), (**d**), (**P**), (**c**) and (**q**) are known, then its Quoted Longterm Debt Planned is:

Q= **/**(**U**+{**$**[1-**f**]-**V**-**$'i**[1+**s**]}[1-**t**][1-**d**])-**P**/[**c**-**q**]

Rule-13578:

If both (**Q**), (**U**), (**$**), (**f**), (**V**), (**$'**), (**i**), (**s**), (**t**), (**d**), (**P**), (**c**) and (**q**) are known, then its Leverage or Gearing Ratio Planned is:

/= {**Q**+**P**/[**c**-**q**]}/(**U**+{**$**[1-**f**]-**V**-**$'i**[1+**s**]}[1-**t**][1-**d**])

Steve Asikin ISBN 14: 978-1514685136, ISBN 10: 1514685132

Rule-13579:

If both (**/**), (**Q**), (**$**), (**f**), (**V**), (**$'**), (**i**), (**s**), (**t**), (**d**), (**P**), (**c**) and (**q**) are known, then its Utilized or Starting Capital must be:

U= {**Q**+**P**/[**c**-**q**]}/**/**-{**$**[1-**f**]-**V**-**$'i**[1+**s**]}[1-**t**][1-**d**]

Rule-13580:

If both (**/**), (**U**), (**Q**), (**f**), (**V**), (**$'**), (**i**), (**s**), (**t**), (**d**), (**P**), (**c**) and (**q**) are known, then its Sales or Revenue Planned is:

$**= **[V**+**$'i[1+**s**]+({**Q**+**P**/[**c**-**q**]}/**/**-**U**)
/{[1-**t**][1-**d**]}**]**/[1-**f**]

Rule-13581:

If both (**/**), (**U**), (**$**), (**Q**), (**V**), (**$'**), (**i**), (**s**), (**t**), (**d**), (**P**), (**c**) and (**q**) are known, then its Fixed Portion Planned is:

f= 1-**[V**+**$'i**[1+**s**]+({**Q**+**P**/[**c**-**q**]}/**/**-**U**)
/{[1-**t**][1-**d**]}**]**/**$**

Rule-13582:

If both (**/**), (**U**), (**$**), (**f**), (**Q**), (**$'**), (**i**), (**s**), (**t**), (**d**), (**P**), (**c**) and (**q**) are known, then its Variable Cost Planned is:

V= **$**[1-**f**]-**$'i**[1+**s**]-({**Q**+**P**/[**c**-**q**]}/**/**-**U**)/{[1-**t**][1-**d**]}

Rule-13583:

If both (**/**), (**U**), (**$**), (**f**), (**V**), (**Q**), (**i**), (**s**), (**t**), (**d**), (**P**), (**c**) and (**q**) are known, then its Sales Past must be:

$'**= **[$[1-**f**]-**V**-({**Q**+**P**/[**c**-**q**]}/**/**-**U**)/{[1-**t**][1-**d**]}**]**
/{**i**[1+**s**]}

Steve Asikin ISBN 14: 978-1514685136, ISBN 10: 1514685132

Rule-13584:

If both (/), (U), (S), (f), (V), (S'), (Q), (s), (t), (d), (P), (c) and (q) are known, then its Interest Portion Planned is:

i= [S[1-f]-V-({Q+P/[c-q]}/ℓU)/{[1-t][1-d]}]
/{S'[1+s]}

Rule-13585:

If both (/), (U), (S), (f), (V), (S'), (i), (Q), (t), (d), (P), (c) and (q) are known, then its Sales Growth Planned is:

s= S/S'-1

or it can also be found as

s= [S[1-f]-V-({Q+P/[c-q]}/ℓU)/{[1-t][1-d]}]
/[S'i]-1

Rule-13586:

If both (/), (U), (S), (f), (V), (S'), (i), (s), (Q), (d), (P), (c) and (q) are known, then its Tax Rate Planned is:

t= 1-({Q+P/[c-q]}/ℓU)/({S[1-f]-V-S'i[1+s]}[1-d])

Rule-13587:

If both (/), (U), (S), (f), (V), (S'), (i), (s), (t), (Q), (P), (c) and (q) are known, then its Dividend Portion Planned is:

d= 1-({Q+P/[c-q]}/ℓU)/({S[1-f]-V-S'i[1+s]}[1-t])

Steve Asikin ISBN 14: 978-1514685136, ISBN 10: 1514685132

Rule-13588:

If both (**/**), (**U**), (**$**), (**f**), (**V**), (**$'**), (**i**), (**s**), (**t**), (**d**), (**Q**), (**c**) and (**q**) are known, then its Procured Inventory Planned is:

P= [**c**-**q**][**/**(**U**+{**$**[1-**f**]-**V**-**$'i**[1+**s**]}[1-**t**][1-**d**])-**Q**]

Rule-13589:

If both (**/**), (**U**), (**$**), (**f**), (**V**), (**$'**), (**i**), (**s**), (**t**), (**d**), (**P**), (**Q**) and (**q**) are known, then its Current Ratio Planned is:

c= **q**+**P**/[**/**(**U**+{**$**[1-**f**]-**V**-**$'i**[1+**s**]}[1-**t**][1-**d**])-**Q**]

Rule-13590:

If both (**/**), (**U**), (**$**), (**f**), (**V**), (**$'**), (**i**), (**s**), (**t**), (**d**), (**P**), (**c**) and (**Q**) are known, then its Quick or Acid Test Ratio Planned is:

q= **c**-**P**/[**/**(**U**+{**$**[1-**f**]-**V**-**$'i**[1+**s**]}[1-**t**][1-**d**])-**Q**]

Rule-13591:

If both (**/**), (**U**), (**$**), (**f**), (**V**), (**$'**), (**i**), (**s**), (**t**), (**d**), (**p**), (**c**) and (**q**) are known, then its Quoted Longterm Debt Planned is:

Q= **/**(**U**+{**$**[1-**f**]-**V**-**$'i**[1+**s**]}[1-**t**][1-**d**])
-**Vp**/{360[**c**-**q**]}

Rule-13592:

If both (**Q**), (**U**), (**$**), (**f**), (**V**), (**$'**), (**i**), (**s**), (**t**), (**d**), (**p**), (**c**) and (**q**) are known, then its Leverage or Gearing Ratio Planned is:

/= (**Q**+**Vp**/{360[**c**-**q**]})
/(**U**+{**$**[1-**f**]-**V**-**$'i**[1+**s**]}[1-**t**][1-**d**])

Steve Asikin ISBN 14: 978-1514685136, ISBN 10: 1514685132

Rule-13593:

If both (**/**), (**Q**), (**$**), (**f**), (**V**), (**$'**), (**i**), (**s**), (**t**), (**d**), (**p**), (**c**) and (**q**) are known, then its Utilized or Starting Capital must be :

U= (**Q**+**Vp**/{360[**c**-**q**]})/**/**
-{**$**[1-**f**]-**V**-**$'i**[1+**s**]}[1-**t**][1-**d**]

Rule-13594:

If both (**/**), (**U**), (**Q**), (**f**), (**V**), (**$'**), (**i**), (**s**), (**t**), (**d**), (**p**), (**c**) and (**q**) are known, then its Sales or Revenue Planned is:

$**= {**V**+**$'i[1+**s**]+[(**Q**+**Vp**/{360[**c**-**q**]})/**/**-**U**]
/{[1-**t**][1-**d**]}}/ [1-**f**]

Rule-13595:

If both (**/**), (**U**), (**$**), (**Q**), (**V**), (**$'**), (**i**), (**s**), (**t**), (**d**), (**p**), (**c**) and (**q**) are known, then its Fixed Portion Planned is:

f= 1-{**V**+**$'i**[1+**s**]+[(**Q**+**Vp**/{360[**c**-**q**]})/**/**-**U**]
/{[1-**t**][1-**d**]}}/**$**

Rule-13596:

If both (**/**), (**U**), (**$**), (**f**), (**Q**), (**$'**), (**i**), (**s**), (**t**), (**d**), (**p**), (**c**) and (**q**) are known, then its Variable Cost Planned is:

V= [**/**(**U**+{**$**[1-**f**]-**$'i**[1+**s**]}[1-**t**][1-**d**])-**Q**]
/(**p**/{360[**c**-**q**]}+**/**[1-**t**][1-**d**])

Steve Asikin ISBN 14: 978-1514685136, ISBN 10: 1514685132

Rule-13597:

If both (**/**), (**U**), (**$**), (**f**), (**V**), (**Q**), (**i**), (**s**), (**t**), (**d**), (**p**), (**c**) and (**q**) are known, then its Sales Past must be:

$'**= {**$[1-**f**]-**V**-[(**Q**+**Vp**/{360[**c**-**q**]})/**/-U**]
/{[1-**t**][1-**d**]}}/{**i**[1+**s**]}

Rule-13598:

If both (**/**), (**U**), (**$**), (**f**), (**V**), (**$'**), (**Q**), (**s**), (**t**), (**d**), (**p**), (**c**) and (**q**) are known, then its Interest Portion Planned is:

i= {**$**[1-**f**]-**V**-[(**Q**+**Vp**/{360[**c**-**q**]})/**/-U**]
/{[1-**t**][1-**d**]}}/{**$'**[1+**s**]}

Rule-13599:

If both (**/**), (**U**), (**$**), (**f**), (**V**), (**$'**), (**i**), (**Q**), (**t**), (**d**), (**p**), (**c**) and (**q**) are known, then its Sales Growth Planned is:

s= {**$**[1-**f**]-**V**-[(**Q**+**Vp**/{360[**c**-**q**]})/**/-U**]
/{[1-**t**][1-**d**]}}/ [**$'i**]-1

Rule-13600:

If both (**/**), (**U**), (**$**), (**f**), (**V**), (**$'**), (**i**), (**s**), (**Q**), (**d**), (**p**), (**c**) and (**q**) are known, then its Tax Rate Planned is:

t= 1-[(**Q**+**Vp**/{360[**c**-**q**]})/**/-U**]
/({**$**[1-**f**]-**V**-**$'i**[1+**s**]}[1-**d**])

Steve Asikin ISBN 14: 978-1514685136, ISBN 10: 1514685132

Rule-13601:

If both (/), (U), ($), (f), (V), ($'), (i), (s), (t), (Q), (p), (c) and (q) are known, then its Dividend Payout Planned is:

d= 1-[(Q+Vp/{360[c-q]})//U]
/({$[1-f]-V-$'i[1+s]}[1-t])

Rule-13602:

If both (/), (U), ($), (f), (V), ($'), (i), (s), (t), (d), (Q), (c) and (q) are known, then its Procured Inventory Days planned:

p= 360[c-q]
[/U+{$[1-f]-V-$'i[1+s]}[1-t][1-d])-Q]/V

Rule-13603:

If both (/), (U), ($), (f), (V), ($'), (i), (s), (t), (d), (p), (Q) and (q) are known, then its Current Ratio Planned is:

c= q+Vp/{360[/U+{$[1-f]-V-$'i[1+s]}[1-t][1-d])
-Q]}

Rule-13604:

If both (/), (U), ($), (f), (V), ($'), (i), (s), (t), (d), (p), (c) and (Q) are known, then its Quick or Acid Test Ratio Planned is:

q= c-Vp/{360[/U+{$[1-f]-V-$'i[1+s]}[1-t][1-d])
-Q]}

Steve Asikin ISBN 14: 978-1514685136, ISBN 10: 1514685132

Rule-13605:

If both (**/**), (**U**), (**S**), (**f**), (**V**), (**S'**), (**i**), (**s**), (**t**), (**d**), (**v**), (**p**), (**c**) and (**q**) are known, then its Quoted Longterm Debt Planned is:

Q= **/**(**U**+{**S**[1-**f**]-**V**-**S'i**[1+**s**]}[1-**t**][1-**d**])
-**Svp**/{360[**c**-**q**]}

Rule-13606:

If both (**Q**), (**U**), (**S**), (**f**), (**V**), (**S'**), (**i**), (**s**), (**t**), (**d**), (**v**), (**p**), (**c**) and (**q**) are known, then its Leverage or Gearing Ratio Planned is:

/= (**Q**+**Svp**/{360[**c**-**q**]})
/(**U**+{**S**[1-**f**]-**V**-**S'i**[1+**s**]}[1-**t**][1-**d**])

Rule-13607:

If both (**/**), (**Q**), (**S**), (**f**), (**V**), (**S'**), (**i**), (**s**), (**t**), (**d**), (**v**), (**p**), (**c**) and (**q**) are known, then its Utilized or Starting Capital must be:

U= (**Q**+**Svp**/{360[**c**-**q**]})/**/**
-{**S**[1-**f**]-**V**-**S'i**[1+**s**]}[1-**t**][1-**d**]

Rule-13608:

If both (**/**), (**U**), (**Q**), (**f**), (**V**), (**S'**), (**i**), (**s**), (**t**), (**d**), (**v**), (**p**), (**c**) and (**q**) are known, then its Sales or Revenue Planned is:

S= [**/**(**U**+{-**V**-**S'i**[1+**s**]}[1-**t**][1-**d**])-**Q**]
/(**vp**/{360[**c**-**q**]}-**/**[1-**f**] [1-**t**][1-**d**])

Steve Asikin ISBN 14: 978-1514685136; ISBN 10: 1514685132

Rule-13609:

If both (/), (U), (S), (Q), (V), (S'), (i), (s), (t), (d), (v), (p), (c) and (q) are known, then its Fixed Portion Planned is:

f= 1-{V+S'i[1+s]+[(Q+Svp/{360[c-q]})//U]
/{[1-t][1-d]}}/S

Rule-13610:

If both (/), (U), (S), (f), (Q), (S'), (i), (s), (t), (d), (v), (p), (c) and (q) are known, then its Variable Portion Planned is:

V= [/U+{S[1-f]-S'i[1+s]}[1-t][1-d])
-Svp/{360[c-q]}-Q]/{/[1-t][1-d]}

Rule-13611:

If both (/), (U), (S), (f), (V), (Q), (i), (s), (t), (d), (v), (p), (c) and (q) are known, then its Sales Past must be:

S'= {S[1-f]-V-[(Q+Svp/{360[c-q]})//U]
/{[1-t][1-d]}}/{i[1+s]}

Rule-13612:

If both (/), (U), (S), (f), (V), (S'), (Q), (s), (t), (d), (v), (p), (c) and (q) are known, then its Interest Portion Planned is:

i= {S[1-f]-V-[(Q+Svp/{360[c-q]})//U]
/{[1-t][1-d]}}/{S'[1+s]}

Steve Asikin ISBN 14: 978-1514685136, ISBN 10: 1514685132

Rule-13613:

If both (**/**), (**U**), (**$**), (**f**), (**V**), (**$'**), (**i**), (**Q**), (**t**), (**d**), (**v**), (**p**), (**c**) and (**q**) are known, then its Sales Growth Planned is:

s= $/$'-1

or it can also be found as

s= {$[1-f]-V-[(Q+$vp/{360[c-q]})/Ⅎ-U]/{[1-t][1-d]}}/ [$'i]-1

Rule-13614:

If both (**/**), (**U**), (**$**), (**f**), (**V**), (**$'**), (**i**), (**s**), (**Q**), (**d**), (**v**), (**p**), (**c**) and (**q**) are known, then its Tax Rate Planned is:

t= 1-[(Q+$vp/{360[c-q]})/Ⅎ-U]/({$[1-f]-V-$'i[1+s]}[1-d])

Rule-13615:

If both (**/**), (**U**), (**$**), (**f**), (**V**), (**$'**), (**i**), (**s**), (**t**), (**Q**), (**v**), (**p**), (**c**) and (**q**) are known, then its Dividend Payout Planned is:

d= 1-[(Q+$vp/{360[c-q]})/Ⅎ-U]/({$[1-f]-V-$'i[1+s]}[1-t])

Rule-13616:

If both (**/**), (**U**), (**$**), (**f**), (**V**), (**$'**), (**i**), (**s**), (**t**), (**d**), (**Q**), (**p**), (**c**) and (**q**) are known, then its Variable Portion Planned is:

v= V/$

or it can also be found as

v= 360[c-q][ⅎU+{$[1-f]-V-$'i[1+s]}[1-t][1-d])-Q]/[$p]

Steve Asikin ISBN 14: 978-1514685136, ISBN 10: 1514685132

Rule-13617:

If both (/), (U), (S), (f), (V), (S'), (i), (s), (t), (d), (v), (Q), (c) and (q) are known, then its Procured Inventory Days Planned:

p= 360[c-q]
[/(U+{S[1-f]-V-S'i[1+s]}[1-t][1-d])-Q]/[Sv]

Rule-13618:

If both (/), (U), (S), (f), (V), (S'), (i), (s), (t), (d), (v), (p), (Q) and (q) are known, then its Current Ratio Planned is:

c= q+Svp/{360[/(U+{S[1-f]-V-S'i[1+s]}[1-t][1-d])
-Q]}

Rule-13619:

If both (/), (U), (S), (f), (V), (S'), (i), (s), (t), (d), (v), (p), (c) and (Q) are known, then its Quick or Acid Test Ratio Planned is:

q= c-Svp/{360
[/(U+{S[1-f]-V-S'i[1+s]}[1-t][1-d])-Q]}

Rule-13620:

If both (/), (U), (S), (f), (V), (S'), (i), (s), (t), (d), (v), (p), (c) and (q) are known, then its Quoted Longterm Debt Planned is:

Q= /(U+{S[1-f]-V-S'i[1+s]}[1-t][1-d])
-S'vp[1+s]/{360[c-q]}

Steve Asikin ISBN 14: 978-1514685136, ISBN 10: 1514685132

Rule-13621:

If both (**Q**), (**U**), (**$**), (**f**), (**V**), (**$'**), (**i**), (**s**), (**t**), (**d**), (**v**), (**p**), (**c**) and (**q**) are known, then its Leverage or Gearing Ratio Planned is:

/= (Q+$'vp[1+s]/{360[c-q]})
/(U+{$[1-f]-V-$'i[1+s]}[1-t][1-d])

Rule-13622:

If both (**/**), (**Q**), (**$**), (**f**), (**V**), (**$'**), (**i**), (**s**), (**t**), (**d**), (**v**), (**p**), (**c**) and (**q**) are known, then its Utilized or Starting Capital must be:

U= (Q+$'vp[1+s]/{360[c-q]})//-{$[1-f]-V
-$'i[1+s]}[1-t][1-d]

Rule-13623:

If both (**/**), (**U**), (**Q**), (**f**), (**V**), (**$'**), (**i**), (**s**), (**t**), (**d**), (**v**), (**p**), (**c**) and (**q**) are known, then its Sales or Revenue Planned is:

$= [Q-/(U-{V+$'i[1+s]}[1-t][1-d])
+$'vp[1+s]/{360[c-q]}]/[1-f]

Rule-13624:

If both (**/**), (**U**), (**$**), (**Q**), (**V**), (**$'**), (**i**), (**s**), (**t**), (**d**), (**v**), (**p**), (**c**) and (**q**) are known, then its Fixed Portion Planned is:

f= 1-{V+$'i[1+s]+[(Q+$'vp[1+s]/{360[c-q]})//
-U]/{[1-t][1-d]}}/$

Steve Asikin ISBN 14: 978-1514685136, ISBN 10: 1514685132

Rule-13625:

If both (**/**), (**U**), (**S**), (**f**), (**Q**), (**S'**), (**i**), (**s**), (**t**), (**d**), (**v**), (**p**), (**c**) and (**q**) are known, then its Variable Cost Planned is:

V= [/U+{S[1-f]-S'i[1+s]}[1-t][1-d])
-S'vp[1+s]/{360[c-q]}-Q]/{/[1-t][1-d]}

Rule-13626:

If both (**/**), (**U**), (**S**), (**f**), (**V**), (**Q**), (**i**), (**s**), (**t**), (**d**), (**v**), (**p**), (**c**) and (**q**) are known, then its Sales Past must be:

S'= [/U+{S[1-f]-V}[1-t][1-d])-Q]
/[[1+s](vp/{360[c-q]}+i/[1-t][1-d])]

Rule-13627:

If both (**/**), (**U**), (**S**), (**f**), (**V**), (**S'**), (**Q**), (**s**), (**t**), (**d**), (**v**), (**p**), (**c**) and (**q**) are known, then its Interest Portion Planned is:

i= {S[1-f]-V-[(Q+S'vp[1+s]/{360[c-q]})//-U]
/{[1-t][1-d]}}/{S'[1+s]}

Rule-13628:

If both (**/**), (**U**), (**S**), (**f**), (**V**), (**S'**), (**i**), (**Q**), (**t**), (**d**), (**v**), (**p**), (**c**) and (**q**) are known, then its Sales Growth Planned is:

s= S/S'-1

or it can also be found as

s= [/U+{S[1-f]-V}[1-t][1-d])-Q]
/[S'(vp/{360[c-q]}+i/[1-t][1-d])]-1

Steve Asikin ISBN 14: 978-1514685136, ISBN 10: 1514685132

Rule-13629:

If both (**/**), (**U**), (**$**), (**f**), (**V**), (**$'**), (**i**), (**s**), (**Q**), (**d**), (**v**), (**p**), (**c**) and (**q**) are known, then its Tax Rate Planned is:

t= 1-[(**Q**+**$'vp**[1+**s**]/{360[**c**-**q**]})/**/**-**U**]
/({**$**[1-**f**]-**V**-**$'i**[1+**s**]}[1-**d**])

Rule-13630:

If both (**/**), (**U**), (**$**), (**f**), (**V**), (**$'**), (**i**), (**s**), (**t**), (**Q**), (**v**), (**p**), (**c**) and (**q**) are known, then its Dividend Payout Planned is:

d= 1-[(**Q**+**$'vp**[1+**s**]/{360[**c**-**q**]})/**/**-**U**]
/({**$**[1-**f**]-**V**-**$'i**[1+**s**]}[1-**t**])

Rule-13631:

If both (**/**), (**U**), (**$**), (**f**), (**V**), (**$'**), (**i**), (**s**), (**t**), (**d**), (**Q**), (**p**), (**c**) and (**q**) are known, then its Variable Portion Planned is:

v= **V**/**$**

or it can also be found as

v= 360[**c**-**q**][**/U**+{**$**[1-**f**]-**V**-**$'i**[1+**s**]}
[1-**t**][1-**d**])-**Q**]/{**$'p**[1+**s**]}

Rule-13632:

If both (**/**), (**U**), (**$**), (**f**), (**V**), (**$'**), (**i**), (**s**), (**t**), (**d**), (**v**), (**Q**), (**c**) and (**q**) are known, then its Procured Inventory Days planned:

p= 360[**c**-**q**][**/U**+{**$**[1-**f**]-**V**-**$'i**[1+**s**]}
[1-**t**][1-**d**])-**Q**]/{**$'v**[1+**s**]}

Steve Asikin ISBN 14: 978-1514685136, ISBN 10: 1514685132

Rule-13633:

If both (/), (U), (S), (f), (V), (S'), (i), (s), (t), (d), (v), (p), (Q) and (q) are known, then its Current Ratio Planned is:

c= q+S'vp[1+s]/{360[/(U+{S[1-f]-V-S'i[1+s]}[1-t][1-d])-Q]}

Rule-13634:

If both (/), (U), (S), (f), (V), (S'), (i), (s), (t), (d), (v), (p), (c) and (Q) are known, then its Quick or Acid Test Ratio Planned is:

q= c-S'vp[1+s]/{360[/(U+{S[1-f]-V-S'i[1+s]}[1-t][1-d])-Q]}

Rule-13635:

If both (/), (U), (S), (V), (S'), (f), (s), (I), (T), (D) and (X) are known, then its Quoted Longterm Debt Planned is:

Q= /{U+S-V-S'f[1+s]-I-T-D}-X

Rule-13636:

If both (/), (U), (S), (V), (S'), (f), (s), (I), (T), (D) and (X) are known, then its Quoted Longterm Debt Planned is:

/= [Q+X]/{U+S-V-S'f[1+s]-I-T-D}

Rule-13637:

If both (/), (Q), (S), (V), (S'), (f), (s), (I), (T), (D) and (X) are known, then its Utiliuzed or Starting Capital must be:

U= [Q+X]/ /-{S-V-S'f[1+s]-I-T-D}

Steve Asikin ISBN 14: 978-1514685136, ISBN 10: 1514685132

Rule-13638:

If both (/), (U), (Q), (V), ($'), (f), (s), (I), (T), (D) and (X) are known, then its Sales or Revenue Planned is:

$= V+$'f[1+s]+I+T+D+[Q+X]/ / U

Rule-13639:

If both (/), (U), ($), (Q), ($'), (f), (s), (I), (T), (D) and (X) are known, then its Variable Cost Planned is:

V= $-$'f[1+s]-I-T-D}-[Q+X]/ / U

Rule-13640:

If both (/), (U), ($), (V), (Q), (f), (s), (I), (T), (D) and (X) are known, then its Sales Past must be:

$'= {$-V-I-T-D+[Q+X]/ / U}/{f[1+s]}

Rule-13641:

If both (/), (U), ($), (V), ($'), (Q), (s), (I), (T), (D) and (X) are known, then its Fixed Portion Planned is:

f= {$-V-I-T-D+[Q+X]/ / U}/{$'[1+s]}

Rule-13642:

If both (/), (U), ($), (V), ($'), (f), (Q), (I), (T), (D) and (X) are known, then its Sales Growth Planned is:

s= $/$'-1

or it can also be found as

s= {$-V-I-T-D+[Q+X]/ / U}/[$'f]-1

Rule-13643:

If both (/), (U), ($), (V), ($'), (f), (s), (Q), (T), (D) and (X) are known, then its Interest Expense Planned is:

I= $-V-$'f[1+s]-T-D}-[Q+X]/ / U

Steve Asikin ISBN 14: 978-1514685136, ISBN 10: 1514685132

Rule-13644:

If both (**/**), (**U**), (**$**), (**V**), (**$'**), (**f**), (**s**), (**I**), (**Q**), (**D**) and (**X**) are known, then its Tax Planned is:

T= **$**-**V**-**$'f**[1+**s**]-**I**- **D**}-[**Q**+**X**]/**/**-**U**

Rule-13645:

If both (**/**), (**U**), (**$**), (**V**), (**$'**), (**f**), (**s**), (**I**), (**T**), (**Q**) and (**X**) are known, then its Dividend Planned is:

D= **$**-**V**-**$'f**[1+**s**]-**I**-**T**-[**Q**+**X**]/**/**-**U**

Rule-13646:

If both (**/**), (**U**), (**$**), (**V**), (**$'**), (**f**), (**s**), (**I**), (**T**), (**D**) and (**Q**) are known, then its Xpress or Current Debt Planned is:

X= **/**{**U**+**$**-**V**-**$'f**[1+**s**]-**I**-**T**-**D**}-**Q**

Rule-13647:

If both (**/**), (**U**), (**$**), (**V**), (**$'**), (**f**), (**s**), (**I**), (**T**), (**D**), (**P**), (**c**) and (**q**) are known, then its Quoted Longterm Debt Planned is:

Q= **/**{**U**+**$**-**V**-**$'f**[1+**s**]-**I**-**T**-**D**}-**P**/[**c**-**q**]

Rule-13648:

If both (**Q**), (**U**), (**$**), (**V**), (**$'**), (**f**), (**s**), (**I**), (**T**), (**D**), (**P**), (**c**) and (**q**) are known, then its Leverage or Gearing Ratio Planned is:

/= {**Q**+**P**/[**c**-**q**]}/{**U**+**$**-**V**-**$'f**[1+**s**]-**I**-**T**-**D**}

Steve Asikin ISBN 14: 978-1514685136, ISBN 10: 1514685132

Rule-13649:

If both (**/**), (**Q**), (**$**), (**V**), (**$'**), (**f**), (**s**), (**I**), (**T**), (**D**), (**P**), (**c**) and (**q**) are known, then its Utilized or Starting Capital must be:

U={**Q**+**P**/[**c**-**q**]}/**/**-{**$**-**V**-**$'f**[1+**s**]-**I**-**T**-**D**}

Rule-13650:

If both (**/**), (**U**), (**Q**), (**V**), (**$'**), (**f**), (**s**), (**I**), (**T**), (**D**), (**P**), (**c**) and (**q**) are known, then its Sales or Revenue Planned is:

$**= **V**+**$'f[1+**s**]+**I**+**T**+**D**+{**Q**+**P**/[**c**-**q**]}/**/**-**U**

Rule-13651:

If both (**/**), (**U**), (**$**), (**Q**), (**$'**), (**f**), (**s**), (**I**), (**T**), (**D**), (**P**), (**c**) and (**q**) are known, then its Variable Cost Planned is:

V= **$**-**$'f**[1+**s**]-**I**-**T**-**D**}-{**Q**+**P**/[**c**-**q**]}/**/**-**U**

Rule-13652:

If both (**/**), (**U**), (**$**), (**V**), (**Q**), (**f**), (**s**), (**I**), (**T**), (**D**), (**P**), (**c**) and (**q**) are known, then its Sales Past must be:

$'**= (**$-**V**-**I**-**T**-**D**+{**Q**+**P**/[**c**-**q**]}/**/**-**U**)/{**f**[1+**s**]}

Rule-13653:

If both (**/**), (**U**), (**$**), (**V**), (**$'**), (**Q**), (**s**), (**I**), (**T**), (**D**), (**P**), (**c**) and (**q**) are known, then its Fixed Portion Planned is:

f= (**$**-**V**-**I**-**T**-**D**+{**Q**+**P**/[**c**-**q**]}/**/**-**U**)/{**$'**[1+**s**]}

Steve Asikin ISBN 14: 978-1514685136, ISBN 10: 1514685132

Rule-13654:

If both (**/**), (**U**), (**$**), (**V**), (**$'**), (**f**), (**Q**), (**I**), (**T**), (**D**), (**P**), (**c**) and (**q**) are known, then its Sales Growth Planned is:

s= $/$'-1

or it can also be found as

s= ($-V-I-T-D+{Q+P/[c-q]}//-U)/[$'f]-1

Rule-13655:

If both (**/**), (**U**), (**$**), (**V**), (**$'**), (**f**), (**s**), (**Q**), (**T**), (**D**), (**P**), (**c**) and (**q**) are known, then its Interest Expense Planned is:

I= $-V-$'f[1+s]-T-D}-{Q+P/[c-q]}//-U

Rule-13656:

If both (**/**), (**U**), (**$**), (**V**), (**$'**), (**f**), (**s**), (**I**), (**Q**), (**D**), (**P**), (**c**) and (**q**) are known, then its Tax Planned is:

T= $-V-$'f[1+s]-I- D}-{Q+P/[c-q]}//-U

Rule-13657:

If both (**/**), (**U**), (**$**), (**V**), (**$'**), (**f**), (**s**), (**I**), (**T**), (**Q**), (**P**), (**c**) and (**q**) are known, then its Dividend Planned is:

D= $-V-$'f[1+s]-I-T-{Q+P/[c-q]}//-U

Rule-13658:

If both (**/**), (**U**), (**$**), (**V**), (**$'**), (**f**), (**s**), (**I**), (**T**), (**D**), (**Q**), (**c**) and (**q**) are known, then its Procured Inventory Planned is:

P= [c-q](/{U+$-V-$'f[1+s]-I-T-D}-Q)

Steve Asikin ISBN 14: 978-1514685136, ISBN 10: 1514685132

Rule-13659:

If both (**/**), (**U**), (**$**), (**V**), (**$'**), (**f**), (**s**), (**I**), (**T**), (**D**), (**P**), (**Q**) and (**q**) are known, then its Current Ratio Planned is:

c= **q**+**P**/(**/**{**U**+**$**-**V**-**$'f**[1+**s**]-**I**-**T**-**D**}-**Q**)

Rule-13660:

If both (**/**), (**U**), (**$**), (**V**), (**$'**), (**f**), (**s**), (**I**), (**T**), (**D**), (**P**), (**c**) and (**Q**) are known, then its Quick or Acid Test Ratio Planned is:

q= **c**-**P**/(**/**{**U**+**$**-**V**-**$'f**[1+**s**]-**I**-**T**-**D**}-**Q**)

Rule-13661:

If both (**/**), (**U**), (**$**), (**V**), (**$'**), (**f**), (**s**), (**I**), (**T**), (**D**), (**p**), (**c**) and (**q**) are known, then its Quoted Longterm Debt Planned is:

Q= **/**{**U**+**$**-**V**-**$'f**[1+**s**]-**I**-**T**-**D**}-**Vp**/{360[**c**-**q**]}

Rule-13662:

If both (**Q**), (**U**), (**$**), (**V**), (**$'**), (**f**), (**s**), (**I**), (**T**), (**D**), (**p**), (**c**) and (**q**) are known, then its Leverage or Gearing Ratio Planned is:

/= (**Q**+**Vp**/{360[**c**-**q**]})/{**U**+**$**-**V**-**$'f**[1+**s**]-**I**-**T**-**D**}

Rule-13663:

If both (**/**), (**Q**), (**$**), (**V**), (**$'**), (**f**), (**s**), (**I**), (**T**), (**D**), (**p**), (**c**) and (**q**) are known, then its Utilized or Starting Capital Planned is:

U=(**Q**+**Vp**/{360[**c**-**q**]})/**/**-{**$**-**V**-**$'f**[1+**s**]-**I**-**T**-**D**}

Steve Asikin ISBN 14: 978-1514685136, ISBN 10: 1514685132

Rule-13664:

If both (/), (U), (Q), (V), (S'), (f), (s), (I), (T), (D), (p), (c) and (q) are known, then its Sales or Revenue Planned is:

S= V+S'f[1+s]+I+T+D+(Q+Vp/{360[c-q]})/ /-U

Rule-13665:

If both (/), (U), (S), (Q), (S'), (f), (s), (I), (T), (D), (p), (c) and (q) are known, then its Variable Cost Planned is:

V= (/{U+S-S'f[1+s]-I-T-D}-Q)/(p/{360[c-q]}+/)

Rule-13666:

If both (/), (U), (S), (V), (Q), (f), (s), (I), (T), (D), (p), (c) and (q) are known, then its Sales Past must be:

S'= [S-V-I-T-D+(Q+Vp/{360[c-q]})/ /-U]/{f[1+s]}

Rule-13667:

If both (/), (U), (S), (V), (S'), (Q), (s), (I), (T), (D), (p), (c) and (q) are known, then its Fixed Cost Planned is:

f= [S-V-I-T-D+(Q+Vp/{360[c-q]})/ /-U]/{S'[1+s]}

Rule-13668:

If both (/), (U), (S), (V), (S'), (f), (Q), (I), (T), (D), (p), (c) and (q) are known, then its Sales Growth Planned is:

s= S/S'-1

or it can also be found as

s=[S-V-I-T-D+(Q+Vp/{360[c-q]})/ /-U]/[S'f]-1

Steve Asikin ISBN 14: 978-1514685136, ISBN 10: 1514685132

Rule-13669:

If both (**/**), (**U**), (**$**), (**V**), (**$'**), (**f**), (**s**), (**Q**), (**T**), (**D**), (**p**), (**c**) and (**q**) are known, then its Interest Expense Planned is:

I= **$-V-$'f**[1+**s**]-**T-D**}-(**Q**+**Vp**/{360[**c-q**]})/**/-U**

Rule-13670:

If both (**/**), (**U**), (**$**), (**V**), (**$'**), (**f**), (**s**), (**I**), (**Q**), (**D**), (**p**), (**c**) and (**q**) are known, then its Tax Planned is:

T= **$-V-$'f**[1+**s**]-**I**- **D**}-(**Q**+**Vp**/{360[**c-q**]})/**/-U**

Rule-13671:

If both (**/**), (**U**), (**$**), (**V**), (**$'**), (**f**), (**s**), (**I**), (**T**), (**Q**), (**p**), (**c**) and (**q**) are known, then its Dividend Payout Planned is:

D= **$-V-$'f**[1+**s**]-**I-T**-(**Q**+**Vp**/{360[**c-q**]})/**/-U**

Rule-13672:

If both (**/**), (**U**), (**$**), (**V**), (**$'**), (**f**), (**s**), (**I**), (**T**), (**D**), (**Q**), (**c**) and (**q**) are known, then its Procured Inventory Days Planned:

p= 360[**c-q**](**/**{**U**+**$-V-$'f**[1+**s**]-**I-T-D**}-**Q**)/**V**

Rule-13673:

If both (**/**), (**U**), (**$**), (**V**), (**$'**), (**f**), (**s**), (**I**), (**T**), (**D**), (**p**), (**Q**) and (**q**) are known, then its Current Ratio Planned is:

c= **q**+**Vp**/[360(**/**{**U**+**$-V-$'f**[1+**s**]-**I-T-D**}-**Q**)]

Steve Asikin ISBN 14: 978-1514685136, ISBN 10: 1514685132

Rule-13674:

If both (**/**), (**U**), (**$**), (**V**), (**$'**), (**f**), (**s**), (**I**), (**T**), (**D**), (**p**), (**c**) and (**Q**) are known, then its Quick or Acid Test Ratio Planned is:

q= c-Vp/[360(/{U+$-V-$'f[1+s]-I-T-D}-Q)]

Rule-13675:

If both (**/**), (**U**), (**$**), (**V**), (**$'**), (**f**), (**s**), (**I**), (**T**), (**D**), (**v**), (**p**), (**c**) and (**q**) are known, then its Quoted Longterm Debt Planned is:

Q= /{U+$-V-$'f[1+s]-I-T-D}-$vp/{360[c-q]}

Rule-13676:

If both (**Q**), (**U**), (**$**), (**V**), (**$'**), (**f**), (**s**), (**I**), (**T**), (**D**), (**v**), (**p**), (**c**) and (**q**) are known, then its Leverage or Gearing Ratio Planned is:

/= (Q+$vp/{360[c-q]})/{U+$-V-$'f[1+s]-I-T-D}

Rule-13677:

If both (**/**), (**Q**), (**$**), (**V**), (**$'**), (**f**), (**s**), (**I**), (**T**), (**D**), (**v**), (**p**), (**c**) and (**q**) are known, then its Utilized or Starting Capital must be:

U=(Q+$vp/{360[c-q]})//-{$-V-$'f[1+s]-I-T-D}

Rule-13678:

If both (**/**), (**U**), (**Q**), (**V**), (**$'**), (**f**), (**s**), (**I**), (**T**), (**D**), (**v**), (**p**), (**c**) and (**q**) are known, then its Sales or Revenue Planned is:

$= (/{U-V-$'f[1+s]-I-T-D}-Q)/(vp/{360[c-q]}-/)

Steve Asikin ISBN 14: 978-1514685136, ISBN 10: 1514685132

Rule-13679:

If both (**/**), (**U**), (**Q**), (**Q**), (**$'**), (**f**), (**s**), (**I**), (**T**), (**D**), (**v**), (**p**), (**c**) and (**q**) are known, then its Variable Cost Planned is:

V= (**/**{**U**+**$** -**$'f**[1+**s**]-**I**-**T**-**D**}-**$vp**/{360[**c**-**q**]}–**Q**)/**/**

Rule-13680:

If both (**/**), (**U**), (**Q**), (**V**), (**Q**), (**f**), (**s**), (**I**), (**T**), (**D**), (**v**), (**p**), (**c**) and (**q**) are known, then its Sales Past must be:

$'**= [**$-**V**-**I**-**T**-**D**+(**Q**+**$vp**/{360[**c**-**q**]})/**/**-**U**]/{**f**[1+**s**]}

Rule-13681:

If both (**/**), (**U**), (**Q**), (**V**), (**$'**), (**Q**), (**s**), (**I**), (**T**), (**D**), (**v**), (**p**), (**c**) and (**q**) are known, then its Fixed Portion Planned is:

f= [**$**-**V**-**I**-**T**-**D**+(**Q**+**$vp**/{360[**c**-**q**]})/**/**-**U**]/{**$'**[1+**s**]}

Rule-13682:

If both (**/**), (**U**), (**Q**), (**V**), (**$'**), (**f**), (**Q**), (**I**), (**T**), (**D**), (**v**), (**p**), (**c**) and (**q**) are known, then its Sales Growth Planned is:

s= **$**/**$'**-1

or it can also be found as

s=[**$**-**V**-**I**-**T**-**D**+(**Q**+**$vp**/{360[**c**-**q**]})/**/**-**U**]/[**$'f**]-1

Rule-13683:

If both (**/**), (**U**), (**Q**), (**V**), (**$'**), (**f**), (**s**), (**Q**), (**T**), (**D**), (**v**), (**p**), (**c**) and (**q**) are known, then its Interest Expense Planned is:

I= **$**-**V**-**$'f**[1+**s**]-**T**-**D**}-(**Q**+**$vp**/{360[**c**-**q**]})/**/**-**U**

Steve Asikin ISBN 14: 978-1514685136, ISBN 10: 1514685132

Rule-13684:

If both (**/**), (**U**), (**Q**), (**V**), (**S'**), (**f**), (**s**), (**I**), (**Q**), (**D**), (**v**), (**p**), (**c**) and (**q**) are known, then its Tax Planned is:

T= **S-V-S'f**[1+**s**]-**I**- **D**}-(**Q**+**Svp**/{360[**c-q**]})/**/-U**

Rule-13685:

If both (**/**), (**U**), (**Q**), (**V**), (**S'**), (**f**), (**s**), (**I**), (**T**), (**Q**), (**v**), (**p**), (**c**) and (**q**) are known, then its Dividend Planned is:

D= **S-V-S'f**[1+**s**]-**I-T**-(**Q**+**Svp**/{360[**c-q**]})/**/-U**

Rule-13686:

If both (**/**), (**U**), (**Q**), (**V**), (**S'**), (**f**), (**s**), (**I**), (**T**), (**D**), (**Q**), (**p**), (**c**) and (**q**) are known, then its Variable Portion Planned is:

v= **V**/**S**

or it can also be found as

v= 360[**c-q**](**/**{**U**+**S-V-S'f**[1+**s**]-**I-T-D**}-**Q**)/[**Sp**]

Rule-13687:

If both (**/**), (**U**), (**Q**), (**V**), (**S'**), (**f**), (**s**), (**I**), (**T**), (**D**), (**v**), (**Q**), (**c**) and (**q**) are known, then its Procured Inventory Planned is:

p= 360[**c-q**](**/**{**U**+**S-V-S'f**[1+**s**]-**I-T-D**}-**Q**)/[**Sv**]

Rule-13688:

If both (**/**), (**U**), (**Q**), (**V**), (**S'**), (**f**), (**s**), (**I**), (**T**), (**D**), (**v**), (**p**), (**Q**) and (**q**) are known, then its Current Ratio Planned is:

c= **q**+**Svp**/[360(**/**{**U**+**S-V-S'f**[1+**s**]-**I-T-D**}-**Q**)]

Steve Asikin ISBN 14: 978-1514685136, ISBN 10: 1514685132

Rule-13689:

If both (**/**), (**U**), (**Q**), (**V**), (**$'**), (**f**), (**s**), (**I**), (**T**), (**D**), (**v**), (**p**), (**c**) and (**Q**) are known, then its Quick or Acid Test Ratio Planned is:

q= **c-$vp**/[360(**/**{**U**+**$-V-$'f**[1+**s**]-**I-T-D**}-**Q**)]

Rule-13690:

If both (**/**), (**U**), (**$**), (**V**), (**$'**), (**f**), (**s**), (**I**), (**T**), (**D**), (**v**), (**p**), (**c**) and (**q**) are known, then its Sales or Revenue Planned is:

Q= **/**{**U**+**$-V-$'f**[1+**s**]-**I-T-D**}-**$'vp**[1+**s**]/{360[**c-q**]}

Rule-13691:

If both (**Q**), (**U**), (**$**), (**V**), (**$'**), (**f**), (**s**), (**I**), (**T**), (**D**), (**v**), (**p**), (**c**) and (**q**) are known, then its Leverage or Gearing Ratio Planned is:

/= (**Q**+**$'vp**[1+**s**]/{360[**c-q**]})
/{**U**+**$-V-$'f**[1+**s**]-**I-T-D**}

Rule-13692:

If both (**/**), (**Q**), (**$**), (**V**), (**$'**), (**f**), (**s**), (**I**), (**T**), (**D**), (**v**), (**p**), (**c**) and (**q**) are known, then its Utilized or Starting Capital must be:

U=(**Q**+**$'vp**[1+**s**]/{360[**c-q**]})/**/**
-{**$-V-$'f**[1+**s**]-**I-T-D**}

Steve Asikin ISBN 14: 978-1514685136, ISBN 10: 1514685132

Rule-13693:

If both (**/**), (**U**), (**Q**), (**V**), (**S'**), (**f**), (**s**), (**I**), (**T**), (**D**), (**v**), (**p**), (**c**) and (**q**) are known, then its Sales or Revenue Planned is:

S= (**Q**-**/**{**U**-**V**-**S'f**[1+**s**]-**I**-**T**-**D**}
+**S'vp**[1+**s**]/{360[**c**-**q**]})/**/**

Rule-13694:

If both (**/**), (**U**), (**S**), (**Q**), (**S'**), (**f**), (**s**), (**I**), (**T**), (**D**), (**v**), (**p**), (**c**) and (**q**) are known, then its Variable Cost Planned is:

V= (**/**{**U**+**S** -**S'f**[1+**s**]-**I**-**T**-**D**}
-**S'vp**[1+**s**]/{360[**c**-**q**]}–**Q**)/**/**

Rule-13695:

If both (**/**), (**U**), (**S**), (**V**), (**Q**), (**f**), (**s**), (**I**), (**T**), (**D**), (**v**), (**p**), (**c**) and (**q**) are known, then its Sales Past must be:

S'= (**/**{**U**+**S**-**V**-**I**-**T**-**D**}-**Q**)
/**[**[1+**s**](**vp**/{360[**c**-**q**]}+ **f/**)**]**

Rule-13696:

If both (**/**), (**U**), (**S**), (**V**), (**S'**), (**Q**), (**s**), (**I**), (**T**), (**D**), (**v**), (**p**), (**c**) and (**q**) are known, then its Fixed Portion Planned is:

f= **[S**-**V**-**I**-**T**-**D**+(**Q**+**S'vp**[1+**s**]/{360[**c**-**q**]})/**/**-**U]**
/{**S'**[1+**s**]}

Steve Asikin ISBN 14: 978-1514685136, ISBN 10: 1514685132

Rule-13697:

If both (**/**), (**U**), (**$**), (**V**), (**$'**), (**f**), (**Q**), (**I**), (**T**), (**D**), (**v**), (**p**), (**c**) and (**q**) are known, then its Sales Growth Planned is:

v= $/$'-1

or it can also be found as

s=(/{U+$-V-I-T-D}-Q)/[$'(vp/{360[c-q]}+ f/)]-1

Rule-13698:

If both (**/**), (**U**), (**$**), (**V**), (**$'**), (**f**), (**s**), (**Q**), (**T**), (**D**), (**v**), (**p**), (**c**) and (**q**) are known, then its Interest Expense Planned is:

I= $-V-$'f[1+s]-T-D}

-(Q+$'vp[1+s]/{360[c-q]})/-U

Rule-13699:

If both (**/**), (**U**), (**$**), (**V**), (**$'**), (**f**), (**s**), (**I**), (**Q**), (**D**), (**v**), (**p**), (**c**) and (**q**) are known, then its Tax Planned is:

T= $-V-$'f[1+s]-I- D}

-(Q+$'vp[1+s]/{360[c-q]})/-U

Rule-13700:

If both (**/**), (**U**), (**$**), (**V**), (**$'**), (**f**), (**s**), (**I**), (**T**), (**Q**), (**v**), (**p**), (**c**) and (**q**) are known, then its Dividend Planned is:

D= $-V-$'f[1+s]-I-T-(Q+$'vp[1+s]/{360[c-q]})/-U

Steve Asikin ISBN 14: 978-1514685136, ISBN 10: 1514685132

Rule-13701:

If both (**/**), (**U**), (**S**), (**V**), (**S'**), (**f**), (**s**), (**I**), (**T**), (**D**), (**Q**), (**p**), (**c**) and (**q**) are known, then its Variable Portion Planned is:

v= **V**/**S**

or it can also be found as

v= 360[**c**-**q**](**/**{**U**+**S**-**V**-**S'f**[1+**s**]-**I**-**T**-**D**}-**Q**)
/{**S'p**[1+**s**]}

Rule-13702:

If both (**/**), (**U**), (**S**), (**V**), (**S'**), (**f**), (**s**), (**I**), (**T**), (**D**), (**v**), (**Q**), (**c**) and (**q**) are known, then its Procured Inventory Days Planned:

p= 360[**c**-**q**](**/**{**U**+**S**-**V**-**S'f**[1+**s**]-**I**-**T**-**D**}-**Q**)
/{**S'v**[1+**s**]}

Rule-13703:

If both (**/**), (**U**), (**S**), (**V**), (**S'**), (**f**), (**s**), (**I**), (**T**), (**D**), (**v**), (**p**), (**Q**) and (**q**) are known, then its Current Ratio Planned is:

c= **q**+**S'vp**[1+**s**]/[360(**/**{**U**+**S**-**V**-**S'f**[1+**s**]-**I**-**T**-**D**}-**Q**)]

Rule-13704:

If both (**/**), (**U**), (**S**), (**V**), (**S'**), (**f**), (**s**), (**I**), (**T**), (**D**), (**v**), (**p**), (**c**) and (**Q**) are known, then its Quick or Acid Test Ratio Planned is:

q= **c**-**S'vp**[1+**s**]/[360(**/**{**U**+**S**-**V**-**S'f**[1+**s**]-**I**-**T**-**D**}-**Q**)]

Steve Asikin ISBN 14: 978-1514685136, ISBN 10: 1514685132

Rule-13705:

If both (**/**), (**U**), (**$**), (**V**), (**$'**), (**f**), (**s**), (**I**), (**T**), (**d**) and (**X**) are known, then its Sales or Revenue Planned is:

Q= **/**(**U**+{**$**-**V**-**$'f**[1+**s**]-**I**-**T**}[1-**d**])-**X**

Rule-13706:

If both (**Q**), (**U**), (**$**), (**V**), (**$'**), (**f**), (**s**), (**I**), (**T**), (**d**) and (**X**) are known, then its Leverage or Gearing Ratio Planned is:

/= [**Q**+**X**]/(**U**+{**$**-**V**-**$'f**[1+**s**]-**I**-**T**}[1-**d**])

Rule-13707:

If both (**/**), (**Q**), (**$**), (**V**), (**$'**), (**f**), (**s**), (**I**), (**T**), (**d**) and (**X**) are known, then its Utilized or Starting Capital must be:

U= [**Q**+**X**]/**/**-{**$**-**V**-**$'f**[1+**s**]-**I**-**T**}[1-**d**]

Rule-13708:

If both (**/**), (**U**), (**$**), (**V**), (**$'**), (**f**), (**s**), (**I**), (**T**), (**d**) and (**X**) are known, then its Sales or Revenue Planned is:

$**= **V**+**$'f[1+**s**]+**I**+**T**+{[**Q**+**X**]/**/**-**U**}/[1-**d**]

Rule-13709:

If both (**/**), (**U**), (**$**), (**Q**), (**$'**), (**f**), (**s**), (**I**), (**T**), (**d**) and (**X**) are known, then its Variable Cost Planned is:

V= {**$**-**$'f**[1+**s**]-**I**-**T**}-{[**Q**+**X**]/**/**-**U**}/[1-**d**]

Rule-13710:

If both (**/**), (**U**), (**$**), (**V**), (**Q**), (**f**), (**s**), (**I**), (**T**), (**d**) and (**X**) are known, then its Sales Past must be:

$'**= ([**$-**V**-**I**-**T**]-{[**Q**+**X**]/**/**-**U**}/[1-**d**])/{**f**[1+**s**]}

Steve Asikin ISBN 14: 978-1514685136, ISBN 10: 1514685132

Rule-13711:

If both (/), (U), (S), (V), (S'), (Q), (s), (I), (T), (d) and (X) are known, then its Fixed Portion Planned is:

f= ([S-V-I-T]-{[Q+X]/ /-U}/[1-d])/{S'[1+s]}

Rule-13712:

If both (/), (U), (S), (V), (S'), (f), (Q), (I), (T), (d) and (X) are known, then its Sales Growth Planned is:

s= S/S'-1

or it can also be found as

s= ([S-V-I-T]-{[Q+X]/ /-U}/[1-d])/[S'f]-1

Rule-13713:

If both (/), (U), (S), (V), (S'), (f), (s), (Q), (T), (d) and (X) are known, then its Interest Expense Planned is:

I= {S-V-S'f[1+s]-T}-{[Q+X]/ /-U}/[1-d]

Rule-13714:

If both (/), (U), (S), (V), (S'), (f), (s), (I), (Q), (d) and (X) are known, then its Tax Planned is:

T= {S-V-S'f[1+s]-I}-{[Q+X]/ /-U}/[1-d]

Rule-13715:

If both (/), (U), (S), (V), (S'), (f), (s), (I), (T), (Q) and (X) are known, then its Dividend Payout Planned is:

d= 1-{[Q+X]/ /-U}/{S-V-S'f[1+s]-I-T}

Steve Asikin ISBN 14: 978-1514685136, ISBN 10: 1514685132

Rule-13716:

If both (**/**), (**U**), (**$**), (**V**), (**$'**), (**f**), (**s**), (**I**), (**T**), (**d**) and (**Q**) are known, then its Xpress or Current Debt Planned is:

X= **/**(**U**+{**$**-**V**-**$'f**[1+**s**]-**I**-**T**}[1-**d**])-**Q**

Rule-13717:

If both (**/**), (**U**), (**$**), (**V**), (**$'**), (**f**), (**s**), (**I**), (**T**), (**d**), (**P**), (**c**) and (**q**) are known, then its Quoted Longterm Debt Planned is:

Q= **/**(**U**+{**$**-**V**-**$'f**[1+**s**]-**I**-**T**}[1-**d**])-**P**/[**c**-**q**]

Rule-13718:

If both (**Q**), (**U**), (**$**), (**V**), (**$'**), (**f**), (**s**), (**I**), (**T**), (**d**), (**P**), (**c**) and (**q**) are known, then its Leverage or Gearing Ratio Planned is:

/= {**Q**+**P**/[**c**-**q**]}/(**U**+{**$**-**V**-**$'f**[1+**s**]-**I**-**T**}[1-**d**])

Rule-13719:

If both (**/**), (**Q**), (**$**), (**V**), (**$'**), (**f**), (**s**), (**I**), (**T**), (**d**), (**P**), (**c**) and (**q**) are known, then its Utilized or Starting Capital Planned is:

U= {**Q**+**P**/[**c**-**q**]}/**/**-{**$**-**V**-**$'f**[1+**s**]-**I**-**T**}[1-**d**]

Rule-13720:

If both (**/**), (**U**), (**Q**), (**V**), (**$'**), (**f**), (**s**), (**I**), (**T**), (**d**), (**P**), (**c**) and (**q**) are known, then its Sales or Revenue Planned is:

$**= **V**+**$'f[1+**s**]+**I**+**T**+({**Q**+**P**/[**c**-**q**]}/**/**-**U**)/[1-**d**]

Steve Asikin ISBN 14: 978-1514685136, ISBN 10: 1514685132

Rule-13721:

If both (/), (U), ($), (Q), ($'), (f), (s), (I), (T), (d), (P), (c) and (q) are known, then its Variable Cost Planned is:

V= {$-$'f[1+s]-I-T}-({Q+P/[c-q]}//U)/[1-d]

Rule-13722:

If both (/), (U), ($), (V), (Q), (f), (s), (I), (T), (d), (P), (c) and (q) are known, then its Sales Past must be:

$'= [[$-V-I-T]-({Q+P/[c-q]}//U)/[1-d]]/{f[1+s]}

Rule-13723:

If both (/), (U), ($), (V), ($'), (Q), (s), (I), (T), (d), (P), (c) and (q) are known, then its Fixed Portion Planned is:

f= [[$-V-I-T]-({Q+P/[c-q]}//U)/[1-d]]/{$'[1+s]}

Rule-13724:

If both (/), (U), ($), (V), ($'), (f), (Q), (I), (T), (d), (P), (c) and (q) are known, then its Sales Growth Planned is:

s= $/$'-1

or it can also be found as

s= [[$-V-I-T]-({Q+P/[c-q]}//U)/[1-d]]/[$'f]-1

Rule-13725:

If both (/), (U), ($), (V), ($'), (f), (s), (Q), (T), (d), (P), (c) and (q) are known, then its Interest Expense Planned is:

I= {$-V-$'f[1+s]-T}-({Q+P/[c-q]}//U)/[1-d]

Steve Asikin ISBN 14: 978-1514685136, ISBN 10: 1514685132

Rule-13726:

If both (**/**), (**U**), (**S**), (**V**), (**S'**), (**f**), (**s**), (**I**), (**Q**), (**d**), (**P**), (**c**) and (**q**) are known, then its Tax Planned is:

T= {**S**-**V**-**S'f**[1+**s**]-**I**}-({**Q**+**P**/[**c**-**q**]}/**/U**)/[1-**d**]

Rule-13727:

If both (**/**), (**U**), (**S**), (**V**), (**S'**), (**f**), (**s**), (**I**), (**T**), (**Q**), (**P**), (**c**) and (**q**) are known, then its Dividend Payout Planned is:

d= 1-({**Q**+**P**/[**c**-**q**]}/**/U**)/{**S**-**V**-**S'f**[1+**s**]-**I**-**T**}

Rule-13728:

If both (**/**), (**U**), (**S**), (**V**), (**S'**), (**f**), (**s**), (**I**), (**T**), (**d**), (**Q**), (**c**) and (**q**) are known, then its Procured Inventory Planned is:

P= [**c**-**q**][**/**(**U**+{**S**-**V**-**S'f**[1+**s**]-**I**-**T**}[1-**d**])-**Q**]

Rule-13729:

If both (**/**), (**U**), (**S**), (**V**), (**S'**), (**f**), (**s**), (**I**), (**T**), (**d**), (**P**), (**c**) and (**q**) are known, then its Current Ratio Planned is:

c= **q**+**P**/[**/**(**U**+{**S**-**V**-**S'f**[1+**s**]-**I**-**T**}[1-**d**])-**Q**]

Rule-13730:

If both (**/**), (**U**), (**S**), (**V**), (**S'**), (**f**), (**s**), (**I**), (**T**), (**d**), (**P**), (**c**) and (**Q**) are known, then its Quick or Acid Test Ratio Planned is:

q= **c**-**P**/[**/**(**U**+{**S**-**V**-**S'f**[1+**s**]-**I**-**T**}[1-**d**])-**Q**]

Steve Asikin ISBN 14: 978-1514685136, ISBN 10: 1514685132

Rule-13731:

If both (**/**), (**U**), (**$**), (**V**), (**$'**), (**f**), (**s**), (**I**), (**T**), (**d**), (**p**), (**c**) and (**q**) are known, then its Quoted Longterm Debt Planned is:

Q= /(U+{$-V-$'f[1+s]-I-T}[1-d])-Vp/{360[c-q]}

Rule-13732:

If both (**Q**), (**U**), (**$**), (**V**), (**$'**), (**f**), (**s**), (**I**), (**T**), (**d**), (**p**), (**c**) and (**q**) are known, then its Leverage or Gearing Ratio Planned is:

/= (Q+Vp/{360[c-q]})
/(U+{$-V-$'f[1+s]-I-T}[1-d])

Rule-13733:

If both (**/**), (**Q**), (**$**), (**V**), (**$'**), (**f**), (**s**), (**I**), (**T**), (**d**), (**p**), (**c**) and (**q**) are known, then its Utilized or Starting Capital must be:

U= (Q+Vp/{360[c-q]})//-{$-V-$'f[1+s]-I-T}[1-d]

Rule-13734:

If both (**/**), (**U**), (**Q**), (**V**), (**$'**), (**f**), (**s**), (**I**), (**T**), (**d**), (**p**), (**c**) and (**q**) are known, then its Sales or Revenue Planned is:

$= V+$'f[1+s]+I+T+[(Q+Vp/{360[c-q]})//-U]
/[1-d]

Steve Asikin ISBN 14: 978-1514685136, ISBN 10: 1514685132

Rule-13735:

If both (**/**), (**U**), (**$**), (**Q**), (**$'**), (**f**), (**s**), (**I**), (**T**), (**d**), (**p**), (**c**) and (**q**) are known, then its Variable Cost Planned is:

V= [**/**(**U**+{**$**-**$'f**[1+**s**]-**I**-**T**}[1-**d**])-**Q**]
/(**p**/{360[**c**-**q**]}+**/**[1-**d**])

Rule-13736:

If both (**/**), (**U**), (**$**), (**V**), (**Q**), (**f**), (**s**), (**I**), (**T**), (**d**), (**p**), (**c**) and (**q**) are known, then its Sales Past must be:

$'**= {[**$-**V**-**I**-**T**]-[(**Q**+**Vp**/{360[**c**-**q**]})/**/**-**U**]
/[1-**d**]}/{**f**[1+**s**]}

Rule-13737:

If both (**/**), (**U**), (**$**), (**V**), (**$'**), (**Q**), (**s**), (**I**), (**T**), (**d**), (**p**), (**c**) and (**q**) are known, then its Fixed Portion Planned is:

f= {[**$**-**V**-**I**-**T**]- [(**Q**+**Vp**/{360[**c**-**q**]})/**/**-**U**]/[1-**d**]}
/{**$'**[1+**s**]}

Rule-13738:

If both (**/**), (**U**), (**$**), (**V**), (**$'**), (**f**), (**Q**), (**I**), (**T**), (**d**), (**p**), (**c**) and (**q**) are known, then its Sales Growth Planned is:

s= **$**/**$'**-1

or it can also be found as

s= {[**$**-**V**-**I**-**T**]- [(**Q**+**Vp**/{360[**c**-**q**]})/**/**-**U**]/[1-**d**]}
/[**$'f**]-1

Steve Asikin ISBN 14: 978-1514685136, ISBN 10: 1514685132

Rule-13739:

If both (/), (U), (S), (V), (S'), (f), (s), (Q), (T), (d), (p), (c) and (q) are known, then its Interest Expense Planned is:

I= {S-V-S'f[1+s]-T}
-[(Q+Vp/{360[c-q]})//-U]/[1-d]

Rule-13740:

If both (/), (U), (S), (V), (S'), (f), (s), (I), (Q), (d), (p), (c) and (q) are known, then its Tax Planned is:

T= {S-V-S'f[1+s]-I}
-[(Q+Vp/{360[c-q]})//-U]/[1-d]

Rule-13741:

If both (/), (U), (S), (V), (S'), (f), (s), (I), (T), (Q), (p), (c) and (q) are known, then its Dividend Payout Planned is:

d= 1-[(Q+Vp/{360[c-q]})//-U]/{S-V-S'f[1+s]-I-T}

Rule-13742:

If both (/), (U), (S), (V), (S'), (f), (s), (I), (T), (d), (Q), (c) and (q) are known, then its Procured Inventory Days planned:

p= 360[c-q][/(U+{S-V-S'f[1+s]-I-T}[1-d])-Q]/V

Rule-13743:

If both (/), (U), (S), (V), (S'), (f), (s), (I), (T), (d), (p), (Q) and (q) are known, then its Current Ratio Planned is:

c= q+Vp/{360[/(U+{S-V-S'f[1+s]-I-T}[1-d])-Q]}

Steve Asikin ISBN 14: 978-1514685136, ISBN 10: 1514685132

Rule-13744:

If both (/), (U), ($), (V), ($'), (f), (s), (I), (T), (d), (p), (c) and (Q) are known, then its Quick or Acid Test Ratio Planned is:

q= c-Vp/{360[/(U+{$-V-$'f[1+s]-I-T}[1-d])-Q]}

Rule-13745:

If both (/), (U), ($), (V), ($'), (f), (s), (I), (T), (d), (v), (p), (c) and (q) are known, then its Quoted Longterm Debt Planned is:

Q= /(U+{$-V-$'f[1+s]-I-T}[1-d])-$vp/{360[c-q]}

Rule-13746:

If both (Q), (U), ($), (V), ($'), (f), (s), (I), (T), (d), (v), (p), (c) and (q) are known, then its Leverage or Gearing Ratio Planned is:

/= (Q+$vp/{360[c-q]})
/(U+{$-V-$'f[1+s]-I-T}[1-d])

Rule-13747:

If both (/), (Q), ($), (V), ($'), (f), (s), (I), (T), (d), (v), (p), (c) and (q) are known, then its Utilized or Starting Capital must be:

U= (Q+$vp/{360[c-q]})//-{$-V-$'f[1+s]-I-T}[1-d]

Rule-13748:

If both (/), (U), (Q), (V), ($'), (f), (s), (I), (T), (d), (v), (p), (c) and (q) are known, then its Sales or Revenue Planned is:

$= [/(U+{$-V-$'f[1+s]-I-T}[1-d])-Q]
/(vp/{360[c-q]}-/[1-d])

Steve Asikin ISBN 14: 978-1514685136, ISBN 10: 1514685132

Rule-13749:

If both (**/**), (**U**), (**$**), (**Q**), (**$'**), (**f**), (**s**), (**I**), (**T**), (**d**), (**v**), (**p**), (**c**) and (**q**) are known, then its Variable Cost Planned is:

V= [/U+{$-$'f[1+s]-I-T}[1-d])
-$vp/{360[c-q]}-Q]/{/[1-d]}

Rule-13750:

If both (**/**), (**U**), (**$**), (**V**), (**Q**), (**f**), (**s**), (**I**), (**T**), (**d**), (**v**), (**p**), (**c**) and (**q**) are known, then its Sales Past must be:

**$'**= {[$-V-I-T]- [(Q+$vp/{360[c-q]})/-U]/[1-d]}
/{f[1+s]}

Rule-13751:

If both (**/**), (**U**), (**$**), (**V**), (**$'**), (**Q**), (**s**), (**I**), (**T**), (**d**), (**v**), (**p**), (**c**) and (**q**) are known, then its Fixed Portion Planned is:

f= {[$-V-I-T]-[(Q+$vp/{360[c-q]})/-U]/[1-d]}
/{$'[1+s]}

Rule-13752:

If both (**/**), (**U**), (**$**), (**V**), (**$'**), (**f**), (**Q**), (**I**), (**T**), (**d**), (**v**), (**p**), (**c**) and (**q**) are known, then its Sales Growth Planned is:

s= $/$'-1

or it can also be found as

s= {[$-V-I-T]
-[(Q+$vp/{360[c-q]})/-U]/[1-d]}/[$'f]-1

Steve Asikin ISBN 14: 978-1514685136, ISBN 10: 1514685132

Rule-13753:

If both (**/**), (**U**), (**S**), (**V**), (**S'**), (**f**), (**s**), (**Q**), (**T**), (**d**), (**v**), (**p**), (**c**) and (**q**) are known, then its Interest Expense Planned is:

I= {**S**-**V**-**S'f**[1+**s**]-**T**}
-[(**Q**+**Svp**/{360[**c**-**q**]})/**/**-**U**]/[1-**d**]

Rule-13754:

If both (**/**), (**U**), (**S**), (**V**), (**S'**), (**f**), (**s**), (**I**), (**Q**), (**d**), (**v**), (**p**), (**c**) and (**q**) are known, then its Tax Planned is:

T= {**S**-**V**-**S'f**[1+**s**]-**I**}
-[(**Q**+**Svp**/{360[**c**-**q**]})/**/**-**U**]/[1-**d**]

Rule-13755:

If both (**/**), (**U**), (**S**), (**V**), (**S'**), (**f**), (**s**), (**I**), (**T**), (**Q**), (**v**), (**p**), (**c**) and (**q**) are known, then its Dividend Payout Planned is:

d= 1-[(**Q**+**Svp**/{360[**c**-**q**]})/**/**-**U**]/{**S**-**V**-**S'f**[1+**s**]-**I**-**T**}

Rule-13756:

If both (**/**), (**U**), (**S**), (**V**), (**S'**), (**f**), (**s**), (**I**), (**T**), (**d**), (**Q**), (**p**), (**c**) and (**q**) are known, then its Variable Portion Planned is:

v= 360[**c**-**q**][**/**(**U**+{**S**-**V**-**S'f**[1+**s**]-**I**-**T**}[1-**d**])-**Q**]/[**Sp**]

Rule-13757:

If both (**/**), (**U**), (**S**), (**V**), (**S'**), (**f**), (**s**), (**I**), (**T**), (**d**), (**v**), (**Q**), (**c**) and (**q**) are known, then its Procured Inventory Days planned:

p= 360[**c**-**q**][**/**(**U**+{**S**-**V**-**S'f**[1+**s**]-**I**-**T**}[1-**d**])-**Q**]/[**Sv**]

Steve Asikin ISBN 14: 978-1514685136, ISBN 10: 1514685132

Rule-13758:

If both (**/**), (**U**), (**$**), (**V**), (**$'**), (**f**), (**s**), (**I**), (**T**), (**d**), (**v**), (**p**), (**Q**) and (**q**) are known, then its Current Ratio Planned is:

c= q+$vp/{360[/(U+{$-V-$'f[1+s]-I-T}[1-d])-Q]}

Rule-13759:

If both (**/**), (**U**), (**$**), (**V**), (**$'**), (**f**), (**s**), (**I**), (**T**), (**d**), (**v**), (**p**), (**c**) and (**Q**) are known, then its Quick or Acid Test Ratio Planned is:

q= c-$vp/{360[/(U+{$-V-$'f[1+s]-I-T}[1-d])-Q]}

Rule-13760:

If both (**/**), (**U**), (**$**), (**V**), (**$'**), (**f**), (**s**), (**I**), (**T**), (**d**), (**v**), (**p**), (**c**) and (**q**) are known, then its Quoted Longterm Debt Planned is:

Q= /(U+{$-V-$'f[1+s]-I-T}[1-d])
-$'vp[1+s]/{360[c-q]}

Rule-13761:

If both (**Q**), (**U**), (**$**), (**V**), (**$'**), (**f**), (**s**), (**I**), (**T**), (**d**), (**v**), (**p**), (**c**) and (**q**) are known, then its Leverage or Gearing Ratio Planned is:

/= (Q+$'vp[1+s]/{360[c-q]})
/(U+{$-V-$'f[1+s]-I-T}[1-d])

Steve Asikin ISBN 14: 978-1514685136, ISBN 10: 1514685132

Rule-13762:

If both (/), (Q), (S), (V), (S'), (f), (s), (I), (T), (d), (v), (p), (c) and (q) are known, then its Utilized or Starting Capital must be:

U= (Q+S'vp[1+s]/{360[c-q]})//
-{S-V-S'f[1+s]-I-T}[1-d]

Rule-13763:

If both (/), (U), (Q), (V), (S'), (f), (s), (I), (T), (d), (v), (p), (c) and (q) are known, then its Sales or Revenue Planned is:

S= [Q-/(U-{V+S'f[1+s]+I+T}[1-d])
+S'vp[1+s]/{360[c-q]}]/{/[1-d]}

Rule-13764:

If both (/), (U), (S), (Q), (S'), (f), (s), (I), (T), (d), (v), (p), (c) and (q) are known, then its Variable Cost Planned is:

V= [/(U+{S-S'f[1+s]-I-T}[1-d])
-S'vp[1+s]/{360[c-q]}-Q]/{/[1-d]}

Rule-13765:

If both (/), (U), (S), (V), (Q), (f), (s), (I), (T), (d), (v), (p), (c) and (q) are known, then its Sales Past must be:

S'= [/(U+{S-V-I-T}[1-d])-Q]
/[[1+s](vp/{360[c-q]}+f/)]

Steve Asikin ISBN 14: 978-1514685136, ISBN 10: 1514685132

Rule-13766:

If both (**/**), (**U**), (**$**), (**V**), (**$'**), (**Q**), (**s**), (**I**), (**T**), (**d**), (**v**), (**p**), (**c**) and (**q**) are known, then its Fixed Portion Planned is:

f= {[$-V-I-T]-[(Q+$'vp[1+s]/{360[c-q]})//-U]/[1-d]}/{$'[1+s]}

Rule-13767:

If both (**/**), (**U**), (**$**), (**V**), (**$'**), (**f**), (**Q**), (**I**), (**T**), (**d**), (**v**), (**p**), (**c**) and (**q**) are known, then its Sales Growth Planned is:

s= $/$'-1

or it can also be found as

s= [/U+{$-V-I-T}[1-d])-Q]/[$'(vp/{360[c-q]}+f/)]-1

Rule-13768:

If both (**/**), (**U**), (**$**), (**V**), (**$'**), (**f**), (**s**), (**Q**), (**T**), (**d**), (**v**), (**p**), (**c**) and (**q**) are known, then its Interest Expense Planned is:

I= {$-V-$'f[1+s]-T}-[(Q+$'vp[1+s]/{360[c-q]})//-U]/[1-d]

Rule-13769:

If both (**/**), (**U**), (**$**), (**V**), (**$'**), (**f**), (**s**), (**I**), (**Q**), (**d**), (**v**), (**p**), (**c**) and (**q**) are known, then its Tax Planned is:

T= {$-V-$'f[1+s]-I}-[(Q+$'vp[1+s]/{360[c-q]})//-U]/[1-d]

Steve Asikin ISBN 14: 978-1514685136, ISBN 10: 1514685132

Rule-13770:

If both (**/**), (**U**), (**$**), (**V**), (**$'**), (**f**), (**s**), (**I**), (**T**), (**Q**), (**v**), (**p**), (**c**) and (**q**) are known, then its Dividend Payout Planned is:

d= 1-[(**Q**+**$'vp**[1+**s**]/{360[**c**-**q**]})/**/-U**]
/{**$**-**V**-**$'f**[1+**s**]-**I**-**T**}

Rule-13771:

If both (**/**), (**U**), (**$**), (**V**), (**$'**), (**f**), (**s**), (**I**), (**T**), (**d**), (**Q**), (**p**), (**c**) and (**q**) are known, then its Variable Portion Planned is:

v= **V**/**$**

or it can also be found as

v= 360[**c**-**q**][**/U**+{**$**-**V**-**$'f**[1+**s**]-**I**-**T**}[1-**d**])-**Q**]
/{**$'p**[1+**s**]}

Rule-13772:

If both (**/**), (**U**), (**$**), (**V**), (**$'**), (**f**), (**s**), (**I**), (**T**), (**d**), (**v**), (**Q**), (**c**) and (**q**) are known, then its Procured Inventory Days Planned:

p= 360[**c**-**q**][**/U**+{**$**-**V**-**$'f**[1+**s**]-**I**-**T**}[1-**d**])-**Q**]
/{**$'v**[1+**s**]}

Rule-13773:

If both (**/**), (**U**), (**$**), (**V**), (**$'**), (**f**), (**s**), (**I**), (**T**), (**d**), (**v**), (**p**), (**Q**) and (**q**) are known, then its Current Ratio Planned is:

c= **q**+**$'vp**[1+**s**]
/{360[**/U**+{**$**-**V**-**$'f**[1+**s**]-**I**-**T**}[1-**d**])-**Q**]}

Steve Asikin ISBN 14: 978-1514685136, ISBN 10: 1514685132

Rule-13774:

If both (**/**), (**U**), (**$**), (**V**), (**$'**), (**f**), (**s**), (**I**), (**T**), (**d**), (**v**), (**p**), (**c**) and (**Q**) are known, then its Quick or Acid Test Ratio Planned is:

q= **c**-**$'vp**[1+**s**]
/{360[**/**(**U**+{**$**-**V**-**$'f**[1+**s**]-**I**-**T**}[1-**d**])-**Q**]}

Rule-13775:

If both (**/**), (**U**), (**$**), (**V**), (**$'**), (**f**), (**s**), (**I**), (**T**), (**A**), (**d**) and (**X**) are known, then its Quoted Longterm Debt Planned is:

Q= **/**{**U**+**$**-**V**-**$'f**[1+**s**]-**I**-**T**-**Ad**}-**X**

Rule-13776:

If both (**Q**), (**U**), (**$**), (**V**), (**$'**), (**f**), (**s**), (**I**), (**T**), (**A**), (**d**) and (**X**) are known, then its Leverage or Gearing Ratio Planned is:

/= [**Q**+**X**]/{**U**+**$**-**V**-**$'f**[1+**s**]-**I**-**T**-**Ad**}

Rule-13777:

If both (**/**), (**Q**), (**$**), (**V**), (**$'**), (**f**), (**s**), (**I**), (**T**), (**A**), (**d**) and (**X**) are known, then its Utilized or Starting Capital must be:

U= [**Q**+**X**]/**/**-{**$**-**V**-**$'f**[1+**s**]-**I**-**T**-**Ad**}

Rule-13778:

If both (**/**), (**U**), (**Q**), (**V**), (**$'**), (**f**), (**s**), (**I**), (**T**), (**A**), (**d**) and (**X**) are known, then its Sales or Revenue Planned is:

$**= [**Q**+**X**]/**/**-{**U**-**V**-**$'f[1+**s**]-**I**-**T**-**Ad**}

Steve Asikin ISBN 14: 978-1514685136, ISBN 10: 1514685132

Rule-13779:

If both (**/**), (**U**), (**$**), (**Q**), (**$'**), (**f**), (**s**), (**I**), (**T**), (**A**), (**d**) and (**X**) are known, then its Variable Cost Planned is:

V= **U**+**$**-**$'f**[1+**s**]-**I**-**T**-**Ad**-[**Q**+**X**]/**/**

Rule-13780:

If both (**/**), (**U**), (**$**), (**V**), (**Q**), (**f**), (**s**), (**I**), (**T**), (**A**), (**d**) and (**X**) are known, then its Sales Past must be:

$'**= {**U**+**$-**V**-**I**-**T**-**Ad**-[**Q**+**X**]/**/**}/{**f**[1+**s**]}

Rule-13781:

If both (**/**), (**U**), (**$**), (**V**), (**$'**), (**Q**), (**s**), (**I**), (**T**), (**A**), (**d**) and (**X**) are known, then its Fixed Portion Planned is:

f= {**U**+**$**-**V**-**I**-**T**-**Ad**-[**Q**+**X**]/**/**}/{**$'**[1+**s**]}

Rule-13782:

If both (**/**), (**U**), (**$**), (**V**), (**$'**), (**f**), (**Q**), (**I**), (**T**), (**A**), (**d**) and (**X**) are known, then its Sales Growth Planned is:

s= **$**/**$'**-1

or it can also be found as

s= {**U**+**$**-**V**-**I**-**T**-**Ad**-[**Q**+**X**]/**/**}/[**$'f**]-1

Rule-13783:

If both (**/**), (**U**), (**$**), (**V**), (**$'**), (**f**), (**s**), (**Q**), (**T**), (**A**), (**d**) and (**X**) are known, then its Interest Expense Planned is:

I= {**U**+**$**-**V**-**$'f**[1+**s**]-**T**-**Ad**}-[**Q**+**X**]/**/**

Rule-13784:

If both (**/**), (**U**), (**$**), (**V**), (**$'**), (**f**), (**s**), (**I**), (**Q**), (**A**), (**d**) and (**X**) are known, then its Tax Planned is:

T= **U**+**$**-**V**-**$'f**[1+**s**]-**I**-**Ad**-[**Q**+**X**]/**/**

Steve Asikin ISBN 14: 978-1514685136, ISBN 10: 1514685132

Rule-13785:

If both (**/**), (**U**), (**$**), (**V**), (**$'**), (**f**), (**s**), (**I**), (**T**), (**Q**), (**d**) and (**X**) are known, then its After Tax Income Planned is:

A= {**U**+**$**-**V**-**$'f**[1+**s**]-**I**-[**Q**+**X**]/**/**}/**d**

Rule-13786:

If both (**/**), (**U**), (**$**), (**V**), (**$'**), (**f**), (**s**), (**I**), (**T**), (**A**), (**Q**) and (**X**) are known, then its Dividend Payout Planned is:

d= {**U**+**$**-**V**-**$'f**[1+**s**]-**I**-[**Q**+**X**]/**/**}/**A**

Rule-13787:

If both (**/**), (**U**), (**$**), (**V**), (**$'**), (**f**), (**s**), (**I**), (**T**), (**A**), (**d**) and (**Q**) are known, then its Xpress or Current Debt Planned is:

X= **/**{**U**+**$**-**V**-**$'f**[1+**s**]-**I**-**T**-**Ad**}-**Q**

Rule-13788:

If both (**/**), (**U**), (**$**), (**V**), (**$'**), (**f**), (**s**), (**I**), (**T**), (**A**), (**d**), (**P**), (**c**) and (**q**) are known, then its Quoted Longterm Debt Planned is:

Q= **/**{**U**+**$**-**V**-**$'f**[1+**s**]-**I**-**T**-**Ad**}-**P**/[**c**-**q**]

Rule-13789:

If both (**Q**), (**U**), (**$**), (**V**), (**$'**), (**f**), (**s**), (**I**), (**T**), (**A**), (**d**), (**P**), (**c**) and (**q**) are known, then its Leverage or Gearing Ratio Planned is:

/= {**Q**+**P**/[**c**-**q**]}/{**U**+**$**-**V**-**$'f**[1+**s**]-**I**-**T**-**Ad**}

Steve Asikin ISBN 14: 978-1514685136, ISBN 10: 1514685132

Rule-13790:

If both (/), (Q), (S), (V), (S'), (f), (s), (I), (T), (A), (d), (P), (c) and (q) are known, then its Utilized or Starting Capital must be:

U= {**Q**+**P**/[**c**-**q**]}//-{**S**-**V**-**S'f**[1+**s**]-**I**-**T**-**Ad**}

Rule-13791:

If both (/), (U), (Q), (V), (S'), (f), (s), (I), (T), (A), (d), (P), (c) and (q) are known, then its Sales or Revenue Planned is:

S= {**Q**+**P**/[**c**-**q**]}//-{**U**-**V**-**S'f**[1+**s**]-**I**-**T**-**Ad**}

Rule-13792:

If both (/), (U), (S), (Q), (S'), (f), (s), (I), (T), (A), (d), (P), (c) and (q) are known, then its Variable Cost Planned is:

V= **U**+**S**-**S'f**[1+**s**]-**I**-**T**-**Ad**-{**Q**+**P**/[**c**-**q**]}//

Rule-13793:

If both (/), (U), (S), (V), (Q), (f), (s), (I), (T), (A), (d), (P), (c) and (q) are known, then its Sales Past must be:

S'= (**U**+**S**-**V**-**I**-**T**-**Ad**-{**Q**+**P**/[**c**-**q**]}//)/{**f**[1+**s**]}

Rule-13794:

If both (/), (U), (S), (V), (S'), (Q), (s), (I), (T), (A), (d), (P), (c) and (q) are known, then its Fixed Portion Planned is:

f= {(**U**+**S**-**V**-**I**-**T**-**Ad**-{**Q**+**P**/[**c**-**q**]}//)/{**S'**[1+**s**]}

Steve Asikin ISBN 14: 978-1514685136, ISBN 10: 1514685132

Rule-13795:

If both (/), (**U**), (**$**), (**V**), (**$'**), (**f**), (**Q**), (**I**), (**T**), (**A**), (**d**), (**P**), (**c**) and (**q**) are known, then its Sales Growth Planned is:

s= $/$'-1

or itcan also be found as

s= (U+$-V-I-T-Ad-{Q+P/[c-q]}//)/ [$'f]-1

Rule-13796:

If both (/), (**U**), (**$**), (**V**), (**$'**), (**f**), (**s**), (**Q**), (**T**), (**A**), (**d**), (**P**), (**c**) and (**q**) are known, then its Interest Expense Planned is:

I= {U+$-V-$'f[1+s]-T-Ad}-{Q+P/[c-q]}//

Rule-13797:

If both (/), (**U**), (**$**), (**V**), (**$'**), (**f**), (**s**), (**I**), (**Q**), (**A**), (**d**), (**P**), (**c**) and (**q**) are known, then its Tax Planned is:

T= U+$-V-$'f[1+s]-I-Ad-{Q+P/[c-q]}//

Rule-13798:

If both (/), (**U**), (**$**), (**V**), (**$'**), (**f**), (**s**), (**I**), (**T**), (**Q**), (**d**), (**P**), (**c**) and (**q**) are known, then its After Tax Income Planned is:

A= (U+$-V-$'f[1+s]-I-{Q+P/[c-q]}//)/d

Rule-13799:

If both (/), (**U**), (**$**), (**V**), (**$'**), (**f**), (**s**), (**I**), (**T**), (**A**), (**Q**), (**P**), (**c**) and (**q**) are known, then its Dividend Payout Planned is:

d= (U+$-V-$'f[1+s]-I-{Q+P/[c-q]}//)/A

Steve Asikin ISBN 14: 978-1514685136, ISBN 10: 1514685132

Rule-13800:

If both (/), (U), ($), (V), ($'), (f), (s), (I), (T), (A), (d), (Q), (c) and (q) are known, then its Procured Inventory Planned is:

P= [c-q](/{U+$-V-$'f[1+s]-I-T-Ad}-Q)

Rule-13801:

If both (/), (U), ($), (V), ($'), (f), (s), (I), (T), (A), (d), (P), (Q) and (q) are known, then its Current Ratio Planned is:

c= q+P/(/{U+$-V-$'f[1+s]-I-T-Ad}-Q)

Rule-13802:

If both (/), (U), ($), (V), ($'), (f), (s), (I), (T), (A), (d), (P), (c) and (Q) are known, then its Quick or Acid Test Ratio Planned is:

q= c-P/(/{U+$-V-$'f[1+s]-I-T-Ad}-Q)

Rule-13803:

If both (/), (U), ($), (V), ($'), (f), (s), (I), (T), (A), (d), (p), (c) and (q) are known, then its Quoted Longterm Debt Planned is:

Q= /{U+$-V-$'f[1+s]-I-T-Ad}-Vp/{360[c-q]}

Rule-13804:

If both (Q), (U), ($), (V), ($'), (f), (s), (I), (T), (A), (d), (p), (c) and (q) are known, then its Leverage or Gearing Ratio Planned is:

/= (Q+Vp/{360[c-q]})/{U+$-V-$'f[1+s]-I-T-Ad}

Steve Asikin ISBN 14: 978-1514685136, ISBN 10: 1514685132

Rule-13805:

If both (/), (Q), ($), (V), ($'), (f), (s), (I), (T), (A), (d), (p), (c) and (q) are known, then its Utilized or Starting Capital must be:

U= (Q+Vp/{360[c-q]})//-{$-V-$'f[1+s]-I-T-Ad}

Rule-13806:

If both (/), (U), (Q), (V), ($'), (f), (s), (I), (T), (A), (d), (p), (c) and (q) are known, then its Sales or Revenue Planned is:

$= (Q+Vp/{360[c-q]})//-{U-V-$'f[1+s]-I-T-Ad}

Rule-13807:

If both (/), (U), ($), (Q), ($'), (f), (s), (I), (T), (A), (d), (p), (c) and (q) are known, then its Variable Cost Planned is:

V= (/{U+$-$'f[1+s]-I-T-Ad}-Q)/(p/{360[c-q]}+/)

Rule-13808:

If both (/), (U), ($), (V), (Q), (f), (s), (I), (T), (A), (d), (p), (c) and (q) are known, then its Sales Past must be:

$'= [U+$-V-I-T-Ad-(Q+Vp/{360[c-q]})//]/{f[1+s]}

Rule-13809:

If both (/), (U), ($), (V), ($'), (Q), (s), (I), (T), (A), (d), (p), (c) and (q) are known, then its Fixed Portion Planned is:

f= [U+$-V-I-T-Ad-(Q+Vp/{360[c-q]})//]/{$'[1+s]}

Steve Asikin ISBN 14: 978-1514685136, ISBN 10: 1514685132

Rule-13810:

If both (**/**), (**U**), (**S**), (**V**), (**S'**), (**f**), (**Q**), (**I**), (**T**), (**A**), (**d**), (**p**), (**c**) and (**q**) are known, then its Sales Growth Planned is:

s= S/S'-1

or it can also be found as

s= [U+S-V-I-T-Ad-(Q+Vp/{360[c-q]})/**/**]/[S'f]-1

Rule-13811:

If both (**/**), (**U**), (**S**), (**V**), (**S'**), (**f**), (**s**), (**Q**), (**T**), (**A**), (**d**), (**p**), (**c**) and (**q**) are known, then its Interest Expense Planned is:

I= {U+S-V-S'f[1+s]-T-Ad}-(Q+Vp/{360[c-q]})/**/**

Rule-13812:

If both (**/**), (**U**), (**S**), (**V**), (**S'**), (**f**), (**s**), (**I**), (**Q**), (**A**), (**d**), (**p**), (**c**) and (**q**) are known, then its Tax Planned is:

T= U+S-V-S'f[1+s]-I-Ad-(Q+Vp/{360[c-q]})/**/**

Rule-13813:

If both (**/**), (**U**), (**S**), (**V**), (**S'**), (**f**), (**s**), (**I**), (**T**), (**Q**), (**d**), (**p**), (**c**) and (**q**) are known, then its After Tax Income Planned is:

A= [U+S-V-S'f[1+s]-I-(Q+Vp/{360[c-q]})/**/**]/d

Rule-13814:

If both (**/**), (**U**), (**S**), (**V**), (**S'**), (**f**), (**s**), (**I**), (**T**), (**A**), (**Q**), (**p**), (**c**) and (**q**) are known, then its Dividend Payout Planned is:

d= [U+S-V-S'f[1+s]-I-(Q+Vp/{360[c-q]})/**/**]/A

Steve Asikin ISBN 14: 978-1514685136, ISBN 10: 1514685132

Rule-13815:

If both (**/**), (**U**), (**$**), (**V**), (**$'**), (**f**), (**s**), (**I**), (**T**), (**A**), (**d**), (**Q**), (**c**) and (**q**) are known, then its Procured Inventory Planned is:

p= 360[**c-q**](**/**{**U+$-V-$'f**[1+**s**]-**I-T-Ad**}-**Q**)/**V**

Rule-13816:

If both (**/**), (**U**), (**$**), (**V**), (**$'**), (**f**), (**s**), (**I**), (**T**), (**A**), (**d**), (**p**), (**Q**) and (**q**) are known, then its Current Ratio Planned is:

c= **q**+**Vp**/[360(**/**{**U+$-V-$'f**[1+**s**]-**I-T-Ad**}-**Q**)]

Rule-13817:

If both (**/**), (**U**), (**$**), (**V**), (**$'**), (**f**), (**s**), (**I**), (**T**), (**A**), (**d**), (**p**), (**c**) and (**Q**) are known, then its Quick or Acid Test Ratio Planned is:

q= **c-Vp**/[360(**/**{**U+$-V-$'f**[1+**s**]-**I-T-Ad**}-**Q**)]

Rule-13818:

If both (**/**), (**U**), (**$**), (**V**), (**$'**), (**f**), (**s**), (**I**), (**T**), (**A**), (**d**), (**v**), (**p**), (**c**) and (**q**) are known, then its Quoted Longterm Debt Planned is:

Q= **/**{**U+$-V-$'f**[1+**s**]-**I-T-Ad**}-**$vp**/{360[**c-q**]}

Rule-13819:

If both (**Q**), (**U**), (**$**), (**V**), (**$'**), (**f**), (**s**), (**I**), (**T**), (**A**), (**d**), (**v**), (**p**), (**c**) and (**q**) are known, then its Leverage or Gearing Ratio Planned is:

/= (**Q**+**$vp**/{360[**c-q**]})/{**U+$-V-$'f**[1+**s**]-**I-T-Ad**}

Steve Asikin ISBN 14: 978-1514685136, ISBN 10: 1514685132

Rule-13820:

If both (**/**), (**Q**), (**$**), (**V**), (**$'**), (**f**), (**s**), (**I**), (**T**), (**A**), (**d**), (**v**), (**p**), (**c**) and (**q**) are known, then its Utilized or Starting Capital must be:

U= (**Q**+**$vp**/{360[**c-q**]})/**/**-{**$-V-$'f**[1+**s**]-**I-T-Ad**}

Rule-13821:

If both (**/**), (**U**), (**Q**), (**V**), (**$'**), (**f**), (**s**), (**I**), (**T**), (**A**), (**d**), (**v**), (**p**), (**c**) and (**q**) are known, then its Sales or Revenue Planned is:

$**= (**/**{**U-V-$'f[1+**s**]-**I-T-Ad**}-**Q**)/(**vp**/{360[**c-q**]}-**/**)

Rule-13822:

If both (**/**), (**U**), (**$**), (**Q**), (**$'**), (**f**), (**s**), (**I**), (**T**), (**A**), (**d**), (**v**), (**p**), (**c**) and (**q**) are known, then its Variable Cost Planned is:

V= (**/**{**U**+**$-$'f**[1+**s**]-**I-T-Ad**}-**$vp**/{360[**c-q**]}-**Q**)/**/**

Rule-13823:

If both (**/**), (**U**), (**$**), (**V**), (**Q**), (**f**), (**s**), (**I**), (**T**), (**A**), (**d**), (**v**), (**p**), (**c**) and (**q**) are known, then its Sales Past must be:

$'**= **[U**+**$-V-I-T-Ad-(**Q**+**$vp**/{360[**c-q**]})/**/]**
/{**f**[1+**s**]}

Rule-13824:

If both (**/**), (**U**), (**$**), (**V**), (**$'**), (**Q**), (**s**), (**I**), (**T**), (**A**), (**d**), (**v**), (**p**), (**c**) and (**q**) are known, then its Fixed Rate Planned is:

f= **[U**+**$-V-I-T-Ad**-(**Q**+**$vp**/{360[**c-q**]})/**/]**
/{**$'**[1+**s**]}

Steve Asikin ISBN 14: 978-1514685136, ISBN 10: 1514685132

Rule-13825:

If both (**/**), (**U**), (**$**), (**V**), (**$'**), (**f**), (**Q**), (**I**), (**T**), (**A**), (**d**), (**v**), (**p**), (**c**) and (**q**) are known, then its Sales Growth Planned is:

s= [U+$-V-I-T-Ad-(Q+$vp/{360[c-q]})/**/**]/[$'f]-1

Rule-13826:

If both (**/**), (**U**), (**$**), (**V**), (**$'**), (**f**), (**s**), (**Q**), (**T**), (**A**), (**d**), (**v**), (**p**), (**c**) and (**q**) are known, then its Interest Expense Planned is:

I= {U+$-V-$'f[1+s]-T-Ad}-(Q+$vp/{360[c-q]})/**/**

Rule-13827:

If both (**/**), (**U**), (**$**), (**V**), (**$'**), (**f**), (**s**), (**I**), (**Q**), (**A**), (**d**), (**v**), (**p**), (**c**) and (**q**) are known, then its Tax Planned is:

T= U+$-V-$'f[1+s]-I-Ad-(Q+$vp/{360[c-q]})/**/**

Rule-13828:

If both (**/**), (**U**), (**$**), (**V**), (**$'**), (**f**), (**s**), (**I**), (**T**), (**Q**), (**d**), (**v**), (**p**), (**c**) and (**q**) are known, then its After Tax Income Planned is:

A= [U+$-V-$'f[1+s]-I-(Q+$vp/{360[c-q]})/**/**]/d

Rule-13829:

If both (**/**), (**U**), (**$**), (**V**), (**$'**), (**f**), (**s**), (**I**), (**T**), (**A**), (**Q**), (**v**), (**p**), (**c**) and (**q**) are known, then its Dividend Payout Planned is:

d= [U+$-V-$'f[1+s]-I-(Q+$vp/{360[c-q]})/**/**]/A

Steve Asikin ISBN 14: 978-1514685136, ISBN 10: 1514685132

Rule-13830:

If both (**/**), (**U**), (**$**), (**V**), (**$'**), (**f**), (**s**), (**I**), (**T**), (**A**), (**d**), (**Q**), (**p**), (**c**) and (**q**) are known, then its Variable Portion Planned is:

v= **V/$**

or it can also be found as

v= 360[**c-q**](**/**{**U**+**$-V-$'f**[1+**s**]-**I-T-Ad**}-**Q**)/[**$p**]

Rule-13831:

If both (**/**), (**U**), (**$**), (**V**), (**$'**), (**f**), (**s**), (**I**), (**T**), (**A**), (**d**), (**v**), (**Q**), (**c**) and (**q**) are known, then its Quoted Procured Inventory Days planned:

p= 360[**c-q**](**/**{**U**+**$-V-$'f**[1+**s**]-**I-T-Ad**}-**Q**)/[**$v**]

Rule-13832:

If both (**/**), (**U**), (**$**), (**V**), (**$'**), (**f**), (**s**), (**I**), (**T**), (**A**), (**d**), (**v**), (**p**), (**Q**) and (**q**) are known, then its Current Ratio Planned is:

c= **q**+**$vp**/[360(**/**{**U**+**$-V-$'f**[1+**s**]-**I-T-Ad**}-**Q**)]

Rule-13833:

If both (**/**), (**U**), (**$**), (**V**), (**$'**), (**f**), (**s**), (**I**), (**T**), (**A**), (**d**), (**v**), (**p**), (**c**) and (**Q**) are known, then its Quick or Acid Test Ratio Planned is:

q= **c-$vp**/[360(**/**{**U**+**$-V-$'f**[1+**s**]-**I-T-Ad**}-**Q**)]

Steve Asikin ISBN 14: 978-1514685136, ISBN 10: 1514685132

Rule-13834:

If both (**/**), (**U**), (**$**), (**V**), (**$'**), (**f**), (**s**), (**I**), (**T**), (**A**), (**d**), (**v**), (**p**), (**c**) and (**q**) are known, then its Quoted Longterm Debt Planned is:

Q= **/**{**U**+**$**-**V**-**$'f**[1+**s**]-**I**-**T**-**Ad**}
-**$'vp**[1+**s**]/{360[**c**-**q**]}

Rule-13835:

If both (**Q**), (**U**), (**$**), (**V**), (**$'**), (**f**), (**s**), (**I**), (**T**), (**A**), (**d**), (**v**), (**p**), (**c**) and (**q**) are known, then its Leverage or Gearing Ratio Planned is:

/= (**Q**+**$'vp**[1+**s**]/{360[**c**-**q**]})
/{**U**+**$**-**V**-**$'f**[1+**s**]-**I**-**T**-**Ad**}

Rule-13836:

If both (**/**), (**Q**), (**$**), (**V**), (**$'**), (**f**), (**s**), (**I**), (**T**), (**A**), (**d**), (**v**), (**p**), (**c**) and (**q**) are known, then its Utilized or Starting Capital must be:

U= (**Q**+**$'vp**[1+**s**]/{360[**c**-**q**]})/**/**
-{**$**-**V**-**$'f**[1+**s**]-**I**-**T**-**Ad**}

Rule-13837:

If both (**/**), (**U**), (**Q**), (**V**), (**$'**), (**f**), (**s**), (**I**), (**T**), (**A**), (**d**), (**v**), (**p**), (**c**) and (**q**) are known, then its Sales or Revenue Planned is:

$**= (**Q**-**/**{**U**-**V**-**$'f[1+**s**]-**I**-**T**-**Ad**}
+**$'vp**[1+**s**]/{360[**c**-**q**]})/**/**

Steve Asikin ISBN 14: 978-1514685136, ISBN 10: 1514685132

Rule-13838:

If both (**/**), (**U**), (**$**), (**Q**), (**$'**), (**f**), (**s**), (**I**), (**T**), (**A**), (**d**), (**v**), (**p**), (**c**) and (**q**) are known, then its Variable Cost Planned is:

V= (**/**{**U**+**$**-**$'f**[1+**s**]-**I**-**T**-**Ad**} -**$'vp**[1+**s**]/{360[**c**-**q**]}-**Q**)/**/**

Rule-13839:

If both (**/**), (**U**), (**$**), (**V**), (**Q**), (**f**), (**s**), (**I**), (**T**), (**A**), (**d**), (**v**), (**p**), (**c**) and (**q**) are known, then its Sales Past must be:

$'**= [**/**{**U**+**$-**V**-**I**-**T**-**Ad**}-**Q**] /[[1+**s**](**vp**/{360[**c**-**q**]}+**f/**)]

Rule-13840:

If both (**/**), (**U**), (**$**), (**V**), (**$'**), (**Q**), (**s**), (**I**), (**T**), (**A**), (**d**), (**v**), (**p**), (**c**) and (**q**) are known, then its Fixed Portion Planned is:

f= [**U**+**$**-**V**-**I**-**T**-**Ad**-(**Q**+**$'vp**[1+**s**]/{360[**c**-**q**]})/**/**] /{**$'**[1+**s**]}

Rule-13841:

If both (**/**), (**U**), (**$**), (**V**), (**$'**), (**f**), (**Q**), (**I**), (**T**), (**A**), (**d**), (**v**), (**p**), (**c**) and (**q**) are known, then its Sales Growth Planned is:

s= **$**/**$'**-1

or it can also be found as

s= [**/**{**U**+**$**-**V**-**I**-**T**-**Ad**}-**Q**]/[**$'**(**vp**/{360[**c**-**q**]}+**f/**)]-1

Steve Asikin ISBN 14: 978-1514685136, ISBN 10: 1514685132

Rule-13842:

If both (**/**), (**U**), (**$**), (**V**), (**$'**), (**f**), (**s**), (**Q**), (**T**), (**A**), (**d**), (**v**), (**p**), (**c**) and (**q**) are known, then its Interest Expense Planned is:

I= {**U**+**$**-**V**-**$'f**[1+**s**]-**T**-**Ad**}
-(**Q**+**$'vp**[1+**s**]/{360[**c**-**q**]})/**/**

Rule-13843:

If both (**/**), (**U**), (**$**), (**V**), (**$'**), (**f**), (**s**), (**I**), (**Q**), (**A**), (**d**), (**v**), (**p**), (**c**) and (**q**) are known, then its Tax Planned is:

T= **U**+**$**-**V**-**$'f**[1+**s**]-**I**-**Ad**
-(**Q**+**$'vp**[1+**s**]/{360[**c**-**q**]})/**/**

Rule-13844:

If both (**/**), (**U**), (**$**), (**V**), (**$'**), (**f**), (**s**), (**I**), (**T**), (**Q**), (**d**), (**v**), (**p**), (**c**) and (**q**) are known, then its After Tax Income Planned is:

A= [**U**+**$**-**V**-**$'f**[1+**s**]-**I**
-(**Q**+**$'vp**[1+**s**]/{360[**c**-**q**]})/**/**]/**d**

Rule-13845:

If both (**/**), (**U**), (**$**), (**V**), (**$'**), (**f**), (**s**), (**I**), (**T**), (**A**), (**Q**), (**v**), (**p**), (**c**) and (**q**) are known, then its Dividend Payout Planned is:

d= [**U**+**$**-**V**-**$'f**[1+**s**]-**I**
-(**Q**+**$'vp**[1+**s**]/{360[**c**-**q**]})/**/**]/**A**

Steve Asikin ISBN 14: 978-1514685136, ISBN 10: 1514685132

Rule-13846:

If both (**/**), (**U**), (**$**), (**V**), (**$'**), (**f**), (**s**), (**I**), (**T**), (**A**), (**d**), (**Q**), (**p**), (**c**) and (**q**) are known, then its Variable Portion Planned is:

v= 360[**c-q**](**/**{**U**+**$-V-$'f**[1+**s**]-**I-T-Ad**}-**Q**)
/{**$'p**[1+**s**]}

Rule-13847:

If both (**/**), (**U**), (**$**), (**V**), (**$'**), (**f**), (**s**), (**I**), (**T**), (**A**), (**d**), (**v**), (**Q**), (**c**) and (**q**) are known, then its Procured Inventory Days planned:

p= 360[**c-q**](**/**{**U**+**$-V-$'f**[1+**s**]-**I-T-Ad**}-**Q**)
/{**$'v**[1+**s**]}

Rule-13848:

If both (**/**), (**U**), (**$**), (**V**), (**$'**), (**f**), (**s**), (**I**), (**T**), (**A**), (**d**), (**v**), (**p**), (**Q**) and (**q**) are known, then its Current Ratio Planned is:

c= **q**+**$'vp**[1+**s**]
/[360(**/**{**U**+**$-V-$'f**[1+**s**]-**I-T-Ad**}-**Q**)]

Rule-13849:

If both (**/**), (**U**), (**$**), (**V**), (**$'**), (**f**), (**s**), (**I**), (**T**), (**A**), (**d**), (**v**), (**p**), (**c**) and (**Q**) are known, then its Quick or Acid Test Ratio Planned is:

q= **c-$'vp**[1+**s**]
/[360(**/**{**U**+**$-V-$'f**[1+**s**]-**I-T-Ad**}-**Q**)]

Steve Asikin ISBN 14: 978-1514685136, ISBN 10: 1514685132

Rule-13850:

If both (**/**), (**U**), (**$**), (**V**), (**$'**), (**f**), (**s**), (**I**), (**t**), (**D**) and (**X**) are known, then its Quoted Longterm Debt Planned is:

Q= **/**(**U**+{**$**-**V**-**$'f**[1+**s**]-**I**}[1-**t**]-**D**)-**X**

Rule-13851:

If both (**Q**), (**U**), (**$**), (**V**), (**$'**), (**f**), (**s**), (**I**), (**t**), (**D**) and (**X**) are known, then its Leverage or Gearing Ratio Planned is:

/= [**Q**+**X**]/(**U**+{**$**-**V**-**$'f**[1+**s**]-**I**}[1-**t**]-**D**)

Rule-13852:

If both (**/**), (**Q**), (**$**), (**V**), (**$'**), (**f**), (**s**), (**I**), (**t**), (**D**) and (**X**) are known, then its Utilized or Starting Capital must be:

U= [**Q**+**X**]/**/**-{**$**-**V**-**$'f**[1+**s**]-**I**}[1-**t**]+**D**

Rule-13853:

If both (**/**), (**U**), (**Q**), (**V**), (**$'**), (**f**), (**s**), (**I**), (**t**), (**D**) and (**X**) are known, then its Sales or Revenue Planned is:

$**= **V**+**$'f[1+**s**]+**I**+{**D**+[**Q**+**X**]/**/**-**U**}/[1-**t**]

Rule-13854:

If both (**/**), (**U**), (**$**), (**Q**), (**$'**), (**f**), (**s**), (**I**), (**t**), (**D**) and (**X**) are known, then its Variable Cost Planned is:

V= {**$**-**V**-**$'f**[1+**s**]-**I**}-{**D**+[**Q**+**X**]/**/**-**U**}/[1-**t**]

Steve Asikin ISBN 14: 978-1514685136, ISBN 10: 1514685132

Rule-13855:

If both (/), (U), ($), (V), (Q), (f), (s), (I), (t), (D) and (X) are known, then its Sales Past must be:

$'= ($-V-I-{D+[Q+X]/-U}/[1-t])/{f[1+s]}

Rule-13856:

If both (/), (U), ($), (V), ($'), (Q), (s), (I), (t), (D) and (X) are known, then its Fixed Portion Planned is:

f= ($-V-I-{D+[Q+X]/-U}/[1-t])/{$'[1+s]}

Rule-13857:

If both (/), (U), ($), (V), ($'), (f), (Q), (I), (t), (D) and (X) are known, then its Sales Growth Planned is:

s= $/$'-1

or it can also be found as

s= ($-V-I-{D+[Q+X]/-U}/[1-t])/[$'f]-1

Rule-13858:

If both (/), (U), ($), (V), ($'), (f), (s), (Q), (t), (D) and (X) are known, then its Interest Expense Planned is:

I= {$-V-$'f[1+s]-I}-{D+[Q+X]/-U}/[1-t]

Rule-13859:

If both (/), (U), ($), (V), ($'), (f), (s), (I), (Q), (D) and (X) are known, then its Tax Rate Planned is:

t= 1-{D+[Q+X]/-U}/{$-V-$'f[1+s]-I}

Rule-13860:

If both (/), (U), ($), (V), ($'), (f), (s), (I), (t), (Q) and (X) are known, then its Dividend Planned is:

D= U+{$-V-$'f[1+s]-I}[1-t]-[Q+X]//

Steve Asikin ISBN 14: 978-1514685136, ISBN 10: 1514685132

Rule-13861:

If both (/), (U), ($), (V), ($'), (f), (s), (I), (t), (D) and (Q) are known, then its Xpress or Current Debt Planned is:

X= /(U+{$-V-$'f[1+s]-I}[1-t]-D)-Q

Rule-13862:

If both (/), (U), ($), (V), ($'), (f), (s), (I), (t), (D), (P), (c) and (q) are known, then its Quoted Longterm Debt Planned is:

Q= /(U+{$-V-$'f[1+s]-I}[1-t]-D)-P/[c-q]

Rule-13863:

If both (Q), (U), ($), (V), ($'), (f), (s), (I), (t), (D), (P), (c) and (q) are known, then its Leverage or Gearing Ratio Planned is:

/= {Q+P/[c-q]}/(U+{$-V-$'f[1+s]-I}[1-t]-D)

Rule-13864:

If both (/), (Q), ($), (V), ($'), (f), (s), (I), (t), (D), (P), (c) and (q) are known, then its Utilized or Starting Capital Planned is:

U= {Q+P/[c-q]}//-{$-V-$'f[1+s]-I}[1-t]+D

Rule-13865:

If both (/), (U), (Q), (V), ($'), (f), (s), (I), (t), (D), (P), (c) and (q) are known, then its Sales or Revenue Planned is:

$= V+$'f[1+s]+I+(D+{Q+P/[c-q]}//-U)/[1-t]

Steve Asikin ISBN 14: 978-1514685136, ISBN 10: 1514685132

Rule-13866:

If both (**/**), (**U**), (**$**), (**Q**), (**$'**), (**f**), (**s**), (**I**), (**t**), (**D**), (**P**), (**c**) and (**q**) are known, then its Variable Cost Planned is:

V= {$-V-$'f[1+s]-I}-(D+{Q+P/[c-q]}/ /-U)/[1-t]

Rule-13867:

If both (**/**), (**U**), (**$**), (**V**), (**Q**), (**f**), (**s**), (**I**), (**t**), (**D**), (**P**), (**c**) and (**q**) are known, then its Sales Past must be:

$'= [$-V-I-(D+{Q+P/[c-q]}/ /-U)/[1-t]]/{f[1+s]}

Rule-13868:

If both (**/**), (**U**), (**$**), (**V**), (**$'**), (**Q**), (**s**), (**I**), (**t**), (**D**), (**P**), (**c**) and (**q**) are known, then its Fixed Portion Planned is:

f= [$-V-I-(D+{Q+P/[c-q]}/ /-U)/[1-t]]/{$'[1+s]}

Rule-13869:

If both (**/**), (**U**), (**$**), (**V**), (**$'**), (**f**), (**Q**), (**I**), (**t**), (**D**), (**P**), (**c**) and (**q**) are known, then its Sales Growth Planned is:

s= $/$'-1

or it can also be found as

s= [$-V-I-(D+{Q+P/[c-q]}/ /-U)/[1-t]]/[$'f]-1

Rule-13870:

If both (**/**), (**U**), (**$**), (**V**), (**$'**), (**f**), (**s**), (**Q**), (**t**), (**D**), (**P**), (**c**) and (**q**) are known, then its Interest Expense Planned is:

I= {$-V-$'f[1+s]-I}-(D+{Q+P/[c-q]}/ /-U)/[1-t]

Steve Asikin ISBN 14: 978-1514685136, ISBN 10: 1514685132

Rule-13871:

If both (**/**), (**U**), (**$**), (**V**), (**$'**), (**f**), (**s**), (**I**), (**Q**), (**D**), (**P**), (**c**) and (**q**) are known, then its Tax Rate Planned is:

t= 1-(**D**+{**Q**+**P**/[**c**-**q**]}/**/**-**U**)/[1-**t**]]/**$**-**V**-**$'f**[1+**s**]-**I**}

Rule-13872:

If both (**/**), (**U**), (**$**), (**V**), (**$'**), (**f**), (**s**), (**I**), (**t**), (**Q**), (**P**), (**c**) and (**q**) are known, then its Dividend Planned is:

D= **U**+{**$**-**V**-**$'f**[1+**s**]-**I**}[1-**t**]-{**Q**+**P**/[**c**-**q**]}/**/**

Rule-13873:

If both (**/**), (**U**), (**$**), (**V**), (**$'**), (**f**), (**s**), (**I**), (**t**), (**D**), (**Q**), (**c**) and (**q**) are known, then its Procured Inventory Planned is:

P= [**c**-**q**](**/**(**U**+{**$**-**V**-**$'f**[1+**s**]-**I**}[1-**t**]-**D**)-**Q**)

Rule-13874:

If both (**/**), (**U**), (**$**), (**V**), (**$'**), (**f**), (**s**), (**I**), (**t**), (**D**), (**P**), (**Q**) and (**q**) are known, then its Current Ratio Planned is:

c= **q**+**P**/(**/**(**U**+{**$**-**V**-**$'f**[1+**s**]-**I**}[1-**t**]-**D**)-**Q**)

Rule-13875:

If both (**/**), (**U**), (**$**), (**V**), (**$'**), (**f**), (**s**), (**I**), (**t**), (**D**), (**P**), (**c**) and (**Q**) are known, then its Quick or Acid Test Ratio Planned is:

q= **c**-**P**/(**/**(**U**+{**$**-**V**-**$'f**[1+**s**]-**I**}[1-**t**]-**D**)-**Q**)

Steve Asikin ISBN 14: 978-1514685136, ISBN 10: 1514685132

Rule-13876:

If both (/), (U), (S), (V), (S'), (f), (s), (I), (t), (D), (p), (c) and (q) are known, then its Quoted Longterm Debt Planned is:

Q= /(U+{S-V-S'f[1+s]-I}[1-t]-D)-Vp/{360[c-q]}

Rule-13877:

If both (Q), (U), (S), (V), (S'), (f), (s), (I), (t), (D), (p), (c) and (q) are known, then its Leverage or Gearing Ratio Planned is:

/= (Q+Vp/{360[c-q]})
/(U+{S-V-S'f[1+s]-I}[1-t]-D)

Rule-13878:

If both (/), (U), (S), (V), (S'), (f), (s), (I), (t), (D), (p), (c) and (q) are known, then its Quoted Longterm Debt Planned is:

U= (Q+Vp/{360[c-q]})//-{S-V-S'f[1+s]-I}[1-t]+D

Rule-13879:

If both (/), (U), (Q), (V), (S'), (f), (s), (I), (t), (D), (p), (c) and (q) are known, then its Sales or Revenue Planned is:

S= V+S'f[1+s]+I
+[D+(Q+Vp/{360[c-q]})//-U]/[1-t]

Steve Asikin ISBN 14: 978-1514685136, ISBN 10: 1514685132

Rule-13880:

If both (/), (U), (Q), (Q), ($'), (f), (s), (I), (t), (D), (p), (c) and (q) are known, then its Variable Cost Planned is:

V= [/(U+{$-$'f[1+s]-I}[1-t]-D)-Q]
/(p/{360[c-q]}+/[1-t])

Rule-13881:

If both (/), (U), (Q), (V), (Q), (f), (s), (I), (t), (D), (p), (c) and (q) are known, then its Sales Past must be:

$'= {$-V-I-[D+(Q+Vp/{360[c-q]})/-U]/[1-t]}
/{f[1+s]}

Rule-13882:

If both (/), (U), (Q), (V), ($'), (Q), (s), (I), (t), (D), (p), (c) and (q) are known, then its Fixed Portion Planned is:

f= {$-V-I-[D+(Q+Vp/{360[c-q]})/-U]/[1-t]}
/{$'[1+s]}

Rule-13883:

If both (/), (U), (Q), (V), ($'), (f), (Q), (I), (t), (D), (p), (c) and (q) are known, then its Sales Growth Planned is:

s= $/$'-1

or it can also be found as

s={$-V-I-[D+(Q+Vp/{360[c-q]})/-U]/[1-t]}
/[$'f]-1

Steve Asikin ISBN 14: 978-1514685136, ISBN 10: 1514685132

Rule-13884:

If both (**A**), (**U**), (**Q**), (**V**), (**S'**), (**f**), (**s**), (**Q**), (**t**), (**D**), (**p**), (**c**) and (**q**) are known, then its Interest Expense Planned is:

I= {**S**-**V**-**S'f**[1+**s**]-**I**}-[**D**+(**Q**+**Vp**/{360[**c**-**q**]})/**A**-**U**]
/[1-**t**]

Rule-13885:

If both (**A**), (**U**), (**Q**), (**V**), (**S'**), (**f**), (**s**), (**I**), (**Q**), (**D**), (**p**), (**c**) and (**q**) are known, then its Tax Rate Planned is:

t= 1-[**D**+(**Q**+**Vp**/{360[**c**-**q**]})/**A**-**U**]/[1-**t**]]/**S**
-**V**-**S'f**[1+**s**]-**I**}

Rule-13886:

If both (**A**), (**U**), (**Q**), (**V**), (**S'**), (**f**), (**s**), (**I**), (**t**), (**Q**), (**p**), (**c**) and (**q**) are known, then its Dividend Planned is:

D= **U**+{**S**-**V**-**S'f**[1+**s**]-**I**}[1-**t**]-(**Q**+**Vp**/{360[**c**-**q**]})/**A**

Rule-13887:

If both (**A**), (**U**), (**Q**), (**V**), (**S'**), (**f**), (**s**), (**I**), (**t**), (**D**), (**Q**), (**c**) and (**q**) are known, then its Procured Inventory Days Planned:

p= 360[**c**-**q**](**AU**+{**S**-**V**-**S'f**[1+**s**]-**I**}[1-**t**]-**D**)-**Q**)/**V**

Rule-13888:

If both (**A**), (**U**), (**Q**), (**V**), (**S'**), (**f**), (**s**), (**I**), (**t**), (**D**), (**p**), (**Q**) and (**q**) are known, then its Current Ratio Planned is:

c= **q**+**Vp**/[360(**AU**+{**S**-**V**-**S'f**[1+**s**]-**I**}[1-**t**]-**D**)-**Q**)]

Steve Asikin ISBN 14: 978-1514685136, ISBN 10: 1514685132

Rule-13889:

If both (/), (U), (Q), (V), (S'), (f), (s), (I), (t), (D), (p), (c) and (Q) are known, then its Quick or Acid Test Ratio Planned is:

q= c-Vp/[360(/(U+{S-V-S'f[1+s]-I}[1-t]-D)-Q)]

Rule-13890:

If both (/), (U), (Q), (V), (S'), (f), (s), (I), (t), (D), (p), (v), (c) and (q) are known, then its Quoted Longterm Debt Planned is:

Q= /(U+{S-V-S'f[1+s]-I}[1-t]-D)-Svp/{360[c-q]}

Rule-13891:

If both (Q), (U), (S), (V), (S'), (f), (s), (I), (t), (D), (p), (v), (c) and (q) are known, then its Leverage or Gearing Ratio Planned is:

/= (Q+Svp/{360[c-q]})
/(U+{S-V-S'f[1+s]-I}[1-t]-D)

Rule-13892:

If both (/), (Q), (S), (V), (S'), (f), (s), (I), (t), (D), (p), (v), (c) and (q) are known, then its Utilized or Starting Capital must be:

U= (Q+Svp/{360[c-q]})//-{S-V-S'f[1+s]-I}[1-t]+D

Rule-13893:

If both (/), (U), (Q), (V), (S'), (f), (s), (I), (t), (D), (p), (v), (c) and (q) are known, then its Sales or Revenue Planned is:

S= [/(U-{V+S'f[1+s]+I}[1-t]-D)-Q]
/(vp/{360[c-q]}-/[1-t])

Steve Asikin ISBN 14: 978-1514685136, ISBN 10: 1514685132

Rule-13894:

If both (/), (U), (S), (Q), (S'), (f), (s), (I), (t), (D), (p), (v), (c) and (q) are known, then its Quoted Variable Cost Planned is:

V= [/U+{S-S'f[1+s]-I}[1-t]-D)
-Svp/{360[c-q]}-Q]/{/[1-t]}

Rule-13895:

If both (/), (U), (S), (V), (Q), (f), (s), (I), (t), (D), (p), (v), (c) and (q) are known, then its Sales Past must be:

S'= {S-V-I-[D+(Q+Svp/{360[c-q]})/-U]
/[1-t]}/{f[1+s]}

Rule-13896:

If both (/), (U), (S), (V), (S'), (Q), (s), (I), (t), (D), (p), (v), (c) and (q) are known, then its Fixed Portion Planned is:

f= {S-V-I-[D+(Q+Svp/{360[c-q]})/-U]/[1-t]}
/{S'[1+s]}

Rule-13897:

If both (/), (U), (S), (V), (S'), (f), (Q), (I), (t), (D), (p), (v), (c) and (q) are known, then its Sales Growth Planned is:

s= S/S'-1

or it can also be found as

s={S-V-I-[D+(Q+Svp/{360[c-q]})/-U]/[1-t]}
/[S'f]-1

Steve Asikin ISBN 14: 978-1514685136, ISBN 10: 1514685132

Rule-13898:

If both (/), (U), (S), (V), (S'), (f), (s), (Q), (t), (D), (p), (v), (c) and (q) are known, then its Interest Expense Planned is:

I= {S-V-S'f[1+s]-I}
-[D+(Q+Svp/{360[c-q]})/-U]/[1-t]

Rule-13899:

If both (/), (U), (S), (V), (S'), (f), (s), (I), (Q), (D), (p), (v), (c) and (q) are known, then its Tax Rate Planned is:

t= 1-[D+(Q+Svp/{360[c-q]})/-U]/[1-t]]/S
-V-S'f[1+s]-I}

Rule-13900:

If both (/), (U), (S), (V), (S'), (f), (s), (I), (t), (Q), (p), (v), (c) and (q) are known, then its Dividend Planned is:

D= U+{S-V-S'f[1+s]-I}[1-t]-(Q+Svp/{360[c-q]})//

Rule-13901:

If both (/), (U), (S), (V), (S'), (f), (s), (I), (t), (D), (Q), (p), (c) and (q) are known, then its Variable Portion Planned is:

v= V/S

or it can also be found as

v= 360[c-q](/U+{S-V-S'f[1+s]-I}[1-t]-D)-Q)/[Sp]

Steve Asikin ISBN 14: 978-1514685136, ISBN 10: 1514685132

Rule-13902:

If both (/), (U), (S), (V), (S'), (f), (s), (I), (t), (D), (Q), (v), (c) and (q) are known, then its Procured Inventory Days Planned:

p= 360[c-q](/U+{S-V-S'f[1+s]-I}[1-t]-D)-Q)/[Sv]

Rule-13903:

If both (/), (U), (S), (V), (S'), (f), (s), (I), (t), (D), (p), (v), (Q) and (q) are known, then its Current Ratio Planned is:

c= q+Svp/[360(/U+{S-V-S'f[1+s]-I}[1-t]-D)-Q)]

Rule-13904:

If both (/), (U), (S), (V), (S'), (f), (s), (I), (t), (D), (p), (v), (c) and (Q) are known, then its Quick or Acid Test Ratio Planned is:

q= c-Svp/[360(/U+{S-V-S'f[1+s]-I}[1-t]-D)-Q)]

Rule-13905:

If both (/), (U), (S), (V), (S'), (f), (s), (I), (t), (D), (v), (p), (c) and (q) are known, then its Quoted Longterm Debt Planned is:

Q= /U+{S-V-S'f[1+s]-I}[1-t]-D)
-S'vp[1+s]/{360[c-q]}

Rule-13906:

If both (Q), (U), (Q), (V), (S'), (f), (s), (I), (t), (D), (v), (p), (c) and (q) are known, then its Leverage or Gearing Ratio Planned is:

/= (Q+S'vp[1+s]/{360[c-q]})
/(U+{S-V-S'f[1+s]-I}[1-t]-D)

Steve Asikin ISBN 14: 978-1514685136, ISBN 10: 1514685132

Rule-13907:

If both (**/**), (**Q**), (**$**), (**V**), (**$'**), (**f**), (**s**), (**I**), (**t**), (**D**), (**v**), (**p**), (**c**) and (**q**) are known, then its Utilized or Starting Capital must be:

U= (**Q**+**$'vp**[1+**s**]/{360[**c-q**]})/**/**
-{**$-V-$'f**[1+**s**]-**I**}[1-**t**]+**D**

Rule-13908:

If both (**/**), (**U**), (**Q**), (**V**), (**$'**), (**f**), (**s**), (**I**), (**t**), (**D**), (**v**), (**p**), (**c**) and (**q**) are known, then its Sales or Revenue Planned is:

$**= **[Q-/(U**-{**V**+**$'f[1+**s**]+**I**}[1-**t**]-**D**)
+**$'vp**[1+**s**]/{360[**c-q**]}/{**/**[1-**t**]}

Rule-13909:

If both (**/**), (**U**), (**$**), (**Q**), (**$'**), (**f**), (**s**), (**I**), (**t**), (**D**), (**v**), (**p**), (**c**) and (**q**) are known, then its Variable Cost Planned is:

V= **[/(U**+{**$-$'f**[1+**s**]-**I**}[1-**t**]-**D**)
-**$'vp**[1+**s**]/{360[**c-q**]}-**Q]**/ {**/**[1-**t**]}

Rule-13910:

If both (**/**), (**U**), (**$**), (**V**), (**Q**), (**f**), (**s**), (**I**), (**t**), (**D**), (**v**), (**p**), (**c**) and (**q**) are known, then its Sales Past must be:

$'**= **[/(U**+{**$-V-I}[1-**t**]-**D**)-**Q]**
/**[**[1+**s**](**vp**/{360[**c-q**]}-**f/**[1-**t**])**]**

Steve Asikin ISBN 14: 978-1514685136, ISBN 10: 1514685132

Rule-13911:

If both (**/**), (**U**), (**S**), (**V**), (**S'**), (**Q**), (**s**), (**I**), (**t**), (**D**), (**v**), (**p**), (**c**) and (**q**) are known, then its Fixed Portion Planned is:

f= {**S**-**V**-**I**-[**D**+(**Q**+**S'vp**[1+**s**]/{360[**c**-**q**]})/**/**-**U**]/[1-**t**]}/{**S'**[1+**s**]}

Rule-13912:

If both (**/**), (**U**), (**S**), (**V**), (**S'**), (**f**), (**Q**), (**I**), (**t**), (**D**), (**v**), (**p**), (**c**) and (**q**) are known, then its Sales Growth Planned is:

s= **S**/**S'**-1

or it can also be found as

s=[**/U**+{**S**-**V**-**I**}[1-**t**]-**D**)-**Q**]/[**S'**(**vp**/{360[**c**-**q**]}-**f/**[1-**t**])]-1

Rule-13913:

If both (**/**), (**U**), (**S**), (**V**), (**S'**), (**f**), (**s**), (**Q**), (**t**), (**D**), (**v**), (**p**), (**c**) and (**q**) are known, then its Interest Expense Planned is:

I= {**S**-**V**-**S'f**[1+**s**]-**I**}-[**D**+(**Q**+**S'vp**[1+**s**]/{360[**c**-**q**]})/**/**-**U**]/[1-**t**]

Rule-13914:

If both (**/**), (**U**), (**S**), (**V**), (**S'**), (**f**), (**s**), (**I**), (**Q**), (**D**), (**v**), (**p**), (**c**) and (**q**) are known, then its Tax Rate Planned is:

t= 1-[**D**+(**Q**+**S'vp**[1+**s**]/{360[**c**-**q**]})/**/**-**U**]/[1-**t**]]/**S**-**V**-**S'f**[1+**s**]-**I**}

Steve Asikin ISBN 14: 978-1514685136, ISBN 10: 1514685132

Rule-13915:

If both (**/**), (**U**), (**$**), (**V**), (**$'**), (**f**), (**s**), (**I**), (**t**), (**Q**), (**v**), (**p**), (**c**) and (**q**) are known, then its Dividend Planned is:

D= **U**+{**$**-**V**-**$'f**[1+**s**]-**I**}[1-**t**]
-(**Q**+**$'vp**[1+**s**]/{360[**c**-**q**]})/**/**

Rule-13916:

If both (**/**), (**U**), (**$**), (**V**), (**$'**), (**f**), (**s**), (**I**), (**t**), (**D**), (**Q**), (**p**), (**c**) and (**q**) are known, then its Variable Portion Planned is:

v=**V**/**$**

or it can also be found as

v= 360[**c**-**q**](**/U**+{**$**-**V**-**$'f**[1+**s**]-**I**}[1-**t**]-**D**)-**Q**)
/{**$'p**[1+**s**]}

Rule-13917:

If both (**/**), (**U**), (**$**), (**V**), (**$'**), (**f**), (**s**), (**I**), (**t**), (**D**), (**v**), (**Q**), (**c**) and (**q**) are known, then its Procured Inventory Days Planned:

p= 360[**c**-**q**](**/U**+{**$**-**V**-**$'f**[1+**s**]-**I**}[1-**t**]-**D**)-**Q**)
/{**$'v**[1+**s**]}

Rule-13918:

If both (**/**), (**U**), (**$**), (**V**), (**$'**), (**f**), (**s**), (**I**), (**t**), (**D**), (**p**), (**v**), (**Q**) and (**q**) are known, then its Current Ratio Planned is:

c= **q**+**$'vp**[1+**s**]
/[360(**/U**+{**$**-**V**-**$'f**[1+**s**]-**I**}[1-**t**]-**D**)-**Q**)]

Steve Asikin ISBN 14: 978-1514685136, ISBN 10: 1514685132

Rule-13919:

If both (**/**), (**U**), (**$**), (**V**), (**$'**), (**f**), (**s**), (**I**), (**t**), (**D**), (**p**), (**v**), (**c**) and (**Q**) are known, then its Quick or Acid Test Ratio Planned is:

q= **c**-**$'vp**[1+**s**]
/[360(**/U**+{**$**-**V**-**$'f**[1+**s**]-**I**}[1-**t**]-**D**)-**Q**)]

Rule-13920:

If both (**/**), (**U**), (**$**), (**V**), (**$'**), (**f**), (**s**), (**I**), (**t**), (**A**), (**d**) and (**X**) are known, then its Quoted Longterm Debt Planned is:

Q= **/**(**U**+{**$**-**V**-**$'f**[1+**s**]-**I**}[1-**t**]-**Ad**)-**X**

Rule-13921:

If both (**Q**), (**U**), (**$**), (**V**), (**$'**), (**f**), (**s**), (**I**), (**t**), (**D**), (**p**), (**v**), (**c**) and (**q**) are known, then its Leverage or Gearing Ratio Planned is:

/= [**Q**+**X**]/(**U**+{**$**-**V**-**$'f**[1+**s**]-**I**}[1-**t**]-**Ad**)

Rule-13922:

If both (**/**), (**Q**), (**$**), (**V**), (**$'**), (**f**), (**s**), (**I**), (**t**), (**A**), (**d**) and (**X**) are known, then its Utilized or Starting Capital must be:

U= **Ad**+[**Q**+**X**]/**/**-{**$**-**V**-**$'f**[1+**s**]-**I**}[1-**t**]

Rule-13923:

If both (**/**), (**U**), (**Q**), (**V**), (**$'**), (**f**), (**s**), (**I**), (**t**), (**A**), (**d**) and (**X**) are known, then its Sales or Revenue Planned is:

$**= **V**+**$'f[1+**s**]+**I**+{**Ad**+[**Q**+**X**]/**/**-**U**}/[1-**t**]

Steve Asikin ISBN 14: 978-1514685136, ISBN 10: 1514685132

Rule-13924:

If both (/), (U), (S), (Q), (S'), (f), (s), (I), (t), (A), (d) and (X) are known, then its Variable Cost Planned is:

V= {S-S'f[1+s]-I}-{Ad+[Q+X]/ /U}/[1-t]

Rule-13925:

If both (/), (U), (S), (V), (Q), (f), (s), (I), (t), (A), (d) and (X) are known, then its Sales Past must be:

S'= (S-V-I-{Ad+[Q+X]/ /U}/[1-t])/{f[1+s]}

Rule-13926:

If both (/), (U), (S), (V), (S'), (Q), (s), (I), (t), (A), (d) and (X) are known, then its Fixed Cost Planned is:

f= (S-V-I-{Ad+[Q+X]/ /U}/[1-t])/{S'[1+s]}

Rule-13927:

If both (/), (U), (S), (V), (S'), (f), (Q), (I), (t), (A), (d) and (X) are known, then its Sales Growth Planned is:

s= S/S'-1

or it can also be found as

s= (S-V-I-{Ad+[Q+X]/ /U}/[1-t])/[S'f]-1

Rule-13928:

If both (/), (U), (S), (V), (S'), (f), (s), (Q), (t), (A), (d) and (X) are known, then its Interest Expense Planned is:

I= {S-V-S'f[1+s]}-{Ad+[Q+X]/ /U}/[1-t]

Steve Asikin ISBN 14: 978-1514685136, ISBN 10: 1514685132

Rule-13929:

If both (/), (U), ($), (V), ($'), (f), (s), (I), (Q), (A), (d) and (X) are known, then its Tax Planned is:

t= 1-{Ad+[Q+X]/ /-U}/{$-V-$'f[1+s]-I}

Rule-13930:

If both (/), (U), ($), (V), ($'), (f), (s), (I), (t), (Q), (d) and (X) are known, then its After Tax Income Planned is:

A= (U+{$-V-$'f[1+s]-I}[1-t]-[Q+X]/ /)/d

Rule-13931:

If both (/), (U), ($), (V), ($'), (f), (s), (I), (t), (A), (Q) and (X) are known, then its Dividend Payout Planned is:

d= (U+{$-V-$'f[1+s]-I}[1-t]-[Q+X]/ /)/A

Rule-13932:

If both (/), (U), ($), (V), ($'), (f), (s), (I), (t), (A), (d) and (Q) are known, then its Xpress or Current Debt Planned is:

X= /(U+{$-V-$'f[1+s]-I}[1-t]-Ad)-Q

Rule-13933:

If both (/), (U), ($), (V), ($'), (f), (s), (I), (t), (A), (d), (P), (c) and (q) are known, then its Quoted Longterm Debt Planned is:

Q= /(U+{$-V-$'f[1+s]-I}[1-t]-Ad)-P/[c-q]

Steve Asikin ISBN 14: 978-1514685136, ISBN 10: 1514685132

Rule-13934:

If both (**Q**), (**U**), (**$**), (**V**), (**$'**), (**f**), (**s**), (**I**), (**t**), (**A**), (**d**), (**P**), (**c**) and (**q**) are known, then its Leverage or Gearing Ratio Planned is:

/= {Q+P/[c-q]}/(U+{$-V-$'f[1+s]-I}[1-t]-Ad)

Rule-13935:

If both (**/**), (**Q**), (**$**), (**V**), (**$'**), (**f**), (**s**), (**I**), (**t**), (**A**), (**d**), (**P**), (**c**) and (**q**) are known, then its Utilized or Starting Capital must be:

U= Ad+{Q+P/[c-q]}/ /-{$-V-$'f[1+s]-I}[1-t]

Rule-13936:

If both (**/**), (**U**), (**Q**), (**V**), (**$'**), (**f**), (**s**), (**I**), (**t**), (**A**), (**d**), (**P**), (**c**) and (**q**) are known, then its Sales or Revenue Planned is:

$= V+$'f[1+s]+I+(Ad+{Q+P/[c-q]}/ /-U)/[1-t]

Rule-13937:

If both (**/**), (**U**), (**$**), (**Q**), (**$'**), (**f**), (**s**), (**I**), (**t**), (**A**), (**d**), (**P**), (**c**) and (**q**) are known, then its Variable Cost Planned is:

V= {$-$'f[1+s]-I}-(Ad+{Q+P/[c-q]}/ /-U)/[1-t]

Rule-13938:

If both (**/**), (**U**), (**$**), (**V**), (**Q**), (**f**), (**s**), (**I**), (**t**), (**A**), (**d**), (**P**), (**c**) and (**q**) are known, then its Sales Past must be:

$'= [$-V-I-(Ad+{Q+P/[c-q]}/ /-U)/[1-t]]/{f[1+s]}

Steve Asikin ISBN 14: 978-1514685136, ISBN 10: 1514685132

Rule-13939:

If both (**/**), (**U**), (**$**), (**V**), (**$'**), (**Q**), (**s**), (**I**), (**t**), (**A**), (**d**), (**P**), (**c**) and (**q**) are known, then its Fixed Portion Planned is:

f= [**$**-**V**-**I**-(**Ad**+{**Q**+**P**/[**c**-**q**]}/**/U**)/[1-**t**]]/{**$'**[1+**s**]}

Rule-13940:

If both (**/**), (**U**), (**$**), (**V**), (**$'**), (**f**), (**Q**), (**I**), (**t**), (**A**), (**d**), (**P**), (**c**) and (**q**) are known, then its Sales Growth Planned is:

s= **$**/**$'**-1

or it can also be found as

s= [**$**-**V**-**I**-(**Ad**+{**Q**+**P**/[**c**-**q**]}/**/U**)/[1-**t**]]/[**$'f**]-1

Rule-13941:

If both (**/**), (**U**), (**$**), (**V**), (**$'**), (**f**), (**s**), (**Q**), (**t**), (**A**), (**d**), (**P**), (**c**) and (**q**) are known, then its Interest Expense Planned is:

I= {**$**-**V**-**$'f**[1+**s**]}-(**Ad**+{**Q**+**P**/[**c**-**q**]}/**/U**)/[1-**t**]

Rule-13942:

If both (**/**), (**U**), (**$**), (**V**), (**$'**), (**f**), (**s**), (**I**), (**Q**), (**A**), (**d**), (**P**), (**c**) and (**q**) are known, then its Tax Rate Planned is:

t= 1-(**Ad**+{**Q**+**P**/[**c**-**q**]}/**/U**)/{**$**-**V**-**$'f**[1+**s**]-**I**}

Rule-13943:

If both (**/**), (**U**), (**$**), (**V**), (**$'**), (**f**), (**s**), (**I**), (**t**), (**Q**), (**d**), (**P**), (**c**) and (**q**) are known, then its After Tax Income Planned is:

A= [**U**+{**$**-**V**-**$'f**[1+**s**]-**I**}[1-**t**]-{**Q**+**P**/[**c**-**q**]}/**/**]/**d**

Steve Asikin ISBN 14: 978-1514685136, ISBN 10: 1514685132

Rule-13944:

If both (**/**), (**U**), (**\$**), (**V**), (**\$'**), (**f**), (**s**), (**I**), (**t**), (**A**), (**Q**), (**P**), (**c**) and (**q**) are known, then its Dividend Payout Planned is:

d= [**U**+{**\$**-**V**-**\$'f**[1+**s**]-**I**}[1-**t**]-{**Q**+**P**/[**c**-**q**]}/**/**]/**A**

Rule-13945:

If both (**/**), (**U**), (**\$**), (**V**), (**\$'**), (**f**), (**s**), (**I**), (**t**), (**A**), (**d**), (**Q**), (**c**) and (**q**) are known, then its Procured Inventory Days planned:

P= [**c**-**q**][**/**(**U**+{**\$**-**V**-**\$'f**[1+**s**]-**I**}[1-**t**]-**Ad**)-**Q**]

Rule-13946:

If both (**/**), (**U**), (**\$**), (**V**), (**\$'**), (**f**), (**s**), (**I**), (**t**), (**A**), (**d**), (**P**), (**Q**) and (**q**) are known, then its Current Ratio Planned is:

c= **q**+**P**/[**/**(**U**+{**\$**-**V**-**\$'f**[1+**s**]-**I**}[1-**t**]-**Ad**)-**Q**]

Rule-13947:

If both (**/**), (**U**), (**\$**), (**V**), (**\$'**), (**f**), (**s**), (**I**), (**t**), (**A**), (**d**), (**P**), (**c**) and (**Q**) are known, then its Quick or Acid Test Ratio Planned is:

q= **c**-**P**/[**/**(**U**+{**\$**-**V**-**\$'f**[1+**s**]-**I**}[1-**t**]-**Ad**)-**Q**]

Rule-13948:

If both (**/**), (**U**), (**\$**), (**V**), (**\$'**), (**f**), (**s**), (**I**), (**t**), (**A**), (**d**), (**p**), (**c**) and (**q**) are known, then its Quoted Longterm Debt Planned is:

Q= **/**(**U**+{**\$**-**V**-**\$'f**[1+**s**]-**I**}[1-**t**]-**Ad**)-**Vp**/{360[**c**-**q**]}

Steve Asikin ISBN 14: 978-1514685136, ISBN 10: 1514685132

Rule-13949:

If both (**Q**), (**U**), (**$**), (**V**), (**$'**), (**f**), (**s**), (**I**), (**t**), (**A**), (**d**), (**p**), (**c**) and (**q**) are known, then its Leverage or Gearing Ratio Planned is:

/= (Q+Vp/{360[c-q]})
/(U+{$-V-$'f[1+s]-I}[1-t]-Ad)

Rule-13950:

If both (**/**), (**Q**), (**$**), (**V**), (**$'**), (**f**), (**s**), (**I**), (**t**), (**A**), (**d**), (**p**), (**c**) and (**q**) are known, then its Utilized or Starting Capital must be:

U= Ad+(Q+Vp/{360[c-q]})/ /-{$-V-$'f[1+s]-I}[1-t]

Rule-13951:

If both (**/**), (**U**), (**Q**), (**V**), (**$'**), (**f**), (**s**), (**I**), (**t**), (**A**), (**d**), (**p**), (**c**) and (**q**) are known, then its Sales or Revenue Planned is:

$= V+$'f[1+s]+I+[Ad+(Q+Vp/{360[c-q]})/ /-U]
/[1-t]

Rule-13952:

If both (**/**), (**U**), (**$**), (**Q**), (**$'**), (**f**), (**s**), (**I**), (**t**), (**A**), (**d**), (**p**), (**c**) and (**q**) are known, then its Variable Cost Planned is:

V= [/(U+{$-$'f[1+s]-I}[1-t]-Ad)-Q]
/(p/{360[c-q]}+/[1-t])

Steve Asikin ISBN 14: 978-1514685136, ISBN 10: 1514685132

Rule-13953:

If both (/), (U), (S), (V), (Q), (f), (s), (I), (t), (A), (d), (p), (c) and (q) are known, then its Sales Past must be:

S'= {S-V-I-[Ad+(Q+Vp/{360[c-q]})/-U]/[1-t]} /{f[1+s]}

Rule-13954:

If both (/), (U), (S), (V), (S'), (Q), (s), (I), (t), (A), (d), (p), (c) and (q) are known, then its Fixed Portion Planned is:

f= {S-V-I-[Ad+(Q+Vp/{360[c-q]})/-U]/[1-t]} /{S'[1+s]}

Rule-13955:

If both (/), (U), (S), (V), (S'), (f), (Q), (I), (t), (A), (d), (p), (c) and (q) are known, then its Sales Growth Planned is:

s= {S-V-I-[Ad+(Q+Vp/{360[c-q]})/-U]/[1-t]} /[S'f]-1

Rule-13956:

If both (/), (U), (S), (V), (S'), (f), (s), (Q), (t), (A), (d), (p), (c) and (q) are known, then its Interest Expense Planned is:

I= {S-V-S'f[1+s]}-[Ad+(Q+Vp/{360[c-q]})/-U] /[1-t]

Steve Asikin ISBN 14: 978-1514685136, ISBN 10: 1514685132

Rule-13957:

If both (**/**), (**U**), (**S**), (**V**), (**S'**), (**f**), (**s**), (**I**), (**Q**), (**A**), (**d**), (**p**), (**c**) and (**q**) are known, then its Tax Rate Planned is:

t= 1-[**Ad**+(**Q**+**Vp**/{360[**c**-**q**]})/**/**-**U**]
/{**S**-**V**-**S'f**[1+**s**]-**I**}

Rule-13958:

If both (**/**), (**U**), (**S**), (**V**), (**S'**), (**f**), (**s**), (**I**), (**t**), (**Q**), (**d**), (**p**), (**c**) and (**q**) are known, then its After Tax Income Planned is:

A= [**U**+{**S**-**V**-**S'f**[1+**s**]-**I**}[1-**t**]
-(**Q**+**Vp**/{360[**c**-**q**]})/**/**]/**d**

Rule-13959:

If both (**/**), (**U**), (**S**), (**V**), (**S'**), (**f**), (**s**), (**I**), (**t**), (**A**), (**Q**), (**p**), (**c**) and (**q**) are known, then its Dividend Payout Planned is:

d= [**U**+{**S**-**V**-**S'f**[1+**s**]-**I**}[1-**t**]
-(**Q**+**Vp**/{360[**c**-**q**]})/**/**]/**A**

Rule-13960:

If both (**/**), (**U**), (**S**), (**V**), (**S'**), (**f**), (**s**), (**I**), (**t**), (**A**), (**d**), (**Q**), (**c**) and (**q**) are known, then its Procured Inventory Days Planned:

p= 360[**c**-**q**][**/**(**U**+{**S**-**V**-**S'f**[1+**s**]-**I**}[1-**t**]-**Ad**)-**Q**]/**V**

Steve Asikin ISBN 14: 978-1514685136, ISBN 10: 1514685132

Rule-13961:

If both (**/**), (**U**), (**$**), (**V**), (**$'**), (**f**), (**s**), (**I**), (**t**), (**A**), (**d**), (**p**), (**Q**) and (**q**) are known, then its Current Ratio Planned is:

c= **q**+**Vp**/{360[**/**(**U**+{**$**-**V**-**$'f**[1+**s**]-**I**}[1-**t**]-**Ad**)-**Q**]}

Rule-13962:

If both (**/**), (**U**), (**$**), (**V**), (**$'**), (**f**), (**s**), (**I**), (**t**), (**A**), (**d**), (**p**), (**c**) and (**Q**) are known, then its Quick or Acid Test Ratio Planned is:

q= **c**-**Vp**/{360[**/**(**U**+{**$**-**V**-**$'f**[1+**s**]-**I**}[1-**t**]-**Ad**)-**Q**]}

Rule-13963:

If both (**/**), (**U**), (**$**), (**V**), (**$'**), (**f**), (**s**), (**I**), (**t**), (**A**), (**d**), (**v**), (**p**), (**c**) and (**q**) are known, then its Quoted Longterm Debt Planned is:

Q= **/**(**U**+{**$**-**V**-**$'f**[1+**s**]-**I**}[1-**t**]-**Ad**)-**$vp**/{360[**c**-**q**]}

Rule-13964:

If both (**Q**), (**U**), (**$**), (**V**), (**$'**), (**f**), (**s**), (**I**), (**t**), (**A**), (**d**), (**v**), (**p**), (**c**) and (**q**) are known, then its Leverage or Gearing Ratio Planned is:

/= (**Q**+**$vp**/{360[**c**-**q**]})
/(**U**+{**$**-**V**-**$'f**[1+**s**]-**I**}[1-**t**]-**Ad**)

Rule-13965:

If both (**/**), (**Q**), (**$**), (**V**), (**$'**), (**f**), (**s**), (**I**), (**t**), (**A**), (**d**), (**v**), (**p**), (**c**) and (**q**) are known, then its Utilized or Sdtarting Capital must be:

U= **Ad**+(**Q**+**$vp**/{360[**c**-**q**]})/**/**
-{**$**-**V**-**$'f**[1+**s**]-**I**}[1-**t**]

Steve Asikin ISBN 14: 978-1514685136, ISBN 10: 1514685132

Rule-13966:

If both (**/**), (**U**), (**Q**), (**V**), (**$'**), (**f**), (**s**), (**I**), (**t**), (**A**), (**d**), (**v**), (**p**), (**c**) and (**q**) are known, then its Sales or Revenue Planned is:

$**= [**/U**-{**V**+**$'f[1+**s**]+**I**}[1-**t**]-**Ad**)-**Q**]
/(**vp**/{360[**c-q**]}-**/**[1-**t**])

Rule-13967:

If both (**/**), (**U**), (**$**), (**Q**), (**$'**), (**f**), (**s**), (**I**), (**t**), (**A**), (**d**), (**v**), (**p**), (**c**) and (**q**) are known, then its Variable Cost Planned is:

V= [**/U**+{**$**-**$'f**[1+**s**]-**I**}[1-**t**]-**Ad**)
-**$vp**/{360[**c-q**]}-**Q**]/{**/**[1-**t**]}

Rule-13968:

If both (**/**), (**U**), (**$**), (**V**), (**Q**), (**f**), (**s**), (**I**), (**t**), (**A**), (**d**), (**v**), (**p**), (**c**) and (**q**) are known, then its Sales Past must be:

$'**= {**$-V-I-[**Ad**+(**Q**+**$vp**/{360[**c-q**]})/**/-U**]/[1-**t**]}
/{**f**[1+**s**]}

Rule-13969:

If both (**/**), (**U**), (**$**), (**V**), (**$'**), (**Q**), (**s**), (**I**), (**t**), (**A**), (**d**), (**v**), (**p**), (**c**) and (**q**) are known, then its Quoted Fixed Portion Planned is:

f= {**$-V-I-**[**Ad**+(**Q**+**$vp**/{360[**c-q**]})/**/-U**]/[1-**t**]}
/{**$'**[1+**s**]}

Steve Asikin ISBN 14: 978-1514685136, ISBN 10: 1514685132

Rule-13970:

If both (/), (U), ($), (V), ($'), (f), (Q), (I), (t), (A), (d), (v), (p), (c) and (q) are known, then its Sales Growth Planned is:

s= $/$'-1

or it can also be found as

s= {$-V-I-[Ad+(Q+$vp/{360[c-q]})/IU]/[1-t]}/[$'f]-1

Rule-13971:

If both (/), (U), ($), (V), ($'), (f), (s), (Q), (t), (A), (d), (v), (p), (c) and (q) are known, then its Interest Expense Planned is:

I= {$-V-$'f[1+s]}-[Ad+(Q+$vp/{360[c-q]})/IU]/[1-t]

Rule-13972:

If both (/), (U), ($), (V), ($'), (f), (s), (I), (Q), (A), (d), (v), (p), (c) and (q) are known, then its Tax Rate Planned is:

t= 1-[Ad+(Q+$vp/{360[c-q]})/IU]/{$-V-$'f[1+s]-I}

Rule-13973:

If both (/), (U), ($), (V), ($'), (f), (s), (I), (t), (Q), (d), (v), (p), (c) and (q) are known, then its After Tax Income Planned is:

A= [U+{$-V-$'f[1+s]-I}[1-t]-(Q+$vp/{360[c-q]})/I]/d

Steve Asikin ISBN 14: 978-1514685136, ISBN 10: 1514685132

Rule-13974:

If both (**/**), (**U**), (**$**), (**V**), (**$'**), (**f**), (**s**), (**I**), (**t**), (**A**), (**Q**), (**v**), (**p**), (**c**) and (**q**) are known, then its Dividend Payout Planned is:

d= [**U**+{**$**-**V**-**$'f**[1+**s**]-**I**}[1-**t**]
-(**Q**+**$vp**/{360[**c**-**q**]})/**/**]/**A**

Rule-13975:

If both (**/**), (**U**), (**$**), (**V**), (**$'**), (**f**), (**s**), (**I**), (**t**), (**A**), (**d**), (**Q**), (**p**), (**c**) and (**q**) are known, then its Variabl;e Portion Planned is:

v= **V**/**$**

or it can also be found as

v= 360[**c**-**q**][**/**(**U**+{**$**-**V**-**$'f**[1+**s**]-**I**}[1-**t**]-**Ad**)-**Q**]
/[**$p**]

Rule-13976:

If both (**/**), (**U**), (**$**), (**V**), (**$'**), (**f**), (**s**), (**I**), (**t**), (**A**), (**d**), (**v**), (**Q**), (**c**) and (**q**) are known, then its Procured Inventory Days Planned:

p= 360[**c**-**q**][**/**(**U**+{**$**-**V**-**$'f**[1+**s**]-**I**}[1-**t**]-**Ad**)-**Q**]
/[**$v**]

Rule-13977:

If both (**/**), (**U**), (**$**), (**V**), (**$'**), (**f**), (**s**), (**I**), (**t**), (**A**), (**d**), (**v**), (**p**), (**Q**) and (**q**) are known, then its Current Ratio Planned is:

c= **q**+**$vp**/{360[**/**(**U**+{**$**-**V**-**$'f**[1+**s**]-**I**}[1-**t**]-**Ad**)-**Q**]}

Steve Asikin ISBN 14: 978-1514685136, ISBN 10: 1514685132

Rule-13978:

If both (**/**), (**U**), (**$**), (**V**), (**$'**), (**f**), (**s**), (**I**), (**t**), (**A**), (**d**), (**v**), (**p**), (**c**) and (**Q**) are known, then its Quick or Acid Test Ratio Planned is:

q= **c**-**$vp**/{360[**/**(**U**+{**$**-**V**-**$'f**[1+**s**]-**I**}[1-**t**]-**Ad**)-**Q**]}

Rule-13979:

If both (**/**), (**U**), (**$**), (**V**), (**$'**), (**f**), (**s**), (**I**), (**t**), (**A**), (**d**), (**v**), (**p**), (**c**) and (**q**) are known, then its Quoted Longterm Debt Planned is:

Q= **/**(**U**+{**$**-**V**-**$'f**[1+**s**]-**I**}[1-**t**]-**Ad**)
-**$'vp**[1+**s**]/{360[**c**-**q**]}

Rule-13980:

If both (**Q**), (**U**), (**$**), (**V**), (**$'**), (**f**), (**s**), (**I**), (**t**), (**A**), (**d**), (**v**), (**p**), (**c**) and (**q**) are known, then its Leverage or Gearing Ratio Planned is:

/= (**Q**+**$'vp**[1+**s**]/{360[**c**-**q**]})
/(**U**+{**$**-**V**-**$'f**[1+**s**]-**I**}[1-**t**]-**Ad**)

Rule-13981:

If both (**/**), (**Q**), (**$**), (**V**), (**$'**), (**f**), (**s**), (**I**), (**t**), (**A**), (**d**), (**v**), (**p**), (**c**) and (**q**) are known, then its Utilized or Starting Capital must be:

U= **Ad**+(**Q**+**$'vp**[1+**s**]/{360[**c**-**q**]})/**/**
-{**$**-**V**-**$'f**[1+**s**]-**I**}[1-**t**]

Steve Asikin ISBN 14: 978-1514685136, ISBN 10: 1514685132

Rule-13982:

If both (**/**), (**U**), (**Q**), (**V**), (**S'**), (**f**), (**s**), (**I**), (**t**), (**A**), (**d**), (**v**), (**p**), (**c**) and (**q**) are known, then its Sales or Revenue Planned is:

S= **[Q-/(U-**{**V**+**S'f**[1+**s**]+**I**}[1-**t**]-**Ad**)
+**S'vp**[1+**s**]/{360[**c-q**]}**]**/{**/**[1-**t**]}

Rule-13983:

If both (**/**), (**U**), (**S**), (**Q**), (**S'**), (**f**), (**s**), (**I**), (**t**), (**A**), (**d**), (**v**), (**p**), (**c**) and (**q**) are known, then its Variable Cost Planned is:

V= **[/(U**+{**S-S'f**[1+**s**]-**I**}[1-**t**]-**Ad**)
-**S'vp**[1+**s**]/{360[**c-q**]}-**Q]**/{**/**[1-**t**]}

Rule-13984:

If both (**/**), (**U**), (**S**), (**V**), (**Q**), (**f**), (**s**), (**I**), (**t**), (**A**), (**d**), (**v**), (**p**), (**c**) and (**q**) are known, then its Sales Past mmust be:

S'= **[/(U**+{**S-V-I**}[1-**t**]-**Ad**)-**Q]**
/**[**[1+**s**](**vp**/{360[**c-q**]}+**f/**[1-**t**])**]**

Rule-13985:

If both (**/**), (**U**), (**S**), (**V**), (**S'**), (**Q**), (**s**), (**I**), (**t**), (**A**), (**d**), (**v**), (**p**), (**c**) and (**q**) are known, then its Fixed Portion Planned is:

f= **{S-V-I-[Ad**+(**Q**+**S'vp**[1+**s**]/{360[**c-q**]})/**/-U]**
/[1-**t**]**}**/{**S'**[1+**s**]}

Steve Asikin ISBN 14: 978-1514685136, ISBN 10: 1514685132

Rule-13986:

If both (/), (U), (S), (V), (S'), (f), (Q), (I), (t), (A), (d), (v), (p), (c) and (q) are known, then its Sales Growth Planned is:

s= [/(U+{S-V-I}[1-t]-Ad)-Q]
/[S'(vp/{360[c-q]}+f/[1-t])]-1

Rule-13987:

If both (/), (U), (S), (V), (S'), (f), (s), (Q), (t), (A), (d), (v), (p), (c) and (q) are known, then its Interest Expense Planned is:

I= {S-V-S'f[1+s]}
-[Ad+(Q+S'vp[1+s]/{360[c-q]})/-U]/[1-t]

Rule-13988:

If both (/), (U), (S), (V), (S'), (f), (s), (I), (Q), (A), (d), (v), (p), (c) and (q) are known, then its Tax Rate Planned is:

t= 1-[Ad+(Q+S'vp[1+s]/{360[c-q]})/-U]
/{S-V-S'f[1+s]-I}

Rule-13989:

If both (/), (U), (S), (V), (S'), (f), (s), (I), (t), (Q), (d), (v), (p), (c) and (q) are known, then its After Tax Income Planned is:

A= [U+{S-V-S'f[1+s]-I}[1-t]
-(Q+S'vp[1+s]/{360[c-q]})/]/d

Steve Asikin ISBN 14: 978-1514685136, ISBN 10: 1514685132

Rule-13990:

If both (**/**), (**U**), (**$**), (**V**), (**$'**), (**f**), (**s**), (**I**), (**t**), (**A**), (**Q**), (**v**), (**p**), (**c**) and (**q**) are known, then its Dividend Payout Planned is:

d= [**U**+{**$**-**V**-**$'f**[1+**s**]-**I**}[1-**t**]
-(**Q**+**$'vp**[1+**s**]/{360[**c**-**q**]})/**/**/**A**

Rule-13991:

If both (**/**), (**U**), (**$**), (**V**), (**$'**), (**f**), (**s**), (**I**), (**t**), (**A**), (**d**), (**Q**), (**p**), (**c**) and (**q**) are known, then its Variable Portion Planned is:

v= **V**/**$**

or it can also be found as

v= 360[**c**-**q**][**/**(**U**+{**$**-**V**-**$'f**[1+**s**]-**I**}[1-**t**]-**Ad**)-**Q**]
/{**$'p**[1+**s**]}

Rule-13992:

If both (**/**), (**U**), (**$**), (**V**), (**$'**), (**f**), (**s**), (**I**), (**t**), (**A**), (**d**), (**v**), (**Q**), (**c**) and (**q**) are known, then its Procured Inventory Days Planned:

p= 360[**c**-**q**][**/**(**U**+{**$**-**V**-**$'f**[1+**s**]-**I**}[1-**t**]-**Ad**)-**Q**]
/{**$'v**[1+**s**]}

Rule-13993:

If both (**/**), (**U**), (**$**), (**V**), (**$'**), (**f**), (**s**), (**I**), (**t**), (**A**), (**d**), (**v**), (**p**), (**Q**) and (**q**) are known, then its Quick or Gearing Ratio Planned is:

c= **q**+**$'vp**[1+**s**]
/{360[**/**(**U**+{**$**-**V**-**$'f**[1+**s**]-**I**}[1-**t**]-**Ad**)-**Q**]}

Steve Asikin ISBN 14: 978-1514685136, ISBN 10: 1514685132

Rule-13994:

If both (**/**), (**U**), (**$**), (**V**), (**$'**), (**f**), (**s**), (**I**), (**t**), (**A**), (**d**), (**v**), (**p**), (**c**) and (**Q**) are known, then its Quick or Acid Test Ratio Planned is:

q= **c**-**$'vp**[1+**s**]
/{360[**/U**+{**$**-**V**-**$'f**[1+**s**]-**I**}[1-**t**]-**Ad**)-**Q**]}

Rule-13995:

If both (**/**), (**U**), (**$**), (**V**), (**$'**), (**f**), (**s**), (**I**), (**t**), (**d**) and (**X**) are known, then its Quoted Longterm Debt Planned is:

Q= **/**(**U**+{**$**-**V**-**$'f**[1+**s**]-**I**}[1-**t**][1-**d**])-**X**

Rule-13996:

If both (**Q**), (**U**), (**$**), (**V**), (**$'**), (**f**), (**s**), (**I**), (**t**), (**d**) and (**X**) are known, then its Leverage or Gearing Ratio Planned is:

/= [**Q**+**X**]/(**U**+{**$**-**V**-**$'f**[1+**s**]-**I**}[1-**t**][1-**d**])

Rule-13997:

If both (**/**), (**Q**), (**$**), (**V**), (**$'**), (**f**), (**s**), (**I**), (**t**), (**d**) and (**X**) are known, then its Utilized or Starting Capital must be:

U= [**Q**+**X**]/**/**-{**$**-**V**-**$'f**[1+**s**]-**I**}[1-**t**][1-**d**]

Rule-13998:

If both (**/**), (**U**), (**Q**), (**V**), (**$'**), (**f**), (**s**), (**I**), (**t**), (**d**) and (**X**) are known, then its Sales or Revenue Planned is:

$**= **V**+**$'f[1+**s**]+**I**+{[**Q**+**X**]/**/**-**U**}/{[1-**t**][1-**d**]}

Steve Asikin ISBN 14: 978-1514685136, ISBN 10: 1514685132

Rule-13999:

If both (**/**), (**U**), (**$**), (**Q**), (**$'**), (**f**), (**s**), (**I**), (**t**), (**d**) and (**X**) are known, then its Variable Cost Planned is:

V= {**$**-**$'f**[1+**s**]-**I**}-{[**Q**+**X**]/**/-U**}/{[1-**t**][1-**d**]}

Rule-14000:

If both (**/**), (**U**), (**$**), (**V**), (**Q**), (**f**), (**s**), (**I**), (**t**), (**d**) and (**X**) are known, then its Sales Past must be:

$'**= (**$-**V**-**I**-{[**Q**+**X**]/**/-U**}/{[1-**t**][1-**d**]})/{**f**[1+**s**]}

Rule-14001:

If both (**/**), (**U**), (**$**), (**V**), (**$'**), (**Q**), (**s**), (**I**), (**t**), (**d**) and (**X**) are known, then its Fixed Portion Planned is:

f= (**$**-**V**-**I**-{[**Q**+**X**]/**/-U**}/{[1-**t**][1-**d**]})/{**$'**[1+**s**]}

Rule-14002:

If both (**/**), (**U**), (**$**), (**V**), (**$'**), (**f**), (**Q**), (**I**), (**t**), (**d**) and (**X**) are known, then its Sales Growth Planned is:

s= (**$**-**V**-**I**-{[**Q**+**X**]/**/-U**}/{[1-**t**][1-**d**]})/[**$'f**]-1

Rule-14003:

If both (**/**), (**U**), (**$**), (**V**), (**$'**), (**f**), (**s**), (**Q**), (**t**), (**d**) and (**X**) are known, then its Interest Expense Planned is:

I= {**$**-**V**-**$'f**[1+**s**]}-{[**Q**+**X**]/**/-U**}/{[1-**t**][1-**d**]}

Rule-14004:

If both (**/**), (**U**), (**$**), (**V**), (**$'**), (**f**), (**s**), (**I**), (**Q**), (**d**) and (**X**) are known, then its Tax Planned is:

t= 1-{[**Q**+**X**]/**/-U**}/([1-**d**]{**$**-**V**-**$'f**[1+**s**]-**I**})

Steve Asikin ISBN 14: 978-1514685136, ISBN 10: 1514685132

Rule-14005:

If both (/), (U), (S), (V), (S'), (f), (s), (I), (t), (Q) and (X) are known, then its Dividend Payout Planned is:

d= 1-{[Q+X]/ /U}/([1-t]{S-V-S'f[1+s]-I})

Rule-14006:

If both (/), (U), (S), (V), (S'), (f), (s), (I), (t), (d) and (Q) are known, then its Xpress or Current Debt Planned is:

X= /(U+{S-V-S'f[1+s]-I}[1-t][1-d])-Q

Rule-14007:

If both (/), (U), (S), (V), (S'), (f), (s), (I), (t), (d), (P), (c) and (q) are known, then its Quoted Longterm Debt Planned is:

Q= /(U+{S-V-S'f[1+s]-I}[1-t][1-d])-P/[c-q]

Rule-14008:

If both (Q), (U), (S), (V), (S'), (f), (s), (I), (t), (d), (P), (c) and (q) are known, then its Leverage or Gearing Ratio Planned is:

/= {Q+P/[c-q]}/(U+{S-V-S'f[1+s]-I}[1-t][1-d])

Rule-14009:

If both (/), (Q), (S), (V), (S'), (f), (s), (I), (t), (d), (P), (c) and (q) are known, then its Utilized or Starting Capital must be:

U= {Q+P/[c-q]}/ /-{S-V-S'f[1+s]-I}[1-t][1-d]

Steve Asikin ISBN 14: 978-1514685136, ISBN 10: 1514685132

Rule-14010:

If both (/), (U), (Q), (V), (S'), (f), (s), (I), (t), (d), (P), (c) and (q) are known, then its Sales or Revenue Planned is:

S= V+S'f[1+s]+I+({Q+P/[c-q]}//-U)/{[1-t][1-d]}

Rule-14011:

If both (/), (U), (S), (Q), (S'), (f), (s), (I), (t), (d), (P), (c) and (q) are known, then its Variable Cost Planned is:

V= {S-S'f[1+s]-I}-({Q+P/[c-q]}/-U)/{[1-t][1-d]}

Rule-14012:

If both (/), (U), (S), (V), (Q), (f), (s), (I), (t), (d), (P), (c) and (q) are known, then its Sales Past must be:

S'= [S-V-I-({Q+P/[c-q]}/-U)
/{[1-t][1-d]}]/{f[1+s]}

Rule-14013:

If both (/), (U), (S), (V), (S'), (Q), (s), (I), (t), (d), (P), (c) and (q) are known, then its Fixed Portion Planned is:

f= [S-V-I-({Q+P/[c-q]}/-U)/{[1-t][1-d]}]
/{S'[1+s]}

Steve Asikin ISBN 14: 978-1514685136, ISBN 10: 1514685132

Rule-14014:

If both (/), (**U**), (**S**), (**V**), (**S'**), (**f**), (**Q**), (**I**), (**t**), (**d**), (**P**), (**c**) and (**q**) are known, then its Sales Growth Planned is:

s= S/S'-1

or it can also be found as

s= [S-V-I-({Q+P/[c-q]}//U)/{[1-t][1-d]}]/[S'f]-1

Rule-14015:

If both (/), (**U**), (**S**), (**V**), (**S'**), (**f**), (**s**), (**Q**), (**t**), (**d**), (**P**), (**c**) and (**q**) are known, then its Interest Expense Ratio Planned is:

I= {S-V-S'f[1+s]}-({Q+P/[c-q]}//U)/{[1-t][1-d]}

Rule-14016:

If both (/), (**U**), (**S**), (**V**), (**S'**), (**f**), (**s**), (**I**), (**Q**), (**d**), (**P**), (**c**) and (**q**) are known, then its Tax Rate Planned is:

t= 1-({Q+P/[c-q]}//U)/([1-d]{S-V-S'f[1+s]-I})

Rule-14017:

If both (/), (**U**), (**S**), (**V**), (**S'**), (**f**), (**s**), (**I**), (**t**), (**Q**), (**P**), (**c**) and (**q**) are known, then its Dividend Payout Planned is:

d= 1-({Q+P/[c-q]}//U)/([1-t]{S-V-S'f[1+s]-I})

Rule-14018:

If both (/), (**U**), (**S**), (**V**), (**S'**), (**f**), (**s**), (**I**), (**t**), (**d**), (**Q**), (**c**) and (**q**) are known, then its Procured Inventory Planned is:

P= [c-q][/U+{S-V-S'f[1+s]-I}[1-t][1-d])-Q]

Steve Asikin ISBN 14: 978-1514685136, ISBN 10: 1514685132

Rule-14019:

If both (**/**), (**U**), (**$**), (**V**), (**$'**), (**f**), (**s**), (**I**), (**t**), (**d**), (**P**), (**Q**) and (**q**) are known, then its Current Ratio Planned is:

c= q+P/[/(U+{$-V-$'f[1+s]-I}[1-t][1-d])-Q]

Rule-14020:

If both (**/**), (**U**), (**$**), (**V**), (**$'**), (**f**), (**s**), (**I**), (**t**), (**d**), (**P**), (**c**) and (**Q**) are known, then its Quick or Acid Test Ratio Planned is:

q= c-P/[/(U+{$-V-$'f[1+s]-I}[1-t][1-d])-Q]

Rule-14021:

If both (**/**), (**U**), (**$**), (**V**), (**$'**), (**f**), (**s**), (**I**), (**t**), (**d**), (**p**), (**c**) and (**q**) are known, then its Quoted Longterm Debt Planned is:

Q= /(U+{$-V-$'f[1+s]-I}[1-t][1-d])-Vp/{360[c-q]}

Rule-14022:

If both (**Q**), (**U**), (**$**), (**V**), (**$'**), (**f**), (**s**), (**I**), (**t**), (**d**), (**p**), (**c**) and (**q**) are known, then its Leverage or Gearing Ratio Planned is:

/= (Q+Vp/{360[c-q]})
/(U+{$-V-$'f[1+s]-I}[1-t][1-d])

Rule-14023:

If both (**/**), (**Q**), (**$**), (**V**), (**$'**), (**f**), (**s**), (**I**), (**t**), (**d**), (**p**), (**c**) and (**q**) are known, then its Utilized or Starting Capital must be:

U= (Q+Vp/{360[c-q]})//
-{$-V-$'f[1+s]-I}[1-t][1-d]

Steve Asikin ISBN 14: 978-1514685136, ISBN 10: 1514685132

Rule-14024:

If both (**/**), (**U**), (**Q**), (**V**), (**S'**), (**f**), (**s**), (**I**), (**t**), (**d**), (**p**), (**c**) and (**q**) are known, then its Sales or Revenue Planned is:

S= **V**+**S'f**[1+**s**]+**I**+**[**(**Q**+**Vp**/{360[**c**-**q**]})/**/-U]**
/{[1-**t**][1-**d**]}

Rule-14025:

If both (**/**), (**U**), (**S**), (**Q**), (**S'**), (**f**), (**s**), (**I**), (**t**), (**d**), (**p**), (**c**) and (**q**) are known, then its Variable Cost Planned is:

V= **[/U**+{**S**-**S'f**[1+**s**]-**I**}[1-**t**][1-**d**])-**Q]**
/(**p**/{360[**c**-**q**]}+**/**[1-**t**][1-**d**])

Rule-14026:

If both (**/**), (**U**), (**S**), (**V**), (**Q**), (**f**), (**s**), (**I**), (**t**), (**d**), (**p**), (**c**) and (**q**) are known, then its Sales Past must be:

S'= **{S**-**V**-**I**-**[**(**Q**+**Vp**/{360[**c**-**q**]})/**/-U]**
/{[1-**t**][1-**d**]}**}**/{**f**[1+**s**]}

Rule-14027:

If both (**/**), (**U**), (**S**), (**V**), (**S'**), (**Q**), (**s**), (**I**), (**t**), (**d**), (**p**), (**c**) and (**q**) are known, then its Fixed Portion Planned is:

f= **{S**-**V**-**I**-**[**(**Q**+**Vp**/{360[**c**-**q**]})/**/-U]**
/{[1-**t**][1-**d**]}**}**/{**S'**[1+**s**]}

Steve Asikin ISBN 14: 978-1514685136, ISBN 10: 1514685132

Rule-14028:

If both (/), (U), ($), (V), ($'), (f), (s), (I), (t), (d), (p), (c) and (q) are known, then its Quoted Longterm Debt Planned is:

s= {$-V-I-[(Q+Vp/{360[c-q]})/*l*U]/{[1-t][1-d]}}
/ [$'f]-1

Rule-14029:

If both (/), (U), ($), (V), ($'), (f), (s), (Q), (t), (d), (p), (c) and (q) are known, then its Interest Expense Planned is:

I= {$-V-$'f[1+s]}-[(Q+Vp/{360[c-q]})/*l*U]
/{[1-t][1-d]}

Rule-14030:

If both (/), (U), ($), (V), ($'), (f), (s), (I), (Q), (d), (p), (c) and (q) are known, then its Tax Rate Planned is:

t= 1-[(Q+Vp/{360[c-q]})/*l*U]
/([1-d]{$-V-$'f[1+s]-I})

Rule-14031:

If both (/), (U), ($), (V), ($'), (f), (s), (I), (t), (Q), (p), (c) and (q) are known, then its Dividend Payout Planned is:

d= 1-[(Q+Vp/{360[c-q]})/*l*U]
/([1-t]{$-V-$'f[1+s]-I})

Steve Asikin ISBN 14: 978-1514685136, ISBN 10: 1514685132

Rule-14032:

If both (**/**), (**U**), (**$**), (**V**), (**$'**), (**f**), (**s**), (**I**), (**t**), (**d**), (**Q**), (**c**) and (**q**) are known, then its Procured Inventory Days Planned:

p= 360[**c**-**q**][**/**(**U**+{**$**-**V**-**$'f**[1+**s**]-**I**}[1-**t**][1-**d**])-**Q**]/**V**

Rule-14033:

If both (**/**), (**U**), (**$**), (**V**), (**$'**), (**f**), (**s**), (**I**), (**t**), (**d**), (**p**), (**Q**) and (**q**) are known, then its Current Ratio Planned is:

c= **q**+**Vp**
/{360[**/**(**U**+{**$**-**V**-**$'f**[1+**s**]-**I**}[1-**t**][1-**d**])-**Q**]}

Rule-14034:

If both (**/**), (**U**), (**$**), (**V**), (**$'**), (**f**), (**s**), (**I**), (**t**), (**d**), (**p**), (**c**) and (**Q**) are known, then its Quick or Acid Test Ratio Planned is:

q= **c**- **Vp**
/{360[**/**(**U**+{**$**-**V**-**$'f**[1+**s**]-**I**}[1-**t**][1-**d**])-**Q**]}

Rule-14035:

If both (**/**), (**U**), (**$**), (**V**), (**$'**), (**f**), (**s**), (**I**), (**t**), (**d**), (**v**), (**p**), (**c**) and (**q**) are known, then its Quoted Longterm Debt Planned is:

Q= **/**(**U**+{**$**-**V**-**$'f**[1+**s**]-**I**}[1-**t**][1-**d**])
-**$vp**/{360[**c**-**q**]}

Rule-14036:

If both (**Q**), (**U**), (**$**), (**V**), (**$'**), (**f**), (**s**), (**I**), (**t**), (**d**), (**v**), (**p**), (**c**) and (**q**) are known, then its Leverage or Gearing Ratio Planned is:

/= (Q+$vp/{360[c-q]})
/(U+{$-V-$'f[1+s]-I}[1-t][1-d])

Rule-14037:

If both (**/**), (**Q**), (**$**), (**V**), (**$'**), (**f**), (**s**), (**I**), (**t**), (**d**), (**v**), (**p**), (**c**) and (**q**) are known, then its Utilized or Starting Capital must be:

U= (Q+$vp/{360[c-q]})//
-{$-V-$'f[1+s]-I}[1-t][1-d]

Rule-14038:

If both (**/**), (**U**), (**Q**), (**V**), (**$'**), (**f**), (**s**), (**I**), (**t**), (**d**), (**v**), (**p**), (**c**) and (**q**) are known, then its Sales or Revenue Planned is:

$= [/(U-{V+$'f[1+s]+I}[1-t][1-d])-V]
/(vp/{360[c-q]}-/[1-t][1-d])

Rule-14039:

If both (**/**), (**U**), (**$**), (**Q**), (**$'**), (**f**), (**s**), (**I**), (**t**), (**d**), (**v**), (**p**), (**c**) and (**q**) are known, then its Variable Cost Planned is:

V= [/(U+{$-$'f[1+s]-I}[1-t][1-d])
-$vp/{360[c-q]}-Q]/{/[1-t][1-d]}

Steve Asikin ISBN 14: 978-1514685136, ISBN 10: 1514685132

Rule-14040:

If both (/), (U), (S), (V), (Q), (f), (s), (I), (t), (d), (v), (p), (c) and (q) are known, then its Sales Past must be:

S'= {S-V-I-[(Q+Svp/{360[c-q]})//-U]
/{[1-t][1-d]}}/{f[1+s]}

Rule-14041:

If both (/), (U), (S), (V), (S'), (Q), (s), (I), (t), (d), (v), (p), (c) and (q) are known, then its Fixed Portion Planned is:

f= {S-V-I-[(Q+Svp/{360[c-q]})//-U]
/{[1-t][1-d]}}/{S'[1+s]}

Rule-14042:

If both (/), (U), (S), (V), (S'), (f), (Q), (I), (t), (d), (v), (p), (c) and (q) are known, then its Sales Growth Planned is:

s= {S-V-I-[(Q+Svp/{360[c-q]})//-U]/{[1-t][1-d]}}
/[S'f]-1

Rule-14043:

If both (/), (U), (S), (V), (S'), (f), (s), (Q), (t), (d), (v), (p), (c) and (q) are known, then its Interest Expense Planned is:

I= {S-V-S'f[1+s]}-[(Q+Svp/{360[c-q]})//-U]
/{[1-t][1-d]}

Steve Asikin ISBN 14: 978-1514685136, ISBN 10: 1514685132

Rule-14044:

If both (**/**), (**U**), (**$**), (**V**), (**$'**), (**f**), (**s**), (**I**), (**Q**), (**d**), (**v**), (**p**), (**c**) and (**q**) are known, then its Tax Rate Planned is:

t= 1-**[**(**Q**+**$vp**/{360[**c-q**]})/**/U]** /([1-**d**]{**$-V-$'f**[1+**s**]-**I**})

Rule-14045:

If both (**/**), (**U**), (**$**), (**V**), (**$'**), (**f**), (**s**), (**I**), (**t**), (**Q**), (**v**), (**p**), (**c**) and (**q**) are known, then its Dividend Payout Planned is:

d= 1-**[**(**Q**+**$vp**/{360[**c-q**]})/**/U]** /([1-**t**]{**$-V-$'f**[1+**s**]-**I**})

Rule-14046:

If both (**/**), (**U**), (**$**), (**V**), (**$'**), (**f**), (**s**), (**I**), (**t**), (**d**), (**Q**), (**p**), (**c**) and (**q**) are known, then its Variable Portion Planned is:

v= **V/$**

or it can also be found as

v= 360[**c-q**]**[/U**+{**$-V-$'f**[1+**s**]-**I**}[1-**t**][1-**d**])-**Q]** /[**$p**]

Rule-14047:

If both (**/**), (**U**), (**$**), (**V**), (**$'**), (**f**), (**s**), (**I**), (**t**), (**d**), (**v**), (**Q**), (**c**) and (**q**) are known, then its Procured Inventory Days Planned:

p= 360[**c-q**]**[/U**+{**$-V-$'f**[1+**s**]-**I**}[1-**t**][1-**d**])-**Q]** /[**$v**]

Steve Asikin ISBN 14: 978-1514685136, ISBN 10: 1514685132

Rule-14048:

If both (/), (**U**), (**$**), (**V**), (**$'**), (**f**), (**s**), (**I**), (**t**), (**d**), (**v**), (**p**), (**Q**) and (**q**) are known, then its Current Ratio Planned is:

c= q+$vp
/{360[/(U+{$-V-$'f[1+s]-I}[1-t][1-d])-Q]}

Rule-14049:

If both (/), (**U**), (**$**), (**V**), (**$'**), (**f**), (**s**), (**I**), (**t**), (**d**), (**v**), (**p**), (**c**) and (**Q**) are known, then its Quick or Acid Test Ratio Planned is:

q= c- $vp
/{360[/(U+{$-V-$'f[1+s]-I}[1-t][1-d])-Q]}

Rule-14050:

If both (/), (**U**), (**$**), (**V**), (**$'**), (**f**), (**s**), (**I**), (**t**), (**d**), (**v**), (**p**), (**c**) and (**q**) are known, then its Quoted Longterm Debt Planned is:

Q= /(U+{$-V-$'f[1+s]-I}[1-t][1-d])
-$'vp[1+s]/{360[c-q]}

Rule-14051:

If both (**Q**), (**U**), (**$**), (**V**), (**$'**), (**f**), (**s**), (**I**), (**t**), (**d**), (**v**), (**p**), (**c**) and (**q**) are known, then its Leverage or Gearing Ratio Planned is:

/= (Q+$'vp[1+s]/{360[c-q]})
/(U+{$-V-$'f[1+s]-I}[1-t][1-d])

Steve Asikin ISBN 14: 978-1514685136, ISBN 10: 1514685132

Rule-14052:

If both (**/**), (**Q**), (**S**), (**V**), (**S'**), (**f**), (**s**), (**I**), (**t**), (**d**), (**v**), (**p**), (**c**) and (**q**) are known, then its Utilized or Starting Capital must be:

U= (**Q**+**S'vp**[1+**s**]/{360[**c**-**q**]})/**/**
-{**S**-**V**-**S'f**[1+**s**]-**I**}[1-**t**][1-**d**]

Rule-14053:

If both (**/**), (**U**), (**Q**), (**V**), (**S'**), (**f**), (**s**), (**I**), (**t**), (**d**), (**v**), (**p**), (**c**) and (**q**) are known, then its Sales or Revenue Planned is:

S= [**Q**-**/**(**U**-{**V**+**S'f**[1+**s**]-**I**}[1-**t**][1-**d**])
+**S'vp**[1+**s**]/{360[**c**-**q**]}]/{**/**[1-**t**][1-**d**]}

Rule-14054:

If both (**/**), (**U**), (**S**), (**Q**), (**S'**), (**f**), (**s**), (**I**), (**t**), (**d**), (**v**), (**p**), (**c**) and (**q**) are known, then its Variable Cost Planned is:

V= [**/**(**U**+{**S**-**S'f**[1+**s**]-**I**}[1-**t**][1-**d**])
-**S'vp**[1+**s**]/{360[**c**-**q**]}-**Q**]/{**/**[1-**t**][1-**d**]}

Rule-14055:

If both (**/**), (**U**), (**S**), (**V**), (**Q**), (**f**), (**s**), (**I**), (**t**), (**d**), (**v**), (**p**), (**c**) and (**q**) are known, then its Sales Past must be:

S'= [**/**(**U**+{**S**-**V**-**I**}[1-**t**][1-**d**])-**Q**]
/[[1+**s**](**vp**/{360[**c**-**q**]}+**f/**[1-**t**][1-**d**])

Steve Asikin ISBN 14: 978-1514685136, ISBN 10: 1514685132

Rule-14056:

If both (**/**), (**U**), (**$**), (**V**), (**$'**), (**Q**), (**s**), (**I**), (**t**), (**d**), (**v**), (**p**), (**c**) and (**q**) are known, then its Fixed Portion Planned is:

f= {**$**-**V**-**I**-[(**Q**+**$'vp**[1+**s**]/{360[**c**-**q**]})/**/U**]
/{[1-**t**][1-**d**]}}/{**$'**[1+**s**]}

Rule-14057:

If both (**/**), (**U**), (**$**), (**V**), (**$'**), (**f**), (**Q**), (**I**), (**t**), (**d**), (**v**), (**p**), (**c**) and (**q**) are known, then its Sales Growth Planned is:

s= **$**/**$'**-1

or it can also be found as

s= [**/U**+{**$**-**V**-**I**}[1-**t**][1-**d**])-**Q**]
/[**$'**(**vp**/{360[**c**-**q**]}+**f**/[1-**t**][1-**d**])-1

Rule-14058:

If both (**/**), (**U**), (**$**), (**V**), (**$'**), (**f**), (**s**), (**Q**), (**t**), (**d**), (**v**), (**p**), (**c**) and (**q**) are known, then its Interest Expense Planned is:

I= {**$**-**V**-**$'f**[1+**s**]}-[(**Q**+**$'vp**[1+**s**]/{360[**c**-**q**]})/**/U**]
/{[1-**t**][1-**d**]}

Rule-14059:

If both (**/**), (**U**), (**$**), (**V**), (**$'**), (**f**), (**s**), (**I**), (**Q**), (**d**), (**v**), (**p**), (**c**) and (**q**) are known, then its Tax Rate Planned is:

t= 1-[(**Q**+**$'vp**[1+**s**]/{360[**c**-**q**]})/**/U**]
/([1-**d**]{**$**-**V**-**$'f**[1+**s**]-**I**})

Steve Asikin ISBN 14: 978-1514685136, ISBN 10: 1514685132

Rule-14060:

If both (**/**), (**U**), (**$**), (**V**), (**$'**), (**f**), (**s**), (**I**), (**t**), (**Q**), (**v**), (**p**), (**c**) and (**q**) are known, then its Dividend Payout Planned is:

d= 1-[(**Q**+**$'vp**[1+**s**]/{360[**c**-**q**]})/**/**-**U**]
/([1-**t**]{**$**-**V**-**$'f**[1+**s**]-**I**})

Rule-14061:

If both (**/**), (**U**), (**$**), (**V**), (**$'**), (**f**), (**s**), (**I**), (**t**), (**d**), (**Q**), (**p**), (**c**) and (**q**) are known, then its Variable Portion Planned is:

v= 360[**c**-**q**][**/**(**U**+{**$**-**V**-**$'f**[1+**s**]-**I**}[1-**t**][1-**d**])-**Q**]
/{**$'p**[1+**s**]}

Rule-14062:

If both (**/**), (**U**), (**$**), (**V**), (**$'**), (**f**), (**s**), (**I**), (**t**), (**d**), (**v**), (**Q**), (**c**) and (**q**) are known, then its Procured Inventory Days Planned:

p= 360[**c**-**q**][**/**(**U**+{**$**-**V**-**$'f**[1+**s**]-**I**}[1-**t**][1-**d**])-**Q**]
/{**$'v**[1+**s**]}

Rule-14063:

If both (**/**), (**U**), (**$**), (**V**), (**$'**), (**f**), (**s**), (**I**), (**t**), (**d**), (**v**), (**p**), (**Q**) and (**q**) are known, then its Current Ratio Planned is:

c= **q**+**$'vp**[1+**s**]
/{360[**/**(**U**+{**$**-**V**-**$'f**[1+**s**]-**I**}[1-**t**][1-**d**])-**Q**]}

Steve Asikin ISBN 14: 978-1514685136, ISBN 10: 1514685132

Rule-14064:

If both (**/**), (**U**), (**$**), (**V**), (**$'**), (**f**), (**s**), (**I**), (**t**), (**d**), (**v**), (**p**), (**c**) and (**Q**) are known, then its Quick or Acid Test Ratio Planned is:

q= **c**-**$'vp**[1+**s**]
/{360[**/**(**U**+{**$**-**V**-**$'f**[1+**s**]-**I**}[1-**t**][1-**d**])-**Q**]}

Rule-14065:

If both (**/**), (**U**), (**$**), (**i**), (**V**), (**$'**), (**f**), (**s**), (**T**), (**D**) and (**X**) are known, then its Quoted Longterm Debt Planned is:

Q= **/**{**U**+**$**[1-**i**]-**V**-**$'f**[1+**s**]-**T**-**D**}-**X**

Rule-14066:

If both (**Q**), (**U**), (**$**), (**i**), (**V**), (**$'**), (**f**), (**s**), (**T**), (**D**) and (**X**) are known, then its Leverage or Gearing Ratio Planned is:

/= [**Q**+**X**]/{**U**+**$**[1-**i**]-**V**-**$'f**[1+**s**]-**T**-**D**}

Rule-14067:

If both (**/**), (**Q**), (**$**), (**i**), (**V**), (**$'**), (**f**), (**s**), (**T**), (**D**) and (**X**) are known, then its Utilized or Starting Capital must be:

U= [**Q**+**X**]/**/**-{**$**[1-**i**]-**V**-**$'f**[1+**s**]-**T**-**D**}

Rule-14068:

If both (**/**), (**U**), (**Q**), (**i**), (**V**), (**$'**), (**f**), (**s**), (**T**), (**D**) and (**X**) are known, then its Sales or Revenue Planned is:

$**= {**V**+**$'f[1+**s**]+**T**+**D**+[**Q**+**X**]/**/**-**U**}/[1-**i**]

Steve Asikin ISBN 14: 978-1514685136, ISBN 10: 1514685132

Rule-14069:

If both (Λ), (U), ($), (Q), (V), ($'), (f), (s), (T), (D) and (X) are known, then its Interest Portion Planned is:

i= 1-{V+$'f[1+s]+T+D+[Q+X]/Λ-U}/$

Rule-14070:

If both (Λ), (U), ($), (i), (Q), ($'), (f), (s), (T), (D) and (X) are known, then its Variable Cost Planned is:

V= U-$[1-i]-$'f[1+s]-T-D-[Q+X]/Λ

Rule-14071:

If both (Λ), (U), ($), (i), (V), (Q), (f), (s), (T), (D) and (X) are known, then its Sales Past must be:

$'= {U-$[1-i]-V-T-D-[Q+X]/Λ}/{f[1+s]}

Rule-14072:

If both (Λ), (U), ($), (i), (V), ($'), (Q), (s), (T), (D) and (X) are known, then its Fixed Cost Planned is:

f= {U-$[1-i]-V-T-D-[Q+X]/Λ}/{$'[1+s]}

Rule-14073:

If both (Λ), (U), ($), (i), (V), ($'), (f), (Q), (T), (D) and (X) are known, then its Sales Growth Planned is:

s= $/$'-1

or it can also be found as

s= {U-$[1-i]-V-T-D-[Q+X]/Λ}/[$'f]-1

Steve Asikin ISBN 14: 978-1514685136, ISBN 10: 1514685132

Rule-14074:

If both (**/**), (**U**), (**$**), (**i**), (**V**), (**$'**), (**f**), (**s**), (**Q**), (**D**) and (**X**) are known, then its Tax Planned is:

T= **U**-**$**[1-**i**]-**V**-**$'f**[1+**s**]-**D**-[**Q**+**X**]/**/**

Rule-14075:

If both (**/**), (**U**), (**$**), (**i**), (**V**), (**$'**), (**f**), (**s**), (**T**), (**Q**) and (**X**) are known, then its Dividend Planned is:

D= **U**-**$**[1-**i**]-**V**-**$'f**[1+**s**]-**T**-[**Q**+**X**]/**/**

Rule-14076:

If both (**/**), (**U**), (**$**), (**i**), (**V**), (**$'**), (**f**), (**s**), (**T**), (**D**) and (**Q**) are known, then its Xpress or Current Debt Planned is:

X= **/**{**U**+**$**[1-**i**]-**V**-**$'f**[1+**s**]-**T**-**D**}-**Q**

Rule-14077:

If both (**/**), (**U**), (**$**), (**i**), (**V**), (**$'**), (**f**), (**s**), (**T**), (**D**), (**P**), (**c**) and (**q**) are known, then its Quoted Longterm Debt Planned is:

Q= **/**{**U**+**$**[1-**i**]-**V**-**$'f**[1+**s**]-**T**-**D**}-**P**/[**c**-**q**]

Rule-14078:

If both (**Q**), (**U**), (**$**), (**i**), (**V**), (**$'**), (**f**), (**s**), (**T**), (**D**), (**P**), (**c**) and (**q**) are known, then its Leverage or Gearing Ratio Planned is:

/= {**Q**+**P**/[**c**-**q**]}/{**U**+**$**[1-**i**]-**V**-**$'f**[1+**s**]-**T**-**D**}

Steve Asikin ISBN 14: 978-1514685136, ISBN 10: 1514685132

Rule-14079:

If both (**/**), (**Q**), (**$**), (**i**), (**V**), (**$'**), (**f**), (**s**), (**T**), (**D**), (**P**), (**c**) and (**q**) are known, then its Utilized or Starting Capital must be:

U= {**Q**+**P**/[**c**-**q**]}/**/**-{**$**[1-**i**]-**V**-**$'f**[1+**s**]-**T**-**D**}

Rule-14080:

If both (**/**), (**U**), (**Q**), (**i**), (**V**), (**$'**), (**f**), (**s**), (**T**), (**D**), (**P**), (**c**) and (**q**) are known, then its Sales or Revenue Planned is:

$**= (**V**+**$'f[1+**s**]+**T**+**D**+{**Q**+**P**/[**c**-**q**]}/**/**-**U**)/[1-**i**]

Rule-14081:

If both (**/**), (**U**), (**$**), (**Q**), (**V**), (**$'**), (**f**), (**s**), (**T**), (**D**), (**P**), (**c**) and (**q**) are known, then its Interest Portion Planned is:

i= 1-(**V**+**$'f**[1+**s**]+**T**+**D**+{**Q**+**P**/[**c**-**q**]}/**/**-**U**)/**$**

Rule-14082:

If both (**/**), (**U**), (**$**), (**i**), (**Q**), (**$'**), (**f**), (**s**), (**T**), (**D**), (**P**), (**c**) and (**q**) are known, then its Variable Cost Planned is:

V= **U**-**$**[1-**i**]-**$'f**[1+**s**]-**T**-**D**-{**Q**+**P**/[**c**-**q**]}/**/**

Rule-14083:

If both (**/**), (**U**), (**$**), (**i**), (**V**), (**Q**), (**f**), (**s**), (**T**), (**D**), (**P**), (**c**) and (**q**) are known, then its Sales Past must be:

$'**= (**U**-**$[1-**i**]-**V**-**T**-**D**-{**Q**+**P**/[**c**-**q**]}/**/**)/{**f**[1+**s**]}

Steve Asikin ISBN 14: 978-1514685136, ISBN 10: 1514685132

Rule-14084:

If both (I), (U), (S), (i), (V), (S'), (Q), (s), (T), (D), (P), (c) and (q) are known, then its Fixed Portion Planned is:

$$f = (U - S[1-i] - V - T - D - \{Q + P/[c-q]\}/I)/\{S'[1+s]\}$$

Rule-14085:

If both (I), (U), (S), (i), (V), (S'), (f), (Q), (T), (D), (P), (c) and (q) are known, then its Sales Growth Planned is:

$$s = S/S' - 1$$

or it can also be found as

$$s = (U - S[1-i] - V - T - D - \{Q + P/[c-q]\}/I)/[S'f] - 1$$

Rule-14086:

If both (I), (U), (S), (i), (V), (S'), (f), (s), (Q), (D), (P), (c) and (q) are known, then its Tax Planned is:

$$T = U - S[1-i] - V - S'f[1+s] - D - \{Q + P/[c-q]\}/I$$

Rule-14087:

If both (I), (U), (S), (i), (V), (S'), (f), (s), (T), (Q), (P), (c) and (q) are known, then its Dividend Planned is:

$$D = U - S[1-i] - V - S'f[1+s] - T - \{Q + P/[c-q]\}/I$$

Rule-14088:

If both (I), (U), (S), (i), (V), (S'), (f), (s), (T), (D), (Q), (c) and (q) are known, then its Procured Inventory Planned is:

$$P = [c-q](I\{U + S[1-i] - V - S'f[1+s] - T - D\} - Q)$$

Steve Asikin ISBN 14: 978-1514685136, ISBN 10: 1514685132

Rule-14089:

If both (**/**), (**U**), (**$**), (**i**), (**V**), (**$'**), (**f**), (**s**), (**T**), (**D**), (**P**), (**Q**) and (**q**) are known, then its Current Ratio Planned is:

c= q+P/(/{U+$[1-i]-V-$'f[1+s]-T-D}-Q)

Rule-14090:

If both (**/**), (**U**), (**$**), (**i**), (**V**), (**$'**), (**f**), (**s**), (**T**), (**D**), (**P**), (**c**) and (**Q**) are known, then its Quick Ratio Planned is:

q= c-P/(/{U+$[1-i]-V-$'f[1+s]-T-D}-Q)

Rule-14091:

If both (**/**), (**U**), (**$**), (**i**), (**V**), (**$'**), (**f**), (**s**), (**T**), (**D**), (**p**), (**c**) and (**q**) are known, then its Quoted Longterm Debt Planned is:

Q= /{U+$[1-i]-V-$'f[1+s]-T-D}-Vp/{360[c-q]}

Rule-14092:

If both (**Q**), (**U**), (**$**), (**i**), (**V**), (**$'**), (**f**), (**s**), (**T**), (**D**), (**p**), (**c**) and (**q**) are known, then its Leverage or Gearing Ratio Planned is:

/= (Q+Vp/{360[c-q])/{U+$[1-i]-V-$'f[1+s]-T-D}

Rule-14093:

If both (**/**), (**Q**), (**$**), (**i**), (**V**), (**$'**), (**f**), (**s**), (**T**), (**D**), (**p**), (**c**) and (**q**) are known, then its Utilized or Starting Capital must be:

U= (Q+Vp/{360[c-q])//-{$[1-i]-V-$'f[1+s]-T-D}

Steve Asikin ISBN 14: 978-1514685136, ISBN 10: 1514685132

Rule-14094:

If both (**/**), (**U**), (**Q**), (**i**), (**V**), (**S'**), (**f**), (**s**), (**T**), (**D**), (**p**), (**c**) and (**q**) are known, then its Sales or Revenue Planned is:

S= [**V**+**S'f**[1+**s**]+**T**+**D**+(**Q**+**Vp**/{360[**c**-**q**])/**/**-**U**]/[1-**i**]

Rule-14095:

If both (**/**), (**U**), (**S**), (**Q**), (**V**), (**S'**), (**f**), (**s**), (**T**), (**D**), (**p**), (**c**) and (**q**) are known, then its Interest Portion Planned is:

i= 1-[**V**+**S'f**[1+**s**]+**T**+**D**+(**Q**+**Vp**/{360[**c**-**q**])/**/**-**U**]/**S**

Rule-14096:

If both (**/**), (**U**), (**S**), (**i**), (**Q**), (**S'**), (**f**), (**s**), (**T**), (**D**), (**p**), (**c**) and (**q**) are known, then its Variable Cost Planned is:

V= (**/**{**U**+**S**[1-**i**]-**S'f**[1+**s**]-**T**-**D**}-**Q**)/(**p**/{360[**c**-**q**]}+**/**)

Rule-14097:

If both (**/**), (**U**), (**S**), (**i**), (**V**), (**Q**), (**f**), (**s**), (**T**), (**D**), (**p**), (**c**) and (**q**) are known, then its Sales Past must be:

S'= [**U**-**S**[1-**i**]-**V**-**T**-**D**-(**Q**+**Vp**/{360[**c**-**q**])/**/**]/{**f**[1+**s**]}

Rule-14098:

If both (**/**), (**U**), (**S**), (**i**), (**V**), (**S'**), (**Q**), (**s**), (**T**), (**D**), (**p**), (**c**) and (**q**) are known, then its Fixed Portion Planned is:

f= [**U**-**S**[1-**i**]-**V**-**T**-**D**-(**Q**+**Vp**/{360[**c**-**q**])/**/**]/{**S'**[1+**s**]}

Steve Asikin ISBN 14: 978-1514685136, ISBN 10: 1514685132

Rule-14099:

If both (**/**), (**U**), (**$**), (**i**), (**V**), (**$'**), (**f**), (**Q**), (**T**), (**D**), (**p**), (**c**) and (**q**) are known, then its Sales Growth Planned is:

s= **$**/**$'**-1

or it can also be found as

s= [**U**-**$**[1-**i**]-**V**-**T**-**D**-(**Q**+**Vp**/{360[**c**-**q**])/**/**]/[**$'f**]-1

Rule-14100:

If both (**/**), (**U**), (**$**), (**i**), (**V**), (**$'**), (**f**), (**s**), (**Q**), (**D**), (**p**), (**c**) and (**q**) are known, then its Tax Planned is:

T= **U**-**$**[1-**i**]-**V**-**$'f**[1+**s**]-**D**-(**Q**+**Vp**/{360[**c**-**q**])/**/**

Rule-14101:

If both (**/**), (**U**), (**$**), (**i**), (**V**), (**$'**), (**f**), (**s**), (**T**), (**Q**), (**p**), (**c**) and (**q**) are known, then its Dividend Payout Planned is:

D= **U**-**$**[1-**i**]-**V**-**$'f**[1+**s**]-**T**-(**Q**+**Vp**/{360[**c**-**q**])/**/**

Rule-14102:

If both (**/**), (**U**), (**$**), (**i**), (**V**), (**$'**), (**f**), (**s**), (**T**), (**D**), (**Q**), (**c**) and (**q**) are known, then its Procured Inventory Days planned:

p= 360[**c**-**q**](**/**{**U**+**$**[1-**i**]-**V**-**$'f**[1+**s**]-**T**-**D**}-**Q**)/**V**

Rule-14103:

If both (**/**), (**U**), (**$**), (**i**), (**V**), (**$'**), (**f**), (**s**), (**T**), (**D**), (**p**), (**Q**) and (**q**) are known, then its Current Ratio Planned is:

c= **q**+**Vp**/[360(**/**{**U**+**$**[1-**i**]-**V**-**$'f**[1+**s**]-**T**-**D**}-**Q**)]

Steve Asikin ISBN 14: 978-1514685136, ISBN 10: 1514685132

Rule-14094:

If both (/), (**U**), (**$**), (**i**), (**V**), (**$'**), (**f**), (**s**), (**T**), (**D**), (**p**), (**c**) and (**q**) are known, then its Quick or Acid Test Ratio Planned is:

q= **c**-**Vp**/[360(/{**U**+**$**[1-**i**]-**V**-**$'f**[1+**s**]-**T**-**D**}-**Q**)]

Rule-14105:

If both (/), (**U**), (**$**), (**i**), (**V**), (**$'**), (**f**), (**s**), (**T**), (**D**), (**v**), (**p**), (**c**) and (**q**) are known, then its Quoted Longterm Debt Planned is:

Q= /{**U**+**$**[1-**i**]-**V**-**$'f**[1+**s**]-**T**-**D**}-**$vp**/{360[**c**-**q**]}

Rule-14106:

If both (**Q**), (**U**), (**$**), (**i**), (**V**), (**$'**), (**f**), (**s**), (**T**), (**D**), (**v**), (**p**), (**c**) and (**q**) are known, then its Leverage or Gearing Ratio Planned is:

/= (**Q**+**$vp**/{360[**c**-**q**])/{**U**+**$**[1-**i**]-**V**-**$'f**[1+**s**]-**T**-**D**}

Rule-14107:

If both (/), (**Q**), (**$**), (**i**), (**V**), (**$'**), (**f**), (**s**), (**T**), (**D**), (**v**), (**p**), (**c**) and (**q**) are known, then its Utilized or Starting Capital must be:

U= (**Q**+**$vp**/{360[**c**-**q**])//-{**$**[1-**i**]-**V**-**$'f**[1+**s**]-**T**-**D**}

Rule-14108:

If both (/), (**U**), (**Q**), (**i**), (**V**), (**$'**), (**f**), (**s**), (**T**), (**D**), (**v**), (**p**), (**c**) and (**q**) are known, then its Sales or Revenue Planned is:

$**= (/{**U**-**V**-**$'f[1+**s**]-**T**-**D**}-**Q**)/(**vp**/{360[**c**-**q**]}-/[1-**i**])

Steve Asikin ISBN 14: 978-1514685136, ISBN 10: 1514685132

Rule-14109:

If both (**/**), (**U**), (**$**), (**Q**), (**V**), (**$'**), (**f**), (**s**), (**T**), (**D**), (**v**), (**p**), (**c**) and (**q**) are known, then its Interest Portion Planned is:

i= 1-[V+$'f[1+s]+T+D+(Q+$vp/{360[c-q]})/-U]/$

Rule-14110:

If both (**/**), (**U**), (**$**), (**i**), (**Q**), (**$'**), (**f**), (**s**), (**T**), (**D**), (**v**), (**p**), (**c**) and (**q**) are known, then its Variable Portion Planned is:

V= (/{U+$[1-i]-$'f[1+s]-T-D}-$vp/{360[c-q]}-Q)//

Rule-14111:

If both (**/**), (**U**), (**$**), (**i**), (**V**), (**Q**), (**f**), (**s**), (**T**), (**D**), (**v**), (**p**), (**c**) and (**q**) are known, then its Sales Past must be:

$'= [U-$[1-i]-V-T-D-(Q+$vp/{360[c-q]})//]
/{f[1+s]}

Rule-14112:

If both (**/**), (**U**), (**$**), (**i**), (**V**), (**$'**), (**Q**), (**s**), (**T**), (**D**), (**v**), (**p**), (**c**) and (**q**) are known, then its Fixed Portion Planned is:

f= [U-$[1-i]-V-T-D-(Q+$vp/{360[c-q]})//]
/{$'[1+s]}

Steve Asikin ISBN 14: 978-1514685136, ISBN 10: 1514685132

Rule-14113:

If both (**I**), (**U**), (**S**), (**i**), (**V**), (**S'**), (**f**), (**Q**), (**T**), (**D**), (**v**), (**p**), (**c**) and (**q**) are known, then its Sales Growth Planned is:

$s = S/S'-1$

or it can also be found as

$s = [U-S[1-i]-V-T-D-(Q+Svp/\{360[c-q])/I]/[S'f]-1$

Rule-14114:

If both (**I**), (**U**), (**S**), (**i**), (**V**), (**S'**), (**f**), (**s**), (**Q**), (**D**), (**v**), (**p**), (**c**) and (**q**) are known, then its Tax Planned is:

$T = U-S[1-i]-V-S'f[1+s]-D-(Q+Svp/\{360[c-q])/I$

Rule-14115:

If both (**I**), (**U**), (**S**), (**i**), (**V**), (**S'**), (**f**), (**s**), (**T**), (**Q**), (**v**), (**p**), (**c**) and (**q**) are known, then its Dividend Planned is:

$D = U-S[1-i]-V-S'f[1+s]-T-(Q+Svp/\{360[c-q])/I$

Rule-14116:

If both (**I**), (**U**), (**S**), (**i**), (**V**), (**S'**), (**f**), (**s**), (**T**), (**D**), (**Q**), (**p**), (**c**) and (**q**) are known, then its Variable Portion Planned is:

$v = 360[c-q](I\{U+S[1-i]-V-S'f[1+s]-T-D\}-Q)/[Sp]$

Rule-14117:

If both (**I**), (**U**), (**S**), (**i**), (**V**), (**S'**), (**f**), (**s**), (**T**), (**D**), (**v**), (**Q**), (**c**) and (**q**) are known, then its Procured Inventory Days Planned:

$p = 360[c-q](I\{U+S[1-i]-V-S'f[1+s]-T-D\}-Q)/[Sv]$

Steve Asikin ISBN 14: 978-1514685136, ISBN 10: 1514685132

Rule-14118:

If both (**/**), (**U**), (**$**), (**i**), (**V**), (**$'**), (**f**), (**s**), (**T**), (**D**), (**v**), (**p**), (**Q**) and (**q**) are known, then its Current Ratio Planned is:

c= **q**+**$vp**/[360(**/**{**U**+**$**[1-**i**]-**V**-**$'f**[1+**s**]-**T**-**D**}-**Q**)]

Rule-14119:

If both (**/**), (**U**), (**$**), (**i**), (**V**), (**$'**), (**f**), (**s**), (**T**), (**D**), (**v**), (**p**), (**c**) and (**Q**) are known, then its Quick or Acid Test Ratio Planned is:

q= **c**-**$vp**/[360(**/**{**U**+**$**[1-**i**]-**V**-**$'f**[1+**s**]-**T**-**D**}-**Q**)]

Rule-14120:

If both (**/**), (**U**), (**$**), (**i**), (**V**), (**$'**), (**f**), (**s**), (**T**), (**D**), (**v**), (**p**), (**c**) and (**q**) are known, then its Quoted Longterm Debt Planned is:

Q= **/**{**U**+**$**[1-**i**]-**V**-**$'f**[1+**s**]-**T**-**D**}
-**$'vp**[1+**s**]/{360[**c**-**q**]}

Rule-14121:

If both (**Q**), (**U**), (**$**), (**i**), (**V**), (**$'**), (**f**), (**s**), (**T**), (**D**), (**v**), (**p**), (**c**) and (**q**) are known, then its Leverage or Gearing Ratio Planned is:

/= (**Q**+**$'vp**[1+**s**]/{360[**c**-**q**]})/{**U**+**$**[1-**i**]-**V**
-**$'f**[1+**s**]-**T**-**D**}

Rule-14122:

If both (**/**), (**Q**), (**$**), (**i**), (**V**), (**$'**), (**f**), (**s**), (**T**), (**D**), (**v**), (**p**), (**c**) and (**q**) are known, then its Utilized or Starting Capital must be:

U= (**Q**+**$'vp**[1+**s**]/{360[**c**-**q**]})/**/**
-{**$**[1-**i**]-**V**-**$'f**[1+**s**]-**T**-**D**}

Steve Asikin ISBN 14: 978-1514685136, ISBN 10: 1514685132

Rule-14123:

If both (/), (U), (Q), (i), (V), (S'), (f), (s), (T), (D), (v), (p), (c) and (q) are known, then its Sales or Revenue Planned is:

S= (Q-/{U-V-S'f[1+s]-T-D}+S'vp[1+s]/{360[c-q]})
/{/[1-i]}

Rule-14124:

If both (/), (U), (S), (Q), (V), (S'), (f), (s), (T), (D), (v), (p), (c) and (q) are known, then its Interest Portion Planned is:

i= 1-[V+S'f[1+s]+T+D
+(Q+S'vp[1+s]/{360[c-q]})/ /-U]/S

Rule-14125:

If both (/), (U), (S), (i), (Q), (S'), (f), (s), (T), (D), (v), (p), (c) and (q) are known, then its Variable Cost Planned is:

V= (/{U+S[1-i]-S'f[1+s]-T-D}
-S'vp[1+s]/{360[c-q]}-Q)//

Rule-14126:

If both (/), (U), (S), (i), (V), (Q), (f), (s), (T), (D), (v), (p), (c) and (q) are known, then its Sales Past must be:

S'= (/{U+S[1-i]-V-T-D}-Q)
/[[1+s](vp/{360[c-q]}-f/)]

Steve Asikin ISBN 14: 978-1514685136, ISBN 10: 1514685132

Rule-14127:

If both (**/**), (**U**), (**S**), (**i**), (**V**), (**S'**), (**Q**), (**s**), (**T**), (**D**), (**v**), (**p**), (**c**) and (**q**) are known, then its Fixed Portion Planned is:

f= [**U**-**S**[1-**i**]-**V**-**T**-**D**-(**Q**+**S'vp**[1+**s**]/{360[**c**-**q**])/**/**]
/{**S'**[1+**s**]}

Rule-14128:

If both (**/**), (**U**), (**S**), (**i**), (**V**), (**S'**), (**f**), (**Q**), (**T**), (**D**), (**v**), (**p**), (**c**) and (**q**) are known, then its Sales Growth Planned is:

s= **S**/**S'**-1

or it can also be found as

s= (**/**{**U**+**S**[1-**i**]-**V**-**T**-**D**}-**Q**)/[**S'**(**vp**/{360[**c**-**q**]}-**f/**)]-1

Rule-14129:

If both (**/**), (**U**), (**S**), (**i**), (**V**), (**S'**), (**f**), (**s**), (**Q**), (**D**), (**v**), (**p**), (**c**) and (**q**) are known, then its Tax Planned is:

T= **U**-**S**[1-**i**]-**V**-**S'f**[1+**s**]-**D**
-(**Q**+**S'vp**[1+**s**]/{360[**c**-**q**])/**/**

Rule-14130:

If both (**/**), (**U**), (**S**), (**i**), (**V**), (**S'**), (**f**), (**s**), (**T**), (**Q**), (**v**), (**p**), (**c**) and (**q**) are known, then its Dividend Planned is:

D= **U**-**S**[1-**i**]-**V**-**S'f**[1+**s**]-**T**
-(**Q**+**S'vp**[1+**s**]/{360[**c**-**q**])/**/**

Steve Asikin ISBN 14: 978-1514685136, ISBN 10: 1514685132

Rule-14131:

If both (**/**), (**U**), (**$**), (**i**), (**V**), (**$'**), (**f**), (**s**), (**T**), (**D**), (**Q**), (**p**), (**c**) and (**q**) are known, then its Variable Portion Planned is:

v= 360[**c-q**](**/**{**U**+**$**[1-**i**]-**V**-**$'f**[1+**s**]-**T**-**D**}-**Q**)
/{**$'p**[1+**s**]}

Rule-14132:

If both (**/**), (**U**), (**$**), (**i**), (**V**), (**$'**), (**f**), (**s**), (**T**), (**D**), (**v**), (**Q**), (**c**) and (**q**) are known, then its Procured Inventory Days planned:

p= 360[**c-q**](**/**{**U**+**$**[1-**i**]-**V**-**$'f**[1+**s**]-**T**-**D**}-**Q**)
/{**$'v**[1+**s**]}

Rule-14133:

If both (**/**), (**U**), (**$**), (**i**), (**V**), (**$'**), (**f**), (**s**), (**T**), (**D**), (**v**), (**p**), (**Q**) and (**q**) are known, then its Current Ratio Planned is:

c= **q**+**$'vp**[1+**s**]
/[360(**/**{**U**+**$**[1-**i**]-**V**-**$'f**[1+**s**]-**T**-**D**}-**Q**)]

Rule-14134:

If both (**/**), (**U**), (**$**), (**i**), (**V**), (**$'**), (**f**), (**s**), (**T**), (**D**), (**v**), (**p**), (**c**) and (**Q**) are known, then its Quick or Acid Test Ratio Planned is:

q= **c-$'vp**[1+**s**]
/[360(**/**{**U**+**$**[1-**i**]-**V**-**$'f**[1+**s**]-**T**-**D**}-**Q**)]

Steve Asikin ISBN 14: 978-1514685136, ISBN 10: 1514685132

Rule-14135:

If both (**/**), (**U**), (**S**), (**i**), (**V**), (**S'**), (**f**), (**s**), (**T**), (**d**) and (**X**) are known, then its Quoted Longterm Debt Planned is:

Q= **/**(**U**+{**S**[1-**i**]-**V**-**S'f**[1+**s**]-**T**}[1-**d**])-**X**

Rule-14136:

If both (**Q**), (**U**), (**S**), (**i**), (**V**), (**S'**), (**f**), (**s**), (**T**), (**d**) and (**X**) are known, then its Leverage or Gearing Ratio Planned is:

/= [**Q**+**X**]/(**U**+{**S**[1-**i**]-**V**-**S'f**[1+**s**]-**T**}[1-**d**])

Rule-14137:

If both (**/**), (**Q**), (**S**), (**i**), (**V**), (**S'**), (**f**), (**s**), (**T**), (**d**) and (**X**) are known, then its Utilized or Starting Capital must be:

U= [**Q**+**X**]/**/**-{**S**[1-**i**]-**V**-**S'f**[1+**s**]-**T**}[1-**d**]

Rule-14138:

If both (**/**), (**U**), (**Q**), (**i**), (**V**), (**S'**), (**f**), (**s**), (**T**), (**d**) and (**X**) are known, then its Sales or Revenue Planned is:

S= (**V**+**S'f**[1+**s**]+**T**+{[**Q**+**X**]/**/**-**U**}/[1-**d**])/[1-**i**]

Rule-14139:

If both (**/**), (**U**), (**S**), (**Q**), (**V**), (**S'**), (**f**), (**s**), (**T**), (**d**) and (**X**) are known, then its Interest Portion Planned is:

i= 1-(**V**+**S'f**[1+**s**]+**T**+{[**Q**+**X**]/**/**-**U**}/[1-**d**])/**S**

Steve Asikin ISBN 14: 978-1514685136, ISBN 10: 1514685132

Rule-14140:

If both (**/**), (**U**), (**$**), (**i**), (**Q**), (**$'**), (**f**), (**s**), (**T**), (**d**) and (**X**) are known, then its Variable Cost Planned is:

V= **$**[1-**i**]-**$'f**[1+**s**]-**T**-{[**Q**+**X**]/**/**-**U**}/[1-**d**]

Rule-14141:

If both (**/**), (**U**), (**$**), (**i**), (**V**), (**Q**), (**f**), (**s**), (**T**), (**d**) and (**X**) are known, then its Sales Past must be:

$'**= (**$[1-**i**]-**T**-{[**Q**+**X**]/**/**-**U**}/[1-**d**])/{**f**[1+**s**]}

Rule-14142:

If both (**/**), (**U**), (**$**), (**i**), (**V**), (**$'**), (**Q**), (**s**), (**T**), (**d**) and (**X**) are known, then its Fixed Portion Planned is:

f= (**$**[1-**i**]-**T**-{[**Q**+**X**]/**/**-**U**}/[1-**d**])/{**$'**[1+**s**]}

Rule-14143:

If both (**/**), (**U**), (**$**), (**i**), (**V**), (**$'**), (**f**), (**Q**), (**T**), (**d**) and (**X**) are known, then its Sales Growth Planned is:

s= **$**/**$'**-1

or it can also be found as

s= (**$**[1-**i**]-**T**-{[**Q**+**X**]/**/**-**U**}/[1-**d**])/[**$'f**]-1

Rule-14144:

If both (**/**), (**U**), (**$**), (**i**), (**V**), (**$'**), (**f**), (**s**), (**Q**), (**d**) and (**X**) are known, then its Tax Planned is:

T= **$**[1-**i**]-**V**-**$'f**[1+**s**]-{[**Q**+**X**]/**/**-**U**}/[1-**d**]

Rule-14145:

If both (**/**), (**U**), (**$**), (**i**), (**V**), (**$'**), (**f**), (**s**), (**T**), (**D**) and (**X**) are known, then its Dividend Payout Planned is:

d= 1-{[**Q**+**X**]/**/**-**U**}/{**$**[1-**i**]-**V**-**$'f**[1+**s**]-**T**}

Steve Asikin ISBN 14: 978-1514685136, ISBN 10: 1514685132

Rule-14146:

If both (/), (U), ($), (i), (V), ($'), (f), (s), (T), (d) and (Q) are known, then its Xpress or Current Debt Planned is:

X= /(U+{$[1-i]-V-$'f[1+s]-T}[1-d])-Q

Rule-14147:

If both (/), (U), ($), (i), (V), ($'), (f), (s), (T), (d), (P), (c) and (q) are known, then its Quoted Longterm Debt Planned is:

Q= /(U+{$[1-i]-V-$'f[1+s]-T}[1-d])-P/[c-q]

Rule-14148:

If both (Q), (U), ($), (i), (V), ($'), (f), (s), (T), (d), (P), (c) and (q) are known, then its Leverage or Gearing Ratio Planned is:

/= {Q+P/[c-q]}/(U+{$[1-i]-V-$'f[1+s]-T}[1-d])

Rule-14149:

If both (/), (Q), ($), (i), (V), ($'), (f), (s), (T), (d), (P), (c) and (q) are known, then its Utilized or Starting Capital must be:

U= {Q+P/[c-q]}/-/-{$[1-i]-V-$'f[1+s]-T}[1-d]

Rule-14150:

If both (/), (U), (Q), (i), (V), ($'), (f), (s), (T), (d), (P), (c) and (q) are known, then its Sales or Revenue Planned is:

$= [V+$'f[1+s]+T+({Q+P/[c-q]}//-U)/[1-d]]/[1-i]

Steve Asikin ISBN 14: 978-1514685136, ISBN 10: 1514685132

Rule-14151:

If both (**/**), (**U**), (**S**), (**Q**), (**V**), (**S'**), (**f**), (**s**), (**T**), (**d**), (**P**), (**c**) and (**q**) are known, then its Interest Portion Planned is:

i= 1-[**V**+**S'f**[1+**s**]+**T**+({**Q**+**P**/[**c**-**q**]}/**/U**)/[1-**d**]]/**S**

Rule-14152:

If both (**/**), (**U**), (**S**), (**i**), (**Q**), (**S'**), (**f**), (**s**), (**T**), (**d**), (**P**), (**c**) and (**q**) are known, then its Variable Cost Planned is:

V= **S**[1-**i**]-**S'f**[1+**s**]-**T**-({**Q**+**P**/[**c**-**q**]}/**/U**)/[1-**d**]

Rule-14153:

If both (**/**), (**U**), (**S**), (**i**), (**V**), (**Q**), (**f**), (**s**), (**T**), (**d**), (**P**), (**c**) and (**q**) are known, then its Sales Past must be:

S'= [**S**[1-**i**]-**T**-({**Q**+**P**/[**c**-**q**]}/**/U**)/[1-**d**]]/{**f**[1+**s**]}

Rule-14154:

If both (**/**), (**U**), (**S**), (**i**), (**V**), (**S'**), (**Q**), (**s**), (**T**), (**d**), (**P**), (**c**) and (**q**) are known, then its Fixed Portion Planned is:

f= [**S**[1-**i**]-**T**-({**Q**+**P**/[**c**-**q**]}/**/U**)/[1-**d**]]/{**S'**[1+**s**]}

Rule-14155:

If both (**/**), (**U**), (**S**), (**i**), (**V**), (**S'**), (**f**), (**Q**), (**T**), (**d**), (**P**), (**c**) and (**q**) are known, then its Sales Growth Planned is:

s= **S**/**S'**-1

or it can also be found as

s= [**S**[1-**i**]-**T**-({**Q**+**P**/[**c**-**q**]}/**/U**)/[1-**d**]]/[**S'f**]-1

Steve Asikin ISBN 14: 978-1514685136, ISBN 10: 1514685132

Rule-14156:

If both (**/**), (**U**), (**$**), (**i**), (**V**), (**$'**), (**f**), (**s**), (**Q**), (**d**), (**P**), (**c**) and (**q**) are known, then its Tax Planned is:

T= **$**[1-**i**]-**V**-**$'f**[1+**s**]-({**Q**+**P**/[**c**-**q**]}/**/U**)/[1-**d**]

Rule-14157:

If both (**/**), (**U**), (**$**), (**i**), (**V**), (**$'**), (**f**), (**s**), (**T**), (**Q**), (**P**), (**c**) and (**q**) are known, then its Dividend Payout Planned is:

d= 1-({**Q**+**P**/[**c**-**q**]}/**/U**)/{**$**[1-**i**]-**V**-**$'f**[1+**s**]-**T**}

Rule-14158:

If both (**/**), (**U**), (**$**), (**i**), (**V**), (**$'**), (**f**), (**s**), (**T**), (**d**), (**Q**), (**c**) and (**q**) are known, then its Procured Inventory Planned is:

P= [**c**-**q**][**/**(**U**+{**$**[1-**i**]-**V**-**$'f**[1+**s**]-**T**}[1-**d**])-**Q**]

Rule-14159:

If both (**/**), (**U**), (**$**), (**i**), (**V**), (**$'**), (**f**), (**s**), (**T**), (**d**), (**P**), (**Q**) and (**q**) are known, then its Current Ratio Planned is:

c= **q**+**P**/[**/**(**U**+{**$**[1-**i**]-**V**-**$'f**[1+**s**]-**T**}[1-**d**])-**Q**]

Rule-14160:

If both (**/**), (**U**), (**$**), (**i**), (**V**), (**$'**), (**f**), (**s**), (**T**), (**d**), (**P**), (**c**) and (**Q**) are known, then its Quick or Acid Test Ratio Planned is:

q= **c**-**P**/[**/**(**U**+{**$**[1-**i**]-**V**-**$'f**[1+**s**]-**T**}[1-**d**])-**Q**]

Steve Asikin ISBN 14: 978-1514685136, ISBN 10: 1514685132

Rule-14161:

If both (**/**), (**U**), (**$**), (**i**), (**V**), (**$'**), (**f**), (**s**), (**T**), (**d**), (**p**), (**c**) and (**q**) are known, then its Quoted Longterm Debt Planned is:

Q= **/**(**U**+{**$**[1-**i**]-**V**-**$'f**[1+**s**]-**T**}[1-**d**])
-**Vp**/{360[**c**-**q**]}

Rule-14162:

If both (**Q**), (**U**), (**$**), (**i**), (**V**), (**$'**), (**f**), (**s**), (**T**), (**d**), (**p**), (**c**) and (**q**) are known, then its Leverage or Gearing Ratio Planned is:

/= (**Q**+**Vp**/{360[**c**-**q**]})
/(**U**+{**$**[1-**i**]-**V**-**$'f**[1+**s**]-**T**}[1-**d**])

Rule-14163:

If both (**/**), (**Q**), (**$**), (**i**), (**V**), (**$'**), (**f**), (**s**), (**T**), (**d**), (**p**), (**c**) and (**q**) are known, then its Utilized or Starting Capital must be:

U= (**Q**+**Vp**/{360[**c**-**q**]})/**/**
-{**$**[1-**i**]-**V**-**$'f**[1+**s**]-**T**}[1-**d**]

Rule-14164:

If both (**/**), (**U**), (**Q**), (**i**), (**V**), (**$'**), (**f**), (**s**), (**T**), (**d**), (**p**), (**c**) and (**q**) are known, then its Sales or Revenue Planned is:

$**= {**V**+**$'f[1+**s**]+**T**
+[(**Q**+**Vp**/{360[**c**-**q**]})/**/**-**U**]/[1-**d**]}/[1-**i**]

Steve Asikin ISBN 14: 978-1514685136, ISBN 10: 1514685132

Rule-14165:

If both (/), (U), (S), (Q), (V), (S'), (f), (s), (T), (d), (p), (c) and (q) are known, then its Interest Portion Planned is:

i= 1-{V+S'f[1+s]+T
+[(Q+Vp/{360[c-q]})//U]/[1-d]}/S

Rule-14166:

If both (/), (U), (S), (i), (Q), (S'), (f), (s), (T), (d), (p), (c) and (q) are known, then its Variable Cost Planned is:

V= [/U+{S[1-i]-S'f[1+s]-T}[1-d])-Q]
/(p/{360[c-q]}+/[1-d])

Rule-14167:

If both (/), (U), (S), (i), (V), (Q), (f), (s), (T), (d), (p), (c) and (q) are known, then its Sales Past must be:

S'= {S[1-i]-T-[(Q+Vp/{360[c-q]})//U]/[1-d]}
/{f[1+s]}

Rule-14168:

If both (/), (U), (S), (i), (V), (S'), (Q), (s), (T), (d), (p), (c) and (q) are known, then its Fixed Portion Planned is:

f= {S[1-i]-T-[(Q+Vp/{360[c-q]})//U]/[1-d]}
/{S'[1+s]}

Steve Asikin ISBN 14: 978-1514685136, ISBN 10: 1514685132

Rule-14169:

If both (**/**), (**U**), (**$**), (**i**), (**V**), (**$'**), (**f**), (**Q**), (**T**), (**d**), (**p**), (**c**) and (**q**) are known, then its Sales Growth Planned is:

s= **$**/**$'**-1

or it can also be found as

s= {**$**[1-**i**]-**T**-[(**Q**+**Vp**/{360[**c**-**q**]})/**/**-**U**]/[1-**d**]}
/ [**$'f**]-1

Rule-14170:

If both (**/**), (**U**), (**$**), (**i**), (**V**), (**$'**), (**f**), (**s**), (**Q**), (**d**), (**p**), (**c**) and (**q**) are known, then its Tax Planned is:

T= **$**[1-**i**]-**V**-**$'f**[1+**s**]-[(**Q**+**Vp**/{360[**c**-**q**]})/**/**-**U**]
/[1-**d**]

Rule-14171:

If both (**/**), (**U**), (**$**), (**i**), (**V**), (**$'**), (**f**), (**s**), (**T**), (**Q**), (**p**), (**c**) and (**q**) are known, then its Dividend Payout Planned is:

d= 1-[(**Q**+**Vp**/{360[**c**-**q**]})/**/**-**U**]
/{**$**[1-**i**]-**V**-**$'f**[1+**s**]-**T**}

Rule-14172:

If both (**/**), (**U**), (**$**), (**i**), (**V**), (**$'**), (**f**), (**s**), (**T**), (**d**), (**Q**), (**c**) and (**q**) are known, then its Procured Inventory Days Planned:

p= 360[**c**-**q**][**/**(**U**+{**$**[1-**i**]-**V**-**$'f**[1+**s**]-**T**}[1-**d**])-**Q**]/**V**

Steve Asikin ISBN 14: 978-1514685136, ISBN 10: 1514685132

Rule-14173:

If both (/), (U), (S), (i), (V), (S'), (f), (s), (T), (d), (p), (Q) and (q) are known, then its Current Ratio Planned is:

c= q+Vp
/{360[/(U+{S[1-i]-V-S'f[1+s]-T}[1-d])-Q]}

Rule-14174:

If both (/), (U), (S), (i), (V), (S'), (f), (s), (T), (d), (p), (c) and (Q) are known, then its Quick or Acid test Ratio Planned is:

q= c-Vp/{360[/(U+{S[1-i]-V-S'f[1+s]-T}[1-d])-Q]}

Rule-14175:

If both (/), (U), (S), (i), (V), (S'), (f), (s), (T), (d), (v), (p), (c) and (q) are known, then its Quoted Longterm Debt Planned is:

Q= /(U+{S[1-i]-V-S'f[1+s]-T}[1-d])
-Svp/{360[c-q]}

Rule-14176:

If both (Q), (U), (S), (i), (V), (S'), (f), (s), (T), (d), (v), (p), (c) and (q) are known, then its Leverage or Gearing Ratio Planned is:

/= (Q+Svp/{360[c-q]})
/(U+{S[1-i]-V-S'f[1+s]-T}[1-d])

Steve Asikin ISBN 14: 978-1514685136, ISBN 10: 1514685132

Rule-14177:

If both (**/**), (**Q**), (**$**), (**i**), (**V**), (**$'**), (**f**), (**s**), (**T**), (**d**), (**v**), (**p**), (**c**) and (**q**) are known, then its Utilized or Starting Capital must be:

U= (**Q**+**$vp**/{360[**c**-**q**]})/**/**
-{**$**[1-**i**]-**V**-**$'f**[1+**s**]-**T**}[1-**d**]

Rule-14178:

If both (**/**), (**U**), (**Q**), (**i**), (**V**), (**$'**), (**f**), (**s**), (**T**), (**d**), (**v**), (**p**), (**c**) and (**q**) are known, then its Sales or Revenue Planned is:

$**= **[/(U**-{**V**+**$'f[1+**s**]+**T**}[1-**d**])-**Q]**
/(**vp**/{360[**c**-**q**]}-**/**[1-**i**] [1-**d**])

Rule-14179:

If both (**/**), (**U**), (**$**), (**Q**), (**V**), (**$'**), (**f**), (**s**), (**T**), (**d**), (**v**), (**p**), (**c**) and (**q**) are known, then its Interest Portion Planned is:

i= 1-**{V**+**$'f**[1+**s**]+**T**
+**[(Q**+**$vp**/{360[**c**-**q**]})/**/-U]**/[1-**d**]**}/$**

Rule-14180:

If both (**/**), (**U**), (**$**), (**i**), (**Q**), (**$'**), (**f**), (**s**), (**T**), (**d**), (**v**), (**p**), (**c**) and (**q**) are known, then its Variable Cost Planned is:

V= **[/(U**+{**$**[1-**i**]-**$'f**[1+**s**]-**T**}[1-**d**])
-**$vp**/{360[**c**-**q**]}-**Q]**/{**/**[1-**d**]}

Steve Asikin ISBN 14: 978-1514685136, ISBN 10: 1514685132

Rule-14181:

If both (Λ), (U), ($), (i), (V), (Q), (f), (s), (T), (d), (v), (p), (c) and (q) are known, then its Sales Past must be:

$'= {$[1-i]-T-[(Q+$vp/{360[c-q]})/ΛU]/[1-d]}
/{f[1+s]}

Rule-14182:

If both (Λ), (U), ($), (i), (V), ($'), (Q), (s), (T), (d), (v), (p), (c) and (q) are known, then its Fixed Portion Planned is:

f= {$[1-i]-T-[(Q+$vp/{360[c-q]})/ΛU]/[1-d]}
/{$'[1+s]}

Rule-14183:

If both (Λ), (U), ($), (i), (V), ($'), (f), (Q), (T), (d), (v), (p), (c) and (q) are known, then its Sales Growth Planned is:

s= $/$'-1

or it can also be found as

s= {$[1-i]-T-[(Q+$vp/{360[c-q]})/ΛU]/[1-d]}
/[$'f]-1

Rule-14184:

If both (Λ), (U), ($), (i), (V), ($'), (f), (s), (Q), (d), (v), (p), (c) and (q) are known, then its Tax Planned is:

T= $[1-i]-V-$'f[1+s]-[(Q+$vp/{360[c-q]})/ΛU]
/[1-d]

Steve Asikin ISBN 14: 978-1514685136, ISBN 10: 1514685132

Rule-14185:

If both (/), (**U**), (**$**), (**i**), (**V**), (**$'**), (**f**), (**s**), (**T**), (**Q**), (**v**), (**p**), (**c**) and (**q**) are known, then its Dividend Payout Planned is:

d= 1-[(Q+$vp/{360[c-q]})//U]
/{$[1-i]-V-$'f[1+s]-T}

Rule-14186:

If both (/), (**U**), (**$**), (**i**), (**V**), (**$'**), (**f**), (**s**), (**T**), (**d**), (**Q**), (**p**), (**c**) and (**q**) are known, then its Variable Portion Planned is:

v= V/$

or it can also be found as

v= 360[c-q][/U+{$[1-i]-V-$'f[1+s]-T}[1-d])-Q]
/[$p]

Rule-14187:

If both (/), (**U**), (**$**), (**i**), (**V**), (**$'**), (**f**), (**s**), (**T**), (**d**), (**v**), (**Q**), (**c**) and (**q**) are known, then its Procured Inventory Days Planned:

p= 360[c-q][/U+{$[1-i]-V-$'f[1+s]-T}[1-d])-Q]
/[$v]

Rule-14188:

If both (/), (**U**), (**$**), (**i**), (**V**), (**$'**), (**f**), (**s**), (**T**), (**d**), (**v**), (**p**), (**Q**) and (**q**) are known, then its Current Ratio Planned is:

c= q+$vp
/{360[/U+{$[1-i]-V-$'f[1+s]-T}[1-d])-Q]}

Steve Asikin ISBN 14: 978-1514685136, ISBN 10: 1514685132

Rule-14189:

If both (**/**), (**U**), (**$**), (**i**), (**V**), (**$'**), (**f**), (**s**), (**T**), (**d**), (**v**), (**p**), (**c**) and (**Q**) are known, then its Quick or Acid Test Ratio Planned is:

q= **c**-**$vp**
/{360[**/**(**U**+{**$**[1-**i**]-**V**-**$'f**[1+**s**]-**T**}[1-**d**])-**Q**]}

Rule-14190:

If both (**/**), (**U**), (**$**), (**i**), (**V**), (**$'**), (**f**), (**s**), (**T**), (**d**), (**v**), (**p**), (**c**) and (**q**) are known, then its Quoted Longterm Debt Planned is:

Q= **/**(**U**+{**$**[1-**i**]-**V**-**$'f**[1+**s**]-**T**}[1-**d**])
-**$'vp**[1+**s**]/{360[**c**-**q**]}

Rule-14191:

If both (**Q**), (**U**), (**$**), (**i**), (**V**), (**$'**), (**f**), (**s**), (**T**), (**d**), (**v**), (**p**), (**c**) and (**q**) are known, then its Leverage or Gearing Ratio Planned is:

/= (**Q**+**$'vp**[1+**s**]/{360[**c**-**q**]})
/(**U**+{**$**[1-**i**]-**V**-**$'f**[1+**s**]-**T**}[1-**d**])

Rule-14192:

If both (**/**), (**Q**), (**$**), (**i**), (**V**), (**$'**), (**f**), (**s**), (**T**), (**d**), (**v**), (**p**), (**c**) and (**q**) are known, then its Utilized or Starting Capital must be:

U= (**Q**+**$'vp**[1+**s**]/{360[**c**-**q**]})/**/**
-{**$**[1-**i**]-**V**-**$'f**[1+**s**]-**T**}[1-**d**]

Steve Asikin ISBN 14: 978-1514685136, ISBN 10: 1514685132

Rule-14193:

If both (**/**), (**U**), (**Q**), (**i**), (**V**), (**S'**), (**f**), (**s**), (**T**), (**d**), (**v**), (**p**), (**c**) and (**q**) are known, then its Sales or Revenue Planned is:

S= [**Q**-**/**(**U**-{**V**+**S'f**[1+**s**]+**T**}[1-**d**])
+**S'vp**[1+**s**]/{360[**c**-**q**]}]/{**/**[1-**i**][1-**d**]}

Rule-14194:

If both (**/**), (**U**), (**S**), (**Q**), (**V**), (**S'**), (**f**), (**s**), (**T**), (**d**), (**v**), (**p**), (**c**) and (**q**) are known, then its Interest Portion Planned is:

i= 1-{**V**+**S'f**[1+**s**]+**T**
+[(**Q**+**S'vp**[1+**s**]/{360[**c**-**q**]})/**/**-**U**]/[1-**d**]}/**S**

Rule-14195:

If both (**/**), (**U**), (**S**), (**i**), (**Q**), (**S'**), (**f**), (**s**), (**T**), (**d**), (**v**), (**p**), (**c**) and (**q**) are known, then its Variable Cost Planned is:

V= [**/**(**U**+{**S**[1-**i**]-**S'f**[1+**s**]-**T**}[1-**d**])
-**S'vp**[1+**s**]/{360[**c**-**q**]}-**Q**]/{**/**[1-**d**]}

Rule-14196:

If both (**/**), (**U**), (**S**), (**i**), (**V**), (**Q**), (**f**), (**s**), (**T**), (**d**), (**v**), (**p**), (**c**) and (**q**) are known, then its Sales Past must be:

S'= [**/**(**U**+{**S**[1-**i**]-**V**-**T**}[1-**d**])-**Q**]
/[[1+**s**](**vp**/{360[**c**-**q**]}+**f**/**/**[1-**d**])]

Steve Asikin ISBN 14: 978-1514685136, ISBN 10: 1514685132

Rule-14197:

If both (**/**), (**U**), (**S**), (**i**), (**V**), (**S'**), (**Q**), (**s**), (**T**), (**d**), (**v**), (**p**), (**c**) and (**q**) are known, then its Fixed Portion Planned is:

f= {**S**[1-**i**]-**T**-[(**Q**+**S'vp**[1+**s**]/{360[**c-q**]})/**/-U**] /[1-**d**]}/{**S'**[1+**s**]}

Rule-14198:

If both (**/**), (**U**), (**S**), (**i**), (**V**), (**S'**), (**f**), (**Q**), (**T**), (**d**), (**v**), (**p**), (**c**) and (**q**) are known, then its Sales Growth Planned is:

s= [**/**(**U**+{**S**[1-**i**]-**V-T**}[1-**d**])-**Q**] /[**S'**(**vp**/{360[**c-q**]}+**f/**[1-**d**])-1

Rule-14199:

If both (**/**), (**U**), (**S**), (**i**), (**V**), (**S'**), (**f**), (**s**), (**Q**), (**d**), (**v**), (**p**), (**c**) and (**q**) are known, then its Tax Planned is:

T= **S**[1-**i**]-**V-S'f**[1+**s**] -[(**Q**+**S'vp**[1+**s**]/{360[**c-q**]})/**/-U**]/[1-**d**]

Rule-14200:

If both (**/**), (**U**), (**S**), (**i**), (**V**), (**S'**), (**f**), (**s**), (**T**), (**Q**), (**v**), (**p**), (**c**) and (**q**) are known, then its Dividend Payout Planned is:

d= 1-[(**Q**+**S'vp**[1+**s**]/{360[**c-q**]})/**/-U**] /{**S**[1-**i**]-**V-S'f**[1+**s**]-**T**}

Steve Asikin ISBN 14: 978-1514685136, ISBN 10: 1514685132

Rule-14201:

If both (/), (U), ($), (i), (V), ($'), (f), (s), (T), (d), (Q), (p), (c) and (q) are known, then its Variable Portion Planned is:

$v= V/\$$

or it can also be found as

$v= 360[c-q][/(U+\{\$[1-i]-V-\$'f[1+s]-T\}[1-d])-Q]/\{\$'p[1+s]\}$

Rule-14202:

If both (/), (U), ($), (i), (V), ($'), (f), (s), (T), (d), (v), (Q), (c) and (q) are known, then its Procured Inventory Days Planned:

$p= 360[c-q][/(U+\{\$[1-i]-V-\$'f[1+s]-T\}[1-d])-Q]/\{\$'v[1+s]\}$

Rule-14203:

If both (/), (U), ($), (i), (V), ($'), (f), (s), (T), (d), (v), (p), (Q) and (q) are known, then its Current Ratio Planned is:

$c= q+\$'vp[1+s]/\{360[/(U+\{\$[1-i]-V-\$'f[1+s]-T\}[1-d])-Q]\}$

Rule-14204:

If both (/), (U), ($), (i), (V), ($'), (f), (s), (T), (d), (v), (p), (c) and (Q) are known, then its Quick or Acid Test Ratio Planned is:

$q= c-\$'vp[1+s]/\{360[/(U+\{\$[1-i]-V-\$'f[1+s]-T\}[1-d])-Q]\}$

Steve Asikin ISBN 14: 978-1514685136, ISBN 10: 1514685132

Rule-14205:

If both (/), (U), ($), (i), (V), ($'), (f), (s), (T), (A), (d) and (X) are known, then its Quoted Longterm Debt Planned is:

Q= /{U+$[1-i]-V-$'f[1+s]-T-Ad}-X

Rule-14206:

If both (Q), (U), ($), (i), (V), ($'), (f), (s), (T), (A), (d) and (X) are known, then its Leverage or Gearing Ratio Planned is:

/= [Q+X]/{U+$[1-i]-V-$'f[1+s]-T-Ad}

Rule-14207:

If both (/), (Q), ($), (i), (V), ($'), (f), (s), (T), (A), (d) and (X) are known, then its Utilized or Starting Capital must be:

U= [Q+X]/ /-{$[1-i]-V-$'f[1+s]-T-Ad}

Rule-14208:

If both (/), (U), (Q), (i), (V), ($'), (f), (s), (T), (A), (d) and (X) are known, then its Sales or Revenue Planned is:

$= {V+$'f[1+s]+T+Ad+[Q+X]/ /-U}/[1-i]

Rule-14209:

If both (/), (U), ($), (Q), (V), ($'), (f), (s), (T), (A), (d) and (X) are known, then its Interest Portion Planned is:

i= 1-{V+$'f[1+s]+T+Ad+[Q+X]/ /-U}/$

Steve Asikin ISBN 14: 978-1514685136, ISBN 10: 1514685132

Rule-14210:

If both (**/**), (**U**), (**$**), (**i**), (**Q**), (**$'**), (**f**), (**s**), (**T**), (**A**), (**d**) and (**X**) are known, then its Variable Cost Planned is:

V= {**U**+**$**[1-**i**]-**$'f**[1+**s**]-**T**-**Ad**}-[**Q**+**X**]/**/**

Rule-14211:

If both (**/**), (**U**), (**$**), (**i**), (**V**), (**Q**), (**f**), (**s**), (**T**), (**A**), (**d**) and (**X**) are known, then its Sales Past must be:

$'**= {**U**+**$[1-**i**]-**V**-**T**-**Ad**-[**Q**+**X**]/**/**}/{**f**[1+**s**]}

Rule-14212:

If both (**/**), (**U**), (**$**), (**i**), (**V**), (**$'**), (**Q**), (**s**), (**T**), (**A**), (**d**) and (**X**) are known, then its Fixed Portion Planned is:

f= {**U**+**$**[1-**i**]-**V**-**T**-**Ad**-[**Q**+**X**]/**/**}/{**$'**[1+**s**]}

Rule-14213:

If both (**/**), (**U**), (**$**), (**i**), (**V**), (**$'**), (**f**), (**Q**), (**T**), (**A**), (**d**) and (**X**) are known, then its Sales Growth Planned is:

s= **$**/**$'**-1

or it can also be found as

s= {**U**+**$**[1-**i**]-**V**-**T**-**Ad**-[**Q**+**X**]/**/**}/[**$'f**]-1

Rule-14214:

If both (**/**), (**U**), (**$**), (**i**), (**V**), (**$'**), (**f**), (**s**), (**Q**), (**A**), (**d**) and (**X**) are known, then its Tax Planned is:

T= {**U**+**$**[1-**i**]-**V**-**$'f**[1+**s**]-**Ad**}-[**Q**+**X**]/**/**

Rule-14215:

If both (**/**), (**U**), (**$**), (**i**), (**V**), (**$'**), (**f**), (**s**), (**T**), (**Q**), (**d**) and (**X**) are known, then its After Tax Income Planned is:

A= {**U**+**$**[1-**i**]-**V**-**$'f**[1+**s**]-**T**-[**Q**+**X**]/**/**}/**d**

Steve Asikin ISBN 14: 978-1514685136, ISBN 10: 1514685132

Rule-14216:

If both (**/**), (**U**), (**$**), (**i**), (**V**), (**$'**), (**f**), (**s**), (**T**), (**A**), (**Q**) and (**X**) are known, then its Dividend Payout Planned is:

d= {**U**+**$**[1-**i**]-**V**-**$'f**[1+**s**]-**T**-[**Q**+**X**]/**/**}/**A**

Rule-14217:

If both (**/**), (**U**), (**$**), (**i**), (**V**), (**$'**), (**f**), (**s**), (**T**), (**A**), (**d**) and (**Q**) are known, then its Xpress or Current Debt Planned is:

X= **/**{**U**+**$**[1-**i**]-**V**-**$'f**[1+**s**]-**T**-**Ad**}-**Q**

Rule-14218:

If both (**/**), (**U**), (**$**), (**i**), (**V**), (**$'**), (**f**), (**s**), (**T**), (**A**), (**d**), (**P**), (**c**) and (**q**) are known, then its Quoted Longterm Debt Planned is:

Q= **/**{**U**+**$**[1-**i**]-**V**-**$'f**[1+**s**]-**T**-**Ad**}-**P**/[**c**-**q**]

Rule-14219:

If both (**Q**), (**U**), (**$**), (**i**), (**V**), (**$'**), (**f**), (**s**), (**T**), (**A**), (**d**), (**P**), (**c**) and (**q**) are known, then its Leverage or Gearing Ratio Planned is:

/= {**Q**+**P**/[**c**-**q**]}/{**U**+**$**[1-**i**]-**V**-**$'f**[1+**s**]-**T**-**Ad**}

Rule-14220:

If both (**/**), (**Q**), (**$**), (**i**), (**V**), (**$'**), (**f**), (**s**), (**T**), (**A**), (**d**), (**P**), (**c**) and (**q**) are known, then its Utilized or Starting Capital must be:

U= {**Q**+**P**/[**c**-**q**]}/**/**-{**$**[1-**i**]-**V**-**$'f**[1+**s**]-**T**-**Ad**}

Steve Asikin ISBN 14: 978-1514685136, ISBN 10: 1514685132

Rule-14221:

If both (/), (U), (Q), (i), (V), (S'), (f), (s), (T), (A), (d), (P), (c) and (q) are known, then its Sales or Revenue Planned is:

S= (V+S'f[1+s]+T+Ad+{Q+P/[c-q]}//-U)/[1-i]

Rule-14222:

If both (/), (U), (S), (Q), (V), (S'), (f), (s), (T), (A), (d), (P), (c) and (q) are known, then its Interest Portion Planned is:

i= 1-(V+S'f[1+s]+T+Ad+{Q+P/[c-q]}//-U)/S

Rule-14223:

If both (/), (U), (S), (i), (Q), (S'), (f), (s), (T), (A), (d), (P), (c) and (q) are known, then its Variable Cost Planned is:

V= {U+S[1-i]-S'f[1+s]-T-Ad}-{Q+P/[c-q]}//

Rule-14224:

If both (/), (U), (S), (i), (V), (Q), (f), (s), (T), (A), (d), (P), (c) and (q) are known, then its Sales Past must be:

S'= (U+S[1-i]-V-T-Ad-{Q+P/[c-q]}//)/{f[1+s]}

Rule-14225:

If both (/), (U), (S), (i), (V), (S'), (Q), (s), (T), (A), (d), (P), (c) and (q) are known, then its Fixed Portion Planned is:

f= (U+S[1-i]-V-T-Ad-{Q+P/[c-q]}//)/{S'[1+s]}

Steve Asikin ISBN 14: 978-1514685136, ISBN 10: 1514685132

Rule-14226:

If both (**/**), (**U**), (**$**), (**i**), (**V**), (**$'**), (**f**), (**Q**), (**T**), (**A**), (**d**), (**P**), (**c**) and (**q**) are known, then its Sales Growth Planned is:

s= $/$'-1

or it can also be found as

s= (U+$[1-i]-V-T-Ad-{Q+P/[c-q]}//)/ [$'f]-1

Rule-14227:

If both (**/**), (**U**), (**$**), (**i**), (**V**), (**$'**), (**f**), (**s**), (**Q**), (**A**), (**d**), (**P**), (**c**) and (**q**) are known, then its Tax Planned is:

T= {U+$[1-i]-V-$'f[1+s]-Ad}-{Q+P/[c-q]}//

Rule-14228:

If both (**/**), (**U**), (**$**), (**i**), (**V**), (**$'**), (**f**), (**s**), (**T**), (**Q**), (**d**), (**P**), (**c**) and (**q**) are known, then its After Tax Income Planned is:

A= (U+$[1-i]-V-$'f[1+s]-T-{Q+P/[c-q]}//)/d

Rule-14229:

If both (**/**), (**U**), (**$**), (**i**), (**V**), (**$'**), (**f**), (**s**), (**T**), (**A**), (**Q**), (**P**), (**c**) and (**q**) are known, then its Dividend Payout Planned is:

d= (U+$[1-i]-V-$'f[1+s]-T-{Q+P/[c-q]}//)/A

Rule-14230:

If both (**/**), (**U**), (**$**), (**i**), (**V**), (**$'**), (**f**), (**s**), (**T**), (**A**), (**d**), (**Q**), (**c**) and (**q**) are known, then its Procured Inventory Planned is:

P= [c-q](/{U+$[1-i]-V-$'f[1+s]-T-Ad}-Q)

Steve Asikin ISBN 14: 978-1514685136, ISBN 10: 1514685132

Rule-14231:

If both (**/**), (**U**), (**$**), (**i**), (**V**), (**$'**), (**f**), (**s**), (**T**), (**A**), (**d**), (**P**), (**Q**) and (**q**) are known, then its Current Ratio Planned is:

c= **q**+**P**/(**/**{**U**+**$**[1-**i**]-**V**-**$'f**[1+**s**]-**T**-**Ad**}-**Q**)

Rule-14232:

If both (**/**), (**U**), (**$**), (**i**), (**V**), (**$'**), (**f**), (**s**), (**T**), (**A**), (**d**), (**P**), (**c**) and (**Q**) are known, then its Quick or Acid Test Ratio Planned is:

q= **c**-**P**/(**/**{**U**+**$**[1-**i**]-**V**-**$'f**[1+**s**]-**T**-**Ad**}-**Q**)

Rule-14233:

If both (**/**), (**U**), (**$**), (**i**), (**V**), (**$'**), (**f**), (**s**), (**T**), (**A**), (**d**), (**p**), (**c**) and (**q**) are known, then its Quoted Longterm Debt Planned is:

Q= **/**{**U**+**$**[1-**i**]-**V**-**$'f**[1+**s**]-**T**-**Ad**}-**Vp**/{360[**c**-**q**]}

Rule-14234:

If both (**Q**), (**U**), (**$**), (**i**), (**V**), (**$'**), (**f**), (**s**), (**T**), (**A**), (**d**), (**p**), (**c**) and (**q**) are known, then its Leverage or Gearing Ratio Planned is:

/= (**Q**+**Vp**/{360[**c**-**q**]})
/{**U**+**$**[1-**i**]-**V**-**$'f**[1+**s**]-**T**-**Ad**}

Rule-14235:

If both (**/**), (**Q**), (**$**), (**i**), (**V**), (**$'**), (**f**), (**s**), (**T**), (**A**), (**d**), (**p**), (**c**) and (**q**) are known, then its Utilized or Starting Capital must be:

U= (**Q**+**Vp**/{360[**c**-**q**]})/**/**-{**$**[1-**i**]-**V**-**$'f**[1+**s**]-**T**-**Ad**}

Steve Asikin ISBN 14: 978-1514685136, ISBN 10: 1514685132

Rule-14236:

If both (**/**), (**U**), (**Q**), (**i**), (**V**), (**$'**), (**f**), (**s**), (**T**), (**A**), (**d**), (**p**), (**c**) and (**q**) are known, then its Sales or Revenue Planned is:

$**= [**V**+**$'f[1+**s**]+**T**+**Ad**+(**Q**+**Vp**/{360[**c**-**q**]})/**/**-**U**]
/[1-**i**]

Rule-14237:

If both (**/**), (**U**), (**$**), (**Q**), (**V**), (**$'**), (**f**), (**s**), (**T**), (**A**), (**d**), (**p**), (**c**) and (**q**) are known, then its Interest Portion Planned is:

i= 1-[**V**+**$'f**[1+**s**]+**T**+**Ad**
+(**Q**+**Vp**/{360[**c**-**q**]})/**/**-**U**]/**$**

Rule-14238:

If both (**/**), (**U**), (**$**), (**i**), (**Q**), (**$'**), (**f**), (**s**), (**T**), (**A**), (**d**), (**p**), (**c**) and (**q**) are known, then its Variable Cost Planned is:

V= (**/**{**U**+**$**[1-**i**]-**$'f**[1+**s**]-**T**-**Ad**}-**Q**)
/(**p**/{360[**c**-**q**]}+**/**)

Rule-14239:

If both (**/**), (**U**), (**$**), (**i**), (**V**), (**Q**), (**f**), (**s**), (**T**), (**A**), (**d**), (**p**), (**c**) and (**q**) are known, then its Sales Past must be:

$'**= [**U**+**$[1-**i**]-**V**-**T**-**Ad**
-(**Q**+**Vp**/{360[**c**-**q**]})/**/**]/{**f**[1+**s**]}

Steve Asikin ISBN 14: 978-1514685136, ISBN 10: 1514685132

Rule-14240:

If both (**/**), (**U**), (**$**), (**i**), (**V**), (**$'**), (**Q**), (**s**), (**T**), (**A**), (**d**), (**p**), (**c**) and (**q**) are known, then its Fixed Portion Planned is:

f= [**U**+**$**[1-**i**]-**V**-**T**-**Ad**-(**Q**+**Vp**/{360[**c**-**q**]})/**/**]
/{**$'**[1+**s**]}

Rule-14241:

If both (**/**), (**U**), (**$**), (**i**), (**V**), (**$'**), (**f**), (**Q**), (**T**), (**A**), (**d**), (**p**), (**c**) and (**q**) are known, then its Sales Growth Planned is:

s= **$**/**$'**-1

or it can also be found as

s= [**U**+**$**[1-**i**]-**V**-**T**-**Ad**-(**Q**+**Vp**/{360[**c**-**q**]})/**/**]/[**$'f**]-1

Rule-14242:

If both (**/**), (**U**), (**$**), (**i**), (**V**), (**$'**), (**f**), (**s**), (**Q**), (**A**), (**d**), (**p**), (**c**) and (**q**) are known, then its Tax Planned is:

T= {**U**+**$**[1-**i**]-**V**-**$'f**[1+**s**]-**Ad**}
-(**Q**+**Vp**/{360[**c**-**q**]})/**/**

Rule-14243:

If both (**/**), (**U**), (**$**), (**i**), (**V**), (**$'**), (**f**), (**s**), (**T**), (**Q**), (**d**), (**p**), (**c**) and (**q**) are known, then its After Tax Income Planned is:

A= [**U**+**$**[1-**i**]-**V**-**$'f**[1+**s**]-**T**
-(**Q**+**Vp**/{360[**c**-**q**]})/**/**]/**d**

Steve Asikin ISBN 14: 978-1514685136, ISBN 10: 1514685132

Rule-14244:

If both (**/**), (**U**), (**$**), (**i**), (**V**), (**$'**), (**f**), (**s**), (**T**), (**A**), (**Q**), (**p**), (**c**) and (**q**) are known, then its Dividend Payout Planned is:

d= [U+$[1-i]-V-$'f[1+s]-T
-(Q+Vp/{360[c-q]})/**/**]/A

Rule-14245:

If both (**/**), (**U**), (**$**), (**i**), (**V**), (**$'**), (**f**), (**s**), (**T**), (**A**), (**d**), (**Q**), (**c**) and (**q**) are known, then its Procured Inventory Days Planned:

p= 360[c-q](/{U+$[1-i]-V-$'f[1+s]-T-Ad}-Q)/V

Rule-14246:

If both (**/**), (**U**), (**$**), (**i**), (**V**), (**$'**), (**f**), (**s**), (**T**), (**A**), (**d**), (**p**), (**Q**) and (**q**) are known, then its Current Ratio Planned is:

c= q+Vp/[360(/{U+$[1-i]-V-$'f[1+s]-T-Ad}-Q)]

Rule-14247:

If both (**/**), (**U**), (**$**), (**i**), (**V**), (**$'**), (**f**), (**s**), (**T**), (**A**), (**d**), (**p**), (**c**) and (**Q**) are known, then its Quick or Acid Test Ratio Planned is:

q= c-Vp/[360(/{U+$[1-i]-V-$'f[1+s]-T-Ad}-Q)]

Rule-14248:

If both (**/**), (**U**), (**$**), (**i**), (**V**), (**$'**), (**f**), (**s**), (**T**), (**A**), (**d**), (**v**), (**p**), (**c**) and (**q**) are known, then its Quoted Longterm Debt Planned is:

Q= /{U+$[1-i]-V-$'f[1+s]-T-Ad}-$vp/{360[c-q]}

Steve Asikin ISBN 14: 978-1514685136, ISBN 10: 1514685132

Rule-14249:

If both (**Q**), (**U**), (**$**), (**i**), (**V**), (**$'**), (**f**), (**s**), (**T**), (**A**), (**d**), (**v**), (**p**), (**c**) and (**q**) are known, then its Leverage ior Gearing Ratio Planned is:

/= (**Q**+**$vp**/{360[**c-q**]})
/{**U**+**$**[1-**i**]-**V-$'f**[1+**s**]-**T-Ad**}

Rule-14250:

If both (**/**), (**Q**), (**$**), (**i**), (**V**), (**$'**), (**f**), (**s**), (**T**), (**A**), (**d**), (**v**), (**p**), (**c**) and (**q**) are known, then its Utilized or Starting Capital must be:

U= (**Q**+**$vp**/{360[**c-q**]})/**/**
-{**$**[1-**i**]-**V-$'f**[1+**s**]-**T-Ad**}

Rule-14251:

If both (**/**), (**U**), (**Q**), (**i**), (**V**), (**$'**), (**f**), (**s**), (**T**), (**A**), (**d**), (**v**), (**p**), (**c**) and (**q**) are known, then its Sales or Revenue Planned is:

$**= (**/**{**U-V-$'f[1+**s**]-**T-Ad**}-**Q**)
/(**vp**/{360[**c-q**]}-**/**[1-**i**])

Rule-14252:

If both (**/**), (**U**), (**$**), (**Q**), (**V**), (**$'**), (**f**), (**s**), (**T**), (**A**), (**d**), (**v**), (**p**), (**c**) and (**q**) are known, then its Interest Portion Planned is:

i= 1-[**V**+**$'f**[1+**s**]+**T**+**Ad**
+(**Q**+**$vp**/{360[**c-q**]})/**/**-**U**]/**$**

Steve Asikin ISBN 14: 978-1514685136, ISBN 10: 1514685132

Rule-14253:

If both (**/**), (**U**), (**$**), (**i**), (**Q**), (**$'**), (**f**), (**s**), (**T**), (**A**), (**d**), (**v**), (**p**), (**c**) and (**q**) are known, then its Variable Cost Planned is:

V= (**/**{**U**+**$**[1-**i**]-**$'f**[1+**s**]-**T**-**Ad**}
-**$vp**/{360[**c**-**q**]}-**Q**)/**/**

Rule-14254:

If both (**/**), (**U**), (**$**), (**i**), (**V**), (**Q**), (**f**), (**s**), (**T**), (**A**), (**d**), (**v**), (**p**), (**c**) and (**q**) are known, then its Sales Past must be:

$'**= [**U**+**$[1-**i**]-**V**-**T**-**Ad**-(**Q**+**$vp**/{360[**c**-**q**]})/**/**]
/{**f**[1+**s**]}

Rule-14255:

If both (**/**), (**U**), (**$**), (**i**), (**V**), (**$'**), (**Q**), (**s**), (**T**), (**A**), (**d**), (**v**), (**p**), (**c**) and (**q**) are known, then its Fixed Portion Planned is:

f= [**U**+**$**[1-**i**]-**V**-**T**-**Ad**-(**Q**+**$vp**/{360[**c**-**q**]})/**/**]
/{**$'**[1+**s**]}

Rule-14256:

If both (**/**), (**U**), (**$**), (**i**), (**V**), (**$'**), (**f**), (**Q**), (**T**), (**A**), (**d**), (**v**), (**p**), (**c**) and (**q**) are known, then its Sales Growth Planned is:

s= **$**/**$'**-1

or it can also be found as

s= [**U**+**$**[1-**i**]-**V**-**T**-**Ad**-(**Q**+**$vp**/{360[**c**-**q**]})/**/**]
/[**$'f**]-1

Steve Asikin ISBN 14: 978-1514685136, ISBN 10: 1514685132

Rule-14257:

If both (**/**), (**U**), (**$**), (**i**), (**V**), (**$'**), (**f**), (**s**), (**Q**), (**A**), (**d**), (**v**), (**p**), (**c**) and (**q**) are known, then its Tax Planned is:

T= {**U**+**$**[1-**i**]-**V**-**$'f**[1+**s**]-**Ad**}
-(**Q**+**$vp**/{360[**c-q**]})/**/**

Rule-14258:

If both (**/**), (**U**), (**$**), (**i**), (**V**), (**$'**), (**f**), (**s**), (**T**), (**Q**), (**d**), (**v**), (**p**), (**c**) and (**q**) are known, then its After Tax Income Planned is:

A= [**U**+**$**[1-**i**]-**V**-**$'f**[1+**s**]-**T**
-(**Q**+**$vp**/{360[**c-q**]})/**/**]/**d**

Rule-14259:

If both (**/**), (**U**), (**$**), (**i**), (**V**), (**$'**), (**f**), (**s**), (**T**), (**A**), (**Q**), (**v**), (**p**), (**c**) and (**q**) are known, then its Dividend Payout Planned is:

d= [**U**+**$**[1-**i**]-**V**-**$'f**[1+**s**]-**T**
-(**Q**+**$vp**/{360[**c-q**]})/**/**]/**A**

Rule-14260:

If both (**/**), (**U**), (**$**), (**i**), (**V**), (**$'**), (**f**), (**s**), (**T**), (**A**), (**d**), (**Q**), (**p**), (**c**) and (**q**) are known, then its Variable Portion Planned is:

v= **V**/**$**

or it can also be found as

v= 360[**c-q**](**/**{**U**+**$**[1-**i**]-**V**-**$'f**[1+**s**]-**T**-**Ad**}-**Q**)/[**$p**]

Steve Asikin ISBN 14: 978-1514685136, ISBN 10: 1514685132

Rule-14261:

If both (**/**), (**U**), (**$**), (**i**), (**V**), (**$'**), (**f**), (**s**), (**T**), (**A**), (**d**), (**v**), (**Q**), (**c**) and (**q**) are known, then its Procured Inventory Days Planned:

p= 360[**c-q**](**/**{**U**+**$**[1-**i**]-**V**-**$'f**[1+**s**]-**T**-**Ad**}-**Q**)/[**$v**]

Rule-14262:

If both (**/**), (**U**), (**$**), (**i**), (**V**), (**$'**), (**f**), (**s**), (**T**), (**A**), (**d**), (**v**), (**p**), (**Q**) and (**q**) are known, then its Current Ratio Planned is:

c= **q**+**$vp**/[360(**/**{**U**+**$**[1-**i**]-**V**-**$'f**[1+**s**]-**T**-**Ad**}-**Q**)]

Rule-14263:

If both (**/**), (**U**), (**$**), (**i**), (**V**), (**$'**), (**f**), (**s**), (**T**), (**A**), (**d**), (**v**), (**p**), (**c**) and (**Q**) are known, then its Quick or Acid Test Ratio Planned is:

q= **c**-**$vp**/[360(**/**{**U**+**$**[1-**i**]-**V**-**$'f**[1+**s**]-**T**-**Ad**}-**Q**)]

Rule-14264:

If both (**/**), (**U**), (**$**), (**i**), (**V**), (**$'**), (**f**), (**s**), (**T**), (**A**), (**d**), (**v**), (**p**), (**c**) and (**q**) are known, then its Quoted Longterm Debt Planned is:

Q= **/**{**U**+**$**[1-**i**]-**V**-**$'f**[1+**s**]-**T**-**Ad**}
-**$'vp**[1+**s**]/{360[**c-q**]}

Steve Asikin ISBN 14: 978-1514685136, ISBN 10: 1514685132

Rule-14265:

If both (**Q**), (**U**), (**S**), (**i**), (**V**), (**S'**), (**f**), (**s**), (**T**), (**A**), (**d**), (**v**), (**p**), (**c**) and (**q**) are known, then its Leverage or Gearing Ratio Planned is:

$$/= (Q+S'vp[1+s]/\{360[c-q]\})/\{U+S[1-i]-V-S'f[1+s]-T-Ad\}$$

Rule-14266:

If both (**/**), (**Q**), (**S**), (**i**), (**V**), (**S'**), (**f**), (**s**), (**T**), (**A**), (**d**), (**v**), (**p**), (**c**) and (**q**) are known, then its Utilized or Starting Capital must be:

$$U= (Q+S'vp[1+s]/\{360[c-q]\})// -\{S[1-i]-V-S'f[1+s]-T-Ad\}$$

Rule-14267:

If both (**/**), (**U**), (**Q**), (**i**), (**V**), (**S'**), (**f**), (**s**), (**T**), (**A**), (**d**), (**v**), (**p**), (**c**) and (**q**) are known, then its Sales or Revenue Planned is:

$$S= (Q-/\{U-V-S'f[1+s]-T-Ad\} +S'vp[1+s]/\{360[c-q]\})/\{/[1-i]\}$$

Rule-14268:

If both (**/**), (**U**), (**S**), (**Q**), (**V**), (**S'**), (**f**), (**s**), (**T**), (**A**), (**d**), (**v**), (**p**), (**c**) and (**q**) are known, then its Interest Portion Planned is:

$$i= 1-[V+S'f[1+s]+T+Ad +(Q+S'vp[1+s]/\{360[c-q]\})//-U]/S$$

Steve Asikin ISBN 14: 978-1514685136, ISBN 10: 1514685132

Rule-14269:

If both (**/**), (**U**), (**$**), (**i**), (**Q**), (**$'**), (**f**), (**s**), (**T**), (**A**), (**d**), (**v**), (**p**), (**c**) and (**q**) are known, then its Variable Cost Planned is:

V= (**/**{**U**+**$**[1-**i**]-**$'f**[1+**s**]-**T**-**Ad**}
-**$'vp**[1+**s**]/{360[**c**-**q**]}-**Q**)/**/**

Rule-14270:

If both (**/**), (**U**), (**$**), (**i**), (**V**), (**Q**), (**f**), (**s**), (**T**), (**A**), (**d**), (**v**), (**p**), (**c**) and (**q**) are known, then its Sales Past must be:

$'**= (**/**{**U**+**$[1-**i**]-**V**-**T**-**Ad**}-**Q**)
/**[**[1+**s**](**vp**/{360[**c**-**q**]}-**f/**)

Rule-14271:

If both (**/**), (**U**), (**$**), (**i**), (**V**), (**$'**), (**Q**), (**s**), (**T**), (**A**), (**d**), (**v**), (**p**), (**c**) and (**q**) are known, then its Fixed Portion Planned is:

f= **[U**+**$**[1-**i**]-**V**-**T**-**Ad**
-(**Q**+**$'vp**[1+**s**]/{360[**c**-**q**]})/**/]**/{**$'**[1+**s**]}

Rule-14272:

If both (**/**), (**U**), (**$**), (**i**), (**V**), (**$'**), (**f**), (**Q**), (**T**), (**A**), (**d**), (**v**), (**p**), (**c**) and (**q**) are known, then its Sales Growth Planned is:

s= **$**/**$'**-1

or it can also be found as

s= (**/**{**U**+**$**[1-**i**]-**V**-**T**-**Ad**}-**Q**)
/**[$'**(**vp**/{360[**c**-**q**]}-**f/**)-1

Steve Asikin ISBN 14: 978-1514685136, ISBN 10: 1514685132

Rule-14273:

If both (**/**), (**U**), (**$**), (**i**), (**V**), (**$'**), (**f**), (**s**), (**Q**), (**A**), (**d**), (**v**), (**p**), (**c**) and (**q**) are known, then its Tax Planned is:

T= {**U**+**$**[1-**i**]-**V**-**$'f**[1+**s**]-**Ad**}
-(**Q**+**$'vp**[1+**s**]/{360[**c**-**q**]})/**/**

Rule-14274:

If both (**/**), (**U**), (**$**), (**i**), (**V**), (**$'**), (**f**), (**s**), (**T**), (**Q**), (**d**), (**v**), (**p**), (**c**) and (**q**) are known, then its After Tax Income Planned is:

A= [**U**+**$**[1-**i**]-**V**-**$'f**[1+**s**]-**T**
-(**Q**+**$'vp**[1+**s**]/{360[**c**-**q**]})/**/]**/**d**

Rule-14275:

If both (**/**), (**U**), (**$**), (**i**), (**V**), (**$'**), (**f**), (**s**), (**T**), (**A**), (**Q**), (**v**), (**p**), (**c**) and (**q**) are known, then its Dividend Payout Planned is:

d= [**U**+**$**[1-**i**]-**V**-**$'f**[1+**s**]-**T**
-(**Q**+**$'vp**[1+**s**]/{360[**c**-**q**]})/**/]**/**A**

Rule-14276:

If both (**/**), (**U**), (**$**), (**i**), (**V**), (**$'**), (**f**), (**s**), (**T**), (**A**), (**d**), (**Q**), (**p**), (**c**) and (**q**) are known, then its Variable Portion Planned is:

v= **V**/**$**

or it can also be found as

v= 360[**c**-**q**](**/**{**U**+**$**[1-**i**]-**V**-**$'f**[1+**s**]-**T**-**Ad**}-**Q**)
/{**$'p**[1+**s**]}

Steve Asikin ISBN 14: 978-1514685136, ISBN 10: 1514685132

Rule-14277:

If both (/), (U), (S), (i), (V), (S'), (f), (s), (T), (A), (d), (v), (Q), (c) and (q) are known, then its Procured Inventory Days Planned:

p= 360[c-q](/{U+S[1-i]-V-S'f[1+s]-T-Ad}-Q) /{S'v[1+s]}

Rule-14278:

If both (/), (U), (S), (i), (V), (S'), (f), (s), (T), (A), (d), (v), (p), (Q) and (q) are known, then its Current Ratio Planned is:

c= q+S'vp[1+s] /[360(/{U+S[1-i]-V-S'f[1+s]-T-Ad}-Q)]

Rule-14279:

If both (/), (U), (S), (i), (V), (S'), (f), (s), (T), (A), (d), (v), (p), (c) and (Q) are known, then its Quick or Acid Test Ratio Planned is:

q= c-S'vp[1+s] /[360(/{U+S[1-i]-V-S'f[1+s]-T-Ad}-Q)]

Rule-14280:

If both (/), (U), (S), (i), (V), (S'), (f), (s), (t), (D) and (X) are known, then its Quoted Longterm Debt Planned is:

Q= /(U+{S[1-i]-V-S'f[1+s]}[1-t]-D)-X

Steve Asikin ISBN 14: 978-1514685136, ISBN 10: 1514685132

Rule-14281:

If both (**Q**), (**U**), (**$**), (**i**), (**V**), (**$'**), (**f**), (**s**), (**t**), (**D**) and (**X**) are known, then its Leverage or Gearing Ratio Planned is:

/= [Q+X]/(U+{$[1-i]-V-$'f[1+s]}[1-t]-D)

Rule-14282:

If both (/), (**Q**), (**$**), (**i**), (**V**), (**$'**), (**f**), (**s**), (**t**), (**D**) and (**X**) are known, then its Utilized or Starting Capital must be:

U= [Q+X]//({$[1-i]-V-$'f[1+s]}[1-t]-D)

Rule-14283:

If both (/), (**U**), (**Q**), (**i**), (**V**), (**$'**), (**f**), (**s**), (**t**), (**D**) and (**X**) are known, then its Sales or Revenue Planned is:

$= (V+$'f[1+s]+{D+[Q+X]//U}/[1-t])/[1-i]

Rule-14284:

If both (/), (**U**), (**$**), (**Q**), (**V**), (**$'**), (**f**), (**s**), (**t**), (**D**) and (**X**) are known, then its Interest Portion Planned is:

i= 1-(V+$'f[1+s]+{D+[Q+X]//U}/[1-t])/$

Rule-14285:

If both (/), (**U**), (**$**), (**i**), (**Q**), (**$'**), (**f**), (**s**), (**t**), (**D**) and (**X**) are known, then its Variable Cost Planned is:

V= $[1-i]-$'f[1+s]}-{D+[Q+X]//U}/[1-t]

Rule-14286:

If both (/), (**U**), (**$**), (**i**), (**V**), (**Q**), (**f**), (**s**), (**t**), (**D**) and (**X**) are known, then its Sales Past must be:

$'= ($[1-i]-V-{D+[Q+X]//U}/[1-t])/{f[1+s]}

Steve Asikin ISBN 14: 978-1514685136, ISBN 10: 1514685132

Rule-14287:

If both (**/**), (**U**), (**$**), (**i**), (**V**), (**$'**), (**Q**), (**s**), (**t**), (**D**) and (**X**) are known, then its Fixed Portion Planned is:

f= (**$**[1-**i**]-**V**-{**D**+[**Q**+**X**]/**/U**}/[1-**t**])/{**$'**[1+**s**]}

Rule-14288:

If both (**/**), (**U**), (**$**), (**i**), (**V**), (**$'**), (**f**), (**s**), (**t**), (**D**) and (**X**) are known, then its Sales Growth Planned is:

s= **$**/**$'**-1

or it can also be found as

s= (**$**[1-**i**]-**V**-{**D**+[**Q**+**X**]/**/U**}/[1-**t**])/[**$'f**]-1

Rule-14289:

If both (**/**), (**U**), (**$**), (**i**), (**V**), (**$'**), (**f**), (**s**), (**Q**), (**D**) and (**X**) are known, then its Tax Rate Planned is:

t= 1-{**D**+[**Q**+**X**]/**/U**}/{**$**[1-**i**]-**V**-**$'f**[1+**s**]}

Rule-14290:

If both (**/**), (**U**), (**$**), (**i**), (**V**), (**$'**), (**f**), (**s**), (**t**), (**Q**) and (**X**) are known, then its Dividend Planned is:

D= {**$**[1-**i**]-**V**-**$'f**[1+**s**]}[1-**t**]-[**Q**+**X**]/**/U**

Rule-14291:

If both (**/**), (**U**), (**$**), (**i**), (**V**), (**$'**), (**f**), (**s**), (**t**), (**D**) and (**Q**) are known, then its Xpress or Current Debt Planned is:

X= **/**(**U**+{**$**[1-**i**]-**V**-**$'f**[1+**s**]}[1-**t**]-**D**)-**Q**

Steve Asikin ISBN 14: 978-1514685136, ISBN 10: 1514685132

Rule-14292:

If both (/), (U), ($), (i), (V), ($'), (f), (s), (t), (D), (P), (c) and (q) are known, then its Quoted Longterm Debt Planned is:

Q= /(U+{$[1-i]-V-$'f[1+s]}[1-t]-D)-P/[c-q]

Rule-14293:

If both (Q), (U), ($), (i), (V), ($'), (f), (s), (t), (D), (P), (c) and (q) are known, then its Leverage or Gearing Ratio Planned is:

/= {Q+P/[c-q]}/(U+{$[1-i]-V-$'f[1+s]}[1-t]-D)

Rule-14294:

If both (/), (Q), ($), (i), (V), ($'), (f), (s), (t), (D), (P), (c) and (q) are known, then its Utilized or Starting Capital must be:

U={Q+P/[c-q]}//-({$[1-i]-V-$'f[1+s]}[1-t]-D)

Rule-14295:

If both (/), (U), (Q), (i), (V), ($'), (f), (s), (t), (D), (P), (c) and (q) are known, then its Sales or Revenue Planned is:

$= [V+$'f[1+s]+(D+{Q+P/[c-q]}//-U)/[1-t]]/[1-i]

Rule-14296:

If both (/), (U), ($), (Q), (V), ($'), (f), (s), (t), (D), (P), (c) and (q) are known, then its Interest Portion Planned is:

i= 1-[V+$'f[1+s]+(D+{Q+P/[c-q]}//-U)/[1-t]]/$

Steve Asikin ISBN 14: 978-1514685136, ISBN 10: 1514685132

Rule-14297:

If both (**/**), (**U**), (**$**), (**i**), (**Q**), (**$'**), (**f**), (**s**), (**t**), (**D**), (**P**), (**c**) and (**q**) are known, then its Variable Cost Planned is:

V= **$**[1-**i**]-**$'f**[1+**s**]}-(**D**+{**Q**+**P**/[**c**-**q**]}/**/U**)/ [1-**t**]

Rule-14298:

If both (**/**), (**U**), (**$**), (**i**), (**V**), (**Q**), (**f**), (**s**), (**t**), (**D**), (**P**), (**c**) and (**q**) are known, then its Sales Past must be:

$'**= [**$[1-**i**]-**V**-(**D**+{**Q**+**P**/[**c**-**q**]}/**/U**)/[1-**t**]]/{**f**[1+**s**]}

Rule-14299:

If both (**/**), (**U**), (**$**), (**i**), (**V**), (**$'**), (**Q**), (**s**), (**t**), (**D**), (**P**), (**c**) and (**q**) are known, then its Fixed Portion Planned is:

f= [**$**[1-**i**]-**V**-(**D**+{**Q**+**P**/[**c**-**q**]}/**/U**)/[1-**t**]]/{**$'**[1+**s**]}

Rule-14300:

If both (**/**), (**U**), (**$**), (**i**), (**V**), (**$'**), (**f**), (**Q**), (**t**), (**D**), (**P**), (**c**) and (**q**) are known, then its Sales Growth Planned is:

s= **$**/**$'**-1

or it can also be found as

s= [**$**[1-**i**]-**V**-(**D**+{**Q**+**P**/[**c**-**q**]}/**/U**)/[1-**t**]]/[**$'f**]-1

Rule-14301:

If both (**/**), (**U**), (**$**), (**i**), (**V**), (**$'**), (**f**), (**s**), (**Q**), (**D**), (**P**), (**c**) and (**q**) are known, then its Tax Rate Planned is:

t= 1-(**D**+{**Q**+**P**/[**c**-**q**]}/**/U**)/{**$**[1-**i**]-**V**-**$'f**[1+**s**]}

Steve Asikin ISBN 14: 978-1514685136, ISBN 10: 1514685132

Rule-14302:

If both (**/**), (**U**), (**$**), (**i**), (**V**), (**$'**), (**f**), (**s**), (**t**), (**Q**), (**P**), (**c**) and (**q**) are known, then its Dividend Planned is:

D= {**$**[1-**i**]-**V**-**$'f**[1+**s**]}[1-**t**]-{**Q**+**P**/[**c**-**q**]}/**/U**

Rule-14303:

If both (**/**), (**U**), (**$**), (**i**), (**V**), (**$'**), (**f**), (**s**), (**t**), (**D**), (**Q**), (**c**) and (**q**) are known, then its Procured Inventory Planned is:

P= [**c**-**q**][**/**(**U**+{**$**[1-**i**]-**V**-**$'f**[1+**s**]}[1-**t**]-**D**)-**Q**]

Rule-14304:

If both (**/**), (**U**), (**$**), (**i**), (**V**), (**$'**), (**f**), (**s**), (**t**), (**D**), (**P**), (**Q**) and (**q**) are known, then its Current Ratio Planned is:

c= **q**+**P**/[**/**(**U**+{**$**[1-**i**]-**V**-**$'f**[1+**s**]}[1-**t**]-**D**)-**Q**]

Rule-14305:

If both (**/**), (**U**), (**$**), (**i**), (**V**), (**$'**), (**f**), (**s**), (**t**), (**D**), (**P**), (**c**) and (**Q**) are known, then its Quick or Acid Test Ratio Planned is:

q= **c**-**P**/[**/**(**U**+{**$**[1-**i**]-**V**-**$'f**[1+**s**]}[1-**t**]-**D**)-**Q**]

Rule-14306:

If both (**/**), (**U**), (**$**), (**i**), (**V**), (**$'**), (**f**), (**s**), (**t**), (**D**), (**p**), (**c**) and (**q**) are known, then its Quoted Longterm Debt Planned is:

Q= **/**(**U**+{**$**[1-**i**]-**V**-**$'f**[1+**s**]}[1-**t**]-**D**)-**Vp**/{360[**c**-**q**]}

Steve Asikin ISBN 14: 978-1514685136, ISBN 10: 1514685132

Rule-14307:

If both (**Q**), (**U**), (**S**), (**i**), (**V**), (**S'**), (**f**), (**s**), (**t**), (**D**), (**p**), (**c**) and (**q**) are known, then its Leverage or Gearing Ratio Planned is:

/= (Q+Vp/{360[c-q]})
/(U+{S[1-i]-V-S'f[1+s]}[1-t]-D)

Rule-14308:

If both (**/**), (**Q**), (**S**), (**i**), (**V**), (**S'**), (**f**), (**s**), (**t**), (**D**), (**p**), (**c**) and (**q**) are known, then its Utilized or Starting Capital must be:

U=(Q+Vp/{360[c-q]})//
-({S[1-i]-V-S'f[1+s]}[1-t]-D)

Rule-14309:

If both (**/**), (**U**), (**Q**), (**i**), (**V**), (**S'**), (**f**), (**s**), (**t**), (**D**), (**p**), (**c**) and (**q**) are known, then its Sales or Revenue Planned is:

S= {V+S'f[1+s]
+[D+(Q+Vp/{360[c-q]})//-U]/[1-t]}/[1-i]

Rule-14310:

If both (**/**), (**U**), (**S**), (**Q**), (**V**), (**S'**), (**f**), (**s**), (**t**), (**D**), (**p**), (**c**) and (**q**) are known, then its Interest Portion Planned is:

i= 1-{V+S'f[1+s]
+[D+(Q+Vp/{360[c-q]})//-U]/[1-t]}/S

Steve Asikin ISBN 14: 978-1514685136, ISBN 10: 1514685132

Rule-14311:

If both (/), (U), ($), (i), (Q), ($'), (f), (s), (t), (D), (p), (c) and (q) are known, then its Variable Cost Planned is:

V= [/(U+{$[1-i]-$'f[1+s]}[1-t]-D)-Q] /(p/{360[c-q]}+/[1-t])

Rule-14312:

If both (/), (U), ($), (i), (V), (Q), (f), (s), (t), (D), (p), (c) and (q) are known, then its Sales Past must be:

$'= {$[1-i]-V-[D+(Q+Vp/{360[c-q]})/-U]/[1-t]} /{f[1+s]}

Rule-14313:

If both (/), (U), ($), (i), (V), ($'), (Q), (s), (t), (D), (p), (c) and (q) are known, then its Fixed Portion Planned is:

f= {$[1-i]-V-[D+(Q+Vp/{360[c-q]})/-U]/[1-t]} /{$'[1+s]}

Rule-14314:

If both (/), (U), ($), (i), (V), ($'), (f), (Q), (t), (D), (p), (c) and (q) are known, then its Sales Growth Planned is:

s= $/$'-1

or it can also be found as

s= {$[1-i]-V-[D+(Q+Vp/{360[c-q]})/-U]/[1-t]} /[$'f]-1

Steve Asikin ISBN 14: 978-1514685136, ISBN 10: 1514685132

Rule-14315:

If both (**/**), (**U**), (**$**), (**i**), (**V**), (**$'**), (**f**), (**s**), (**Q**), (**D**), (**p**), (**c**) and (**q**) are known, then its Tax Rate Planned is:

t= 1-[**D**+(**Q**+**Vp**/{360[**c**-**q**]})/**/U**]
/{**$**[1-**i**]-**V**-**$'f**[1+**s**]}

Rule-14316:

If both (**/**), (**U**), (**$**), (**i**), (**V**), (**$'**), (**f**), (**s**), (**t**), (**Q**), (**p**), (**c**) and (**q**) are known, then its Dividend Planned is:

D= {**$**[1-**i**]-**V**-**$'f**[1+**s**]}[1-**t**]
-(**Q**+**Vp**/{360[**c**-**q**]})/**/U**

Rule-14317:

If both (**/**), (**U**), (**$**), (**i**), (**V**), (**$'**), (**f**), (**s**), (**t**), (**D**), (**Q**), (**c**) and (**q**) are known, then its Procured Inventory Days Planned:

p= 360[**c**-**q**][**/**(**U**+{**$**[1-**i**]-**V**-**$'f**[1+**s**]}[1-**t**]-**D**)-**Q**]/**V**

Rule-14318:

If both (**/**), (**U**), (**$**), (**i**), (**V**), (**$'**), (**f**), (**s**), (**t**), (**D**), (**p**), (**Q**) and (**q**) are known, then its Current Ratio Planned is:

c= **q**+**Vp**
/{360[**/**(**U**+{**$**[1-**i**]-**V**-**$'f**[1+**s**]}[1-**t**]-**D**)-**Q**]}

Rule-14319:

If both (**/**), (**U**), (**$**), (**i**), (**V**), (**$'**), (**f**), (**s**), (**t**), (**D**), (**p**), (**c**) and (**Q**) are known, then its Quick or Acid Test Ratio Planned is:

q= **c**-**Vp**/{360[**/**(**U**+{**$**[1-**i**]-**V**-**$'f**[1+**s**]}[1-**t**]-**D**)-**Q**]}

Steve Asikin ISBN 14: 978-1514685136, ISBN 10: 1514685132

Rule-14320:

If both (/), (U), ($), (i), (V), ($'), (f), (s), (t), (D), (p), (v), (c) and (q) are known, then its Quoted Longterm Debt Planned is:

Q= /(U+{$[1-i]-V-$'f[1+s]}[1-t]-D)
-$vp/{360[c-q]}

Rule-14321:

If both (Q), (U), ($), (i), (V), ($'), (f), (s), (t), (D), (p), (v), (c) and (q) are known, then its Leverage or Gearing Ratio Planned is:

/= (Q+$vp/{360[c-q]})
/(U+{$[1-i]-V-$'f[1+s]}[1-t]-D)

Rule-14322:

If both (/), (Q), ($), (i), (V), ($'), (f), (s), (t), (D), (p), (v), (c) and (q) are known, then its Utilized or Starting Capital must be:

U=(Q+$vp/{360[c-q]})//
-({$[1-i]-V-$'f[1+s]}[1-t]-D)

Rule-14323:

If both (/), (U), (Q), (i), (V), ($'), (f), (s), (t), (D), (p), (v), (c) and (q) are known, then its Sales or Revenue Planned is:

$= [/(U-{V+$'f[1+s]}[1-t]-D)-Q]
/(vp/{360[c-q]}-$/[1-i] [1-t])

Steve Asikin ISBN 14: 978-1514685136, ISBN 10: 1514685132

Rule-14324:

If both (**/**), (**U**), (**$**), (**Q**), (**V**), (**$'**), (**f**), (**s**), (**t**), (**D**), (**p**), (**v**), (**c**) and (**q**) are known, then its Interest Portion Planned is:

i= 1-{**V**+**$'f**[1+**s**]+[**D**+(**Q**+**$vp**/{360[**c**-**q**]})/**/U**]
/[1-**t**]}/**$**

Rule-14325:

If both (**/**), (**U**), (**$**), (**i**), (**Q**), (**$'**), (**f**), (**s**), (**t**), (**D**), (**p**), (**v**), (**c**) and (**q**) are known, then its Variable Cost Planned is:

V= [**/U**+{**$**[1-**i**]-**$'f**[1+**s**]}[1-**t**]-**D**)
-**$vp**/{360[**c**-**q**]}-**Q**]/{**/**[1-**t**]}

Rule-14326:

If both (**/**), (**U**), (**$**), (**i**), (**V**), (**Q**), (**f**), (**s**), (**t**), (**D**), (**p**), (**v**), (**c**) and (**q**) are known, then its Sales Past must be:

$'**= {**$[1-**i**]-**V**-[**D**+(**Q**+**$vp**/{360[**c**-**q**]})/**/U**]/[1-**t**]}
/{**f**[1+**s**]}

Rule-14327:

If both (**/**), (**U**), (**$**), (**i**), (**V**), (**$'**), (**Q**), (**s**), (**t**), (**D**), (**p**), (**v**), (**c**) and (**q**) are known, then its Fixed Portion Planned is:

f= {**$**[1-**i**]-**V**-[**D**+(**Q**+**$vp**/{360[**c**-**q**]})/**/U**]/[1-**t**]}
/{**$'**[1+**s**]}

Steve Asikin ISBN 14: 978-1514685136, ISBN 10: 1514685132

Rule-14328:

If both (/), (U), (S), (i), (V), (S'), (f), (Q), (t), (D), (p), (v), (c) and (q) are known, then its Sales Growth Planned is:

s= {S[1-i]-V-[D+(Q+Svp/{360[c-q]})/⁄U]/[1-t]}/[S'f]-1

Rule-14329:

If both (/), (U), (S), (i), (V), (S'), (f), (s), (Q), (D), (p), (v), (c) and (q) are known, then its Tax Rate Planned is:

t= 1-[D+(Q+Svp/{360[c-q]})/⁄U]/{S[1-i]-V-S'f[1+s]}

Rule-14330:

If both (/), (U), (S), (i), (V), (S'), (f), (s), (t), (Q), (p), (v), (c) and (q) are known, then its Dividend Planned is:

D= {S[1-i]-V-S'f[1+s]}[1-t]-(Q+Svp/{360[c-q]})/⁄U

Rule-14331:

If both (/), (U), (S), (i), (V), (S'), (f), (s), (t), (D), (p), (Q), (c) and (q) are known, then its Variable Portion Planned is:

v= V/S

or it can also be found as

v= 360[c-q][⁄U+{S[1-i]-V-S'f[1+s]}[1-t]-D)-Q]/[Sp]

Steve Asikin ISBN 14: 978-1514685136, ISBN 10: 1514685132

Rule-14332:

If both (**/**), (**U**), (**$**), (**i**), (**V**), (**$'**), (**f**), (**s**), (**t**), (**D**), (**Q**), (**v**), (**c**) and (**q**) are known, then its Procured Inventory Days Planned:

p= 360[**c**-**q**][**/**(**U**+{**$**[1-**i**]-**V**-**$'f**[1+**s**]}[1-**t**]-**D**)-**Q**]
/[**$v**]

Rule-14333:

If both (**/**), (**U**), (**$**), (**i**), (**V**), (**$'**), (**f**), (**s**), (**t**), (**D**), (**p**), (**v**), (**Q**) and (**q**) are known, then its Current Ratio Planned is:

c= **q**+**$vp**
/{360[**/**(**U**+{**$**[1-**i**]-**V**-**$'f**[1+**s**]}[1-**t**]-**D**)-**Q**]}

Rule-14334:

If both (**/**), (**U**), (**$**), (**i**), (**V**), (**$'**), (**f**), (**s**), (**t**), (**D**), (**p**), (**v**), (**c**) and (**q**) are known, then its Quoted Longterm Debt Planned is:

q= **c**-**$vp**
/{360[**/**(**U**+{**$**[1-**i**]-**V**-**$'f**[1+**s**]}[1-**t**]-**D**)-**Q**]}

Rule-14335:

If both (**/**), (**U**), (**$**), (**i**), (**V**), (**$'**), (**f**), (**s**), (**t**), (**D**), (**p**), (**v**), (**c**) and (**q**) are known, then its Quoted Longterm Debt Planned is:

Q= **/**(**U**+{**$**[1-**i**]-**V**-**$'f**[1+**s**]}[1-**t**]-**D**)
-**$'vp**[1+**s**]/{360[**c**-**q**]}

Steve Asikin ISBN 14: 978-1514685136, ISBN 10: 1514685132

Rule-14336:

If both (**Q**), (**U**), (**$**), (**i**), (**V**), (**$'**), (**f**), (**s**), (**t**), (**D**), (**p**), (**v**), (**c**) and (**q**) are known, then its Leverage or Gearing Ratio Planned is:

/= (Q+$'vp[1+s]/{360[c-q]})
/(U+{$[1-i]-V-$'f[1+s]}[1-t]-D)

Rule-14337:

If both (**/**), (**Q**), (**$**), (**i**), (**V**), (**$'**), (**f**), (**s**), (**t**), (**D**), (**p**), (**v**), (**c**) and (**q**) are known, then its Utilized or Starting Capital must be:

U=(Q+$'vp[1+s]/{360[c-q]})//
-({$[1-i]-V-$'f[1+s]}[1-t]-D)

Rule-14338:

If both (**/**), (**U**), (**Q**), (**i**), (**V**), (**$'**), (**f**), (**s**), (**t**), (**D**), (**p**), (**v**), (**c**) and (**q**) are known, then its Sales or Revenue Planned is:

$= [Q-/(U-{V+$'f[1+s]}[1-t]-D)
+$'vp[1+s]/{360[c-q]}]/{/[1-i] [1-t]}

Rule-14339:

If both (**/**), (**U**), (**$**), (**Q**), (**V**), (**$'**), (**f**), (**s**), (**t**), (**D**), (**p**), (**v**), (**c**) and (**q**) are known, then its Interest Portion Planned is:

i= 1-{V+$'f[1+s]+[D
+(Q+$'vp[1+s]/{360[c-q]})//-U]/[1-t]}/$

Steve Asikin ISBN 14: 978-1514685136, ISBN 10: 1514685132

Rule-14340:

If both (**/**), (**U**), (**$**), (**i**), (**V**), (**$'**), (**f**), (**s**), (**t**), (**D**), (**p**), (**v**), (**c**) and (**q**) are known, then its Sales or Revenue Planned is:

V= [**/**(**U**+{**$**[1-**i**]-**$'f**[1+**s**]}[1-**t**]-**D**)
-**$'vp**[1+**s**]/{360[**c**-**q**]}-**Q**]/{**/**[1-**t**]}

Rule-14341:

If both (**/**), (**U**), (**$**), (**i**), (**V**), (**Q**), (**f**), (**s**), (**t**), (**D**), (**p**), (**v**), (**c**) and (**q**) are known, then its Sales Past must be:

$'**= [**/**(**U**+{**$[1-**i**]-**V**}[1-**t**]-**D**)-**Q**]
/[[1+**s**](**vp**/{360[**c**-**q**]}+**f/**[1-**t**])

Rule-14342:

If both (**/**), (**U**), (**$**), (**i**), (**V**), (**$'**), (**Q**), (**s**), (**t**), (**D**), (**p**), (**v**), (**c**) and (**q**) are known, then its Fixed Portion Planned is:

f= {**$**[1-**i**]-**V**-[**D**+(**Q**+**$'vp**[1+**s**]/{360[**c**-**q**]})/**/**-**U**]
/[1-**t**]}//{**$'**[1+**s**]}

Rule-14343:

If both (**/**), (**U**), (**$**), (**i**), (**V**), (**$'**), (**f**), (**Q**), (**t**), (**D**), (**p**), (**v**), (**c**) and (**q**) are known, then its Sales Growth Planned is:

s= **$**/**$'**-1

or it can also be found as

s= [**/**(**U**+{**$**[1-**i**]-**V**}[1-**t**]-**D**)-**Q**]
/[**$'**(**vp**/{360[**c**-**q**]}+**f/**[1-**t**])-1

Steve Asikin ISBN 14: 978-1514685136, ISBN 10: 1514685132

Rule-14344:

If both (/), (U), (S), (i), (V), (S'), (f), (Q), (Q), (D), (p), (v), (c) and (q) are known, then its Tax Rate Planned is:

t= 1-[D+(Q+S'vp[1+s]/{360[c-q]})/ℓU]
/{S[1-i]-V-S'f[1+s]}

Rule-14345:

If both (/), (U), (S), (i), (V), (S'), (f), (Q), (t), (Q), (p), (v), (c) and (q) are known, then its Dividend Planned is:

D= {S[1-i]-V-S'f[1+s]}[1-t]
-(Q+S'vp[1+s]/{360[c-q]})/ℓU

Rule-14346:

If both (/), (U), (S), (i), (V), (S'), (f), (Q), (t), (D), (p), (Q), (c) and (q) are known, then its Variable Portion Planned is:

v= 360[c-q][/(U+{S[1-i]-V-S'f[1+s]}[1-t]-D)-Q]
/{S'p[1+s]}

Rule-14347:

If both (/), (U), (S), (i), (V), (S'), (f), (Q), (t), (D), (Q), (v), (c) and (q) are known, then its Procured Inventory Days Planned:

p= 360[c-q][/(U+{S[1-i]-V-S'f[1+s]}[1-t]-D)-Q]
/{S'v[1+s]}

Steve Asikin ISBN 14: 978-1514685136, ISBN 10: 1514685132

Rule-14348:

If both (**/**), (**U**), (**S**), (**i**), (**V**), (**S'**), (**f**), (**Q**), (**t**), (**D**), (**p**), (**v**), (**Q**) and (**q**) are known, then its Current Ratio Planned is:

c= **q**+**S'vp**[1+**s**]
/{360[**/**(**U**+{**S**[1-**i**]-**V**-**S'f**[1+**s**]}[1-**t**]-**D**)-**Q**]}

Rule-14349:

If both (**/**), (**U**), (**S**), (**i**), (**V**), (**S'**), (**f**), (**Q**), (**t**), (**D**), (**p**), (**v**), (**c**) and (**Q**) are known, then its Quick or Acid Test Ratio Planned is:

q= **c**-**S'vp**[1+**s**]
/{360[**/**(**U**+{**S**[1-**i**]-**V**-**S'f**[1+**s**]}[1-**t**]-**D**)-**Q**]}

Rule-14350:

If both (**/**), (**U**), (**S**), (**i**), (**V**), (**S'**), (**f**), (**Q**), (**t**), (**A**), (**d**) and (**X**) are known, then its Quoted Longterm Debt Planned is:

Q= **/**(**U**+{**S**[1-**i**]-**V**-**S'f**[1+**s**]}[1-**t**]-**Ad**)-**X**

Rule-14351:

If both (**Q**), (**U**), (**S**), (**i**), (**V**), (**S'**), (**f**), (**Q**), (**t**), (**A**), (**d**) and (**X**) are known, then its Leverage or Gearing Ratio Planned is:

/= [**Q**+**X**]/(**U**+{**S**[1-**i**]-**V**-**S'f**[1+**s**]}[1-**t**]-**Ad**)

Rule-14352:

If both (**/**), (**Q**), (**S**), (**i**), (**V**), (**S'**), (**f**), (**Q**), (**t**), (**A**), (**d**) and (**X**) are known, then its Utilized or Starting Capital must be:

U= [**Q**+**X**]/**/**({**S**[1-**i**]-**V**-**S'f**[1+**s**]}[1-**t**]-**Ad**)

Steve Asikin ISBN 14: 978-1514685136, ISBN 10: 1514685132

Rule-14353:

If both (/), (U), (Q), (i), (V), (S'), (f), (Q), (t), (A), (d) and (X) are known, then its Sales or Revenue Planned is:

S= (V+S'f[1+s]+{Ad+[Q+X]/-U}/[1-t])/[1-i]

Rule-14354:

If both (/), (U), (S), (Q), (V), (S'), (f), (Q), (t), (A), (d) and (X) are known, then its Interest Portion Planned is:

i= 1-(V+S'f[1+s]+{Ad+[Q+X]/-U}/[1-t])/S

Rule-14355:

If both (/), (U), (S), (i), (Q), (S'), (f), (Q), (t), (A), (d) and (X) are known, then its Variable Cost Planned is:

V= S[1-i]-S'f[1+s]-{Ad+[Q+X]/-U}/[1-t]

Rule-14356:

If both (/), (U), (S), (i), (V), (Q), (f), (Q), (t), (A), (d) and (X) are known, then its Sales Past must be:

S'= (S[1-i]-V-{Ad+[Q+X]/-U}/[1-t])/{f[1+s]}

Rule-14357:

If both (/), (U), (S), (i), (V), (S'), (Q), (Q), (t), (A), (d) and (X) are known, then its Fixed Portion Planned is:

f= (S[1-i]-V-{Ad+[Q+X]/-U}/[1-t])/{S'[1+s]}

Steve Asikin ISBN 14: 978-1514685136, ISBN 10: 1514685132

Rule-14358:

If both (/), (U), ($), (i), (V), ($'), (f), (Q), (t), (A), (d) and (X) are known, then its Sales Growth Planned is:

s= $/$'-1

or it can also be found as

s= 1-($[1-i]-V-{Ad+[Q+X]//-U}/[1-t])/[$'f]-1

Rule-14359:

If both (/), (U), ($), (i), (V), ($'), (f), (s), (Q), (A), (d) and (X) are known, then its Tax Rate Planned is:

t= 1-{Ad+[Q+X]/-U}/{$[1-i]-V-$'f[1+s]}

Rule-14360:

If both (/), (U), ($), (i), (V), ($'), (f), (s), (t), (Q), (d) and (X) are known, then its After Tax Income Planned is:

A= (U+{$[1-i]-V-$'f[1+s]}[1-t]-[Q+X]/ /)/d

Rule-14361:

If both (/), (U), ($), (i), (V), ($'), (f), (s), (t), (A), (Q) and (X) are known, then its Dividend Payout Planned is:

d= (U+{$[1-i]-V-$'f[1+s]}[1-t]-[Q+X]/ /)/A

Rule-14362:

If both (/), (U), ($), (i), (V), ($'), (f), (s), (t), (A), (d) and (Q) are known, then its Xpress or Current Debt Planned is:

X= /(U+{$[1-i]-V-$'f[1+s]}[1-t]-Ad)-Q

Steve Asikin ISBN 14: 978-1514685136, ISBN 10: 1514685132

Rule-14363:

If both (**/**), (**U**), (**$**), (**i**), (**V**), (**$'**), (**f**), (**s**), (**t**), (**A**), (**d**), (**P**), (**c**) and (**q**) are known, then its Quoted Longterm Debt Planned is:

Q= **/**(**U**+{**$**[1-**i**]-**V**-**$'f**[1+**s**]}[1-**t**]-**Ad**)-**P**/[**c**-**q**]

Rule-14364:

If both (**Q**), (**U**), (**$**), (**i**), (**V**), (**$'**), (**f**), (**s**), (**t**), (**A**), (**d**), (**P**), (**c**) and (**q**) are known, then its Leverage or Gearing Ratio Planned is:

/= {**Q**+**P**/[**c**-**q**]}/(**U**+{**$**[1-**i**]-**V**-**$'f**[1+**s**]}[1-**t**]-**Ad**)

Rule-14365:

If both (**/**), (**Q**), (**$**), (**i**), (**V**), (**$'**), (**f**), (**s**), (**t**), (**A**), (**d**), (**P**), (**c**) and (**q**) are known, then its Utilized or Starting Capital must be:

U= {**Q**+**P**/[**c**-**q**]}/**/**-({**$**[1-**i**]-**V**-**$'f**[1+**s**]}[1-**t**]-**Ad**)

Rule-14366:

If both (**/**), (**U**), (**Q**), (**i**), (**V**), (**$'**), (**f**), (**s**), (**t**), (**A**), (**d**), (**P**), (**c**) and (**q**) are known, then its Sales or Revenue Planned is:

$**= (**V**+**$'f[1+**s**]+(**Ad**+{**Q**+**P**/[**c**-**q**]}/**/**-**U**)/[1-**t**])/[1-**i**]

Rule-14367:

If both (**/**), (**U**), (**$**), (**Q**), (**V**), (**$'**), (**f**), (**s**), (**t**), (**A**), (**d**), (**P**), (**c**) and (**q**) are known, then its Interest Portion Planned is:

i= 1-(**V**+**$'f**[1+**s**]+(**Ad**+{**Q**+**P**/[**c**-**q**]}/**/**-**U**)/[1-**t**])/**$**

Steve Asikin ISBN 14: 978-1514685136, ISBN 10: 1514685132

Rule-14368:

If both (**/**), (**U**), (**$**), (**i**), (**Q**), (**$'**), (**f**), (**s**), (**t**), (**A**), (**d**), (**P**), (**c**) and (**q**) are known, then its Variable Cost Planned is:

V= **$**[1-**i**]-**$'f**[1+**s**]- (**Ad**+{**Q**+**P**/[**c**-**q**]}/**/-U**)/[1-**t**]

Rule-14369:

If both (**/**), (**U**), (**$**), (**i**), (**V**), (**Q**), (**f**), (**s**), (**t**), (**A**), (**d**), (**P**), (**c**) and (**q**) are known, then its Sales Past must be:

$'**= [**$[1-**i**]-**V**-(**Ad**+{**Q**+**P**/[**c**-**q**]}/**/-U**)/[1-**t**]]
/{**f**[1+**s**]}

Rule-14370:

If both (**/**), (**U**), (**$**), (**i**), (**V**), (**$'**), (**Q**), (**s**), (**t**), (**A**), (**d**), (**P**), (**c**) and (**q**) are known, then its Fixed Portion Planned is:

f= [**$**[1-**i**]-**V**-(**Ad**+{**Q**+**P**/[**c**-**q**]}/**/-U**)/[1-**t**]]
/{**$'**[1+**s**]}

Rule-14371:

If both (**/**), (**U**), (**$**), (**i**), (**V**), (**$'**), (**f**), (**Q**), (**t**), (**A**), (**d**), (**P**), (**c**) and (**q**) are known, then its Sales Growth Planned is:

s= **$**/**$'**-1

or it can also be found as

s= 1-[**$**[1-**i**]-**V**-(**Ad**+{**Q**+**P**/[**c**-**q**]}/**/-U**)/[1-**t**]]/[**$'f**]-1

Steve Asikin ISBN 14: 978-1514685136, ISBN 10: 1514685132

Rule-14372:

If both (**/**), (**U**), (**$**), (**i**), (**V**), (**$'**), (**f**), (**s**), (**Q**), (**A**), (**d**), (**P**), (**c**) and (**q**) are known, then its Tax Rate Planned is:

t= 1-(**Ad**+{**Q**+**P**/[**c**-**q**]}/**/**-**U**)/{**$**[1-**i**]-**V**-**$'f**[1+**s**]}

Rule-14373:

If both (**/**), (**U**), (**$**), (**i**), (**V**), (**$'**), (**f**), (**s**), (**t**), (**Q**), (**d**), (**P**), (**c**) and (**q**) are known, then its After Tax Income Planned is:

A= (**U**+{**$**[1-**i**]-**V**-**$'f**[1+**s**]}[1-**t**]-{**Q**+**P**/[**c**-**q**]}/**/**)/**d**

Rule-14374:

If both (**/**), (**U**), (**$**), (**i**), (**V**), (**$'**), (**f**), (**s**), (**t**), (**A**), (**Q**), (**P**), (**c**) and (**q**) are known, then its Dividend Payout Planned is:

d= (**U**+{**$**[1-**i**]-**V**-**$'f**[1+**s**]}[1-**t**]-{**Q**+**P**/[**c**-**q**]}/**/**)/**A**

Rule-14375:

If both (**/**), (**U**), (**$**), (**i**), (**V**), (**$'**), (**f**), (**s**), (**t**), (**A**), (**d**), (**Q**), (**c**) and (**q**) are known, then its Procured Inventory Planned is:

P= [**c**-**q**][**/**(**U**+{**$**[1-**i**]-**V**-**$'f**[1+**s**]}[1-**t**]-**Ad**)-**Q**]

Rule-14376:

If both (**/**), (**U**), (**$**), (**i**), (**V**), (**$'**), (**f**), (**s**), (**t**), (**A**), (**d**), (**P**), (**Q**) and (**q**) are known, then its Current Ratio Planned is:

c= **q**+**P**/[**/**(**U**+{**$**[1-**i**]-**V**-**$'f**[1+**s**]}[1-**t**]-**Ad**)-**Q**]

Steve Asikin ISBN 14: 978-1514685136, ISBN 10: 1514685132

Rule-14377:

If both (**/**), (**U**), (**$**), (**i**), (**V**), (**$'**), (**f**), (**s**), (**t**), (**A**), (**d**), (**P**), (**c**) and (**Q**) are known, then its Quick or Acid Test Ratio Planned is:

q= **c**-**P**/[**/**(**U**+{**$**[1-**i**]-**V**-**$'f**[1+**s**]}[1-**t**]-**Ad**)-**Q**]

Rule-14378:

If both (**/**), (**U**), (**$**), (**i**), (**V**), (**$'**), (**f**), (**s**), (**t**), (**A**), (**d**), (**p**), (**c**) and (**q**) are known, then its Quoted Longterm Debt Planned is:

Q= **/**(**U**+{**$**[1-**i**]-**V**-**$'f**[1+**s**]}[1-**t**]-**Ad**)
-**Vp**/{360[**c**-**q**]}

Rule-14379:

If both (**Q**), (**U**), (**$**), (**i**), (**V**), (**$'**), (**f**), (**s**), (**t**), (**A**), (**d**), (**p**), (**c**) and (**q**) are known, then its Leverage or Gearing Ratio Planned is:

/= (**Q**+**Vp**/{360[**c**-**q**]})
/(**U**+{**$**[1-**i**]-**V**-**$'f**[1+**s**]}[1-**t**]-**Ad**)

Rule-14380:

If both (**/**), (**Q**), (**$**), (**i**), (**V**), (**$'**), (**f**), (**s**), (**t**), (**A**), (**d**), (**p**), (**c**) and (**q**) are known, then its Utilized or Starting Capital must be:

U= (**Q**+**Vp**/{360[**c**-**q**]})/**/**
-({**$**[1-**i**]-**V**-**$'f**[1+**s**]}[1-**t**]-**Ad**)

Steve Asikin ISBN 14: 978-1514685136, ISBN 10: 1514685132

Rule-14381:

If both (**/**), (**U**), (**Q**), (**i**), (**V**), (**$'**), (**f**), (**s**), (**t**), (**A**), (**d**), (**p**), (**c**) and (**q**) are known, then its Sales or Revenue Planned is:

$= {V+$'f[1+s]
+[Ad+(Q+Vp/{360[c-q]})/-U]/[1-t]}/[1-i]

Rule-14382:

If both (**/**), (**U**), (**$**), (**Q**), (**V**), (**$'**), (**f**), (**s**), (**t**), (**A**), (**d**), (**p**), (**c**) and (**q**) are known, then its Interest Portion Planned is:

i= 1-{V+$'f[1+s]
+[Ad+(Q+Vp/{360[c-q]})/-U]/[1-t]}/$

Rule-14383:

If both (**/**), (**U**), (**$**), (**i**), (**Q**), (**$'**), (**f**), (**s**), (**t**), (**A**), (**d**), (**p**), (**c**) and (**q**) are known, then its Variable Cost Planned is:

V= [/(U+{$[1-i]-$'f[1+s]}[1-t]-Ad)-Q]
/(p/{360[c-q]} /[1-t])

Rule-14384:

If both (**/**), (**U**), (**$**), (**i**), (**V**), (**Q**), (**f**), (**s**), (**t**), (**A**), (**d**), (**p**), (**c**) and (**q**) are known, then its Sales Past must be:

$'= {$[1-i]-V
-[Ad+(Q+Vp/{360[c-q]})/-U]/[1-
t]}/{f[1+s]}

Steve Asikin ISBN 14: 978-1514685136, ISBN 10: 1514685132

Rule-14385:

If both (**/**), (**U**), (**$**), (**i**), (**V**), (**$'**), (**f**), (**s**), (**t**), (**A**), (**d**), (**p**), (**c**) and (**q**) are known, then its Quoted Longterm Debt Planned is:

f= {**$**[1-**i**]-**V**-[**Ad**+(**Q**+**Vp**/{360[**c**-**q**]})/**/**-**U**]/[1-**t**]}
/{**$'**[1+**s**]}

Rule-14386:

If both (**/**), (**U**), (**$**), (**i**), (**V**), (**$'**), (**f**), (**Q**), (**t**), (**A**), (**d**), (**p**), (**c**) and (**q**) are known, then its Sales Growth Planned is:

s= **$**/**$'**-1

or it can also be found as

s= {**$**[1-**i**]-**V**-[**Ad**+(**Q**+**Vp**/{360[**c**-**q**]})/**/**-**U**]/[1-**t**]}
/ [**$'f**]-1

Rule-14387:

If both (**/**), (**U**), (**$**), (**i**), (**V**), (**$'**), (**f**), (**s**), (**Q**), (**A**), (**d**), (**p**), (**c**) and (**q**) are known, then its Tax Rate Planned is:

t= 1-[**Ad**+(**Q**+**Vp**/{360[**c**-**q**]})/**/**-**U**]
/{**$**[1-**i**]-**V**-**$'f**[1+**s**]}

Rule-14388:

If both (**/**), (**U**), (**$**), (**i**), (**V**), (**$'**), (**f**), (**s**), (**t**), (**Q**), (**d**), (**p**), (**c**) and (**q**) are known, then its After Tax Income Planned is:

A= [**U**+{**$**[1-**i**]-**V**-**$'f**[1+**s**]}[1-**t**]
-(**Q**+**Vp**/{360[**c**-**q**]})/**/**]/**d**

Steve Asikin ISBN 14: 978-1514685136, ISBN 10: 1514685132

Rule-14389:

If both (**/**), (**U**), (**S**), (**i**), (**V**), (**S'**), (**f**), (**s**), (**t**), (**A**), (**Q**), (**p**), (**c**) and (**q**) are known, then its Dividend Payout Planned is:

d= [**U**+{**S**[1-**i**]-**V**-**S'f**[1+**s**]}[1-**t**] -(**Q**+**Vp**/{360[**c**-**q**]})/**/**]/**A**

Rule-14390:

If both (**/**), (**U**), (**S**), (**i**), (**V**), (**S'**), (**f**), (**s**), (**t**), (**A**), (**d**), (**Q**), (**c**) and (**q**) are known, then its Procured Inventory Days Planned:

p= 360[**c**-**q**] [**/**(**U**+{**S**[1-**i**]-**V**-**S'f**[1+**s**]}[1-**t**]-**Ad**)-**Q**]/**V**

Rule-14391:

If both (**/**), (**U**), (**S**), (**i**), (**V**), (**S'**), (**f**), (**s**), (**t**), (**A**), (**d**), (**p**), (**Q**) and (**q**) are known, then its Current Ratio Planned is:

c= **q**+**Vp** /{360[**/**(**U**+{**S**[1-**i**]-**V**-**S'f**[1+**s**]}[1-**t**]-**Ad**)-**Q**]}

Rule-14392:

If both (**/**), (**U**), (**S**), (**i**), (**V**), (**S'**), (**f**), (**s**), (**t**), (**A**), (**d**), (**p**), (**c**) and (**Q**) are known, then its Quick or Acid Test Ratio Planned is:

q= **c**-**Vp** /{360[**/**(**U**+{**S**[1-**i**]-**V**-**S'f**[1+**s**]}[1-**t**]-**Ad**)-**Q**]}

Steve Asikin ISBN 14: 978-1514685136, ISBN 10: 1514685132

Rule-14393:

If both (/), (U), ($), (i), (V), ($'), (f), (s), (t), (A), (d), (v), (p), (c) and (q) are known, then its Quoted Longterm Debt Planned is:

Q= /(U+{$[1-i]-V-$'f[1+s]}[1-t]-Ad)
-$vp/{360[c-q]}

Rule-14394:

If both (Q), (U), ($), (i), (V), ($'), (f), (s), (t), (A), (d), (v), (p), (c) and (q) are known, then its Leverage or Gearing Ratio Planned is:

/= (Q+$vp/{360[c-q]})
/(U+{$[1-i]-V-$'f[1+s]}[1-t]-Ad)

Rule-14395:

If both (/), (Q), ($), (i), (V), ($'), (f), (s), (t), (A), (d), (v), (p), (c) and (q) are known, then its Utilized or Starting Capital must be:

U= (Q+$vp/{360[c-q]})//
-({$[1-i]-V-$'f[1+s]}[1-t]-Ad)

Rule-14396:

If both (/), (U), (Q), (i), (V), ($'), (f), (s), (t), (A), (d), (v), (p), (c) and (q) are known, then its Sales or Revenue Planned is:

$= [/(U-{V+$'f[1+s]}[1-t]-Ad)-Q]
/(vp/{360[c-q]}-/[1-i] [1-t])

Steve Asikin ISBN 14: 978-1514685136, ISBN 10: 1514685132

Rule-14397:

If both (**/**), (**U**), (**S**), (**Q**), (**V**), (**S'**), (**f**), (**s**), (**t**), (**A**), (**d**), (**v**), (**p**), (**c**) and (**q**) are known, then its Interest Portion Planned is:

i= 1-{**V**+**S'f**[1+**s**]
+[**Ad**+(**Q**+**Svp**/{360[**c**-**q**]})/**/**-**U**]/[1-**t**]}/**S**

Rule-14398:

If both (**/**), (**U**), (**S**), (**i**), (**Q**), (**S'**), (**f**), (**s**), (**t**), (**A**), (**d**), (**v**), (**p**), (**c**) and (**q**) are known, then its Variable Cost Planned is:

V= [**/**(**U**+{**S**[1-**i**]-**S'f**[1+**s**]}[1-**t**]-**Ad**)
-**Svp**/{360[**c**-**q**]}-**Q**]/{**/**[1-**t**]}

Rule-14399:

If both (**/**), (**U**), (**S**), (**i**), (**V**), (**Q**), (**f**), (**s**), (**t**), (**A**), (**d**), (**v**), (**p**), (**c**) and (**q**) are known, then its Sales Past must be:

S'= {**S**[1-**i**]-**V**-[**Ad**+(**Q**+**Svp**/{360[**c**-**q**]})/**/**-**U**]/[1-**t**]}
/{**f**[1+**s**]}

Rule-14400:

If both (**/**), (**U**), (**S**), (**i**), (**V**), (**S'**), (**Q**), (**s**), (**t**), (**A**), (**d**), (**v**), (**p**), (**c**) and (**q**) are known, then its Fixed Portion Planned is:

f= {**S**[1-**i**]-**V**-[**Ad**+(**Q**+**Svp**/{360[**c**-**q**]})/**/**-**U**]/[1-**t**]}
/{**S'**[1+**s**]}

Steve Asikin ISBN 14: 978-1514685136, ISBN 10: 1514685132

Rule-14401:

If both (**/**), (**U**), (**$**), (**i**), (**V**), (**$'**), (**f**), (**Q**), (**t**), (**A**), (**d**), (**v**), (**p**), (**c**) and (**q**) are known, then its Sales Growth Planned is:

s= **$**/**$'**-1

or it can also be found as

s= {**$**[1-**i**]-**V**-[**Ad**+(**Q**+**$vp**/{360[**c**-**q**]})/**/**-**U**]/[1-**t**]}
/ [**$'f**]-1

Rule-14402:

If both (**/**), (**U**), (**$**), (**i**), (**V**), (**$'**), (**f**), (**s**), (**Q**), (**A**), (**d**), (**v**), (**p**), (**c**) and (**q**) are known, then its Tax Rate Planned is:

t= 1-[**Ad**+(**Q**+**$vp**/{360[**c**-**q**]})/**/**-**U**]
/{**$**[1-**i**]-**V**-**$'f**[1+**s**]}

Rule-14403:

If both (**/**), (**U**), (**$**), (**i**), (**V**), (**$'**), (**f**), (**s**), (**t**), (**Q**), (**d**), (**v**), (**p**), (**c**) and (**q**) are known, then its After Tax Income Planned is:

A= [**U**+{**$**[1-**i**]-**V**-**$'f**[1+**s**]}[1-**t**]
-(**Q**+**$vp**/{360[**c**-**q**]})/**/**]/**d**

Rule-14404:

If both (**/**), (**U**), (**$**), (**i**), (**V**), (**$'**), (**f**), (**s**), (**t**), (**A**), (**Q**), (**v**), (**p**), (**c**) and (**q**) are known, then its Dividend Payout Planned is:

d= [**U**+{**$**[1-**i**]-**V**-**$'f**[1+**s**]}[1-**t**]
-(**Q**+**$vp**/{360[**c**-**q**]})/**/**]/**A**

Steve Asikin ISBN 14: 978-1514685136, ISBN 10: 1514685132

Rule-14405:

If both (/), (U), (S), (i), (V), (S'), (f), (s), (t), (A), (d), (v), (p), (c) and (q) are known, then its Variable Portion Planned is:

v= 360[c-q][/(U+{S[1-i]-V-S'f[1+s]}[1-t]-Ad)-Q]
/[Sp]

Rule-14406:

If both (/), (U), (S), (i), (V), (S'), (f), (s), (t), (A), (d), (v), (p), (c) and (q) are known, then its Procured Inventory Days Planned:

p= 360[c-q][/(U+{S[1-i]-V-S'f[1+s]}[1-t]-Ad)-Q]
/[Sv]

Rule-14407:

If both (/), (U), (S), (i), (V), (S'), (f), (s), (t), (A), (d), (v), (p), (Q) and (q) are known, then its Current Ratio Planned is:

c= q+Svp
/{360[/(U+{S[1-i]-V-S'f[1+s]}[1-t]-Ad)-Q]}

Rule-14408:

If both (/), (U), (S), (i), (V), (S'), (f), (s), (t), (A), (d), (v), (p), (c) and (Q) are known, then its Quick or Acid Test Ratio Planned is:

q= c-Svp
/{360[/(U+{S[1-i]-V-S'f[1+s]}[1-t]-Ad)-Q]}

Steve Asikin ISBN 14: 978-1514685136, ISBN 10: 1514685132

Rule-14409:

If both (/), (U), (S), (i), (V), (S'), (f), (s), (t), (A), (d), (v), (p), (c) and (q) are known, then its Quoted Longterm Debt Planned is:

Q= /(U+{S[1-i]-V-S'f[1+s]}[1-t]-Ad)
-S'vp[1+s]/{360[c-q]}

Rule-14410:

If both (Q), (U), (S), (i), (V), (S'), (f), (s), (t), (A), (d), (v), (p), (c) and (q) are known, then its Leverage or Gearing Ratio Planned is:

/= (Q+S'vp[1+s]/{360[c-q]})
/(U+{S[1-i]-V-S'f[1+s]}[1-t]-Ad)

Rule-14411:

If both (/), (Q), (S), (i), (V), (S'), (f), (s), (t), (A), (d), (v), (p), (c) and (q) are known, then its Utilized or Starting capital must be:

U= (Q+S'vp[1+s]/{360[c-q]})//
-({S[1-i]-V-S'f[1+s]}[1-t]-Ad)

Rule-14412:

If both (/), (U), (Q), (i), (V), (S'), (f), (s), (t), (A), (d), (v), (p), (c) and (q) are known, then its Sales or Revenue Planned is:

S= [Q-/(U+{ -V-S'f[1+s]}[1-t]-Ad)
+S'vp[1+s]/{360[c-q]}]/{/[1-i][1-t]}

Steve Asikin ISBN 14: 978-1514685136, ISBN 10: 1514685132

Rule-14413:

If both (**/**), (**U**), (**$**), (**Q**), (**V**), (**$'**), (**f**), (**s**), (**t**), (**A**), (**d**), (**v**), (**p**), (**c**) and (**q**) are known, then its Interest Portion Planned is:

i= 1-**[V**+**$'f**[1+**s**]+**[Ad**
+(**Q**+**$'vp**[1+**s**]/{360[**c-q**]})/**/-U]**/[1-**t**]]/**$**

Rule-14414:

If both (**/**), (**U**), (**$**), (**i**), (**Q**), (**$'**), (**f**), (**s**), (**t**), (**A**), (**d**), (**v**), (**p**), (**c**) and (**q**) are known, then its Variable Cost Planned is:

V= **[/(U**+{**$**[1-**i**]-**$'f**[1+**s**]}[1-**t**]-**Ad**)
-**$'vp**[1+**s**]/{360[**c-q**]}-**Q]**/{**/**[1-**t**]}

Rule-14415:

If both (**/**), (**U**), (**$**), (**i**), (**V**), (**Q**), (**f**), (**s**), (**t**), (**A**), (**d**), (**v**), (**p**), (**c**) and (**q**) are known, then its Sales Past must be:

$'**= **[/(U**+{**$[1-**i**]-**V**}[1-**t**]-**Ad**)-**Q]**
/**[**[1+**s**](**vp**/{360[**c-q**]}+**f/**[1-**t**])

Rule-14416:

If both (**/**), (**U**), (**$**), (**i**), (**V**), (**$'**), (**Q**), (**s**), (**t**), (**A**), (**d**), (**v**), (**p**), (**c**) and (**q**) are known, then its Variable Cost Planned is:

f= **{$**[1-**i**]-**V-[Ad**+(**Q**+**$'vp**[1+**s**]/{360[**c-q**]})/**/-U]**
/[1-**t**]**}**/{**$'**[1+**s**]}

Steve Asikin ISBN 14: 978-1514685136, ISBN 10: 1514685132

Rule-14417:

If both (/), (U), (S), (i), (V), (S'), (f), (Q), (t), (A), (d), (v), (p), (c) and (q) are known, then its Sales Growth Planned is:

s= S/S'-1

or it can also be found as

s= [/(U+{S[1-i]-V}[1-t]-Ad)-Q]
/[S'(vp/{360[c-q]}+f/[1-t])-1

Rule-14418:

If both (/), (U), (S), (i), (V), (S'), (f), (s), (Q), (A), (d), (v), (p), (c) and (q) are known, then its Tax Rate Planned is:

t= 1-[Ad+(Q+S'vp[1+s]/{360[c-q]})/-U]
/{S[1-i]-V-S'f[1+s]}

Rule-14419:

If both (/), (U), (S), (i), (V), (S'), (f), (s), (t), (Q), (d), (v), (p), (c) and (q) are known, then its After Tax Income Planned is:

A= [U+{S[1-i]-V-S'f[1+s]}[1-t]
-(Q+S'vp[1+s]/{360[c-q]})/]/d

Rule-14420:

If both (/), (U), (S), (i), (V), (S'), (f), (s), (t), (A), (Q), (v), (p), (c) and (q) are known, then its Dividend Payout Planned is:

d= [U+{S[1-i]-V-S'f[1+s]}[1-t]
-(Q+S'vp[1+s]/{360[c-q]})/]/A

Steve Asikin ISBN 14: 978-1514685136, ISBN 10: 1514685132

Rule-14421:

If both (/), (U), ($), (i), (V), ($'), (f), (s), (t), (A), (d), (Q), (p), (c) and (q) are known, then its Variable Portion Planned is:

v= V/$

or it can also be found as

v= 360[c-q][/(U+{$[1-i]-V-$'f[1+s]}[1-t]-Ad)-Q] /{$'p[1+s]}

Rule-14422:

If both (/), (U), ($), (i), (V), ($'), (f), (s), (t), (A), (d), (v), (Q), (c) and (q) are known, then its Procured Inventory Days Planned:

p= 360[c-q][/(U+{$[1-i]-V-$'f[1+s]}[1-t]-Ad)-Q] /{$'v[1+s]}

Rule-14423:

If both (/), (U), ($), (i), (V), ($'), (f), (s), (t), (A), (d), (v), (p), (Q) and (q) are known, then its Current Ratio Planned is:

c= q+$'vp[1+s] /{360[/(U+{$[1-i]-V-$'f[1+s]}[1-t]-Ad)-Q]}

Rule-14424:

If both (/), (U), ($), (i), (V), ($'), (f), (s), (t), (A), (d), (v), (p), (c) and (Q) are known, then its Quick or Acid Test Ratio Planned is:

q= c-$'vp[1+s] /{360[/(U+{$[1-i]-V-$'f[1+s]}[1-t]-Ad)-Q]}

Steve Asikin ISBN 14: 978-1514685136, ISBN 10: 1514685132

Rule-14425:

If both (/), (**U**), ($), (i), (**V**), ($'), (**f**), (s), (**t**), (**d**) and (**X**) are known, then its Quoted Longterm Debt Planned is:

Q= /(**U**+{$[1-**i**]-**V**-$'**f**[1+s]}[1-**t**][1-**d**])-**X**

Rule-14426:

If both (**Q**), (**U**), ($), (i), (**V**), ($'), (**f**), (s), (**t**), (**d**) and (**X**) are known, then its Leverage or Gearing Ratio Planned is:

/= [**Q**+**X**]/(**U**+{$[1-**i**]-**V**-$'**f**[1+s]}[1-**t**][1-**d**])

Rule-14427:

If both (/), (**Q**), ($), (i), (**V**), ($'), (**f**), (s), (**t**), (**d**) and (**X**) are known, then its Utilized or Starting capital must be:

U= [**Q**+**X**]//-{$[1-**i**]-**V**-$'**f**[1+s]}[1-**t**][1-**d**]

Rule-14428:

If both (/), (**U**), (**Q**), (i), (**V**), ($'), (**f**), (s), (**t**), (**d**) and (**X**) are known, then its Sales or Revenue Planned is:

$= (**V**+$'**f**[1+s]+{[**Q**+**X**]//-**U**}/{[1-**t**][1-**d**]})/[1-**i**]

Rule-14429:

If both (/), (**U**), ($), (**Q**), (**V**), ($'), (**f**), (s), (**t**), (**d**) and (**X**) are known, then its Interest Portion Planned is:

i= 1-(**V**+$'**f**[1+s]+{[**Q**+**X**]//-**U**}/{[1-**t**][1-**d**]})/$

Steve Asikin ISBN 14: 978-1514685136, ISBN 10: 1514685132

Rule-14430:
> If both (I), (U), (S), (i), (Q), (S'), (f), (s), (t), (d) and
> (X) are known, then its Variable Cost Planned is:
> $$V = S[1-i]-S'f[1+s]-\{[Q+X]/I-U\}/\{[1-t][1-d]\}$$

Rule-14431:
> If both (I), (U), (S), (i), (V), (Q), (f), (s), (t), (d) and
> (X) are known, then its Sales Past must be:
> $$S' = (S[1-i]-V-\{[Q+X]/I-U\}/\{[1-t][1-d]\})/\{f[1+s]\}$$

Rule-14432:
> If both (I), (U), (S), (i), (V), (S'), (Q), (s), (t), (d) and
> (X) are known, then its Fixed Portion Planned is:
> $$f = (S[1-i]-V-\{[Q+X]/I-U\}/\{[1-t][1-d]\})/\{S'[1+s]\}$$

Rule-14433:
> If both (I), (U), (S), (i), (V), (S'), (f), (Q), (t), (d) and
> (X) are known, then its Sales Growth Planned is:
> $$s = S/S'-1$$
>> or it can also be found as
> $$s = (S[1-i]-V-\{[Q+X]/I-U\}/\{[1-t][1-d]\})/[S'f]-1$$

Rule-14434:
> If both (I), (U), (S), (i), (V), (S'), (f), (s), (Q), (d) and
> (X) are known, then its Tax Rate Planned is:
> $$t = 1-\{[Q+X]/I-U\}/([1-d]\ \{S[1-i]-V-S'f[1+s]\})$$

Rule-14435:
> If both (I), (U), (S), (i), (V), (S'), (f), (s), (t), (Q) and
> (X) are known, then its Dividend Payout Planned is:
> $$d = 1-\{[Q+X]/I-U\}/([1-t]\ \{S[1-i]-V-S'f[1+s]\})$$

Steve Asikin ISBN 14: 978-1514685136, ISBN 10: 1514685132

Rule-14436:

If both (/), (U), ($), (i), (V), ($'), (f), (s), (t), (d) and (Q) are known, then its Xpress or Current Debt Planned is:

X= /(U+{$[1-i]-V-$'f[1+s]}[1-t][1-d])-Q

Rule-14437:

If both (/), (U), ($), (i), (V), ($'), (f), (s), (t), (d), (P), (c) and (q) are known, then its Quoted Longterm Debt Planned is:

Q= /(U+{$[1-i]-V-$'f[1+s]}[1-t][1-d])-P/[c-q]

Rule-14438:

If both (Q), (U), ($), (i), (V), ($'), (f), (s), (t), (d), (P), (c) and (q) are known, then its Leverage or Gearing Ratio Planned is:

/= {Q+P/[c-q]}/(U+{$[1-i]-V-$'f[1+s]}[1-t][1-d])

Rule-14439:

If both (/), (Q), ($), (i), (V), ($'), (f), (s), (t), (d), (P), (c) and (q) are known, then its Utilized or Starting Capital must be:

U= {Q+P/[c-q]}/ /-{$[1-i]-V-$'f[1+s]}[1-t][1-d]

Rule-14440:

If both (/), (U), (Q), (i), (V), ($'), (f), (s), (t), (d), (P), (c) and (q) are known, then its Sales or Revenue Planned is:

$= [V+$'f[1+s]+({Q+P/[c-q]}/ /–U)/{[1-t][1-d]}]
/[1-i]

Steve Asikin ISBN 14: 978-1514685136, ISBN 10: 1514685132

Rule-14441:

If both (**/**), (**U**), (**$**), (**Q**), (**V**), (**$'**), (**f**), (**s**), (**t**), (**d**), (**P**), (**c**) and (**q**) are known, then its Interest Portion Planned is:

i= 1-[**V**+**$'f**[1+**s**]+({**Q**+**P**/[**c**-**q**]}/**/**–**U**)
/{[1-**t**][1-**d**]}]/**$**

Rule-14442:

If both (**/**), (**U**), (**$**), (**i**), (**Q**), (**$'**), (**f**), (**s**), (**t**), (**d**), (**P**), (**c**) and (**q**) are known, then its Variable Cost Planned is:

V= **$**[1-**i**]-**$'f**[1+**s**]-({**Q**+**P**/[**c**-**q**]}/**/**–**U**)/{[1-**t**][1-**d**]}

Rule-14443:

If both (**/**), (**U**), (**$**), (**i**), (**V**), (**Q**), (**f**), (**s**), (**t**), (**d**), (**P**), (**c**) and (**q**) are known, then its Sales Past must be:

$'**= [**$[1-**i**]-**V**-({**Q**+**P**/[**c**-**q**]}/**/**–**U**)/{[1-**t**][1-**d**]}]
/{**f**[1+**s**]}

Rule-14444:

If both (**/**), (**U**), (**$**), (**i**), (**V**), (**$'**), (**Q**), (**s**), (**t**), (**d**), (**P**), (**c**) and (**q**) are known, then its Fixed Portion Planned is:

f= [**$**[1-**i**]-**V**-({**Q**+**P**/[**c**-**q**]}/**/**–**U**)/{[1-**t**][1-**d**]}]
/{**$'**[1+**s**]}

Steve Asikin ISBN 14: 978-1514685136, ISBN 10: 1514685132

Rule-14445:

If both (**/**), (**U**), (**$**), (**i**), (**V**), (**$'**), (**f**), (**Q**), (**t**), (**d**), (**P**), (**c**) and (**q**) are known, then its Sales Growth Planned is:

s= **$**/**$'**-1

or it can also be found as

s= [**$**[1-**i**]-**V**-({**Q**+**P**/[**c**-**q**]}/**/**–**U**)/{[1-**t**][1-**d**]}]
/[**$'f**]-1

Rule-14446:

If both (**/**), (**U**), (**$**), (**i**), (**V**), (**$'**), (**f**), (**s**), (**Q**), (**d**), (**P**), (**c**) and (**q**) are known, then its Tax Rate Planned is:

t = 1-({**Q**+**P**/[**c**-**q**]}/**/**–**U**)
/([1-**d**] {**$**[1-**i**]-**V**-**$'f**[1+**s**]})

Rule-14447:

If both (**/**), (**U**), (**$**), (**i**), (**V**), (**$'**), (**f**), (**s**), (**t**), (**Q**), (**P**), (**c**) and (**q**) are known, then its Dividend Payout Planned is:

d = 1-({**Q**+**P**/[**c**-**q**]}/**/**–**U**)
/([1-**t**]{**$**[1-**i**]-**V**-**$'f**[1+**s**]})

Rule-14448:

If both (**/**), (**U**), (**$**), (**i**), (**V**), (**$'**), (**f**), (**s**), (**t**), (**d**), (**Q**), (**c**) and (**q**) are known, then its Procured Inventory Planned is:

P= [**c**-**q**][**/**(**U**+{**$**[1-**i**]-**V**-**$'f**[1+**s**]}[1-**t**][1-**d**])-**Q**]

Steve Asikin ISBN 14: 978-1514685136, ISBN 10: 1514685132

Rule-14449:

If both (**/**), (**U**), (**$**), (**i**), (**V**), (**$'**), (**f**), (**s**), (**t**), (**d**), (**P**), (**Q**) and (**q**) are known, then its Current Ratio Planned is:

c= q+P/[/(U+{$[1-i]-V-$'f[1+s]}[1-t][1-d])-Q]

Rule-14450:

If both (**/**), (**U**), (**$**), (**i**), (**V**), (**$'**), (**f**), (**s**), (**t**), (**d**), (**P**), (**c**) and (**Q**) are known, then its Quick or Acid Test Ratio Planned is:

q= c-P/[/(U+{$[1-i]-V-$'f[1+s]}[1-t][1-d])-Q]

Rule-14451:

If both (**/**), (**U**), (**$**), (**i**), (**V**), (**$'**), (**f**), (**s**), (**t**), (**d**), (**p**), (**c**) and (**q**) are known, then its Quoted Longterm Debt Planned is:

Q= /(U+{$[1-i]-V-$'f[1+s]}[1-t][1-d])
-Vp/{360[c-q]}

Rule-14452:

If both (**Q**), (**U**), (**$**), (**i**), (**V**), (**$'**), (**f**), (**s**), (**t**), (**d**), (**p**), (**c**) and (**q**) are known, then its Leverage or Gearing Ratio Planned is:

/= (Q+Vp/{360[c-q]})
/(U+{$[1-i]-V-$'f[1+s]}[1-t][1-d])

Rule-14453:

If both (**/**), (**Q**), (**$**), (**i**), (**V**), (**$'**), (**f**), (**s**), (**t**), (**d**), (**p**), (**c**) and (**q**) are known, then its Utilized or Starting Capital must be:

U= (Q+Vp/{360[c-q]})//
-{$[1-i]-V-$'f[1+s]}[1-t][1-d]

Steve Asikin ISBN 14: 978-1514685136, ISBN 10: 1514685132

Rule-14454:

If both (**/**), (**U**), (**Q**), (**i**), (**V**), (**$'**), (**f**), (**s**), (**t**), (**d**), (**p**), (**c**) and (**q**) are known, then its Sales or Revenue Planned is:

$= {V+$'f[1+s]+[(Q+Vp/{360[c-q]})//–U]
/{[1-t][1-d]}}/[1-i]

Rule-14455:

If both (**/**), (**U**), (**$**), (**Q**), (**V**), (**$'**), (**f**), (**s**), (**t**), (**d**), (**p**), (**c**) and (**q**) are known, then its Interest Portion Planned is:

i= 1-{V+$'f[1+s]+[(Q+Vp/{360[c-q]})//–U]
/{[1-t][1-d]}}/$

Rule-14456:

If both (**/**), (**U**), (**$**), (**i**), (**Q**), (**$'**), (**f**), (**s**), (**t**), (**d**), (**p**), (**c**) and (**q**) are known, then its Variable Cost Planned is:

V= [/U+{$[1-i]-$'f[1+s]}[1-t][1-d])-Q]
/(p/{360[c-q]}+/[1-t][1-d])

Rule-14457:

If both (**/**), (**U**), (**$**), (**i**), (**V**), (**Q**), (**f**), (**s**), (**t**), (**d**), (**p**), (**c**) and (**q**) are known, then its Sales Past must be:

$'= {$[1-i]-V-[(Q+Vp/{360[c-q]})//–U]
/{[1-t][1-d]}}/{f[1+s]}

Steve Asikin ISBN 14: 978-1514685136, ISBN 10: 1514685132

Rule-14458:

If both (I), (U), (S), (i), (V), (S'), (Q), (s), (t), (d), (p), (c) and (q) are known, then its Fixed Portion Planned is:

$$f = \{S[1\text{-}i]\text{-}V\text{-}[(Q+Vp/\{360[c\text{-}q]\})/I-U]$$
$$/\{[1\text{-}t][1\text{-}d]\}\}/\{S'[1+s]\}$$

Rule-14459:

If both (I), (U), (S), (i), (V), (S'), (f), (Q), (t), (d), (p), (c) and (q) are known, then its Sales Growth Planned is:

$$s = S/S'\text{-}1$$

or it can also be found as

$$s = \{S[1\text{-}i]\text{-}V\text{-}[(Q+Vp/\{360[c\text{-}q]\})/I-U]$$
$$/\{[1\text{-}t][1\text{-}d]\}\}/[S'f]\text{-}1$$

Rule-14460:

If both (I), (U), (S), (i), (V), (S'), (f), (s), (Q), (d), (p), (c) and (q) are known, then its Tax Rate Planned is:

$$t = 1\text{-}[(Q+Vp/\{360[c\text{-}q]\})/I-U]$$
$$/ ([1\text{-}d]\{S[1\text{-}i]\text{-}V\text{-}S'f[1+s]\})$$

Rule-14461:

If both (I), (U), (S), (i), (V), (S'), (f), (s), (t), (Q), (p), (c) and (q) are known, then its Dividend Payout Planned is:

$$d = 1\text{-}[(Q+Vp/\{360[c\text{-}q]\})/I-U]$$
$$/([1\text{-}t] \{S[1\text{-}i]\text{-}V\text{-}S'f[1+s]\})$$

Steve Asikin ISBN 14: 978-1514685136, ISBN 10: 1514685132

Rule-14462:

If both (**/**), (**U**), (**$**), (**i**), (**V**), (**$'**), (**f**), (**s**), (**t**), (**d**), (**Q**), (**c**) and (**q**) are known, then its Procured Inventory Days Planned:

p= 360[**c**-**q**]
[**/**(**U**+{**$**[1-**i**]-**V**-**$'f**[1+**s**]}[1-**t**][1-**d**])-**Q**]/**V**

Rule-14463:

If both (**/**), (**U**), (**$**), (**i**), (**V**), (**$'**), (**f**), (**s**), (**t**), (**d**), (**p**), (**Q**) and (**q**) are known, then its Current Ratio Planned is:

c= **q**+**Vp**/{360
[**/**(**U**+{**$**[1-**i**]-**V**-**$'f**[1+**s**]}[1-**t**][1-**d**])-**Q**]}

Rule-14464:

If both (**/**), (**U**), (**$**), (**i**), (**V**), (**$'**), (**f**), (**s**), (**t**), (**d**), (**p**), (**c**) and (**Q**) are known, then its Quick or Acid Test Ratio Planned is:

q= **c**-**Vp**/{360
[**/**(**U**+{**$**[1-**i**]-**V**-**$'f**[1+**s**]}[1-**t**][1-**d**])-**Q**]}

Rule-14465:

If both (**/**), (**U**), (**$**), (**i**), (**V**), (**$'**), (**f**), (**s**), (**t**), (**d**), (**v**), (**p**), (**c**) and (**q**) are known, then its Quoted Longterm Debt Planned is:

Q= **/**(**U**+{**$**[1-**i**]-**V**-**$'f**[1+**s**]}[1-**t**][1-**d**])
-**$vp**/{360[**c**-**q**]}

Steve Asikin ISBN 14: 978-1514685136, ISBN 10: 1514685132

Rule-14466:

If both (**Q**), (**U**), (**$**), (**i**), (**V**), (**$'**), (**f**), (**s**), (**t**), (**d**), (**v**), (**p**), (**c**) and (**q**) are known, then its Leverage or Gearing Ratio Planned is:

/= (**Q**+**$vp**/{360[**c**-**q**]})
/(**U**+{**$**[1-**i**]-**V**-**$'f**[1+**s**]}[1-**t**][1-**d**])

Rule-14467:

If both (**/**), (**Q**), (**$**), (**i**), (**V**), (**$'**), (**f**), (**s**), (**t**), (**d**), (**v**), (**p**), (**c**) and (**q**) are known, then its Utilized or Starting capital must be:

U= (**Q**+**$vp**/{360[**c**-**q**]})/**/**
-{**$**[1-**i**]-**V**-**$'f**[1+**s**]}[1-**t**][1-**d**]

Rule-14468:

If both (**/**), (**U**), (**Q**), (**i**), (**V**), (**$'**), (**f**), (**s**), (**t**), (**d**), (**v**), (**p**), (**c**) and (**q**) are known, then its Sales or Revenue Planned is:

$**= **[/**(**U**-{**V**+**$'f[1+**s**]}[1-**t**][1-**d**])-**Q]**
/(**vp**/{360[**c**-**q**]}-**/**[1-**i**][1-**t**][1-**d**])

Rule-14469:

If both (**/**), (**U**), (**$**), (**Q**), (**V**), (**$'**), (**f**), (**s**), (**t**), (**d**), (**v**), (**p**), (**c**) and (**q**) are known, then its Interest Portion Planned is:

i= 1-**{V**+**$'f**[1+**s**]+**[(Q**+**$vp**/{360[**c**-**q**]})/**/**–**U]**
/{[1-**t**][1-**d**]}**}**/**$**

Steve Asikin ISBN 14: 978-1514685136, ISBN 10: 1514685132

Rule-14470:

If both (**/**), (**U**), (**$**), (**i**), (**Q**), (**$'**), (**f**), (**s**), (**t**), (**d**), (**v**), (**p**), (**c**) and (**q**) are known, then its Variable Cost Planned is:

V= [**/**(**U**+{**$**[1-**i**] -**$'f**[1+**s**]}[1-**t**][1-**d**])
-**$vp**/{360[**c-q**]}-**Q**]/{**/**[1-**t**][1-**d**]}

Rule-14471:

If both (**/**), (**U**), (**$**), (**i**), (**V**), (**Q**), (**f**), (**s**), (**t**), (**d**), (**v**), (**p**), (**c**) and (**q**) are known, then its Sales Past must be:

$'**= {**$[1-**i**]-**V**-[(**Q**+**$vp**/{360[**c-q**]})/**/**–**U**]
/{[1-**t**][1-**d**]}}/{**f**[1+**s**]}

Rule-14472:

If both (**/**), (**U**), (**$**), (**i**), (**V**), (**$'**), (**f**), (**s**), (**t**), (**d**), (**v**), (**p**), (**c**) and (**q**) are known, then its Quoted Longterm Debt Planned is:

f= {**$**[1-**i**]-**V**-[(**Q**+**$vp**/{360[**c-q**]})/**/**–**U**]
/{[1-**t**][1-**d**]}}/{**$'**[1+**s**]}

Rule-14473:

If both (**/**), (**U**), (**$**), (**i**), (**V**), (**$'**), (**f**), (**Q**), (**t**), (**d**), (**v**), (**p**), (**c**) and (**q**) are known, then its Sales Growth Planned is:

s= **$**/**$'**-1

or it can also be found as

s= {**$**[1-**i**]-**V**-[(**Q**+**$vp**/{360[**c-q**]})/**/**–**U**]
/{[1-**t**][1-**d**]}}/[**$'f**]-1

Steve Asikin ISBN 14: 978-1514685136, ISBN 10: 1514685132

Rule-14474:

If both (/), (U), (S), (i), (V), (S'), (f), (s), (Q), (d), (v), (p), (c) and (q) are known, then its Tax Rate Planned is:

t = 1-[(Q+Svp/{360[c-q]})//–U]
/([1-d] {S[1-i]-V-S'f[1+s]})

Rule-14475:

If both (/), (U), (S), (i), (V), (S'), (f), (s), (t), (Q), (v), (p), (c) and (q) are known, then its Dividend Payout Planned is:

d = 1-[(Q+Svp/{360[c-q]})//–U]
/([1-t] {S[1-i]-V-S'f[1+s]})

Rule-14476:

If both (/), (U), (S), (i), (V), (S'), (f), (s), (t), (d), (Q), (p), (c) and (q) are known, then its Variable Portion Planned is:

v= V/S

or it can also be found as

v= 360[c-q]
[/(U+{S[1-i]-V-S'f[1+s]}[1-t][1-d])-Q]/[Sp]

Rule-14477:

If both (/), (U), (S), (i), (V), (S'), (f), (s), (t), (d), (v), (Q), (c) and (q) are known, then its Procured Inventory Planned is:

p= 360[c-q]
[/(U+{S[1-i]-V-S'f[1+s]}[1-t][1-d])-Q]/[Sv]

Steve Asikin ISBN 14: 978-1514685136, ISBN 10: 1514685132

Rule-14478:

If both (**/**), (**U**), (**$**), (**i**), (**V**), (**$'**), (**f**), (**s**), (**t**), (**d**), (**v**), (**p**), (**Q**) and (**q**) are known, then its Current Ratio Planned is:

c= q+$vp/{360
[/(U+{$[1-i]-V-$'f[1+s]}[1-t][1-d])-Q]}

Rule-14479:

If both (**/**), (**U**), (**$**), (**i**), (**V**), (**$'**), (**f**), (**s**), (**t**), (**d**), (**v**), (**p**), (**c**) and (**Q**) are known, then its Quick or Acid Test Ratio Planned is:

q= c-$vp/{360
[/(U+{$[1-i]-V-$'f[1+s]}[1-t][1-d])-Q]}

Rule-14480:

If both (**/**), (**U**), (**$**), (**i**), (**V**), (**$'**), (**f**), (**s**), (**t**), (**d**), (**v**), (**p**), (**c**) and (**q**) are known, then its Quoted Longterm Debt Planned is:

Q= /(U+{$[1-i]-V-$'f[1+s]}[1-t][1-d])
-$'vp[1+s]/{360[c-q]}

Rule-14481:

If both (**Q**), (**U**), (**$**), (**i**), (**V**), (**$'**), (**f**), (**s**), (**t**), (**d**), (**v**), (**p**), (**c**) and (**q**) are known, then its Leverage or Gearing Ratio Planned is:

/= (Q+$'vp[1+s]/{360[c-q]})
/(U+{$[1-i]-V-$'f[1+s]}[1-t][1-d])

Steve Asikin ISBN 14: 978-1514685136, ISBN 10: 1514685132

Rule-14482:

If both (**/**), (**Q**), (**$**), (**i**), (**V**), (**$'**), (**f**), (**s**), (**t**), (**d**), (**v**), (**p**), (**c**) and (**q**) are known, then its Utilized or Starting capital must be:

U= (**Q**+**$'vp**[1+**s**]/{360[**c-q**]})/**/**
-{**$**[1-**i**]-**V-$'f**[1+**s**]}[1-**t**][1-**d**]

Rule-14483:

If both (**/**), (**U**), (**Q**), (**i**), (**V**), (**$'**), (**f**), (**s**), (**t**), (**d**), (**v**), (**p**), (**c**) and (**q**) are known, then its Sales or Revenue Planned is:

$**= **[Q-/(U**-{**V**+**$'f[1+**s**]}[1-**t**][1-**d**])
+**$'vp**[1+**s**]/{360[**c-q**]}**]**/{**/**[1-**i**] [1-**t**][1-**d**]}

Rule-14484:

If both (**/**), (**U**), (**$**), (**Q**), (**V**), (**$'**), (**f**), (**s**), (**t**), (**d**), (**v**), (**p**), (**c**) and (**q**) are known, then its Interest Portion Planned is:

i= 1-**{V**+**$'f**[1+**s**]+**[**(**Q**+**$'vp**[1+**s**]/{360[**c-q**]})/**/**
--**U]**/{[1-**t**][1-**d**]}**}**/**$**

Rule-14485:

If both (**/**), (**U**), (**$**), (**i**), (**Q**), (**$'**), (**f**), (**s**), (**t**), (**d**), (**v**), (**p**), (**c**) and (**q**) are known, then its Variable Cost Planned is:

V= **[/(U**+{**$**[1-**i**] -**$'f**[1+**s**]}[1-**t**][1-**d**])
-**$'vp**[1+**s**]/{360[**c-q**]}-**Q]**/**{/**[1-**t**][1-**d**]}

Steve Asikin ISBN 14: 978-1514685136, ISBN 10: 1514685132

Rule-14486:

If both (**/**), (**U**), (**$**), (**i**), (**V**), (**Q**), (**f**), (**s**), (**t**), (**d**), (**v**), (**p**), (**c**) and (**q**) are known, then its Sales Past must be:

$'= [/U+{$[1-i]-V}[1-t][1-d])-Q]
/[[1+s](vp/{360[c-q]}+f/[1-t][1-d])]

Rule-14487:

If both (**/**), (**U**), (**$**), (**i**), (**V**), (**$'**), (**Q**), (**s**), (**t**), (**d**), (**v**), (**p**), (**c**) and (**q**) are known, then its Fixed Portion Planned is:

f= {$[1-i]-V-[(Q+$'vp[1+s]/{360[c-q]})//–U]
/{[1-t][1-d]}}/{$'[1+s]}

Rule-14488:

If both (**/**), (**U**), (**$**), (**i**), (**V**), (**$'**), (**f**), (**Q**), (**t**), (**d**), (**v**), (**p**), (**c**) and (**q**) are known, then its Sales Growth Planned is:

s= $/$'-1

or it can also be found as

s= [/U+{$[1-i]-V}[1-t][1-d])-Q]
/[$'(vp/{360[c-q]}+f/[1-t][1-d])]/[$'f]-1

Rule-14489:

If both (**/**), (**U**), (**$**), (**i**), (**V**), (**$'**), (**f**), (**s**), (**Q**), (**d**), (**v**), (**p**), (**c**) and (**q**) are known, then its Tax Rate Planned is:

t = 1-[(Q+$'vp[1+s]/{360[c-q]})//–U]
/([1-d] {$[1-i]-V-$'f[1+s]})

Steve Asikin ISBN 14: 978-1514685136, ISBN 10: 1514685132

Rule-14490:

If both (/), (U), ($), (i), (V), ($'), (f), (s), (t), (Q), (v), (p), (c) and (q) are known, then its Dividend Payout Planned is:

d = 1-[(Q+$'vp[1+s]/{360[c-q]})//–U]
/([1-t] {$[1-i]-V-$'f[1+s]})

Rule-14491:

If both (/), (U), ($), (i), (V), ($'), (f), (s), (t), (d), (Q), (p), (c) and (q) are known, then its Variable Portion Planned is:

v= V/$

or it can also be found as

v= 360[c-q][/(U+{$[1-i]-V-$'f[1+s]}[1-t][1-d])-Q]
/{$'p[1+s]}

Rule-14492:

If both (/), (U), ($), (i), (V), ($'), (f), (s), (t), (d), (v), (Q), (c) and (q) are known, then its Procured Inventory Days Planned:

p= 360[c-q][/(U+{$[1-i]-V-$'f[1+s]}[1-t][1-d])-Q]
/{$'v[1+s]}

Rule-14493:

If both (/), (U), ($), (i), (V), ($'), (f), (s), (t), (d), (v), (p), (Q) and (q) are known, then its Current Ratio Planned is:

c= q+$'vp[1+s]/{360[/(U+{$[1-i]-V-$'f[1+s]}
[1-t][1-d])-Q]}

Steve Asikin ISBN 14: 978-1514685136, ISBN 10: 1514685132

Rule-14494:

If both (**/**), (**U**), (**$**), (**i**), (**V**), (**$'**), (**f**), (**s**), (**t**), (**d**), (**v**), (**p**), (**c**) and (**Q**) are known, then its Quick or Acid Test Ratio Planned is:

q= **c**-**$'vp**[1+**s**]/{360[**/**{**U**+{**$**[1-**i**]-**V**-**$'f**[1+**s**]}[1-**t**][1-**d**])-**Q**]}

Rule-14495:

If both (**/**), (**U**), (**$**), (**V**), (**$'**), (**s**), (**f**), (**i**), (**T**), (**D**) and (**X**) are known, then its Quoted Longterm Debt Planned is:

Q= **/**{**U**+**$**-**V**-**$'**[1+**s**][**f**+**i**]-**T**-**D**}-**X**

Rule-14496:

If both (**Q**), (**U**), (**$**), (**V**), (**$'**), (**s**), (**f**), (**i**), (**T**), (**D**) and (**X**) are known, then its Leverage or Gearing Ratio Planned is:

/= [**Q**+**X**]/{**U**+**$**-**V**-**$'**[1+**s**][**f**+**i**]-**T**-**D**}

Rule-14497:

If both (**/**), (**Q**), (**$**), (**V**), (**$'**), (**s**), (**f**), (**i**), (**T**), (**D**) and (**X**) are known, then its Utilized or Starting Capital must be:

U= [**Q**+**X**]/**/**-{**$**-**V**-**$'**[1+**s**][**f**+**i**]-**T**-**D**}

Rule-14498:

If both (**/**), (**U**), (**Q**), (**V**), (**$'**), (**s**), (**f**), (**i**), (**T**), (**D**) and (**X**) are known, then its Sales or Revenue Planned is:

$**= [**Q**+**X**]/**/**-{**U**-**V**-**$'[1+**s**][**f**+**i**]-**T**-**D**}

Steve Asikin ISBN 14: 978-1514685136, ISBN 10: 1514685132

Rule-14499:

If both (/), (U), (S), (Q), (S'), (s), (f), (i), (T), (D) and (X) are known, then its Variable Cost Planned is:

V= {U+S-S'[1+s][f+i]-T-D}-[Q+X]//

Rule-14500:

If both (/), (U), (S), (V), (Q), (s), (f), (i), (T), (D) and (X) are known, then its Sales Past must be:

S'= {U+S-V-T-D-[Q+X]//}/{[1+s][f+i]}

Rule-14501:

If both (/), (U), (S), (V), (S'), (Q), (f), (i), (T), (D) and (X) are known, then its Sales Growth Planned is:

s= S/S'-1

or it can also be found as

s= {U+S-V-T-D-[Q+X]//}/{S'[f+i]}-1

Rule-14502:

If both (/), (U), (S), (V), (S'), (s), (Q), (i), (T), (D) and (X) are known, then its Fixed Portion Planned is:

f= {U+S-V-T-D-[Q+X]//}/{S'[1+s]}-i

Rule-14503:

If both (/), (U), (S), (V), (S'), (s), (f), (Q), (T), (D) and (X) are known, then its Interest Portion Planned is:

i= {U+S-V-T-D-[Q+X]//}/{S'[1+s]}-f

Rule-14504:

If both (/), (U), (S), (V), (S'), (s), (f), (i), (Q), (D) and (X) are known, then its Tax Planned is:

T= {U+S-V-S'[1+s][f+i]-D}-[Q+X]//

Steve Asikin ISBN 14: 978-1514685136, ISBN 10: 1514685132

Rule-14505:

If both (**/**), (**U**), (**$**), (**V**), (**$'**), (**s**), (**f**), (**i**), (**T**), (**Q**) and (**X**) are known, then its Dividend Planned is:

D= {**U**+**$**-**V**-**$'**[1+**s**][**f**+**i**]-**T**}-[**Q**+**X**]/**/**

Rule-14506:

If both (**/**), (**U**), (**$**), (**V**), (**$'**), (**s**), (**f**), (**i**), (**T**), (**D**) and (**Q**) are known, then its Xpress or Current Debt Planned is:

X= **/**{**U**+**$**-**V**-**$'**[1+**s**][**f**+**i**]-**T**-**D**}-**Q**

Rule-14507:

If both (**/**), (**U**), (**$**), (**V**), (**$'**), (**s**), (**f**), (**i**), (**T**), (**D**), (**P**), (**c**) and (**q**) are known, then its Quoted Longterm Debt Planned is:

Q= **/**{**U**+**$**-**V**-**$'**[1+**s**][**f**+**i**]-**T**-**D**}-**P**/[**c**-**q**]

Rule-14508:

If both (**I**), (**U**), (**$**), (**V**), (**$'**), (**s**), (**f**), (**i**), (**T**), (**D**), (**P**), (**c**) and (**q**) are known, then its Leverage or Gearing Ratio Planned is:

/= {**Q**+**P**/[**c**-**q**]}/{**U**+**$**-**V**-**$'**[1+**s**][**f**+**i**]-**T**-**D**}

Rule-14509:

If both (**/**), (**Q**), (**$**), (**V**), (**$'**), (**s**), (**f**), (**i**), (**T**), (**D**), (**P**), (**c**) and (**q**) are known, then its Utilized or Starting Capital must be:

U= {**Q**+**P**/[**c**-**q**]}/**/**-{**$**-**V**-**$'**[1+**s**][**f**+**i**]-**T**-**D**}

Steve Asikin ISBN 14: 978-1514685136, ISBN 10: 1514685132

Rule-14510:

If both (**/**), (**U**), (**Q**), (**V**), (**S'**), (**s**), (**f**), (**i**), (**T**), (**D**), (**P**), (**c**) and (**q**) are known, then its Sales or Revenue Planned is:

S= {**Q**+**P**/[**c**-**q**]}/**/**-{**U**-**V**-**S'**[1+**s**][**f**+**i**]-**T**-**D**}

Rule-14511:

If both (**/**), (**U**), (**S**), (**Q**), (**S'**), (**s**), (**f**), (**i**), (**T**), (**D**), (**P**), (**c**) and (**q**) are known, then its Variable Cost Planned is:

V= {**U**+**S**-**S'**[1+**s**][**f**+**i**]-**T**-**D**}-{**Q**+**P**/[**c**-**q**]}/**/**

Rule-14512:

If both (**/**), (**U**), (**S**), (**V**), (**Q**), (**s**), (**f**), (**i**), (**T**), (**D**), (**P**), (**c**) and (**q**) are known, then its Sales Past must be:

S'= (**U**+**S**-**V**-**T**-**D**-{**Q**+**P**/[**c**-**q**]}/**/**)/{[1+**s**][**f**+**i**]}

Rule-14513:

If both (**/**), (**U**), (**S**), (**V**), (**S'**), (**Q**), (**f**), (**i**), (**T**), (**D**), (**P**), (**c**) and (**q**) are known, then its Sales Growth Planned is:

s= **S**/**S'**-1

or it can also be found as

s= (**U**+**S**-**V**-**T**-**D**-{**Q**+**P**/[**c**-**q**]}/**/**)/{**S'**[**f**+**i**]}-1

Rule-14514:

If both (**/**), (**U**), (**S**), (**V**), (**S'**), (**s**), (**Q**), (**i**), (**T**), (**D**), (**P**), (**c**) and (**q**) are known, then its Fixed Portion Planned is:

f= (**U**+**S**-**V**-**T**-**D**-{**Q**+**P**/[**c**-**q**]}/**/**)/{**S'**[1+**s**]}-**i**

Steve Asikin ISBN 14: 978-1514685136, ISBN 10: 1514685132

Rule-14515:

If both (**/**), (**U**), (**$**), (**V**), (**$'**), (**s**), (**f**), (**Q**), (**T**), (**D**), (**P**), (**c**) and (**q**) are known, then its Interest Portion Planned is:

i= (**U**+**$**-**V**-**T**-**D**-{**Q**+**P**/[**c**-**q**]}/**/**)/{**$'**[1+**s**]}-**f**

Rule-14516:

If both (**/**), (**U**), (**$**), (**V**), (**$'**), (**s**), (**f**), (**i**), (**Q**), (**D**), (**P**), (**c**) and (**q**) are known, then its Tax Planned is:

T= {**U**+**$**-**V**-**$'**[1+**s**][**f**+**i**]-**D**}-{**Q**+**P**/[**c**-**q**]}/**/**

Rule-14517:

If both (**/**), (**U**), (**$**), (**V**), (**$'**), (**s**), (**f**), (**i**), (**T**), (**Q**), (**P**), (**c**) and (**q**) are known, then its Dividend Planned is:

D= {**U**+**$**-**V**-**$'**[1+**s**][**f**+**i**]-**T**}-{**Q**+**P**/[**c**-**q**]}/**/**

Rule-14518:

If both (**/**), (**U**), (**$**), (**V**), (**$'**), (**s**), (**f**), (**i**), (**T**), (**D**), (**Q**), (**c**) and (**q**) are known, then its Procured Inventory Planned is:

P= [**c**-**q**](**/**{**U**+**$**-**V**-**$'**[1+**s**][**f**+**i**]-**T**-**D**}-**Q**)

Rule-14519:

If both (**/**), (**U**), (**$**), (**V**), (**$'**), (**s**), (**f**), (**i**), (**T**), (**D**), (**P**), (**Q**) and (**q**) are known, then its Current Ratio Planned is:

c= **q**+**P**/(**/**{**U**+**$**-**V**-**$'**[1+**s**][**f**+**i**]-**T**-**D**}-**Q**)

Steve Asikin ISBN 14: 978-1514685136, ISBN 10: 1514685132

Rule-14520:

If both (**/**), (**U**), (**$**), (**V**), (**$'**), (**s**), (**f**), (**i**), (**T**), (**D**), (**P**), (**c**) and (**Q**) are known, then its Quick or Acid Test Ratio Planned is:

q= **c**-**P**/(**/**{**U**+**$**-**V**-**$'**[1+**s**][**f**+**i**]-**T**-**D**}-**Q**)

Rule-14521:

If both (**/**), (**U**), (**$**), (**V**), (**$'**), (**s**), (**f**), (**i**), (**T**), (**D**), (**p**), (**c**) and (**q**) are known, then its Quoted Longterm Debt Planned is:

Q= **/**{**U**+**$**-**V**-**$'**[1+**s**][**f**+**i**]-**T**-**D**}-**Vp**/{360[**c**-**q**]}

Rule-14522:

If both (**Q**), (**U**), (**$**), (**V**), (**$'**), (**s**), (**f**), (**i**), (**T**), (**D**), (**p**), (**c**) and (**q**) are known, then its Leverage or Gearing Ratio Planned is:

/= (**Q**+**Vp**/{360[**c**-**q**]})/{**U**+**$**-**V**-**$'**[1+**s**][**f**+**i**]-**T**-**D**}

Rule-14523:

If both (**/**), (**Q**), (**$**), (**V**), (**$'**), (**s**), (**f**), (**i**), (**T**), (**D**), (**p**), (**c**) and (**q**) are known, then its Utilized or Starting Capital must be:

U= (**Q**+**Vp**/{360[**c**-**q**]})/**/**-{**$**-**V**-**$'**[1+**s**][**f**+**i**]-**T**-**D**}

Rule-14524:

If both (**/**), (**U**), (**Q**), (**V**), (**$'**), (**s**), (**f**), (**i**), (**T**), (**D**), (**p**), (**c**) and (**q**) are known, then its Sales or Revenue Planned is:

$**= (**Q**+**Vp**/{360[**c**-**q**]})/**/**-{**U**-**V**-**$'[1+**s**][**f**+**i**]-**T**-**D**}

Steve Asikin ISBN 14: 978-1514685136, ISBN 10: 1514685132

Rule-14525:

If both (Λ), (**U**), (**$**), (**Q**), (**$'**), (**s**), (**f**), (**i**), (**T**), (**D**), (**p**), (**c**) and (**q**) are known, then its Variable Cost Planned is:

V= (Λ{**U**+**$**-**$'**[1+**s**][**f**+**i**]-**T**-**D**}-**Q**)/(**p**/{360[**c**-**q**]}+Λ)

Rule-14526:

If both (Λ), (**U**), (**$**), (**V**), (**Q**), (**s**), (**f**), (**i**), (**T**), (**D**), (**p**), (**c**) and (**q**) are known, then its Sales Past must be:

$'**= [**U**+**$-**V**-**T**-**D**-(**Q**+**Vp**/{360[**c**-**q**]})/Λ]
/{[1+**s**][**f**+**i**]}

Rule-14527:

If both (Λ), (**U**), (**$**), (**V**), (**$'**), (**Q**), (**f**), (**i**), (**T**), (**D**), (**p**), (**c**) and (**q**) are known, then its Sales Growth Planned is:

s= **$**/**$'**-1

or it can also be found as

s= [**U**+**$**-**V**-**T**-**D**-(**Q**+**Vp**/{360[**c**-**q**]})/Λ]/{**$'**[**f**+**i**]}-1

Rule-14528:

If both (Λ), (**U**), (**$**), (**V**), (**$'**), (**s**), (**Q**), (**i**), (**T**), (**D**), (**p**), (**c**) and (**q**) are known, then its Fixed Portion Planned is:

f= [**U**+**$**-**V**-**T**-**D**-(**Q**+**Vp**/{360[**c**-**q**]})/Λ]/{**$'**[1+**s**]}-**i**

Rule-14529:

If both (Λ), (**U**), (**$**), (**V**), (**$'**), (**s**), (**f**), (**Q**), (**T**), (**D**), (**p**), (**c**) and (**q**) are known, then its Interest Portion Planned is:

i= [**U**+**$**-**V**-**T**-**D**-(**Q**+**Vp**/{360[**c**-**q**]})/Λ]/{**$'**[1+**s**]}-**f**

Steve Asikin ISBN 14: 978-1514685136, ISBN 10: 1514685132

Rule-14530:

If both (/), (U), (S), (V), (S'), (s), (f), (i), (Q), (D), (p), (c) and (q) are known, then its Tax Planned is:

T= {U+S-V-S'[1+s][f+i]-D}-(Q+Vp/{360[c-q]})//

Rule-14531:

If both (/), (U), (S), (V), (S'), (s), (f), (i), (T), (Q), (p), (c) and (q) are known, then its Dividend Planned is:

D= {U+S-V-S'[1+s][f+i]-T}-(Q+Vp/{360[c-q]})//

Rule-14532:

If both (/), (U), (S), (V), (S'), (s), (f), (i), (T), (D), (Q), (c) and (q) are known, then its Procured Inventory Days Planned:

p= 360[c-q](/{U+S-V-S'[1+s][f+i]-T-D}-Q)/V

Rule-14533:

If both (/), (U), (S), (V), (S'), (s), (f), (i), (T), (D), (p), (Q) and (q) are known, then its Current Ratio Planned is:

c= q+Vp/[360(/{U+S-V-S'[1+s][f+i]-T-D}-Q)]

Rule-14534:

If both (/), (U), (S), (V), (S'), (s), (f), (i), (T), (D), (p), (c) and (Q) are known, then its Quick or Acid Test Ratio Planned is:

q= c-Vp/[360(/{U+S-V-S'[1+s][f+i]-T-D}-Q)]

Steve Asikin ISBN 14: 978-1514685136, ISBN 10: 1514685132

Rule-14535:

If both (**/**), (**U**), (**$**), (**V**), (**$'**), (**s**), (**f**), (**i**), (**T**), (**D**), (**v**), (**p**), (**c**) and (**q**) are known, then its Quoted Longterm Debt Planned is:

Q= **/**{**U**+**$**-**V**-**$'**[1+**s**][**f**+**i**]-**T**-**D**}-**$vp**/{360[**c**-**q**]}

Rule-14536:

If both (**Q**), (**U**), (**$**), (**V**), (**$'**), (**s**), (**f**), (**i**), (**T**), (**D**), (**v**), (**p**), (**c**) and (**q**) are known, then its Leverage or Gearing Ratio Planned is:

/= (**Q**+**$vp**/{360[**c**-**q**]})/{**U**+**$**-**V**-**$'**[1+**s**][**f**+**i**]-**T**-**D**}

Rule-14537:

If both (**/**), (**Q**), (**$**), (**V**), (**$'**), (**s**), (**f**), (**i**), (**T**), (**D**), (**v**), (**p**), (**c**) and (**q**) are known, then its Utilized or Starting Capital must be:

U= (**Q**+**$vp**/{360[**c**-**q**]})/**/**-{**$**-**V**-**$'**[1+**s**][**f**+**i**]-**T**-**D**}

Rule-14538:

If both (**/**), (**U**), (**Q**), (**V**), (**$'**), (**s**), (**f**), (**i**), (**T**), (**D**), (**v**), (**p**), (**c**) and (**q**) are known, then its Sales or Revenue Planned is:

$**= **/**{**U**-**V**-**$'[1+**s**][**f**+**i**]-**T**-**D**}-**Q**)/(**vp**/{360[**c**-**q**]}-**/**)

Rule-14539:

If both (**/**), (**U**), (**$**), (**Q**), (**$'**), (**s**), (**f**), (**i**), (**T**), (**D**), (**v**), (**p**), (**c**) and (**q**) are known, then its Variable Cost Planned is:

V= (**/**{**U**+**$**-**$'**[1+**s**][**f**+**i**]-**T**-**D**}-**$vp**/{360[**c**-**q**]}-**Q**)/**/**

Steve Asikin ISBN 14: 978-1514685136, ISBN 10: 1514685132

Rule-14540:

If both (**/**), (**U**), (**$**), (**V**), (**Q**), (**s**), (**f**), (**i**), (**T**), (**D**), (**v**), (**p**), (**c**) and (**q**) are known, then its Sales Past must be:

$'**= [**U**+**$-**V**-**T**-**D**-(**Q**+**$vp**/{360[**c**-**q**]})/**/**]
/{[1+**s**][**f**+**i**]}

Rule-14541:

If both (**/**), (**U**), (**$**), (**V**), (**$'**), (**Q**), (**f**), (**i**), (**T**), (**D**), (**v**), (**p**), (**c**) and (**q**) are known, then its Sales Growth Planned is:

s= **$**/**$'**-1

or it can also be found as

s= [**U**+**$**-**V**-**T**-**D**-(**Q**+**$vp**/{360[**c**-**q**]})/**/**]/{**$'**[**f**+**i**]}-1

Rule-14542:

If both (**/**), (**U**), (**$**), (**V**), (**$'**), (**s**), (**Q**), (**i**), (**T**), (**D**), (**v**), (**p**), (**c**) and (**q**) are known, then its Fixed Portion Planned is:

f= [**U**+**$**-**V**-**T**-**D**-(**Q**+**$vp**/{360[**c**-**q**]})/**/**]/{**$'**[1+**s**]}-**i**

Rule-14543:

If both (**/**), (**U**), (**$**), (**V**), (**$'**), (**s**), (**f**), (**Q**), (**T**), (**D**), (**v**), (**p**), (**c**) and (**q**) are known, then its Interest Portion Planned is:

i= [**U**+**$**-**V**-**T**-**D**-(**Q**+**$vp**/{360[**c**-**q**]})/**/**]/{**$'**[1+**s**]}-**f**

Rule-14544:

If both (**/**), (**U**), (**$**), (**V**), (**$'**), (**s**), (**f**), (**i**), (**Q**), (**D**), (**v**), (**p**), (**c**) and (**q**) are known, then its Tax Planned is:

T= {**U**+**$**-**V**-**$'**[1+**s**][**f**+**i**]-**D**}-(**Q**+**$vp**/{360[**c**-**q**]})/**/**

Steve Asikin ISBN 14: 978-1514685136, ISBN 10: 1514685132

Rule-14545:

If both (**/**), (**U**), (**$**), (**V**), (**$'**), (**s**), (**f**), (**i**), (**T**), (**D**), (**v**), (**p**), (**c**) and (**q**) are known, then its Quoted Longterm Debt Planned is:

D= {U+$-V-$'[1+s][f+i]-T}-(Q+$vp/{360[c-q]})//

Rule-14546:

If both (**/**), (**U**), (**$**), (**V**), (**$'**), (**s**), (**f**), (**i**), (**T**), (**D**), (**Q**), (**p**), (**c**) and (**q**) are known, then its Variable Portion Planned is:

v= V/$

or it can also be found as

v= 360[c-q](/{U+$-V-$'[1+s][f+i]-T-D}-Q)/[$p]

Rule-14547:

If both (**/**), (**U**), (**$**), (**V**), (**$'**), (**s**), (**f**), (**i**), (**T**), (**D**), (**v**), (**Q**), (**c**) and (**q**) are known, then its Procured Inventory Days Planned:

p= 360[c-q](/{U+$-V-$'[1+s][f+i]-T-D}-Q)/[$v]

Rule-14548:

If both (**/**), (**U**), (**$**), (**V**), (**$'**), (**s**), (**f**), (**i**), (**T**), (**D**), (**v**), (**p**), (**Q**) and (**q**) are known, then its Current Ratio Planned is:

c= q+$vp/[360(/{U+$-V-$'[1+s][f+i]-T-D}-Q)]

Rule-14549:

If both (**/**), (**U**), (**$**), (**V**), (**$'**), (**s**), (**f**), (**i**), (**T**), (**D**), (**v**), (**p**), (**c**) and (**Q**) are known, then its Quick or Acid Test Ratio Planned is:

q= c-$vp/[360(/{U+$-V-$'[1+s][f+i]-T-D}-Q)]

Steve Asikin ISBN 14: 978-1514685136, ISBN 10: 1514685132

Rule-14550:

If both (/), (U), ($), (V), ($'), (s), (f), (i), (T), (D), (v), (p), (c) and (q) are known, then its Quoted Longterm Debt Planned is:

Q= /{U+$-V-$'[1+s][f+i]-T-D}
-$'vp[1+s]/{360[c-q]}

Rule-14551:

If both (Q), (U), ($), (V), ($'), (s), (f), (i), (T), (D), (v), (p), (c) and (q) are known, then its Leverage or Gearing Ratio Planned is:

/= (Q+$'vp[1+s]/{360[c-q]})
/{U+$-V-$'[1+s][f+i]-T-D}

Rule-14552:

If both (/), (Q), ($), (V), ($'), (s), (f), (i), (T), (D), (v), (p), (c) and (q) are known, then its Utilized or Starting Capital must be:

U= (Q+$'vp[1+s]/{360[c-q]})//
-{$-V-$'[1+s][f+i]-T-D}

Rule-14553:

If both (/), (U), (Q), (V), ($'), (s), (f), (i), (T), (D), (v), (p), (c) and (q) are known, then its Sales or Revenue Planned is:

$= (Q-/{U-V-$'[1+s][f+i]-T-D}
+$'vp[1+s]/{360[c-q]})//

Steve Asikin ISBN 14: 978-1514685136, ISBN 10: 1514685132

Rule-14554:

If both (**/**), (**U**), (**$**), (**Q**), (**$'**), (**s**), (**f**), (**i**), (**T**), (**D**), (**v**), (**p**), (**c**) and (**q**) are known, then its Variable Cost Planned is:

V= (**/**{**U**+**$**-**$'**[1+**s**][**f**+**i**]-**T**-**D**}
-**$'vp**[1+**s**]/{360[**c**-**q**]}-**Q**)/**/**

Rule-14555:

If both (**/**), (**U**), (**$**), (**V**), (**Q**), (**s**), (**f**), (**i**), (**T**), (**D**), (**v**), (**p**), (**c**) and (**q**) are known, then its Sales Past must be:

$'**= (**/**{**U**+**$-**V**-**T**-**D**}-**Q**)
/**[**[1+**s**](**vp**/{360[**c**-**q**]}+**/**[**f**+**i**])**]**

Rule-14556:

If both (**/**), (**U**), (**$**), (**V**), (**$'**), (**Q**), (**f**), (**i**), (**T**), (**D**), (**v**), (**p**), (**c**) and (**q**) are known, then its Sales Growth Planned is:

s= **$**/**$'**-1

or it can also be found as

s= (**/**{**U**+**$**-**V**-**T**-**D**}-**Q**)/**[$'**(**vp**/{360[**c**-**q**]}+**/**[**f**+**i**])**]**-1

Rule-14557:

If both (**/**), (**U**), (**$**), (**V**), (**$'**), (**s**), (**Q**), (**i**), (**T**), (**D**), (**v**), (**p**), (**c**) and (**q**) are known, then its Fixed Portion Planned is:

f= **[U**+**$**-**V**-**T**-**D**-(**Q**+**$'vp**[1+**s**]/{360[**c**-**q**]})/**/]**
/{**$'**[1+**s**]}-**i**

Steve Asikin ISBN 14: 978-1514685136, ISBN 10: 1514685132

Rule-14558:

If both (/), (U), ($), (V), ($'), (s), (f), (Q), (T), (D), (v), (p), (c) and (q) are known, then its Interest Portion Debt Planned is:

i= [U+$-V-T-D-(Q+$'vp[1+s]/{360[c-q]})//]
/{$'[1+s]}-f

Rule-14559:

If both (/), (U), ($), (V), ($'), (s), (f), (i), (Q), (D), (v), (p), (c) and (q) are known, then its Tax Planned is:

T= {U+$-V-$'[1+s][f+i]-D}
-(Q+$'vp[1+s]/{360[c-q]})//

Rule-14560:

If both (/), (U), ($), (V), ($'), (s), (f), (i), (T), (Q), (v), (p), (c) and (q) are known, then its Dividend Planned is:

D= {U+$-V-$'[1+s][f+i]-T}
-(Q+$'vp[1+s]/{360[c-q]})//

Rule-14561:

If both (/), (U), ($), (V), ($'), (s), (f), (i), (T), (D), (Q), (p), (c) and (q) are known, then its Variable Portion Planned is:

v= V/$

or it can also be found as

v= 360[c-q](/{U+$-V-$'[1+s][f+i]-T-D}-Q)
/{$'p[1+s]}

Steve Asikin ISBN 14: 978-1514685136, ISBN 10: 1514685132

Rule-14562:

If both (**/**), (**U**), (**$**), (**V**), (**$'**), (**s**), (**f**), (**i**), (**T**), (**D**), (**v**), (**Q**), (**c**) and (**q**) are known, then its Procured Inventory Days Planned:

p= 360[**c-q**](**/**{**U**+**$**-**V**-**$'**[1+**s**][**f**+**i**]-**T**-**D**}-**Q**)
/{**$'v**[1+**s**]}

Rule-14563:

If both (**/**), (**U**), (**$**), (**V**), (**$'**), (**s**), (**f**), (**i**), (**T**), (**D**), (**v**), (**p**), (**Q**) and (**q**) are known, then its Current Ratio Planned is:

c= **q**+**$'vp**[1+**s**]
/[360(**/**{**U**+**$**-**V**-**$'**[1+**s**][**f**+**i**]-**T**-**D**}-**Q**)]

Rule-14564:

If both (**/**), (**U**), (**$**), (**V**), (**$'**), (**s**), (**f**), (**i**), (**T**), (**D**), (**v**), (**p**), (**c**) and (**Q**) are known, then its Quick or Acid Test Ratio Planned is:

q= **c**-**$'vp**[1+**s**]
/[360(**/**{**U**+**$**-**V**-**$'**[1+**s**][**f**+**i**]-**T**-**D**}-**Q**)]

Rule-14565:

If both (**/**), (**U**), (**$**), (**V**), (**$'**), (**s**), (**f**), (**i**), (**T**), (**d**) and (**X**) are known, then its Quoted Longterm Debt Planned is:

Q= **/**(**U**+{**$**-**V**-**$'**[1+**s**][**f**+**i**]-**T**}[1-**d**])-**X**

Steve Asikin ISBN 14: 978-1514685136, ISBN 10: 1514685132

Rule-14566:

If both (**Q**), (**U**), (**$**), (**V**), (**$'**), (**s**), (**f**), (**i**), (**T**), (**d**) and (**X**) are known, then its Leverage or Gearing Ratio Planned is:

/= [Q+X]/(U+{$-V-$'[1+s][f+i]-T}[1-d])

Rule-14567:

If both (**/**), (**Q**), (**$**), (**V**), (**$'**), (**s**), (**f**), (**i**), (**T**), (**d**) and (**X**) are known, then its Utilized or Starting Capital must be:

U= [Q+X]//{$-V-$'[1+s][f+i]-T}[1-d]

Rule-14568:

If both (**/**), (**U**), (**Q**), (**V**), (**$'**), (**s**), (**f**), (**i**), (**T**), (**d**) and (**X**) are known, then its Sales or Revenue Planned is:

$= V+$'[1+s][f+i]+T+{[Q+X]//U}/[1-d]

Rule-14569:

If both (**/**), (**U**), (**$**), (**Q**), (**$'**), (**s**), (**f**), (**i**), (**T**), (**d**) and (**X**) are known, then its Variable Cost Planned is:

V= {$-$'[1+s][f+i]-T}-{[Q+X]//U}/[1-d]

Rule-14570:

If both (**/**), (**U**), (**$**), (**V**), (**Q**), (**s**), (**f**), (**i**), (**T**), (**d**) and (**X**) are known, then its Sales Past must be:

$'= {$-V-T}-{[Q+X]//U}/{[1+s][f+i] [1-d]}

Steve Asikin ISBN 14: 978-1514685136, ISBN 10: 1514685132

Rule-14571:

If both (**/**), (**U**), (**$**), (**V**), (**$'**), (**Q**), (**f**), (**i**), (**T**), (**d**) and (**X**) are known, then its Sales Growth Planned is:

s= **$**/**$'**-1

or it can also be found as

s= {**$**-**V**-**T**}-{[**Q**+**X**]/**/-U**}/{**$'**[**f**+**i**] [1-**d**]}-1

Rule-14572:

If both (**/**), (**U**), (**$**), (**V**), (**$'**), (**s**), (**Q**), (**i**), (**T**), (**d**) and (**X**) are known, then its Fixed Portion Planned is:

f= {**$**-**V**-**T**}-{[**Q**+**X**]/**/-U**}/{**$'**[1+**s**][1-**d**]}-**i**

Rule-14573:

If both (**/**), (**U**), (**$**), (**V**), (**$'**), (**s**), (**f**), (**Q**), (**T**), (**d**) and (**X**) are known, then its Interest Portion Planned is:

i= {**$**-**V**-**T**}-{[**Q**+**X**]/**/-U**}/{**$'**[1+**s**][1-**d**]}-**f**

Rule-14574:

If both (**/**), (**U**), (**$**), (**V**), (**$'**), (**s**), (**f**), (**i**), (**Q**), (**d**) and (**X**) are known, then its Tax Planned is:

T= {**$**-**V**-**$'**[1+**s**][**f**+**i**]}-{[**Q**+**X**]/**/-U**}/[1-**d**]

Rule-14575:

If both (**/**), (**U**), (**$**), (**V**), (**$'**), (**s**), (**f**), (**i**), (**T**), (**Q**) and (**X**) are known, then its Dividend Payout Planned is:

d= 1-{[**Q**+**X**]/**/-U**}/{**$**-**V**-**$'**[1+**s**][**f**+**i**]-**T**}

Steve Asikin ISBN 14: 978-1514685136, ISBN 10: 1514685132

Rule-14576:

If both (/), (U), (S), (V), (S'), (s), (f), (i), (T), (d) and (Q) are known, then its Xpress or Current Debt Planned is:

X= /(U+{S-V-S'[1+s][f+i]-T}[1-d])-Q

Rule-14577:

If both (/), (U), (S), (V), (S'), (s), (f), (i), (T), (d), (P), (c) and (q) are known, then its Quoted Longterm Debt Planned is:

Q= /(U+{S-V-S'[1+s][f+i]-T}[1-d])-P/[c-q]

Rule-14578:

If both (Q), (U), (S), (V), (S'), (s), (f), (i), (T), (d), (P), (c) and (q) are known, then its Leverage or Gearing Ratio Planned is:

/= {Q+P/[c-q]}/(U+{S-V-S'[1+s][f+i]-T}[1-d])

Rule-14579:

If both (/), (Q), (S), (V), (S'), (s), (f), (i), (T), (d), (P), (c) and (q) are known, then its Utilized or Starting Capital must be:

U= {Q+P/[c-q]}/ /-{S-V-S'[1+s][f+i]-T}[1-d]

Rule-14580:

If both (/), (U), (Q), (V), (S'), (s), (f), (i), (T), (d), (P), (c) and (q) are known, then its Sales or Revenue Planned is:

S= V+S'[1+s][f+i]+T+({Q+P/[c-q]}/ /-U)/[1-d]

Steve Asikin ISBN 14: 978-1514685136, ISBN 10: 1514685132

Rule-14581:

If both (**/**), (**U**), (**$**), (**Q**), (**$'**), (**s**), (**f**), (**i**), (**T**), (**d**), (**P**), (**c**) and (**q**) are known, then its Variable Cost Planned is:

V= {**$**-**$'**[1+**s**][**f**+**i**]-**T**}-({**Q**+**P**/[**c**-**q**]}/**/-U**)/[1-**d**]

Rule-14582:

If both (**/**), (**U**), (**$**), (**V**), (**Q**), (**s**), (**f**), (**i**), (**T**), (**d**), (**P**), (**c**) and (**q**) are known, then its Sales Past must be:

$'**= {**$-**V**-**T**}-({**Q**+**P**/[**c**-**q**]}/**/-U**)/{[1+**s**][**f**+**i**] [1-**d**]}

Rule-14583:

If both (**/**), (**U**), (**$**), (**V**), (**$'**), (**Q**), (**f**), (**i**), (**T**), (**d**), (**P**), (**c**) and (**q**) are known, then its Sales Growth Planned is:

s= **$**/**$'**-1

or it can also be found as

s= {**$**-**V**-**T**}-({**Q**+**P**/[**c**-**q**]}/**/-U**)/{**$'**[**f**+**i**] [1-**d**]}-1

Rule-14584:

If both (**/**), (**U**), (**$**), (**V**), (**$'**), (**s**), (**Q**), (**i**), (**T**), (**d**), (**P**), (**c**) and (**q**) are known, then its Fixed Portion Planned is:

f= {**$**-**V**-**T**}-({**Q**+**P**/[**c**-**q**]}/**/-U**)/{**$'**[1+**s**][1-**d**]}-**i**

Rule-14585:

If both (**/**), (**U**), (**$**), (**V**), (**$'**), (**s**), (**f**), (**Q**), (**T**), (**d**), (**P**), (**c**) and (**q**) are known, then its Interest Portion Planned is:

i= {**$**-**V**-**T**}-({**Q**+**P**/[**c**-**q**]}/**/-U**)/{**$'**[1+**s**][1-**d**]}-**f**

Steve Asikin ISBN 14: 978-1514685136, ISBN 10: 1514685132

Rule-14586:

If both (**/**), (**U**), (**$**), (**V**), (**$'**), (**s**), (**f**), (**i**), (**Q**), (**d**), (**P**), (**c**) and (**q**) are known, then its Tax Planned is:

T= {**$**-**V**-**$'**[1+**s**][**f**+**i**]}-({**Q**+**P**/[**c**-**q**]}/**/**-**U**)/[1-**d**]

Rule-14587:

If both (**/**), (**U**), (**$**), (**V**), (**$'**), (**s**), (**f**), (**i**), (**T**), (**Q**), (**P**), (**c**) and (**q**) are known, then its Dividend Payout Planned is:

d= 1-({**Q**+**P**/[**c**-**q**]}/**/**-**U**)/{**$**-**V**-**$'**[1+**s**][**f**+**i**]-**T**}

Rule-14588:

If both (**/**), (**U**), (**$**), (**V**), (**$'**), (**s**), (**f**), (**i**), (**T**), (**d**), (**Q**), (**c**) and (**q**) are known, then its Procured Inventory Planned is:

P= [**c**-**q**][**/**(**U**+{**$**-**V**-**$'**[1+**s**][**f**+**i**]-**T**}[1-**d**])-**Q**]

Rule-14589:

If both (**/**), (**U**), (**$**), (**V**), (**$'**), (**s**), (**f**), (**i**), (**T**), (**d**), (**P**), (**Q**) and (**q**) are known, then its Current Ratio Planned is:

c= **q**+**P**/[**/**(**U**+{**$**-**V**-**$'**[1+**s**][**f**+**i**]-**T**}[1-**d**])-**Q**]

Rule-14590:

If both (**/**), (**U**), (**$**), (**V**), (**$'**), (**s**), (**f**), (**i**), (**T**), (**d**), (**P**), (**c**) and (**Q**) are known, then its Quick or Acid Test Ratio Planned is:

q= **c**-**P**/[**/**(**U**+{**$**-**V**-**$'**[1+**s**][**f**+**i**]-**T**}[1-**d**])-**Q**]

Steve Asikin ISBN 14: 978-1514685136, ISBN 10: 1514685132

Rule-14591:

If both (/), (U), (S), (V), (S'), (s), (f), (i), (T), (d), (p), (c) and (q) are known, then its Quoted Longterm Debt Planned is:

Q= /(U+{S-V-S'[1+s][f+i]-T}[1-d])-Vp/{360[c-q]}

Rule-14592:

If both (Q), (U), (S), (V), (S'), (s), (f), (i), (T), (d), (p), (c) and (q) are known, then its Leverage or Gearing Ratio Planned is:

/= (Q+Vp/{360[c-q]})
/(U+{S-V-S'[1+s][f+i]-T}[1-d])

Rule-14593:

If both (/), (Q), (S), (V), (S'), (s), (f), (i), (T), (d), (p), (c) and (q) are known, then its Utilized or Starting Capital must be:

U= (Q+Vp/{360[c-q]})//
-{S-V-S'[1+s][f+i]-T}[1-d]

Rule-14594:

If both (/), (U), (Q), (V), (S'), (s), (f), (i), (T), (d), (p), (c) and (q) are known, then its Sales or Revenue Planned is:

S= V+S'[1+s][f+i]+T+[(Q+Vp/{360[c-q]})//-U]
/[1-d]

Steve Asikin ISBN 14: 978-1514685136, ISBN 10: 1514685132

Rule-14595:

If both (**/**), (**U**), (**$**), (**Q**), (**$'**), (**s**), (**f**), (**i**), (**T**), (**d**), (**p**), (**c**) and (**q**) are known, then its Variable Cost Planned is:

V= [/(U+{$-$'[1+s][f+i]-T}[1-d])-Q]
/(p/{360[c-q]}+/[1-d])

Rule-14596:

If both (**/**), (**U**), (**$**), (**V**), (**Q**), (**s**), (**f**), (**i**), (**T**), (**d**), (**p**), (**c**) and (**q**) are known, then its Sales Past must be:

$'= {$-V-T}-[(Q+Vp/{360[c-q]})/-U]
/{[1+s][f+i] [1-d]}

Rule-14597:

If both (**/**), (**U**), (**$**), (**V**), (**$'**), (**Q**), (**f**), (**i**), (**T**), (**d**), (**p**), (**c**) and (**q**) are known, then its Sales Growth Planned is:

s= $/$'-1

or it can also be found as

s= {$-V-T}-[(Q+Vp/{360[c-q]})/-U]
/{$'[f+i] [1-d]}-1

Rule-14598:

If both (**/**), (**U**), (**$**), (**V**), (**$'**), (**s**), (**Q**), (**i**), (**T**), (**d**), (**p**), (**c**) and (**q**) are known, then its Fixed Portion Planned is:

f= {$-V-T}-[(Q+Vp/{360[c-q]})/-U]
/{$'[1+s][1-d]}-i

Steve Asikin ISBN 14: 978-1514685136, ISBN 10: 1514685132

Rule-14599:

If both (**/**), (**U**), (**$**), (**V**), (**$'**), (**s**), (**f**), (**Q**), (**T**), (**d**), (**p**), (**c**) and (**q**) are known, then its Interest Portion Planned is:

i= {$-V-T}-[(Q+Vp/{360[c-q]})/**/**U]
/{$'[1+s][1-d]}-f

Rule-14600:

If both (**/**), (**U**), (**$**), (**V**), (**$'**), (**s**), (**f**), (**i**), (**Q**), (**d**), (**p**), (**c**) and (**q**) are known, then its Tax Planned is:

T= {$-V-$'[1+s][f+i]}
-[(Q+Vp/{360[c-q]})/**/**U]/[1-d]

Rule-14601:

If both (**/**), (**U**), (**$**), (**V**), (**$'**), (**s**), (**f**), (**i**), (**T**), (**Q**), (**p**), (**c**) and (**q**) are known, then its Dividend Portion Planned is:

d= 1-[(Q+Vp/{360[c-q]})/**/**U]
/{$-V-$'[1+s][f+i]-T}

Rule-14602:

If both (**/**), (**U**), (**$**), (**V**), (**$'**), (**s**), (**f**), (**i**), (**T**), (**d**), (**Q**), (**c**) and (**q**) are known, then its Procured Inventory Days Planned:

p= 360[c-q][**/**(U+{$-V-$'[1+s][f+i]-T}[1-d])-Q]/V

Rule-14603:

If both (**/**), (**U**), (**$**), (**V**), (**$'**), (**s**), (**f**), (**i**), (**T**), (**d**), (**p**), (**Q**) and (**q**) are known, then its Current Ratio Planned is:

c= q+Vp/{360[**/**(U+{$-V-$'[1+s][f+i]-T}[1-d])-Q]}

Steve Asikin ISBN 14: 978-1514685136, ISBN 10: 1514685132

Rule-14604:

If both (**/**), (**U**), (**$**), (**V**), (**$'**), (**s**), (**f**), (**i**), (**T**), (**d**), (**p**), (**c**) and (**Q**) are known, then its Quick or Acid Test Ratio Planned is:

q= **c**-**Vp**/{360[**/**(**U**+{**$**-**V**-**$'**[1+**s**][**f**+**i**]-**T**}[1-**d**])-**Q**]}

Rule-14605:

If both (**/**), (**U**), (**$**), (**V**), (**$'**), (**s**), (**f**), (**i**), (**T**), (**d**), (**v**), (**p**), (**c**) and (**q**) are known, then its Quoted Longterm Debt Planned is:

Q= **/**(**U**+{**$**-**V**-**$'**[1+**s**][**f**+**i**]-**T**}[1-**d**])
-**$vp**/{360[**c**-**q**]}

Rule-14606:

If both (**Q**), (**U**), (**$**), (**V**), (**$'**), (**s**), (**f**), (**i**), (**T**), (**d**), (**v**), (**p**), (**c**) and (**q**) are known, then its Leverage or Gearing Ratio Planned is:

/= (**Q**+**$vp**/{360[**c**-**q**]})
/(**U**+{**$**-**V**-**$'**[1+**s**][**f**+**i**]-**T**}[1-**d**])

Rule-14607:

If both (**/**), (**Q**), (**$**), (**V**), (**$'**), (**s**), (**f**), (**i**), (**T**), (**d**), (**v**), (**p**), (**c**) and (**q**) are known, then its Utilized or Starting Capital must be:

U= (**Q**+**$vp**/{360[**c**-**q**]})/**/**
-{**$**-**V**-**$'**[1+**s**][**f**+**i**]-**T**}[1-**d**]

Steve Asikin ISBN 14: 978-1514685136, ISBN 10: 1514685132

Rule-14608:

If both (**/**), (**U**), (**Q**), (**V**), (**$'**), (**s**), (**f**), (**i**), (**T**), (**d**), (**v**), (**p**), (**c**) and (**q**) are known, then its Sales or Revenue Planned is:

$= [/(U+{ -V-$'[1+s][f+i]-T}[1-d])-Q]
/(vp/{360[c-q]}-/[1-d])

Rule-14609:

If both (**/**), (**U**), (**$**), (**Q**), (**$'**), (**s**), (**f**), (**i**), (**T**), (**d**), (**v**), (**p**), (**c**) and (**q**) are known, then its Variable Cost Planned is:

V= [/(U+{$-$'[1+s][f+i]-T}[1-d])
-$vp/{360[c-q]}-Q]/{/[1-d]}

Rule-14610:

If both (**/**), (**U**), (**$**), (**V**), (**Q**), (**s**), (**f**), (**i**), (**T**), (**d**), (**v**), (**p**), (**c**) and (**q**) are known, then its Sales Past must be:

$'= {$-V-T}-[(Q+$vp/{360[c-q]})/-U]
/{[1+s][f+i] [1-d]}

Rule-14611:

If both (**/**), (**U**), (**$**), (**V**), (**$'**), (**Q**), (**f**), (**i**), (**T**), (**d**), (**v**), (**p**), (**c**) and (**q**) are known, then its Sales Growth Planned is:

s= $/$'-1

or it can also be found as

s= {$-V-T}-[(Q+$vp/{360[c-q]})/-U]
/{$'[f+i] [1-d]}-1

Steve Asikin ISBN 14: 978-1514685136, ISBN 10: 1514685132

Rule-14612:

If both $(\textbf{\textit{l}})$, $(\textbf{U})$, $(\textbf{S})$, $(\textbf{V})$, $(\textbf{S'})$, $(\textbf{s})$, $(\textbf{Q})$, $(\textbf{i})$, $(\textbf{T})$, $(\textbf{d})$, $(\textbf{v})$, $(\textbf{p})$, $(\textbf{c})$ and $(\textbf{q})$ are known, then its Fixed Portion Planned is:

$$f = \{S\text{-}V\text{-}T\}\text{-}[(Q+Svp/\{360[c\text{-}q]\})/\textit{l}U]$$
$$/\{S'[1+s][1\text{-}d]\}\text{-}i$$

Rule-14613:

If both $(\textbf{\textit{l}})$, $(\textbf{U})$, $(\textbf{S})$, $(\textbf{V})$, $(\textbf{S'})$, $(\textbf{s})$, $(\textbf{f})$, $(\textbf{Q})$, $(\textbf{T})$, $(\textbf{d})$, $(\textbf{v})$, $(\textbf{p})$, $(\textbf{c})$ and $(\textbf{q})$ are known, then its Interest Portion Planned is:

$$i = \{S\text{-}V\text{-}T\}\text{-}[(Q+Svp/\{360[c\text{-}q]\})/\textit{l}U]$$
$$/\{S'[1+s][1\text{-}d]\}\text{-}f$$

Rule-14614:

If both $(\textbf{\textit{l}})$, $(\textbf{U})$, $(\textbf{S})$, $(\textbf{V})$, $(\textbf{S'})$, $(\textbf{s})$, $(\textbf{f})$, $(\textbf{i})$, $(\textbf{Q})$, $(\textbf{d})$, $(\textbf{v})$, $(\textbf{p})$, $(\textbf{c})$ and $(\textbf{q})$ are known, then its Tax Planned is:

$$T = \{S\text{-}V\text{-}S'[1+s][f+i]\}$$
$$\text{-}[(Q+Svp/\{360[c\text{-}q]\})/\textit{l}U]/[1\text{-}d]$$

Rule-14615:

If both $(\textbf{\textit{l}})$, $(\textbf{U})$, $(\textbf{S})$, $(\textbf{V})$, $(\textbf{S'})$, $(\textbf{s})$, $(\textbf{f})$, $(\textbf{i})$, $(\textbf{T})$, $(\textbf{Q})$, $(\textbf{v})$, $(\textbf{p})$, $(\textbf{c})$ and $(\textbf{q})$ are known, then its Dividend Payout Planned is:

$$d = 1\text{-}[(Q+Svp/\{360[c\text{-}q]\})/\textit{l}U]$$
$$/\{S\text{-}V\text{-}S'[1+s][f+i]\text{-}T\}$$

Steve Asikin ISBN 14: 978-1514685136, ISBN 10: 1514685132

Rule-14616:

If both (**/**), (**U**), (**$**), (**V**), (**$'**), (**s**), (**f**), (**i**), (**T**), (**d**), (**Q**), (**p**), (**c**) and (**q**) are known, then its Variable Portion Planned is:

v= **V**/**$**

or it can also be found

v= 360[**c**-**q**]
[**/**(**U**+{**$**-**V**-**$'**[1+**s**][**f**+**i**]-**T**}[1-**d**])-**Q**]/[**$p**]

Rule-14617:

If both (**/**), (**U**), (**$**), (**V**), (**$'**), (**s**), (**f**), (**i**), (**T**), (**d**), (**v**), (**Q**), (**c**) and (**q**) are known, then its Procured Inventory Days Planned is:

p= 360[**c**-**q**]
[**/**(**U**+{**$**-**V**-**$'**[1+**s**][**f**+**i**]-**T**}[1-**d**])-**Q**]/[**$v**]

Rule-14618:

If both (**/**), (**U**), (**$**), (**V**), (**$'**), (**s**), (**f**), (**i**), (**T**), (**d**), (**v**), (**p**), (**Q**) and (**q**) are known, then its Current Ratio Planned is:

c= **q**+**$vp**
/{360[**/**(**U**+{**$**-**V**-**$'**[1+**s**][**f**+**i**]-**T**}[1-**d**])-**Q**]}

Rule-14619:

If both (**/**), (**U**), (**$**), (**V**), (**$'**), (**s**), (**f**), (**i**), (**T**), (**d**), (**v**), (**p**), (**c**) and (**Q**) are known, then its Quick or Acid Test Ratio Planned is:

q= **c**-**$vp**
/{360[**/**(**U**+{**$**-**V**-**$'**[1+**s**][**f**+**i**]-**T**}[1-**d**])-**Q**]}

Steve Asikin ISBN 14: 978-1514685136, ISBN 10: 1514685132

Rule-14620:

If both (**/**), (**U**), (**$**), (**V**), (**$'**), (**s**), (**f**), (**i**), (**T**), (**d**), (**v**), (**p**), (**c**) and (**q**) are known, then its Quoted Longterm Debt Planned is:

Q= **/**(**U**+{**$**-**V**-**$'**[1+**s**][**f**+**i**]-**T**}[1-**d**])
-**$'vp**[1+**s**]/{360[**c**-**q**]}

Rule-14621:

If both (**Q**), (**U**), (**$**), (**V**), (**$'**), (**s**), (**f**), (**i**), (**T**), (**d**), (**v**), (**p**), (**c**) and (**q**) are known, then its Leverage or Gearing Ratio Planned is:

/= (**Q**+**$'vp**[1+**s**]/{360[**c**-**q**]})
/(**U**+{**$**-**V**-**$'**[1+**s**][**f**+**i**]-**T**}[1-**d**])

Rule-14622:

If both (**/**), (**Q**), (**$**), (**V**), (**$'**), (**s**), (**f**), (**i**), (**T**), (**d**), (**v**), (**p**), (**c**) and (**q**) are known, then its Utilized or Starting Capital must be:

U= (**Q**+**$'vp**[1+**s**]/{360[**c**-**q**]})/**/**-{**$**-**V**
-**$'**[1+**s**][**f**+**i**]-**T**}[1-**d**]

Rule-14623:

If both (**/**), (**U**), (**Q**), (**V**), (**$'**), (**s**), (**f**), (**i**), (**T**), (**d**), (**v**), (**p**), (**c**) and (**q**) are known, then its Sales or Revenue Planned is:

$**= [**Q**-**/**(**U**-{**V**+**$'[1+**s**][**f**+**i**]-**T**}[1-**d**])
+**$'vp**[1+**s**]/{360[**c**-**q**]}]/{**/**[1-**d**]}

Steve Asikin ISBN 14: 978-1514685136, ISBN 10: 1514685132

Rule-14624:

If both (**/**), (**U**), (**$**), (**Q**), (**$'**), (**s**), (**f**), (**i**), (**T**), (**d**), (**v**), (**p**), (**c**) and (**q**) are known, then its Variable Cost Planned is:

V= [**/**(**U**+{**$**-**$'**[1+**s**][**f**+**i**]-**T**}[1-**d**])
-**$'vp**[1+**s**]/{360[**c**-**q**]}-**Q**]/{**/**[1-**d**]}

Rule-14625:

If both (**/**), (**U**), (**$**), (**V**), Q), (**s**), (**f**), (**i**), (**T**), (**d**), (**v**), (**p**), (**c**) and (**q**) are known, then its Sales Past must be:

$'**= [**/**(**U**+{**$-**V**-**T**}[1-**d**])-**Q**]
/[[1+**s**](**vp**/{360[**c**-**q**]}+**/**[**f**+**i**][1-**d**])]

Rule-14626:

If both (**/**), (**U**), (**$**), (**V**), (**$'**), (**Q**), (**f**), (**i**), (**T**), (**d**), (**v**), (**p**), (**c**) and (**q**) are known, then its Sales Growth Planned is:

s= **$**/**$'**-1

or it can also be found as

s= [**/**(**U**+{**$**-**V**-**T**}[1-**d**])-**Q**]
/[**$'**(**vp**/{360[**c**-**q**]}+**/**[**f**+**i**][1-**d**])]-1

Rule-14627:

If both (**/**), (**U**), (**$**), (**V**), (**$'**), (**s**), (**Q**), (**i**), (**T**), (**d**), (**v**), (**p**), (**c**) and (**q**) are known, then its Fixed Portion Planned is:

f= {**$**-**V**-**T**}-[(**Q**+**$'vp**[1+**s**]/{360[**c**-**q**]})/**/**-**U**]
/{**$'**[1+**s**][1-**d**]}-**i**

Steve Asikin ISBN 14: 978-1514685136, ISBN 10: 1514685132

Rule-14628:

If both (**/**), (**U**), (**$**), (**V**), (**$'**), (**s**), (**f**), (**Q**), (**T**), (**d**), (**v**), (**p**), (**c**) and (**q**) are known, then its Interest Portion Planned is:

i= {$-V-T}-[(Q+$'vp[1+s]/{360[c-q]})/**/**U]
/{$'[1+s][1-d]}-f

Rule-14629:

If both (**/**), (**U**), (**$**), (**V**), (**$'**), (**s**), (**f**), (**i**), (**Q**), (**d**), (**v**), (**p**), (**c**) and (**q**) are known, then its Tax Planned is:

T= {$-V-$'[1+s][f+i]}
-[(Q+$'vp[1+s]/{360[c-q]})/**/**U]/[1-d]

Rule-14630:

If both (**/**), (**U**), (**$**), (**V**), (**$'**), (**s**), (**f**), (**i**), (**T**), (**Q**), (**v**), (**p**), (**c**) and (**q**) are known, then its Dividend Payout Planned is:

d= 1-[(Q+$'vp[1+s]/{360[c-q]})/**/**U]
/{$-V-$'[1+s][f+i]-T}

Rule-14631:

If both (**/**), (**U**), (**$**), (**V**), (**$'**), (**s**), (**f**), (**i**), (**T**), (**d**), (**Q**), (**p**), (**c**) and (**q**) are known, then its Variable Portion Planned is:

v= V/$

or it can also be found as

v= 360[c-q][**/**U+{$-V-$'[1+s][f+i]-T}[1-d])-Q]
/{$'p[1+s]}

Steve Asikin ISBN 14: 978-1514685136, ISBN 10: 1514685132

Rule-14632:

If both (**/**), (**U**), (**S**), (**V**), (**S'**), (**s**), (**f**), (**i**), (**T**), (**d**), (**v**), (**Q**), (**c**) and (**q**) are known, then its Procured Inventory Days Planned:

p= 360[**c**-**q**][**/**(**U**+{**S**-**V**-**S'**[1+**s**][**f**+**i**]-**T**}[1-**d**])-**Q**]
/{**S'v**[1+**s**]}

Rule-14633:

If both (**/**), (**U**), (**S**), (**V**), (**S'**), (**s**), (**f**), (**i**), (**T**), (**d**), (**v**), (**p**), (**Q**) and (**q**) are known, then its Current Ratio Planned is:

c= **q**+**S'vp**[1+**s**]
/{360[**/**(**U**+{**S**-**V**-**S'**[1+**s**][**f**+**i**]-**T**}[1-**d**])-**Q**]}

Rule-14634:

If both (**/**), (**U**), (**S**), (**V**), (**S'**), (**s**), (**f**), (**i**), (**T**), (**d**), (**v**), (**p**), (**c**) and (**Q**) are known, then its Quick or Acid Test Ratio Planned is:

q= **c**-**S'vp**[1+**s**]
/{360[**/**(**U**+{**S**-**V**-**S'**[1+**s**][**f**+**i**]-**T**}[1-**d**])-**Q**]}

Rule-14635:

If both (**/**), (**U**), (**S**), (**V**), (**S'**), (**s**), (**f**), (**i**), (**T**), (**A**), (**d**) and (**X**) are known, then its Quoted Longterm Debt Planned is:

Q= **/**{**U**+**S**-**V**-**S'**[1+**s**][**f**+**i**]-**T**-**Ad**}-**X**

Steve Asikin ISBN 14: 978-1514685136, ISBN 10: 1514685132

Rule-14636:

If both (**Q**), (**U**), (**$**), (**V**), (**$'**), (**s**), (**f**), (**i**), (**T**), (**A**), (**d**) and (**X**) are known, then its Leverage or Gearing Ratio Planned is:

/= [**Q**+**X**]/{**U**+**$**-**V**-**$'**[1+**s**][**f**+**i**]-**T**-**Ad**}

Rule-14637:

If both (**/**), (**Q**), (**$**), (**V**), (**$'**), (**s**), (**f**), (**i**), (**T**), (**A**), (**d**) and (**X**) are known, then its Utilized or Starting Capital must be:

U= [**Q**+**X**]/**/**-{**$**-**V**-**$'**[1+**s**][**f**+**i**]-**T**-**Ad**}

Rule-14638:

If both (**/**), (**U**), (**Q**), (**V**), (**$'**), (**s**), (**f**), (**i**), (**T**), (**A**), (**d**) and (**X**) are known, then its Sales or Revenue Planned is:

$**= [**Q**+**X**]/**/**-{**U**-**V**-**$'[1+**s**][**f**+**i**]-**T**-**Ad**}

Rule-14639:

If both (**/**), (**U**), (**$**), (**Q**), (**$'**), (**s**), (**f**), (**i**), (**T**), (**A**), (**d**) and (**X**) are known, then its Variable Cost Planned is:

V= {**U**+**$**-**$'**[1+**s**][**f**+**i**]-**T**-**Ad**}-[**Q**+**X**]/**/**

Rule-14640:

If both (**/**), (**U**), (**$**), (**V**), (**Q**), (**s**), (**f**), (**i**), (**T**), (**A**), (**d**) and (**X**) are known, then its Sales Past must be:

$'**= {**U**+**$-**V**-**T**-**Ad**-[**Q**+**X**]/**/**}/{[1+**s**][**f**+**i**]}

Steve Asikin ISBN 14: 978-1514685136, ISBN 10: 1514685132

Rule-14641:

If both (/), (U), ($), (V), ($'), (Q), (f), (i), (T), (A), (d) and (X) are known, then its Sales Growth Planned is:

s= $/$'-1

or it can also be found as

s= {U+$-V-T-Ad-[Q+X]//}/{$'[f+i]}-1

Rule-14642:

If both (/), (U), ($), (V), ($'), (s), (Q), (i), (T), (A), (d) and (X) are known, then its Fixed Portion Planned is:

f= {U+$-V-T-Ad-[Q+X]//}/{$'[1+s]}-i

Rule-14643:

If both (/), (U), ($), (V), ($'), (s), (f), (Q), (T), (A), (d) and (X) are known, then its Interest Portion Planned is:

i= {U+$-V-T-Ad-[Q+X]//}/{$'[1+s]}-f

Rule-14644:

If both (/), (U), ($), (V), ($'), (s), (f), (i), (Q), (A), (d) and (X) are known, then its Tax Planned is:

T= {U+$-V-$'[1+s][f+i]-Ad}-[Q+X]//

Rule-14645:

If both (/), (U), ($), (V), ($'), (s), (f), (i), (T), (Q), (d) and (X) are known, then its After Tax Income Planned is:

A= {U+$-V-$'[1+s][f+i]-T-[Q+X]//}/d

Steve Asikin ISBN 14: 978-1514685136, ISBN 10: 1514685132

Rule-14646:

If both (/), (**U**), (**S**), (**V**), (**S'**), (**s**), (**f**), (**i**), (**T**), (**A**), (**Q**) and (**X**) are known, then its Dividend Payout Planned is:

d= {**U**+**S**-**V**-**S'**[1+**s**][**f**+**i**]-**T**-[**Q**+**X**]//}/**A**

Rule-14647:

If both (/), (**U**), (**S**), (**V**), (**S'**), (**s**), (**f**), (**i**), (**T**), (**A**), (**d**) and (**Q**) are known, then its Xpress or Current Debt Planned is:

X= /{**U**+**S**-**V**-**S'**[1+**s**][**f**+**i**]-**T**-**Ad**}-**Q**

Rule-14648:

If both (/), (**U**), (**S**), (**V**), (**S'**), (**s**), (**f**), (**i**), (**T**), (**A**), (**d**), (**P**), (**c**) and (**q**) are known, then its Quoted Longterm Debt Planned is:

Q= /{**U**+**S**-**V**-**S'**[1+**s**][**f**+**i**]-**T**-**Ad**}-**P**/[**c**-**q**]

Rule-14649:

If both (**Q**), (**U**), (**S**), (**V**), (**S'**), (**s**), (**f**), (**i**), (**T**), (**A**), (**d**), (**P**), (**c**) and (**q**) are known, then its Leverage or Gearing Ratio Planned is:

/= {**Q**+**P**/[**c**-**q**]}/{**U**+**S**-**V**-**S'**[1+**s**][**f**+**i**]-**T**-**Ad**}

Rule-14650:

If both (/), (**Q**), (**S**), (**V**), (**S'**), (**s**), (**f**), (**i**), (**T**), (**A**), (**d**), (**P**), (**c**) and (**q**) are known, then its Utilized or Starting Capital must be:

U= {**Q**+**P**/[**c**-**q**]}//-{**S**-**V**-**S'**[1+**s**][**f**+**i**]-**T**-**Ad**}

Steve Asikin ISBN 14: 978-1514685136, ISBN 10: 1514685132

Rule-14651:

If both (**/**), (**U**), (**Q**), (**V**), (**S'**), (**s**), (**f**), (**i**), (**T**), (**A**), (**d**), (**P**), (**c**) and (**q**) are known, then its Sales or Revenue Planned is:

S= {**Q**+**P**/[**c**-**q**]}/**/**-{**U**-**V**-**S'**[1+**s**][**f**+**i**]-**T**-**Ad**}

Rule-14652:

If both (**/**), (**U**), (**S**), (**Q**), (**S'**), (**s**), (**f**), (**i**), (**T**), (**A**), (**d**), (**P**), (**c**) and (**q**) are known, then its Variable Cost Planned is:

V= {**U**+**S**-**S'**[1+**s**][**f**+**i**]-**T**-**Ad**}-{**Q**+**P**/[**c**-**q**]}/**/**

Rule-14653:

If both (**/**), (**U**), (**S**), (**V**), (**Q**), (**s**), (**f**), (**i**), (**T**), (**A**), (**d**), (**P**), (**c**) and (**q**) are known, then its Sales Past Planned is:

S'= (**U**+**S**-**V**-**T**-**Ad**-{**Q**+**P**/[**c**-**q**]}/**/**)/{[1+**s**][**f**+**i**]}

Rule-14654:

If both (**/**), (**U**), (**S**), (**V**), (**S'**), (**s**), (**f**), (**i**), (**T**), (**A**), (**d**), (**P**), (**c**) and (**q**) are known, then its Sales Growth Planned is:

s= **S**/**S'**-1

or it can also be found as

s= (**U**+**S**-**V**-**T**-**Ad**-{**Q**+**P**/[**c**-**q**]}/**/**)/{**S'**[**f**+**i**]}-1

Rule-14655:

If both (**/**), (**U**), (**S**), (**V**), (**S'**), (**s**), (**Q**), (**i**), (**T**), (**A**), (**d**), (**P**), (**c**) and (**q**) are known, then its Fixed Portion Planned is:

f= (**U**+**S**-**V**-**T**-**Ad**-{**Q**+**P**/[**c**-**q**]}/**/**)/{**S'**[1+**s**]}-**i**

Steve Asikin ISBN 14: 978-1514685136, ISBN 10: 1514685132

Rule-14656:

If both (**/**), (**U**), (**$**), (**V**), (**$'**), (**s**), (**f**), (**Q**), (**T**), (**A**), (**d**), (**P**), (**c**) and (**q**) are known, then its Interest Portion Planned is:

i= (**U**+**$**-**V**-**T**-**Ad**-{**Q**+**P**/[**c**-**q**]}/**/**)/{**$'**[1+**s**]}-**f**

Rule-14657:

If both (**/**), (**U**), (**$**), (**V**), (**$'**), (**s**), (**f**), (**i**), (**Q**), (**A**), (**d**), (**P**), (**c**) and (**q**) are known, then its Tax Planned is:

T= {**U**+**$**-**V**-**$'**[1+**s**][**f**+**i**]-**Ad**}-{**Q**+**P**/[**c**-**q**]}/**/**

Rule-14658:

If both (**/**), (**U**), (**$**), (**V**), (**$'**), (**s**), (**f**), (**i**), (**T**), (**Q**), (**d**), (**P**), (**c**) and (**q**) are known, then its After Tax Income Planned is:

A= (**U**+**$**-**V**-**$'**[1+**s**][**f**+**i**]-**T**-{**Q**+**P**/[**c**-**q**]}/**/**)/**d**

Rule-14659:

If both (**/**), (**U**), (**$**), (**V**), (**$'**), (**s**), (**f**), (**i**), (**T**), (**A**), (**Q**), (**P**), (**c**) and (**q**) are known, then its Dividend Payout Planned is:

d= (**U**+**$**-**V**-**$'**[1+**s**][**f**+**i**]-**T**-{**Q**+**P**/[**c**-**q**]}/**/**)/**A**

Rule-14660:

If both (**/**), (**U**), (**$**), (**V**), (**$'**), (**s**), (**f**), (**i**), (**T**), (**A**), (**d**), (**Q**), (**c**) and (**q**) are known, then its Procured Inventory Planned is:

P= [**c**-**q**](**/**{**U**+**$**-**V**-**$'**[1+**s**][**f**+**i**]-**T**-**Ad**}-**Q**)

Steve Asikin ISBN 14: 978-1514685136, ISBN 10: 1514685132

Rule-14661:

If both (/), (U), (S), (V), (S'), (s), (f), (i), (T), (A), (d), (P), (Q) and (q) are known, then its Current Ratio Planned is:

c= q+P/(/{U+S-V-S'[1+s][f+i]-T-Ad}-Q)

Rule-14662:

If both (/), (U), (S), (V), (S'), (s), (f), (i), (T), (A), (d), (P), (c) and (Q) are known, then its Quick or Acid Test Ratio Planned is:

q= c-P/(/{U+S-V-S'[1+s][f+i]-T-Ad}-Q)

Rule-14663:

If both (/), (U), (S), (V), (S'), (s), (f), (i), (T), (A), (d), (p), (c) and (q) are known, then its Quoted Longterm Debt Planned is:

Q= /{U+S-V-S'[1+s][f+i]-T-Ad}-Vp/{360[c-q]}

Rule-14664:

If both (Q), (U), (S), (V), (S'), (s), (f), (i), (T), (A), (d), (p), (c) and (q) are known, then its Leverage or Gearing Ratio Planned is:

/= (Q+Vp/{360[c-q]})/{U+S-V-S'[1+s][f+i]-T-Ad}

Rule-14665:

If both (/), (Q), (S), (V), (S'), (s), (f), (i), (T), (A), (d), (p), (c) and (q) are known, then its Utilized or Starting Capital must be:

U= (Q+Vp/{360[c-q]})//-{S-V-S'[1+s][f+i]-T-Ad}

Steve Asikin ISBN 14: 978-1514685136, ISBN 10: 1514685132

Rule-14666:

If both (/), (U), (Q), (V), (S'), (s), (f), (i), (T), (A), (d), (p), (c) and (q) are known, then its Sales or Revenue Planned is:

S= (Q+Vp/{360[c-q]})/Ⱡ{U-V-S'[1+s][f+i]-T-Ad}

Rule-14667:

If both (/), (U), (S), (Q), (S'), (s), (f), (i), (T), (A), (d), (p), (c) and (q) are known, then its Variable Cost Planned is:

V= (/{U+S-S'[1+s][f+i]-T-Ad}-Q)
/(p/{360[c-q]}+/)

Rule-14668:

If both (/), (U), (S), (V), (Q), (s), (f), (i), (T), (A), (d), (p), (c) and (q) are known, then its Sales Past must be:

S'= [U+S-V-T-Ad-(Q+Vp/{360[c-q]})//]
/{[1+s][f+i]}

Rule-14669:

If both (/), (U), (S), (V), (S'), (Q), (f), (i), (T), (A), (d), (p), (c) and (q) are known, then its Sales Growth Planned is:

s= S/S'-1

or it can also be found as

s=[U+S-V-T-Ad-(Q+Vp/{360[c-q]})//]/{S'[f+i]}-1

Steve Asikin ISBN 14: 978-1514685136, ISBN 10: 1514685132

Rule-14670:

If both (**/**), (**U**), (**$**), (**V**), (**$'**), (**s**), (**Q**), (**i**), (**T**), (**A**), (**d**), (**p**), (**c**) and (**q**) are known, then its Fixed Portion Planned is:

f= [**U**+**$**-**V**-**T**-**Ad**-(**Q**+**Vp**/{360[**c**-**q**]})/**/**]/{**$'**[1+**s**]}-**i**

Rule-14671:

If both (**/**), (**U**), (**$**), (**V**), (**$'**), (**s**), (**f**), (**Q**), (**T**), (**A**), (**d**), (**p**), (**c**) and (**q**) are known, then its Interest Portion Debt Planned is:

i= [**U**+**$**-**V**-**T**-**Ad**-(**Q**+**Vp**/{360[**c**-**q**]})/**/**]/{**$'**[1+**s**]}-**f**

Rule-14672:

If both (**/**), (**U**), (**$**), (**V**), (**$'**), (**s**), (**f**), (**i**), (**Q**), (**A**), (**d**), (**p**), (**c**) and (**q**) are known, then its Tax Planned is:

T= {**U**+**$**-**V**-**$'**[1+**s**][**f**+**i**]-**Ad**}-(**Q**+**Vp**/{360[**c**-**q**]})/**/**

Rule-14673:

If both (**/**), (**U**), (**$**), (**V**), (**$'**), (**s**), (**f**), (**i**), (**T**), (**Q**), (**d**), (**p**), (**c**) and (**q**) are known, then its After Tax Income Planned is:

A= [**U**+**$**-**V**-**$'**[1+**s**][**f**+**i**]-**T**-(**Q**+**Vp**/{360[**c**-**q**]})/**/**]/**d**

Rule-14674:

If both (**/**), (**U**), (**$**), (**V**), (**$'**), (**s**), (**f**), (**i**), (**T**), (**A**), (**Q**), (**p**), (**c**) and (**q**) are known, then its Dividend Payout Planned is:

d= [**U**+**$**-**V**-**$'**[1+**s**][**f**+**i**]-**T**-(**Q**+**Vp**/{360[**c**-**q**]})/**/**]/**A**

Steve Asikin ISBN 14: 978-1514685136, ISBN 10: 1514685132

Rule-14675:

If both (/), (U), (S), (V), (S'), (s), (f), (i), (T), (A), (d), (Q), (c) and (q) are known, then its Procured Inventory Days Planned:

p= 360[c-q](/{U+S-V-S'[1+s][f+i]-T-Ad}-Q)/V

Rule-14676:

If both (/), (U), (S), (V), (S'), (s), (f), (i), (T), (A), (d), (p), (Q) and (q) are known, then its Current Ratio Planned is:

c= q+Vp/[360(/{U+S-V-S'[1+s][f+i]-T-Ad}-Q)]

Rule-14677:

If both (/), (U), (S), (V), (S'), (s), (f), (i), (T), (A), (d), (p), (c) and (Q) are known, then its Quick or Acid Test Ratio Planned is:

q= c-Vp/[360(/{U+S-V-S'[1+s][f+i]-T-Ad}-Q)]

Rule-14678:

If both (/), (U), (S), (V), (S'), (s), (f), (i), (T), (A), (d), (v), (p), (c) and (q) are known, then its Quoted Longterm Debt Planned is:

Q= /{U+S-V-S'[1+s][f+i]-T-Ad}-Svp/{360[c-q]}

Rule-14679:

If both (Q), (U), (S), (V), (S'), (s), (f), (i), (T), (A), (d), (v), (p), (c) and (q) are known, then its Leverage or Gearing Ratio Planned is:

/= (Q+Svp/{360[c-q]})
/{U+S-V-S'[1+s][f+i]-T-Ad}

Steve Asikin ISBN 14: 978-1514685136, ISBN 10: 1514685132

Rule-14680:

If both (/), (Q), ($), (V), ($'), (s), (f), (i), (T), (A), (d), (v), (p), (c) and (q) are known, then its Utilized or Starting Capital must be:

U= (Q+$vp/{360[c-q]})/ℓ-{$-V-$'[1+s][f+i]-T-Ad}

Rule-14681:

If both (/), (U), (Q), (V), ($'), (s), (f), (i), (T), (A), (d), (v), (p), (c) and (q) are known, then its Sales or Revenue Planned is:

$= (/{U-V-$'[1+s][f+i]-T-Ad}-Q)
/(vp/{360[c-q]}-/)

Rule-14682:

If both (/), (U), ($), (Q), ($'), (s), (f), (i), (T), (A), (d), (v), (p), (c) and (q) are known, then its Variable Cost Planned is:

V= (/{U+$-$'[1+s][f+i]-T-Ad}
-$vp/{360[c-q]}+Q)//

Rule-14683:

If both (/), (U), ($), (V), (Q), (s), (f), (i), (T), (A), (d), (v), (p), (c) and (q) are known, then its Sales Past must be:

$'= [U+$-V-T-Ad-(Q+$vp/{360[c-q]})//]
/{[1+s][f+i]}

Steve Asikin ISBN 14: 978-1514685136, ISBN 10: 1514685132

Rule-14684:

If both (/), (U), ($), (V), ($'), (Q), (f), (i), (T), (A), (d), (v), (p), (c) and (q) are known, then its Sales Growth Planned is:

s= $/$'-1

or it can also be found as

s=[U+$-V-T-Ad-(Q+$vp/{360[c-q]})//]
/{$'[f+i]}-1

Rule-14685:

If both (/), (U), ($), (V), ($'), (s), (Q), (i), (T), (A), (d), (v), (p), (c) and (q) are known, then its Fixed Portion Planned is:

f= [U+$-V-T-Ad-(Q+$vp/{360[c-q]})//]
/{$'[1+s]}-i

Rule-14686:

If both (/), (U), ($), (V), ($'), (s), (f), (Q), (T), (A), (d), (v), (p), (c) and (q) are known, then its Interest Portion Planned is:

i= [U+$-V-T-Ad-(Q+$vp/{360[c-q]})//]
/{$'[1+s]}-f

Rule-14687:

If both (/), (U), ($), (V), ($'), (s), (f), (i), (Q), (A), (d), (v), (p), (c) and (q) are known, then its Tax Planned is:

T= {U+$-V-$'[1+s][f+i]-Ad}
-(Q+$vp/{360[c-q]})//

Steve Asikin ISBN 14: 978-1514685136, ISBN 10: 1514685132

Rule-14688:

If both (**/**), (**U**), (**$**), (**V**), (**$'**), (**s**), (**f**), (**i**), (**T**), (**Q**), (**d**), (**v**), (**p**), (**c**) and (**q**) are known, then its After Tax Income Planned is:

A= [U+$-V-$'[1+s][f+i]-T
-(Q+$vp/{360[c-q]})/**/**]/d

Rule-14689:

If both (**/**), (**U**), (**$**), (**V**), (**$'**), (**s**), (**f**), (**i**), (**T**), (**A**), (**Q**), (**v**), (**p**), (**c**) and (**q**) are known, then its Dividend Payout Planned is:

d= [U+$-V-$'[1+s][f+i]-T
-(Q+$vp/{360[c-q]})/**/**]/A

Rule-14690:

If both (**/**), (**U**), (**$**), (**V**), (**$'**), (**s**), (**f**), (**i**), (**T**), (**A**), (**d**), (**Q**), (**p**), (**c**) and (**q**) are known, then its Variable Portion Longterm Debt Planned is:

v= 360[c-q](/{U+$-V-$'[1+s][f+i]-T-Ad}-Q)/[$p]

Rule-14691:

If both (**/**), (**U**), (**$**), (**V**), (**$'**), (**s**), (**f**), (**i**), (**T**), (**A**), (**d**), (**v**), (**Q**), (**c**) and (**q**) are known, then its Procured Inventory Days Planned is:

p= 360[c-q](/{U+$-V-$'[1+s][f+i]-T-Ad}-Q)/[$v]

Rule-14692:

If both (**/**), (**U**), (**$**), (**V**), (**$'**), (**s**), (**f**), (**i**), (**T**), (**A**), (**d**), (**v**), (**p**), (**Q**) and (**q**) are known, then its Current Ratio Planned is:

c= q+$vp/[360(/{U+$-V-$'[1+s][f+i]-T-Ad}-Q)]

Steve Asikin ISBN 14: 978-1514685136, ISBN 10: 1514685132

Rule-14693:

If both (**/**), (**U**), (**$**), (**V**), (**$'**), (**s**), (**f**), (**i**), (**T**), (**A**), (**d**), (**v**), (**p**), (**c**) and (**Q**) are known, then its Quick or Acid Test Ratio Planned is:

q= **c**-**$vp**/[360(**/**{**U**+**$**-**V**-**$'**[1+**s**][**f**+**i**]-**T**-**Ad**}-**Q**)]

Rule-14694:

If both (**/**), (**U**), (**$**), (**V**), (**$'**), (**s**), (**f**), (**i**), (**T**), (**A**), (**d**), (**v**), (**p**), (**c**) and (**q**) are known, then its Quoted Longterm Debt Planned is:

Q= **/**{**U**+**$**-**V**-**$'**[1+**s**][**f**+**i**]-**T**-**Ad**}
-**$'vp**[1+**s**]/{360[**c**-**q**]}

Rule-14695:

If both (**Q**), (**U**), (**$**), (**V**), (**$'**), (**s**), (**f**), (**i**), (**T**), (**A**), (**d**), (**v**), (**p**), (**c**) and (**q**) are known, then its Leverage or Gearing Ratio Planned is:

/= (**Q**+**$'vp**[1+**s**]/{360[**c**-**q**]})
/{**U**+**$**-**V**-**$'**[1+**s**][**f**+**i**]-**T**-**Ad**}

Rule-14696:

If both (**/**), (**Q**), (**$**), (**V**), (**$'**), (**s**), (**f**), (**i**), (**T**), (**A**), (**d**), (**v**), (**p**), (**c**) and (**q**) are known, then its Utilized or Starting Capital must be:

U= (**Q**+**$'vp**[1+**s**]/{360[**c**-**q**]})/**/**
-{**$**-**V**-**$'**[1+**s**][**f**+**i**]-**T**-**Ad**}

Steve Asikin ISBN 14: 978-1514685136, ISBN 10: 1514685132

Rule-14697:

If both (**/**), (**U**), (**Q**), (**V**), (**S'**), (**s**), (**f**), (**i**), (**T**), (**A**), (**d**), (**v**), (**p**), (**c**) and (**q**) are known, then its Sales or Revenue Planned is:

S= (**Q**-**/**{**U**-**V**-**S'**[1+**s**][**f**+**i**]-**T**-**Ad**}
+**S'vp**[1+**s**]/{360[**c**-**q**]})/**/**

Rule-14698:

If both (**/**), (**U**), (**S**), (**Q**), (**S'**), (**s**), (**f**), (**i**), (**T**), (**A**), (**d**), (**v**), (**p**), (**c**) and (**q**) are known, then its Variable Cost Planned is:

V= (**/**{**U**+**S**-**S'**[1+**s**][**f**+**i**]-**T**-**Ad**}
-**S'vp**[1+**s**]/{360[**c**-**q**]}+**Q**)/**/**

Rule-14699:

If both (**/**), (**U**), (**S**), (**V**), (**Q**), (**s**), (**f**), (**i**), (**T**), (**A**), (**d**), (**v**), (**p**), (**c**) and (**q**) are known, then its Sales Past must be:

S'= (**/**{**U**+**S**-**V**-**T**-**Ad**}-**Q**)
/**[**[1+**s**](**vp**/{360[**c**-**q**]}+**/**[**f**+**i**])**]**

Rule-14700:

If both (**/**), (**U**), (**S**), (**V**), (**S'**), (**Q**), (**f**), (**i**), (**T**), (**A**), (**d**), (**v**), (**p**), (**c**) and (**q**) are known, then its Sales Growth Planned is:

s= **S**/**S'**-1

or it can also be found as

s=(**/**{**U**+**S**-**V**-**T**-**Ad**}-**Q**)
/**[S'**(**vp**/{360[**c**-**q**]}+**/**[**f**+**i**])**]**-1

Steve Asikin ISBN 14: 978-1514685136, ISBN 10: 1514685132

Rule-14701:

If both (**/**), (**U**), (**$**), (**V**), (**$'**), (**s**), (**Q**), (**i**), (**T**), (**A**), (**d**), (**v**), (**p**), (**c**) and (**q**) are known, then its Fixed Portion Planned is:

f= [**U**+**$**-**V**-**T**-**Ad**-(**Q**+**$'vp**[1+**s**]/{360[**c**-**q**]})/**/**]
/{**$'**[1+**s**]}-**i**

Rule-14702:

If both (**/**), (**U**), (**$**), (**V**), (**$'**), (**s**), (**f**), (**Q**), (**T**), (**A**), (**d**), (**v**), (**p**), (**c**) and (**q**) are known, then its Interest Portion Planned is:

i= [**U**+**$**-**V**-**T**-**Ad**-(**Q**+**$'vp**[1+**s**]/{360[**c**-**q**]})/**/**]
/{**$'**[1+**s**]}-**f**

Rule-14703:

If both (**/**), (**U**), (**$**), (**V**), (**$'**), (**s**), (**f**), (**i**), (**Q**), (**A**), (**d**), (**v**), (**p**), (**c**) and (**q**) are known, then its Tax Planned is:

T= {**U**+**$**-**V**-**$'**[1+**s**][**f**+**i**]-**Ad**}
-(**Q**+**$'vp**[1+**s**]/{360[**c**-**q**]})/**/**

Rule-14704:

If both (**/**), (**U**), (**$**), (**V**), (**$'**), (**s**), (**f**), (**i**), (**T**), (**Q**), (**d**), (**v**), (**p**), (**c**) and (**q**) are known, then its After Tax Income Planned is:

A= [**U**+**$**-**V**-**$'**[1+**s**][**f**+**i**]-**T**
-(**Q**+**$'vp**[1+**s**]/{360[**c**-**q**]})/**/**]/**d**

Steve Asikin ISBN 14: 978-1514685136, ISBN 10: 1514685132

Rule-14705:

If both (Λ), (U), (S), (V), (S'), (s), (f), (i), (T), (A), (Q), (v), (p), (c) and (q) are known, then its Dividend Payout Planned is:

d= [U+S-V-S'[1+s][f+i]-T
-(Q+S'vp[1+s]/{360[c-q]})/Λ]/A

Rule-14706:

If both (Λ), (U), (S), (V), (S'), (s), (f), (i), (T), (A), (d), (Q), (p), (c) and (q) are known, then its Variable Portion Planned is:

v= V/S

or it can also be found as

v= 360[c-q](Λ{U+S-V-S'[1+s][f+i]-T-Ad}-Q)
/{S'p[1+s]}

Rule-14707:

If both (Λ), (U), (S), (V), (S'), (s), (f), (i), (T), (A), (d), (v), (Q), (c) and (q) are known, then its Procured Inventory Days Planned:

p= 360[c-q](Λ{U+S-V-S'[1+s][f+i]-T-Ad}-Q)
/{S'v[1+s]}

Rule-14708:

If both (Λ), (U), (S), (V), (S'), (s), (f), (i), (T), (A), (d), (v), (p), (Q) and (q) are known, then its Current Ratio Planned is:

c= q+S'vp[1+s]
/[360(Λ{U+S-V-S'[1+s][f+i]-T-Ad}-Q)]

Steve Asikin ISBN 14: 978-1514685136, ISBN 10: 1514685132

Rule-14709:

If both (/), (U), (S), (V), (S'), (s), (f), (i), (T), (A), (d), (v), (p), (c) and (Q) are known, then its Quick or Acid Test Ratio Planned is:

q= c-S'vp[1+s]
/[360(/{U+S-V-S'[1+s][f+i]-T-Ad}-Q)]

Rule-14710:

If both (/), (U), (S), (V), (S'), (s), (f), (i), (t), (D) and (X) are known, then its Quoted Longterm Debt Planned is:

Q= /(U+{S-V-S'[1+s][f+i]}[1-t]-D)-X

Rule-14711:

If both (Q), (U), (S), (V), (S'), (s), (f), (i), (t), (D) and (X) are known, then its Leverage or Gearing Ratio Planned is:

/= [Q+X]/(U+{S-V-S'[1+s][f+i]}[1-t]-D)

Rule-14712:

If both (/), (Q), (S), (V), (S'), (s), (f), (i), (t), (D) and (X) are known, then its Utilized or Starting Capital must be:

U= [Q+X]//-({S-V-S'[1+s][f+i]}[1-t]-D)

Rule-14713:

If both (/), (U), (Q), (V), (S'), (s), (f), (i), (t), (D) and (X) are known, then its Sales or Revenue Planned is:

S= V+S'[1+s][f+i]}+{[Q+X]//-U+D}/[1-t]

Steve Asikin ISBN 14: 978-1514685136, ISBN 10: 1514685132

Rule-14714:

If both (**/**), (**U**), (**$**), (**Q**), (**$'**), (**s**), (**f**), (**i**), (**t**), (**D**) and (**X**) are known, then its Variable Cost Planned is:

V= {**$**-**$'**[1+**s**][**f**+**i**]}-{[**Q**+**X**] /**/**-**U**+**D**}/[1-**t**]

Rule-14715:

If both (**/**), (**U**), (**$**), (**V**), (**Q**), (**s**), (**f**), (**i**), (**t**), (**D**) and (**X**) are known, then its Sales Past must be:

$'**= (**$-**V**-{[**Q**+**X**]/**/**-**U**+**D**}/[1-**t**])/{[1+**s**][**f**+**i**]}

Rule-14716:

If both (**/**), (**U**), (**$**), (**V**), (**$'**), (**Q**), (**f**), (**i**), (**t**), (**D**) and (**X**) are known, then its Sales Growth Planned is:

s= **$**/**$'**-1

or it can also be found as

s= (**$**-**V**-{[**Q**+**X**]/**/**-**U**+**D**}/[1-**t**])/{**$'**[**f**+**i**]}-1

Rule-14717:

If both (**/**), (**U**), (**$**), (**V**), (**$'**), (**s**), (**Q**), (**i**), (**t**), (**D**) and (**X**) are known, then its Fixed Portion Planned is:

f= (**$**-**V**-{[**Q**+**X**]/**/**-**U**+**D**}/[1-**t**])/{**$'**[1+**s**]}-**i**

Rule-14718:

If both (**/**), (**U**), (**$**), (**V**), (**$'**), (**s**), (**f**), (**Q**), (**t**), (**D**) and (**X**) are known, then its Interest Portion Planned is:

i= (**$**-**V**-{[**Q**+**X**]/**/**-**U**+**D**}/[1-**t**])/{**$'**[1+**s**]}-**f**

Steve Asikin ISBN 14: 978-1514685136, ISBN 10: 1514685132

Rule-14719:

If both (**/**), (**U**), (**$**), (**V**), (**$'**), (**s**), (**f**), (**i**), (**Q**), (**D**) and (**X**) are known, then its Tax Rate Planned is:

t= 1-{[**Q**+**X**]/**/**-**U**+**D**}/{**$**-**V**-**$'**[1+**s**][**f**+**i**]}

Rule-14720:

If both (**/**), (**U**), (**$**), (**V**), (**$'**), (**s**), (**f**), (**i**), (**t**), (**Q**) and (**X**) are known, then its Dividend Planned is:

D= **U**+{**$**-**V**-**$'**[1+**s**][**f**+**i**]}[1-**t**]-[**Q**+**X**]/**/**

Rule-14721:

If both (**/**), (**U**), (**$**), (**V**), (**$'**), (**s**), (**f**), (**i**), (**t**), (**D**) and (**Q**) are known, then its Xpress or Current Debt Planned is:

X= **/**(**U**+{**$**-**V**-**$'**[1+**s**][**f**+**i**]}[1-**t**]-**D**)-**Q**

Rule-14722:

If both (**/**), (**U**), (**$**), (**V**), (**$'**), (**s**), (**f**), (**i**), (**t**), (**D**), (**P**), (**c**) and (**q**) are known, then its Quoted Longterm Debt Planned is:

Q= **/**(**U**+{**$**-**V**-**$'**[1+**s**][**f**+**i**]}[1-**t**]-**D**)-**P**/[**c**-**q**]

Rule-14723:

If both (**Q**), (**U**), (**$**), (**V**), (**$'**), (**s**), (**f**), (**i**), (**t**), (**D**), (**P**), (**c**) and (**q**) are known, then its Leverage or Gearing Ratio Planned is:

/= {**Q**+**P**/[**c**-**q**]}/(**U**+{**$**-**V**-**$'**[1+**s**][**f**+**i**]}[1-**t**]-**D**)

Steve Asikin ISBN 14: 978-1514685136, ISBN 10: 1514685132

Rule-14724:

If both (/), (Q), (S), (V), (S'), (s), (f), (i), (t), (D), (P), (c) and (q) are known, then its Utilized or Starting Capital must be:

U= {Q+P/[c-q]}//-({S-V-S'[1+s][f+i]}[1-t]-D)

Rule-14725:

If both (/), (U), (Q), (V), (S'), (s), (f), (i), (t), (D), (P), (c) and (q) are known, then its Sales or Revenue Planned is:

S= V+S'[1+s][f+i]}+({Q+P/[c-q]}//-U+D)/[1-t]

Rule-14726:

If both (/), (U), (S), (Q), (S'), (s), (f), (i), (t), (D), (P), (c) and (q) are known, then its Variable Cost Planned is:

V= {S-S'[1+s][f+i]}-({Q+P/[c-q]}//-U+D)/[1-t]

Rule-14727:

If both (/), (U), (S), (V), (Q), (s), (f), (i), (t), (D), (P), (c) and (q) are known, then its Sales Past must be:

S'= [S-V-({Q+P/[c-q]}//-U+D)/[1-t]]/{[1+s][f+i]}

Rule-14728:

If both (/), (U), (S), (V), (S'), (Q), (f), (i), (t), (D), (P), (c) and (q) are known, then its Sales Growth Planned is:

s= S/S'-1

or it can also be found as

s= [S-V-({Q+P/[c-q]}//-U+D)/[1-t]]/{S'[f+i]}-1

Steve Asikin ISBN 14: 978-1514685136, ISBN 10: 1514685132

Rule-14729:

If both (/), (U), ($), (V), ($'), (s), (Q), (i), (t), (D), (P), (c) and (q) are known, then its Fixed Portion Planned is:

f= [$-V-({Q+P/[c-q]}/ /U+D)/[1-t]]/{$'[1+s]}-i

Rule-14730:

If both (/), (U), ($), (V), ($'), (s), (f), (Q), (t), (D), (P), (c) and (q) are known, then its Interest Portion Planned is:

i= [$-V-({Q+P/[c-q]}/ /U+D)/[1-t]]/{$'[1+s]}-f

Rule-14731:

If both (/), (U), ($), (V), ($'), (s), (f), (i), (Q), (D), (P), (c) and (q) are known, then its Tax Rate Planned is:

t= 1-({Q+P/[c-q]}/ /U+D)/{$-V-$'[1+s][f+i]}

Rule-14732:

If both (/), (U), ($), (V), ($'), (s), (f), (i), (t), (Q), (P), (c) and (q) are known, then its Dividend Planned is:

D= U+{$-V-$'[1+s][f+i]}[1-t]-{Q+P/[c-q]}/ /

Rule-14733:

If both (/), (U), ($), (V), ($'), (s), (f), (i), (t), (D), (Q), (c) and (q) are known, then its Procured Inventory Planned is:

P= [c-q][/(U+{$-V-$'[1+s][f+i]}[1-t]-D)-Q]

Steve Asikin ISBN 14: 978-1514685136, ISBN 10: 1514685132

Rule-14734:

If both (**/**), (**U**), (**$**), (**V**), (**$'**), (**s**), (**f**), (**i**), (**t**), (**D**), (**P**), (**Q**) and (**q**) are known, then its Current Ratio Planned is:

c= q+P/[/(U+{$-V-$'[1+s][f+i]}[1-t]-D)-Q]

Rule-14735:

If both (**/**), (**U**), (**$**), (**V**), (**$'**), (**s**), (**f**), (**i**), (**t**), (**D**), (**P**), (**c**) and (**Q**) are known, then its Quick or Acid test Ratio Planned is:

q= c-P/[/(U+{$-V-$'[1+s][f+i]}[1-t]-D)-Q]

Rule-14736:

If both (**/**), (**U**), (**$**), (**V**), (**$'**), (**s**), (**f**), (**i**), (**t**), (**D**), (**p**), (**c**) and (**q**) are known, then its Quoted Longterm Debt Planned is:

Q= /(U+{$-V-$'[1+s][f+i]}[1-t]-D)-Vp/{360[c-q]}

Rule-14737:

If both (**Q**), (**U**), (**$**), (**V**), (**$'**), (**s**), (**f**), (**i**), (**t**), (**D**), (**p**), (**c**) and (**q**) are known, then its Leverage or Gearing Ratio Planned is:

/= (Q+Vp/{360[c-q]})
/(U+{$-V-$'[1+s][f+i]}[1-t]-D)

Rule-14738:

If both (**/**), (**Q**), (**$**), (**V**), (**$'**), (**s**), (**f**), (**i**), (**t**), (**D**), (**p**), (**c**) and (**q**) are known, then its Utilized or Starting Capital must be:

U= (Q+Vp/{360[c-q]})//
-({$-V-$'[1+s][f+i]}[1-t]-D)

Steve Asikin ISBN 14: 978-1514685136, ISBN 10: 1514685132

Rule-14739:

If both (/), (**U**), (**Q**), (**V**), (**$'**), (**s**), (**f**), (**i**), (**t**), (**D**), (**p**), (**c**) and (**q**) are known, then its Sales or Revenue Planned is:

$= V+$'[1+s][f+i]}
+[(Q+Vp/{360[c-q]})/ /U+D]/[1-t]

Rule-14740:

If both (/), (**U**), (**$**), (**Q**), (**$'**), (**s**), (**f**), (**i**), (**t**), (**D**), (**p**), (**c**) and (**q**) are known, then its Variable Cost Planned is:

V= [/(U+{$-$'[1+s][f+i]}[1-t]-D)-Q]
/(p/{360[c-q]}+/[1-t])

Rule-14741:

If both (/), (**U**), (**$**), (**V**), (**Q**), (**s**), (**f**), (**i**), (**t**), (**D**), (**p**), (**c**) and (**q**) are known, then its Sales Past must be:

$'= {$-V-[(Q+Vp/{360[c-q]})/ /U+D]/[1-t]}
/{[1+s][f+i]}

Rule-14742:

If both (/), (**U**), (**$**), (**V**), (**$'**), (**Q**), (**f**), (**i**), (**t**), (**D**), (**p**), (**c**) and (**q**) are known, then its Sales Growth Planned is:

s= $/$'-1

or it can also be found as

s= {$-V-[(Q+Vp/{360[c-q]})/ /U+D]/[1-t]}
/{$'[f+i]}-1

Steve Asikin ISBN 14: 978-1514685136, ISBN 10: 1514685132

Rule-14743:

If both (**/**), (**U**), (**$**), (**V**), (**$'**), (**s**), (**Q**), (**i**), (**t**), (**D**), (**p**), (**c**) and (**q**) are known, then its Fixed Portion Planned is:

f= {**$**-**V**-[(**Q**+**Vp**/{360[**c**-**q**]})/**/**-**U**+**D**]/[1-**t**]}
/{**$'**[1+**s**]}-**i**

Rule-14744:

If both (**/**), (**U**), (**$**), (**V**), (**$'**), (**s**), (**f**), (**Q**), (**t**), (**D**), (**p**), (**c**) and (**q**) are known, then its Interest Portion Planned is:

i= {**$**-**V**-[(**Q**+**Vp**/{360[**c**-**q**]})/**/**-**U**+**D**]/[1-**t**]}
/{**$'**[1+**s**]}-**f**

Rule-14745:

If both (**/**), (**U**), (**$**), (**V**), (**$'**), (**s**), (**f**), (**i**), (**Q**), (**D**), (**p**), (**c**) and (**q**) are known, then its Tax Rate Planned is:

t= 1-[(**Q**+**Vp**/{360[**c**-**q**]})/**/**-**U**+**D**]
/{**$**-**V**-**$'**[1+**s**][**f**+**i**]}

Rule-14746:

If both (**/**), (**U**), (**$**), (**V**), (**$'**), (**s**), (**f**), (**i**), (**t**), (**Q**), (**p**), (**c**) and (**q**) are known, then its Dividend Planned is:

D= **U**+{**$**-**V**-**$'**[1+**s**][**f**+**i**]}[1-**t**]
-(**Q**+**Vp**/{360[**c**-**q**]})/**/**

Rule-14747:

If both (**/**), (**U**), (**$**), (**V**), (**$'**), (**s**), (**f**), (**i**), (**t**), (**D**), (**Q**), (**c**) and (**q**) are known, then its Procured Inventory Days Planned:

p= 360[**c**-**q**][**/**(**U**+{**$**-**V**-**$'**[1+**s**][**f**+**i**]}[1-**t**]-**D**)-**Q**]/**V**

Steve Asikin ISBN 14: 978-1514685136, ISBN 10: 1514685132

Rule-14748:

If both (**/**), (**U**), (**$**), (**V**), (**$'**), (**s**), (**f**), (**i**), (**t**), (**D**), (**p**), (**Q**) and (**q**) are known, then its Current Ratio Planned is:

c= **q**+**Vp**/{360[**/**(**U**+{**$**-**V**-**$'**[1+**s**][**f**+**i**]}[1-**t**]-**D**)-**Q**]}

Rule-14749:

If both (**/**), (**U**), (**$**), (**V**), (**$'**), (**s**), (**f**), (**i**), (**t**), (**D**), (**p**), (**c**) and (**Q**) are known, then its Quick or Acid Test Ratio Planned is:

q= **c**-**Vp**/{360[**/**(**U**+{**$**-**V**-**$'**[1+**s**][**f**+**i**]}[1-**t**]-**D**)-**Q**]}

Rule-14750:

If both (**/**), (**U**), (**$**), (**V**), (**$'**), (**s**), (**f**), (**i**), (**t**), (**D**), (**v**), (**p**), (**c**) and (**q**) are known, then its Quoted Longterm Debt Planned is:

Q= **/**(**U**+{**$**-**V**-**$'**[1+**s**][**f**+**i**]}[1-**t**]-**D**)
-**$vp**/{360[**c**-**q**]}

Rule-14751:

If both (**Q**), (**U**), (**$**), (**V**), (**$'**), (**s**), (**f**), (**i**), (**t**), (**D**), (**v**), (**p**), (**c**) and (**q**) are known, then its Leverage or Gearing Ratio Planned is:

/= (**Q**+**$vp**/{360[**c**-**q**]})
/(**U**+{**$**-**V**-**$'**[1+**s**][**f**+**i**]}[1-**t**]-**D**)

Steve Asikin ISBN 14: 978-1514685136, ISBN 10: 1514685132

Rule-14752:

If both (**/**), (**Q**), (**$**), (**V**), (**$'**), (**s**), (**f**), (**i**), (**t**), (**D**), (**v**), (**p**), (**c**) and (**q**) are known, then its Utilized or Starting Capital must be:

U= (**Q**+**$vp**/{360[**c**-**q**]})/**/**
-({**$**-**V**-**$'**[1+**s**][**f**+**i**]}[1-**t**]-**D**)

Rule-14753:

If both (**/**), (**U**), (**Q**), (**V**), (**$'**), (**s**), (**f**), (**i**), (**t**), (**D**), (**v**), (**p**), (**c**) and (**q**) are known, then its Sales or Revenue Planned is:

$**= [**/**(**U**-{**V**+**$'[1+**s**][**f**+**i**]}[1-**t**]-**D**)-**Q**]
/(**vp**/{360[**c**-**q**]}-**/**[1-**t**])

Rule-14754:

If both (**/**), (**U**), (**$**), (**Q**), (**$'**), (**s**), (**f**), (**i**), (**t**), (**D**), (**v**), (**p**), (**c**) and (**q**) are known, then its Variable Cost Planned is:

V= [**/**(**U**+{**$**-**$'**[1+**s**][**f**+**i**]}[1-**t**]-**D**)
-**$vp**/{360[**c**-**q**]}-**Q**]/{**/**[1-**t**]}

Rule-14755:

If both (**/**), (**U**), (**$**), (**V**), (**Q**), (**s**), (**f**), (**i**), (**t**), (**D**), (**v**), (**p**), (**c**) and (**q**) are known, then its Sales Past must be:

$'**= {**$-**V**-[(**Q**+**$vp**/{360[**c**-**q**]})/**/**-**U**+**D**]/[1-**t**]}
/{[1+**s**][**f**+**i**]}

Steve Asikin ISBN 14: 978-1514685136, ISBN 10: 1514685132

Rule-14756:

If both (/), (U), ($), (V), ($'), (Q), (f), (i), (t), (D), (v), (p), (c) and (q) are known, then its Sales Growth Planned is:

s= $/$'-1

or it can also be found as

s= {$-V-[(Q+$vp/{360[c-q]})//-U+D]/[1-t]}
/{$'[f+i]}-1

Rule-14757:

If both (/), (U), ($), (V), ($'), (s), (Q), (i), (t), (D), (v), (p), (c) and (q) are known, then its Fixed Portion Planned is:

f= {$-V-[(Q+$vp/{360[c-q]})/ /-U+D]/[1-t]}
/{$'[1+s]}-i

Rule-14758:

If both (/), (U), ($), (V), ($'), (s), (f), (Q), (t), (D), (v), (p), (c) and (q) are known, then its Interest Portion Planned is:

i= {$-V-[(Q+$vp/{360[c-q]})/ /-U+D]/[1-t]}
/{$'[1+s]}-f

Rule-14759:

If both (/), (U), ($), (V), ($'), (s), (f), (i), (Q), (D), (v), (p), (c) and (q) are known, then its Tax Rate Planned is:

t= 1-[(Q+$vp/{360[c-q]})/ /-U+D]
/{$-V-$'[1+s][f+i]}

Steve Asikin ISBN 14: 978-1514685136, ISBN 10: 1514685132

Rule-14760:

If both (**/**), (**U**), (**$**), (**V**), (**$'**), (**s**), (**f**), (**i**), (**t**), (**Q**), (**v**), (**p**), (**c**) and (**q**) are known, then its Dividend Planned is:

D= **U**+{**$**-**V**-**$'**[1+**s**][**f**+**i**]}[1-**t**]
-(**Q**+**$vp**/{360[**c**-**q**]})/**/**

Rule-14761:

If both (**/**), (**U**), (**$**), (**V**), (**$'**), (**s**), (**f**), (**i**), (**t**), (**D**), (**Q**), (**p**), (**c**) and (**q**) are known, then its Variable Portion Planned is:

v= **V**/**$**

or it can also be found as

v= 360[**c**-**q**][**/**(**U**+{**$**-**V**-**$'**[1+**s**][**f**+**i**]}[1-**t**]-**D**)-**Q**]
/[**$p**]

Rule-14762:

If both (**/**), (**U**), (**$**), (**V**), (**$'**), (**s**), (**f**), (**i**), (**t**), (**D**), (**v**), (**Q**), (**c**) and (**q**) are known, then its Procured Inventory Days Planned:

p= 360[**c**-**q**][**/**(**U**+{**$**-**V**-**$'**[1+**s**][**f**+**i**]}[1-**t**]-**D**)-**Q**]
/[**$v**]

Rule-14763:

If both (**/**), (**U**), (**$**), (**V**), (**$'**), (**s**), (**f**), (**i**), (**t**), (**D**), (**v**), (**p**), (**Q**) and (**q**) are known, then its Current Ratio Planned is:

c= **q**+**$vp**
/{360[**/**(**U**+{**$**-**V**-**$'**[1+**s**][**f**+**i**]}[1-**t**]-**D**)-**Q**]}

Steve Asikin ISBN 14: 978-1514685136, ISBN 10: 1514685132

Rule-14764:

If both (/), (U), (S), (V), (S'), (s), (f), (i), (t), (D), (v), (p), (c) and (Q) are known, then its Quick or Acid Test Ratio Planned is:

q= c-Svp/{360[/(U+{S-V-S'[1+s][f+i]}[1-t]-D)-Q]}

Rule-14765:

If both (/), (U), (S), (V), (S'), (s), (f), (i), (t), (D), (v), (p), (c) and (q) are known, then its Quoted Longterm Debt Planned is:

Q= /(U+{S-V-S'[1+s][f+i]}[1-t]-D)
-S'vp[1+s]/{360[c-q]}

Rule-14766:

If both (Q), (U), (S), (V), (S'), (s), (f), (i), (t), (D), (v), (p), (c) and (q) are known, then its Leverage or Gearing Ratio Planned is:

/= (Q+S'vp[1+s]/{360[c-q]})
/(U+{S-V-S'[1+s][f+i]}[1-t]-D)

Rule-14767:

If both (/), (Q), (S), (V), (S'), (s), (f), (i), (t), (D), (v), (p), (c) and (q) are known, then its Utilized or Starting capital must be:

U= (Q+S'vp[1+s]/{360[c-q]})//
-({S-V-S'[1+s][f+i]}[1-t]-D)

Steve Asikin ISBN 14: 978-1514685136, ISBN 10: 1514685132

Rule-14768:

If both (/), (U), (Q), (V), ($'), (s), (f), (i), (t), (D), (v), (p), (c) and (q) are known, then its Sales or Revenue Planned is:

$= [Q-/(U+{ -V-$'[1+s][f+i]}[1-t]-D)
+$'vp[1+s]/{360[c-q]}]/{/[1-t]}

Rule-14769:

If both (/), (U), ($), (Q), ($'), (s), (f), (i), (t), (D), (v), (p), (c) and (q) are known, then its Variable Cost Planned is:

V= [/(U+{$-$'[1+s][f+i]}[1-t]-D)
-$'vp[1+s]/{360[c-q]}-Q]/{/[1-t]}

Rule-14770:

If both (/), (U), ($), (V), (Q), (s), (f), (i), (t), (D), (v), (p), (c) and (q) are known, then its Sales Pastmust be:

$'= [/(U+{$-V}[1-t]-D)-Q]
/[[1+s](vp/{360[c-q]}+/[f+i][1-t])]

Rule-14771:

If both (/), (U), ($), (V), ($'), (Q), (f), (i), (t), (D), (v), (p), (c) and (q) are known, then its Sales Growth Planned is:

s= $/$'-1

or it can also be found as

s= [/(U+{$-V}[1-t]-D)-Q]
/[$'(vp/{360[c-q]}+/[f+i] [1-t])]-1

Steve Asikin ISBN 14: 978-1514685136, ISBN 10: 1514685132

Rule-14772:

If both (/), (U), (S), (V), (S'), (s), (Q), (i), (t), (D), (v), (p), (c) and (q) are known, then its Fixed Portion Planned is:

f= {S-V-[(Q+S'vp[1+s]/{360[c-q]})//-U+D]/[1-t]}
/{S'[1+s]}-i

Rule-14773:

If both (/), (U), (S), (V), (S'), (s), (f), (Q), (t), (D), (v), (p), (c) and (q) are known, then its Interest Portion Planned is:

i= {S-V-[(Q+S'vp[1+s]/{360[c-q]})//-U+D]/[1-t]}
/{S'[1+s]}-f

Rule-14774:

If both (/), (U), (S), (V), (S'), (s), (f), (i), (Q), (D), (v), (p), (c) and (q) are known, then its Tax Rate Planned is:

t= 1-[(Q+S'vp[1+s]/{360[c-q]})//-U+D]
/{S-V-S'[1+s][f+i]}

Rule-14775:

If both (/), (U), (S), (V), (S'), (s), (f), (i), (t), (Q), (v), (p), (c) and (q) are known, then its Dividend Planned is:

D= U+{S-V-S'[1+s][f+i]}[1-t]
-(Q+S'vp[1+s]/{360[c-q]})//

Steve Asikin ISBN 14: 978-1514685136, ISBN 10: 1514685132

Rule-14776:

If both (**/**), (**U**), (**$**), (**V**), (**$'**), (**s**), (**f**), (**i**), (**t**), (**D**), (**Q**), (**p**), (**c**) and (**q**) are known, then its Variable Portion Planned is:

v= **V**/**$**

or it can also be found as

v= 360[**c**-**q**][**/**(**U**+{**$**-**V**-**$'**[1+**s**][**f**+**i**]}[1-**t**]-**D**)-**Q**]/{**$'p**[1+**s**]}

Rule-14777:

If both (**/**), (**U**), (**$**), (**V**), (**$'**), (**s**), (**f**), (**i**), (**t**), (**D**), (**v**), (**Q**), (**c**) and (**q**) are known, then its Procured Inventory Days Planned:

p= 360[**c**-**q**][**/**(**U**+{**$**-**V**-**$'**[1+**s**][**f**+**i**]}[1-**t**]-**D**)-**Q**]/{**$'v**[1+**s**]}

Rule-14778:

If both (**/**), (**U**), (**$**), (**V**), (**$'**), (**s**), (**f**), (**i**), (**t**), (**D**), (**v**), (**p**), (**Q**) and (**q**) are known, then its Current Ratio Planned is:

c= **q**+**$'vp**[1+**s**]/{360[**/**(**U**+{**$**-**V**-**$'**[1+**s**][**f**+**i**]}[1-**t**]-**D**)-**Q**]}

Rule-14779:

If both (**/**), (**U**), (**$**), (**V**), (**$'**), (**s**), (**f**), (**i**), (**t**), (**D**), (**v**), (**p**), (**c**) and (**Q**) are known, then its Quick or Acid Test Ratio Planned is:

q= **c**-**$'vp**[1+**s**]/{360[**/**(**U**+{**$**-**V**-**$'**[1+**s**][**f**+**i**]}[1-**t**]-**D**)-**Q**]}

Steve Asikin ISBN 14: 978-1514685136, ISBN 10: 1514685132

Rule-14780:

If both (**/**), (**U**), (**$**), (**V**), (**$'**), (**s**), (**f**), (**i**), (**t**), (**A**), (**d**) and (**X**) are known, then its Quoted Longterm Debt Planned is:

Q= **/**(**U**+{**$**-**V**-**$'**[1+**s**][**f**+**i**]}[1-**t**]-**Ad**)-**X**

Rule-14781:

If both (**Q**), (**U**), (**$**), (**V**), (**$'**), (**s**), (**f**), (**i**), (**t**), (**A**), (**d**) and (**X**) are known, then its Leverage or Gearing Ratio Planned is:

/= [**Q**+**X**]/(**U**+{**$**-**V**-**$'**[1+**s**][**f**+**i**]}[1-**t**]-**Ad**)

Rule-14782:

If both (**/**), (**Q**), (**$**), (**V**), (**$'**), (**s**), (**f**), (**i**), (**t**), (**A**), (**d**) and (**X**) are known, then its Utilized or Starting Capital must be:

U= **Ad**+[**Q**+**X**]/**/**-{**$**-**V**-**$'**[1+**s**][**f**+**i**]}[1-**t**]

Rule-14783:

If both (**/**), (**U**), (**Q**), (**V**), (**$'**), (**s**), (**f**), (**i**), (**t**), (**A**), (**d**) and (**X**) are known, then its Sales or Revenue Planned is:

$**= **V**+**$'[1+**s**][**f**+**i**]}+{**Ad**+[**Q**+**X**]/**/**-**U**}/[1-**t**]

Rule-14784:

If both (**/**), (**U**), (**$**), (**Q**), (**$'**), (**s**), (**f**), (**i**), (**t**), (**A**), (**d**) and (**X**) are known, then its Variable Cost Planned is:

V= {**$**-**$'**[1+**s**][**f**+**i**]}-{**Ad**+[**Q**+**X**]/**/**-**U**}/[1-**t**]

Steve Asikin ISBN 14: 978-1514685136, ISBN 10: 1514685132

Rule-14785:

If both (**/**), (**U**), (**$**), (**V**), (**Q**), (**s**), (**f**), (**i**), (**t**), (**A**), (**d**) and (**X**) are known, then its Sales Past must be:

$$ \$' = (\$ - V - \{Ad + [Q+X]/{/}U\}/[1-t])/\{[1+s][f+i]\} $$

Rule-14786:

If both (**/**), (**U**), (**$**), (**V**), (**$'**), (**Q**), (**f**), (**i**), (**t**), (**A**), (**d**) and (**X**) are known, then its Sales Growth Planned is:

$$ s = \$/\$' - 1 $$

or it can also be found as

$$ s = (\$ - V - \{Ad + [Q+X]/{/}U\}/[1-t])/\{\$'[f+i]\} - 1 $$

Rule-14787:

If both (**/**), (**U**), (**$**), (**V**), (**$'**), (**s**), (**Q**), (**i**), (**t**), (**A**), (**d**) and (**X**) are known, then its Fixed Portion Planned is:

$$ f = (\$ - V - \{Ad + [Q+X]/{/}U\}/[1-t])/\{\$'[1+s]\} - i $$

Rule-14788:

If both (**/**), (**U**), (**$**), (**V**), (**$'**), (**s**), (**f**), (**Q**), (**t**), (**A**), (**d**) and (**X**) are known, then its Interest Portion Planned is:

$$ i = (\$ - V - \{Ad + [Q+X]/{/}U\}/[1-t])/\{\$'[1+s]\} - f $$

Rule-14789:

If both (**/**), (**U**), (**$**), (**V**), (**$'**), (**s**), (**f**), (**i**), (**Q**), (**A**), (**d**) and (**X**) are known, then its Tax Rate Planned is:

$$ t = 1 - \{Ad + [Q+X]/{/}U\}/\{\$ - V - \$'[1+s][f+i]\} $$

Steve Asikin ISBN 14: 978-1514685136, ISBN 10: 1514685132

Rule-14790:

If both (/), (U), ($), (V), ($'), (s), (f), (i), (t), (Q), (d) and (X) are known, then its After Tax Income Planned is:

A= (U+{$-V-$'[1+s][f+i]}[1-t]-[Q+X]/ /)/d

Rule-14791:

If both (/), (U), ($), (V), ($'), (s), (f), (i), (t), (A), (Q) and (X) are known, then its Dividend Payout Planned is:

d= (U+{$-V-$'[1+s][f+i]}[1-t]-[Q+X]/ /)/A

Rule-14792:

If both (/), (U), ($), (V), ($'), (s), (f), (i), (t), (A), (d) and (Q) are known, then its Xpress or Current Debt Planned is:

X= /(U+{$-V-$'[1+s][f+i]}[1-t]-Ad)-Q

Rule-14793:

If both (/), (U), ($), (V), ($'), (s), (f), (i), (t), (A), (d), (P), (c) and (q) are known, then its Quoted Longterm Debt Planned is:

Q= /(U+{$-V-$'[1+s][f+i]}[1-t]-Ad)-P/[c-q]

Rule-14794:

If both (Q), (U), ($), (V), ($'), (s), (f), (i), (t), (A), (d), (P), (c) and (q) are known, then its Leverage or Gearing Ratio Planned is:

/= {Q+P/[c-q]}/(U+{$-V-$'[1+s][f+i]}[1-t]-Ad)

Steve Asikin ISBN 14: 978-1514685136, ISBN 10: 1514685132

Rule-14795:

If both (**/**), (**Q**), (**$**), (**V**), (**$'**), (**s**), (**f**), (**i**), (**t**), (**A**), (**d**), (**P**), (**c**) and (**q**) are known, then its Utilized or Starting Capital must be:

U= **Ad**+{**Q**+**P**/[**c**-**q**]}/**/**-{**$**-**V**-**$'**[1+**s**][**f**+**i**]}[1-**t**]

Rule-14796:

If both (**/**), (**U**), (**Q**), (**V**), (**$'**), (**s**), (**f**), (**i**), (**t**), (**A**), (**d**), (**P**), (**c**) and (**q**) are known, then its Sales or Revenue Planned is:

$**= **V**+**$'[1+**s**][**f**+**i**]}+ (**Ad**+{**Q**+**P**/[**c**-**q**]}/**/**-**U**)/[1-**t**]

Rule-14797:

If both (**/**), (**U**), (**$**), (**Q**), (**$'**), (**s**), (**f**), (**i**), (**t**), (**A**), (**d**), (**P**), (**c**) and (**q**) are known, then its Variable Cost Planned is:

V= {**$**-**$'**[1+**s**][**f**+**i**]}-(**Ad**+{**Q**+**P**/[**c**-**q**]}/**/**-**U**)/[1-**t**]

Rule-14798:

If both (**/**), (**U**), (**$**), (**V**), (**Q**), (**s**), (**f**), (**i**), (**t**), (**A**), (**d**), (**P**), (**c**) and (**q**) are known, then its Sales Past must be:

$'**= [**$-**V**-(**Ad**+{**Q**+**P**/[**c**-**q**]}/**/**-**U**)/[1-**t**]]
/{[1+**s**][**f**+**i**]}

Steve Asikin ISBN 14: 978-1514685136, ISBN 10: 1514685132

Rule-14799:

If both (**/**), (**U**), (**$**), (**V**), (**$'**), (**Q**), (**f**), (**i**), (**t**), (**A**), (**d**), (**P**), (**c**) and (**q**) are known, then its Sales Growth Planned is:

s= $/$'-1

or it can also be found as

s= [$-V-(Ad+{Q+P/[c-q]}//-U)/[1-t]]/{$'[f+i]}-1

Rule-14800:

If both (**/**), (**U**), (**$**), (**V**), (**$'**), (**s**), (**Q**), (**i**), (**t**), (**A**), (**d**), (**P**), (**c**) and (**q**) are known, then its Fixed Portion Planned is:

f= [$-V-(Ad+{Q+P/[c-q]}//-U)/[1-t]]/{$'[1+s]}-i

Rule-14801:

If both (**/**), (**U**), (**$**), (**V**), (**$'**), (**s**), (**f**), (**Q**), (**t**), (**A**), (**d**), (**P**), (**c**) and (**q**) are known, then its Interest Portion Planned is:

i= [$-V-(Ad+{Q+P/[c-q]}//-U)/[1-t]]/{$'[1+s]}-f

Rule-14802:

If both (**/**), (**U**), (**$**), (**V**), (**$'**), (**s**), (**f**), (**i**), (**Q**), (**A**), (**d**), (**P**), (**c**) and (**q**) are known, then its Tax Rate Planned is:

t= 1-(Ad+{Q+P/[c-q]}//-U)/{$-V-$'[1+s][f+i]}

Rule-14803:

If both (**/**), (**U**), (**$**), (**V**), (**$'**), (**s**), (**f**), (**i**), (**t**), (**Q**), (**d**), (**P**), (**c**) and (**q**) are known, then its After Tax Income Planned is:

A= (U+{$-V-$'[1+s][f+i]}[1-t]-{Q+P/[c-q]}//)/d

Steve Asikin ISBN 14: 978-1514685136, ISBN 10: 1514685132

Rule-14804:

If both (**/**), (**U**), (**S**), (**V**), (**S'**), (**s**), (**f**), (**i**), (**t**), (**A**), (**Q**), (**P**), (**c**) and (**q**) are known, then its Dividend Payout Planned is:

d= (**U**+{**S**-**V**-**S'**[1+**s**][**f**+**i**]}[1-**t**]-{**Q**+**P**/[**c**-**q**]}/**/**)/**A**

Rule-14805:

If both (**/**), (**U**), (**S**), (**V**), (**S'**), (**s**), (**f**), (**i**), (**t**), (**A**), (**d**), (**Q**), (**c**) and (**q**) are known, then its Procured Inventory Planned is:

P= [**c**-**q**][**/**(**U**+{**S**-**V**-**S'**[1+**s**][**f**+**i**]}[1-**t**]-**Ad**)-**Q**]

Rule-14806:

If both (**/**), (**U**), (**S**), (**V**), (**S'**), (**s**), (**f**), (**i**), (**t**), (**A**), (**d**), (**P**), (**Q**) and (**q**) are known, then its Current Ratio Planned is:

c= **q**+**P**/[**/**(**U**+{**S**-**V**-**S'**[1+**s**][**f**+**i**]}[1-**t**]-**Ad**)-**Q**]

Rule-14807:

If both (**/**), (**U**), (**S**), (**V**), (**S'**), (**s**), (**f**), (**i**), (**t**), (**A**), (**d**), (**P**), (**c**) and (**Q**) are known, then its Quick or Acid Test Ratio Planned is:

q= **c**-**P**/[**/**(**U**+{**S**-**V**-**S'**[1+**s**][**f**+**i**]}[1-**t**]-**Ad**)-**Q**]

Rule-14808:

If both (**/**), (**U**), (**S**), (**V**), (**S'**), (**s**), (**f**), (**i**), (**t**), (**A**), (**d**), (**p**), (**c**) and (**q**) are known, then its Quoted Longterm Debt Planned is:

Q= **/**(**U**+{**S**-**V**-**S'**[1+**s**][**f**+**i**]}[1-**t**]-**Ad**)
-**Vp**/{360[**c**-**q**]}

Steve Asikin ISBN 14: 978-1514685136, ISBN 10: 1514685132

Rule-14809:

If both (**Q**), (**U**), (**$**), (**V**), (**$'**), (**s**), (**f**), (**i**), (**t**), (**A**), (**d**), (**p**), (**c**) and (**q**) are known, then its Leverage or Gearing Ratio Planned is:

/= (Q+Vp/{360[c-q]})
/(U+{$-V-$'[1+s][f+i]}[1-t]-Ad)

Rule-14810:

If both (**/**), (**Q**), (**$**), (**V**), (**$'**), (**s**), (**f**), (**i**), (**t**), (**A**), (**d**), (**p**), (**c**) and (**q**) are known, then its Utilized or Starting Capital must be:

U= Ad+(Q+Vp/{360[c-q]})//
-{$-V-$'[1+s][f+i]}[1-t]

Rule-14811:

If both (**/**), (**U**), (**Q**), (**V**), (**$'**), (**s**), (**f**), (**i**), (**t**), (**A**), (**d**), (**p**), (**c**) and (**q**) are known, then its Sales or Revenue Planned is:

$= V+$'[1+s][f+i]}+[Ad+(Q+Vp/{360[c-q]})//-U]
/[1-t]

Rule-14812:

If both (**/**), (**U**), (**$**), (**Q**), (**$'**), (**s**), (**f**), (**i**), (**t**), (**A**), (**d**), (**p**), (**c**) and (**q**) are known, then its Variable Cost Debt Planned is:

V= [/(U+{$-$'[1+s][f+i]}[1-t]-Ad)-Q]
/(p/{360[c-q]}+/[1-t])

Steve Asikin ISBN 14: 978-1514685136, ISBN 10: 1514685132

Rule-14813:

If both (/), (U), ($), (V), (Q), (s), (f), (i), (t), (A), (d), (p), (c) and (q) are known, then its Sales Past must be:

$'= {$-V-[Ad+(Q+Vp/{360[c-q]})/ /U]/[1-t]}
/{[1+s][f+i]}

Rule-14814:

If both (/), (U), ($), (V), ($'), (Q), (f), (i), (t), (A), (d), (p), (c) and (q) are known, then its Sales Growth Planned is:

s= $/$'-1

or it can also be found as

s= {$-V-[Ad+(Q+Vp/{360[c-q]})/ /U]/[1-t]}
/{$'[f+i]}-1

Rule-14815:

If both (/), (U), ($), (V), ($'), (s), (Q), (i), (t), (A), (d), (p), (c) and (q) are known, then its Fixed Portion Planned is:

f= {$-V-[Ad+(Q+Vp/{360[c-q]})/ /U]/[1-t]}
/{$'[1+s]}-i

Rule-14816:

If both (/), (U), ($), (V), ($'), (s), (f), (Q), (t), (A), (d), (p), (c) and (q) are known, then its Interest Portion Planned is:

i= {$-V-[Ad+(Q+Vp/{360[c-q]})/ /U]/[1-t]}
/{$'[1+s]}-f

Steve Asikin ISBN 14: 978-1514685136, ISBN 10: 1514685132

Rule-14817:

If both (/), (U), ($), (V), ($'), (s), (f), (i), (Q), (A), (d), (p), (c) and (q) are known, then its Tax Rate Planned is:

t= 1-[Ad+(Q+Vp/{360[c-q]})/ /-U]
/{$-V-$'[1+s][f+i]}

Rule-14818:

If both (/), (U), ($), (V), ($'), (s), (f), (i), (t), (Q), (d), (p), (c) and (q) are known, then its After Tax Income Planned is:

A= [U+{$-V-$'[1+s][f+i]}[1-t
]-(Q+Vp/{360[c-q]})/ /]/d

Rule-14819:

If both (/), (U), ($), (V), ($'), (s), (f), (i), (t), (A), (Q), (p), (c) and (q) are known, then its Dividend Payout Planned is:

d= [U+{$-V-$'[1+s][f+i]}[1-t]
-(Q+Vp/{360[c-q]})/ /]/A

Rule-14820:

If both (/), (U), ($), (V), ($'), (s), (f), (i), (t), (A), (d), (Q), (c) and (q) are known, then its Procured Inventory Days Planned:

p= 360[c-q]
[/(U+{$-V-$'[1+s][f+i]}[1-t]-Ad)-Q]/V

Steve Asikin ISBN 14: 978-1514685136, ISBN 10: 1514685132

Rule-14821:

If both (**/**), (**U**), (**$**), (**V**), (**$'**), (**s**), (**f**), (**i**), (**t**), (**A**), (**d**), (**p**), (**Q**) and (**q**) are known, then its Current Ratio Planned is:

c= **q**+**Vp**
/{360[**/**(**U**+{**$**-**V**-**$'**[1+**s**][**f**+**i**]}[1-**t**]-**Ad**)-**Q**]}

Rule-14822:

If both (**/**), (**U**), (**$**), (**V**), (**$'**), (**s**), (**f**), (**i**), (**t**), (**A**), (**d**), (**p**), (**c**) and (**Q**) are known, then its Quick or Acid Test Ratio Planned is:

q= **c**-**Vp**
/{360[**/**(**U**+{**$**-**V**-**$'**[1+**s**][**f**+**i**]}[1-**t**]-**Ad**)-**Q**]}

Rule-14823:

If both (**/**), (**U**), (**$**), (**V**), (**$'**), (**s**), (**f**), (**i**), (**t**), (**A**), (**d**), (**v**), (**p**), (**c**) and (**q**) are known, then its Quoted Longterm Debt Planned is:

Q= **/**(**U**+{**$**-**V**-**$'**[1+**s**][**f**+**i**]}[1-**t**]-**Ad**)
-**$vp**/{360[**c**-**q**]}

Rule-14824:

If both (**Q**), (**U**), (**$**), (**V**), (**$'**), (**s**), (**f**), (**i**), (**t**), (**A**), (**d**), (**v**), (**p**), (**c**) and (**q**) are known, then its Leverage or Gearing Ratio Planned is:

/= (**Q**+**$vp**/{360[**c**-**q**]})
/(**U**+{**$**-**V**-**$'**[1+**s**][**f**+**i**]}[1-**t**]-**Ad**)

Steve Asikin ISBN 14: 978-1514685136, ISBN 10: 1514685132

Rule-14825:

If both (**/**), (**Q**), (**$**), (**V**), (**$'**), (**s**), (**f**), (**i**), (**t**), (**A**), (**d**), (**v**), (**p**), (**c**) and (**q**) are known, then its Utilized or Starting Capital must be:

U= Ad+(Q+$vp/{360[c-q]})//
-{$-V-$'[1+s][f+i]}[1-t]

Rule-14826:

If both (**/**), (**U**), (**$**), (**V**), (**Q**), (**s**), (**f**), (**i**), (**t**), (**A**), (**d**), (**v**), (**p**), (**c**) and (**q**) are known, then its Sales Past must be:

$= [/(U+{$-V-$'[1+s][f+i]}[1-t]-Ad)-Q]
/(vp/{360[c-q]}-/[1-t])

Rule-14827:

If both (**/**), (**U**), (**$**), (**Q**), (**$'**), (**s**), (**f**), (**i**), (**t**), (**A**), (**d**), (**v**), (**p**), (**c**) and (**q**) are known, then its Variable Cost Planned is:

V= [/(U+{$-V-$'[1+s][f+i]}[1-t]-Ad)
-$vp/{360[c-q]}-Q]/{/[1-t]}

Rule-14828:

If both (**/**), (**U**), (**$**), (**V**), (**Q**), (**s**), (**f**), (**i**), (**t**), (**A**), (**d**), (**v**), (**p**), (**c**) and (**q**) are known, then its Sales Past must be:

$'= {$-V-[Ad+(Q+$vp/{360[c-q]})/-U]/[1-t]}
/{[1+s][f+i]}

Steve Asikin ISBN 14: 978-1514685136, ISBN 10: 1514685132

Rule-14829:

If both (Ⱡ), (**U**), (**$**), (**V**), (**$'**), (**Q**), (**f**), (**i**), (**t**), (**A**), (**d**), (**v**), (**p**), (**c**) and (**q**) are known, then its Sales Growth Planned is:

s= $/$'-1

or it can also be found as

s= {$-V-[Ad+(Q+$vp/{360[c-q]})/ⱠU]/[1-t]}
/{$'[f+i]}-1

Rule-14830:

If both (Ⱡ), (**U**), (**$**), (**V**), (**$'**), (**s**), (**Q**), (**i**), (**t**), (**A**), (**d**), (**v**), (**p**), (**c**) and (**q**) are known, then its Fixed Portion Planned is:

f= {$-V-[Ad+(Q+$vp/{360[c-q]})/ⱠU]/[1-t]}
/{$'[1+s]}-i

Rule-14831:

If both (Ⱡ), (**U**), (**$**), (**V**), (**$'**), (**s**), (**f**), (**Q**), (**t**), (**A**), (**d**), (**v**), (**p**), (**c**) and (**q**) are known, then its Interest Portion Planned is:

i= {$-V-[Ad+(Q+$vp/{360[c-q]})/ⱠU]/[1-t]}
/{$'[1+s]}-f

Rule-14832:

If both (Ⱡ), (**U**), (**$**), (**V**), (**$'**), (**s**), (**f**), (**i**), (**Q**), (**A**), (**d**), (**v**), (**p**), (**c**) and (**q**) are known, then its Tax Rate Planned is:

t= 1-[Ad+(Q+$vp/{360[c-q]})/ⱠU]
/{$-V-$'[1+s][f+i]}

Steve Asikin ISBN 14: 978-1514685136, ISBN 10: 1514685132

Rule-14833:

If both (**/**), (**U**), (**$**), (**V**), (**$'**), (**s**), (**f**), (**i**), (**t**), (**Q**), (**d**), (**v**), (**p**), (**c**) and (**q**) are known, then its After Tax Income Planned is:

A= [**U**+{**$**-**V**-**$'**[1+**s**][**f**+**i**]}[1-**t**]
-(**Q**+**$vp**/{360[**c**-**q**]})/**/**]/**d**

Rule-14834:

If both (**/**), (**U**), (**$**), (**V**), (**$'**), (**s**), (**f**), (**i**), (**t**), (**A**), (**Q**), (**v**), (**p**), (**c**) and (**q**) are known, then its Dividend Payout Planned is:

d= [**U**+{**$**-**V**-**$'**[1+**s**][**f**+**i**]}[1-**t**]
-(**Q**+**$vp**/{360[**c**-**q**]})/**/**]/**A**

Rule-14835:

If both (**/**), (**U**), (**$**), (**V**), (**$'**), (**s**), (**f**), (**i**), (**t**), (**A**), (**d**), (**Q**), (**p**), (**c**) and (**q**) are known, then its Variable Portion Planned is:

v= **V**/**$**

or it can also be found as

v= 360[**c**-**q**][**/**(**U**+{**$**-**V**-**$'**[1+**s**][**f**+**i**]}[1-**t**]-**Ad**)-**Q**]
/[**$p**]

Rule-14836:

If both (**/**), (**U**), (**$**), (**V**), (**$'**), (**s**), (**f**), (**i**), (**t**), (**A**), (**d**), (**v**), (**Q**), (**c**) and (**q**) are known, then its Procured Inventory Days Planned:

p= 360[**c**-**q**][**/**(**U**+{**$**-**V**-**$'**[1+**s**][**f**+**i**]}[1-**t**]-**Ad**)-**Q**]
/[**$v**]

Steve Asikin ISBN 14: 978-1514685136, ISBN 10: 1514685132

Rule-14837:

If both (**/**), (**U**), (**$**), (**V**), (**$'**), (**s**), (**f**), (**i**), (**t**), (**A**), (**d**), (**v**), (**p**), (**Q**) and (**q**) are known, then its Current Ratio Planned is:

c= q+$vp
/{360[/(U+{$-V-$'[1+s][f+i]}[1-t]-Ad)-Q]}

Rule-14838:

If both (**/**), (**U**), (**$**), (**V**), (**$'**), (**s**), (**f**), (**i**), (**t**), (**A**), (**d**), (**v**), (**p**), (**c**) and (**Q**) are known, then its Quick or Acid Test Ratio Planned is:

q= c-$vp
/{360[/(U+{$-V-$'[1+s][f+i]}[1-t]-Ad)-Q]}

Rule-14839:

If both (**/**), (**U**), (**$**), (**V**), (**$'**), (**s**), (**f**), (**i**), (**t**), (**A**), (**d**), (**v**), (**p**), (**c**) and (**q**) are known, then its Quoted Longterm Debt Planned is:

Q= /(U+{$-V-$'[1+s][f+i]}[1-t]-Ad)
-$'vp[1+s]/{360[c-q]}

Rule-14840:

If both (**Q**), (**U**), (**$**), (**V**), (**$'**), (**s**), (**f**), (**i**), (**t**), (**A**), (**d**), (**v**), (**p**), (**c**) and (**q**) are known, then its Leverage or Gearing Ratio Planned is:

/= (Q+$'vp[1+s]/{360[c-q]})
/(U+{$-V-$'[1+s][f+i]}[1-t]-Ad)

Steve Asikin ISBN 14: 978-1514685136, ISBN 10: 1514685132

Rule-14841:

If both (/), (Q), (S), (V), (S'), (s), (f), (i), (t), (A), (d), (v), (p), (c) and (q) are known, then its Utilized or Starting Capital must be:

U= Ad+(Q+S'vp[1+s]/{360[c-q]})//
-{S-V-S'[1+s][f+i]}[1-t]

Rule-14842:

If both (/), (U), (Q), (V), (S'), (s), (f), (i), (t), (A), (d), (v), (p), (c) and (q) are known, then its Sales or Revenue Planned is:

S= [Q-/(U+{S-V-S'[1+s][f+i]}[1-t]-Ad)
+S'vp[1+s]/{360[c-q]}]/{/[1-t]}

Rule-14843:

If both (/), (U), (S), (Q), (S'), (s), (f), (i), (t), (A), (d), (v), (p), (c) and (q) are known, then its Variable Cost Planned is:

V= [/(U+{S-V-S'[1+s][f+i]}[1-t]-Ad)
-S'vp[1+s]/{360[c-q]}-Q]/{/[1-t]}

Rule-14844:

If both (/), (U), (S), (V), (Q), (s), (f), (i), (t), (A), (d), (v), (p), (c) and (q) are known, then its Sales Past must be:

S'= [/(U+{S-V-S'[1+s][f+i]}[1-t]-Ad)-Q]
/[[1+s](vp/{360[c-q]}+/[f+i][1-t])]

Steve Asikin ISBN 14: 978-1514685136, ISBN 10: 1514685132

Rule-14845:

If both (/), (U), (S), (V), (S'), (Q), (f), (i), (t), (A), (d), (v), (p), (c) and (q) are known, then its Sales Growth Planned is:

s= S/S'-1

or it can also be found as

s= [/U+{S-V-S'[1+s][f+i]}[1-t]-Ad)-Q]
/[S'(vp/{360[c-q]}+/[f+i][1-t])]-1

Rule-14846:

If both (/), (U), (S), (V), (S'), (s), (Q), (i), (t), (A), (d), (v), (p), (c) and (q) are known, then its Fixed Portion Planned is:

f= {S-V-[Ad+(Q+S'vp[1+s]/{360[c-q]})/-U]
/[1-t]}/{S'[1+s]}-i

Rule-14847:

If both (/), (U), (S), (V), (S'), (s), (f), (Q), (t), (A), (d), (v), (p), (c) and (q) are known, then its Interest Portion Planned is:

i= {S-V-[Ad+(Q+S'vp[1+s]/{360[c-q]})/-U]
/[1-t]}/{S'[1+s]}-f

Rule-14848:

If both (/), (U), (S), (V), (S'), (s), (f), (i), (Q), (A), (d), (v), (p), (c) and (q) are known, then its Tax Rate Planned is:

t= 1-[Ad+(Q+S'vp[1+s]/{360[c-q]})/-U]
/{S-V-S'[1+s][f+i]}

Steve Asikin ISBN 14: 978-1514685136, ISBN 10: 1514685132

Rule-14849:

If both (/), (U), ($), (V), ($'), (s), (f), (i), (t), (Q), (d), (v), (p), (c) and (q) are known, then its After Tax Income Planned is:

A= [U+{$-V-$'[1+s][f+i]}[1-t]
-(Q+$'vp[1+s]/{360[c-q]})//]/d

Rule-14850:

If both (/), (U), ($), (V), ($'), (s), (f), (i), (t), (A), (Q), (v), (p), (c) and (q) are known, then its Dividend Payout Planned is:

d= [U+{$-V-$'[1+s][f+i]}[1-t]
-(Q+$'vp[1+s]/{360[c-q]})//]/A

Rule-14851:

If both (/), (U), ($), (V), ($'), (s), (f), (i), (t), (A), (d), (Q), (p), (c) and (q) are known, then its Variable Portion Planned is:

v= V/$

or it can also be found as

v= 360[c-q][/(U+{$-V-$'[1+s][f+i]}[1-t]-Ad)-Q]
/{$'p[1+s]}

Rule-14852:

If both (/), (U), ($), (V), ($'), (s), (f), (i), (t), (A), (d), (v), (Q), (c) and (q) are known, then its Procured Inventory Days Planned:

p= 360[c-q][/(U+{$-V-$'[1+s][f+i]}[1-t]-Ad)-Q]
/{$'v[1+s]}

Steve Asikin ISBN 14: 978-1514685136, ISBN 10: 1514685132

Rule-14853:

If both (**/**), (**U**), (**S**), (**V**), (**S'**), (**s**), (**f**), (**i**), (**t**), (**A**), (**d**), (**v**), (**p**), (**Q**) and (**q**) are known, then its Current Ratio Planned is:

c= **q**+**S'vp**[1+**s**]
/{360[**/**(**U**+{**S**-**V**-**S'**[1+**s**][**f**+**i**]}[1-**t**]-**Ad**)-**Q**]}

Rule-14854:

If both (**/**), (**U**), (**S**), (**V**), (**S'**), (**s**), (**f**), (**i**), (**t**), (**A**), (**d**), (**v**), (**p**), (**c**) and (**Q**) are known, then its Quick or Acid Test Ratio Planned is:

q= **c**-**S'vp**[1+**s**]
/{360[**/**(**U**+{**S**-**V**-**S'**[1+**s**][**f**+**i**]}[1-**t**]-**Ad**)-**Q**]}

Rule-14855:

If both (**/**), (**U**), (**S**), (**V**), (**S'**), (**s**), (**f**), (**i**), (**t**), (**d**) and (**X**) are known, then its Quoted Longterm Debt Planned is:

Q= **/**(**U**+{**S**-**V**-**S'**[1+**s**][**f**+**i**]}[1-**t**][1-**d**])-**X**

Rule-14856:

If both (**Q**), (**U**), (**S**), (**V**), (**S'**), (**s**), (**f**), (**i**), (**t**), (**d**) and (**X**) are known, then its Leverage or Gearing Ratio Planned is:

/= [**Q**+**X**]/(**U**+{**S**-**V**-**S'**[1+**s**][**f**+**i**]}[1-**t**][1-**d**])

Rule-14857:

If both (**/**), (**Q**), (**S**), (**V**), (**S'**), (**s**), (**f**), (**i**), (**t**), (**d**) and (**X**) are known, then its Utilized or Starting Capital must be:

U= [**Q**+**X**]/**/**-{**S**-**V**-**S'**[1+**s**][**f**+**i**]}[1-**t**][1-**d**]

Steve Asikin ISBN 14: 978-1514685136, ISBN 10: 1514685132

Rule-14858:

If both (/), (**U**), (**Q**), (**V**), (**$'**), (**s**), (**f**), (**i**), (**t**), (**d**) and (**X**) are known, then its Sales or Revenue Planned is:

$= V+$'[1+s][f+i]+{[Q+X]//-U}/{[1-t][1-d]}

Rule-14859:

If both (/), (**U**), (**$**), (**Q**), (**$'**), (**s**), (**f**), (**i**), (**t**), (**d**) and (**X**) are known, then its Variable Cost Planned is:

V= $-$'[1+s][f+i]-{[Q+X]//-U}/{[1-t][1-d]}

Rule-14860:

If both (/), (**U**), (**$**), (**V**), (**Q**), (**s**), (**f**), (**i**), (**t**), (**d**) and (**X**) are known, then its Sales Past must be:

$'= ($-V-{[Q+X]//-U}/{[1-t][1-d]})/{[1+s][f+i]}

Rule-14861:

If both (/), (**U**), (**$**), (**V**), (**$'**), (**Q**), (**f**), (**i**), (**t**), (**d**) and (**X**) are known, then its Sales Growth Planned is:

s= $/$'-1

or it can also be found as

s= ($-V-{[Q+X]//-U}/{[1-t][1-d]})/{$'[f+i]}-1

Rule-14862:

If both (/), (**U**), (**$**), (**V**), (**$'**), (**s**), (**Q**), (**i**), (**t**), (**d**) and (**X**) are known, then its Fixed Portion Planned is:

f= ($-V-{[Q+X]//-U}/{[1-t][1-d]})/{$'[1+s]}-i

Steve Asikin ISBN 14: 978-1514685136, ISBN 10: 1514685132

Rule-14863:

If both (**/**), (**U**), (**$**), (**V**), (**$'**), (**s**), (**f**), (**Q**), (**t**), (**d**) and (**X**) are known, then its Interest Portion Planned is:

i= (**$**-**V**-{[**Q**+**X**]/**/**-**U**}/{[1-**t**][1-**d**]})/{**$'**[1+**s**]}-**f**

Rule-14864:

If both (**/**), (**U**), (**$**), (**V**), (**$'**), (**s**), (**f**), (**i**), (**Q**), (**d**) and (**X**) are known, then its Tax Rate Planned is:

t= 1-{[**Q**+**X**]/**/**-**U**}/([1-**d**] {**$**-**V**-**$'**[1+**s**][**f**+**i**]})

Rule-14865:

If both (**/**), (**U**), (**$**), (**V**), (**$'**), (**s**), (**f**), (**i**), (**t**), (**Q**) and (**X**) are known, then its Dividend Payout Planned is:

d= 1-{[**Q**+**X**]/**/**-**U**}/([1-**t**] {**$**-**V**-**$'**[1+**s**][**f**+**i**]})

Rule-14866:

If both (**/**), (**U**), (**$**), (**V**), (**$'**), (**s**), (**f**), (**i**), (**t**), (**d**) and (**Q**) are known, then its Xpress or Current Debt Planned is:

X= **/**(**U**+{**$**-**V**-**$'**[1+**s**][**f**+**i**]}[1-**t**][1-**d**])-**Q**

Rule-14867:

If both (**/**), (**U**), (**$**), (**V**), (**$'**), (**s**), (**f**), (**i**), (**t**), (**d**), (**P**), (**c**) and (**q**) are known, then its Quoted Longterm Debt Planned is:

Q= **/**(**U**+{**$**-**V**-**$'**[1+**s**][**f**+**i**]}[1-**t**][1-**d**])-**P**/[**c**-**q**]

Steve Asikin ISBN 14: 978-1514685136, ISBN 10: 1514685132

Rule-14868:

If both (**Q**), (**U**), (**$**), (**V**), (**$'**), (**s**), (**f**), (**i**), (**t**), (**d**), (**P**), (**c**) and (**q**) are known, then its Leverage or Gearing Ratio Planned is:

/= {Q+P/[c-q]}/(U+{$-V-$'[1+s][f+i]}[1-t][1-d])

Rule-14869:

If both (**/**), (**Q**), (**$**), (**V**), (**$'**), (**s**), (**f**), (**i**), (**t**), (**d**), (**P**), (**c**) and (**q**) are known, then its Utilized or Starting Capital must be:

U= {Q+P/[c-q]}/ /-{$-V-$'[1+s][f+i]}[1-t][1-d]

Rule-14870:

If both (**/**), (**U**), (**Q**), (**V**), (**$'**), (**s**), (**f**), (**i**), (**t**), (**d**), (**P**), (**c**) and (**q**) are known, then its Sales or Revenue Planned is:

$= V+$'[1+s][f+i]+({Q+P/[c-q]}/ /-U)/{[1-t][1-d]}

Rule-14871:

If both (**/**), (**U**), (**$**), (**Q**), (**$'**), (**s**), (**f**), (**i**), (**t**), (**d**), (**P**), (**c**) and (**q**) are known, then its Variable Cost Planned is:

V= $-$'[1+s][f+i]-({Q+P/[c-q]}/ /-U)/{[1-t][1-d]}

Rule-14872:

If both (**/**), (**U**), (**$**), (**V**), (**Q**), (**s**), (**f**), (**i**), (**t**), (**d**), (**P**), (**c**) and (**q**) are known, then its Sales Past must be:

$'= [$-V-({Q+P/[c-q]}/ /-U)/{[1-t][1-d]}]
/{[1+s][f+i]}

Steve Asikin ISBN 14: 978-1514685136, ISBN 10: 1514685132

Rule-14873:

If both (**/**), (**U**), (**$**), (**V**), (**$'**), (**Q**), (**f**), (**i**), (**t**), (**d**), (**P**), (**c**) and (**q**) are known, then its Sales Growth Planned is:

s= **$**/**$'**-1

or it can also be found as

s= [**$**-**V**-({**Q**+**P**/[**c**-**q**]}/**/**-**U**)/{[1-**t**][1-**d**]}]
/{**$'**[**f**+**i**]}-1

Rule-14874:

If both (**/**), (**U**), (**$**), (**V**), (**$'**), (**s**), (**Q**), (**i**), (**t**), (**d**), (**P**), (**c**) and (**q**) are known, then its Fixed Portion Planned is:

f= [**$**-**V**-({**Q**+**P**/[**c**-**q**]}/**/**-**U**)/{[1-**t**][1-**d**]}]
/{**$'**[1+**s**]}-**i**

Rule-14875:

If both (**/**), (**U**), (**$**), (**V**), (**$'**), (**s**), (**f**), (**Q**), (**t**), (**d**), (**P**), (**c**) and (**q**) are known, then its Interest Portion Planned is:

i= [**$**-**V**-({**Q**+**P**/[**c**-**q**]}/**/**-**U**)/{[1-**t**][1-**d**]}]
/{**$'**[1+**s**]}-**f**

Rule-14876:

If both (**/**), (**U**), (**$**), (**V**), (**$'**), (**s**), (**f**), (**i**), (**Q**), (**d**), (**P**), (**c**) and (**q**) are known, then its Tax Rate Planned is:

t= 1-({**Q**+**P**/[**c**-**q**]}/**/**-**U**)/ ([1-**d**] {**$**-**V**-**$'**[1+**s**][**f**+**i**]})

Steve Asikin ISBN 14: 978-1514685136, ISBN 10: 1514685132

Rule-14877:

If both (/), (U), (S), (V), (S'), (s), (f), (i), (t), (Q), (P), (c) and (q) are known, then its Dividend Payout Planned is:

d= 1-({Q+P/[c-q]}/AU)/([1-t] {S-V-S'[1+s][f+i]})

Rule-14878:

If both (/), (U), (S), (V), (S'), (s), (f), (i), (t), (d), (Q), (c) and (q) are known, then its Procured Inventory Planned is:

P= [c-q][AU+{S-V-S'[1+s][f+i]}[1-t][1-d])-Q]

Rule-14879:

If both (/), (U), (S), (V), (S'), (s), (f), (i), (t), (d), (P), (Q) and (q) are known, then its Current Ratio Planned is:

c= q+P/[AU+{S-V-S'[1+s][f+i]}[1-t][1-d])-Q]

Rule-14880:

If both (/), (U), (S), (V), (S'), (s), (f), (i), (t), (d), (P), (c) and (Q) are known, then its Quick or Acid test Ratio Planned is:

q= c-P/[AU+{S-V-S'[1+s][f+i]}[1-t][1-d])-Q]

Rule-14881:

If both (/), (U), (S), (V), (S'), (s), (f), (i), (t), (d), (p), (c) and (q) are known, then its Quoted Longterm Debt Planned is:

Q= AU+{S-V-S'[1+s][f+i]}[1-t][1-d])
-Vp/{360[c-q]}

Steve Asikin ISBN 14: 978-1514685136, ISBN 10: 1514685132

Rule-14882:

If both (**Q**), (**U**), (**$**), (**V**), (**$'**), (**s**), (**f**), (**i**), (**t**), (**d**), (**p**), (**c**) and (**q**) are known, then its Leverage or Gearing Ratio Planned is:

/= (**Q**+**Vp**/{360[**c**-**q**]})
/(**U**+{**$**-**V**-**$'**[1+**s**][**f**+**i**]}[1-**t**][1-**d**])

Rule-14883:

If both (**/**), (**Q**), (**$**), (**V**), (**$'**), (**s**), (**f**), (**i**), (**t**), (**d**), (**p**), (**c**) and (**q**) are known, then its Utilized or Starting capital must be:

U= (**Q**+**Vp**/{360[**c**-**q**]})/**/**
-{**$**-**V**-**$'**[1+**s**][**f**+**i**]}[1-**t**][1-**d**]

Rule-14884:

If both (**/**), (**U**), (**Q**), (**V**), (**$'**), (**s**), (**f**), (**i**), (**t**), (**d**), (**p**), (**c**) and (**q**) are known, then its Sales or Revenue Planned is:

$**= **V**+**$'[1+**s**][**f**+**i**]+[(**Q**+**Vp**/{360[**c**-**q**]})/**/**-**U**]
/{[1-**t**][1-**d**]}

Rule-14885:

If both (**/**), (**U**), (**$**), (**Q**), (**$'**), (**s**), (**f**), (**i**), (**t**), (**d**), (**p**), (**c**) and (**q**) are known, then its Variable Cost Planned is:

V= [**/**(**U**+{**$**-**$'**[1+**s**][**f**+**i**]}[1-**t**][1-**d**])-**Q**]
/(**p**/{360[**c**-**q**]}+**/**[1-**t**][1-**d**])

Steve Asikin ISBN 14: 978-1514685136, ISBN 10: 1514685132

Rule-14886:

If both (**/**), (**U**), (**$**), (**V**), (**Q**), (**s**), (**f**), (**i**), (**t**), (**d**), (**p**), (**c**) and (**q**) are known, then its Sales Past must be:

$'**={**$-**V**-[(**Q**+**Vp**/{360[**c**-**q**]})/**/**-**U**]/{[1-**t**][1-**d**]}}
/{[1+**s**][**f**+**i**]}

Rule-14887:

If both (**/**), (**U**), (**$**), (**V**), (**$'**), (**Q**), (**f**), (**i**), (**t**), (**d**), (**p**), (**c**) and (**q**) are known, then its Sales Growth Planned is:

s= **$**/**$'**-1

or it can also be found as

s= {**$**-**V**-[(**Q**+**Vp**/{360[**c**-**q**]})/**/**-**U**]/{[1-**t**][1-**d**]}}
/{**$'**[**f**+**i**]}-1

Rule-14888:

If both (**/**), (**U**), (**$**), (**V**), (**$'**), (**s**), (**Q**), (**i**), (**t**), (**d**), (**p**), (**c**) and (**q**) are known, then its Fixed Portion Planned is:

f= {**$**-**V**-[(**Q**+**Vp**/{360[**c**-**q**]})/**/**-**U**]/{[1-**t**][1-**d**]}}
/{**$'**[1+**s**]}-**i**

Rule-14889:

If both (**/**), (**U**), (**$**), (**V**), (**$'**), (**s**), (**f**), (**Q**), (**t**), (**d**), (**p**), (**c**) and (**q**) are known, then its Interest Portion Planned is:

i= {**$**-**V**-[(**Q**+**Vp**/{360[**c**-**q**]})/**/**-**U**]/{[1-**t**][1-**d**]}}
/{**$'**[1+**s**]}-**f**

Steve Asikin ISBN 14: 978-1514685136, ISBN 10: 1514685132

Rule-14890:

If both (**/**), (**U**), (**$**), (**V**), (**$'**), (**s**), (**f**), (**i**), (**Q**), (**d**), (**p**), (**c**) and (**q**) are known, then its Tax Rate Planned is:

t= 1-[(**Q**+**Vp**/{360[**c**-**q**]})/**/**-**U**]
/([1-**d**] {**$**-**V**-**$'**[1+**s**][**f**+**i**]})

Rule-14891:

If both (**/**), (**U**), (**$**), (**V**), (**$'**), (**s**), (**f**), (**i**), (**t**), (**Q**), (**p**), (**c**) and (**q**) are known, then its Dividend Payout Planned is:

d= 1-[(**Q**+**Vp**/{360[**c**-**q**]})/**/**-**U**]
/([1-**t**] {**$**-**V**-**$'**[1+**s**][**f**+**i**]})

Rule-14892:

If both (**/**), (**U**), (**$**), (**V**), (**$'**), (**s**), (**f**), (**i**), (**t**), (**d**), (**Q**), (**c**) and (**q**) are known, then its Procured Inventory Days Planned:

p= 360[**c**-**q**]
[**/U**+{**$**-**V**-**$'**[1+**s**][**f**+**i**]}[1-**t**][1-**d**])-**Q**]/**V**

Rule-14893:

If both (**/**), (**U**), (**$**), (**V**), (**$'**), (**s**), (**f**), (**i**), (**t**), (**d**), (**p**), (**Q**) and (**q**) are known, then its Current Ratio Planned is:

c= **q**+**Vp**
/{360[**/U**+{**$**-**V**-**$'**[1+**s**][**f**+**i**]}[1-**t**][1-**d**])-**Q**]}

Steve Asikin ISBN 14: 978-1514685136, ISBN 10: 1514685132

Rule-14894:

If both (**/**), (**U**), (**$**), (**V**), (**$'**), (**s**), (**f**), (**i**), (**t**), (**d**), (**p**), (**c**) and (**Q**) are known, then its Quick or Acid test Ratio Planned is:

q= c-Vp /{360[/(U+{$-V-$'[1+s][f+i]}[1-t][1-d])-Q]}

Rule-14895:

If both (**/**), (**U**), (**$**), (**V**), (**$'**), (**s**), (**f**), (**i**), (**t**), (**d**), (**v**), (**p**), (**c**) and (**q**) are known, then its Quoted Longterm Debt Planned is:

Q= /(U+{$-V-$'[1+s][f+i]}[1-t][1-d]) -$vp/{360[c-q]}

Rule-14896:

If both (**Q**), (**U**), (**$**), (**V**), (**$'**), (**s**), (**f**), (**i**), (**t**), (**d**), (**v**), (**p**), (**c**) and (**q**) are known, then its Leverage or Gearing Ratio Planned is:

/= (Q+$vp/{360[c-q]}) /(U+{$-V-$'[1+s][f+i]}[1-t][1-d])

Rule-14897:

If both (**/**), (**Q**), (**$**), (**V**), (**$'**), (**s**), (**f**), (**i**), (**t**), (**d**), (**v**), (**p**), (**c**) and (**q**) are known, then its Utilized or Starting Capital must be:

U= (Q+$vp/{360[c-q]})// -{$-V-$'[1+s][f+i]}[1-t][1-d]

Steve Asikin ISBN 14: 978-1514685136, ISBN 10: 1514685132

Rule-14898:

If both (**/**), (**U**), (**Q**), (**V**), (**$'**), (**s**), (**f**), (**i**), (**t**), (**d**), (**v**), (**p**), (**c**) and (**q**) are known, then its Sales or Revenue Planned is:

$= [/(U-{V+$'[1+s][f+i]}[1-t][1-d])-Q]
/(vp/{360[c-q]}-/[1-t][1-d])

Rule-14899:

If both (**/**), (**U**), (**$**), (**Q**), (**$'**), (**s**), (**f**), (**i**), (**t**), (**d**), (**v**), (**p**), (**c**) and (**q**) are known, then its Variable Cost Planned is:

V= [/(U+{$-$'[1+s][f+i]}[1-t][1-d])-$vp/{360[c-q]}-Q]/{/[1-t][1-d]}

Rule-14900:

If both (**/**), (**U**), (**$**), (**V**), (**Q**), (**s**), (**f**), (**i**), (**t**), (**d**), (**v**), (**p**), (**c**) and (**q**) are known, then its Sales Past must be:

$'={$-V-[(Q+$vp/{360[c-q]})//-U]/{[1-t][1-d]}}
/{[1+s][f+i]}

Rule-14901:

If both (**/**), (**U**), (**$**), (**V**), (**$'**), (**Q**), (**f**), (**i**), (**t**), (**d**), (**v**), (**p**), (**c**) and (**q**) are known, then its Sales Growth Planned is:

s= $/$'-1

or it can also be found as

s= {$-V-[(Q+$vp/{360[c-q]})//-U]/{[1-t][1-d]}}
/{$'[f+i]}-1

Steve Asikin ISBN 14: 978-1514685136, ISBN 10: 1514685132

Rule-14902:

If both (/), (U), (S), (V), (S'), (s), (Q), (i), (t), (d), (v), (p), (c) and (q) are known, then its Fixed Portion Planned is:

f= {S-V-[(Q+Svp/{360[c-q]})/-U]/{[1-t][1-d]}}
/{S'[1+s]}-i

Rule-14903:

If both (/), (U), (S), (V), (S'), (s), (f), (Q), (t), (d), (v), (p), (c) and (q) are known, then its Interest Portion Planned is:

i= {S-V-[(Q+Svp/{360[c-q]})/-U]/{[1-t][1-d]}}
/{S'[1+s]}-f

Rule-14904:

If both (/), (U), (S), (V), (S'), (s), (f), (i), (Q), (d), (v), (p), (c) and (q) are known, then its Tax Rate Planned is:

t= 1-[(Q+Svp/{360[c-q]})/-U]
/([1-d]{S-V-S'[1+s][f+i]})

Rule-14905:

If both (/), (U), (S), (V), (S'), (s), (f), (i), (t), (Q), (v), (p), (c) and (q) are known, then its Dividend Payout Planned is:

d= 1-[(Q+Svp/{360[c-q]})/-U]
/([1-t] {S-V-S'[1+s][f+i]})

Steve Asikin ISBN 14: 978-1514685136, ISBN 10: 1514685132

Rule-14906:

If both (/), (U), ($), (V), ($'), (s), (f), (i), (t), (d), (Q), (p), (c) and (q) are known, then its Variable Portion Planned is:

v= V/$

or it can also be found as

v= 360[c-q][/(U+{$-V-$'[1+s][f+i]}[1-t][1-d])-Q]/[$p]

Rule-14907:

If both (/), (U), ($), (V), ($'), (s), (f), (i), (t), (d), (v), (Q), (c) and (q) are known, then its Procured Inventory Days Planned:

p= 360[c-q][/(U+{$-V-$'[1+s][f+i]}[1-t][1-d])-Q]/[$v]

Rule-14908:

If both (/), (U), ($), (V), ($'), (s), (f), (i), (t), (d), (v), (p), (Q) and (q) are known, then its Current Ratio Planned is:

c= q+$vp/{360[/(U+{$-V-$'[1+s][f+i]}[1-t][1-d])-Q]}

Rule-14909:

If both (/), (U), ($), (V), ($'), (s), (f), (i), (t), (d), (v), (p), (c) and (Q) are known, then its Quick or Acid Test Ratio Planned is:

q= c-$vp/{360[/(U+{$-V-$'[1+s][f+i]}[1-t][1-d])-Q]}

Steve Asikin ISBN 14: 978-1514685136, ISBN 10: 1514685132

Rule-14910:

If both (/), (U), (S), (V), (S'), (s), (f), (i), (t), (d), (v), (p), (c) and (q) are known, then its Quoted Longterm Debt Planned is:

Q= /(U+{S-V-S'[1+s][f+i]}[1-t][1-d])
-S'vp[1+s]/{360[c-q]}

Rule-14911:

If both (Q), (U), (S), (V), (S'), (s), (f), (i), (t), (d), (v), (p), (c) and (q) are known, then its Leverage or Gearing Ratio Planned is:

/= (Q+S'vp[1+s]/{360[c-q]})
/(U+{S-V-S'[1+s][f+i]}[1-t][1-d])

Rule-14912:

If both (/), (Q), (S), (V), (S'), (s), (f), (i), (t), (d), (v), (p), (c) and (q) are known, then its Utilized or Starting Capital must be:

U= (Q+S'vp[1+s]/{360[c-q]})//
-{S-V-S'[1+s][f+i]}[1-t][1-d]

Rule-14913:

If both (/), (U), (Q), (V), (S'), (s), (f), (i), (t), (d), (v), (p), (c) and (q) are known, then its Sales or Revenue Planned is:

S= [Q-/(U+{S-V-S'[1+s][f+i]}[1-t][1-d]
+ S'vp[1+s]/{360[c-q]}/{/[1-t][1-d]}

Steve Asikin ISBN 14: 978-1514685136, ISBN 10: 1514685132

Rule-14914:

If both (**/**), (**U**), (**S**), (**Q**), (**S'**), (**s**), (**f**), (**i**), (**t**), (**d**), (**v**), (**p**), (**c**) and (**q**) are known, then its Variable Cost Planned is:

V= **[/U**+{**S**-**S'**[1+**s**][**f**+**i**]}[1-**t**][1-**d**])
-**S'vp**[1+**s**]/{360[**c**-**q**]}-**Q]**/{**/**[1-**t**][1-**d**]}

Rule-14915:

If both (**/**), (**U**), (**S**), (**V**), (**Q**), (**s**), (**f**), (**i**), (**t**), (**d**), (**v**), (**p**), (**c**) and (**q**) are known, then its Sales Past must be:

S'= **[/U**+{**S**-**V**-**S'**[1+**s**][**f**+**i**]}[1-**t**][1-**d**])-**Q]**
/**[**[1+**s**](**vp**/{360[**c**-**q**]}+**/**[**f**+**i**][1-**t**][1-**d**])**]**

Rule-14916:

If both (**/**), (**U**), (**S**), (**V**), (**S'**), (**Q**), (**f**), (**i**), (**t**), (**d**), (**v**), (**p**), (**c**) and (**q**) are known, then its Sales Growth Planned is:

s= **S**/**S'**-1

or it can also be found as

s= **[/U**+{**S**-**V**-**S'**[1+**s**][**f**+**i**]}[1-**t**][1-**d**])-**Q]**
/**[S'**(**vp**/{360[**c**-**q**]}+**/**[**f**+**i**][1-**t**][1-**d**])**]**-1

Rule-14917:

If both (**/**), (**U**), (**S**), (**V**), (**S'**), (**s**), (**Q**), (**i**), (**t**), (**d**), (**v**), (**p**), (**c**) and (**q**) are known, then its Fixed Portion Planned is:

f= **{S**-**V**-**[**(**Q**+**S'vp**[1+**s**]/{360[**c**-**q**]})/**/**-**U]**
/{[1-**t**][1-**d**]}**}**/{**S'**[1+**s**]}-**i**

Steve Asikin ISBN 14: 978-1514685136, ISBN 10: 1514685132

Rule-14918:

If both (**/**), (**U**), (**$**), (**V**), (**$'**), (**s**), (**f**), (**Q**), (**t**), (**d**), (**v**), (**p**), (**c**) and (**q**) are known, then its Interest Portion Planned is:

i= {**$**-**V**-[(**Q**+**$'vp**[1+**s**]/{360[**c**-**q**]})/**/**-**U**]
/{[1-**t**][1-**d**]}}/{**$'**[1+**s**]}-**f**

Rule-14919:

If both (**/**), (**U**), (**$**), (**V**), (**$'**), (**s**), (**f**), (**i**), (**Q**), (**d**), (**v**), (**p**), (**c**) and (**q**) are known, then its Tax Rate Planned is:

t= 1-[(**Q**+**$'vp**[1+**s**]/{360[**c**-**q**]})/**/**-**U**]
/([1-**d**]{**$**-**V**-**$'**[1+**s**][**f**+**i**]})

Rule-14920:

If both (**/**), (**U**), (**$**), (**V**), (**$'**), (**s**), (**f**), (**i**), (**t**), (**Q**), (**v**), (**p**), (**c**) and (**q**) are known, then its Dividend Payout Planned is:

d= 1-[(**Q**+**$'vp**[1+**s**]/{360[**c**-**q**]})/**/**-**U**]
/([1-**t**]{**$**-**V**-**$'**[1+**s**][**f**+**i**]})

Rule-14921:

If both (**/**), (**U**), (**$**), (**V**), (**$'**), (**s**), (**f**), (**i**), (**t**), (**d**), (**Q**), (**p**), (**c**) and (**q**) are known, then its Variable Cost Planned is:

v= **V**/**$**

or it can also be found as

v= 360[**c**-**q**][**/**(**U**+{**$**-**V**-**$'**[1+**s**][**f**+**i**]}[1-**t**][1-**d**])-**Q**]
/{**$'p**[1+**s**]}

Steve Asikin ISBN 14: 978-1514685136, ISBN 10: 1514685132

Rule-14922:

If both (/), (**U**), (**S**), (**V**), (**S'**), (**s**), (**f**), (**i**), (**t**), (**d**), (**v**), (**Q**), (**c**) and (**q**) are known, then its Procured Inventory Days Planned:

$$p= 360[c-q][/(U+\{S-V-S'[1+s][f+i]\}[1-t][1-d])-Q]/\{S'v[1+s]\}$$

Rule-14923:

If both (/), (**U**), (**S**), (**V**), (**S'**), (**s**), (**f**), (**i**), (**t**), (**d**), (**v**), (**p**), (**Q**) and (**q**) are known, then its Current Ratio Planned is:

$$c= q+S'vp[1+s]/\{360[/(U+\{S-V-S'[1+s][f+i]\}[1-t][1-d])-Q]\}$$

Rule-14924:

If both (/), (**U**), (**S**), (**V**), (**S'**), (**s**), (**f**), (**i**), (**t**), (**d**), (**v**), (**p**), (**c**) and (**Q**) are known, then its Quick or Acid Test Ratio Planned is:

$$q= c-S'vp[1+s]/\{360[/(U+\{S-V-S'[1+s][f+i]\}[1-t][1-d])-Q]\}$$

Rule-14925:

If both (/), (**U**), (**S**), (**v**), (**F**), (**I**), (**T**), (**D**) and (**X**) are known, then its Quoted Longterm Debt Planned is:

$$Q= /\{U+S[1-v]-F-I-T-D\}-X$$

Rule-14926:

If both (**Q**), (**U**), (**S**), (**v**), (**F**), (**I**), (**T**), (**D**) and (**X**) are known, then its Leverage or Gearing Ratio Planned is:

$$/= [Q+X]/\{U+S[1-v]-F-I-T-D\}$$

Steve Asikin ISBN 14: 978-1514685136, ISBN 10: 1514685132

Rule-14927:

If both (/), (Q), ($), (v), (F), (I), (T), (D) and (X) are known, then its Utilized or Starting Capital must be:

U= F+I+T+D+[Q+X]//+$[1-v]

Rule-14928:

If both (/), (U), (Q), (v), (F), (I), (T), (D) and (X) are known, then its Sales or Revenue Planned is:

$= {F+I+T+D+[Q+X]//-U}/$

Rule-14929:

If both (/), (U), ($), (Q), (F), (I), (T), (D) and (X) are known, then its Variable Portion Planned is:

v= 1-{F+I+T+D+[Q+X]//-U}/$

Rule-14930:

If both (/), (U), ($), (v), (Q), (I), (T), (D) and (X) are known, then its Fixed Cost Planned is:

F= {U+$[1-v]-I-T-D}-[Q+X]//

Rule-14931:

If both (/), (U), ($), (v), (F), (Q), (T), (D) and (X) are known, then its Interest Expense Planned is:

I= {U+$[1-v]-F-T-D}-[Q+X]//

Rule-14932:

If both (/), (U), ($), (v), (F), (I), (Q), (D) and (X) are known, then its Tax Planned is:

T= {U+$[1-v]-F-I-D}-[Q+X]//

Steve Asikin ISBN 14: 978-1514685136, ISBN 10: 1514685132

Rule-14933:

If both (**/**), (**U**), (**$**), (**v**), (**F**), (**I**), (**T**), (**Q**) and (**X**) are known, then its Dividend Planned is:

D= {**U**+**$**[1-**v**]-**F**-**I**-**T**}-[**Q**+**X**]/**/**

Rule-14934:

If both (**/**), (**U**), (**$**), (**v**), (**F**), (**I**), (**T**), (**D**) and (**Q**) are known, then its Xpress or Current Debt Planned is:

X= **/**{**U**+**$**[1-**v**]-**F**-**I**-**T**-**D**}-**Q**

Rule-14935:

If both (**/**), (**U**), (**$**), (**v**), (**F**), (**I**), (**T**), (**D**), (**P**), (**c**) and (**q**) are known, then its Quoted Longterm Debt Planned is:

Q= **/**{**U**+**$**[1-**v**]-**F**-**I**-**T**-**D**}-**P**/[**c**-**q**]

Rule-14936:

If both (**Q**), (**U**), (**$**), (**v**), (**F**), (**I**), (**T**), (**D**), (**P**), (**c**) and (**q**) are known, then its Leverage or Gearing Ratio Planned is:

/= {**Q**+**P**/[**c**-**q**]}/{**U**+**$**[1-**v**]-**F**-**I**-**T**-**D**}

Rule-14937:

If both (**/**), (**Q**), (**$**), (**v**), (**F**), (**I**), (**T**), (**D**), (**P**), (**c**) and (**q**) are known, then its Utilized or Starting Capital must be:

U= **F**+**I**+**T**+**D**+{**Q**+**P**/[**c**-**q**]}/**/**+**$**[1-**v**]

Steve Asikin ISBN 14: 978-1514685136, ISBN 10: 1514685132

Rule-14938:

If both (**/**), (**U**), (**Q**), (**v**), (**F**), (**I**), (**T**), (**D**), (**P**), (**c**) and (**q**) are known, then its Sales or Revenue Planned is:

$**= (**F**+**I**+**T**+**D**+{**Q**+**P**/[**c**-**q**]}/**/-U**)/**$

Rule-14939:

If both (**/**), (**U**), (**$**), (**Q**), (**F**), (**I**), (**T**), (**D**), (**P**), (**c**) and (**q**) are known, then its Variable Portion Planned is:

v= 1-(**F**+**I**+**T**+**D**+{**Q**+**P**/[**c**-**q**]}/**/-U**)/**$**

Rule-14940:

If both (**/**), (**U**), (**$**), (**v**), (**Q**), (**I**), (**T**), (**D**), (**P**), (**c**) and (**q**) are known, then its Fixed Cost Planned is:

F= {**U**+**$**[1-**v**]-**I**-**T**-**D**}-{**Q**+**P**/[**c**-**q**]}/**/**

Rule-14941:

If both (**/**), (**U**), (**$**), (**v**), (**F**), (**Q**), (**T**), (**D**), (**P**), (**c**) and (**q**) are known, then its Interest Expense Planned is:

I= {**U**+**$**[1-**v**]-**F**-**T**-**D**}-{**Q**+**P**/[**c**-**q**]}/**/**

Rule-14942:

If both (**/**), (**U**), (**$**), (**v**), (**F**), (**I**), (**Q**), (**D**), (**P**), (**c**) and (**q**) are known, then its Tax Planned is:

T= {**U**+**$**[1-**v**]-**F**-**I**-**D**}-{**Q**+**P**/[**c**-**q**]}/**/**

Rule-14943:

If both (**/**), (**U**), (**$**), (**v**), (**F**), (**I**), (**T**), (**Q**), (**P**), (**c**) and (**q**) are known, then its Dividend Planned is:

D= {**U**+**$**[1-**v**]-**F**-**I**-**T**}-{**Q**+**P**/[**c**-**q**]}/**/**

Steve Asikin ISBN 14: 978-1514685136, ISBN 10: 1514685132

Rule-14944:

If both (/), (U), ($), (v), (F), (I), (T), (D), (Q), (c) and (q) are known, then its Procured Inventory Planned is:

P= [c-q](/{U+$[1-v]-F-I-T-D}-Q)

Rule-14945:

If both (/), (U), ($), (v), (F), (I), (T), (D), (P), (Q) and (q) are known, then its Current Ratio Planned is:

c= q+P/(/{U+$[1-v]-F-I-T-D}-Q)

Rule-14946:

If both (/), (U), ($), (v), (F), (I), (T), (D), (P), (c) and (Q) are known, then its Quick or Acid Test Ratio Planned is:

q= c-P/(/{U+$[1-v]-F-I-T-D}-Q)

Rule-14947:

If both (/), (U), ($), (v), (F), (I), (T), (D), (V), (p), (c) and (q) are known, then its Quoted Longterm Debt Planned is:

Q= /{U+$[1-v]-F-I-T-D}-Vp/{360[c-q]}

Rule-14948:

If both (Q), (U), ($), (v), (F), (I), (T), (D), (V), (p), (c) and (q) are known, then its Leverage or Gearing Ratio Planned is:

/= (Q+Vp/{360[c-q]})/{U+$[1-v]-F-I-T-D}

Steve Asikin ISBN 14: 978-1514685136, ISBN 10: 1514685132

Rule-14949:

If both (**/**), (**Q**), (**$**), (**v**), (**F**), (**I**), (**T**), (**D**), (**V**), (**p**), (**c**) and (**q**) are known, then its Utilized or Starting Capital must be:

U= **F**+**I**+**T**+**D**+(**Q**+**Vp**/{360[**c**-**q**]})/**/**+**$**[1-**v**]

Rule-14950:

If both (**/**), (**U**), (**Q**), (**v**), (**F**), (**I**), (**T**), (**D**), (**V**), (**p**), (**c**) and (**q**) are known, then its Sales or Revenue Planned is:

$= **[F**+**I**+**T**+**D**+(**Q**+**Vp**/{360[**c**-**q**]})/**/-U]**/[1-**v**]

Rule-14951:

If both (**/**), (**U**), (**$**), (**Q**), (**F**), (**I**), (**T**), (**D**), (**V**), (**p**), (**c**) and (**q**) are known, then its Variable Portion Planned is:

v= **V**/**$**

or it can also be found as

v= 1-**[F**+**I**+**T**+**D**+(**Q**+**Vp**/{360[**c**-**q**]})/**/-U]**/**$**

Rule-14952:

If both (**/**), (**U**), (**$**), (**v**), (**Q**), (**I**), (**T**), (**D**), (**V**), (**p**), (**c**) and (**q**) are known, then its Fixed Cost Planned is:

F= {**U**+**$**[1-**v**]-**I**-**T**-**D**}-(**Q**+**Vp**/{360[**c**-**q**]})/**/**

Rule-14953:

If both (**/**), (**U**), (**$**), (**v**), (**F**), (**Q**), (**T**), (**D**), (**V**), (**p**), (**c**) and (**q**) are known, then its Interest Expense Planned is:

I= {**U**+**$**[1-**v**]-**F**-**T**-**D**}-(**Q**+**Vp**/{360[**c**-**q**]})/**/**

Steve Asikin ISBN 14: 978-1514685136, ISBN 10: 1514685132

Rule-14954:

If both (/), (U), (S), (v), (F), (I), (Q), (D), (V), (p), (c) and (q) are known, then its Tax Planned is:

$$T= \{U+S[1-v]-F-I-D\}-(Q+Vp/\{360[c-q]\})//$$

Rule-14955:

If both (/), (U), (S), (v), (F), (I), (T), (Q), (V), (p), (c) and (q) are known, then its Dividend Planned is:

$$D= \{U+S[1-v]-F-I-T\}-(Q+Vp/\{360[c-q]\})//$$

Rule-14956:

If both (/), (U), (S), (v), (F), (I), (T), (D), (Q), (p), (c) and (q) are known, then its Variable Cost Planned is:

$$V= 360[c-q](/\{U+S[1-v]-F-I-T-D\}-Q)/p$$

Rule-14957:

If both (/), (U), (S), (v), (F), (I), (T), (D), (V), (Q), (c) and (q) are known, then its Procured Inventory Days Planned:

$$p= 360[c-q](/\{U+S[1-v]-F-I-T-D\}-Q)/V$$

Rule-14958:

If both (/), (U), (S), (v), (F), (I), (T), (D), (V), (p), (Q) and (q) are known, then its Current Ratio Planned is:

$$c= q+Vp/[360(/\{U+S[1-v]-F-I-T-D\}-Q)]$$

Rule-14959:

If both (/), (U), (S), (v), (F), (I), (T), (D), (V), (p), (c) and (Q) are known, then its Quick or Acid Test Ratio Planned is:

$$q= c-Vp/[360(/\{U+S[1-v]-F-I-T-D\}-Q)]$$

Steve Asikin ISBN 14: 978-1514685136, ISBN 10: 1514685132

Rule-14960:

If both $(/)$, (U), $(\$)$, (v), (F), (I), (T), (D), (p), (c) and (q) are known, then its Quoted Longterm Debt Planned is:

$$Q= /\{U+\$[1-v]-F-I-T-D\}-\$vp/\{360[c-q]\}$$

Rule-14961:

If both (Q), (U), $(\$)$, (v), (F), (I), (T), (D), (p), (c) and (q) are known, then its Leverage or Gearing Ratio Planned is:

$$/= (Q+\$vp/\{360[c-q]\})/\{U+\$[1-v]-F-I-T-D\}$$

Rule-14962:

If both $(/)$, (Q), $(\$)$, (v), (F), (I), (T), (D), (p), (c) and (q) are known, then its Utilized or Starting capital must be:

$$U= F+I+T+D+(Q+\$vp/\{360[c-q]\})//-\$[1-v]$$

Rule-14963:

If both $(/)$, (U), (Q), (v), (F), (I), (T), (D), (p), (c) and (q) are known, then its Sales or Revenue Planned is:

$$\$= (/\{U-F-I-T-D\}-Q)/(vp/\{360[c-q]\}-/[1-v])$$

Rule-14964:

If both $(/)$, (U), $(\$)$, (Q), (F), (I), (T), (D), (p), (c) and (q) are known, then its Variable Portion Planned is:

$$v= (/\{U+\$-F-I-T-D\}-Q)/[\$(p/\{360[c-q]\}+/)]$$

Rule-14965:

If both $(/)$, (U), $(\$)$, (v), (Q), (I), (T), (D), (p), (c) and (q) are known, then its Fixed Cost Planned is:

$$F= \{U+\$[1-v]-I-T-D\}-(Q+\$vp/\{360[c-q]\})//$$

Steve Asikin ISBN 14: 978-1514685136, ISBN 10: 1514685132

Rule-14966:

If both (**/**), (**U**), (**$**), (**v**), (**F**), (**Q**), (**T**), (**D**), (**p**), (**c**) and (**q**) are known, then its Interest Expense Planned is:

I= {**U**+**$**[1-**v**]-**F**-**T**-**D**}-(**Q**+**$vp**/{360[**c**-**q**]})/**/**

Rule-14967:

If both (**/**), (**U**), (**$**), (**v**), (**F**), (**I**), (**Q**), (**D**), (**p**), (**c**) and (**q**) are known, then its Tax Planned is:

T= {**U**+**$**[1-**v**]-**F**-**I**-**D**}-(**Q**+**$vp**/{360[**c**-**q**]})/**/**

Rule-14968:

If both (**/**), (**U**), (**$**), (**v**), (**F**), (**I**), (**T**), (**Q**), (**p**), (**c**) and (**q**) are known, then its Dividend Planned is:

D= {**U**+**$**[1-**v**]-**F**-**I**-**T**}-(**Q**+**$vp**/{360[**c**-**q**]})/**/**

Rule-14969:

If both (**/**), (**U**), (**$**), (**v**), (**F**), (**I**), (**T**), (**D**), (**Q**), (**c**) and (**q**) are known, then its Procured Inventory Days Planned is:

p= 360[**c**-**q**](**/**{**U**+**$**[1-**v**]-**F**-**I**-**T**-**D**}-**Q**)/[**$v**]

Rule-14970:

If both (**/**), (**U**), (**$**), (**v**), (**F**), (**I**), (**T**), (**D**), (**p**), (**Q**) and (**q**) are known, then its Current Ratio Planned is:

c= **q**+**$vp**/[360(**/**{**U**+**$**[1-**v**]-**F**-**I**-**T**-**D**}-**Q**)]

Rule-14971:

If both (**/**), (**U**), (**$**), (**v**), (**F**), (**I**), (**T**), (**D**), (**p**), (**c**) and (**Q**) are known, then its Quick or Acid Test Ratio Planned is:

q= **c**-**$vp**/[360(**/**{**U**+**$**[1-**v**]-**F**-**I**-**T**-**D**}-**Q**)]

Steve Asikin ISBN 14: 978-1514685136, ISBN 10: 1514685132

Rule-14972:

If both (**/**), (**U**), (**$**), (**v**), (**F**), (**I**), (**T**), (**D**), (**$'**), (**p**), (**s**), (**c**) and (**q**) are known, then its Quoted Longterm Debt Planned is:

Q= **/**{**U**+**$**[1-**v**]-**F**-**I**-**T**-**D**}-**$'vp**[1+**s**]/{360[**c**-**q**]}

Rule-14973:

If both (**/**), (**U**), (**$**), (**v**), (**F**), (**I**), (**T**), (**D**), (**$'**), (**p**), (**s**), (**c**) and (**q**) are known, then its Quoted Longterm Debt Planned is:

/= (**Q**+**$'vp**[1+**s**]/{360[**c**-**q**]})/{**U**+**$**[1-**v**]-**F**-**I**-**T**-**D**}

Rule-14974:

If both (**/**), (**Q**), (**$**), (**v**), (**F**), (**I**), (**T**), (**D**), (**$'**), (**p**), (**s**), (**c**) and (**q**) are known, then its Utilized or Starting Capital must be:

U= **F**+**I**+**T**+**D**+(**Q**+**$'vp**[1+**s**]/{360[**c**-**q**]})/**/**-**$**[1-**v**]

Rule-14975:

If both (**/**), (**U**), (**Q**), (**v**), (**F**), (**I**), (**T**), (**D**), (**$'**), (**p**), (**s**), (**c**) and (**q**) are known, then its Sales or Revenue Planned is:

$**= (**Q**-**/**{**U**-**F**-**I**-**T**-**D**}+**$'[1+**s**]**vp**/{360[**c**-**q**]})
/{**/**[1-**v**]}

Rule-14976:

If both (**/**), (**U**), (**$**), (**Q**), (**F**), (**I**), (**T**), (**D**), (**$'**), (**p**), (**s**), (**c**) and (**q**) are known, then its Variable Portion Planned is:

v= (**/**{**U**+**$**-**F**-**I**-**T**-**D**}-**Q**)/(**$'p**[1+**s**]/{360[**c**-**q**]}+**$/**)

Steve Asikin ISBN 14: 978-1514685136, ISBN 10: 1514685132

Rule-14977:

If both (**/**), (**U**), (**$**), (**v**), (**Q**), (**I**), (**T**), (**D**), (**$'**), (**p**), (**s**), (**c**) and (**q**) are known, then its Fixed Cost Planned is:

F= {**U**+**$**[1-**v**]-**I**-**T**-**D**}-(**Q**+**$'vp**[1+**s**]/{360[**c**-**q**]})/**/**

Rule-14978:

If both (**/**), (**U**), (**$**), (**v**), (**F**), (**Q**), (**T**), (**D**), (**$'**), (**p**), (**s**), (**c**) and (**q**) are known, then its Interest Expense Planned is:

I= {**U**+**$**[1-**v**]-**F**-**T**-**D**}-(**Q**+**$'vp**[1+**s**]/{360[**c**-**q**]})/**/**

Rule-14979:

If both (**/**), (**U**), (**$**), (**v**), (**F**), (**I**), (**Q**), (**D**), (**$'**), (**p**), (**s**), (**c**) and (**q**) are known, then its Tax Planned is:

T= {**U**+**$**[1-**v**]-**F**-**I**-**D**}-(**Q**+**$'vp**[1+**s**]/{360[**c**-**q**]})/**/**

Rule-14980:

If both (**/**), (**U**), (**$**), (**v**), (**F**), (**I**), (**T**), (**Q**), (**$'**), (**p**), (**s**), (**c**) and (**q**) are known, then its Dividend Planned is:

D= {**U**+**$**[1-**v**]-**F**-**I**-**T**}-(**Q**+**$'vp**[1+**s**]/{360[**c**-**q**]})/**/**

Rule-14981:

If both (**/**), (**U**), (**$**), (**v**), (**F**), (**I**), (**T**), (**D**), (**Q**), (**p**), (**s**), (**c**) and (**q**) are known, then its Sales Past must be:

$'**= (**/**{**U**+**$[1-**v**]-**F**-**I**-**T**-**D**}-**Q**)/([1+**s**]**vp**/{360[**c**-**q**]})

Rule-14982:

If both (**/**), (**U**), (**$**), (**v**), (**F**), (**I**), (**T**), (**D**), (**$'**), (**Q**), (**s**), (**c**) and (**q**) are known, then its Procured Inventory Days Planned:

p= 360[**c**-**q**](**/**{**U**+**$**[1-**v**]-**F**-**I**-**T**-**D**}-**Q**)/{**$'v**[1+**s**]}

Steve Asikin ISBN 14: 978-1514685136, ISBN 10: 1514685132

Rule-14983:

If both (**/**), (**U**), (**$**), (**v**), (**F**), (**I**), (**T**), (**D**), (**$'**), (**Q**), (**s**), (**c**) and (**q**) are known, then its Sales Growth Planned:

s= $/$'-1

or it could also be found as

s= 1-(/{U+$[1-v]-F-I-T-D}-Q)/($'vp /360[c-q]})

Rule-14984:

If both (**/**), (**U**), (**$**), (**v**), (**F**), (**I**), (**T**), (**D**), (**$'**), (**p**), (**s**), (**Q**) and (**q**) are known, then its Current Ratio Planned is:

c= q+$'vp[1+s]/[360(/{U+$[1-v]-F-I-T-D}-Q)]

Rule-14985:

If both (**/**), (**U**), (**$**), (**v**), (**F**), (**I**), (**T**), (**D**), (**$'**), (**p**), (**s**), (**c**) and (**Q**) are known, then its Quick or Acid Test Ratio Planned is:

q= c-$'vp[1+s]/[360(/{U+$[1-v]-F-I-T-D}-Q)]

Rule-14986:

If both (**/**), (**U**), (**$**), (**v**), (**F**), (**I**), (**T**), (**d**) and (**X**) are known, then its Quoted Longterm Debt Planned is:

Q= /(U+{$[1-v]-F-I-T}[1-d])-X

Rule-14987:

If both (**Q**), (**U**), (**$**), (**v**), (**F**), (**I**), (**T**), (**d**) and (**X**) are known, then its Leverage or Gearing Ratio Planned is:

/= [Q+X]/(U+{$[1-v]-F-I-T}[1-d])

Steve Asikin ISBN 14: 978-1514685136, ISBN 10: 1514685132

Rule-14988:

If both (/), (Q), ($), (v), (F), (I), (T), (d) and (X) are known, then its Utilized or Starting Capital must be:

U= [Q+X]/ℓ-{$[1-v]-F-I-T}[1-d]

Rule-14989:

If both (/), (U), (Q), (v), (F), (I), (T), (d) and (X) are known, then its Sales or Revenue Planned is:

$= (F+I+T+{[Q+X]/ℓ-U}/[1-d])/[1-v]

Rule-14990:

If both (/), (U), ($), (Q), (F), (I), (T), (d) and (X) are known, then its Variable Portion Planned is:

v= 1-(F+I+T+{[Q+X]/ℓ-U}/[1-d])/$

Rule-14991:

If both (/), (U), ($), (v), (Q), (I), (T), (d) and (X) are known, then its Fixed Cost Planned is:

F= {$[1-v]-I-T}-({[Q+X]//}-U)/[1-d]

Rule-14992:

If both (/), (U), ($), (v), (F), (Q), (T), (d) and (X) are known, then its Interest Expense Planned is:

I= {$[1-v]-F-T}-({[Q+X]//}-U)/[1-d]

Rule-14993:

If both (/), (U), ($), (v), (F), (I), (Q), (d) and (X) are known, then its Tax Planned is:

T= {$[1-v]-F-I}-({[Q+X]//}-U)/[1-d]

Steve Asikin ISBN 14: 978-1514685136, ISBN 10: 1514685132

Rule-14994:

If both (**/**), (**U**), (**$**), (**v**), (**F**), (**I**), (**T**), (**Q**) and (**X**) are known, then its Dividend Payout Planned is:

d= 1-({[**Q**+**X**]/**/**}-**U**)/{**$**[1-**v**]-**F**-**I**-**T**}

Rule-14995:

If both (**/**), (**U**), (**$**), (**v**), (**F**), (**I**), (**T**), (**d**) and (**Q**) are known, then its Xpress or Current Debt Planned is:

X= **/**(**U**+{**$**[1-**v**]-**F**-**I**-**T**}[1-**d**])-**Q**

Rule-14996:

If both (**/**), (**U**), (**$**), (**v**), (**F**), (**I**), (**T**), (**d**), (**P**), (**c**) and (**q**) are known, then its Quoted Longterm Debt Planned is:

Q= **/**(**U**+{**$**[1-**v**]-**F**-**I**-**T**}[1-**d**])-**P**/[**c**-**q**]

Rule-14997:

If both (**Q**), (**U**), (**$**), (**v**), (**F**), (**I**), (**T**), (**d**), (**P**), (**c**) and (**q**) are known, then its Leverage or Gearing Ratio Planned is:

/= {**Q**+**P**/[**c**-**q**]}/(**U**+{**$**[1-**v**]-**F**-**I**-**T**}[1-**d**])

Rule-14998:

If both (**/**), (**Q**), (**$**), (**v**), (**F**), (**I**), (**T**), (**d**), (**P**), (**c**) and (**q**) are known, then its Utilized or Starting Capital must be:

U= {**Q**+**P**/[**c**-**q**]}/**/**-{**$**[1-**v**]-**F**-**I**-**T**}[1-**d**]

Steve Asikin ISBN 14: 978-1514685136, ISBN 10: 1514685132

Rule-14999:

If both (**/**), (**U**), (**Q**), (**v**), (**F**), (**I**), (**T**), (**d**), (**P**), (**c**) and (**q**) are known, then its Sales or Revenue Planned is:

$= (**F**+**I**+**T**+{**Q**+**P**/[**c**-**q**]}/**/**-**U**)/[1-**d**])/[1-**v**]

Rule-15000:

If both (**/**), (**U**), (**$**), (**Q**), (**F**), (**I**), (**T**), (**d**), (**P**), (**c**) and (**q**) are known, then its Variable Portion Planned is:

v= 1-(**F**+**I**+**T**+{**Q**+**P**/[**c**-**q**]}/**/**-**U**)/[1-**d**])/**$**

Rule-15001:

If both (**/**), (**U**), (**$**), (**v**), (**Q**), (**I**), (**T**), (**d**), (**P**), (**c**) and (**q**) are known, then its Fixed Cost Planned is:

F= {**$**[1-**v**]-**I**-**T**}-{**Q**+**P**/[**c**-**q**]}/**/**-**U**)/[1-**d**]

Rule-15002:

If both (**/**), (**U**), (**$**), (**v**), (**F**), (**Q**), (**T**), (**d**), (**P**), (**c**) and (**q**) are known, then its Interest Expense Planned is:

I= {**$**[1-**v**]-**F**-**T**}-{**Q**+**P**/[**c**-**q**]}/**/**-**U**)/[1-**d**]

Rule-15003:

If both (**/**), (**U**), (**$**), (**v**), (**F**), (**I**), (**Q**), (**d**), (**P**), (**c**) and (**q**) are known, then its Tax Planned is:

T= {**$**[1-**v**]-**F**-**I**}-{**Q**+**P**/[**c**-**q**]}/**/**-**U**)/[1-**d**]

Rule-15004:

If both (**/**), (**U**), (**$**), (**v**), (**F**), (**I**), (**T**), (**Q**), (**P**), (**c**) and (**q**) are known, then its Dividend Payout Planned is:

d= 1-({**Q**+**P**/[**c**-**q**]}/**/**-**U**)/{**$**[1-**v**]-**F**-**I**-**T**}

Steve Asikin ISBN 14: 978-1514685136, ISBN 10: 1514685132

Rule-15005:

If both (**/**), (**U**), (**$**), (**v**), (**F**), (**I**), (**T**), (**d**), (**Q**), (**c**) and (**q**) are known, then its Procured Inventory Planned is:

P= [**c**-**q**][**/**(**U**+{**$**[1-**v**]-**F**-**I**-**T**}[1-**d**])-**Q**]

Rule-15006:

If both (**/**), (**U**), (**$**), (**v**), (**F**), (**I**), (**T**), (**d**), (**P**), (**Q**) and (**q**) are known, then its Current Ratio Planned is:

c= **q**+**P**/[**/**(**U**+{**$**[1-**v**]-**F**-**I**-**T**}[1-**d**])-**Q**]

Rule-15007:

If both (**/**), (**U**), (**$**), (**v**), (**F**), (**I**), (**T**), (**d**), (**P**), (**c**) and (**Q**) are known, then its Quick or Acid Test Ratio Planned is:

q= **c**-**P**/[**/**(**U**+{**$**[1-**v**]-**F**-**I**-**T**}[1-**d**])-**Q**]

Rule-15008:

If both (**/**), (**U**), (**$**), (**v**), (**F**), (**I**), (**T**), (**d**), (**V**), (**p**), (**c**) and (**q**) are known, then its Quoted Longterm Debt Planned is:

Q= **/**(**U**+{**$**[1-**v**]-**F**-**I**-**T**}[1-**d**])-**Vp**/{360[**c**-**q**]}

Rule-15009:

If both (**Q**), (**U**), (**$**), (**v**), (**F**), (**I**), (**T**), (**d**), (**V**), (**p**), (**c**) and (**q**) are known, then its Leverage or Gearing Ratio Planned is:

/= (**Q**+**Vp**/{360[**c**-**q**]})/(**U**+{**$**[1-**v**]-**F**-**I**-**T**}[1-**d**])

Steve Asikin ISBN 14: 978-1514685136, ISBN 10: 1514685132

Rule-15010:

If both (**/**), (**Q**), (**$**), (**v**), (**F**), (**I**), (**T**), (**d**), (**V**), (**p**), (**c**) and (**q**) are known, then its Utilized or Starting Capital must be:

U= (Q+Vp/{360[c-q]})//-{$[1-v]-F-I-T}[1-d]

Rule-15011:

If both (**/**), (**U**), (**Q**), (**v**), (**F**), (**I**), (**T**), (**d**), (**V**), (**p**), (**c**) and (**q**) are known, then its Sales or Revenue Planned is:

$= [F+I+T+(Q+Vp/{360[c-q]})//-U]/[1-d])/[1-v]

Rule-15012:

If both (**/**), (**U**), (**$**), (**Q**), (**F**), (**I**), (**T**), (**d**), (**V**), (**p**), (**c**) and (**q**) are known, then its Variable Portion Planned is:

v= V/$

or it can also be found as

v= 1-[F+I+T+(Q+Vp/{360[c-q]})//-U]/[1-d])/$

Rule-15013:

If both (**/**), (**U**), (**$**), (**v**), (**Q**), (**I**), (**T**), (**d**), (**V**), (**p**), (**c**) and (**q**) are known, then its Fixed Cost Planned is:

F= [$[1-v]-I-T}-(Q+Vp/{360[c-q]})//-U]/[1-d]

Rule-15014:

If both (**/**), (**U**), (**$**), (**v**), (**F**), (**Q**), (**T**), (**d**), (**V**), (**p**), (**c**) and (**q**) are known, then its Interest Expense Planned is:

I= [$[1-v]-F-T}-(Q+Vp/{360[c-q]})//-U]/[1-d]

Steve Asikin ISBN 14: 978-1514685136, ISBN 10: 1514685132

Rule-15015:

If both (/), (U), ($), (v), (F), (I), (Q), (d), (V), (p), (c) and (q) are known, then its Tax Planned is:

T= [$[1-v]-F-I}-(Q+Vp/{360[c-q]})/ł-U]/[1-d]

Rule-15016:

If both (/), (U), ($), (v), (F), (I), (T), (Q), (V), (p), (c) and (q) are known, then its Dividend Payout Planned is:

d= 1-[(Q+Vp/{360[c-q]})/ł-U]/{$[1-v]-F-I-T}

Rule-15017:

If both (/), (U), ($), (v), (F), (I), (T), (d), (Q), (p), (c) and (q) are known, then its Variable Cost Planned is:

V= 360[c-q][ł(U+{$[1-v]-F-I-T}[1-d])-Q]/p

Rule-15018:

If both (/), (U), ($), (v), (F), (I), (T), (d), (V), (Q), (c) and (q) are known, then its Procured Inventory Days Planned:

p= 360[c-q][ł(U+{$[1-v]-F-I-T}[1-d])-Q]/V

Rule-15019:

If both (/), (U), ($), (v), (F), (I), (T), (d), (V), (p), (Q) and (q) are known, then its Current Ratio Planned is:

c= q+Vp/{360[ł(U+{$[1-v]-F-I-T}[1-d])-Q]}

Steve Asikin ISBN 14: 978-1514685136, ISBN 10: 1514685132

Rule-15020:

If both (**/**), (**U**), (**S**), (**v**), (**F**), (**I**), (**T**), (**d**), (**V**), (**p**), (**c**) and (**Q**) are known, then its Quick or Acid Test Ratio Planned is:

q= **c**-**Vp**/{360[**/**(**U**+{**S**[1-**v**]-**F**-**I**-**T**}[1-**d**])-**Q**]}

Rule-15021:

If both (**/**), (**U**), (**S**), (**v**), (**F**), (**I**), (**T**), (**d**), (**p**), (**c**) and (**q**) are known, then its Quoted Longterm Debt Planned is:

Q= **/**(**U**+{**S**[1-**v**]-**F**-**I**-**T**}[1-**d**])-**Svp**/{360[**c**-**q**]}

Rule-15022:

If both (**Q**), (**U**), (**S**), (**v**), (**F**), (**I**), (**T**), (**d**), (**p**), (**c**) and (**q**) are known, then its Leverage or Gearing Ratio Planned is:

/= (**Q**+**Svp**/{360[**c**-**q**]})/(**U**+{**S**[1-**v**]-**F**-**I**-**T**}[1-**d**])

Rule-15023:

If both (**/**), (**Q**), (**S**), (**v**), (**F**), (**I**), (**T**), (**d**), (**p**), (**c**) and (**q**) are known, then its Utilized or Starting Capital must be:

U= (**Q**+**Svp**/{360[**c**-**q**]})/**/**-{**S**[1-**v**]-**F**-**I**-**T**}[1-**d**]

Rule-15024:

If both (**/**), (**U**), (**Q**), (**v**), (**F**), (**I**), (**T**), (**d**), (**p**), (**c**) and (**q**) are known, then its Sales or Revenue Planned is:

S= (**/**{**U**-[**F**+**I**+**T**][1-**d**]}-**Q**)
/(**vp**/{360[**c**-**q**]}-**/**[1-**v**][1-**d**])

Steve Asikin ISBN 14: 978-1514685136, ISBN 10: 1514685132

Rule-15025:

If both (**/**), (**U**), (**$**), (**Q**), (**F**), (**I**), (**T**), (**d**), (**p**), (**c**) and (**q**) are known, then its Variable Portion Planned is:

v= **/**{**U**+[**$**-**F**-**I**-**T**][1-**d**])-**Q**}
/[**$**(**p**/{360[**c**-**q**]}+**I**[1-**d**])]

Rule-15026:

If both (**/**), (**U**), (**$**), (**v**), (**Q**), (**I**), (**T**), (**d**), (**p**), (**c**) and (**q**) are known, then its Fixed Cost Planned is:

F= [**$**[1-**v**]-**I**-**T**}-(**Q**+**$vp**/{360[**c**-**q**]})/**/**-**U**]/[1-**d**]

Rule-15027:

If both (**/**), (**U**), (**$**), (**v**), (**F**), (**Q**), (**T**), (**d**), (**p**), (**c**) and (**q**) are known, then its Interest Expense Planned is:

I= [**$**[1-**v**]-**F**-**T**}-(**Q**+**$vp**/{360[**c**-**q**]})/**/**-**U**]/[1-**d**]

Rule-15028:

If both (**/**), (**U**), (**$**), (**v**), (**F**), (**I**), (**Q**), (**d**), (**p**), (**c**) and (**q**) are known, then its Tax Planned is:

T= [**$**[1-**v**]-**F**-**I**}-(**Q**+**$vp**/{360[**c**-**q**]})/**/**-**U**]/[1-**d**]

Rule-15029:

If both (**/**), (**U**), (**$**), (**v**), (**F**), (**I**), (**T**), (**Q**), (**p**), (**c**) and (**q**) are known, then its Dividend Payout Planned is:

d= 1-[(**Q**+**$vp**/{360[**c**-**q**]})/**/**-**U**]/{**$**[1-**v**]-**F**-**I**-**T**}

Rule-15030:

If both (**/**), (**U**), (**$**), (**v**), (**F**), (**I**), (**T**), (**d**), (**Q**), (**c**) and (**q**) are known, then its Procured Inventory Days Planned:

p= 360[**c**-**q**][**/**(**U**+{**$**[1-**v**]-**F**-**I**-**T**}[1-**d**])-**Q**]/[**$v**]

Steve Asikin ISBN 14: 978-1514685136, ISBN 10: 1514685132

Rule-15031:

If both (/), (U), ($), (v), (F), (I), (T), (d), (p), (Q) and (q) are known, then its Current Ratio Planned is:

c= q+$vp/{360[/(U+{$[1-v]-F-I-T}[1-d])-Q]}

Rule-15032:

If both (/), (U), ($), (v), (F), (I), (T), (d), (p), (c) and (Q) are known, then its Quick or Acid Test Ratio Planned is:

q= c-$vp/{360[/(U+{$[1-v]-F-I-T}[1-d])-Q]}

Rule-15033:

If both (/), (U), ($), (v), (F), (I), (T), (d), ($'), (p), (s), (c) and (q) are known, then its Quoted Longterm Debt Planned is:

Q= /(U+{$[1-v]-F-I-T}[1-d])
-$'vp[1+s]/{360[c-q]}

Rule-15034:

If both (Q), (U), ($), (v), (F), (I), (T), (d), ($'), (p), (s), (c) and (q) are known, then its Leverage or Gearing Ratio Planned is:

/= (Q+$'vp[1+s]/{360[c-q]})
/(U+{$[1-v]-F-I-T}[1-d])

Rule-15035:

If both (/), (Q), ($), (v), (F), (I), (T), (d), ($'), (p), (s), (c) and (q) are known, then its Utilized or Starting Capital must be:

U= (Q+$'vp[1+s]/{360[c-q]})//
-{$[1-v]-F-I-T}[1-d]

Steve Asikin ISBN 14: 978-1514685136, ISBN 10: 1514685132

Rule-15036:

If both (/), (U), (Q), (v), (F), (I), (T), (d), (S'), (p), (s), (c) and (q) are known, then its Sales or Revenue Planned is:

S= [Q-/{U-[F+I+T][1-d]}+S'vp[1+s]/{360[c-q]}}
/{/[1-v][1-d]}

Rule-15037:

If both (/), (U), (S), (Q), (F), (I), (T), (d), (S'), (p), (s), (c) and (q) are known, then its Variable Portion Planned is:

v= {/[U+S-F-I-T][1-d]-Q}
/(S'[1+s]p/{360[c-q]}+S/[1-d])

Rule-15038:

If both (/), (U), (S), (v), (Q), (I), (T), (d), (S'), (p), (s), (c) and (q) are known, then its Fixed Cost Planned is:

F= [S[1-v]-I-T}-(Q+S'vp[1+s]/{360[c-q]})//U]
/[1-d]

Rule-15039:

If both (/), (U), (S), (v), (F), (Q), (T), (d), (S'), (p), (s), (c) and (q) are known, then its Interest Expense Planned is:

I= [S[1-v]-F-T}-(Q+S'vp[1+s]/{360[c-q]})//U]
/[1-d]

Steve Asikin ISBN 14: 978-1514685136, ISBN 10: 1514685132

Rule-15040:

If both (**/**), (**U**), (**$**), (**v**), (**F**), (**I**), (**Q**), (**d**), (**$'**), (**p**), (**s**), (**c**) and (**q**) are known, then its Tax Planned is:

T= [**$**[1-**v**]-**F**-**I**}-(**Q**+**$'vp**[1+**s**]/{360[**c**-**q**]})/**/**-**U**]
/[1-**d**]

Rule-15041:

If both (**/**), (**U**), (**$**), (**v**), (**F**), (**I**), (**T**), (**Q**), (**$'**), (**p**), (**s**), (**c**) and (**q**) are known, then its Dividend Payout Planned is:

d= 1-[(**Q**+**$'vp**[1+**s**]/{360[**c**-**q**]})/**/**-**U**]
/{**$**[1-**v**]-**F**-**I**-**T**}

Rule-15042:

If both (**/**), (**U**), (**$**), (**v**), (**F**), (**I**), (**T**), (**d**), (**Q**), (**p**), (**s**), (**c**) and (**q**) are known, then its Sales Past must be:

$'**= (**/U**+{**$[1-**v**]-**F**-**I**-**T**}[1-**d**])-**Q**)
/([1+**s**]**vp**/{360[**c**-**q**]})

Rule-15043:

If both (**/**), (**U**), (**$**), (**v**), (**F**), (**I**), (**T**), (**d**), (**$'**), (**Q**), (**s**), (**c**) and (**q**) are known, then its Procured Inventory Days Planned:

p= 360[**c**-**q**][**/U**+{**$**[1-**v**]-**F**-**I**-**T**}[1-**d**])-**Q**]
/{**$'v**[1+**s**]}

Steve Asikin ISBN 14: 978-1514685136, ISBN 10: 1514685132

Rule-15044:

If both (**/**), (**U**), (**$**), (**v**), (**F**), (**I**), (**T**), (**d**), (**$'**), (**p**), (**Q**), (**c**) and (**q**) are known, then its Current Ratio Planned is:

s= $/$'-1

or it can also be found as

s= /(U+{$[1-v]-F-I-T}[1-d]-Q)
/($'vp/{360[c-q]})-1

Rule-15045:

If both (**/**), (**U**), (**$**), (**v**), (**F**), (**I**), (**T**), (**d**), (**$'**), (**p**), (**s**), (**Q**) and (**q**) are known, then its Current Ratio Planned is:

c= q+$'vp[1+s]
/{360[/(U+{$[1-v]-F-I-T}[1-d])-Q]}

Rule-15046:

If both (**/**), (**U**), (**$**), (**v**), (**F**), (**I**), (**T**), (**d**), (**$'**), (**p**), (**s**), (**c**) and (**Q**) are known, then its Quick or Acid Test Ratio Planned is:

q= c-$'vp[1+s]/{360[/(U+{$[1-v]-F-I-T}[1-d])-Q]}

Rule-15047:

If both (**/**), (**U**), (**$**), (**v**), (**F**), (**I**), (**T**), (**A**), (**d**) and (**X**) are known, then its Quoted Longterm Debt Planned is:

Q= /{U+$[1-v]-F-I-T-Ad}-X

Steve Asikin ISBN 14: 978-1514685136, ISBN 10: 1514685132

Rule-15048:

If both (**Q**), (**U**), (**$**), (**v**), (**F**), (**I**), (**T**), (**A**), (**d**) and (**X**) are known, then its Leverage or Gearing Ratio Planned is:

/= [**Q**+**X**]/{**U**+**$**[1-**v**]-**F**-**I**-**T**-**Ad**}

Rule-15049:

If both (**/**), (**U**), (**Q**), (**v**), (**F**), (**I**), (**T**), (**A**), (**d**) and (**X**) are known, then its Sales or Revenue Planned is:

U= **F**+**I**+**T**+**Ad**+[**Q**+**X**]/**/**-**$**[1-**v**]

Rule-15050:

If both (**/**), (**U**), (**Q**), (**v**), (**F**), (**I**), (**T**), (**A**), (**d**) and (**X**) are known, then its Sales or Revenue Planned is:

$= {**F**+**I**+**T**+**Ad**+[**Q**+**X**]/**/**-**U**}/[1-**v**]

Rule-15051:

If both (**/**), (**U**), (**$**), (**Q**), (**F**), (**I**), (**T**), (**A**), (**d**) and (**X**) are known, then its Variable Portion Planned is:

v= 1-{**F**+**I**+**T**+**Ad**+[**Q**+**X**]/**/**-**U**}/**$**

Rule-15052:

If both (**/**), (**U**), (**$**), (**v**), (**Q**), (**I**), (**T**), (**A**), (**d**) and (**X**) are known, then its Fixed Cost Planned is:

F= {**U**+**$**[1-**v**]-**I**-**T**-**Ad**}-[**Q**+**X**]/**/**

Rule-15053:

If both (**/**), (**U**), (**$**), (**v**), (**F**), (**Q**), (**T**), (**A**), (**d**) and (**X**) are known, then its Interest Expense Planned is:

I= {**U**+**$**[1-**v**]-**F**-**T**-**Ad**}-[**Q**+**X**]/**/**

Steve Asikin ISBN 14: 978-1514685136, ISBN 10: 1514685132

Rule-15054:

If both (**/**), (**U**), (**S**), (**v**), (**F**), (**I**), (**Q**), (**A**), (**d**) and (**X**) are known, then its Tax Planned is:

T= {**U**+**S**[1-**v**]-**F**-**I**-**Ad**}-[**Q**+**X**]/**/**

Rule-15055:

If both (**/**), (**U**), (**S**), (**v**), (**F**), (**I**), (**T**), (**Q**), (**d**) and (**X**) are known, then its After Tax Income Planned is:

A= {**U**+**S**[1-**v**]-**F**-**I**-**T**-[**Q**+**X**]/**/**}/**d**

Rule-15056:

If both (**/**), (**U**), (**S**), (**v**), (**F**), (**I**), (**T**), (**A**), (**Q**) and (**X**) are known, then its Dividend Payout Planned is:

d= {**U**+**S**[1-**v**]-**F**-**I**-**T**-[**Q**+**X**]/**/**}/**A**

Rule-15057:

If both (**/**), (**U**), (**S**), (**v**), (**F**), (**I**), (**T**), (**A**), (**d**) and (**Q**) are known, then its Xpress or Current Debt Planned is:

X= **/**{**U**+**S**[1-**v**]-**F**-**I**-**T**-**Ad**}-**Q**

Rule-15058:

If both (**/**), (**U**), (**S**), (**v**), (**F**), (**I**), (**T**), (**A**), (**d**), (**P**), (**c**) and (**q**) are known, then its Quoted Longterm Debt Planned is:

Q= **/**{**U**+**S**[1-**v**]-**F**-**I**-**T**-**Ad**}-**P**/[**c**-**q**]

Rule-15059:

If both (**Q**), (**U**), (**S**), (**v**), (**F**), (**I**), (**T**), (**A**), (**d**), (**P**), (**c**) and (**q**) are known, then its Leverage or Gearing Ratio Planned is:

/= {**Q**+**P**/[**c**-**q**]}/{**U**+**S**[1-**v**]-**F**-**I**-**T**-**Ad**}

Steve Asikin ISBN 14: 978-1514685136, ISBN 10: 1514685132

Rule-15060:

If both (**/**), (**Q**), (**$**), (**v**), (**F**), (**I**), (**T**), (**A**), (**d**), (**P**), (**c**) and (**q**) are known, then its Utilized or Starting Capital must be:

U= **F**+**I**+**T**+**Ad**+{**Q**+**P**/[**c**-**q**]}/**/**-**$**[1-**v**]

Rule-15061:

If both (**/**), (**U**), (**Q**), (**v**), (**F**), (**I**), (**T**), (**A**), (**d**), (**P**), (**c**) and (**q**) are known, then its Sales or Revenue Planned is:

$= (**F**+**I**+**T**+**Ad**+{**Q**+**P**/[**c**-**q**]}/**/**-**U**)/[1-**v**]

Rule-15062:

If both (**/**), (**U**), (**$**), (**Q**), (**F**), (**I**), (**T**), (**A**), (**d**), (**P**), (**c**) and (**q**) are known, then its Variable Portion Planned is:

v= 1-(**F**+**I**+**T**+**Ad**+{**Q**+**P**/[**c**-**q**]}/**/**-**U**)/**$**

Rule-15063:

If both (**/**), (**U**), (**$**), (**v**), (**Q**), (**I**), (**T**), (**A**), (**d**), (**P**), (**c**) and (**q**) are known, then its Fixed Cost Planned is:

F= {**U**+**$**[1-**v**]-**I**-**T**-**Ad**}-{**Q**+**P**/[**c**-**q**]}/**/**

Rule-15064:

If both (**/**), (**U**), (**$**), (**v**), (**F**), (**Q**), (**T**), (**A**), (**d**), (**P**), (**c**) and (**q**) are known, then its Interest Portion Planned is:

I= {**U**+**$**[1-**v**]-**F**-**T**-**Ad**}-{**Q**+**P**/[**c**-**q**]}/**/**

Steve Asikin ISBN 14: 978-1514685136, ISBN 10: 1514685132

Rule-15065:

If both (**/**), (**U**), (**$**), (**v**), (**F**), (**I**), (**Q**), (**A**), (**d**), (**P**), (**c**) and (**q**) are known, then its Tax Planned is:

T= {**U**+**$**[1-**v**]-**F**-**I**-**Ad**}-{**Q**+**P**/[**c**-**q**]}/**/**

Rule-15066:

If both (**/**), (**U**), (**$**), (**v**), (**F**), (**I**), (**T**), (**Q**), (**d**), (**P**), (**c**) and (**q**) are known, then its After Tax Income Planned is:

A= (**U**+**$**[1-**v**]-**F**-**I**-**T**-{**Q**+**P**/[**c**-**q**]}/**/**)/**d**

Rule-15067:

If both (**/**), (**U**), (**$**), (**v**), (**F**), (**I**), (**T**), (**A**), (**Q**), (**P**), (**c**) and (**q**) are known, then its Dividend Payout Planned is:

d= (**U**+**$**[1-**v**]-**F**-**I**-**T**-{**Q**+**P**/[**c**-**q**]}/**/**)/**A**

Rule-15068:

If both (**/**), (**U**), (**$**), (**v**), (**F**), (**I**), (**T**), (**A**), (**d**), (**Q**), (**c**) and (**q**) are known, then its Procured Inventory Days Planned is:

P= [**c**-**q**](**/**{**U**+**$**[1-**v**]-**F**-**I**-**T**-**Ad**}-**Q**)

Rule-15069:

If both (**/**), (**U**), (**$**), (**v**), (**F**), (**I**), (**T**), (**A**), (**d**), (**P**), (**Q**) and (**q**) are known, then its Current Ratio Planned is:

c= **q**+**P**/(**/**{**U**+**$**[1-**v**]-**F**-**I**-**T**-**Ad**}-**Q**)

Steve Asikin ISBN 14: 978-1514685136, ISBN 10: 1514685132

Rule-15070:

If both (**/**), (**U**), (**$**), (**v**), (**F**), (**I**), (**T**), (**A**), (**d**), (**P**), (**c**) and (**Q**) are known, then its Quick or Acid test Ratio Planned is:

q= **c**-**P**/(**/**{**U**+**$**[1-**v**]-**F**-**I**-**T**-**Ad**}-**Q**)

Rule-15071:

If both (**/**), (**U**), (**$**), (**v**), (**F**), (**I**), (**T**), (**A**), (**d**), (**V**), (**p**), (**c**) and (**q**) are known, then its Quoted Longterm Debt Planned is:

Q= **/**{**U**+**$**[1-**v**]-**F**-**I**-**T**-**Ad**}-**Vp**/{360[**c**-**q**]}

Rule-15072:

If both (**Q**), (**U**), (**$**), (**v**), (**F**), (**I**), (**T**), (**A**), (**d**), (**V**), (**p**), (**c**) and (**q**) are known, then its Leverage or Gearing Ratio Planned is:

/= (**Q**+**Vp**/{360[**c**-**q**]})/{**U**+**$**[1-**v**]-**F**-**I**-**T**-**Ad**}

Rule-15073:

If both (**/**), (**Q**), (**$**), (**v**), (**F**), (**I**), (**T**), (**A**), (**d**), (**V**), (**p**), (**c**) and (**q**) are known, then its Utilized or Starting Capital must be:

U= **F**+**I**+**T**+**Ad**+(**Q**+**Vp**/{360[**c**-**q**]})/**/**-**$**[1-**v**]

Rule-15074:

If both (**/**), (**U**), (**Q**), (**v**), (**F**), (**I**), (**T**), (**A**), (**d**), (**V**), (**p**), (**c**) and (**q**) are known, then its Sales or Revenue Planned is:

$= [**F**+**I**+**T**+**Ad**+(**Q**+**Vp**/{360[**c**-**q**]})/**/**-**U**]/[1-**v**]

Steve Asikin ISBN 14: 978-1514685136, ISBN 10: 1514685132

Rule-15075:

If both (/), (U), (S), (Q), (F), (I), (T), (A), (d), (V), (p), (c) and (q) are known, then its Variable Portion Planned is:

v= V/S

or it can also be found as

v= 1-[F+I+T+Ad+(Q+Vp/{360[c-q]})/ /-U]/S

Rule-15076:

If both (/), (U), (S), (v), (Q), (I), (T), (A), (d), (V), (p), (c) and (q) are known, then its Fixed Cost Planned is:

F= {U+S[1-v]-I-T-Ad}-(Q+Vp/{360[c-q]})//

Rule-15077:

If both (/), (U), (S), (v), (F), (Q), (T), (A), (d), (V), (p), (c) and (q) are known, then its Interest Expense Planned is:

I= {U+S[1-v]-F-T-Ad}-(Q+Vp/{360[c-q]})//

Rule-15078:

If both (/), (U), (S), (v), (F), (I), (Q), (A), (d), (V), (p), (c) and (q) are known, then its Tax Planned is:

T= {U+S[1-v]-F-I-Ad}-(Q+Vp/{360[c-q]})//

Rule-15079:

If both (/), (U), (S), (v), (F), (I), (T), (Q), (d), (V), (p), (c) and (q) are known, then its After Tax Income Planned is:

A= [U+S[1-v]-F-I-T-(Q+Vp/{360[c-q]})//]/d

Steve Asikin ISBN 14: 978-1514685136, ISBN 10: 1514685132

Rule-15080:

If both (**/**), (**U**), (**$**), (**v**), (**F**), (**I**), (**T**), (**A**), (**Q**), (**V**), (**p**), (**c**) and (**q**) are known, then its Dividend Payout Planned is:

d= [**U**+**$**[1-**v**]-**F**-**I**-**T**-(**Q**+**Vp**/{360[**c**-**q**]})/**/**]/**A**

Rule-15081:

If both (**/**), (**U**), (**$**), (**v**), (**F**), (**I**), (**T**), (**A**), (**d**), (**Q**), (**p**), (**c**) and (**q**) are known, then its Variable Cost Planned is:

V= 360[**c**-**q**](**/**{**U**+**$**[1-**v**]-**F**-**I**-**T**-**Ad**}-**Q**)/**p**

Rule-15082:

If both (**/**), (**U**), (**$**), (**v**), (**F**), (**I**), (**T**), (**A**), (**d**), (**V**), (**Q**), (**c**) and (**q**) are known, then its Procured Inventory Days Planned:

p= 360[**c**-**q**](**/**{**U**+**$**[1-**v**]-**F**-**I**-**T**-**Ad**}-**Q**)/**V**

Rule-15083:

If both (**/**), (**U**), (**$**), (**v**), (**F**), (**I**), (**T**), (**A**), (**d**), (**V**), (**p**), (**Q**) and (**q**) are known, then its Current Ratio Planned is:

c= **q**+**Vp**/[360(**/**{**U**+**$**[1-**v**]-**F**-**I**-**T**-**Ad**}-**Q**)]

Rule-15084:

If both (**/**), (**U**), (**$**), (**v**), (**F**), (**I**), (**T**), (**A**), (**d**), (**V**), (**p**), (**c**) and (**Q**) are known, then its Quick or Acid Test Ratio Planned is:

q= **c**-**Vp**/[360(**/**{**U**+**$**[1-**v**]-**F**-**I**-**T**-**Ad**}-**Q**)]

Steve Asikin ISBN 14: 978-1514685136, ISBN 10: 1514685132

Rule-15085:

If both (/), (U), ($), (v), (F), (I), (T), (A), (d), (p), (c) and (q) are known, then its Quoted Longterm Debt Planned is:

Q= /{U+$[1-v]-F-I-T-Ad}-$vp/{360[c-q]}

Rule-15086:

If both (Q), (U), ($), (v), (F), (I), (T), (A), (d), (p), (c) and (q) are known, then its Leverage or Gearing Ratio Planned is:

/= (Q+$vp/{360[c-q]})/{U+$[1-v]-F-I-T-Ad}

Rule-15087:

If both (/), (Q), ($), (v), (F), (I), (T), (A), (d), (p), (c) and (q) are known, then its Utilized or Starting Capital must be:

U= F+I+T+Ad+(Q+$vp/{360[c-q]})//-$[1-v]

Rule-15088:

If both (/), (U), (Q), (v), (F), (I), (T), (A), (d), (p), (c) and (q) are known, then its Sales or Revenue Planned is:

$= (/{U-F-I-T-Ad}-Q)/(vp/{360[c-q]}-/[1-v])

Rule-15089:

If both (/), (U), ($), (Q), (F), (I), (T), (A), (d), (p), (c) and (q) are known, then its Variable Portion Planned is:

v= {/[U+$-F-I-T-Ad]-Q}/[$(p/{360[c-q]}+/)]

Steve Asikin ISBN 14: 978-1514685136, ISBN 10: 1514685132

Rule-15090:

If both (**/**), (**U**), (**$**), (**v**), (**Q**), (**I**), (**T**), (**A**), (**d**), (**p**), (**c**) and (**q**) are known, then its Fixed Cost Planned is:

F= {**U**+**$**[1-**v**]-**I**-**T**-**Ad**}-(**Q**+**$vp**/{360[**c**-**q**]})/**/**

Rule-15091:

If both (**/**), (**U**), (**$**), (**v**), (**F**), (**Q**), (**T**), (**A**), (**d**), (**p**), (**c**) and (**q**) are known, then its Interest Expense Planned is:

I= {**U**+**$**[1-**v**]-**F**-**T**-**Ad**}-(**Q**+**$vp**/{360[**c**-**q**]})/**/**

Rule-15092:

If both (**/**), (**U**), (**$**), (**v**), (**F**), (**I**), (**Q**), (**A**), (**d**), (**p**), (**c**) and (**q**) are known, then its Tax Planned is:

T= {**U**+**$**[1-**v**]-**F**-**I**-**Ad**}-(**Q**+**$vp**/{360[**c**-**q**]})/**/**

Rule-15093:

If both (**/**), (**U**), (**$**), (**v**), (**F**), (**I**), (**T**), (**Q**), (**d**), (**p**), (**c**) and (**q**) are known, then its After Tax Income Planned is:

A= **[U**+**$**[1-**v**]-**F**-**I**-**T**-(**Q**+**$vp**/{360[**c**-**q**]})/**/]**/**d**

Rule-15094:

If both (**/**), (**U**), (**$**), (**v**), (**F**), (**I**), (**T**), (**A**), (**Q**), (**p**), (**c**) and (**q**) are known, then its Dividend Payout Planned is:

d= **[U**+**$**[1-**v**]-**F**-**I**-**T**-(**Q**+**$vp**/{360[**c**-**q**]})/**/]**/**A**

Steve Asikin ISBN 14: 978-1514685136, ISBN 10: 1514685132

Rule-15095:

If both (**/**), (**U**), (**$**), (**v**), (**F**), (**I**), (**T**), (**A**), (**d**), (**Q**), (**c**) and (**q**) are known, then its Procured Inventory Days Planned:

p= 360[**c**-**q**](**/**{**U**+**$**[1-**v**]-**F**-**I**-**T**-**Ad**}-**Q**)/[**$v**]

Rule-15096:

If both (**/**), (**U**), (**$**), (**v**), (**F**), (**I**), (**T**), (**A**), (**d**), (**p**), (**Q**) and (**q**) are known, then its Current Ratio Planned is:

c= **q**+**$vp**/[360(**/**{**U**+**$**[1-**v**]-**F**-**I**-**T**-**Ad**}-**Q**)]

Rule-15097:

If both (**/**), (**U**), (**$**), (**v**), (**F**), (**I**), (**T**), (**A**), (**d**), (**p**), (**c**) and (**Q**) are known, then its Quick or Acid Test Ratio Planned is:

q= **c**-**$vp**/[360(**/**{**U**+**$**[1-**v**]-**F**-**I**-**T**-**Ad**}-**Q**)]

Rule-15098:

If both (**/**), (**U**), (**$**), (**v**), (**F**), (**I**), (**T**), (**A**), (**d**), (**$'**), (**p**), (**s**), (**c**) and (**q**) are known, then its Quoted Longterm Debt Planned is:

Q= **/**{**U**+**$**[1-**v**]-**F**-**I**-**T**-**Ad**}-**$'vp**[1+**s**]/{360[**c**-**q**]}

Rule-15099:

If both (**Q**), (**U**), (**$**), (**v**), (**F**), (**I**), (**T**), (**A**), (**d**), (**$'**), (**p**), (**s**), (**c**) and (**q**) are known, then its Leverage or Gearing Ratio Planned is:

/= (**Q**+**$'vp**[1+**s**]/{360[**c**-**q**]})
/{**U**+**$**[1-**v**]-**F**-**I**-**T**-**Ad**}

Steve Asikin ISBN 14: 978-1514685136, ISBN 10: 1514685132

Rule-15100:

If both (**/**), (**Q**), (**$**), (**v**), (**F**), (**I**), (**T**), (**A**), (**d**), (**$'**), (**p**), (**s**), (**c**) and (**q**) are known, then its Utilized or Starting Capital must be:

U= **F**+**I**+**T**+**Ad**+(**Q**+**$'vp**[1+**s**]/{360[**c**-**q**]})/**/**-**$**[1-**v**]

Rule-15101:

If both (**/**), (**U**), (**Q**), (**v**), (**F**), (**I**), (**T**), (**A**), (**d**), (**$'**), (**p**), (**s**), (**c**) and (**q**) are known, then its Sales or Revenue Planned is:

$= (**/**{**U**-**F**-**I**-**T**-**Ad**}-**Q**)/(**vp**/{360[**c**-**q**]}-**/**[1-**v**])

Rule-15102:

If both (**/**), (**U**), (**$**), (**Q**), (**F**), (**I**), (**T**), (**A**), (**d**), (**$'**), (**p**), (**s**), (**c**) and (**q**) are known, then its Variable Portion Planned is:

v= {**/**[**U**+**$**-**F**-**I**-**T**-**Ad**]-**Q**}/[**$**(**p**/{360[**c**-**q**]}+**/**)]

Rule-15103:

If both (**/**), (**U**), (**$**), (**v**), (**Q**), (**I**), (**T**), (**A**), (**d**), (**$'**), (**p**), (**s**), (**c**) and (**q**) are known, then its Fixed Cost Planned is:

F= {**U**+**$**[1-**v**]-**I**-**T**-**Ad**}
-(**Q**+**$'vp**[1+**s**]/{360[**c**-**q**]})/**/**

Rule-15104:

If both (**/**), (**U**), (**$**), (**v**), (**F**), (**Q**), (**T**), (**A**), (**d**), (**$'**), (**p**), (**s**), (**c**) and (**q**) are known, then its Interest Expense Planned is:

I= {**U**+**$**[1-**v**]-**F**-**T**-**Ad**}
-(**Q**+**$'vp**[1+**s**]/{360[**c**-**q**]})/**/**

Steve Asikin ISBN 14: 978-1514685136, ISBN 10: 1514685132

Rule-15105:

If both (**/**), (**U**), (**$**), (**v**), (**F**), (**I**), (**Q**), (**A**), (**d**), (**$'**), (**p**), (**s**), (**c**) and (**q**) are known, then its Tax Planned is:

T= {**U**+**$**[1-**v**]-**F**-**I**-**Ad**}
-(**Q**+**$'vp**[1+**s**]/{360[**c**-**q**]})/**/**

Rule-15106:

If both (**/**), (**U**), (**$**), (**v**), (**F**), (**I**), (**T**), (**Q**), (**d**), (**$'**), (**p**), (**s**), (**c**) and (**q**) are known, then its After Tax Income Planned is:

A= [**U**+**$**[1-**v**]-**F**-**I**-**T**
-(**Q**+**$'vp**[1+**s**]/{360[**c**-**q**]})/**/**]/**d**

Rule-15107:

If both (**/**), (**U**), (**$**), (**v**), (**F**), (**I**), (**T**), (**A**), (**d**), (**Q**), (**p**), (**s**), (**c**) and (**q**) are known, then its Sales Past must be:

$'**= (**/**{**U**+**$[1-**v**]-**F**-**I**-**T**-**Ad**}-**Q**)
/(**vp**[1+**s**]/{360[**c**-**q**]})

Rule-15108:

If both (**/**), (**U**), (**$**), (**v**), (**F**), (**I**), (**T**), (**A**), (**Q**), (**$'**), (**p**), (**s**), (**c**) and (**q**) are known, then its Dividend Payout Planned is:

d= [**U**+**$**[1-**v**]-**F**-**I**-**T**
-(**Q**+**$'vp**[1+**s**]/{360[**c**-**q**]})/**/**]/**A**

Rule-15109:

If both (**/**), (**U**), (**$**), (**v**), (**F**), (**I**), (**T**), (**A**), (**d**), (**$'**), (**Q**), (**s**), (**c**) and (**q**) are known, then its Procured Inventory Days Planned:

p= 360[**c**-**q**](**/**{**U**+**$**[1-**v**]-**F**-**I**-**T**-**Ad**}-**Q**)/{**$'v**[1+**s**]}

Steve Asikin ISBN 14: 978-1514685136, ISBN 10: 1514685132

Rule-15110:

If both (**/**), (**U**), (**$**), (**v**), (**F**), (**I**), (**T**), (**A**), (**d**), (**$'**), (**p**), (**Q**), (**c**) and (**q**) are known, then its Procured Inventory Days Planned:

s= **$**/**$'**-1

or it can also be found as

s= (**/**{**U**+**$**[1-**v**]-**F**-**I**-**T**-**Ad**}-**Q**)/(**$'vp**/{360[**c**-**q**]})-1

Rule-15111:

If both (**/**), (**U**), (**$**), (**v**), (**F**), (**I**), (**T**), (**A**), (**d**), (**$'**), (**p**), (**s**), (**Q**) and (**q**) are known, then its Current Ratio Planned is:

c= **q**+**$'vp**[1+**s**]/[360(**/**{**U**+**$**[1-**v**]-**F**-**I**-**T**-**Ad**}-**Q**)]

Rule-15112:

If both (**/**), (**U**), (**$**), (**v**), (**F**), (**I**), (**T**), (**A**), (**d**), (**$'**), (**p**), (**s**), (**c**) and (**Q**) are known, then its Quick or Acid Test Ratio Planned is:

q= **c**-**$'vp**[1+**s**]/[360(**/**{**U**+**$**[1-**v**]-**F**-**I**-**T**-**Ad**}-**Q**)]

Rule-15113:

If both (**/**), (**U**), (**$**), (**v**), (**F**), (**I**), (**t**), (**D**) and (**X**) are known, then its Quoted Longterm Debt Planned is:

Q= **/**(**U**+{**$**[1-**v**]-**F**-**I**}[1-**t**]-**D**)-**X**

Rule-15114:

If both (**Q**), (**U**), (**$**), (**v**), (**F**), (**I**), (**t**), (**D**) and (**X**) are known, then its Leverage or Gearing Ratio Planned is:

/= [**Q**-**X**]/(**U**+{**$**[1-**v**]-**F**-**I**}[1-**t**]-**D**)

Steve Asikin ISBN 14: 978-1514685136, ISBN 10: 1514685132

Rule-15115:

If both (/), (Q), ($), (v), (F), (I), (t), (D) and (X) are known, then its Utilized or Starting Capital must be:

U= D+[Q-X]/ /-{$[1-v]-F-I}[1-t]

Rule-15116:

If both (/), (U), (Q), (v), (F), (I), (t), (D) and (X) are known, then its Sales or Revenue Planned is:

$= (F+I+{D+[Q-X]/ /-U}/[1-t])/[1-v]

Rule-15117:

If both (/), (U), ($), (Q), (F), (I), (t), (D) and (X) are known, then its Variable Portion Planned is:

v= 1-(F+I+{D+[Q-X]/ /-U}/[1-t])/$

Rule-15118:

If both (/), (U), ($), (v), (Q), (I), (t), (D) and (X) are known, then its Fixed Cost Planned is:

F= $[1-v]-I-{D+[Q-X]/ /-U}/[1-t]

Rule-15119:

If both (/), (U), ($), (v), (F), (Q), (t), (D) and (X) are known, then its Interest Expense Planned is:

I= $[1-v]-F-{D+[Q-X]/ /-U}/[1-t]

Rule-15120:

If both (/), (U), ($), (v), (F), (I), (Q), (D) and (X) are known, then its Tax Rate Planned is:

t= 1-{D+[Q-X]/ /-U}/{$[1-v]-F-I}

Steve Asikin ISBN 14: 978-1514685136, ISBN 10: 1514685132

Rule-15121:

If both (/), (**U**), (**S**), (**v**), (**F**), (**I**), (**t**), (**Q**) and (**X**) are known, then its Dividend Planned is:

D= **U**+{**S**[1-**v**]-**F**-**I**}[1-**t**]-[**Q**-**X**]//

Rule-15122:

If both (/), (**U**), (**S**), (**v**), (**F**), (**I**), (**t**), (**D**) and (**Q**) are known, then its Xpress or Current Debt Planned is:

X= /(**U**+{**S**[1-**v**]-**F**-**I**}[1-**t**]-**D**)-**Q**

Rule-15123:

If both (/), (**U**), (**S**), (**v**), (**F**), (**I**), (**t**), (**D**), (**P**), (**c**) and (**q**) are known, then its Quoted Longterm Debt Planned is:

Q= /(**U**+{**S**[1-**v**]-**F**-**I**}[1-**t**]-**D**)-**P**/[**c**-**q**]

Rule-15124:

If both (**Q**), (**U**), (**S**), (**v**), (**F**), (**I**), (**t**), (**D**), (**P**), (**c**) and (**q**) are known, then its Leverage or Gearing Ratio Planned is:

/= {**Q**-**P**/[**c**-**q**]}/(**U**+{**S**[1-**v**]-**F**-**I**}[1-**t**]-**D**)

Rule-15125:

If both (/), (**Q**), (**S**), (**v**), (**F**), (**I**), (**t**), (**D**), (**P**), (**c**) and (**q**) are known, then its Utilized or Starting Capital must be:

U= **D**+{**Q**-**P**/[**c**-**q**]}//-{**S**[1-**v**]-**F**-**I**}[1-**t**]

Steve Asikin ISBN 14: 978-1514685136, ISBN 10: 1514685132

Rule-15126:

If both (**/**), (**U**), (**Q**), (**v**), (**F**), (**I**), (**t**), (**D**), (**P**), (**c**) and (**q**) are known, then its Sales or Revenue Planned is:

$= [F+I+(D+{Q-P/[c-q]}//U)/[1-t]]/[1-v]

Rule-15127:

If both (**/**), (**U**), (**$**), (**Q**), (**F**), (**I**), (**t**), (**D**), (**P**), (**c**) and (**q**) are known, then its Variable Portion Planned is:

v= 1-[F+I+(D+{Q-P/[c-q]}//U)/[1-t]]/$

Rule-15128:

If both (**/**), (**U**), (**$**), (**v**), (**Q**), (**I**), (**t**), (**D**), (**P**), (**c**) and (**q**) are known, then its Fixed Cost Planned is:

F= $[1-v]-I-(D+{Q-P/[c-q]}//U)/ [1-t]

Rule-15129:

If both (**/**), (**U**), (**$**), (**v**), (**F**), (**Q**), (**t**), (**D**), (**P**), (**c**) and (**q**) are known, then its Interest Expense Planned is:

I= $[1-v]-F-(D+{Q-P/[c-q]}//U)/[1-t]

Rule-15130:

If both (**/**), (**U**), (**$**), (**v**), (**F**), (**I**), (**Q**), (**D**), (**P**), (**c**) and (**q**) are known, then its Tax Rate Planned is:

t= 1-(D+{Q-P/[c-q]}//U)/{$[1-v]-F-I}

Rule-15131:

If both (**/**), (**U**), (**$**), (**v**), (**F**), (**I**), (**t**), (**Q**), (**P**), (**c**) and (**q**) are known, then its Dividend Planned is:

D= U+{$[1-v]-F-I}[1-t]- {Q-P/[c-q]}//

Steve Asikin ISBN 14: 978-1514685136, ISBN 10: 1514685132

Rule-15132:

If both (**/**), (**U**), (**$**), (**v**), (**F**), (**I**), (**t**), (**D**), (**Q**), (**c**) and (**q**) are known, then its Procured Inventory Planned is:

P= [**c**-**q**][**/**(**U**+{**$**[1-**v**]-**F**-**I**}[1-**t**]-**D**)-**Q**]

Rule-15133:

If both (**/**), (**U**), (**$**), (**v**), (**F**), (**I**), (**t**), (**D**), (**P**), (**Q**) and (**q**) are known, then its Current Ratio Planned is:

c= **q**+**P**/[**/**(**U**+{**$**[1-**v**]-**F**-**I**}[1-**t**]-**D**)-**Q**]

Rule-15134:

If both (**/**), (**U**), (**$**), (**v**), (**F**), (**I**), (**t**), (**D**), (**P**), (**c**) and (**Q**) are known, then its Quick or Acid Test Ratio Planned is:

q= **c**-**P**/[**/**(**U**+{**$**[1-**v**]-**F**-**I**}[1-**t**]-**D**)-**Q**]

Rule-15135:

If both (**/**), (**U**), (**$**), (**v**), (**F**), (**I**), (**t**), (**D**), (**V**), (**p**), (**c**) and (**q**) are known, then its Quoted Longterm Debt Planned is:

Q= **/**(**U**+{**$**[1-**v**]-**F**-**I**}[1-**t**]-**D**)-**Vp**/{360[**c**-**q**]}

Rule-15136:

If both (**Q**), (**U**), (**$**), (**v**), (**F**), (**I**), (**t**), (**D**), (**V**), (**p**), (**c**) and (**q**) are known, then its Leverage or Gearing Ratio Planned is:

/= (**Q**-**Vp**/{360[**c**-**q**]})/(**U**+{**$**[1-**v**]-**F**-**I**}[1-**t**]-**D**)

Steve Asikin ISBN 14: 978-1514685136, ISBN 10: 1514685132

Rule-15137:

If both (**/**), (**Q**), (**\$**), (**v**), (**F**), (**I**), (**t**), (**D**), (**V**), (**p**), (**c**) and (**q**) are known, then its Utilized or Starting capital must be:

$$\mathbf{U}= \mathbf{D}+(\mathbf{Q}-\mathbf{Vp}/\{360[\mathbf{c}-\mathbf{q}]\})/\boldsymbol{/}-\{\mathbf{\$}[1-\mathbf{v}]-\mathbf{F}-\mathbf{I}\}[1-\mathbf{t}]$$

Rule-15138:

If both (**/**), (**U**), (**Q**), (**v**), (**F**), (**I**), (**t**), (**D**), (**V**), (**p**), (**c**) and (**q**) are known, then its Sales or Revenue Planned is:

$$\mathbf{\$}= \{\mathbf{F}+\mathbf{I}+[\mathbf{D}+(\mathbf{Q}-\mathbf{Vp}/\{360[\mathbf{c}-\mathbf{q}]\})/\boldsymbol{/}-\mathbf{U}]/[1-\mathbf{t}]\}/[1-\mathbf{v}]$$

Rule-15139:

If both (**/**), (**U**), (**\$**), (**Q**), (**F**), (**I**), (**t**), (**D**), (**V**), (**p**), (**c**) and (**q**) are known, then its Variable Portion Planned is:

$$\mathbf{v}= 1-\{\mathbf{F}+\mathbf{I}+[\mathbf{D}+(\mathbf{Q}-\mathbf{Vp}/\{360[\mathbf{c}-\mathbf{q}]\})/\boldsymbol{/}-\mathbf{U}]/[1-\mathbf{t}]\}/\mathbf{\$}$$

Rule-15140:

If both (**/**), (**U**), (**\$**), (**v**), (**Q**), (**I**), (**t**), (**D**), (**V**), (**p**), (**c**) and (**q**) are known, then its Fixed Cost Planned is:

$$\mathbf{F}= \mathbf{\$}[1-\mathbf{v}]-\mathbf{I}-[\mathbf{D}+(\mathbf{Q}-\mathbf{Vp}/\{360[\mathbf{c}-\mathbf{q}]\})/\boldsymbol{/}-\mathbf{U}]/ [1-\mathbf{t}]$$

Rule-15141:

If both (**/**), (**U**), (**\$**), (**v**), (**F**), (**Q**), (**t**), (**D**), (**V**), (**p**), (**c**) and (**q**) are known, then its Interest Expense Planned is:

$$\mathbf{I}= \mathbf{\$}[1-\mathbf{v}]-\mathbf{F}-[\mathbf{D}+(\mathbf{Q}-\mathbf{Vp}/\{360[\mathbf{c}-\mathbf{q}]\})/\boldsymbol{/}-\mathbf{U}]/ [1-\mathbf{t}]$$

Steve Asikin ISBN 14: 978-1514685136, ISBN 10: 1514685132

Rule-15142:

If both (**/**), (**U**), (**$**), (**v**), (**F**), (**I**), (**Q**), (**D**), (**V**), (**p**), (**c**) and (**q**) are known, then its Tax Rate Planned is:

t= 1-[D+(Q-Vp/{360[c-q]})/**/**-U]/{$[1-v]-F-I}

Rule-15143:

If both (**/**), (**U**), (**$**), (**v**), (**F**), (**I**), (**t**), (**Q**), (**V**), (**p**), (**c**) and (**q**) are known, then its Dividend Planned is:

D= U+{$[1-v]-F-I}[1-t]-(Q-Vp/{360[c-q]})/**/**

Rule-15144:

If both (**/**), (**U**), (**$**), (**v**), (**F**), (**I**), (**t**), (**D**), (**Q**), (**p**), (**c**) and (**q**) are known, then its Variable Cost Planned is:

V= 360[c-q][**/**(U+{$[1-v]-F-I}[1-t]-D)-Q]/p

Rule-15145:

If both (**/**), (**U**), (**$**), (**v**), (**F**), (**I**), (**t**), (**D**), (**V**), (**Q**), (**c**) and (**q**) are known, then its Procured Inventory Days Planned is:

p= 360[c-q][**/**(U+{$[1-v]-F-I}[1-t]-D)-Q]/V

Rule-15146:

If both (**/**), (**U**), (**$**), (**v**), (**F**), (**I**), (**t**), (**D**), (**V**), (**p**), (**Q**) and (**q**) are known, then its Current Ratio Planned is:

c= q+Vp/{360[**/**(U+{$[1-v]-F-I}[1-t]-D)-Q]}

Rule-15147:

If both (**/**), (**U**), (**$**), (**v**), (**F**), (**I**), (**t**), (**D**), (**V**), (**p**), (**c**) and (**Q**) are known, then its Quick or Acid Test Ratio Planned is:

q= c-Vp/{360[**/**(U+{$[1-v]-F-I}[1-t]-D)-Q]}

Steve Asikin ISBN 14: 978-1514685136, ISBN 10: 1514685132

Rule-15148:

If both (**/**), (**U**), (**$**), (**v**), (**F**), (**I**), (**t**), (**D**), (**p**), (**c**) and (**q**) are known, then its Quoted Longterm Debt Planned is:

Q= /(U+{$[1-v]-F-I}[1-t]-D)-$vp/{360[c-q]}

Rule-15149:

If both (**Q**), (**U**), (**$**), (**v**), (**F**), (**I**), (**t**), (**D**), (**p**), (**c**) and (**q**) are known, then its Leverage or Gearing Ratio Planned is:

/= (Q-$vp/{360[c-q]})/(U+{$[1-v]-F-I}[1-t]-D)

Rule-15150:

If both (**/**), (**Q**), (**$**), (**v**), (**F**), (**I**), (**t**), (**D**), (**p**), (**c**) and (**q**) are known, then its Utilized or Starting capital must be:

U= D+(Q-$vp/{360[c-q]})/ /-{$[1-v]-F-I}[1-t]

Rule-15151:

If both (**/**), (**U**), (**Q**), (**v**), (**F**), (**I**), (**t**), (**D**), (**p**), (**c**) and (**q**) are known, then its Sales or Revenue Planned is:

$= (/{U-[F+I][1-t]-D}-Q)/(vp/{360[c-q]}-/[1-v])

Rule-15152:

If both (**/**), (**U**), (**$**), (**Q**), (**F**), (**I**), (**t**), (**D**), (**p**), (**c**) and (**q**) are known, then its Variable Portion Planned is:

v= [/(U+$-F-I}[1-t]-D)-Q]/[$(p/{360[c-q]}+/)]

Steve Asikin ISBN 14: 978-1514685136, ISBN 10: 1514685132

Rule-15153:

If both (/), (U), ($), (v), (Q), (I), (t), (D), (p), (c) and (q) are known, then its Fixed Cost Planned is:

F= $[1-v]-I-[D+(Q-$vp/{360[c-q]})//U]/ [1-t]

Rule-15154:

If both (/), (U), ($), (v), (F), (Q), (t), (D), (p), (c) and (q) are known, then its Interest Expense Planned is:

I= $[1-v]-F-[D+(Q-$vp/{360[c-q]})//U]/ [1-t]

Rule-15155:

If both (/), (U), ($), (v), (F), (I), (Q), (D), (p), (c) and (q) are known, then its Tax Rate Planned is:

t= 1-[D+(Q-$vp/{360[c-q]})//U]/{$[1-v]-F-I}

Rule-15156:

If both (/), (U), ($), (v), (F), (I), (t), (Q), (p), (c) and (q) are known, then its Dividend Planned is:

D= U+{$[1-v]-F-I}[1-t]-(Q-$vp/{360[c-q]})//

Rule-15157:

If both (/), (U), ($), (v), (F), (I), (t), (D), (Q), (c) and (q) are known, then its Procured Inventory Days Planned is:

p= 360[c-q][/(U+{$[1-v]-F-I}[1-t]-D)-Q]/[$v]

Rule-15158:

If both (/), (U), ($), (v), (F), (I), (t), (D), (p), (Q) and (q) are known, then its Current Ratio Planned is:

c= q+$vp/{360[/(U+{$[1-v]-F-I}[1-t]-D)-Q]}

Steve Asikin ISBN 14: 978-1514685136, ISBN 10: 1514685132

Rule-15159:

If both (/), (U), (S), (v), (F), (I), (t), (D), (p), (c) and (Q) are known, then its Quick or Acid test Ratio Planned is:

q= c-Svp/{360[/(U+{S[1-v]-F-I}[1-t]-D)-Q]}

Rule-15160:

If both (/), (U), (S), (v), (F), (I), (t), (D), (S'), (p), (s), (c) and (q) are known, then its Quoted Longterm Debt Planned is:

Q= /(U+{S[1-v]-F-I}[1-t]-D)-S'vp[1+s]/{360[c-q]}

Rule-15161:

If both (Q), (U), (S), (v), (F), (I), (t), (D), (S'), (p), (s), (c) and (q) are known, then its Leverage or Gearing Ratio Planned is:

/= (Q-S'vp[1+s]/{360[c-q]})
/(U+{S[1-v]-F-I}[1-t]-D)

Rule-15162:

If both (/), (Q), (S), (v), (F), (I), (t), (D), (S'), (p), (s), (c) and (q) are known, then its Utilized or Starting Capital must be:

U= D+(Q-S'vp[1+s]/{360[c-q]})//
-{S[1-v]-F-I}[1-t]

Steve Asikin ISBN 14: 978-1514685136, ISBN 10: 1514685132

Rule-15163:

If both (**/**), (**U**), (**Q**), (**v**), (**F**), (**I**), (**t**), (**D**), (**$'**), (**p**), (**s**), (**c**) and (**q**) are known, then its Sales or Revenue Planned is:

$$ \$= (Q-/\{U-[F+I][1-t]-D\}+\$'vp[1+s]/\{360[c-q]\}) /\{/[1-v]\} $$

Rule-15164:

If both (**/**), (**U**), (**$**), (**Q**), (**F**), (**I**), (**t**), (**D**), (**$'**), (**p**), (**s**), (**c**) and (**q**) are known, then its Variable Portion Planned is:

$$ v= /\{U+[\$-F-I][1-t]-D\}-Q) /(\$'[1+s]p/\{360[c-q]\}+\$/[1-t]) $$

Rule-15165:

If both (**/**), (**U**), (**$**), (**v**), (**Q**), (**I**), (**t**), (**D**), (**$'**), (**p**), (**s**), (**c**) and (**q**) are known, then its Fixed Cost Planned is:

$$ F= \$[1-v]-I-[D+(Q-\$'vp[1+s]/\{360[c-q]\})/\text{/-}U] / [1-t] $$

Rule-15166:

If both (**/**), (**U**), (**$**), (**v**), (**F**), (**Q**), (**t**), (**D**), (**$'**), (**p**), (**s**), (**c**) and (**q**) are known, then its Interest Expense Planned is:

$$ I= \$[1-v]-F-[D+(Q-\$'vp[1+s]/\{360[c-q]\})/\text{/-}U] / [1-t] $$

Steve Asikin ISBN 14: 978-1514685136, ISBN 10: 1514685132

Rule-15167:

If both (**/**), (**U**), (**$**), (**v**), (**F**), (**I**), (**Q**), (**D**), (**$'**), (**p**), (**s**), (**c**) and (**q**) are known, then its Tax Rate Planned is:

t= 1-[**D**+(**Q**-**$'vp**[1+**s**]/{360[**c**-**q**]})/**/**-**U**]
/{**$**[1-**v**]-**F**-**I**}

Rule-15168:

If both (**/**), (**U**), (**$**), (**v**), (**F**), (**I**), (**t**), (**Q**), (**$'**), (**p**), (**s**), (**c**) and (**q**) are known, then its Dividend Planned is:

D= **U**+{**$**[1-**v**]-**F**-**I**}[1-**t**]
-(**Q**-**$'vp**[1+**s**]/{360[**c**-**q**]})/**/**

Rule-15169:

If both (**/**), (**U**), (**$**), (**v**), (**F**), (**I**), (**t**), (**D**), (**Q**), (**p**), (**s**), (**c**) and (**q**) are known, then its Sales Past must be:

$'**= [**/**(**U**+{**$[1-**v**]-**F**-**I**}[1-**t**]-**D**)-**Q**]
/([1+**s**]**vp**/{360[**c**-**q**]}

Rule-15170:

If both (**/**), (**U**), (**$**), (**v**), (**F**), (**I**), (**t**), (**D**), (**$'**), (**Q**), (**s**), (**c**) and (**q**) are known, then its Procured Inventory Planned is:

p= 360[**c**-**q**][**/**(**U**+{**$**[1-**v**]-**F**-**I**}[1-**t**]-**D**)-**Q**]
/{**$'p**[1+**s**]}

Steve Asikin ISBN 14: 978-1514685136, ISBN 10: 1514685132

Rule-15171:

If both (**/**), (**U**), (**$**), (**v**), (**F**), (**I**), (**t**), (**D**), (**$'**), (**p**), (**s**), (**c**) and (**q**) are known, then its Quoted Longterm Debt Planned is:

s= $/$'-1

or it can also be found as

s= [/(U+{$[1-v]-F-I}[1-t]-D)-Q]
/($'vp/{360[c-q]}-1

Rule-15172:

If both (**/**), (**U**), (**$**), (**v**), (**F**), (**I**), (**t**), (**D**), (**$'**), (**p**), (**s**), (**Q**) and (**q**) are known, then its Current Ratio Planned is:

c= q+$'vp[1+s]
/{360[/(U+{$[1-v]-F-I}[1-t]-D)-Q]}

Rule-15173:

If both (**/**), (**U**), (**$**), (**v**), (**F**), (**I**), (**t**), (**D**), (**$'**), (**p**), (**s**), (**c**) and (**Q**) are known, then its Quick or Acid Test Ratio Planned is:

q= c-$'vp[1+s]/{360[/(U+{$[1-v]-F-I}[1-t]-D)-Q]}

Rule-15174:

If both (**/**), (**U**), (**$**), (**v**), (**F**), (**I**), (**t**), (**A**), (**d**) and (**X**) are known, then its Quoted Longterm Debt Planned is:

Q= /(U+{$[1-v]-F-I}[1-t]-Ad)-X

Rule-15175:

If both (**Q**), (**U**), (**$**), (**v**), (**F**), (**I**), (**t**), (**A**), (**d**) and (**X**) are known, then its Leverage or Gearing Ratio Planned is:

/= [Q+X]/(U+{$[1-v]-F-I}[1-t]-Ad)

Steve Asikin ISBN 14: 978-1514685136, ISBN 10: 1514685132

Rule-15176:

If both (**/**), (**Q**), (**$**), (**v**), (**F**), (**I**), (**t**), (**A**), (**d**) and (**X**) are known, then its Utilized or Starting capital must be:

U= **Ad**+[**Q**+**X**]/**/**{**$**[1-**v**]-**F**-**I**}[1-**t**]

Rule-15177:

If both (**/**), (**U**), (**Q**), (**v**), (**F**), (**I**), (**t**), (**A**), (**d**) and (**X**) are known, then its Sales or Revenue Planned is:

$= (**F**+**I**+{**Ad**+[**Q**+**X**]/**/U**}/[1-**t**])/[1-**v**]

Rule-15178:

If both (**/**), (**U**), (**$**), (**Q**), (**F**), (**I**), (**t**), (**A**), (**d**) and (**X**) are known, then its Variable Portion Planned is:

v= 1-(**F**+**I**+{**Ad**+[**Q**+**X**]/**/U**}/[1-**t**])/**$**

Rule-15179:

If both (**/**), (**U**), (**$**), (**v**), (**Q**), (**I**), (**t**), (**A**), (**d**) and (**X**) are known, then its Fixed Cost Planned is:

F= **$**[1-**v**]-**I**-{**Ad**+[**Q**+**X**]/**/U**}/[1-**t**]

Rule-15180:

If both (**/**), (**U**), (**$**), (**v**), (**F**), (**Q**), (**t**), (**A**), (**d**) and (**X**) are known, then its Interest Expense Planned is:

I= **$**[1-**v**]-**F**-{**Ad**+[**Q**+**X**]/**/U**}/[1-**t**]

Rule-15181:

If both (**/**), (**U**), (**$**), (**v**), (**F**), (**I**), (**t**), (**Q**), (**d**) and (**X**) are known, then its After Tax Income Planned is:

t= 1-{**Ad**+[**Q**+**X**]/**/U**}/{**$**[1-**v**]-**F**-**I**}

Steve Asikin ISBN 14: 978-1514685136, ISBN 10: 1514685132

Rule-15182:

If both (ƛ), (U), ($), (v), (F), (I), (t), (Q), (d) and (X) are known, then its After Tax Income Planned is:

A= ({$[1-v]-F-I}[1-t]-[Q+X]/ƛU)/d

Rule-15183:

If both (ƛ), (U), ($), (v), (F), (I), (t), (A), (Q) and (X) are known, then its Dividend Payout Planned is:

d= ({$[1-v]-F-I}[1-t]-[Q+X]/ƛU)/A

Rule-15184:

If both (ƛ), (U), ($), (v), (F), (I), (t), (A), (d) and (Q) are known, then its Xpress or Current Debt Planned is:

X= ƛ(U+{$[1-v]-F-I}[1-t]-Ad)-Q

Rule-15185:

If both (ƛ), (U), ($), (v), (F), (I), (t), (A), (d), (P), (c) and (q) are known, then its Quoted Longterm Debt Planned is:

Q= ƛ(U+{$[1-v]-F-I}[1-t]-Ad)-P/[c-q]

Rule-15186:

If both (Q), (U), ($), (v), (F), (I), (t), (A), (d), (P), (c) and (q) are known, then its Leverage or Gearing Ratio Planned is:

ƛ= {Q+P/[c-q]}/(U+{$[1-v]-F-I}[1-t]-Ad)

Steve Asikin ISBN 14: 978-1514685136, ISBN 10: 1514685132

Rule-15187:

If both (/), (Q), ($), (v), (F), (I), (t), (A), (d), (P), (c) and (q) are known, then its Utilized or Starting Capital must be:

U= Ad+{Q+P/[c-q]}/ℓ{$[1-v]-F-I}[1-t]

Rule-15188:

If both (/), (U), (Q), (v), (F), (I), (t), (A), (d), (P), (c) and (q) are known, then its Sales or Revenue Planned is:

$= [F+I+(Ad+{Q+P/[c-q]}/ℓU)/[1-t]]/[1-v]

Rule-15189:

If both (/), (U), ($), (Q), (F), (I), (t), (A), (d), (P), (c) and (q) are known, then its VariablePortion Planned is:

v= 1-[F+I+(Ad+{Q+P/[c-q]}/ℓU)/[1-t]]/$

Rule-15190:

If both (/), (U), ($), (v), (Q), (I), (t), (A), (d), (P), (c) and (q) are known, then its Fixed Cost Planned is:

F= $[1-v]-I-(Ad+{Q+P/[c-q]}/ℓU)/[1-t]

Rule-15191:

If both (/), (U), ($), (v), (F), (Q), (t), (A), (d), (P), (c) and (q) are known, then its Interest Expense Planned is:

I= $[1-v]-F-(Ad+{Q+P/[c-q]}/ℓU)/[1-t]

Steve Asikin ISBN 14: 978-1514685136, ISBN 10: 1514685132

Rule-15192:

If both (/), (U), ($), (v), (F), (I), (t), (Q), (d), (P), (c) and (q) are known, then its After Tax Income Planned is:

t= 1-{**Ad**+{**Q**+**P**/[**c**-**q**]}/**/**-**U**}/{**$**[1-**v**]-**F**-**I**}

Rule-15193:

If both (/), (U), ($), (v), (F), (I), (t), (Q), (d), (P), (c) and (q) are known, then its After Tax Income Planned is:

A= ({**$**[1-**v**]-**F**-**I**}[1-**t**]-{**Q**+**P**/[**c**-**q**]}/**/**-**U**)/**d**

Rule-15194:

If both (/), (U), ($), (v), (F), (I), (t), (A), (Q), (P), (c) and (q) are known, then its Dividend Payout Planned is:

d= ({**$**[1-**v**]-**F**-**I**}[1-**t**]-{**Q**+**P**/[**c**-**q**]}/**/**-**U**)/**A**

Rule-15195:

If both (/), (U), ($), (v), (F), (I), (t), (A), (d), (Q), (c) and (q) are known, then its Procured Inventory Planned is:

P= [**c**-**q**][**/**(**U**+{**$**[1-**v**]-**F**-**I**}[1-**t**]-**Ad**)-**Q**]

Rule-15196:

If both (/), (U), ($), (v), (F), (I), (t), (A), (d), (P), (Q) and (q) are known, then its Current Ratio Planned is:

c= **q**+**P**/[**/**(**U**+{**$**[1-**v**]-**F**-**I**}[1-**t**]-**Ad**)-**Q**]

Steve Asikin ISBN 14: 978-1514685136, ISBN 10: 1514685132

Rule-15197:

If both (**/**), (**U**), (**$**), (**v**), (**F**), (**I**), (**t**), (**A**), (**d**), (**P**), (**c**) and (**Q**) are known, then its Quick or Acid Test Ratio Planned is:

q= **c**-**P**/[**/**(**U**+{**$**[1-**v**]-**F**-**I**}[1-**t**]-**Ad**)-**Q**]

Rule-15198:

If both (**/**), (**U**), (**$**), (**v**), (**F**), (**I**), (**t**), (**A**), (**d**), (**V**), (**p**), (**c**) and (**q**) are known, then its Quoted Longterm Debt Planned is:

Q= **/**(**U**+{**$**[1-**v**]-**F**-**I**}[1-**t**]-**Ad**)-**Vp**/{360[**c**-**q**]}

Rule-15199:

If both (**Q**), (**U**), (**$**), (**v**), (**F**), (**I**), (**t**), (**A**), (**d**), (**V**), (**p**), (**c**) and (**q**) are known, then its Leverage or Gearing Ratio Planned is:

/= (**Q**+**Vp**/{360[**c**-**q**]})/(**U**+{**$**[1-**v**]-**F**-**I**}[1-**t**]-**Ad**)

Rule-15200:

If both (**/**), (**Q**), (**$**), (**v**), (**F**), (**I**), (**t**), (**A**), (**d**), (**V**), (**p**), (**c**) and (**q**) are known, then its Utilized or Starting Capital must be:

U= **Ad**+(**Q**+**Vp**/{360[**c**-**q**]})/**/**-{**$**[1-**v**]-**F**-**I**}[1-**t**]

Rule-15201:

If both (**/**), (**U**), (**Q**), (**v**), (**F**), (**I**), (**t**), (**A**), (**d**), (**V**), (**p**), (**c**) and (**q**) are known, then its Sales or Revenue Planned is:

$= {**F**+**I**+[**Ad**+(**Q**+**Vp**/{360[**c**-**q**]})/**/**-**U**]/[1-**t**]}
/[1-**v**]

Steve Asikin ISBN 14: 978-1514685136, ISBN 10: 1514685132

Rule-15202:

If both (**/**), (**U**), (**$**), (**Q**), (**F**), (**I**), (**t**), (**A**), (**d**), (**V**), (**p**), (**c**) and (**q**) are known, then its Variable Portion Planned is:

v= **V**/**$**

or it can also be found as

v= 1-{**F**+**I**+[**Ad**+(**Q**+**Vp**/{360[**c**-**q**]})/**/**-**U**]/[1-**t**]}/**$**

Rule-15203:

If both (**/**), (**U**), (**$**), (**v**), (**Q**), (**I**), (**t**), (**A**), (**d**), (**V**), (**p**), (**c**) and (**q**) are known, then its Fixed Cost Planned is:

F= **$**[1-**v**]-**I**-[**Ad**+(**Q**+**Vp**/{360[**c**-**q**]})/**/**-**U**]/[1-**t**]

Rule-15204:

If both (**/**), (**U**), (**$**), (**v**), (**F**), (**Q**), (**t**), (**A**), (**d**), (**V**), (**p**), (**c**) and (**q**) are known, then its Interest Expense Planned is:

I= **$**[1-**v**]-**F**-[**Ad**+(**Q**+**Vp**/{360[**c**-**q**]})/**/**-**U**]/[1-**t**]

Rule-15205:

If both (**/**), (**U**), (**$**), (**v**), (**F**), (**I**), (**t**), (**Q**), (**d**), (**V**), (**p**), (**c**) and (**q**) are known, then its After Tax Income Planned is:

t= 1-[**Ad**+(**Q**+**Vp**/{360[**c**-**q**]})/**/**–**U**]/{**$**[1-**v**]-**F**-**I**}

Rule-15206:

If both (**/**), (**U**), (**$**), (**v**), (**F**), (**I**), (**t**), (**Q**), (**d**), (**V**), (**p**), (**c**) and (**q**) are known, then its After Tax Income Planned is:

A= [{**$**[1-**v**]-**F**-**I**}[1-**t**]-(**Q**+**Vp**/{360[**c**-**q**]})/**/**-**U**]/**d**

Steve Asikin ISBN 14: 978-1514685136, ISBN 10: 1514685132

Rule-15207:

If both (**/**), (**U**), (**$**), (**v**), (**F**), (**I**), (**t**), (**A**), (**Q**), (**V**), (**p**), (**c**) and (**q**) are known, then its Dividend Payout Planned is:

d= [{**$**[1-**v**]-**F**-**I**}[1-**t**]-(**Q**+**Vp**/{360[**c**-**q**]})/**/U**]/**A**

Rule-15208:

If both (**/**), (**U**), (**$**), (**v**), (**F**), (**I**), (**t**), (**A**), (**d**), (**Q**), (**p**), (**c**) and (**q**) are known, then its Variable Cost Planned is:

V= 360[**c**-**q**][**/**(**U**+{**$**[1-**v**]-**F**-**I**}[1-**t**]-**Ad**)-**Q**]/**p**

Rule-15209:

If both (**/**), (**U**), (**$**), (**v**), (**F**), (**I**), (**t**), (**A**), (**d**), (**V**), (**Q**), (**c**) and (**q**) are known, then its Procured Inventory Days Planned is:

p= 360[**c**-**q**][**/**(**U**+{**$**[1-**v**]-**F**-**I**}[1-**t**]-**Ad**)-**Q**]/**V**

Rule-15210:

If both (**/**), (**U**), (**$**), (**v**), (**F**), (**I**), (**t**), (**A**), (**d**), (**V**), (**p**), (**Q**) and (**q**) are known, then its Current Ratio Planned is:

c= **q**+**Vp**/{360[**/**(**U**+{**$**[1-**v**]-**F**-**I**}[1-**t**]-**Ad**)-**Q**]}

Rule-15211:

If both (**/**), (**U**), (**$**), (**v**), (**F**), (**I**), (**t**), (**A**), (**d**), (**V**), (**p**), (**c**) and (**Q**) are known, then its Quick or Acid Test Ratio Planned is:

q= **c**-**Vp**/{360[**/**(**U**+{**$**[1-**v**]-**F**-**I**}[1-**t**]-**Ad**)-**Q**]}

Steve Asikin ISBN 14: 978-1514685136, ISBN 10: 1514685132

Rule-15212:

If both (**/**), (**U**), (**$**), (**v**), (**F**), (**I**), (**t**), (**A**), (**d**), (**p**), (**c**) and (**q**) are known, then its Quoted Longterm Debt Planned is:

Q= **/**(**U**+{**$**[1-**v**]-**F**-**I**}[1-**t**]-**Ad**)-**$vp**/{360[**c**-**q**]}

Rule-15213:

If both (**Q**), (**U**), (**$**), (**v**), (**F**), (**I**), (**t**), (**A**), (**d**), (**p**), (**c**) and (**q**) are known, then its Leverage or Gearing Ratio Planned is:

/= (**Q**+**$vp**/{360[**c**-**q**]})/(**U**+{**$**[1-**v**]-**F**-**I**}[1-**t**]-**Ad**)

Rule-15214:

If both (**/**), (**Q**), (**$**), (**v**), (**F**), (**I**), (**t**), (**A**), (**d**), (**p**), (**c**) and (**q**) are known, then its Utilized or Starting Capital must be:

U= **Ad**+(**Q**+**$vp**/{360[**c**-**q**]})/**/**-{**$**[1-**v**]-**F**-**I**}[1-**t**]

Rule-15215:

If both (**/**), (**U**), (**Q**), (**v**), (**F**), (**I**), (**t**), (**A**), (**d**), (**p**), (**c**) and (**q**) are known, then its Sales or Revenue Planned is:

$= (**/**{**U**-[**F**+**I**][1-**t**]-**Ad**}-**Q**)
/(**vp**/{360[**c**-**q**]}-**/**[1-**v**][1-**t**])

Rule-15216:

If both (**/**), (**U**), (**$**), (**Q**), (**F**), (**I**), (**t**), (**A**), (**d**), (**p**), (**c**) and (**q**) are known, then its Variable Portion Planned is:

v= (**/**{**U**+[**$**-**F**-**I**][1-**t**]-**Ad**}-**Q**)
/[**$**(**p**/{360[**c**-**q**]}+**/**[1-**t**])]

Steve Asikin ISBN 14: 978-1514685136, ISBN 10: 1514685132

Rule-15217:

If both (**/**), (**U**), (**$**), (**v**), (**Q**), (**I**), (**t**), (**A**), (**d**), (**p**), (**c**) and (**q**) are known, then its Fixed Cost Planned is:

F= **$**[1-**v**]-**I**-**[Ad**+(**Q**+**$vp**/{360[**c-q**]})/**/U]**/[1-**t**]

Rule-15218:

If both (**/**), (**U**), (**$**), (**v**), (**F**), (**Q**), (**t**), (**A**), (**d**), (**p**), (**c**) and (**q**) are known, then its Interest Expense Planned is:

I= **$**[1-**v**]-**F**-**[Ad**+(**Q**+**$vp**/{360[**c-q**]})/**/U]**/[1-**t**]

Rule-15219:

If both (**/**), (**U**), (**$**), (**v**), (**F**), (**I**), (**Q**), (**A**), (**d**), (**p**), (**c**) and (**q**) are known, then its Tax Planned is:

t= 1-**[Ad**+(**Q**+**$vp**/{360[**c-q**]})/**/U]**/{**$**[1-**v**]-**F**-**I**}

Rule-15220:

If both (**/**), (**U**), (**$**), (**v**), (**F**), (**I**), (**t**), (**Q**), (**d**), (**p**), (**c**) and (**q**) are known, then its After Tax Income Planned is:

A= **[**{**$**[1-**v**]-**F**-**I**}[1-**t**]-(**Q**+**$vp**/{360[**c-q**]})/**/U]**/**d**

Rule-15221:

If both (**/**), (**U**), (**$**), (**v**), (**F**), (**I**), (**t**), (**A**), (**Q**), (**p**), (**c**) and (**q**) are known, then its Dividend Payout Planned is:

d= **[**{**$**[1-**v**]-**F**-**I**}[1-**t**]-(**Q**+**$vp**/{360[**c-q**]})/**/U]**/**A**

Steve Asikin ISBN 14: 978-1514685136, ISBN 10: 1514685132

Rule-15222:

If both (**/**), (**U**), (**$**), (**v**), (**F**), (**I**), (**t**), (**A**), (**d**), (**Q**), (**c**) and (**q**) are known, then its Procured Inventory Days Planned:

p= 360[**c-q**][**/**(**U**+{**$**[1-**v**]-**F-I**}[1-**t**]-**Ad**)-**Q**]/[**$v**]

Rule-15223:

If both (**/**), (**U**), (**$**), (**v**), (**F**), (**I**), (**t**), (**A**), (**d**), (**p**), (**Q**) and (**q**) are known, then its Current Ratio Planned is:

c= **q**+**$vp**/{360[**/**(**U**+{**$**[1-**v**]-**F-I**}[1-**t**]-**Ad**)-**Q**]}

Rule-15224:

If both (**/**), (**U**), (**$**), (**v**), (**F**), (**I**), (**t**), (**A**), (**d**), (**p**), (**c**) and (**Q**) are known, then its Quick or Acid Test Ratio Planned is:

q= **c**-**$vp**/{360[**/**(**U**+{**$**[1-**v**]-**F-I**}[1-**t**]-**Ad**)-**Q**]}

Rule-15225:

If both (**/**), (**U**), (**$**), (**v**), (**F**), (**I**), (**t**), (**A**), (**d**), (**$'**). (**p**), (**s**), (**c**) and (**q**) are known, then its Quoted Longterm Debt Planned is:

Q= **/**(**U**+{**$**[1-**v**]-**F-I**}[1-**t**]-**Ad**)
-**$'vp**[1+**s**]/{360[**c-q**]}

Rule-15226:

If both (**Q**), (**U**), (**$**), (**v**), (**F**), (**I**), (**t**), (**A**), (**d**), (**$'**). (**p**), (**s**), (**c**) and (**q**) are known, then its Leverage or Gearing Ratio Planned is:

/= (**Q**+**$'vp**[1+**s**]/{360[**c-q**]})
/(**U**+{**$**[1-**v**]-**F-I**}[1-**t**]-**Ad**)

Steve Asikin ISBN 14: 978-1514685136, ISBN 10: 1514685132

Rule-15227:

If both (**/**), (**Q**), (**$**), (**v**), (**F**), (**I**), (**t**), (**A**), (**d**), (**$'**). (**p**), (**s**), (**c**) and (**q**) are known, then its Utilized or Starting capital must be:

U= **Ad**+(**Q**+**$'vp**[1+**s**]/{360[**c**-**q**]})/**/**
-{**$**[1-**v**]-**F**-**I**}[1-**t**]

Rule-15228:

If both (**/**), (**U**), (**Q**), (**v**), (**F**), (**I**), (**t**), (**A**), (**d**), (**$'**). (**p**), (**s**), (**c**) and (**q**) are known, then its Sales or Revenue Planned is:

$**= (**Q**-**/**{**U**-[**F**+**I**][1-**t**]-**Ad**}+**$'p[1+**s**]/{360[**c**-**q**]})
/{**/**[1-**v**] [1-**t**]}

Rule-15229:

If both (**/**), (**U**), (**$**), (**Q**), (**F**), (**I**), (**t**), (**A**), (**d**), (**$'**). (**p**), (**s**), (**c**) and (**q**) are known, then its Variable Portion Planned is:

v= (**/**{**U**+[**$**-**F**-**I**][1-**t**]-**Ad**}-**Q**)
/(**$'p**[1+**s**]/{360[**c**-**q**]}+**$/**[1-**t**])

Rule-15230:

If both (**/**), (**U**), (**$**), (**v**), (**Q**), (**I**), (**t**), (**A**), (**d**), (**$'**). (**p**), (**s**), (**c**) and (**q**) are known, then its Fixed Cost Planned is:

F= **$**[1-**v**]-**I**
-[**Ad**+(**Q**+**$'vp**[1+**s**]/{360[**c**-**q**]})/**/**-**U**]/[1-**t**]

Steve Asikin ISBN 14: 978-1514685136, ISBN 10: 1514685132

Rule-15231:

If both (**/**), (**U**), (**$**), (**v**), (**F**), (**Q**), (**t**), (**A**), (**d**), (**$'**). (**p**), (**s**), (**c**) and (**q**) are known, then its Interest Expense Planned is:

I= **$**[1-**v**]-**F**-**[Ad**+(**Q**+**$'vp**[1+**s**]/{360[**c**-**q**]})/**/-U]** /[1-**t**]

Rule-15232:

If both (**/**), (**U**), (**$**), (**v**), (**F**), (**I**), (**Q**), (**A**), (**d**), (**$'**). (**p**), (**s**), (**c**) and (**q**) are known, then its Tax Rate Planned is:

t= 1-**[Ad**+(**Q**+**$'vp**[1+**s**]/{360[**c**-**q**]})/**/-U]** /{**$**[1-**v**]-**F**-**I**}

Rule-15233:

If both (**/**), (**U**), (**$**), (**v**), (**F**), (**I**), (**t**), (**Q**), (**d**), (**$'**). (**p**), (**s**), (**c**) and (**q**) are known, then its After Tax Income Planned is:

A= **[**{**$**[1-**v**]-**F**-**I**}[1-**t**] -(**Q**+**$'vp**[1+**s**]/{360[**c**-**q**]})/**/-U]/d**

Rule-15234:

If both (**/**), (**U**), (**$**), (**v**), (**F**), (**I**), (**t**), (**A**), (**Q**), (**$'**). (**p**), (**s**), (**c**) and (**q**) are known, then its Dividend Payout Planned is:

d= **[**{**$**[1-**v**]-**F**-**I**}[1-**t**] -(**Q**+**$'vp**[1+**s**]/{360[**c**-**q**]})/**/-U]/A**

Steve Asikin ISBN 14: 978-1514685136, ISBN 10: 1514685132

Rule-15235:

If both (/), (U), (S), (v), (F), (I), (t), (A), (d), (Q). (p), (s), (c) and (q) are known, then its Sales Past must be:

S'= [/U+{S[1-v]-F-I}[1-t]-Ad)-Q]
/([1+s]vp/{360[c-q]})

Rule-15236:

If both (/), (U), (S), (v), (F), (I), (t), (A), (d), (S'). (Q), (s), (c) and (q) are known, then its Procured Inventory Days Planned:

p= 360[c-q][/U+{S[1-v]-F-I}[1-t]-Ad)-Q]
/{S'v[1+s]}

Rule-15237:

If both (/), (U), (S), (v), (F), (I), (t), (A), (d), (S'). (p), (Q), (c) and (q) are known, then its Sales Growth Planned is:

s= S/S'-1

or it can al;so be found as

s= /U+{S[1-v]-F-I}[1-t]-Ad)-Q]
/(S'vp/{360[c-q]})-1

Rule-15238:

If both (/), (U), (S), (v), (F), (I), (t), (A), (d), (S'). (p), (s), (Q) and (q) are known, then its Current Ratio Planned is:

c= q+S'vp[1+s]
/{360[/U+{S[1-v]-F-I}[1-t]-Ad)-Q]}

Steve Asikin ISBN 14: 978-1514685136, ISBN 10: 1514685132

Rule-15239:

If both (**/**), (**U**), (**S**), (**v**), (**F**), (**I**), (**t**), (**A**), (**d**), (**S'**). (**p**), (**s**), (**c**) and (**Q**) are known, then its Quick or Acid Test Ratio Planned is:

q= **c**-**S'vp**[1+**s**]
/{360[**/**(**U**+{**S**[1-**v**]-**F**-**I**}[1-**t**]-**Ad**)-**Q**]}

Rule-15240:

If both (**/**), (**U**), (**S**), (**v**), (**F**), (**I**), (**t**), (**d**) and (**X**) are known, then its Quoted Longterm Debt Planned is:

Q= **/**(**U**+{**S**[1-**v**]-**F**-**I**}[1-**t**][1-**d**])-**X**

Rule-15241:

If both (**Q**), (**U**), (**S**), (**v**), (**F**), (**I**), (**t**), (**d**) and (**X**) are known, then its Leverage or Gearing Ratio Planned is:

/= [**Q**+**X**]/(**U**+{**S**[1-**v**]-**F**-**I**}[1-**t**][1-**d**])

Rule-15242:

If both (**/**), (**Q**), (**S**), (**v**), (**F**), (**I**), (**t**), (**d**) and (**X**) are known, then its Utilized or Starting Capital must be:

U= [**Q**+**X**]/**/**-{**S**[1-**v**]-**F**-**I**}[1-**t**][1-**d**]

Rule-15243:

If both (**/**), (**U**), (**Q**), (**v**), (**F**), (**I**), (**t**), (**d**) and (**X**) are known, then its Sales or Revenue Planned is:

S= (**F**+**I**+{[**Q**+**X**]/**/**-**U**}/{[1-**t**][1-**d**]})/[1-**v**]

Rule-15244:

If both (**/**), (**U**), (**S**), (**Q**), (**F**), (**I**), (**t**), (**d**) and (**X**) are known, then its Variable Portion Planned is:

v= 1-(**F**+**I**+{[**Q**+**X**]/**/**-**U**}/{[1-**t**][1-**d**]})/**S**

Steve Asikin ISBN 14: 978-1514685136, ISBN 10: 1514685132

Rule-15245:

If both (**/**), (**U**), (**$**), (**v**), (**Q**), (**I**), (**t**), (**d**) and (**X**) are known, then its Fixed Cost Planned is:

F= **$**[1-**v**]-**I**-{[**Q**+**X**]/**/**-**U**}/{[1-**t**][1-**d**]}

Rule-15246:

If both (**/**), (**U**), (**$**), (**v**), (**F**), (**Q**), (**t**), (**d**) and (**X**) are known, then its Interest Expense Planned is:

I= **$**[1-**v**]-**F**-{[**Q**+**X**]/**/**-**U**}/{[1-**t**][1-**d**]}

Rule-15247:

If both (**/**), (**U**), (**$**), (**v**), (**F**), (**I**), (**Q**), (**d**) and (**X**) are known, then its Tax Rate Planned is:

t = 1-{[**Q**+**X**]/**/**-**U**}/([1-**d**]{**$**[1-**v**]-**F**-**I**})

Rule-15248:

If both (**/**), (**U**), (**$**), (**v**), (**F**), (**I**), (**t**), (**Q**) and (**X**) are known, then its Dividend Payout Planned is:

d = 1-{[**Q**+**X**]/**/**-**U**}/([1-**t**]{**$**[1-**v**]-**F**-**I**})

Rule-15249:

If both (**/**), (**U**), (**$**), (**v**), (**F**), (**I**), (**t**), (**d**) and (**Q**) are known, then its Xpress or Current Debt Planned is:

X= **/**(**U**+{**$**[1-**v**]-**F**-**I**}[1-**t**][1-**d**])-**Q**

Rule-15250:

If both (**/**), (**U**), (**$**), (**v**), (**F**), (**I**), (**t**), (**d**), (**P**), (**c**) and (**q**) are known, then its Quoted Longterm Debt Planned is:

Q= **/**(**U**+{**$**[1-**v**]-**F**-**I**}[1-**t**][1-**d**])-**P**/[**c**-**q**]

Steve Asikin ISBN 14: 978-1514685136, ISBN 10: 1514685132

Rule-15251:

If both (**Q**), (**U**), (**$**), (**v**), (**F**), (**I**), (**t**), (**d**), (**P**), (**c**) and (**q**) are known, then its Leverage or Gearing Ratio Planned is:

/= {**Q**+**P**/[**c**-**q**]}/(**U**+{**$**[1-**v**]-**F**-**I**}[1-**t**][1-**d**])

Rule-15252:

If both (**/**), (**Q**), (**$**), (**v**), (**F**), (**I**), (**t**), (**d**), (**P**), (**c**) and (**q**) are known, then its Utilized or Starting Capital must be:

U= {**Q**+**P**/[**c**-**q**]}/**/**-{**$**[1-**v**]-**F**-**I**}[1-**t**][1-**d**]

Rule-15253:

If both (**/**), (**U**), (**Q**), (**v**), (**F**), (**I**), (**t**), (**d**), (**P**), (**c**) and (**q**) are known, then its Sales or Revenue Planned is:

$= [**F**+**I**+({**Q**+**P**/[**c**-**q**]}/**/**-**U**)/{[1-**t**][1-**d**]}]/[1-**v**]

Rule-15254:

If both (**/**), (**U**), (**$**), (**Q**), (**F**), (**I**), (**t**), (**d**), (**P**), (**c**) and (**q**) are known, then its Variable Portion Planned is:

v= 1-[**F**+**I**+({**Q**+**P**/[**c**-**q**]}/**/**-**U**)/{[1-**t**][1-**d**]}]/**$**

Rule-15255:

If both (**/**), (**U**), (**$**), (**v**), (**Q**), (**I**), (**t**), (**d**), (**P**), (**c**) and (**q**) are known, then its Fixed Cost Planned is:

F= **$**[1-**v**]-**I**-({**Q**+**P**/[**c**-**q**]}/**/**-**U**)/{[1-**t**][1-**d**]}

Steve Asikin ISBN 14: 978-1514685136, ISBN 10: 1514685132

Rule-15256:

If both (/), (U), (S), (v), (F), (Q), (t), (d), (P), (c) and (q) are known, then its Interest Expense Planned is:

I= S[1-v]-F-({Q+P/[c-q]}/ /U)/{[1-t][1-d]}

Rule-15257:

If both (/), (U), (S), (v), (F), (I), (Q), (d), (P), (c) and (q) are known, then its Tax Rate Planned is:

t = 1-({Q+P/[c-q]}/ /U)/ ([1-d]{S[1-v]-F-I})

Rule-15258:

If both (/), (U), (S), (v), (F), (I), (t), (Q), (P), (c) and (q) are known, then its Dividend Payout Planned is:

d = 1-({Q+P/[c-q]}/ /U)/([1-t]{S[1-v]-F-I})

Rule-15259:

If both (/), (U), (S), (v), (F), (I), (t), (d), (Q), (c) and (q) are known, then its Procured Inventory Planned is:

P= [c-q][/U+{S[1-v]-F-I}[1-t][1-d])-Q]

Rule-15260:

If both (/), (U), (S), (v), (F), (I), (t), (d), (P), (Q) and (q) are known, then its Current Ratio Planned is:

c= q+P/[/U+{S[1-v]-F-I}[1-t][1-d])-Q]

Rule-15261:

If both (/), (U), (S), (v), (F), (I), (t), (d), (P), (c) and (Q) are known, then its Quick or Acid test Ratio Planned is:

q= c-P/[/U+{S[1-v]-F-I}[1-t][1-d])-Q]

Steve Asikin ISBN 14: 978-1514685136, ISBN 10: 1514685132

Rule-15262:

If both (Ł), (**U**), (**S**), (**v**), (**F**), (**I**), (**t**), (**d**), (**V**), (**p**), (**c**) and (**q**) are known, then its Quoted Longterm Debt Planned is:

$$Q= \text{Ł}(U+\{S[1-v]-F-I\}[1-t][1-d])-Vp/\{360[c-q]\}$$

Rule-15263:

If both (**Q**), (**U**), (**S**), (**v**), (**F**), (**I**), (**t**), (**d**), (**V**), (**p**), (**c**) and (**q**) are known, then its Leverage or Gearing Ratio Planned is:

$$\text{Ł}= (Q+Vp/\{360[c-q]\})/(U+\{S[1-v]-F-I\}[1-t][1-d])$$

Rule-15264:

If both (Ł), (**Q**), (**S**), (**v**), (**F**), (**I**), (**t**), (**d**), (**V**), (**p**), (**c**) and (**q**) are known, then its Utilized or Starting Capital must be:

$$U= (Q+Vp/\{360[c-q]\})/\text{Ł}-\{S[1-v]-F-I\}[1-t][1-d]$$

Rule-15265:

If both (Ł), (**U**), (**Q**), (**v**), (**F**), (**I**), (**t**), (**d**), (**V**), (**p**), (**c**) and (**q**) are known, then its Sales or Revenue Planned is:

$$S= \{F+I+[(Q+Vp/\{360[c-q]\})/\text{Ł}-U]/\{[1-t][1-d]\}\} / [1-v]$$

Steve Asikin ISBN 14: 978-1514685136, ISBN 10: 1514685132

Rule-15266:

If both (**/**), (**U**), (**$**), (**Q**), (**F**), (**I**), (**t**), (**d**), (**V**), (**p**), (**c**) and (**q**) are known, then its Variable Portion Planned is:

v= **V**/**$**

or it can also be found as

v= 1-{**F**+**I**+[(**Q**+**Vp**/{360[**c**-**q**]})/**/U**]/{[1-**t**][1-**d**]}}/**$**

Rule-15267:

If both (**/**), (**U**), (**$**), (**v**), (**Q**), (**I**), (**t**), (**d**), (**V**), (**p**), (**c**) and (**q**) are known, then its Fixed Cost Planned is:

F= **$**[1-**v**]-**I**-[(**Q**+**Vp**/{360[**c**-**q**]})/**/U**]/{[1-**t**][1-**d**]}

Rule-15268:

If both (**/**), (**U**), (**$**), (**v**), (**F**), (**Q**), (**t**), (**d**), (**V**), (**p**), (**c**) and (**q**) are known, then its Interest Expense Planned is:

I= **$**[1-**v**]-**F**-[(**Q**+**Vp**/{360[**c**-**q**]})/**/U**]/{[1-**t**][1-**d**]}

Rule-15269:

If both (**/**), (**U**), (**$**), (**v**), (**F**), (**I**), (**Q**), (**d**), (**V**), (**p**), (**c**) and (**q**) are known, then its Tax Rate Planned is:

t = 1-[(**Q**+**Vp**/{360[**c**-**q**]})/**/U**]/([1-**d**]{**$**[1-**v**]-**F**-**I**})

Rule-15270:

If both (**/**), (**U**), (**$**), (**v**), (**F**), (**I**), (**t**), (**Q**), (**V**), (**p**), (**c**) and (**q**) are known, then its Dividend Payout Planned is:

d = 1-[(**Q**+**Vp**/{360[**c**-**q**]})/**/U**]/([1-**t**]{**$**[1-**v**]-**F**-**I**})

Steve Asikin ISBN 14: 978-1514685136, ISBN 10: 1514685132

Rule-15271:

If both (**/**), (**U**), (**$**), (**v**), (**F**), (**I**), (**t**), (**d**), (**Q**), (**p**), (**c**) and (**q**) are known, then its Variable Cost Planned is:

V= 360[**c-q**][**/(U**+{**$**[1-**v**]-**F-I**}[1-**t**][1-**d**])-**Q**]/**p**

Rule-15272:

If both (**/**), (**U**), (**$**), (**v**), (**F**), (**I**), (**t**), (**d**), (**V**), (**Q**), (**c**) and (**q**) are known, then its Procured Inventory Days Planned:

p= 360[**c-q**][**/(U**+{**$**[1-**v**]-**F-I**}[1-**t**][1-**d**])-**Q**]/**V**

Rule-15273:

If both (**/**), (**U**), (**$**), (**v**), (**F**), (**I**), (**t**), (**d**), (**V**), (**p**), (**Q**) and (**q**) are known, then its Current Ratio Planned is:

c= **q**+**Vp**/{360[**/(U**+{**$**[1-**v**]-**F-I**}[1-**t**][1-**d**])-**Q**]}

Rule-15274:

If both (**/**), (**U**), (**$**), (**v**), (**F**), (**I**), (**t**), (**d**), (**V**), (**p**), (**c**) and (**Q**) are known, then its Quick or Acid Test Ratio Planned is:

q= **c-Vp**/{360[**/(U**+{**$**[1-**v**]-**F-I**}[1-**t**][1-**d**])-**Q**]}

Rule-15275:

If both (**/**), (**U**), (**$**), (**v**), (**F**), (**I**), (**t**), (**d**), (**p**), (**c**) and (**q**) are known, then its Quoted Longterm Debt Planned is:

Q= **/(U**+{**$**[1-**v**]-**F-I**}[1-**t**][1-**d**])-**$vp**/{360[**c-q**]}

Steve Asikin ISBN 14: 978-1514685136, ISBN 10: 1514685132

Rule-15276:

If both (**Q**), (**U**), (**$**), (**v**), (**F**), (**I**), (**t**), (**d**), (**p**), (**c**) and (**q**) are known, then its Leverage or Gearing Ratio Planned is:

/= (Q+$vp/{360[c-q]})/(U+{$[1-v]-F-I}[1-t][1-d])

Rule-15277:

If both (**/**), (**Q**), (**$**), (**v**), (**F**), (**I**), (**t**), (**d**), (**p**), (**c**) and (**q**) are known, then its Utilized or Starting Capital must be:

U= (Q+$vp/{360[c-q]})//-{$[1-v]-F-I}[1-t][1-d]

Rule-15278:

If both (**/**), (**U**), (**Q**), (**v**), (**F**), (**I**), (**t**), (**d**), (**p**), (**c**) and (**q**) are known, then its Sales or Revenue Planned is:

$= (/{U-[F+I][1-t][1-d]}-Q)
/(vp/{360[c-q]}-/[1-v] [1-t][1-d])

Rule-15279:

If both (**/**), (**U**), (**$**), (**Q**), (**F**), (**I**), (**t**), (**d**), (**p**), (**c**) and (**q**) are known, then its Variable Portion Planned is:

v= (/{U+[$-F-I][1-t][1-d]}-Q)
/[$(p/{360[c-q]}+/[1-t][1-d])]

Rule-15280:

If both (**/**), (**U**), (**$**), (**v**), (**Q**), (**I**), (**t**), (**d**), (**p**), (**c**) and (**q**) are known, then its Fixed Cost Planned is:

F= $[1-v]-I-[(Q+$vp/{360[c-q]})//-U]
/{[1-t][1-d]}

Steve Asikin ISBN 14: 978-1514685136, ISBN 10: 1514685132

Rule-15281:

If both (**/**), (**U**), (**$**), (**v**), (**F**), (**Q**), (**t**), (**d**), (**p**), (**c**) and (**q**) are known, then its Interest Expense Planned is:

I= **$**[1-**v**]-**F**-**[**(**Q**+**$vp**/{360[**c**-**q**]})//**U]**
/{[1-**t**][1-**d**]}

Rule-15282:

If both (**/**), (**U**), (**$**), (**v**), (**F**), (**I**), (**Q**), (**d**), (**p**), (**c**) and (**q**) are known, then its Tax Rate Planned is:

t = 1-**[**(**Q**+**$vp**/{360[**c**-**q**]})//**U]**
/([1-**d**]{**$**[1-**v**]-**F**-**I**})

Rule-15283:

If both (**/**), (**U**), (**$**), (**v**), (**F**), (**I**), (**t**), (**Q**), (**p**), (**c**) and (**q**) are known, then its Dividend Payout Planned is:

d = 1-**[**(**Q**+**$vp**/{360[**c**-**q**]})//**U]**
/([1-**t**]{**$**[1-**v**]-**F**-**I**})

Rule-15284:

If both (**/**), (**U**), (**$**), (**v**), (**F**), (**I**), (**t**), (**d**), (**Q**), (**c**) and (**q**) are known, then its Procured Inventory Days Planned:

p= 360[**c**-**q**]**[/U**+{**$**[1-**v**]-**F**-**I**}[1-**t**][1-**d**])-**Q]**/[**$v**]

Rule-15285:

If both (**/**), (**U**), (**$**), (**v**), (**F**), (**I**), (**t**), (**d**), (**p**), (**Q**) and (**q**) are known, then its Current Ratio Planned is:

c= **q**+**$vp**/**{**360**[/U**+{**$**[1-**v**]-**F**-**I**}[1-**t**][1-**d**])-**Q]}**

Steve Asikin ISBN 14: 978-1514685136, ISBN 10: 1514685132

Rule-15286:

If both (/), (**U**), (**S**), (**v**), (**F**), (**I**), (**t**), (**d**), (**p**), (**c**) and (**Q**) are known, then its Quick or Acid test Ratio Planned is:

q= **c**-**Svp**/{360[/(**U**+{**S**[1-**v**]-**F**-**I**}[1-**t**][1-**d**])-**Q**]}

Rule-15287:

If both (/), (**U**), (**S**), (**v**), (**F**), (**I**), (**t**), (**d**), (**S'**), (**p**), (**s**), (**c**) and (**q**) are known, then its Quoted Longterm Debt Planned is:

Q= /(**U**+{**S**[1-**v**]-**F**-**I**}[1-**t**][1-**d**])
-**S'vp**[1+**s**]/{360[**c**-**q**]}

Rule-15288:

If both (/), (**U**), (**S**), (**v**), (**F**), (**I**), (**t**), (**d**), (**S'**), (**p**), (**s**), (**c**) and (**q**) are known, then its Quoted Longterm Debt Planned is:

/= (**Q**+**Svp**/{360[**c**-**q**]})/(**U**+{**S**[1-**v**]-**F**-**I**}[1-**t**][1-**d**])

Rule-15289:

If both (/), (**Q**), (**S**), (**v**), (**F**), (**I**), (**t**), (**d**), (**S'**), (**p**), (**s**), (**c**) and (**q**) are known, then its Utilized or Starting Capital must be:

U= (**Q**+**Svp**/{360[**c**-**q**]})/**/**-{**S**[1-**v**]-**F**-**I**}[1-**t**][1-**d**]

Rule-15290:

If both (/), (**U**), (**Q**), (**v**), (**F**), (**I**), (**t**), (**d**), (**S'**), (**p**), (**s**), (**c**) and (**q**) are known, then its Sales or Revenue Planned is:

S= (**Q**-/{**U**-[**F**+**I**][1-**t**][1-**d**]}+**S'vp**[1+**s**]/{360[**c**-**q**]})
/{/[1-**v**] [1-**t**][1-**d**]}

Steve Asikin ISBN 14: 978-1514685136, ISBN 10: 1514685132

Rule-15291:

If both (**/**), (**U**), (**$**), (**Q**), (**F**), (**I**), (**t**), (**d**), (**$'**), (**p**), (**s**), (**c**) and (**q**) are known, then its Variable Portion Planned is:

v= (**/**{**U**+[**$-F-I**][1-**t**][1-**d**]}-**Q**)
/(**$'p**[1+**s**]/{360[**c-q**]}+**$/**[1-**t**][1-**d**])

Rule-15292:

If both (**/**), (**U**), (**$**), (**v**), (**Q**), (**I**), (**t**), (**d**), (**$'**), (**p**), (**s**), (**c**) and (**q**) are known, then its Fixed Cost Planned is:

F= **$**[1-**v**]-**I**-[(**Q**+**$vp**/{360[**c-q**]})/**/U**]
/{[1-**t**][1-**d**]}

Rule-15293:

If both (**/**), (**U**), (**$**), (**v**), (**F**), (**Q**), (**t**), (**d**), (**$'**), (**p**), (**s**), (**c**) and (**q**) are known, then its Interest Expense Planned is:

I= **$**[1-**v**]-**F**-[(**Q**+**$vp**/{360[**c-q**]})
/**/U**]/{[1-**t**][1-**d**]}

Rule-15294:

If both (**/**), (**U**), (**$**), (**v**), (**F**), (**I**), (**Q**), (**d**), (**$'**), (**p**), (**s**), (**c**) and (**q**) are known, then its Tax Rate Planned is:

t = 1-[(**Q**+**$vp**/{360[**c-q**]})/**/U**]
/([1-**d**]{**$**[1-**v**]-**F-I**})

Steve Asikin ISBN 14: 978-1514685136, ISBN 10: 1514685132

Rule-15295:

If both (**/**), (**U**), (**$**), (**v**), (**F**), (**I**), (**t**), (**Q**), (**$'**), (**p**), (**s**), (**c**) and (**q**) are known, then its Dividend Payout Planned is:

d = 1-[(**Q**+**$vp**/{360[**c**-**q**]})/**/-U**]
/([1-**t**]{**$**[1-**v**]-**F**-**I**})

Rule-15296:

If both (**/**), (**U**), (**$**), (**v**), (**F**), (**I**), (**t**), (**d**), (**Q**), (**p**), (**s**), (**c**) and (**q**) are known, then its Sales Past must be:

$'**= [**/U**+{**$[1-**v**]-**F**-**I**}[1-**t**][1-**d**])-**Q**]
/(**vp**[1+**s**]/{360[**c**-**q**]}

Rule-15297:

If both (**/**), (**U**), (**$**), (**v**), (**F**), (**I**), (**t**), (**d**), (**$'**), (**Q**), (**s**), (**c**) and (**q**) are known, then its Procured Inventory Days Planned:

p= 360[**c**-**q**][**/U**+{**$**[1-**v**]-**F**-**I**}[1-**t**][1-**d**])-**Q**]/[**$v**]

Rule-15298:

If both (**/**), (**U**), (**$**), (**v**), (**F**), (**I**), (**t**), (**d**), (**$'**), (**p**), (**Q**), (**c**) and (**q**) are known, then its Sales Growth Planned is:

s= **$**/**$'**-1

or it can also be found as

s= [**/U**+{**$**[1-**v**]-**F**-**I**}[1-**t**][1-**d**])-**Q**]
/(**$'vp**/{360[**c**-**q**]}-1

Steve Asikin ISBN 14: 978-1514685136, ISBN 10: 1514685132

Rule-15299:

If both (**/**), (**U**), (**$**), (**v**), (**F**), (**I**), (**t**), (**d**), (**$'**), (**p**), (**s**), (**Q**) and (**q**) are known, then its Current Ratio Planned is:

c= q+$vp/{360[/(U+{$[1-v]-F-I}[1-t][1-d])-Q]}

Rule-15300:

If both (**/**), (**U**), (**$**), (**v**), (**F**), (**I**), (**t**), (**d**), (**$'**), (**p**), (**s**), (**c**) and (**Q**) are known, then its Quick or Acid test Ratio Planned is:

q= c-$vp/{360[/(U+{$[1-v]-F-I}[1-t][1-d])-Q]}

Rule-15301:

If both (**/**), (**U**), (**$**), (**v**), (**i**), (**F**), (**T**), (**D**) and (**X**) are known, then its Quoted Longterm Debt Planned is:

Q= /{U+$[1-v-i]-F-T-D}-X

Rule-15302:

If both (**Q**), (**U**), (**$**), (**v**), (**i**), (**F**), (**T**), (**D**) and (**X**) are known, then its Leverage or Gearing Ratio Planned is:

/= [Q+X]/{U+$[1-v-i]-F-T-D}

Rule-15303:

If both (**/**), (**Q**), (**$**), (**v**), (**i**), (**F**), (**T**), (**D**) and (**X**) are known, then its Utilized or Starting Capital must be:

U= F+T+D+[Q+X]/-/-$[1-v-i]

Rule-15304:

If both (**/**), (**U**), (**Q**), (**v**), (**i**), (**F**), (**T**), (**D**) and (**X**) are known, then its Sales or Revenue Planned is:

$= {F+T+D+[Q+X]/-/-U}/[1-v-i]

Steve Asikin ISBN 14: 978-1514685136, ISBN 10: 1514685132

Rule-15305:

If both (**/**), (**U**), (**$**), (**Q**), (**i**), (**F**), (**T**), (**D**) and (**X**) are known, then its Variable Portion Planned is:

v= 1-**i**-{**F**+**T**+**D**+[**Q**+**X**]/**/**-**U**}/**$**

Rule-15306:

If both (**/**), (**U**), (**$**), (**v**), (**Q**), (**F**), (**T**), (**D**) and (**X**) are known, then its Interest Portion Planned is:

i= 1-**v**-{**F**+**T**+**D**+[**Q**+**X**]/**/**-**U**}/**$**

Rule-15307:

If both (**/**), (**U**), (**$**), (**v**), (**i**), (**Q**), (**T**), (**D**) and (**X**) are known, then its Fixed Cost Planned is:

F= **$**[1-**v**-**i**]-**F**-**T**-**D**-[**Q**+**X**]/**/**-**U**

Rule-15308:

If both (**/**), (**U**), (**$**), (**v**), (**i**), (**F**), (**Q**), (**D**) and (**X**) are known, then its Tax Planned is:

T= **$**[1-**v**-**i**]-**F**-**D**-[**Q**+**X**]/**/**-**U**

Rule-15309:

If both (**/**), (**U**), (**$**), (**v**), (**i**), (**F**), (**T**), (**Q**) and (**X**) are known, then its Dividend Planned is:

D= **$**[1-**v**-**i**]-**F**-**T**-[**Q**+**X**]/**/**-**U**

Rule-15310:

If both (**/**), (**U**), (**$**), (**v**), (**i**), (**F**), (**T**), (**D**) and (**Q**) are known, then its Xpress or Current Debt Planned is:

X= **/**{**U**+**$**[1-**v**-**i**]-**F**-**T**-**D**}-**Q**

Steve Asikin ISBN 14: 978-1514685136, ISBN 10: 1514685132

Rule-15311:

If both (**/**), (**U**), (**$**), (**v**), (**i**), (**F**), (**T**), (**D**), (**P**), (**c**) and (**q**) are known, then its Quoted Longterm Debt Planned is:

Q= /{U+$[1-v-i]-F-T-D}-P/[c-q]

Rule-15312:

If both (**Q**), (**U**), (**$**), (**v**), (**i**), (**F**), (**T**), (**D**), (**P**), (**c**) and (**q**) are known, then its Leverage or Gearing Ratio Planned is:

/= {Q+P/[c-q]}/{U+$[1-v-i]-F-T-D}

Rule-15313:

If both (**/**), (**Q**), (**$**), (**v**), (**i**), (**F**), (**T**), (**D**), (**P**), (**c**) and (**q**) are known, then its Utilized or Starting Capital must be:

U= F+T+D+{Q+P/[c-q]}/ /-$[1-v-i]

Rule-15314:

If both (**/**), (**U**), (**Q**), (**v**), (**i**), (**F**), (**T**), (**D**), (**P**), (**c**) and (**q**) are known, then its Sales or Revenue Planned is:

$= (F+T+D+{Q+P/[c-q]}/ /-U)/ [1-v-i]

Rule-15315:

If both (**/**), (**U**), (**$**), (**Q**), (**i**), (**F**), (**T**), (**D**), (**P**), (**c**) and (**q**) are known, then its Variable Portion Planned is:

v= 1-i-(F+T+D+{Q+P/[c-q]}/ /-U)/$

Steve Asikin ISBN 14: 978-1514685136, ISBN 10: 1514685132

Rule-15316:

If both (**/**), (**U**), (**$**), (**v**), (**Q**), (**F**), (**T**), (**D**), (**P**), (**c**) and (**q**) are known, then its Interest Portion Planned is:

i= 1-**v**-(**F**+**T**+**D**+{**Q**+**P**/[**c**-**q**]}/**/**-**U**)/**$**

Rule-15317:

If both (**/**), (**U**), (**$**), (**v**), (**i**), (**Q**), (**T**), (**D**), (**P**), (**c**) and (**q**) are known, then its Fixed Cost Planned is:

F= **$**[1-**v**-**i**]-**F**-**T**-**D**-{**Q**+**P**/[**c**-**q**]}/**/**-**U**

Rule-15318:

If both (**/**), (**U**), (**$**), (**v**), (**i**), (**F**), (**Q**), (**D**), (**P**), (**c**) and (**q**) are known, then its Tax Planned is:

T= **$**[1-**v**-**i**]-**F**-**D**-{**Q**+**P**/[**c**-**q**]}/**/**-**U**

Rule-15319:

If both (**/**), (**U**), (**$**), (**v**), (**i**), (**F**), (**T**), (**Q**), (**P**), (**c**) and (**q**) are known, then its Dividend Planned is:

D= **$**[1-**v**-**i**]-**F**-**T**-{**Q**+**P**/[**c**-**q**]}/**/**-**U**

Rule-15320:

If both (**/**), (**U**), (**$**), (**v**), (**i**), (**F**), (**T**), (**D**), (**Q**), (**c**) and (**q**) are known, then its Procured Inventory Planned is:

P= [**c**-**q**](**/**{**U**+**$**[1-**v**-**i**]-**F**-**T**-**D**}-**Q**)

Rule-15321:

If both (**/**), (**U**), (**$**), (**v**), (**i**), (**F**), (**T**), (**D**), (**P**), (**Q**) and (**q**) are known, then its Current Ratio Planned is:

c= **q**+**P**/(**/**{**U**+**$**[1-**v**-**i**]-**F**-**T**-**D**}-**Q**)

Steve Asikin ISBN 14: 978-1514685136, ISBN 10: 1514685132

Rule-15322:

If both (**/**), (**U**), (**$**), (**v**), (**i**), (**F**), (**T**), (**D**), (**P**), (**c**) and (**Q**) are known, then its Quick or Acid Test Ratio Planned is:

q= c-P/(/{U+$[1-v-i]-F-T-D}-Q)

Rule-15323:

If both (**/**), (**U**), (**$**), (**v**), (**i**), (**F**), (**T**), (**D**), (**V**), (**p**), (**c**) and (**q**) are known, then its Quoted Longterm Debt Planned is:

Q= /{U+$[1-v-i]-F-T-D}-Vp/{360[c-q]}

Rule-15324:

If both (**Q**), (**U**), (**$**), (**v**), (**i**), (**F**), (**T**), (**D**), (**V**), (**p**), (**c**) and (**q**) are known, then its Leverage or Gearing Ratio Planned is:

/= (Q+Vp/{360[c-q]})/{U+$[1-v-i]-F-T-D}

Rule-15325:

If both (**/**), (**Q**), (**$**), (**v**), (**i**), (**F**), (**T**), (**D**), (**V**), (**p**), (**c**) and (**q**) are known, then its Utilized or Starting capital must be:

U= F+T+D+(Q+Vp/{360[c-q]})/-$[1-v-i]

Rule-15326:

If both (**/**), (**U**), (**Q**), (**v**), (**i**), (**F**), (**T**), (**D**), (**V**), (**p**), (**c**) and (**q**) are known, then its Sales or Revenue Planned is:

$= [F+T+D+(Q+Vp/{360[c-q]})/-U]/[1-v-i]

Steve Asikin ISBN 14: 978-1514685136, ISBN 10: 1514685132

Rule-15327:

If both (/), (U), ($), (Q), (i), (F), (T), (D), (V), (p), (c) and (q) are known, then its Variable Cost Planned is:

v= V/$

or it can also be found as

v= (/{U+$[1-i]-F-T-D}-Vp/{360[c-q]}-Q)/[$/]

Rule-15328:

If both (/), (U), ($), (v), (Q), (F), (T), (D), (V), (p), (c) and (q) are known, then its Interest Portion Planned is:

i= 1-v-[F+T+D+(Q+Vp/{360[c-q]})/-U]/$

Rule-15329:

If both (/), (U), ($), (v), (i), (Q), (T), (D), (V), (p), (c) and (q) are known, then its Fixed Cost Planned is:

F= $[1-v-i]-F-T-D-(Q+Vp/{360[c-q]})/-U

Rule-15330:

If both (/), (U), ($), (v), (i), (F), (T), (D), (V), (p), (c) and (q) are known, then its Tax Planned is:

T= $[1-v-i]-F-D-(Q+Vp/{360[c-q]})/-U

Rule-15331:

If both (/), (U), ($), (v), (i), (F), (T), (D), (V), (p), (c) and (q) are known, then its Dividend Planned is:

D= $[1-v-i]-F-T-(Q+Vp/{360[c-q]})/-U

Steve Asikin ISBN 14: 978-1514685136, ISBN 10: 1514685132

Rule-15332:

If both (**/**), (**U**), (**$**), (**v**), (**i**), (**F**), (**T**), (**D**), (**Q**), (**p**), (**c**) and (**q**) are known, then its Variable Cost Planned is:

V= 360[**c-q**](**/**{**U+$**[1-**v-i**]-**F-T-D**}-**Q**)/**p**

Rule-15333:

If both (**/**), (**U**), (**$**), (**v**), (**i**), (**F**), (**T**), (**D**), (**V**), (**Q**), (**c**) and (**q**) are known, then its Procured Inventory Days Planned is:

p= 360[**c-q**](**/**{**U+$**[1-**v-i**]-**F-T-D**}-**Q**)/**V**

Rule-15334:

If both (**/**), (**U**), (**$**), (**v**), (**i**), (**F**), (**T**), (**D**), (**V**), (**p**), (**Q**) and (**q**) are known, then its Current Ratio Planned is:

c= **q**+**Vp**/[360(**/**{**U+$**[1-**v-i**]-**F-T-D**}-**Q**)]

Rule-15335:

If both (**/**), (**U**), (**$**), (**v**), (**i**), (**F**), (**T**), (**D**), (**V**), (**p**), (**c**) and (**Q**) are known, then its Quick or Acid test Ratio Planned is:

q= **c-Vp**/[360(**/**{**U+$**[1-**v-i**]-**F-T-D**}-**Q**)]

Rule-15336:

If both (**/**), (**U**), (**$**), (**v**), (**i**), (**F**), (**T**), (**D**), (**p**), (**c**) and (**q**) are known, then its Quoted Longterm Debt Planned is:

Q= **/**{**U+$**[1-**v-i**]-**F-T-D**}-**$vp**/{360[**c-q**]}

Steve Asikin ISBN 14: 978-1514685136, ISBN 10: 1514685132

Rule-15337:

If both (**Q**), (**U**), (**S**), (**v**), (**i**), (**F**), (**T**), (**D**), (**p**), (**c**) and (**q**) are known, then its Leverage or Gearing Ratio Planned is:

/= (Q+Svp/{360[c-q]})/{U+S[1-v-i]-F-T-D}

Rule-15338:

If both (**/**), (**Q**), (**S**), (**v**), (**i**), (**F**), (**T**), (**D**), (**p**), (**c**) and (**q**) are known, then its Utilized or Starting Capital must be:

U= F+T+D+(Q+Svp/{360[c-q]})/**/**-S[1-v-i]

Rule-15339:

If both (**/**), (**U**), (**Q**), (**v**), (**i**), (**F**), (**T**), (**D**), (**p**), (**c**) and (**q**) are known, then its Sales or Revenue Planned is:

S= (/{U-F-T-D}-Q)/(vp/{360[c-q]}-/[1-v-i])

Rule-15340:

If both (**/**), (**U**), (**S**), (**Q**), (**i**), (**F**), (**T**), (**D**), (**p**), (**c**) and (**q**) are known, then its Variable Portion Planned is:

v= (/{U+S[1-i]-F-T-D}-Q)/[S(p/{360[c-q]}+/)]

Rule-15341:

If both (**/**), (**U**), (**S**), (**v**), (**Q**), (**F**), (**T**), (**D**), (**p**), (**c**) and (**q**) are known, then its Interest Portion Planned is:

i= 1-v-[F+T+D+(Q+Svp/{360[c-q]})/**/**-U]/S

Rule-15342:

If both (**/**), (**U**), (**S**), (**v**), (**i**), (**Q**), (**T**), (**D**), (**p**), (**c**) and (**q**) are known, then its Fixed Cost Planned is:

F= S[1-v-i]-F-T-D-(Q+Svp/{360[c-q]})/**/**-U

Steve Asikin ISBN 14: 978-1514685136, ISBN 10: 1514685132

Rule-15343:

If both (**/**), (**U**), (**$**), (**v**), (**i**), (**F**), (**Q**), (**D**), (**p**), (**c**) and (**q**) are known, then its Tax Planned is:

T= **$**[1-**v**-**i**]-**F**-**D**-(**Q**+**$vp**/{360[**c**-**q**]})/**/**-**U**

Rule-15344:

If both (**/**), (**U**), (**$**), (**v**), (**i**), (**F**), (**T**), (**Q**), (**p**), (**c**) and (**q**) are known, then its Dividend Planned is:

D= **$**[1-**v**-**i**]-**F**-**T**-(**Q**+**$vp**/{360[**c**-**q**]})/**/**-**U**

Rule-15345:

If both (**/**), (**U**), (**$**), (**v**), (**i**), (**F**), (**T**), (**D**), (**Q**), (**c**) and (**q**) are known, then its Procured Inventory Days Planned:

p= 360[**c**-**q**](**/**{**U**+**$**[1-**v**-**i**]-**F**-**T**-**D**}-**Q**)/[**$v**]

Rule-15346:

If both (**/**), (**U**), (**$**), (**v**), (**i**), (**F**), (**T**), (**D**), (**p**), (**Q**) and (**q**) are known, then its Current Ratio Planned is:

c= **q**+**$vp**/[360(**/**{**U**+**$**[1-**v**-**i**]-**F**-**T**-**D**}-**Q**)]

Rule-15347:

If both (**/**), (**U**), (**$**), (**v**), (**i**), (**F**), (**T**), (**D**), (**p**), (**c**) and (**Q**) are known, then its Quick or Acid test Ratio Planned is:

q= **c**-**$vp**/[360(**/**{**U**+**$**[1-**v**-**i**]-**F**-**T**-**D**}-**Q**)]

Steve Asikin ISBN 14: 978-1514685136, ISBN 10: 1514685132

Rule-15348:

If both (**/**), (**U**), (**$**), (**v**), (**i**), (**F**), (**T**), (**D**), (**$'**), (**p**), (**s**), (**c**) and (**q**) are known, then its Quoted Longterm Debt Planned is:

Q= **/**{**U**+**$**[1-**v**-**i**]-**F**-**T**-**D**}-**$'vp**[1+**s**]/{360[**c**-**q**]}

Rule-15349:

If both (**Q**), (**U**), (**$**), (**v**), (**i**), (**F**), (**T**), (**D**), (**$'**), (**p**), (**s**), (**c**) and (**q**) are known, then its Leverage or Gearing Ratio Planned is:

/= (**Q**+**$'vp**[1+**s**]/{360[**c**-**q**]})/{**U**+**$**[1-**v**-**i**]-**F**-**T**-**D**}

Rule-15350:

If both (**/**), (**Q**), (**$**), (**v**), (**i**), (**F**), (**T**), (**D**), (**$'**), (**p**), (**s**), (**c**) and (**q**) are known, then its Utilized or Starting Capital must be:

U= **F**+**T**+**D**+(**Q**+**$'vp**[1+**s**]/{360[**c**-**q**]})/**/**-**$**[1-**v**-**i**]

Rule-15351:

If both (**/**), (**U**), (**Q**), (**v**), (**i**), (**F**), (**T**), (**D**), (**$'**), (**p**), (**s**), (**c**) and (**q**) are known, then its Sales or Revenue Planned is:

$**= (**Q**-**/**{**U**-**F**-**T**-**D**}+**$'vp[1+**s**]/{360[**c**-**q**]})
/{**/**[1-**v**-**i**]}

Rule-15352:

If both (**/**), (**U**), (**$**), (**Q**), (**i**), (**F**), (**T**), (**D**), (**$'**), (**p**), (**s**), (**c**) and (**q**) are known, then its Variable Portiom Planned is:

v= (**/**{**U**+**$**[1-**i**]-**F**-**T**-**D**}-**Q**)
/(**$'p**[1+**s**]/{360[**c**-**q**]}-**$/**)

Steve Asikin ISBN 14: 978-1514685136, ISBN 10: 1514685132

Rule-15353:

If both (**/**), (**U**), (**$**), (**v**), (**Q**), (**F**), (**T**), (**D**), (**$'**), (**p**), (**s**), (**c**) and (**q**) are known, then its Interest Portion Planned is:

i= 1-**v**-**[F**+**T**+**D**+(**Q**+**$'vp**[1+**s**]/{360[**c**-**q**]})/**/-U]**/**$**

Rule-15354:

If both (**/**), (**U**), (**$**), (**v**), (**i**), (**Q**), (**T**), (**D**), (**$'**), (**p**), (**s**), (**c**) and (**q**) are known, then its Fixed Cost Planned is:

F= **$**[1-**v**-**i**]-**F**-**T**-**D**-(**Q**+**$'vp**[1+**s**]/{360[**c**-**q**]})/**/-U**

Rule-15355:

If both (**/**), (**U**), (**$**), (**v**), (**i**), (**F**), (**Q**), (**D**), (**$'**), (**p**), (**s**), (**c**) and (**q**) are known, then its Tax Planned is:

T= **$**[1-**v**-**i**]-**F**-**D**-(**Q**+**$'vp**[1+**s**]/{360[**c**-**q**]})/**/-U**

Rule-15356:

If both (**/**), (**U**), (**$**), (**v**), (**i**), (**F**), (**T**), (**Q**), (**$'**), (**p**), (**s**), (**c**) and (**q**) are known, then its Dividend Payout Planned is:

D= **$**[1-**v**-**i**]-**F**-**T**-(**Q**+**$'vp**[1+**s**]/{360[**c**-**q**]})/**/-U**

Rule-15357:

If both (**/**), (**U**), (**$**), (**v**), (**i**), (**F**), (**T**), (**D**), (**Q**), (**p**), (**s**), (**c**) and (**q**) are known, then its Sales Past must be:

$'**= (**/**{**U**+**$[1-**v**-**i**]-**F**-**T**-**D**}-**Q**)/(**vp**[1+**s**]/{360[**c**-**q**]})

Steve Asikin ISBN 14: 978-1514685136, ISBN 10: 1514685132

Rule-15358:

If both (**/**), (**U**), (**$**), (**v**), (**i**), (**F**), (**T**), (**D**), (**$'**), (**Q**), (**s**), (**c**) and (**q**) are known, then its Procured Inventory Days Planned:

p= 360[**c**-**q**](**/**{**U**+**$**[1-**v**-**i**]-**F**-**T**-**D**}-**Q**)/{**$'v**[1+**s**]}

Rule-15359:

If both (**/**), (**U**), (**$**), (**v**), (**i**), (**F**), (**T**), (**D**), (**$'**), (**p**), (**Q**), (**c**) and (**q**) are known, then its Sales Growth Planned is:

s= **$**/**$'**-1

or it can also be found as

s= (**/**{**U**+**$**[1-**v**-**i**]-**F**-**T**-**D**}-**Q**)/(**$'vp**/{360[**c**-**q**]})-1

Rule-15360:

If both (**/**), (**U**), (**$**), (**v**), (**i**), (**F**), (**T**), (**D**), (**$'**), (**p**), (**s**), (**Q**) and (**q**) are known, then its Current Ratio Planned is:

c= **q**+**$'vp**[1+**s**]/[360(**/**{**U**+**$**[1-**v**-**i**]-**F**-**T**-**D**}-**Q**)]

Rule-15361:

If both (**/**), (**U**), (**$**), (**v**), (**i**), (**F**), (**T**), (**D**), (**$'**), (**p**), (**s**), (**c**) and (**Q**) are known, then its Quick or Acid test Ratio Planned is:

q= **c**-**$'vp**[1+**s**]/[360(**/**{**U**+**$**[1-**v**-**i**]-**F**-**T**-**D**}-**Q**)]

Rule-15362:

If both (**/**), (**U**), (**$**), (**v**), (**i**), (**F**), (**T**), (**d**) and (**X**) are known, then its Quoted Longterm Debt Planned is:

Q= **/**(**U**+{**$**[1-**v**-**i**]-**F**-**T**}[1-**d**])-**X**

Steve Asikin ISBN 14: 978-1514685136, ISBN 10: 1514685132

Rule-15363:

If both (**Q**), (**U**), (**$**), (**v**), (**i**), (**F**), (**T**), (**d**) and (**X**) are known, then its Leverage or Gearing Ratio Planned is:

/= [**Q**+**X**]/(**U**+{**$**[1-**v**-**i**]-**F**-**T**}[1-**d**])

Rule-15364:

If both (**/**), (**Q**), (**$**), (**v**), (**i**), (**F**), (**T**), (**d**) and (**X**) are known, then its Utilized or Starting Capital must be:

U= [**Q**+**X**]/**/**-{**$**[1-**v**-**i**]-**F**-**T**}[1-**d**]

Rule-15365:

If both (**/**), (**U**), (**Q**), (**v**), (**i**), (**F**), (**T**), (**d**) and (**X**) are known, then its Sales or Revenue Planned is:

$= (**F**+**T**+{[**Q**+**X**]/**/**-**U**}/[1-**d**])/[1-**v**-**i**]

Rule-15366:

If both (**/**), (**U**), (**$**), (**Q**), (**i**), (**F**), (**T**), (**d**) and (**X**) are known, then its Variable Cost Planned is:

v= 1-**i**-(**F**+**T**+{[**Q**+**X**]/**/**-**U**}/[1-**d**])/**$**

Rule-15367:

If both (**/**), (**U**), (**$**), (**v**), (**Q**), (**F**), (**T**), (**d**) and (**X**) are known, then its Interest Portion Planned is:

i= 1-**v**-(**F**+**T**+{[**Q**+**X**]/**/**-**U**}/[1-**d**])/**$**

Rule-15368:

If both (**/**), (**U**), (**$**), (**v**), (**i**), (**Q**), (**T**), (**d**) and (**X**) are known, then its Fixed Cost Planned is:

F= **$**[1-**v**-**i**]-**T**+{[**Q**+**X**]/**/**-**U**}/[1-**d**]

Steve Asikin ISBN 14: 978-1514685136, ISBN 10: 1514685132

Rule-15369:

If both (Ł), (**U**), ($), (**v**), (**i**), (**F**), (**Q**), (**d**) and (**X**) are known, then its Tax Planned is:

T= **$**[1-**v**-**i**]-**F**+{[**Q**+**X**]/**Ł-U**}/[1-**d**]

Rule-15370:

If both (Ł), (**U**), ($), (**v**), (**i**), (**F**), (**T**), (**Q**) and (**X**) are known, then its Dividend Payout Planned is:

d= 1-{[**Q**+**X**]/**Ł-U**}/{**$**[1-**v**-**i**]-**F**-**T**}

Rule-15371:

If both (Ł), (**U**), ($), (**v**), (**i**), (**F**), (**T**), (**d**) and (**Q**) are known, then its Xpress or Current Debt Planned is:

X= **Ł**(**U**+{**$**[1-**v**-**i**]-**F**-**T**}[1-**d**])-**Q**

Rule-15372:

If both (Ł), (**U**), ($), (**v**), (**i**), (**F**), (**T**), (**d**), (**P**), (**c**) and (**q**) are known, then its Quoted Longterm Debt Planned is:

Q= **Ł**(**U**+{**$**[1-**v**-**i**]-**F**-**T**}[1-**d**])-**P**/[**c**-**q**]

Rule-15373:

If both (**Q**), (**U**), ($), (**v**), (**i**), (**F**), (**T**), (**d**), (**P**), (**c**) and (**q**) are known, then its Leverage or Gearing Ratio Planned is:

Ł= {**Q**+**P**/[**c**-**q**]}/(**U**+{**$**[1-**v**-**i**]-**F**-**T**}[1-**d**])

Steve Asikin ISBN 14: 978-1514685136, ISBN 10: 1514685132

Rule-15374:

If both (**/**), (**Q**), (**$**), (**v**), (**i**), (**F**), (**T**), (**d**), (**P**), (**c**) and (**q**) are known, then its Utilized or Starting Capital must be:

U= {**Q**+**P**/[**c**-**q**]}//{**$**[1-**v**-**i**]-**F**-**T**}[1-**d**]

Rule-15375:

If both (**/**), (**U**), (**Q**), (**v**), (**i**), (**F**), (**T**), (**d**), (**P**), (**c**) and (**q**) are known, then its Sales or Revenue Planned is:

$= [**F**+**T**+({**Q**+**P**/[**c**-**q**]}//**U**)/[1-**d**]]/[1-**v**-**i**]

Rule-15376:

If both (**/**), (**U**), (**$**), (**Q**), (**i**), (**F**), (**T**), (**d**), (**P**), (**c**) and (**q**) are known, then its Variable Portion Planned is:

v= 1-**i**-[**F**+**T**+({**Q**+**P**/[**c**-**q**]}//**U**)/[1-**d**]]/**$**

Rule-15377:

If both (**/**), (**U**), (**$**), (**v**), (**Q**), (**F**), (**T**), (**d**), (**P**), (**c**) and (**q**) are known, then its Interest Portion Planned is:

i= 1-**v**-[**F**+**T**+({**Q**+**P**/[**c**-**q**]}//**U**)/[1-**d**]]/**$**

Rule-15378:

If both (**/**), (**U**), (**$**), (**v**), (**i**), (**Q**), (**T**), (**d**), (**P**), (**c**) and (**q**) are known, then its Fixed Cost Planned is:

F= **$**[1-**v**-**i**]-**T**+({**Q**+**P**/[**c**-**q**]}//**U**)/[1-**d**]

Rule-15379:

If both (**/**), (**U**), (**$**), (**v**), (**i**), (**F**), (**Q**), (**d**), (**P**), (**c**) and (**q**) are known, then its Tax Planned is:

T= **$**[1-**v**-**i**]-**F**+({**Q**+**P**/[**c**-**q**]}//**U**)/[1-**d**]

Steve Asikin ISBN 14: 978-1514685136, ISBN 10: 1514685132

Rule-15380:

If both (**/**), (**U**), (**$**), (**v**), (**i**), (**F**), (**T**), (**Q**), (**P**), (**c**) and (**q**) are known, then its Dividend Payout Planned is:

d= 1-({**Q**+**P**/[**c**-**q**]}/**/**-**U**)/{**$**[1-**v**-**i**]-**F**-**T**}

Rule-15381:

If both (**/**), (**U**), (**$**), (**v**), (**i**), (**F**), (**T**), (**d**), (**Q**), (**c**) and (**q**) are known, then its Procured Inventory Days Planned:

P= [**c**-**q**][**/**(**U**+{**$**[1-**v**-**i**]-**F**-**T**}[1-**d**])-**Q**]

Rule-15382:

If both (**/**), (**U**), (**$**), (**v**), (**i**), (**F**), (**T**), (**d**), (**P**), (**Q**) and (**q**) are known, then its Current Ratio Planned is:

c= **q**+**P**/[**/**(**U**+{**$**[1-**v**-**i**]-**F**-**T**}[1-**d**])-**Q**]

Rule-15383:

If both (**/**), (**U**), (**$**), (**v**), (**i**), (**F**), (**T**), (**d**), (**P**), (**c**) and (**Q**) are known, then its Quick or Acid Test Ratio Planned is:

q= **c**-**P**/[**/**(**U**+{**$**[1-**v**-**i**]-**F**-**T**}[1-**d**])-**Q**]

Rule-15384:

If both (**/**), (**U**), (**$**), (**v**), (**i**), (**F**), (**T**), (**d**), (**V**), (**p**), (**c**) and (**q**) are known, then its Quoted Longterm Debt Planned is:

Q= **/**(**U**+{**$**[1-**v**-**i**]-**F**-**T**}[1-**d**])-**Vp**/{360[**c**-**q**]}

Steve Asikin ISBN 14: 978-1514685136, ISBN 10: 1514685132

Rule-15385:

If both (**Q**), (**U**), (**$**), (**v**), (**i**), (**F**), (**T**), (**d**), (**V**), (**p**), (**c**) and (**q**) are known, then its Leverage or Gearing Ratio Planned is:

/= (**Q**+**Vp**/{360[**c**-**q**]})/(**U**+{**$**[1-**v**-**i**]-**F**-**T**}[1-**d**])

Rule-15386:

If both (**/**), (**Q**), (**$**), (**v**), (**i**), (**F**), (**T**), (**d**), (**V**), (**p**), (**c**) and (**q**) are known, then its Utilized or Starting Capital must be:

U= (**Q**+**Vp**/{360[**c**-**q**]})/**/**-{**$**[1-**v**-**i**]-**F**-**T**}[1-**d**]

Rule-15387:

If both (**/**), (**U**), (**Q**), (**v**), (**i**), (**F**), (**T**), (**d**), (**V**), (**p**), (**c**) and (**q**) are known, then its Sales or Revenue Planned is:

$= **{F**+**T**+**[**(**Q**+**Vp**/{360[**c**-**q**]})/**/**-**U]**/[1-**d**]**}**/[1-**v**-**i**]

Rule-15388:

If both (**/**), (**U**), (**$**), (**Q**), (**i**), (**F**), (**T**), (**d**), (**V**), (**p**), (**c**) and (**q**) are known, then its Variable Portion Planned is:

v= **V**/**$**

or it can also be found as

v= 1-**i**-**{F**+**T**+**[**(**Q**+**Vp**/{360[**c**-**q**]})/**/**-**U]**/[1-**d**]**}**/**$**

Rule-15389:

If both (**/**), (**U**), (**$**), (**v**), (**Q**), (**F**), (**T**), (**d**), (**V**), (**p**), (**c**) and (**q**) are known, then its Interest Portion Planned is:

i= 1-**v**-**{F**+**T**+**[**(**Q**+**Vp**/{360[**c**-**q**]})/**/**-**U]**/[1-**d**]**}**/**$**

Steve Asikin ISBN 14: 978-1514685136, ISBN 10: 1514685132

Rule-15390:

If both (**/**), (**U**), (**$**), (**v**), (**i**), (**Q**), (**T**), (**d**), (**V**), (**p**), (**c**) and (**q**) are known, then its Fixed Portion Planned is:

F= **$**[1-**v**-**i**]-**T**+**[**(**Q**+**Vp**/{360[**c**-**q**]})/**/U]**/[1-**d**]

Rule-15391:

If both (**/**), (**U**), (**$**), (**v**), (**i**), (**F**), (**Q**), (**d**), (**V**), (**p**), (**c**) and (**q**) are known, then its Tax Planned is:

T= **$**[1-**v**-**i**]-**F**+**[**(**Q**+**Vp**/{360[**c**-**q**]})/**/U]**/[1-**d**]

Rule-15392:

If both (**/**), (**U**), (**$**), (**v**), (**i**), (**F**), (**T**), (**Q**), (**V**), (**p**), (**c**) and (**q**) are known, then its Dividend Payout Planned is:

d= 1-**[**(**Q**+**Vp**/{360[**c**-**q**]})/**/U]**/{**$**[1-**v**-**i**]-**F**-**T**}

Rule-15393:

If both (**/**), (**U**), (**$**), (**v**), (**i**), (**F**), (**T**), (**d**), (**Q**), (**p**), (**c**) and (**q**) are known, then its Variable Cost Planned is:

V= 360[**c**-**q**]**[/(U**+{**$**[1-**v**-**i**]-**F**-**T**}[1-**d**])-**Q]**/**p**

Rule-15394:

If both (**/**), (**U**), (**$**), (**v**), (**i**), (**F**), (**T**), (**d**), (**V**), (**Q**), (**c**) and (**q**) are known, then its Procured Inventory Days Planned:

p= 360[**c**-**q**]**[/(U**+{**$**[1-**v**-**i**]-**F**-**T**}[1-**d**])-**Q]**/**V**

Steve Asikin ISBN 14: 978-1514685136, ISBN 10: 1514685132

Rule-15395:

If both (**/**), (**U**), (**$**), (**v**), (**i**), (**F**), (**T**), (**d**), (**V**), (**p**), (**Q**) and (**q**) are known, then its Current Ratio Planned is:

c= **q**+**Vp**/{360[**/**(**U**+{**$**[1-**v**-**i**]-**F**-**T**}[1-**d**])-**Q**]}

Rule-15396:

If both (**/**), (**U**), (**$**), (**v**), (**i**), (**F**), (**T**), (**d**), (**V**), (**p**), (**c**) and (**Q**) are known, then its Quick or Acid Test Ratio Planned is:

q= **c**-**Vp**/{360[**/**(**U**+{**$**[1-**v**-**i**]-**F**-**T**}[1-**d**])-**Q**]}

Rule-15397:

If both (**/**), (**U**), (**$**), (**v**), (**i**), (**F**), (**T**), (**d**), (**p**), (**c**) and (**q**) are known, then its Quoted Longterm Debt Planned is:

Q= **/**(**U**+{**$**[1-**v**-**i**]-**F**-**T**}[1-**d**])-**$vp**/{360[**c**-**q**]}

Rule-15398:

If both (**Q**), (**U**), (**$**), (**v**), (**i**), (**F**), (**T**), (**d**), (**p**), (**c**) and (**q**) are known, then its Leverage or Gearing Ratio Planned is:

/= (**Q**+**$vp**/{360[**c**-**q**]})/(**U**+{**$**[1-**v**-**i**]-**F**-**T**}[1-**d**])

Rule-15399:

If both (**/**), (**Q**), (**$**), (**v**), (**i**), (**F**), (**T**), (**d**), (**p**), (**c**) and (**q**) are known, then its Utilized or Starting Capital must be:

U= (**Q**+**$vp**/{360[**c**-**q**]})/**/**-{**$**[1-**v**-**i**]-**F**-**T**}[1-**d**]

Steve Asikin ISBN 14: 978-1514685136, ISBN 10: 1514685132

Rule-15400:

If both (/), (U), (Q), (v), (i), (F), (T), (d), (p), (c) and (q) are known, then its Sales or Revenue Planned is:

S= (/{U-[F+T][1-d]}-Q)
/(vp/{360[c-q]}-/[1-v-i][1-d])

Rule-15401:

If both (/), (U), (S), (Q), (i), (F), (T), (d), (p), (c) and (q) are known, then its Variable Portion Planned is:

v= [/(U+{S[1-i]-F-T}[1-d])-Q]
/[S(p/{360[c-q]}+/[1-d])]

Rule-15402:

If both (/), (U), (S), (v), (Q), (F), (T), (d), (p), (c) and (q) are known, then its Interest Portion Planned is:

i= 1-v-{F+T+[(Q+Svp/{360[c-q]})/U]/[1-d]}/S

Rule-15403:

If both (/), (U), (S), (v), (i), (Q), (T), (d), (p), (c) and (q) are known, then its Fixed Cost Planned is:

F= S[1-v-i]-T+[(Q+Svp/{360[c-q]})/U]/[1-d]

Rule-15404:

If both (/), (U), (S), (v), (i), (F), (Q), (d), (p), (c) and (q) are known, then its Tax Planned is:

T= S[1-v-i]-F+[(Q+Svp/{360[c-q]})/U]/[1-d]

Steve Asikin ISBN 14: 978-1514685136, ISBN 10: 1514685132

Rule-15405:

If both (/), (U), ($), (v), (i), (F), (T), (Q), (p), (c) and (q) are known, then its Dividend Payout Planned is:

d= 1-[(Q+$vp/{360[c-q]})//U]/{$[1-v-i]-F-T}

Rule-15406:

If both (/), (U), ($), (v), (i), (F), (T), (d), (Q), (c) and (q) are known, then its Procured Inventory Days Planned:

p= 360[c-q][/(U+{$[1-v-i]-F-T}[1-d])-Q]/[$v]

Rule-15407:

If both (/), (U), ($), (v), (i), (F), (T), (d), (p), (Q) and (q) are known, then its Current Ratio Planned is:

c= q+$vp/{360[/(U+{$[1-v-i]-F-T}[1-d])-Q]}

Rule-15408:

If both (/), (U), ($), (v), (i), (F), (T), (d), (p), (c) and (Q) are known, then its Quick or Acid Test Ratio Planned is:

q= c-$vp/{360[/(U+{$[1-v-i]-F-T}[1-d])-Q]}

Rule-15409:

If both (/), (U), ($), (v), (i), (F), (T), (d), ($'), (p), (s), (c) and (q) are known, then its Quoted Longterm Debt Planned is:

Q= /(U+{$[1-v-i]-F-T}[1-d])-$'vp[1+s]/{360[c-q]}

Steve Asikin ISBN 14: 978-1514685136, ISBN 10: 1514685132

Rule-15410:

If both (**Q**), (**U**), (**$**), (**v**), (**i**), (**F**), (**T**), (**d**), (**$'**), (**p**), (**s**), (**c**) and (**q**) are known, then its Leverage or Gearing Ratio Planned is:

/= (Q+$'vp[1+s]/{360[c-q]})
/(U+{$[1-v-i]-F-T}[1-d])

Rule-15411:

If both (/), (**Q**), (**$**), (**v**), (**i**), (**F**), (**T**), (**d**), (**$'**), (**p**), (**s**), (**c**) and (**q**) are known, then its Utilized or Starting Capital must be:

U= (Q+$'vp[1+s]/{360[c-q]})//
-{$[1-v-i]-F-T}[1-d]

Rule-15412:

If both (/), (**U**), (**Q**), (**v**), (**i**), (**F**), (**T**), (**d**), (**$'**), (**p**), (**s**), (**c**) and (**q**) are known, then its Sales or Revenue Planned is:

$= [Q-/(U+{ -F-T}[1-d])+$'vp[1+s]/{360[c-q]}]
/{/[1-v-i][1-d]}

Rule-15413:

If both (/), (**U**), (**$**), (**Q**), (**i**), (**F**), (**T**), (**d**), (**$'**), (**p**), (**s**), (**c**) and (**q**) are known, then its Variable Portion Planned is:

v= [/(U+{$[1-i]-F-T}[1-d])-Q]
/($'p[1+s]/{360[c-q]}-$/[1-d])

Steve Asikin ISBN 14: 978-1514685136, ISBN 10: 1514685132

Rule-15414:

If both (**/**), (**U**), (**$**), (**v**), (**Q**), (**F**), (**T**), (**d**), (**$'**), (**p**), (**s**), (**c**) and (**q**) are known, then its Interest Portion Planned is:

i= 1-**v**-{**F**+**T**+[(**Q**+**$'vp**[1+**s**]/{360[**c**-**q**]})/**/-U**] /[1-**d**]}/**$**

Rule-15415:

If both (**/**), (**U**), (**$**), (**v**), (**i**), (**Q**), (**T**), (**d**), (**$'**), (**p**), (**s**), (**c**) and (**q**) are known, then its Fixed Cost Planned is:

F= **$**[1-**v**-**i**]-**T**+[(**Q**+**$'vp**[1+**s**]/{360[**c**-**q**]})/**/-U**] /[1-**d**]

Rule-15416:

If both (**/**), (**U**), (**$**), (**v**), (**i**), (**F**), (**Q**), (**d**), (**$'**), (**p**), (**s**), (**c**) and (**q**) are known, then its Tax Planned is:

T= **$**[1-**v**-**i**]-**F**+[(**Q**+**$'vp**[1+**s**]/{360[**c**-**q**]})/**/-U**] /[1-**d**]

Rule-15417:

If both (**/**), (**U**), (**$**), (**v**), (**i**), (**F**), (**T**), (**Q**), (**$'**), (**p**), (**s**), (**c**) and (**q**) are known, then its Dividend Payout Planned is:

d= 1-[(**Q**+**$'vp**[1+**s**]/{360[**c**-**q**]})/**/-U**] /{**$**[1-**v**-**i**]-**F**-**T**}

Rule-15418:

If both (**/**), (**U**), (**$**), (**v**), (**i**), (**F**), (**T**), (**d**), (**Q**), (**p**), (**s**), (**c**) and (**q**) are known, then its Sales Past must be:

$'**= [**/U**+{**$[1-**v**-**i**]-**F**-**T**}[1-**d**])-**Q**] /(**vp**[1+**s**]/{360[**c**-**q**]})

Steve Asikin ISBN 14: 978-1514685136, ISBN 10: 1514685132

Rule-15419:

If both (**/**), (**U**), (**$**), (**v**), (**i**), (**F**), (**T**), (**d**), (**$'**), (**Q**), (**s**), (**c**) and (**q**) are known, then its Procured Inventory Days Planned:

p= 360[**c**-**q**][**/**(**U**+{**$**[1-**v**-**i**]-**F**-**T**}[1-**d**])-**Q**]
/{**$'v**[1+**s**]}

Rule-15420:

If both (**/**), (**U**), (**$**), (**v**), (**i**), (**F**), (**T**), (**d**), (**$'**), (**p**), (**Q**), (**c**) and (**q**) are known, then its Sales Growth Planned is:

s= **$**/**$'**-1

or it can also be found as

s= [**/**(**U**+{**$**[1-**v**-**i**]-**F**-**T**}[1-**d**])-**Q**]
/(**$'vp**/{360[**c**-**q**]})-1

Rule-15421:

If both (**/**), (**U**), (**$**), (**v**), (**i**), (**F**), (**T**), (**d**), (**$'**), (**p**), (**s**), (**Q**) and (**q**) are known, then its Current Ratio Planned is:

c= **q**+**$'vp**[1+**s**]
/{360[**/**(**U**+{**$**[1-**v**-**i**]-**F**-**T**}[1-**d**])-**Q**]}

Rule-15422:

If both (**/**), (**U**), (**$**), (**v**), (**i**), (**F**), (**T**), (**d**), (**$'**), (**p**), (**s**), (**c**) and (**Q**) are known, then its Quick or Acid Test Ratio Planned is:

q= **c**-**$'vp**[1+**s**]/{360[**/**(**U**+{**$**[1-**v**-**i**]-**F**-**T**}[1-**d**])-**Q**]}

Steve Asikin ISBN 14: 978-1514685136, ISBN 10: 1514685132

Rule-15423:

If both (/), (U), ($), (v), (i), (F), (T), (A), (d) and (X) are known, then its Quoted Longterm Debt Planned is:

Q= /{U+$[1-v-i]-F-T-Ad}-X

Rule-15424:

If both (Q), (U), ($), (v), (i), (F), (T), (A), (d) and (X) are known, then its Leverage or Gearing Ratio Planned is:

/= [Q+X]/{U+$[1-v-i]-F-T-Ad}

Rule-15425:

If both (/), (Q), ($), (v), (i), (F), (T), (A), (d) and (X) are known, then its Utilized or Starting Capital must be:

U= F+T+Ad+[Q+X]/ /-$[1-v-i]

Rule-15426:

If both (/), (U), (Q), (v), (i), (F), (T), (A), (d) and (X) are known, then its Sales or Revenue Planned is:

$= {F+T+Ad+[Q+X]/ /-U}/[1-v-i]

Rule-15427:

If both (/), (U), ($), (Q), (i), (F), (T), (A), (d) and (X) are known, then its Variable Portion Planned is:

v= 1-i-{F+T+Ad+[Q+X]/ /-U}/$

Rule-15428:

If both (/), (U), ($), (v), (Q), (F), (T), (A), (d) and (X) are known, then its Interest Portion Planned is:

i= 1-v-{F+T+Ad+[Q+X]/ /-U}/$

Steve Asikin ISBN 14: 978-1514685136, ISBN 10: 1514685132

Rule-15429:

If both (**/**), (**U**), (**$**), (**v**), (**i**), (**Q**), (**T**), (**A**), (**d**) and (**X**) are known, then its Fixed Cost Planned is:

F= **$**[1-**v**-**i**]-**T**-**Ad**}-[**Q**+**X**]/**/**-**U**

Rule-15430:

If both (**/**), (**U**), (**$**), (**v**), (**i**), (**F**), (**Q**), (**A**), (**d**) and (**X**) are known, then its Tax Planned is:

T= **$**[1-**v**-**i**]-**F**-**Ad**}-[**Q**+**X**]/**/**-**U**

Rule-15431:

If both (**/**), (**U**), (**$**), (**v**), (**i**), (**F**), (**T**), (**Q**), (**d**) and (**X**) are known, then its After Tax Income Planned is:

A= {**$**[1-**v**-**i**]-**F**-**T**-[**Q**+**X**]/**/**-**U**}/**d**

Rule-15432:

If both (**/**), (**U**), (**$**), (**v**), (**i**), (**F**), (**T**), (**A**), (**Q**) and (**X**) are known, then its Dividend Payout Planned is:

d= {**$**[1-**v**-**i**]-**F**-**T**-[**Q**+**X**]/**/**-**U**}/**A**

Rule-15433:

If both (**/**), (**U**), (**$**), (**v**), (**i**), (**F**), (**T**), (**A**), (**d**) and (**Q**) are known, then its Xpress or Cureent Debt Planned is:

X= **/**{**U**+**$**[1-**v**-**i**]-**F**-**T**-**Ad**}-**Q**

Rule-15434:

If both (**/**), (**U**), (**$**), (**v**), (**i**), (**F**), (**T**), (**A**), (**d**), (**P**), (**c**) and (**q**) are known, then its Quoted Longterm Debt Planned is:

Q= **/**{**U**+**$**[1-**v**-**i**]-**F**-**T**-**Ad**}-**P**/[**c**-**q**]

Steve Asikin ISBN 14: 978-1514685136, ISBN 10: 1514685132

Rule-15435:

If both (**Q**), (**U**), (**$**), (**v**), (**i**), (**F**), (**T**), (**A**), (**d**), (**P**), (**c**) and (**q**) are known, then its Leverage or Gearing Ratio Planned is:

/= {Q+P/[c-q]}/{U+$[1-v-i]-F-T-Ad}

Rule-15436:

If both (**/**), (**Q**), (**$**), (**v**), (**i**), (**F**), (**T**), (**A**), (**d**), (**P**), (**c**) and (**q**) are known, then its Utilized or Starting Capital must be:

U= F+T+Ad+{Q+P/[c-q]}//-$[1-v-i]

Rule-15437:

If both (**/**), (**U**), (**Q**), (**v**), (**i**), (**F**), (**T**), (**A**), (**d**), (**P**), (**c**) and (**q**) are known, then its Sales or Revenue Planned is:

$= (F+T+Ad+{Q+P/[c-q]}//-U)/[1-v-i]

Rule-15438:

If both (**/**), (**U**), (**$**), (**Q**), (**i**), (**F**), (**T**), (**A**), (**d**), (**P**), (**c**) and (**q**) are known, then its Variable Portion Planned is:

v= 1-i-(F+T+Ad+{Q+P/[c-q]}//-U)/$

Rule-15439:

If both (**/**), (**U**), (**$**), (**v**), (**Q**), (**F**), (**T**), (**A**), (**d**), (**P**), (**c**) and (**q**) are known, then its Interest Portion Planned is:

i= 1-v-(F+T+Ad+{Q+P/[c-q]}//-U)/$

Steve Asikin ISBN 14: 978-1514685136, ISBN 10: 1514685132

Rule-15440:

If both (**/**), (**U**), (**$**), (**v**), (**i**), (**Q**), (**T**), (**A**), (**d**), (**P**), (**c**) and (**q**) are known, then its Fixed Cost Planned is:

F= **$**[1-**v**-**i**]-**T**-**Ad**}-{**Q**+**P**/[**c**-**q**]}/**/**-**U**

Rule-15441:

If both (**/**), (**U**), (**$**), (**v**), (**i**), (**F**), (**Q**), (**A**), (**d**), (**P**), (**c**) and (**q**) are known, then its Tax Planned is:

T= **$**[1-**v**-**i**]-**F**-**Ad**}-{**Q**+**P**/[**c**-**q**]}/**/**-**U**

Rule-15442:

If both (**/**), (**U**), (**$**), (**v**), (**i**), (**F**), (**T**), (**Q**), (**d**), (**P**), (**c**) and (**q**) are known, then its After Tax Income Planned is:

A= (**$**[1-**v**-**i**]-**F**-**T**-{**Q**+**P**/[**c**-**q**]}/**/**-**U**)/**d**

Rule-15443:

If both (**/**), (**U**), (**$**), (**v**), (**i**), (**F**), (**T**), (**A**), (**Q**), (**P**), (**c**) and (**q**) are known, then its Dividend Payout Planned is:

d= (**$**[1-**v**-**i**]-**F**-**T**-{**Q**+**P**/[**c**-**q**]}/**/**-**U**)/**A**

Rule-15444:

If both (**/**), (**U**), (**$**), (**v**), (**i**), (**F**), (**T**), (**A**), (**d**), (**Q**), (**c**) and (**q**) are known, then its Procured Inventory Planned is:

P= [**c**-**q**](**/**{**U**+**$**[1-**v**-**i**]-**F**-**T**-**Ad**}-**Q**)

Steve Asikin ISBN 14: 978-1514685136, ISBN 10: 1514685132

Rule-15445:

If both (**/**), (**U**), (**S**), (**v**), (**i**), (**F**), (**T**), (**A**), (**d**), (**P**), (**Q**) and (**q**) are known, then its Current Ratio Planned is:

c= **q**+**P**/(**/**{**U**+**S**[1-**v**-**i**]-**F**-**T**-**Ad**}-**Q**)

Rule-15446:

If both (**/**), (**U**), (**S**), (**v**), (**i**), (**F**), (**T**), (**A**), (**d**), (**P**), (**c**) and (**Q**) are known, then its Quick or Acid Test Ratio Planned is:

q= **c**-**P**/(**/**{**U**+**S**[1-**v**-**i**]-**F**-**T**-**Ad**}-**Q**)

Rule-15447:

If both (**/**), (**U**), (**S**), (**v**), (**i**), (**F**), (**T**), (**A**), (**d**), (**V**), (**p**), (**c**) and (**q**) are known, then its Quoted Longterm Debt Planned is:

Q= **/**{**U**+**S**[1-**v**-**i**]-**F**-**T**-**Ad**}-**Vp**/{360[**c**-**q**]}

Rule-15448:

If both (**Q**), (**U**), (**S**), (**v**), (**i**), (**F**), (**T**), (**A**), (**d**), (**V**), (**p**), (**c**) and (**q**) are known, then its Leverage or Gearing Ratio Planned is:

/= (**Q**+**Vp**/{360[**c**-**q**]})/{**U**+**S**[1-**v**-**i**]-**F**-**T**-**Ad**}

Rule-15449:

If both (**/**), (**Q**), (**S**), (**v**), (**i**), (**F**), (**T**), (**A**), (**d**), (**V**), (**p**), (**c**) and (**q**) are known, then its Utilized or Starting Capital must be:

U= **F**+**T**+**Ad**+(**Q**+**Vp**/{360[**c**-**q**]})/**/**-**S**[1-**v**-**i**]

Steve Asikin ISBN 14: 978-1514685136, ISBN 10: 1514685132

Rule-15450:

If both (/), (**U**), (**Q**), (**v**), (**i**), (**F**), (**T**), (**A**), (**d**), (**V**), (**p**), (**c**) and (**q**) are known, then its Sales or Revenue Planned is:

$= [F+T+Ad+(Q+Vp/{360[c-q]})//U]/[1-v-i]

Rule-15451:

If both (/), (**U**), ($), (**Q**), (**i**), (**F**), (**T**), (**A**), (**d**), (**V**), (**p**), (**c**) and (**q**) are known, then its Variable Portion Planned is:

v= V/$

or it can also be found as

v= 1-i-[F+T+Ad+(Q+Vp/{360[c-q]})//U]/$

Rule-15452:

If both (/), (**U**), ($), (**v**), (**Q**), (**F**), (**T**), (**A**), (**d**), (**V**), (**p**), (**c**) and (**q**) are known, then its Interest Portion Planned is:

i= 1-v-[F+T+Ad+(Q+Vp/{360[c-q]})//U]/$

Rule-15453:

If both (/), (**U**), ($), (**v**), (**i**), (**Q**), (**T**), (**A**), (**d**), (**V**), (**p**), (**c**) and (**q**) are known, then its Fixed Portion Planned is:

F= $[1-v-i]-T-Ad}-(Q+Vp/{360[c-q]})//U

Rule-15454:

If both (/), (**U**), ($), (**v**), (**i**), (**F**), (**Q**), (**A**), (**d**), (**V**), (**p**), (**c**) and (**q**) are known, then its Tax Planned is:

T= $[1-v-i]-F-Ad}-(Q+Vp/{360[c-q]})//U

Steve Asikin ISBN 14: 978-1514685136, ISBN 10: 1514685132

Rule-15455:

If both (**/**), (**U**), (**$**), (**v**), (**i**), (**F**), (**T**), (**Q**), (**d**), (**V**), (**p**), (**c**) and (**q**) are known, then its After Tax Income Planned is:

A= [**$**[1-**v**-**i**]-**F**-**T**-(**Q**+**Vp**/{360[**c**-**q**]})/**/**-**U**]/**d**

Rule-15456:

If both (**/**), (**U**), (**$**), (**v**), (**i**), (**F**), (**T**), (**A**), (**Q**), (**V**), (**p**), (**c**) and (**q**) are known, then its Dividend Payout Planned is:

d= [**$**[1-**v**-**i**]-**F**-**T**-(**Q**+**Vp**/{360[**c**-**q**]})/**/**-**U**]/**A**

Rule-15457:

If both (**/**), (**U**), (**$**), (**v**), (**i**), (**F**), (**T**), (**A**), (**d**), (**Q**), (**p**), (**c**) and (**q**) are known, then its Variable Cost Planned is:

V= 360[**c**-**q**](**/**{**U**+**$**[1-**v**-**i**]-**F**-**T**-**Ad**}-**Q**)/**d**

Rule-15458:

If both (**/**), (**U**), (**$**), (**v**), (**i**), (**F**), (**T**), (**A**), (**d**), (**V**), (**Q**), (**c**) and (**q**) are known, then its Procured Inventory Days Planned:

p= 360[**c**-**q**](**/**{**U**+**$**[1-**v**-**i**]-**F**-**T**-**Ad**}-**Q**)/**V**

Rule-15459:

If both (**/**), (**U**), (**$**), (**v**), (**i**), (**F**), (**T**), (**A**), (**d**), (**V**), (**p**), (**Q**) and (**q**) are known, then its Current Ratio Planned is:

c= **q**+**Vp**/[360(**/**{**U**+**$**[1-**v**-**i**]-**F**-**T**-**Ad**}-**Q**)]

Steve Asikin ISBN 14: 978-1514685136, ISBN 10: 1514685132

<u>Rule-15460</u>:
If both (I), (U), (S), (v), (i), (F), (T), (A), (d), (V), (p), (c) and (Q) are known, then its Quick or Acid Test Ratio Planned is:
$$q= c\text{-}Vp/[360(I\{U+S[1\text{-}v\text{-}i]\text{-}F\text{-}T\text{-}Ad\}\text{-}Q)]$$

<u>Rule-15461</u>:
If both (I), (U), (S), (v), (i), (F), (T), (A), (d), (p), (c) and (q) are known, then its Quoted Longterm Debt Planned is:
$$Q= I\{U+S[1\text{-}v\text{-}i]\text{-}F\text{-}T\text{-}Ad\}\text{-}Svp/\{360[c\text{-}q]\}$$

<u>Rule-15462</u>:
If both (Q), (U), (S), (v), (i), (F), (T), (A), (d), (p), (c) and (q) are known, then its Leverage or Gearing Ratio Planned is:
$$I= (Q+Svp/\{360[c\text{-}q]\})/\{U+S[1\text{-}v\text{-}i]\text{-}F\text{-}T\text{-}Ad\}$$

<u>Rule-15463</u>:
If both (I), (Q), (S), (v), (i), (F), (T), (A), (d), (p), (c) and (q) are known, then its Utilized or Starting Capital must be:
$$U= F+T+Ad+(Q+Svp/\{360[c\text{-}q]\})/I\text{-}S[1\text{-}v\text{-}i]$$

<u>Rule-15464</u>:
If both (I), (U), (Q), (v), (i), (F), (T), (A), (d), (p), (c) and (q) are known, then its Sales or Revenue Planned is:
$$S= \{Q\text{-}I[U\text{-}F\text{-}T\text{-}Ad]\text{-}Svp/\{360[c\text{-}q]\})/\{I[1\text{-}v\text{-}i]\}$$

Steve Asikin ISBN 14: 978-1514685136, ISBN 10: 1514685132

Rule-15465:

If both (**/**), (**U**), (**$**), (**Q**), (**i**), (**F**), (**T**), (**A**), (**d**), (**p**), (**c**) and (**q**) are known, then its Variable Portion Planned is:

v= (/{**U**+**$**[1-**i**] -**F**-**T**-**Ad**}-**Q**)/[**$**(**p**/{360[**c**-**q**]}+/)]

Rule-15466:

If both (**/**), (**U**), (**$**), (**v**), (**Q**), (**F**), (**T**), (**A**), (**d**), (**p**), (**c**) and (**q**) are known, then its Interest Portion Planned is:

i= 1-**v**-[**F**+**T**+**Ad**+(**Q**+**$vp**/{360[**c**-**q**]})/**/**-**U**]/**$**

Rule-15467:

If both (**/**), (**U**), (**$**), (**v**), (**i**), (**Q**), (**T**), (**A**), (**d**), (**p**), (**c**) and (**q**) are known, then its Fixed Cost Planned is:

F= **$**[1-**v**-**i**]-**T**-**Ad**}-(**Q**+**$vp**/{360[**c**-**q**]})/**/**-**U**

Rule-15468:

If both (**/**), (**U**), (**$**), (**v**), (**i**), (**F**), (**Q**), (**A**), (**d**), (**p**), (**c**) and (**q**) are known, then its Tax Planned is:

T= **$**[1-**v**-**i**]-**F**-**Ad**}-(**Q**+**$vp**/{360[**c**-**q**]})/**/**-**U**

Rule-15469:

If both (**/**), (**U**), (**$**), (**v**), (**i**), (**F**), (**T**), (**Q**), (**d**), (**p**), (**c**) and (**q**) are known, then its After Tax Income Planned is:

A= [**$**[1-**v**-**i**]-**F**-**T**-(**Q**+**$vp**/{360[**c**-**q**]})/**/**-**U**]/**d**

Steve Asikin ISBN 14: 978-1514685136, ISBN 10: 1514685132

Rule-15470:

If both (**/**), (**U**), (**$**), (**v**), (**i**), (**F**), (**T**), (**A**), (**Q**), (**p**), (**c**) and (**q**) are known, then its Dividend Payout Planned is:

d= [**$**[1-**v**-**i**]-**F**-**T**-(**Q**+**$vp**/{360[**c**-**q**]})/**/**-**U**]/**A**

Rule-15471:

If both (**/**), (**U**), (**$**), (**v**), (**i**), (**F**), (**T**), (**A**), (**d**), (**Q**), (**c**) and (**q**) are known, then its Procured Inventory Days Planned is:

p= 360[**c**-**q**](**/**{**U**+**$**[1-**v**-**i**]-**F**-**T**-**Ad**}-**Q**)/[**$v**]

Rule-15472:

If both (**/**), (**U**), (**$**), (**v**), (**i**), (**F**), (**T**), (**A**), (**d**), (**p**), (**Q**) and (**q**) are known, then its Current Ratio Planned is:

c= **q**+**$vp**/[360(**/**{**U**+**$**[1-**v**-**i**]-**F**-**T**-**Ad**}-**Q**)]

Rule-15473:

If both (**/**), (**U**), (**$**), (**v**), (**i**), (**F**), (**T**), (**A**), (**d**), (**p**), (**c**) and (**Q**) are known, then its Quick or Acid test Ratio Planned is:

q= **c**-**$vp**/[360(**/**{**U**+**$**[1-**v**-**i**]-**F**-**T**-**Ad**}-**Q**)]

Rule-15474:

If both (**/**), (**U**), (**$**), (**v**), (**i**), (**F**), (**T**), (**A**), (**d**), (**$'**), (**p**), (**s**), (**c**) and (**q**) are known, then its Quoted Longterm Debt Planned is:

Q= **/**{**U**+**$**[1-**v**-**i**]-**F**-**T**-**Ad**}-**$'vp**[1+**s**]/{360[**c**-**q**]}

Steve Asikin ISBN 14: 978-1514685136, ISBN 10: 1514685132

Rule-15475:

If both (**Q**), (**U**), (**$**), (**v**), (**i**), (**F**), (**T**), (**A**), (**d**), (**$'**), (**p**), (**s**), (**c**) and (**q**) are known, then its Leverage or Gearing Ratio Planned is:

/= (**Q**+**$'vp**[1+**s**]/{360[**c-q**]})
/{**U**+**$**[1-**v-i**]-**F-T-Ad**}

Rule-15476:

If both (**/**), (**Q**), (**$**), (**v**), (**i**), (**F**), (**T**), (**A**), (**d**), (**$'**), (**p**), (**s**), (**c**) and (**q**) are known, then its Utilized or Starting Capital must be:

U= **F**+**T**+**Ad**+(**Q**+**$'vp**[1+**s**]/{360[**c-q**]})/**/**-**$**[1-**v-i**]

Rule-15477:

If both (**/**), (**U**), (**Q**), (**v**), (**i**), (**F**), (**T**), (**A**), (**d**), (**$'**), (**p**), (**s**), (**c**) and (**q**) are known, then its Sales or Revenue Planned is:

$**= {**Q**-**/**[**U-F-T-Ad**]-**$'vp[1+**s**]/{360[**c-q**]})
/{**/**[1-**v-i**]}

Rule-15478:

If both (**/**), (**U**), (**$**), (**Q**), (**i**), (**F**), (**T**), (**A**), (**d**), (**$'**), (**p**), (**s**), (**c**) and (**q**) are known, then its Variable Portion Planned is:

v= **V**/**$**

or it can also be found as

v= (**/**{**U**+**$**[1-**i**]-**F-T-Ad**}-**Q**)
/(**$'p**[1+**s**]/{360[**c-q**]}+**$/**)

Steve Asikin ISBN 14: 978-1514685136, ISBN 10: 1514685132

Rule-15479:

If both (**/**), (**U**), (**$**), (**v**), (**Q**), (**F**), (**T**), (**A**), (**d**), (**$'**), (**p**), (**s**), (**c**) and (**q**) are known, then its Interest Portion Planned is:

i= 1-**v**-[**F**+**T**+**Ad**+(**Q**+**$'vp**[1+**s**]/{360[**c**-**q**]})/**/U**]/**$**

Rule-15480:

If both (**/**), (**U**), (**$**), (**v**), (**i**), (**Q**), (**T**), (**A**), (**d**), (**$'**), (**p**), (**s**), (**c**) and (**q**) are known, then its Fixed Cost Planned is:

F= **$**[1-**v**-**i**]-**T**-**Ad**}-(**Q**+**$'vp**[1+**s**]/{360[**c**-**q**]})/**/U**

Rule-15481:

If both (**/**), (**U**), (**$**), (**v**), (**i**), (**F**), (**Q**), (**A**), (**d**), (**$'**), (**p**), (**s**), (**c**) and (**q**) are known, then its Tax Planned is:

T= **$**[1-**v**-**i**]-**F**-**Ad**}-(**Q**+**$'vp**[1+**s**]/{360[**c**-**q**]})/**/U**

Rule-15482:

If both (**/**), (**U**), (**$**), (**v**), (**i**), (**F**), (**T**), (**Q**), (**d**), (**$'**), (**p**), (**s**), (**c**) and (**q**) are known, then its After Tax Income Planned is:

A= [**$**[1-**v**-**i**]-**F**-**T**-(**Q**+**$'vp**[1+**s**]/{360[**c**-**q**]})/**/U**]/**d**

Rule-15483:

If both (**/**), (**U**), (**$**), (**v**), (**i**), (**F**), (**T**), (**A**), (**Q**), (**$'**), (**p**), (**s**), (**c**) and (**q**) are known, then its Dividend Payout Planned is:

d= [**$**[1-**v**-**i**]-**F**-**T**-(**Q**+**$'vp**[1+**s**]/{360[**c**-**q**]})/**/U**]/**A**

Steve Asikin ISBN 14: 978-1514685136, ISBN 10: 1514685132

Rule-15484:

If both (**/**), (**U**), (**$**), (**v**), (**i**), (**F**), (**T**), (**A**), (**d**), (**Q**), (**p**), (**s**), (**c**) and (**q**) are known, then its Sales Past must be:

$'**= (**/**{**U**+**$[1-**v**-**i**]-**F**-**T**-**Ad**}-**Q**)
/(**vp**[1+**s**]/{360[**c**-**q**]})

Rule-15485:

If both (**/**), (**U**), (**$**), (**v**), (**i**), (**F**), (**T**), (**A**), (**d**), (**$'**), (**Q**), (**s**), (**c**) and (**q**) are known, then its Procured Inventory Days Planned:

p= 360[**c**-**q**](**/**{**U**+**$**[1-**v**-**i**]-**F**-**T**-**Ad**}-**Q**)/{**$'v**[1+**s**]}

Rule-15486:

If both (**/**), (**U**), (**$**), (**v**), (**i**), (**F**), (**T**), (**A**), (**d**), (**$'**), (**p**), (**Q**), (**c**) and (**q**) are known, then its Sales Growth Planned is:

s= **$**/**$'**-1

or it can also be found as

s= (**/**{**U**+**$**[1-**v**-**i**]-**F**-**T**-**Ad**}-**Q**)/(**$'vp**/{360[**c**-**q**]})-1

Rule-15487:

If both (**/**), (**U**), (**$**), (**v**), (**i**), (**F**), (**T**), (**A**), (**d**), (**$'**), (**p**), (**s**), (**Q**) and (**q**) are known, then its Current Ratio Planned is:

c= **q**+**$'vp**[1+**s**]/[360(**/**{**U**+**$**[1-**v**-**i**]-**F**-**T**-**Ad**}-**Q**)]

Rule-15488:

If both (**/**), (**U**), (**$**), (**v**), (**i**), (**F**), (**T**), (**A**), (**d**), (**$'**), (**p**), (**s**), (**c**) and (**Q**) are known, then its Quick or Acid test Ratio Planned is:

q= **c**-**$'vp**[1+**s**]/[360(**/**{**U**+**$**[1-**v**-**i**]-**F**-**T**-**Ad**}-**Q**)]

Steve Asikin ISBN 14: 978-1514685136, ISBN 10: 1514685132

Rule-15489:

If both (**/**), (**U**), (**S**), (**v**), (**i**), (**F**), (**t**), (**D**) and (**X**) are known, then its Quoted Longterm Debt Planned is:

Q= **/**(**U**+{**S**[1-**v**-**i**]-**F**}[1-**t**]-**D**)-**X**

Rule-15490:

If both (**Q**), (**U**), (**S**), (**v**), (**i**), (**F**), (**t**), (**D**) and (**X**) are known, then its Leverage or Gearing Ratio Planned is:

/= [**Q**+**X**]/(**U**+{**S**[1-**v**-**i**]-**F**}[1-**t**]-**D**)

Rule-15491:

If both (**/**), (**Q**), (**S**), (**v**), (**i**), (**F**), (**t**), (**D**) and (**X**) are known, then its Utilized or Starting Capital must be:

U= **D**+[**Q**+**X**]/**/**-{**S**[1-**v**-**i**]-**F**}[1-**t**]

Rule-15492:

If both (**/**), (**U**), (**Q**), (**v**), (**i**), (**F**), (**t**), (**D**) and (**X**) are known, then its Sales or Revenue Planned is:

S= (**F**+{**D**+[**Q**+**X**]/**/**-**U**}/[1-**t**])[1-**v**-**i**]

Rule-15493:

If both (**/**), (**U**), (**S**), (**Q**), (**i**), (**F**), (**t**), (**D**) and (**X**) are known, then its Variable Portion Planned is:

v= 1-**i**-(**F**+{**D**+[**Q**+**X**]/**/**-**U**}/[1-**t**])/**S**

Rule-15494:

If both (**/**), (**U**), (**S**), (**v**), (**Q**), (**F**), (**t**), (**D**) and (**X**) are known, then its Interest Portion Planned is:

i= 1-**v**-(**F**+{**D**+[**Q**+**X**]/**/**-**U**}/[1-**t**])/**S**

Steve Asikin ISBN 14: 978-1514685136, ISBN 10: 1514685132

Rule-15495:

If both (/), (U), ($), (v), (i), (Q), (t), (D) and (X) are known, then its Fixed Cost Planned is:

F= $[1-v-i]-{D+[Q+X]//-U}/[1-t]

Rule-15496:

If both (/), (U), ($), (v), (i), (F), (Q), (D) and (X) are known, then its Tax Rate Planned is:

t= 1-{D+[Q+X]/-U}/{$[1-v-i]-F}

Rule-15497:

If both (/), (U), ($), (v), (i), (F), (t), (Q) and (X) are known, then its Dividend Planned is:

D= U+{$[1-v-i]-F}[1-t]-[Q+X]//

Rule-15498:

If both (/), (U), ($), (v), (i), (F), (t), (D) and (Q) are known, then its Xpress or Current Debt Planned is:

X= /(U+{$[1-v-i]-F}[1-t]-D)-Q

Rule-15499:

If both (/), (U), ($), (v), (i), (F), (t), (D), (P), (c) and (q) are known, then its Quoted Longterm Debt Planned is:

Q= /(U+{$[1-v-i]-F}[1-t]-D)-P/[c-q]

Rule-15500:

If both (Q), (U), ($), (v), (i), (F), (t), (D), (P), (c) and (q) are known, then its Leverage or Gearing Ratio Planned is:

/= {Q+P/[c-q]}/(U+{$[1-v-i]-F}[1-t]-D)

Steve Asikin ISBN 14: 978-1514685136, ISBN 10: 1514685132

Rule-15501:

If both (/), (**Q**), (**$**), (**v**), (**i**), (**F**), (**t**), (**D**), (**P**), (**c**) and (**q**) are known, then its Utilized or Starting Capital must be:

U= **D**+{**Q**+**P**/[**c**-**q**]}//-{**$**[1-**v**-**i**]-**F**}[1-**t**]

Rule-15502:

If both (/), (**U**), (**Q**), (**v**), (**i**), (**F**), (**t**), (**D**), (**P**), (**c**) and (**q**) are known, then its Sales or Revenue Planned is:

$= **[F**+(**D**+{**Q**+**P**/[**c**-**q**]}//-**U**)/[1-**t**]**]**/[1-**v**-**i**]

Rule-15503:

If both (/), (**U**), (**$**), (**Q**), (**i**), (**F**), (**t**), (**D**), (**P**), (**c**) and (**q**) are known, then its Variable Portion Planned is:

v= 1-**i**-**[F**+(**D**+{**Q**+**P**/[**c**-**q**]}//-**U**)/[1-**t**]**]**/**$**

Rule-15504:

If both (/), (**U**), (**$**), (**v**), (**Q**), (**F**), (**t**), (**D**), (**P**), (**c**) and (**q**) are known, then its Interest Portion Planned is:

i= 1-**v**-**[F**+(**D**+{**Q**+**P**/[**c**-**q**]}//-**U**)/[1-**t**]**]**/**$**

Rule-15505:

If both (/), (**U**), (**$**), (**v**), (**i**), (**Q**), (**t**), (**D**), (**P**), (**c**) and (**q**) are known, then its Fixed Cost Planned is:

F= **$**[1-**v**-**i**]-(**D**+{**Q**+**P**/[**c**-**q**]}//-**U**)/[1-**t**]

Rule-15506:

If both (/), (**U**), (**$**), (**v**), (**i**), (**F**), (**Q**), (**D**), (**P**), (**c**) and (**q**) are known, then its Tax Rate Planned is:

t= 1-(**D**+{**Q**+**P**/[**c**-**q**]}//-**U**)/{**$**[1-**v**-**i**]-**F**}

Steve Asikin ISBN 14: 978-1514685136, ISBN 10: 1514685132

Rule-15507:

If both (**/**), (**U**), (**S**), (**v**), (**i**), (**F**), (**t**), (**Q**), (**P**), (**c**) and (**q**) are known, then its Dividend Planned is:

D= **U**+{**S**[1-**v**-**i**]-**F**}[1-**t**]-{**Q**+**P**/[**c**-**q**]}/**/**

Rule-15508:

If both (**/**), (**U**), (**S**), (**v**), (**i**), (**F**), (**t**), (**D**), (**Q**), (**c**) and (**q**) are known, then its Procured Inventory Planned is:

P= [**c**-**q**][**/**(**U**+{**S**[1-**v**-**i**]-**F**}[1-**t**]-**D**)-**Q**]

Rule-15509:

If both (**/**), (**U**), (**S**), (**v**), (**i**), (**F**), (**t**), (**D**), (**P**), (**Q**) and (**q**) are known, then its Current Ratio Planned is:

c= **q**+**P**/[**/**(**U**+{**S**[1-**v**-**i**]-**F**}[1-**t**]-**D**)-**Q**]

Rule-15510:

If both (**/**), (**U**), (**S**), (**v**), (**i**), (**F**), (**t**), (**D**), (**P**), (**c**) and (**Q**) are known, then its Quick or Acid Test Ratio Planned is:

q= **c**-**P**/[**/**(**U**+{**S**[1-**v**-**i**]-**F**}[1-**t**]-**D**)-**Q**]

Rule-15511:

If both (**/**), (**U**), (**S**), (**v**), (**i**), (**F**), (**t**), (**D**), (**V**), (**p**), (**c**) and (**q**) are known, then its Quoted Longterm Debt Planned is:

Q= **/**(**U**+{**S**[1-**v**-**i**]-**F**}[1-**t**]-**D**)-**Vp**/{360[**c**-**q**]}

Steve Asikin ISBN 14: 978-1514685136, ISBN 10: 1514685132

Rule-15512:

If both (**Q**), (**U**), (**$**), (**v**), (**i**), (**F**), (**t**), (**D**), (**V**), (**p**), (**c**) and (**q**) are known, then its Leverage or Gearing Ratio Planned is:

/= (Q+Vp/{360[c-q]})/(U+{$[1-v-i]-F}[1-t]-D)

Rule-15513:

If both (/), (**Q**), (**$**), (**v**), (**i**), (**F**), (**t**), (**D**), (**V**), (**p**), (**c**) and (**q**) are known, then its Utilized or Starting Capital must be:

U= D+(Q+Vp/{360[c-q]})/ /-{$[1-v-i]-F}[1-t]

Rule-15514:

If both (/), (**U**), (**Q**), (**v**), (**i**), (**F**), (**t**), (**D**), (**V**), (**p**), (**c**) and (**q**) are known, then its Sales or Revenue Planned is:

$= {F+[D+(Q+Vp/{360[c-q]})/ /-U]/[1-t]}/[1-v-i]

Rule-15515:

If both (/), (**U**), (**$**), (**Q**), (**i**), (**F**), (**t**), (**D**), (**V**), (**p**), (**c**) and (**q**) are known, then its Variable Portion Planned is:

v= V/$

or it can also be found as

v= 1-i-{F+[D+(Q+Vp/{360[c-q]})/ /-U]/[1-t]}/$

Steve Asikin ISBN 14: 978-1514685136, ISBN 10: 1514685132

Rule-15516:

If both (/), (U), ($), (v), (Q), (F), (t), (D), (V), (p), (c) and (q) are known, then its Interest Portion Planned is:

i= 1-v-{F+[D+(Q+Vp/{360[c-q]})/-U]/[1-t]}/$

Rule-15517:

If both (/), (U), ($), (v), (i), (Q), (t), (D), (V), (p), (c) and (q) are known, then its Fixed Cost Planned is:

F= $[1-v-i]-[D+(Q+Vp/{360[c-q]})/-U]/[1-t]

Rule-15518:

If both (/), (U), ($), (v), (i), (F), (Q), (D), (V), (p), (c) and (q) are known, then its Tax Rate Planned is:

t= 1-[D+(Q+Vp/{360[c-q]})/-U]/{$[1-v-i]-F}

Rule-15519:

If both (/), (U), ($), (v), (i), (F), (t), (Q), (V), (p), (c) and (q) are known, then its Dividend Planned is:

D= U+{$[1-v-i]-F}[1-t]- (Q+Vp/{360[c-q]})//

Rule-15520:

If both (/), (U), ($), (v), (i), (F), (t), (D), (Q), (p), (c) and (q) are known, then its Variable Portion Planned is:

V= 360[c-q][/(U+{$[1-v-i]-F}[1-t]-D)-Q]/p

Steve Asikin ISBN 14: 978-1514685136, ISBN 10: 1514685132

Rule-15521:

If both (/), (U), ($), (v), (i), (F), (t), (D), (V), (Q), (c) and (q) are known, then its Procured Inventory Days Planned:

p= 360[c-q][/(U+{$[1-v-i]-F}[1-t]-D)-Q]/V

Rule-15522:

If both (/), (U), ($), (v), (i), (F), (t), (D), (V), (p), (Q) and (q) are known, then its Current Ratio Planned is:

c= q+Vp/{360[/(U+{$[1-v-i]-F}[1-t]-D)-Q]}

Rule-15523:

If both (/), (U), ($), (v), (i), (F), (t), (D), (V), (p), (c) and (Q) are known, then its Quick or Acid Test Ratio Planned is:

q= c-Vp/{360[/(U+{$[1-v-i]-F}[1-t]-D)-Q]}

Rule-15524:

If both (/), (U), ($), (v), (i), (F), (t), (D), (p), (c) and (q) are known, then its Quoted Longterm Debt Planned is:

Q= /(U+{$[1-v-i]-F}[1-t]-D)-$vp/{360[c-q]}

Rule-15525:

If both (Q), (U), ($), (v), (i), (F), (t), (D), (p), (c) and (q) are known, then its Leverage or Gearing Ratio Planned is:

/= (Q+$vp/{360[c-q]})/(U+{$[1-v-i]-F}[1-t]-D)

Steve Asikin ISBN 14: 978-1514685136, ISBN 10: 1514685132

Rule-15526:

If both (**/**), (**Q**), (**$**), (**v**), (**i**), (**F**), (**t**), (**D**), (**p**), (**c**) and (**q**) are known, then its Utilized or Starting Capital must be:

U= **D**+(**Q**+**$vp**/{360[**c**-**q**]})/**/**-{**$**[1-**v**-**i**]-**F**}[1-**t**]

Rule-15527:

If both (**/**), (**U**), (**Q**), (**v**), (**i**), (**F**), (**t**), (**D**), (**p**), (**c**) and (**q**) are known, then its Sales or Revenue Planned is:

$= (**/**{**U**-**F**[1-**t**]-**D**}-**Q**)
/(**vp**/{360[**c**-**q**]}-**/**[1-**v**-**i**][1-**t**])

Rule-15528:

If both (**/**), (**U**), (**$**), (**Q**), (**i**), (**F**), (**t**), (**D**), (**p**), (**c**) and (**q**) are known, then its Variable Portion Planned is:

v= **[/**(**U**+{**$**[1-**i**]-**F**}[1-**t**]-**D**)-**Q]**
/**[$**(**p**/{360[**c**-**q**]}+**/**[1-**t**])**]**

Rule-15529:

If both (**/**), (**U**), (**$**), (**v**), (**Q**), (**F**), (**t**), (**D**), (**p**), (**c**) and (**q**) are known, then its Interest Portion Planned is:

i= 1-**v**-**{F**+**[D**+(**Q**+**$vp**/{360[**c**-**q**]})/**/**-**U]**/[1-**t**]**}**/**$**

Rule-15530:

If both (**/**), (**U**), (**$**), (**v**), (**i**), (**Q**), (**t**), (**D**), (**p**), (**c**) and (**q**) are known, then its Fixed Cost Planned is:

F= **$**[1-**v**-**i**]-**[D**+(**Q**+**$vp**/{360[**c**-**q**]})/**/**-**U]**/[1-**t**]

Steve Asikin ISBN 14: 978-1514685136, ISBN 10: 1514685132

Rule-15531:

If both (**/**), (**U**), (**$**), (**v**), (**i**), (**F**), (**Q**), (**D**), (**p**), (**c**) and (**q**) are known, then its Tax Rate Planned is:

t= 1-[**D**+(**Q**+**$vp**/{360[**c-q**]})/**/-U**]/{**$**[1-**v-i**]-**F**}

Rule-15532:

If both (**/**), (**U**), (**$**), (**v**), (**i**), (**F**), (**t**), (**Q**), (**p**), (**c**) and (**q**) are known, then its Dividend Planned is:

D= **U**+{**$**[1-**v-i**]-**F**}[1-**t**]- (**Q**+**$vp**/{360[**c-q**]})/**/**

Rule-15533:

If both (**/**), (**U**), (**$**), (**v**), (**i**), (**F**), (**t**), (**D**), (**Q**), (**c**) and (**q**) are known, then its Procured Inventory Days Planned:

p= 360[**c-q**][**/**(**U**+{**$**[1-**v-i**]-**F**}[1-**t**]-**D**)-**Q**]/[**$v**]

Rule-15534:

If both (**/**), (**U**), (**$**), (**v**), (**i**), (**F**), (**t**), (**D**), (**p**), (**Q**) and (**q**) are known, then its Current Ratio Planned is:

c= **q**+**$vp**/{360[**/**(**U**+{**$**[1-**v-i**]-**F**}[1-**t**]-**D**)-**Q**]}

Rule-15535:

If both (**/**), (**U**), (**$**), (**v**), (**i**), (**F**), (**t**), (**D**), (**p**), (**c**) and (**Q**) are known, then its Quick or Acid Test Ratio Planned is:

q= **c-$vp**/{360[**/**(**U**+{**$**[1-**v-i**]-**F**}[1-**t**]-**D**)-**Q**]}

Steve Asikin ISBN 14: 978-1514685136, ISBN 10: 1514685132

Rule-15536:

If both (**/**), (**U**), (**S**), (**v**), (**i**), (**F**), (**t**), (**D**), (**S'**), (**p**), (**s**), (**c**) and (**q**) are known, then its Quoted Longterm Debt Planned is:

Q= **/**(**U**+{**S**[1-**v**-**i**]-**F**}[1-**t**]-**D**)-**S'vp**[1+**s**]/{360[**c**-**q**]}

Rule-15537:

If both (**Q**), (**U**), (**S**), (**v**), (**i**), (**F**), (**t**), (**D**), (**S'**), (**p**), (**s**), (**c**) and (**q**) are known, then its Leverage or Gearing Ratio Planned is:

/= (**Q**+**S'vp**[1+**s**]/{360[**c**-**q**]})
/(**U**+{**S**[1-**v**-**i**]-**F**}[1-**t**]-**D**)

Rule-15538:

If both (**/**), (**Q**), (**S**), (**v**), (**i**), (**F**), (**t**), (**D**), (**S'**), (**p**), (**s**), (**c**) and (**q**) are known, then its Utilized or Starting Cap[ital must be:

U= **D**+(**Q**+**S'vp**[1+**s**]/{360[**c**-**q**]})/**/**
-{**S**[1-**v**-**i**]-**F**}[1-**t**]

Rule-15539:

If both (**/**), (**U**), (**Q**), (**v**), (**i**), (**F**), (**t**), (**D**), (**S'**), (**p**), (**s**), (**c**) and (**q**) are known, then its Sales or Revenue Planned is:

S= (**Q**-**/**{**U**-**F**[1-**t**]-**D**}+**S'vp**[1+**s**]/{360[**c**-**q**]})
/{**/**[1-**v**-**i**][1-**t**]}

Steve Asikin ISBN 14: 978-1514685136, ISBN 10: 1514685132

Rule-15540:

If both (/), (U), (S), (Q), (i), (F), (t), (D), (S'), (p), (s), (c) and (q) are known, then its Variable Portion Planned is:

v= [/(U+{S[1-i]-F}[1-t]-D)-Q]
/(S''p[1+s]/{360[c-q]}-S/[1-t])

Rule-15541:

If both (/), (U), (S), (v), (Q), (F), (t), (D), (S'), (p), (s), (c) and (q) are known, then its Interest Portion Planned is:

i= 1-v-{F+[D+(Q+S'vp[1+s]/{360[c-q]})//-U]
/[1-t]}/S

Rule-15542:

If both (/), (U), (S), (v), (i), (Q), (t), (D), (S'), (p), (s), (c) and (q) are known, then its Fixed Cost Planned is:

F= S[1-v-i]-[D+(Q+S'vp[1+s]/{360[c-q]})//-U]
/[1-t]

Rule-15543:

If both (/), (U), (S), (v), (i), (F), (Q), (D), (S'), (p), (s), (c) and (q) are known, then its Tax Rate Planned is:

t= 1-[D+(Q+S'vp[1+s]/{360[c-q]})//-U]
/{S[1-v-i]-F}

Rule-15544:

If both (/), (U), (S), (v), (i), (F), (t), (Q), (S'), (p), (s), (c) and (q) are known, then its Dividend Planned is:

D= U+{S[1-v-i]-F}[1-t]
-(Q+S'vp[1+s]/{360[c-q]})//

Steve Asikin ISBN 14: 978-1514685136, ISBN 10: 1514685132

Rule-15545:

If both (**/**), (**U**), (**$**), (**v**), (**i**), (**F**), (**t**), (**D**), (**Q**), (**p**), (**s**), (**c**) and (**q**) are known, then its Sales Past must be:

$'**= [**/**(**U**+{**$[1-**v**-**i**]-**F**}[1-**t**]-**D**)-**Q**]
/(**vp**[1+**s**]/{360[**c**-**q**]})

Rule-15546:

If both (**/**), (**U**), (**$**), (**v**), (**i**), (**F**), (**t**), (**D**), (**$'**), (**Q**), (**s**), (**c**) and (**q**) are known, then its Procured Inventory Days Planned:

p= 360[**c**-**q**][**/**(**U**+{**$**[1-**v**-**i**]-**F**}[1-**t**]-**D**)-**Q**]
/{**$'v**[1+**s**]}

Rule-15547:

If both (**/**), (**U**), (**$**), (**v**), (**i**), (**F**), (**t**), (**D**), (**$'**), (**p**), (**Q**), (**c**) and (**q**) are known, then its Sales Growth Planned is:

s= **$**/**$'**-1

or it can also be found as

s= [**/**(**U**+{**$**[1-**v**-**i**]-**F**}[1-**t**]-**D**)-**Q**]
/(**$'vp**/{360[**c**-**q**]})-1

Rule-15548:

If both (**/**), (**U**), (**$**), (**v**), (**i**), (**F**), (**t**), (**D**), (**$'**), (**p**), (**s**), (**Q**) and (**q**) are known, then its Current Ratio Planned is:

c= **q**+**$'vp**[1+**s**]
/{360[**/**(**U**+{**$**[1-**v**-**i**]-**F**}[1-**t**]-**D**)-**Q**]}

Steve Asikin ISBN 14: 978-1514685136, ISBN 10: 1514685132

Rule-15549:

If both (/), (U), (S), (v), (i), (F), (t), (D), (S'), (p), (s), (c) and (Q) are known, then its Quick or Acid Test Ratio Planned is:

q= c-S'vp[1+s]/{360[/(U+{S[1-v-i]-F}[1-t]-D)-Q]}

Rule-15550:

If both (/), (U), (S), (v), (i), (F), (t), (A), (d) and (X) are known, then its Quoted Longterm Debt Planned is:

Q= /(U+{S[1-v-i]-F}[1-t]-Ad)-X

Rule-15551:

If both (Q), (U), (S), (v), (i), (F), (t), (A), (d) and (X) are known, then its Leverage or Gearing Ratio Planned is:

/= [Q+X]/(U+{S[1-v-i]-F}[1-t]-Ad)

Rule-15552:

If both (/), (Q), (S), (v), (i), (F), (t), (A), (d) and (X) are known, then its Utilized or Starting Capital must be:

U= Ad+[Q+X]//-{S[1-v-i]-F}[1-t]

Rule-15553:

If both (/), (U), (Q), (v), (i), (F), (t), (A), (d) and (X) are known, then its Sales or Revenue Planned is:

S= (F+{Ad+[Q+X]//-U}/[1-t])/[1-v-i]

Steve Asikin ISBN 14: 978-1514685136, ISBN 10: 1514685132

Rule-15554:

If both (/), (U), ($), (Q), (i), (F), (t), (A), (d) and (X) are known, then its Variable Portion Planned is:

v= 1-i-(F+{Ad+[Q+X]/ /U}/[1-t])/$

Rule-15555:

If both (/), (U), ($), (v), (Q), (F), (t), (A), (d) and (X) are known, then its Interest Portion Planned is:

i= 1-v-(F+{Ad+[Q+X]/ /U}/[1-t])/$

Rule-15556:

If both (/), (U), ($), (v), (i), (Q), (t), (A), (d) and (X) are known, then its Fixed Cost Planned is:

F= U+$[1-v-i]-{Ad+[Q+X]/ /}/[1-t]

Rule-15557:

If both (/), (U), ($), (v), (i), (F), (Q), (A), (d) and (X) are known, then its Tax Rate Planned is:

t= 1-{Ad+[Q+X]/ /U}/{$[1-v-i]-F}

Rule-15558:

If both (/), (U), ($), (v), (i), (F), (t), (Q), (d) and (X) are known, then its After Tax Income Planned is:

A= (U+{$[1-v-i]-F}[1-t]-[Q+X]/ /)/d

Rule-15559:

If both (/), (U), ($), (v), (i), (F), (t), (A), (Q) and (X) are known, then its Dividend Payout Planned is:

d= (U+{$[1-v-i]-F}[1-t]-[Q+X]/ /)/A

Steve Asikin ISBN 14: 978-1514685136, ISBN 10: 1514685132

Rule-15560:

If both (**/**), (**U**), (**$**), (**v**), (**i**), (**F**), (**t**), (**A**), (**d**) and (**Q**) are known, then its Xpress or Current Debt Planned is:

X= **/**(**U**+{**$**[1-**v**-**i**]-**F**}[1-**t**]-**Ad**)-**Q**

Rule-15561:

If both (**/**), (**U**), (**$**), (**v**), (**i**), (**F**), (**t**), (**A**), (**d**), (**P**), (**c**) and (**q**) are known, then its Quoted Longterm Debt Planned is:

Q= **/**(**U**+{**$**[1-**v**-**i**]-**F**}[1-**t**]-**Ad**)-**P**/[**c**-**q**]

Rule-15562:

If both (**Q**), (**U**), (**$**), (**v**), (**i**), (**F**), (**t**), (**A**), (**d**), (**P**), (**c**) and (**q**) are known, then its Leverage or Gearing Ratio Planned is:

/= {**Q**+**P**/[**c**-**q**]}/(**U**+{**$**[1-**v**-**i**]-**F**}[1-**t**]-**Ad**)

Rule-15563:

If both (**/**), (**Q**), (**$**), (**v**), (**i**), (**F**), (**t**), (**A**), (**d**), (**P**), (**c**) and (**q**) are known, then its Utilized or Starting Capital must be:

U= **Ad**+{**Q**+**P**/[**c**-**q**]}/**/**-{**$**[1-**v**-**i**]-**F**}[1-**t**]

Rule-15564:

If both (**/**), (**U**), (**Q**), (**v**), (**i**), (**F**), (**t**), (**A**), (**d**), (**P**), (**c**) and (**q**) are known, then its Sales or Revenue Planned is:

$= [**F**+(**Ad**+{**Q**+**P**/[**c**-**q**]}/**/**-**U**)/[1-**t**]]/[1-**v**-**i**]

Steve Asikin ISBN 14: 978-1514685136, ISBN 10: 1514685132

Rule-15565:

If both (/), (**U**), (**S**), (**Q**), (**i**), (**F**), (**t**), (**A**), (**d**), (**P**), (**c**) and (**q**) are known, then its Variable Portion Planned is:

v= 1-i-[F+(Ad+{Q+P/[c-q]}/*l*-U)/[1-t]]/S

Rule-15566:

If both (/), (**U**), (**S**), (**v**), (**Q**), (**F**), (**t**), (**A**), (**d**), (**P**), (**c**) and (**q**) are known, then its Interest Portion Planned is:

i= 1-v-[F+(Ad+{Q+P/[c-q]}/*l*-U)/[1-t]]/S

Rule-15567:

If both (/), (**U**), (**S**), (**v**), (**i**), (**Q**), (**t**), (**A**), (**d**), (**P**), (**c**) and (**q**) are known, then its Fixed Cost Planned is:

F= S[1-v-i]-(Ad+{Q+P/[c-q]}/*l*-U)/[1-t]

Rule-15568:

If both (/), (**U**), (**S**), (**v**), (**i**), (**F**), (**Q**), (**A**), (**d**), (**P**), (**c**) and (**q**) are known, then its Tax Rate Planned is:

t= 1-(Ad+{Q+P/[c-q]}/*l*-U)/{S[1-v-i]-F}

Rule-15569:

If both (/), (**U**), (**S**), (**v**), (**i**), (**F**), (**t**), (**Q**), (**d**), (**P**), (**c**) and (**q**) are known, then its After Tax Income Planned is:

A= (U+{S[1-v-i]-F}[1-t]-{Q+P/[c-q]}/*l*)/d

Steve Asikin ISBN 14: 978-1514685136, ISBN 10: 1514685132

Rule-15570:

If both (**/**), (**U**), (**$**), (**v**), (**i**), (**F**), (**t**), (**A**), (**Q**), (**P**), (**c**) and (**q**) are known, then its Dividend Portion Planned is:

d= (**U**+{**$**[1-**v**-**i**]-**F**}[1-**t**]-{**Q**+**P**/[**c**-**q**]}/**/**)/**A**

Rule-15571:

If both (**/**), (**U**), (**$**), (**v**), (**i**), (**F**), (**t**), (**A**), (**d**), (**Q**), (**c**) and (**q**) are known, then its Procured Inventory Planned is:

P= [**c**-**q**][**/**(**U**+{**$**[1-**v**-**i**]-**F**}[1-**t**]-**Ad**)-**Q**]

Rule-15572:

If both (**/**), (**U**), (**$**), (**v**), (**i**), (**F**), (**t**), (**A**), (**d**), (**P**), (**Q**) and (**q**) are known, then its Current Ratio Planned is:

c= **q**+**P**/[**/**(**U**+{**$**[1-**v**-**i**]-**F**}[1-**t**]-**Ad**)-**Q**]

Rule-15573:

If both (**/**), (**U**), (**$**), (**v**), (**i**), (**F**), (**t**), (**A**), (**d**), (**P**), (**c**) and (**Q**) are known, then its Quick or Acid Test Ratio Planned is:

q= **c**-**P**/[**/**(**U**+{**$**[1-**v**-**i**]-**F**}[1-**t**]-**Ad**)-**Q**]

Rule-15574:

If both (**/**), (**U**), (**$**), (**v**), (**i**), (**F**), (**t**), (**A**), (**d**), (**V**), (**p**), (**c**) and (**q**) are known, then its Quoted Longterm Debt Planned is:

Q= **/**(**U**+{**$**[1-**v**-**i**]-**F**}[1-**t**]-**Ad**)-**Vp**/{360[**c**-**q**]}

Steve Asikin ISBN 14: 978-1514685136, ISBN 10: 1514685132

Rule-15575:

If both (**Q**), (**U**), (**$**), (**v**), (**i**), (**F**), (**t**), (**A**), (**d**), (**V**), (**p**), (**c**) and (**q**) are known, then its Leverage or Gearing Ratio Planned is:

/= (Q+Vp/{360[c-q]})/(U+{$[1-v-i]-F}[1-t]-Ad)

Rule-15576:

If both (**/**), (**Q**), (**$**), (**v**), (**i**), (**F**), (**t**), (**A**), (**d**), (**V**), (**p**), (**c**) and (**q**) are known, then its Utilized or Starting Capital must be:

U= Ad+(Q+Vp/{360[c-q]})//-{$[1-v-i]-F}[1-t]

Rule-15577:

If both (**/**), (**U**), (**Q**), (**v**), (**i**), (**F**), (**t**), (**A**), (**d**), (**V**), (**p**), (**c**) and (**q**) are known, then its Sales or Revenue Planned is:

$= {F+[Ad+(Q+Vp/{360[c-q]})//-U]/[1-t]}/[1-v-i]

Rule-15578:

If both (**/**), (**U**), (**$**), (**Q**), (**i**), (**F**), (**t**), (**A**), (**d**), (**V**), (**p**), (**c**) and (**q**) are known, then its Variable Portion Planned is:

v= V/$

or it can also be found as

v= 1-i-{F+[Ad+(Q+Vp/{360[c-q]})//-U]/[1-t]}/$

Rule-15579:

If both (**/**), (**U**), (**$**), (**v**), (**Q**), (**F**), (**t**), (**A**), (**d**), (**V**), (**p**), (**c**) and (**q**) are known, then its Interest Portion Planned is:

i= 1-v-{F+[Ad+(Q+Vp/{360[c-q]})//-U]/[1-t]}/$

Steve Asikin ISBN 14: 978-1514685136, ISBN 10: 1514685132

Rule-15580:

If both (**/**), (**U**), (**$**), (**v**), (**i**), (**Q**), (**t**), (**A**), (**d**), (**V**), (**p**), (**c**) and (**q**) are known, then its Fixed Cost Planned is:

F= **$**[1-**v**-**i**]-**[Ad**+(**Q**+**Vp**/{360[**c**-**q**]})//**U]**/[1-**t**]

Rule-15581:

If both (**/**), (**U**), (**$**), (**v**), (**i**), (**F**), (**Q**), (**A**), (**d**), (**V**), (**p**), (**c**) and (**q**) are known, then its Tax Rate Planned is:

t= 1-**[Ad**+(**Q**+**Vp**/{360[**c**-**q**]})//**U]**/{**$**[1-**v**-**i**]-**F**}

Rule-15582:

If both (**/**), (**U**), (**$**), (**v**), (**i**), (**F**), (**t**), (**Q**), (**d**), (**V**), (**p**), (**c**) and (**q**) are known, then its After Tax Income Planned is:

A= **[U**+{**$**[1-**v**-**i**]-**F**}[1-**t**]-(**Q**+**Vp**/{360[**c**-**q**]})//**]**/**d**

Rule-15583:

If both (**/**), (**U**), (**$**), (**v**), (**i**), (**F**), (**t**), (**A**), (**Q**), (**V**), (**p**), (**c**) and (**q**) are known, then its Dividend Payout Planned is:

d= **[U**+{**$**[1-**v**-**i**]-**F**}[1-**t**]-(**Q**+**Vp**/{360[**c**-**q**]})//**]**/**A**

Rule-15584:

If both (**/**), (**U**), (**$**), (**v**), (**i**), (**F**), (**t**), (**A**), (**d**), (**Q**), (**p**), (**c**) and (**q**) are known, then its Variable Cost Planned is:

V= 360[**c**-**q**]**[/(U**+{**$**[1-**v**-**i**]-**F**}[1-**t**]-**Ad**)-**Q]**/**p**

Steve Asikin ISBN 14: 978-1514685136, ISBN 10: 1514685132

Rule-15585:

If both (**/**), (**U**), (**S**), (**v**), (**i**), (**F**), (**t**), (**A**), (**d**), (**V**), (**Q**), (**c**) and (**q**) are known, then its Procured Inventory Days Planned is:

p= 360[**c**-**q**][**/**(**U**+{**S**[1-**v**-**i**]-**F**}[1-**t**]-**Ad**)-**Q**]/**V**

Rule-15586:

If both (**/**), (**U**), (**S**), (**v**), (**i**), (**F**), (**t**), (**A**), (**d**), (**V**), (**p**), (**Q**) and (**q**) are known, then its Current Ratio Planned is:

c= **q**+**Vp**/{360[**/**(**U**+{**S**[1-**v**-**i**]-**F**}[1-**t**]-**Ad**)-**Q**]}

Rule-15587:

If both (**/**), (**U**), (**S**), (**v**), (**i**), (**F**), (**t**), (**A**), (**d**), (**V**), (**p**), (**c**) and (**Q**) are known, then its Quick or Acid Test Ratio Planned is:

q= **c**- **Vp**/{360[**/**(**U**+{**S**[1-**v**-**i**]-**F**}[1-**t**]-**Ad**)-**Q**]}

Rule-15588:

If both (**/**), (**U**), (**S**), (**v**), (**i**), (**F**), (**t**), (**A**), (**d**), (**p**), (**c**) and (**q**) are known, then its Quoted Longterm Debt Planned is:

Q= **/**(**U**+{**S**[1-**v**-**i**]-**F**}[1-**t**]-**Ad**)-**Svp**/{360[**c**-**q**]}

Rule-15589:

If both (**Q**), (**U**), (**S**), (**v**), (**i**), (**F**), (**t**), (**A**), (**d**), (**p**), (**c**) and (**q**) are known, then its Leverage or Gearing Ratio Planned is:

/= (**Q**+**Svp**/{360[**c**-**q**]})/(**U**+{**S**[1-**v**-**i**]-**F**}[1-**t**]-**Ad**)

Steve Asikin ISBN 14: 978-1514685136, ISBN 10: 1514685132

Rule-15590:

If both (**/**), (**Q**), (**S**), (**v**), (**i**), (**F**), (**t**), (**A**), (**d**), (**p**), (**c**) and (**q**) are known, then its Utilized or Starting Capital must be:

U= **Ad**+(**Q**+**Svp**/{360[**c**-**q**]})/**/**-{**S**[1-**v**-**i**]-**F**}[1-**t**]

Rule-15591:

If both (**/**), (**U**), (**Q**), (**v**), (**i**), (**F**), (**t**), (**A**), (**d**), (**p**), (**c**) and (**q**) are known, then its Sales or Revenue Planned is:

S= (**/**{**U**-**F**[1-**t**]-**Ad**}-**Q**)
/(**vp**/{360[**c**-**q**]}-**/**[1-**v**-**i**][1-**t**])

Rule-15592:

If both (**/**), (**U**), (**S**), (**Q**), (**i**), (**F**), (**t**), (**A**), (**d**), (**p**), (**c**) and (**q**) are known, then its Variable Portion Planned is:

v= **[/**(**U**+{**S**[1-**i**]-**F**}[1-**t**]-**Ad**)-**Q]**
/**[S**(**p**/{360[**c**-**q**]}+**/**[1-**t**])**]**

Rule-15593:

If both (**/**), (**U**), (**S**), (**v**), (**Q**), (**F**), (**t**), (**A**), (**d**), (**p**), (**c**) and (**q**) are known, then its Interest Portion Planned is:

i= 1-**v**-**{F**+**[Ad**+(**Q**+**Svp**/{360[**c**-**q**]})/**/**-**U]**/[1-**t**]**}**/**S**

Rule-15594:

If both (**/**), (**U**), (**S**), (**v**), (**i**), (**Q**), (**t**), (**A**), (**d**), (**p**), (**c**) and (**q**) are known, then its Fixed Cost Planned is:

F= **S**[1-**v**-**i**]-**[Ad**+(**Q**+**Svp**/{360[**c**-**q**]})/**/**-**U]**/[1-**t**]

Steve Asikin ISBN 14: 978-1514685136, ISBN 10: 1514685132

Rule-15595:

If both (**/**), (**U**), (**$**), (**v**), (**i**), (**F**), (**Q**), (**A**), (**d**), (**p**), (**c**) and (**q**) are known, then its Tax Rate Planned is:

t= 1-[**Ad**+(**Q**+**$vp**/{360[**c**-**q**]})/**/**-**U**]/{**$**[1-**v**-**i**]-**F**}

Rule-15596:

If both (**/**), (**U**), (**$**), (**v**), (**i**), (**F**), (**t**), (**Q**), (**d**), (**p**), (**c**) and (**q**) are known, then its After Tax Income Planned is:

A= [**U**+{**$**[1-**v**-**i**]-**F**}[1-**t**]-(**Q**+**$vp**/{360[**c**-**q**]})/**/**]/**d**

Rule-15597:

If both (**/**), (**U**), (**$**), (**v**), (**i**), (**F**), (**t**), (**A**), (**Q**), (**p**), (**c**) and (**q**) are known, then its Dividend Payout Planned is:

d= [**U**+{**$**[1-**v**-**i**]-**F**}[1-**t**]-(**Q**+**$vp**/{360[**c**-**q**]})/**/**]/**A**

Rule-15598:

If both (**/**), (**U**), (**$**), (**v**), (**i**), (**F**), (**t**), (**A**), (**d**), (**Q**), (**c**) and (**q**) are known, then its Procured Inventory Days Planned:

p= 360[**c**-**q**][**/**(**U**+{**$**[1-**v**-**i**]-**F**}[1-**t**]-**Ad**)-**Q**]/[**$v**]

Rule-15599:

If both (**/**), (**U**), (**$**), (**v**), (**i**), (**F**), (**t**), (**A**), (**d**), (**p**), (**Q**) and (**q**) are known, then its Current Ratio Planned is:

c= **q**+**$vp**/{360[**/**(**U**+{**$**[1-**v**-**i**]-**F**}[1-**t**]-**Ad**)-**Q**]}

Steve Asikin ISBN 14: 978-1514685136, ISBN 10: 1514685132

Rule-15600:

If both (**/**), (**U**), (**$**), (**v**), (**i**), (**F**), (**t**), (**A**), (**d**), (**p**), (**c**) and (**Q**) are known, then its Quick or Acid Test Ratio Planned is:

q= **c**-**$vp**/{360[**/**(**U**+{**$**[1-**v**-**i**]-**F**}[1-**t**]-**Ad**)-**Q**]}

Rule-15601:

If both (**/**), (**U**), (**$**), (**v**), (**i**), (**F**), (**t**), (**A**), (**d**), (**$'**), (**p**), (**s**), (**c**) and (**q**) are known, then its Quoted Longterm Debt Planned is:

Q= **/**(**U**+{**$**[1-**v**-**i**]-**F**}[1-**t**]-**Ad**)
-**$'vp**[1+**s**]/{360[**c**-**q**]}

Rule-15602:

If both (**Q**), (**U**), (**$**), (**v**), (**i**), (**F**), (**t**), (**A**), (**d**), (**$'**), (**p**), (**s**), (**c**) and (**q**) are known, then its Leverage or Gearing Ratio Planned is:

/= (**Q**+**$'vp**[1+**s**]/{360[**c**-**q**]})
/(**U**+{**$**[1-**v**-**i**]-**F**}[1-**t**]-**Ad**)

Rule-15603:

If both (**/**), (**Q**), (**$**), (**v**), (**i**), (**F**), (**t**), (**A**), (**d**), (**$'**), (**p**), (**s**), (**c**) and (**q**) are known, then its Utilized or Starting Capital must be:

U= **Ad**+(**Q**+**$'vp**[1+**s**]/{360[**c**-**q**]})/**/**
-{**$**[1-**v**-**i**]-**F**}[1-**t**]

Steve Asikin ISBN 14: 978-1514685136, ISBN 10: 1514685132

Rule-15604:

If both (/), (U), (Q), (v), (i), (F), (t), (A), (d), (S'), (p), (s), (c) and (q) are known, then its Sales or Revenue Planned is:

S= (Q-/{U-F[1-t]-Ad}+S'vp[1+s]/{360[c-q]})
/[1-v-i]

Rule-15605:

If both (/), (U), (S), (Q), (i), (F), (t), (A), (d), (S'), (p), (s), (c) and (q) are known, then its Variable Portion Planned is:

v= [/(U+{S[1-i]-F}[1-t]-Ad)-Q]
/(S'p[1+s]/{360[c-q]}+S/[1-t])

Rule-15606:

If both (/), (U), (S), (v), (Q), (F), (t), (A), (d), (S'), (p), (s), (c) and (q) are known, then its Interest Portion Planned is:

i= 1-v-{F+[Ad+(Q+S'vp[1+s]/{360[c-q]})//-U]
/[1-t]}/S

Rule-15607:

If both (/), (U), (S), (v), (i), (Q), (t), (A), (d), (S'), (p), (s), (c) and (q) are known, then its Fixed Cost Planned is:

F= S[1-v-i]-[Ad+(Q+S'vp[1+s]/{360[c-q]})//-U]
/[1-t]

Steve Asikin ISBN 14: 978-1514685136, ISBN 10: 1514685132

Rule-15608:

If both (/), (U), ($), (v), (i), (F), (Q), (A), (d), ($'), (p), (s), (c) and (q) are known, then its Tax Rate Planned is:

t= 1-[Ad+(Q+$'vp[1+s]/{360[c-q]})/ /U]
/{$[1-v-i]-F}

Rule-15609:

If both (/), (U), ($), (v), (i), (F), (t), (Q), (d), ($'), (p), (s), (c) and (q) are known, then its After Tax Income Planned is:

A= [U+{$[1-v-i]-F}[1-t]
-(Q+$'vp[1+s]/{360[c-q]})/ /]/d

Rule-15610:

If both (/), (U), ($), (v), (i), (F), (t), (A), (Q), ($'), (p), (s), (c) and (q) are known, then its Dividend Payout Planned is:

d= [U+{$[1-v-i]-F}[1-t]
-(Q+$'vp[1+s]/{360[c-q]})/ /]/A

Rule-15611:

If both (/), (U), ($), (v), (i), (F), (t), (A), (d), (Q), (p), (s), (c) and (q) are known, then its Sales Past must be:

$'= [/(U+{$[1-v-i]-F}[1-t]-Ad)-Q]
/(vp[1+s]/{360[c-q]})

Steve Asikin ISBN 14: 978-1514685136, ISBN 10: 1514685132

Rule-15612:

If both (/), (U), (S), (v), (i), (F), (t), (A), (d), (S'), (Q), (s), (c) and (q) are known, then its Procured Inventory Days Planned:

p= 360[c-q][/(U+{S[1-v-i]-F}[1-t]-Ad)-Q]
/{S'v[1+s]}

Rule-15613:

If both (/), (U), (S), (v), (i), (F), (t), (A), (d), (S'), (p), (Q), (c) and (q) are known, then its Sales Growth Planned is:

s= S/S'-1

or it can also be found as

s= [/(U+{S[1-v-i]-F}[1-t]-Ad)-Q]
/(S'vp/{360[c-q]})-1

Rule-15614:

If both (/), (U), (S), (v), (i), (F), (t), (A), (d), (S'), (p), (s), (Q) and (q) are known, then its Current Ratio Planned is:

c= q+S'vp[1+s]
/{360[/(U+{S[1-v-i]-F}[1-t]-Ad)-Q]}

Rule-15615:

If both (/), (U), (S), (v), (i), (F), (t), (A), (d), (S'), (p), (s), (c) and (Q) are known, then its Quick or Acid Test Ratio Planned is:

q= c-S'vp[1+s]
/{360[/(U+{S[1-v-i]-F}[1-t]-Ad)-Q]}

Steve Asikin ISBN 14: 978-1514685136, ISBN 10: 1514685132

Rule-15616:

If both (**/**), (**U**), (**S**), (**v**), (**i**), (**F**), (**t**), (**d**) and (**X**) are known, then its Quoted Longterm Debt Planned is:

Q= **/**(**U**+{**S**[1-**v**-**i**]-**F**}[1-**t**][1-**d**])-**X**

Rule-15617:

If both (**Q**), (**U**), (**S**), (**v**), (**i**), (**F**), (**t**), (**d**) and (**X**) are known, then its Leverage or Garing Ratio Planned is:

/= [**Q**+**X**]/(**U**+{**S**[1-**v**-**i**]-**F**}[1-**t**][1-**d**])

Rule-15618:

If both (**/**), (**Q**), (**S**), (**v**), (**i**), (**F**), (**t**), (**d**) and (**X**) are known, then its Utilized or Starting Capital must be:

U= [**Q**+**X**]/**/**-{**S**[1-**v**-**i**]-**F**}[1-**t**][1-**d**]

Rule-15619:

If both (**/**), (**U**), (**Q**), (**v**), (**i**), (**F**), (**t**), (**d**) and (**X**) are known, then its Sales or Revenue Planned is:

S= (**F**+{[**Q**+**X**]/**/**-**U**}/{[1-**t**] [1-**d**]})/[1-**v**-**i**]

Rule-15620:

If both (**/**), (**U**), (**S**), (**Q**), (**i**), (**F**), (**t**), (**d**) and (**X**) are known, then its Variable Portion Planned is:

v= 1-**i**-(**F**+{[**Q**+**X**]/**/**-**U**}/{[1-**t**] [1-**d**]})/**S**

Rule-15621:

If both (**/**), (**U**), (**S**), (**v**), (**Q**), (**F**), (**t**), (**d**) and (**X**) are known, then its Interest Portion Planned is:

i= 1-**v**- (**F**+{[**Q**+**X**]/**/**-**U**}/{[1-**t**] [1-**d**]})/**S**

Steve Asikin ISBN 14: 978-1514685136, ISBN 10: 1514685132

Rule-15622:

If both (**L**), (**U**), (**S**), (**v**), (**i**), (**Q**), (**t**), (**d**) and (**X**) are known, then its Fixed Cost Planned is:

F= **S**[1-**v**-**i**]-{[**Q**+**X**]/**LU**}/{[1-**t**] [1-**d**]}

Rule-15623:

If both (**L**), (**U**), (**S**), (**v**), (**i**), (**F**), (**Q**), (**d**) and (**X**) are known, then its Tax Rate Planned is:

t = 1-{[**Q**+**X**]/**LU**}/([1-**d**]{**S**[1-**v**-**i**]-**F**})

Rule-15624:

If both (**L**), (**U**), (**S**), (**v**), (**i**), (**F**), (**t**), (**Q**) and (**X**) are known, then its Dividend Payout Planned is:

d = 1-{[**Q**+**X**]/**LU**}/([1-**t**]{**S**[1-**v**-**i**]-**F**})

Rule-15625:

If both (**L**), (**U**), (**S**), (**v**), (**i**), (**F**), (**t**), (**d**) and (**Q**) are known, then its Xpress or Current Debt Planned is:

X= **L**(**U**+{**S**[1-**v**-**i**]-**F**}[1-**t**][1-**d**])-**Q**

Rule-15626:

If both (**L**), (**U**), (**S**), (**v**), (**i**), (**F**), (**t**), (**d**), (**P**), (**c**) and (**q**) are known, then its Quoted Longterm Debt Planned is:

Q= **L**(**U**+{**S**[1-**v**-**i**]-**F**}[1-**t**][1-**d**])-**P**/[**c**-**q**]

Rule-15627:

If both (**Q**), (**U**), (**S**), (**v**), (**i**), (**F**), (**t**), (**d**), (**P**), (**c**) and (**q**) are known, then its Leverage or Gearing Ratio Planned is:

L= {**Q**+**P**/[**c**-**q**]}/(**U**+{**S**[1-**v**-**i**]-**F**}[1-**t**][1-**d**])

Steve Asikin ISBN 14: 978-1514685136, ISBN 10: 1514685132

Rule-15628:

If both (**/**), (**Q**), (**$**), (**v**), (**i**), (**F**), (**t**), (**d**), (**P**), (**c**) and (**q**) are known, then its Utilized or Starting Capital must be:

U= {**Q**+**P**/[**c**-**q**]}/**/**-{**$**[1-**v**-**i**]-**F**}[1-**t**][1-**d**]

Rule-15629:

If both (**/**), (**U**), (**Q**), (**v**), (**i**), (**F**), (**t**), (**d**), (**P**), (**c**) and (**q**) are known, then its Sales or Revenue Planned is:

$= [**F**+({**Q**+**P**/[**c**-**q**]}/**/**-**U**)/{[1-**t**] [1-**d**]}]/[1-**v**-**i**]

Rule-15630:

If both (**/**), (**U**), (**$**), (**Q**), (**i**), (**F**), (**t**), (**d**), (**P**), (**c**) and (**q**) are known, then its Variable Portion Planned is:

v= 1-**i**-[**F**+({**Q**+**P**/[**c**-**q**]}/**/**-**U**)/{[1-**t**] [1-**d**]}]/**$**

Rule-15631:

If both (**/**), (**U**), (**$**), (**v**), (**Q**), (**F**), (**t**), (**d**), (**P**), (**c**) and (**q**) are known, then its Interest Portion Planned is:

i= 1-**v**- [**F**+({**Q**+**P**/[**c**-**q**]}/**/**-**U**)/{[1-**t**] [1-**d**]}]/**$**

Rule-15632:

If both (**/**), (**U**), (**$**), (**v**), (**i**), (**Q**), (**t**), (**d**), (**P**), (**c**) and (**q**) are known, then its Fixed Cost Planned is:

F= **$**[1-**v**-**i**]-({**Q**+**P**/[**c**-**q**]}/**/**-**U**)/{[1-**t**] [1-**d**]}

Rule-15633:

If both (**/**), (**U**), (**$**), (**v**), (**i**), (**F**), (**Q**), (**d**), (**P**), (**c**) and (**q**) are known, then its Tax Rate Planned is:

t = 1-({**Q**+**P**/[**c**-**q**]}/**/**-**U**)/([1-**d**]{**$**[1-**v**-**i**]-**F**})

Steve Asikin ISBN 14: 978-1514685136, ISBN 10: 1514685132

Rule-15634:

If both (/), (U), (S), (v), (i), (F), (t), (Q), (P), (c) and (q) are known, then its Dividend Payout Planned is:

d = 1-({Q+P/[c-q]}//-U)/([1-t]{S[1-v-i]-F})

Rule-15635:

If both (/), (U), (S), (v), (i), (F), (t), (d), (Q), (c) and (q) are known, then its Procured Inventory Planned is:

P= [c-q][/(U+{S[1-v-i]-F}[1-t][1-d])-Q]

Rule-15636:

If both (/), (U), (S), (v), (i), (F), (t), (d), (P), (Q) and (q) are known, then its Current Ratio Planned is:

c= q+P/[/(U+{S[1-v-i]-F}[1-t][1-d])-Q]

Rule-15637:

If both (/), (U), (S), (v), (i), (F), (t), (d), (P), (c) and (Q) are known, then its Quick or Acid Test Ratio Planned is:

q= c-P/[/(U+{S[1-v-i]-F}[1-t][1-d])-Q]

Rule-15638:

If both (/), (U), (S), (v), (i), (F), (t), (d), (V), (p), (c) and (q) are known, then its Quoted Longterm Debt Planned is:

Q= /(U+{S[1-v-i]-F}[1-t][1-d])-Vp/{360[c-q]}

Steve Asikin ISBN 14: 978-1514685136, ISBN 10: 1514685132

Rule-15639:

If both (**Q**), (**U**), (**$**), (**v**), (**i**), (**F**), (**t**), (**d**), (**V**), (**p**), (**c**) and (**q**) are known, then its Leverage or Gearing Ratio Planned is:

/= (**Q**+**Vp**/{360[**c**-**q**]})/(**U**+{**$**[1-**v**-**i**]-**F**}[1-**t**][1-**d**])

Rule-15640:

If both (**/**), (**Q**), (**$**), (**v**), (**i**), (**F**), (**t**), (**d**), (**V**), (**p**), (**c**) and (**q**) are known, then its Utilized or Starting Capital must be:

U= (**Q**+**Vp**/{360[**c**-**q**]})/**/**-{**$**[1-**v**-**i**]-**F**}[1-**t**][1-**d**]

Rule-15641:

If both (**/**), (**U**), (**Q**), (**v**), (**i**), (**F**), (**t**), (**d**), (**V**), (**p**), (**c**) and (**q**) are known, then its Sales or Revenue Planned is:

$= **{F**+**[**(**Q**+**Vp**/{360[**c**-**q**]})/**/**-**U]**
/{[1-**t**] [1-**d**]}**}**/[1-**v**-**i**]

Rule-15642:

If both (**/**), (**U**), (**$**), (**Q**), (**i**), (**F**), (**t**), (**d**), (**V**), (**p**), (**c**) and (**q**) are known, then its Variable Portion Planned is:

v= 1-**i**-**{F**+**[**(**Q**+**Vp**/{360[**c**-**q**]})/**/**-**U]**
/{[1-**t**] [1-**d**]}**}**/**$**

Steve Asikin ISBN 14: 978-1514685136, ISBN 10: 1514685132

Rule-15643:

If both (/), (U), ($), (v), (Q), (F), (t), (d), (V), (p), (c) and (q) are known, then its Interest Portion Planned is:

i= 1-v-{F+[(Q+Vp/{360[c-q]})/ /U] /{[1-t] [1-d]}}/$

Rule-15644:

If both (/), (U), ($), (v), (i), (Q), (t), (d), (V), (p), (c) and (q) are known, then its Fixed Cost Planned is:

F= $[1-v-i]-[(Q+Vp/{360[c-q]})/ /U]/{[1-t] [1-d]}

Rule-15645:

If both (/), (U), ($), (v), (i), (F), (Q), (d), (V), (p), (c) and (q) are known, then its Tax Rate Planned is:

t = 1-[(Q+Vp/{360[c-q]})/ /U]/([1-d]{$[1-v-i]-F})

Rule-15646:

If both (/), (U), ($), (v), (i), (F), (t), (Q), (V), (p), (c) and (q) are known, then its Dividend Payout Planned is:

d = 1-[(Q+Vp/{360[c-q]})/ /U]/([1-t]{$[1-v-i]-F})

Rule-15647:

If both (/), (U), ($), (v), (i), (F), (t), (d), (Q), (p), (c) and (q) are known, then its Variable Cost Planned is:

V= 360[c-q][/(U+{$[1-v-i]-F}[1-t][1-d])-Q]/p

Steve Asikin ISBN 14: 978-1514685136, ISBN 10: 1514685132

Rule-15648:

If both (**/**), (**U**), (**$**), (**v**), (**i**), (**F**), (**t**), (**d**), (**V**), (**Q**), (**c**) and (**q**) are known, then its Procured Inventory Days Planned:

$$p= 360[c-q][/(U+\{\$[1-v-i]-F\}[1-t][1-d])-Q]/V$$

Rule-15649:

If both (**/**), (**U**), (**$**), (**v**), (**i**), (**F**), (**t**), (**d**), (**V**), (**p**), (**Q**) and (**q**) are known, then its Current Ratio Planned is:

$$c= q+Vp/\{360[/(U+\{\$[1-v-i]-F\}[1-t][1-d])-Q]\}$$

Rule-15650:

If both (**/**), (**U**), (**$**), (**v**), (**i**), (**F**), (**t**), (**d**), (**V**), (**p**), (**c**) and (**Q**) are known, then its Quick or Acid Test Ratio Planned is:

$$q= c-Vp/\{360[/(U+\{\$[1-v-i]-F\}[1-t][1-d])-Q]\}$$

Rule-15651:

If both (**/**), (**U**), (**$**), (**v**), (**i**), (**F**), (**t**), (**d**), (**p**), (**c**) and (**q**) are known, then its Quoted Longterm Debt Planned is:

$$Q= /(U+\{\$[1-v-i]-F\}[1-t][1-d])-\$vp/\{360[c-q]\}$$

Rule-15652:

If both (**Q**), (**U**), (**$**), (**v**), (**i**), (**F**), (**t**), (**d**), (**p**), (**c**) and (**q**) are known, then its Leverage or Gearing Ratio Planned is:

$$/= (Q+\$vp/\{360[c-q]\})/(U+\{\$[1-v-i]-F\}[1-t][1-d])$$

Steve Asikin ISBN 14: 978-1514685136, ISBN 10: 1514685132

Rule-15653:

If both (**Ł**), (**Q**), (**S**), (**v**), (**i**), (**F**), (**t**), (**d**), (**p**), (**c**) and (**q**) are known, then its Utilized or Starting Capital must be:

U= (**Q**+**Svp**/{360[**c**-**q**]})/**Ł**{**S**[1-**v**-**i**]-**F**}[1-**t**][1-**d**]

Rule-15654:

If both (**Ł**), (**U**), (**Q**), (**v**), (**i**), (**F**), (**t**), (**d**), (**p**), (**c**) and (**q**) are known, then its Sales or Revenue Planned is:

S= **Ł**(**U**+{**S**[1-**v**-**i**]-**F**}[1-**t**][1-**d**])-**Svp**/{360[**c**-**q**]}

Rule-15655:

If both (**Ł**), (**U**), (**S**), (**Q**), (**i**), (**F**), (**t**), (**d**), (**p**), (**c**) and (**q**) are known, then its Variable Portion Planned is:

v= **[Ł(U**+{**S**[1-**i**]-**F**}[1-**t**][1-**d**])-**Q]**
/**[S(p**/{360[**c**-**q**]}+**Ł**[1-**t**][1-**d**])**]**

Rule-15656:

If both (**Ł**), (**U**), (**S**), (**v**), (**Q**), (**F**), (**t**), (**d**), (**p**), (**c**) and (**q**) are known, then its Interest Portion Planned is:

i= 1-**v**-**{F+[(Q**+**Svp**/{360[**c**-**q**]})/**ŁU]**
/{[1-**t**] [1-**d**]}**}/S**

Rule-15657:

If both (**Ł**), (**U**), (**S**), (**v**), (**i**), (**Q**), (**t**), (**d**), (**p**), (**c**) and (**q**) are known, then its Fixed Cost Planned is:

F= **S**[1-**v**-**i**]-**[(Q**+**Svp**/{360[**c**-**q**]})/**ŁU]**
/{[1-**t**] [1-**d**]}

Steve Asikin ISBN 14: 978-1514685136, ISBN 10: 1514685132

Rule-15658:

If both (**/**), (**U**), (**$**), (**v**), (**i**), (**F**), (**Q**), (**d**), (**p**), (**c**) and (**q**) are known, then its Tax Rate Planned is:

t = 1-**[**(**Q**+**$vp**/{360[**c-q**]})/**/U]**
/([1-**d**]{**$**[1-**v-i**]-**F**})

Rule-15659:

If both (**/**), (**U**), (**$**), (**v**), (**i**), (**F**), (**t**), (**Q**), (**p**), (**c**) and (**q**) are known, then its Dividend Payout Planned is:

d = 1-**[**(**Q**+**$vp**/{360[**c-q**]})/**/U]**
/([1-**t**]{**$**[1-**v-i**]-**F**})

Rule-15660:

If both (**/**), (**U**), (**$**), (**v**), (**i**), (**F**), (**t**), (**d**), (**Q**), (**c**) and (**q**) are known, then its Procured Inventory Days Planned:

p= 360[**c-q**]**[/U**+{**$**[1-**v-i**]-**F**}[1-**t**][1-**d**])-**Q]**/[**$v**]

Rule-15661:

If both (**/**), (**U**), (**$**), (**v**), (**i**), (**F**), (**t**), (**d**), (**p**), (**Q**) and (**q**) are known, then its Current Ratio Planned is:

c= **q**+**$vp**/{360**[/U**+{**$**[1-**v-i**]-**F**}[1-**t**][1-**d**])-**Q]}**

Rule-15662:

If both (**/**), (**U**), (**$**), (**v**), (**i**), (**F**), (**t**), (**d**), (**p**), (**c**) and (**Q**) are known, then its Quick or Acid Test Ratio Planned is:

q= **c-$vp**/{360**[/U**+{**$**[1-**v-i**]-**F**}[1-**t**][1-**d**])-**Q]}**

Steve Asikin ISBN 14: 978-1514685136, ISBN 10: 1514685132

Rule-15663:

If both (/), (U), (S), (v), (i), (F), (t), (d), (S'), (p), (s), (c) and (q) are known, then its Quoted Longterm Debt Planned is:

$$Q= /(U+\{S[1-v-i]-F\}[1-t][1-d]) -S'vp[1+s]/\{360[c-q]\}$$

Rule-15664:

If both (Q), (U), (S), (v), (i), (F), (t), (d), (S'), (p), (s), (c) and (q) are known, then its Leverage or Gearing Ratio Planned is:

$$/= (Q+S'vp[1+s]/\{360[c-q]\}) /(U+\{S[1-v-i]-F\}[1-t][1-d])$$

Rule-15665:

If both (/), (Q), (S), (v), (i), (F), (t), (d), (S'), (p), (s), (c) and (q) are known, then its Utilized or Starting Capital must be:

$$U= (Q+S'vp[1+s]/\{360[c-q]\})// -\{S[1-v-i]-F\}[1-t][1-d]$$

Rule-15666:

If both (/), (U), (Q), (v), (i), (F), (t), (d), (S'), (p), (s), (c) and (q) are known, then its Sales or Revenue Planned is:

$$S= (Q-/\{U-F[1-t][1-d]\}+S'[1+s]vp/\{360[c-q]\}) /\{/[1-v-i] [1-t][1-d]\}$$

Steve Asikin ISBN 14: 978-1514685136, ISBN 10: 1514685132

Rule-15667:

If both (**/**), (**U**), (**$**), (**Q**), (**i**), (**F**), (**t**), (**d**), (**$'**), (**p**), (**s**), (**c**) and (**q**) are known, then its Variable Cost Planned is:

v= [**/**(**U**+{**$**[1-**i**]-**F**}[1-**t**][1-**d**])-**Q**]
/(**$'p**[1+**s**]/{360[**c**-**q**]}-**$/**[1-**t**][1-**d**])

Rule-15668:

If both (**/**), (**U**), (**$**), (**v**), (**Q**), (**F**), (**t**), (**d**), (**$'**), (**p**), (**s**), (**c**) and (**q**) are known, then its Interest Portion Planned is:

i= 1-**v**-{**F**+[(**Q**+**$'vp**[1+**s**]/{360[**c**-**q**]})/**/**-**U**]
/{[1-**t**] [1-**d**]}}/**$**

Rule-15669:

If both (**/**), (**U**), (**$**), (**v**), (**i**), (**Q**), (**t**), (**d**), (**$'**), (**p**), (**s**), (**c**) and (**q**) are known, then its Fixed Cost Planned is:

F= **$**[1-**v**-**i**]-[(**Q**+**$'vp**[1+**s**]/{360[**c**-**q**]})/**/**-**U**]
/{[1-**t**] [1-**d**]}

Rule-15670:

If both (**/**), (**U**), (**$**), (**v**), (**i**), (**F**), (**Q**), (**d**), (**$'**), (**p**), (**s**), (**c**) and (**q**) are known, then its Tax Rate Planned is:

t = 1-[(**Q**+**$'vp**[1+**s**]/{360[**c**-**q**]})/**/**-**U**]
/([1-**d**]{**$**[1-**v**-**i**]-**F**})

Rule-15671:

If both (**/**), (**U**), (**$**), (**v**), (**i**), (**F**), (**t**), (**Q**), (**$'**), (**p**), (**s**), (**c**) and (**q**) are known, then its Dividend Payout Planned is:

d = 1-[(**Q**+**$'vp**[1+**s**]/{360[**c**-**q**]})/**/**-**U**]
/([1-**t**]{**$**[1-**v**-**i**]-**F**})

Steve Asikin ISBN 14: 978-1514685136, ISBN 10: 1514685132

Rule-15672:

If both (/), (U), ($), (v), (i), (F), (t), (d), (Q), (p), (s), (c) and (q) are known, then its Sales Past must be:

$'= [/(U+{$[1-v-i]-F}[1-t][1-d])-Q]
/([1+s]vp/{360[c-q]}

Rule-15673:

If both (/), (U), ($), (v), (i), (F), (t), (d), ($'), (Q), (s), (c) and (q) are known, then its Procured Inventory Days Planned:

p= 360[c-q][/(U+{$[1-v-i]-F}[1-t][1-d])-Q]
/{$'v[1+s]}

Rule-15674:

If both (/), (U), ($), (v), (i), (F), (t), (d), ($'), (p), (Q), (c) and (q) are known, then its Sales Growth Planned is:

s= $/$'-1

or it can also be found as

s= [/(U+{$[1-v-i]-F}[1-t][1-d])-Q]
/($'vp/{360[c-q]}-1

Rule-15675:

If both (/), (U), ($), (v), (i), (F), (t), (d), ($'), (p), (s), (Q) and (q) are known, then its Current Ratio Planned is:

c= q+$'vp[1+s]
/{360[/(U+{$[1-v-i]-F}[1-t][1-d])-Q]}

Steve Asikin ISBN 14: 978-1514685136, ISBN 10: 1514685132

Rule-15676:

If both (**/**), (**U**), (**$**), (**v**), (**i**), (**F**), (**t**), (**d**), (**$'**), (**p**), (**s**), (**c**) and (**Q**) are known, then its Quick or Acid Test Ratio Planned is:

q= **c**-**$'vp**[1+**s**]
/{360[**/**(**U**+{**$**[1-**v**-**i**]-**F**}[1-**t**][1-**d**])-**Q**]}

Rule-15677:

If both (**/**), (**U**), (**$**), (**v**), (**i**), (**F**), (**$'**), (**i**), (**s**), (**T**), (**D**) and (**X**) are known, then its Quoted Longterm Debt Planned is:

Q= **/**{**U**+**$**[1-**v**]-**F**-**$'i**[1+**s**]-**T**-**D**}-**X**

Rule-15678:

If both (**Q**), (**U**), (**$**), (**v**), (**i**), (**F**), (**$'**), (**i**), (**s**), (**T**), (**D**) and (**X**) are known, then its Leverage or Gearing Ratio Planned is:

/= [**Q**+**X**]/{**U**+**$**[1-**v**]-**F**-**$'i**[1+**s**]-**T**-**D**}

Rule-15679:

If both (**/**), (**Q**), (**$**), (**v**), (**i**), (**F**), (**$'**), (**i**), (**s**), (**T**), (**D**) and (**X**) are known, then its Utilized or Starting Capital must be:

U= **F**+**$'i**[1+**s**]+**T**+**D**+[**Q**+**X**]/**/**-**$**[1-**v**]

Rule-15680:

If both (**/**), (**U**), (**Q**), (**v**), (**i**), (**F**), (**$'**), (**i**), (**s**), (**T**), (**D**) and (**X**) are known, then its Sales or Revenue Planned is:

$**= {**F**+**$'i[1+**s**]+**T**+**D**+[**Q**+**X**]/**/**-**U**}/[1-**v**]

Steve Asikin ISBN 14: 978-1514685136, ISBN 10: 1514685132

Rule-15681:

If both (/), (U), ($), (Q), (i), (F), ($'), (i), (s), (T), (D) and (X) are known, then its Variable Cost Planned is:

v= 1-{F+$'i[1+s]+T+D+[Q+X]/ŁU}/$

Rule-15682:

If both (/), (U), ($), (v), (i), (Q), ($'), (i), (s), (T), (D) and (X) are known, then its Fixed Cost Planned is:

F= $[1-v]-$'i[1+s]-T-D-[Q+X]/ŁU

Rule-15683:

If both (/), (U), ($), (v), (i), (F), (Q), (i), (s), (T), (D) and (X) are known, then its Sales Past must be:

$'= {$[1-v]-F-T-D-[Q+X]/ŁU}/{i[1+s]}

Rule-15684:

If both (/), (U), ($), (v), (i), (F), ($'), (Q), (s), (T), (D) and (X) are known, then its Interest Portion Planned is:

i= {$[1-v]-F-T-D-[Q+X]/ŁU}/{$'[1+s]}

Rule-15685:

If both (/), (U), ($), (v), (i), (F), ($'), (i), (Q), (T), (D) and (X) are known, then its Sales Growth Planned is:

s= $/$'-1

or it can also be found as

s= {$[1-v]-F-T-D-[Q+X]/ŁU}/[$'i]-1

Steve Asikin ISBN 14: 978-1514685136, ISBN 10: 1514685132

Rule-15686:

If both (**/**), (**U**), (**$**), (**v**), (**i**), (**F**), (**$'**), (**i**), (**s**), (**Q**), (**D**) and (**X**) are known, then its Tax Planned is:

T= **$**[1-**v**]-**F**-**$'i**[1+**s**]-**D**-[**Q**+**X**]/**/**-**U**

Rule-15687:

If both (**/**), (**U**), (**$**), (**v**), (**i**), (**F**), (**$'**), (**i**), (**s**), (**T**), (**Q**) and (**X**) are known, then its Dividend Planned is:

D= **$**[1-**v**]-**F**-**T**-**$'i**[1+**s**]-[**Q**+**X**]/**/**-**U**

Rule-15688:

If both (**/**), (**U**), (**$**), (**v**), (**i**), (**F**), (**$'**), (**i**), (**s**), (**T**), (**D**) and (**Q**) are known, then its Xpress or Current Debt Planned is:

X= **/**{**U**+**$**[1-**v**]-**F**-**$'i**[1+**s**]-**T**-**D**}-**Q**

Rule-15689:

If both (**/**), (**U**), (**$**), (**v**), (**F**), (**$'**), (**i**), (**s**), (**T**), (**D**), (**P**), (**c**) and (**q**) are known, then its Quoted Longterm Debt Planned is:

Q= **/**{**U**+**$**[1-**v**]-**F**-**$'i**[1+**s**]-**T**-**D**}-**P**/[**c**-**q**]

Rule-15690:

If both (**Q**), (**U**), (**$**), (**v**), (**F**), (**$'**), (**i**), (**s**), (**T**), (**D**), (**P**), (**c**) and (**q**) are known, then its Leverage or Gearing Ratio Planned is:

/= {**Q**+**P**/[**c**-**q**]}/{**U**+**$**[1-**v**]-**F**-**$'i**[1+**s**]-**T**-**D**}

Steve Asikin ISBN 14: 978-1514685136, ISBN 10: 1514685132

Rule-15691:

If both (**/**), (**Q**), (**$**), (**v**), (**F**), (**$'**), (**i**), (**s**), (**T**), (**D**), (**P**), (**c**) and (**q**) are known, then its Utilized or Starting Capital must be:

U= **F**+**$'i**[1+**s**]+**T**+**D**+{**Q**+**P**/[**c**-**q**]}/**/**-**$**[1-**v**]

Rule-15692:

If both (**/**), (**U**), (**Q**), (**v**), (**F**), (**$'**), (**i**), (**s**), (**T**), (**D**), (**P**), (**c**) and (**q**) are known, then its Sales or Revenue Planned is:

$**= (**F**+**$'i[1+**s**]+**T**+**D**+{**Q**+**P**/[**c**-**q**]}/**/**-**U**)/[1-**v**]

Rule-15693:

If both (**/**), (**U**), (**$**), (**Q**), (**F**), (**$'**), (**i**), (**s**), (**T**), (**D**), (**P**), (**c**) and (**q**) are known, then its Variable Portion Planned is:

v= 1-(**F**+**$'i**[1+**s**]+**T**+**D**+{**Q**+**P**/[**c**-**q**]}/**/**-**U**)/**$**

Rule-15694:

If both (**/**), (**U**), (**$**), (**v**), (**Q**), (**$'**), (**i**), (**s**), (**T**), (**D**), (**P**), (**c**) and (**q**) are known, then its Fixed Cost Planned is:

F= **$**[1-**v**]-**$'i**[1+**s**]-**T**-**D**-{**Q**+**P**/[**c**-**q**]}/**/**-**U**

Rule-15695:

If both (**/**), (**U**), (**$**), (**v**), (**F**), (**Q**), (**i**), (**s**), (**T**), (**D**), (**P**), (**c**) and (**q**) are known, then its Sales Past must be:

$'**= (**$[1-**v**]-**F**-**T**-**D**-{**Q**+**P**/[**c**-**q**]}/**/**-**U**)/{**i**[1+**s**]}

Steve Asikin ISBN 14: 978-1514685136, ISBN 10: 1514685132

Rule-15696:

If both (**/**), (**U**), (**$**), (**v**), (**F**), (**$'**), (**Q**), (**s**), (**T**), (**D**), (**P**), (**c**) and (**q**) are known, then its Interest Portion Planned is:

i= (**$**[1-**v**]-**F**-**T**-**D**-{**Q**+**P**/[**c**-**q**]}/**/**-**U**)/{**$'**[1+**s**]}

Rule-15697:

If both (**/**), (**U**), (**$**), (**v**), (**F**), (**$'**), (**i**), (**Q**), (**T**), (**D**), (**P**), (**c**) and (**q**) are known, then its Sales Growth Planned is:

s= **$**/**$'**-1

or it can also be found as

s= (**$**[1-**v**]-**F**-**T**-**D**-{**Q**+**P**/[**c**-**q**]}/**/**-**U**)/[**$'i**]-1

Rule-15698:

If both (**/**), (**U**), (**$**), (**v**), (**F**), (**$'**), (**i**), (**s**), (**Q**), (**D**), (**P**), (**c**) and (**q**) are known, then its Tax Planned is:

T= **$**[1-**v**]-**F**-**$'i**[1+**s**]-**D**-{**Q**+**P**/[**c**-**q**]}/**/**-**U**

Rule-15699:

If both (**/**), (**U**), (**$**), (**v**), (**F**), (**$'**), (**i**), (**s**), (**T**), (**Q**), (**P**), (**c**) and (**q**) are known, then its Dividend Planned is:

D= **$**[1-**v**]-**F**-**T**-**$'i**[1+**s**]-{**Q**+**P**/[**c**-**q**]}/**/**-**U**

Rule-15700:

If both (**/**), (**U**), (**$**), (**v**), (**F**), (**$'**), (**i**), (**s**), (**T**), (**D**), (**Q**), (**c**) and (**q**) are known, then its Procured Inventory Planned is:

P= [**c**-**q**](**/**{**U**+**$**[1-**v**]-**F**-**$'i**[1+**s**]-**T**-**D**}-**Q**)

Steve Asikin ISBN 14: 978-1514685136, ISBN 10: 1514685132

Rule-15701:

If both (**/**), (**U**), (**$**), (**v**), (**F**), (**$'**), (**i**), (**s**), (**T**), (**D**), (**P**), (**Q**) and (**q**) are known, then its Current Ratio Planned is:

c= **q**+**P**/(**/**{**U**+**$**[1-**v**]-**F**-**$'i**[1+**s**]-**T**-**D**}-**Q**)

Rule-15702:

If both (**/**), (**U**), (**$**), (**v**), (**F**), (**$'**), (**i**), (**s**), (**T**), (**D**), (**P**), (**c**) and (**Q**) are known, then its Quick or Acid test Ratio Planned is:

q= **c**-**P**/(**/**{**U**+**$**[1-**v**]-**F**-**$'i**[1+**s**]-**T**-**D**}-**Q**)

Rule-15703:

If both (**/**), (**U**), (**$**), (**v**), (**F**), (**$'**), (**i**), (**s**), (**T**), (**D**), (**V**), (**p**), (**c**) and (**q**) are known, then its Quoted Longterm Debt Planned is:

Q= **/**{**U**+**$**[1-**v**]-**F**-**$'i**[1+**s**]-**T**-**D**}-**Vp**/{360[**c**-**q**]}

Rule-15704:

If both (**Q**), (**U**), (**$**), (**v**), (**F**), (**$'**), (**i**), (**s**), (**T**), (**D**), (**V**), (**p**), (**c**) and (**q**) are known, then its Leverage or Gearing Ratio Planned is:

/= (**Q**+**Vp**/{360[**c**-**q**]})/{**U**+**$**[1-**v**]-**F**-**$'i**[1+**s**]-**T**-**D**}

Rule-15705:

If both (**/**), (**Q**), (**$**), (**v**), (**F**), (**$'**), (**i**), (**s**), (**T**), (**D**), (**V**), (**p**), (**c**) and (**q**) are known, then its Utilized or Starting Capital must be:

U= **F**+**$'i**[1+**s**]+**T**+**D**+(**Q**+**Vp**/{360[**c**-**q**]})/**/**-**$**[1-**v**]

Steve Asikin ISBN 14: 978-1514685136, ISBN 10: 1514685132

Rule-15706:

If both (/), (U), (Q), (v), (F), (S'), (i), (s), (T), (D), (V), (p), (c) and (q) are known, then its Sales or Revenue Planned is:

S= [F+S'i[1+s]+T+D+(Q+Vp/{360[c-q]})/ƒ-U]
/[1-v]

Rule-15707:

If both (/), (U), (S), (Q), (F), (S'), (i), (s), (T), (D), (V), (p), (c) and (q) are known, then its Variable Portion Planned is:

v= V/S

or it can also be found as

v= 1-[F+S'i[1+s]+T+D+(Q+Vp/{360[c-q]})/ƒ-U]/S

Rule-15708:

If both (/), (U), (S), (v), (Q), (S'), (i), (s), (T), (D), (V), (p), (c) and (q) are known, then its Fixed Cost Planned is:

F= S[1-v]-S'i[1+s]-T-D-(Q+Vp/{360[c-q]})/ƒ-U

Rule-15709:

If both (/), (U), (S), (v), (F), (Q), (i), (s), (T), (D), (V), (p), (c) and (q) are known, then its Sales Past must be:

S'= [S[1-v]-F-T-D-(Q+Vp/{360[c-q]})/ƒ-U]
/{i[1+s]}

Steve Asikin ISBN 14: 978-1514685136, ISBN 10: 1514685132

Rule-15710:

If both (**/**), (**U**), (**$**), (**v**), (**F**), (**$'**), (**Q**), (**s**), (**T**), (**D**), (**V**), (**p**), (**c**) and (**q**) are known, then its Interest Portion Planned is:

i= **[$**[1-**v**]-**F**-**T**-**D**-(**Q**+**Vp**/{360[**c**-**q**]})/**/U]** /{**$'**[1+**s**]}

Rule-15711:

If both (**/**), (**U**), (**$**), (**v**), (**F**), (**$'**), (**i**), (**Q**), (**T**), (**D**), (**V**), (**p**), (**c**) and (**q**) are known, then its Sales Growth Planned is:

s= **$**/**$'**-1

or it can also be found as

s= **[$**[1-**v**]-**F**-**T**-**D**-(**Q**+**Vp**/{360[**c**-**q**]})/**/U]**/[**$'i**]-1

Rule-15712:

If both (**/**), (**U**), (**$**), (**v**), (**F**), (**$'**), (**i**), (**s**), (**Q**), (**D**), (**V**), (**p**), (**c**) and (**q**) are known, then its Tax Planned is:

T= **$**[1-**v**]-**F**-**$'i**[1+**s**]-**D**-(**Q**+**Vp**/{360[**c**-**q**]})/**/U**

Rule-15713:

If both (**/**), (**U**), (**$**), (**v**), (**F**), (**$'**), (**i**), (**s**), (**T**), (**Q**), (**V**), (**p**), (**c**) and (**q**) are known, then its Dividend Planned is:

D= **$**[1-**v**]-**F**-**T**-**$'i**[1+**s**]- (**Q**+**Vp**/{360[**c**-**q**]})/**/U**

Rule-15714:

If both (**/**), (**U**), (**$**), (**v**), (**F**), (**$'**), (**i**), (**s**), (**T**), (**D**), (**Q**), (**p**), (**c**) and (**q**) are known, then its Variable Cost Planned is:

V= 360[**c**-**q**](**/**{**U**+**$**[1-**v**]-**F**-**$'i**[1+**s**]-**T**-**D**}-**Q**)/**p**

Steve Asikin ISBN 14: 978-1514685136, ISBN 10: 1514685132

Rule-15715:

If both (**/**), (**U**), (**$**), (**v**), (**F**), (**$'**), (**i**), (**s**), (**T**), (**D**), (**V**), (**Q**), (**c**) and (**q**) are known, then its Procured Inventory Days Planned:

p= 360[**c-q**](**/**{**U**+**$**[1-**v**]-**F-$'i**[1+**s**]-**T-D**}-**Q**)/**V**

Rule-15716:

If both (**/**), (**U**), (**$**), (**v**), (**F**), (**$'**), (**i**), (**s**), (**T**), (**D**), (**V**), (**p**), (**Q**) and (**q**) are known, then its Current Ratio Planned is:

c= **q**+**Vp**/[360(**/**{**U**+**$**[1-**v**]-**F-$'i**[1+**s**]-**T-D**}-**Q**)]

Rule-15717:

If both (**/**), (**U**), (**$**), (**v**), (**F**), (**$'**), (**i**), (**s**), (**T**), (**D**), (**V**), (**p**), (**c**) and (**q**) are known, then its Quick or Acid test Ratio Planned is:

q= **c-Vp**/[360(**/**{**U**+**$**[1-**v**]-**F-$'i**[1+**s**]-**T-D**}-**Q**)]

Rule-15718:

If both (**/**), (**U**), (**$**), (**v**), (**F**), (**$'**), (**i**), (**s**), (**T**), (**D**), (**p**), (**c**) and (**q**) are known, then its Quoted Longterm Debt Planned is:

Q= **/**{**U**+**$**[1-**v**]-**F-$'i**[1+**s**]-**T-D**}-**$vp**/{360[**c-q**]}

Rule-15719:

If both (**Q**), (**U**), (**$**), (**v**), (**F**), (**$'**), (**i**), (**s**), (**T**), (**D**), (**p**), (**c**) and (**q**) are known, then its Leverage or Gearing Ratio Planned is:

/= (**Q**+**$vp**/{360[**c-q**]})/{**U**+**$**[1-**v**]-**F-$'i**[1+**s**]-**T-D**}

Steve Asikin ISBN 14: 978-1514685136, ISBN 10: 1514685132

Rule-15720:

If both (**/**), (**Q**), (**$**), (**v**), (**F**), (**$'**), (**i**), (**s**), (**T**), (**D**), (**p**), (**c**) and (**q**) are known, then its Utilized or Starting Capital must be:

U= **F**+**$'i**[1+**s**]+**T**+**D**+(**Q**+**$vp**/{360[**c**-**q**]})/**/**-**$**[1-**v**]

Rule-15721:

If both (**/**), (**U**), (**Q**), (**v**), (**F**), (**$'**), (**i**), (**s**), (**T**), (**D**), (**p**), (**c**) and (**q**) are known, then its Sales or Revenue Planned is:

$**= (**/**{**U**-**F**-**$'i[1+**s**]-**T**-**D**}-**Q**)/(**vp**/{360[**c**-**q**]}-[1-**v**])

Rule-15722:

If both (**/**), (**U**), (**$**), (**Q**), (**F**), (**$'**), (**i**), (**s**), (**T**), (**D**), (**p**), (**c**) and (**q**) are known, then its Variable Portion Planned is:

v= (**/**{**U**+**$**-**F**-**$'i**[1+**s**]-**T**-**D**}-**Q**)/[**$**(**p**/{360[**c**-**q**]}-**/**)]

Rule-15723:

If both (**/**), (**U**), (**$**), (**v**), (**Q**), (**$'**), (**i**), (**s**), (**T**), (**D**), (**p**), (**c**) and (**q**) are known, then its Fixed Cost Planned is:

F= **$**[1-**v**]-**$'i**[1+**s**]-**T**-**D**-(**Q**+**$vp**/{360[**c**-**q**]})/**/**-**U**

Rule-15724:

If both (**/**), (**U**), (**$**), (**v**), (**F**), (**Q**), (**i**), (**s**), (**T**), (**D**), (**p**), (**c**) and (**q**) are known, then its Sales Past must be:

$'**= [**$[1-**v**]-**F**-**T**-**D**-(**Q**+**$vp**/{360[**c**-**q**]})/**/**-**U**]
/{**i**[1+**s**]}

Steve Asikin ISBN 14: 978-1514685136, ISBN 10: 1514685132

Rule-15725:

If both (/), (U), ($), (v), (F), ($'), (Q), (s), (T), (D), (p), (c) and (q) are known, then its Interest Portion Planned is:

i= [$[1-v]-F-T-D-(Q+$vp/{360[c-q]})/ / U]
/{$'[1+s]}

Rule-15726:

If both (/), (U), ($), (v), (F), ($'), (i), (Q), (T), (D), (p), (c) and (q) are known, then its Sales Growth Planned is:

s= $/$'-1

or it can also be found as

s= [$[1-v]-F-T-D-(Q+$vp/{360[c-q]})/ / U]/[$'i]-1

Rule-15727:

If both (/), (U), ($), (v), (F), ($'), (i), (s), (Q), (D), (p), (c) and (q) are known, then its Tax Planned is:

T= $[1-v]-F-$'i[1+s]-D-(Q+$vp/{360[c-q]})/ / U

Rule-15728:

If both (/), (U), ($), (v), (F), ($'), (i), (s), (T), (Q), (p), (c) and (q) are known, then its Dividend Planned is:

D= $[1-v]-F-T-$'i[1+s]- (Q+$vp/{360[c-q]})/ / U

Rule-15729:

If both (/), (U), ($), (v), (F), ($'), (i), (s), (T), (D), (Q), (c) and (q) are known, then its Procured Inventory Days Planned:

p= 360[c-q](/{U+$[1-v]-F-$'i[1+s]-T-D}-Q)/[$v]

Steve Asikin ISBN 14: 978-1514685136, ISBN 10: 1514685132

Rule-15730:

If both (**/**), (**U**), (**$**), (**v**), (**F**), (**$'**), (**i**), (**s**), (**T**), (**D**), (**p**), (**Q**) and (**q**) are known, then its Current Ratio Planned is:

c= q+$vp/[360(/{U+$[1-v]-F-$'i[1+s]-T-D}-Q)]

Rule-15731:

If both (**/**), (**U**), (**$**), (**v**), (**F**), (**$'**), (**i**), (**s**), (**T**), (**D**), (**p**), (**c**) and (**Q**) are known, then its Quick or Acid Test Ratio Planned is:

q= c-$vp/[360(/{U+$[1-v]-F-$'i[1+s]-T-D}-Q)]

Rule-15732:

If both (**/**), (**U**), (**$**), (**v**), (**F**), (**$'**), (**i**), (**s**), (**T**), (**D**), (**p**), (**c**) and (**q**) are known, then its Quoted Longterm Debt Planned is:

Q= /{U+$[1-v]-F-$'i[1+s]-T-D}
-$'vp[1+s]/{360[c-q]}

Rule-15733:

If both (**Q**), (**U**), (**$**), (**v**), (**F**), (**$'**), (**i**), (**s**), (**T**), (**D**), (**p**), (**c**) and (**q**) are known, then its Leverage or Gearing Ratio Planned is:

/= (Q+$'vp[1+s]/{360[c-q]})
/{U+$[1-v]-F-$'i[1+s]-T-D}

Steve Asikin ISBN 14: 978-1514685136, ISBN 10: 1514685132

Rule-15734:

If both (**/**), (**Q**), (**S**), (**v**), (**F**), (**S'**), (**i**), (**s**), (**T**), (**D**), (**p**), (**c**) and (**q**) are known, then its Utilized or Starting Capital must be:

U= **F**+**S'i**[1+**s**]+**T**+**D**
+(**Q**+**S'vp**[1+**s**]/{360[**c**-**q**]})/**/**-**S**[1-**v**]

Rule-15735:

If both (**/**), (**U**), (**Q**), (**v**), (**F**), (**S'**), (**i**), (**s**), (**T**), (**D**), (**p**), (**c**) and (**q**) are known, then its Sales or Revenue Planned is:

S= (**Q**-**/**{**U**-**F**-**S'i**[1+**s**]-**T**-**D**}+**S'**[1+**s**]**vp**/{360[**c**-**q**]})
/{**/**[1-**v**]}

Rule-15736:

If both (**/**), (**U**), (**S**), (**Q**), (**F**), (**S'**), (**i**), (**s**), (**T**), (**D**), (**p**), (**c**) and (**q**) are known, then its Variable Portion Planned is:

v= **V**/**S**

or it can also be found as

v= (**/**{**U**+**S**-**F**-**S'i**[1+**s**]-**T**-**D**}-**Q**)
/(**S'**[1+**s**]**vp**/{360[**c**-**q**]}+**S/**)

Rule-15737:

If both (**/**), (**U**), (**S**), (**v**), (**Q**), (**S'**), (**i**), (**s**), (**T**), (**D**), (**p**), (**c**) and (**q**) are known, then its Fixed Cost Planned is:

F= **S**[1-**v**]-**S'i**[1+**s**]-**T**-**D**
-(**Q**+**S'vp**[1+**s**]/{360[**c**-**q**]})/**/**-**U**

Steve Asikin ISBN 14: 978-1514685136, ISBN 10: 1514685132

Rule-15738:

If both (**/**), (**U**), (**$**), (**v**), (**F**), (**Q**), (**i**), (**s**), (**T**), (**D**), (**p**), (**c**) and (**q**) are known, then its Sales Past must be:

$'**= (**/**{**U**+**$[1-**v**]-**F**-**T**-**D**}-**Q**)
/[[1+**s**](**vp**/{360[**c**-**q**]}+**i/**)]

Rule-15739:

If both (**/**), (**U**), (**$**), (**v**), (**F**), (**$'**), (**Q**), (**s**), (**T**), (**D**), (**p**), (**c**) and (**q**) are known, then its Interest Portion Planned is:

i= [**$**[1-**v**]-**F**-**T**-**D**-(**Q**+**$'vp**[1+**s**]/{360[**c**-**q**]})/**/**-**U**]
/{**$'**[1+**s**]}

Rule-15740:

If both (**/**), (**U**), (**$**), (**v**), (**F**), (**$'**), (**i**), (**Q**), (**T**), (**D**), (**p**), (**c**) and (**q**) are known, then its Sales Growth Planned is:

s= **$**/**$'**-1

or it can also be found as

s= (**/**{**U**+**$**[1-**v**]-**F**-**T**-**D**}-**Q**)
/[**$'**(**vp**/{360[**c**-**q**]}+**i/**)]-1

Rule-15741:

If both (**/**), (**U**), (**$**), (**v**), (**F**), (**$'**), (**i**), (**s**), (**Q**), (**D**), (**p**), (**c**) and (**q**) are known, then its Tax Planned is:

T= **$**[1-**v**]-**F**-**$'i**[1+**s**]-**D**
-(**Q**+**$'vp**[1+**s**]/{360[**c**-**q**]})/**/**-**U**

Steve Asikin ISBN 14: 978-1514685136, ISBN 10: 1514685132

Rule-15742:

If both (**/**), (**U**), (**$**), (**v**), (**F**), (**$'**), (**i**), (**s**), (**T**), (**Q**), (**p**), (**c**) and (**q**) are known, then its Dividend Planned is:

D= **$**[1-**v**]-**F**-**T**-**$'i**[1+**s**]
-(**Q**+**$'vp**[1+**s**]/{360[**c**-**q**]})/**/**-**U**

Rule-15743:

If both (**/**), (**U**), (**$**), (**v**), (**F**), (**$'**), (**i**), (**s**), (**T**), (**D**), (**Q**), (**c**) and (**q**) are known, then its Procured Inventory Days Planned:

p= 360[**c**-**q**](**/**{**U**+**$**[1-**v**]-**F**-**$'i**[1+**s**]-**T**-**D**}-**Q**)
/{**$'v**[1+**s**]}

Rule-15744:

If both (**/**), (**U**), (**$**), (**v**), (**F**), (**$'**), (**i**), (**s**), (**T**), (**D**), (**p**), (**Q**) and (**q**) are known, then its Current Ratio Planned is:

c= **q**+**$'vp**[1+**s**]
/[360(**/**{**U**+**$**[1-**v**]-**F**-**$'i**[1+**s**]-**T**-**D**}-**Q**)]

Rule-15745:

If both (**/**), (**U**), (**$**), (**v**), (**F**), (**$'**), (**i**), (**s**), (**T**), (**D**), (**p**), (**c**) and (**Q**) are known, then its Quick or Acid Test Ratio Planned is:

q= **c**-**$'vp**[1+**s**]
/[360(**/**{**U**+**$**[1-**v**]-**F**-**$'i**[1+**s**]-**T**-**D**}-**Q**)]

Steve Asikin ISBN 14: 978-1514685136, ISBN 10: 1514685132

Rule-15746:

If both (**/**), (**U**), (**$**), (**v**), (**F**), (**$'**), (**i**), (**s**), (**T**), (**d**) and (**X**) are known, then its Quoted Longterm Debt Planned is:

Q= **/**(**U**+{**$**[1-**v**]-**F**-**$'i**[1+**s**]-**T**}[1-**d**])-**X**

Rule-15747:

If both (**Q**), (**U**), (**$**), (**v**), (**F**), (**$'**), (**i**), (**s**), (**T**), (**d**) and (**X**) are known, then its Leverage or Gearing Ratio Planned is:

/= [**Q**+**X**]/(**U**+{**$**[1-**v**]-**F**-**$'i**[1+**s**]-**T**}[1-**d**])

Rule-15748:

If both (**/**), (**Q**), (**$**), (**v**), (**F**), (**$'**), (**i**), (**s**), (**T**), (**d**) and (**X**) are known, then its Utilized or Starting Capital must be:

U= [**Q**+**X**]/**/**-{**$**[1-**v**]-**F**-**$'i**[1+**s**]-**T**}[1-**d**]

Rule-15749:

If both (**/**), (**U**), (**Q**), (**v**), (**F**), (**$'**), (**i**), (**s**), (**T**), (**d**) and (**X**) are known, then its Sales or Revenue Planned is:

$**= (**F**+**$'i[1+**s**]+**T**+{[**Q**+**X**]/**/**-**U**}/[1-**d**])/[1-**v**]

Rule-15750:

If both (**/**), (**U**), (**$**), (**Q**), (**F**), (**$'**), (**i**), (**s**), (**T**), (**d**) and (**X**) are known, then its Variable Portion Planned is:

v= 1-(**F**+**$'i**[1+**s**]+**T**+{[**Q**+**X**]/**/**-**U**}/[1-**d**])/**$**

Steve Asikin ISBN 14: 978-1514685136, ISBN 10: 1514685132

Rule-15751:

If both (/), (U), ($), (v), (Q), ($'), (i), (s), (T), (d) and (X) are known, then its Fixed Cost Planned is:

F= $[1-v]-$'i[1+s]-T-{[Q+X]/⁄U}/[1-d]

Rule-15752:

If both (/), (U), ($), (v), (F), (Q), (i), (s), (T), (d) and (X) are known, then its Sales Past must be:

$'= ($[1-v]-F-T-{[Q+X]/⁄U}/[1-d])/{i[1+s]}

Rule-15753:

If both (/), (U), ($), (v), (F), ($'), (Q), (s), (T), (d) and (X) are known, then its Interest Portion Planned is:

i= ($[1-v]-F-T-{[Q+X]/⁄U}/[1-d])/{$'[1+s]}

Rule-15754:

If both (/), (U), ($), (v), (F), ($'), (i), (Q), (T), (d) and (X) are known, then its Sales Growth Planned is:

s= $/$'-1

or it can also be found as

s= ($[1-v]-F-T-{[Q+X]/⁄U}/[1-d])/[$'i]-1

Rule-15755:

If both (/), (U), ($), (v), (F), ($'), (i), (s), (Q), (d) and (X) are known, then its Tax Planned is:

T= $[1-v]-F-$'i[1+s]-{[Q+X]/⁄U}/[1-d]

Rule-15756:

If both (/), (U), ($), (v), (F), ($'), (i), (s), (T), (Q) and (X) are known, then its Dividend Payout Planned is:

d= 1-{[Q+X]/⁄U}/{$[1-v]-F-$'i[1+s]-T}

Steve Asikin ISBN 14: 978-1514685136, ISBN 10: 1514685132

Rule-15757:

If both (**/**), (**U**), (**$**), (**v**), (**F**), (**$'**), (**i**), (**s**), (**T**), (**d**) and (**Q**) are known, then its Xpress or Current Debt Planned is:

X= **/**(**U**+{**$**[1-**v**]-**F**-**$'i**[1+**s**]-**T**}[1-**d**])-**Q**

Rule-15758:

If both (**/**), (**U**), (**$**), (**v**), (**F**), (**$'**), (**i**), (**s**), (**T**), (**d**), (**P**), (**c**) and (**q**) are known, then its Quoted Longterm Debt Planned is:

Q= **/**(**U**+{**$**[1-**v**]-**F**-**$'i**[1+**s**]-**T**}[1-**d**])-**P**/[**c**-**q**]

Rule-15759:

If both (**Q**), (**U**), (**$**), (**v**), (**F**), (**$'**), (**i**), (**s**), (**T**), (**d**), (**P**), (**c**) and (**q**) are known, then its Leverage or Gearing Ratio Planned is:

/= {**Q**+**P**/[**c**-**q**]}/(**U**+{**$**[1-**v**]-**F**-**$'i**[1+**s**]-**T**}[1-**d**])

Rule-15760:

If both (**/**), (**Q**), (**$**), (**v**), (**F**), (**$'**), (**i**), (**s**), (**T**), (**d**), (**P**), (**c**) and (**q**) are known, then its Utilized or Starting Capital must be:

U= {**Q**+**P**/[**c**-**q**]}/**/**-{**$**[1-**v**]-**F**-**$'i**[1+**s**]-**T**}[1-**d**]

Rule-15761:

If both (**/**), (**U**), (**Q**), (**v**), (**F**), (**$'**), (**i**), (**s**), (**T**), (**d**), (**P**), (**c**) and (**q**) are known, then its Sales or Revenue Planned is:

$**= [**F**+**$'i[1+**s**]+**T**+({**Q**+**P**/[**c**-**q**]}/**/**-**U**)/[1-**d**]]/[1-**v**]

Steve Asikin ISBN 14: 978-1514685136, ISBN 10: 1514685132

Rule-15762:

If both (**/**), (**U**), (**$**), (**Q**), (**F**), (**$'**), (**i**), (**s**), (**T**), (**d**), (**P**), (**c**) and (**q**) are known, then its Variable Portion Planned is:

v= 1-[**F**+**$'i**[1+**s**]+**T**+({**Q**+**P**/[**c**-**q**]}/**/**-**U**)/[1-**d**]]/**$**

Rule-15763:

If both (**/**), (**U**), (**$**), (**v**), (**Q**), (**$'**), (**i**), (**s**), (**T**), (**d**), (**P**), (**c**) and (**q**) are known, then its Fixed Cost Planned is:

F= **$**[1-**v**]-**$'i**[1+**s**]-**T**-({**Q**+**P**/[**c**-**q**]}/**/**-**U**)/[1-**d**]

Rule-15764:

If both (**/**), (**U**), (**$**), (**v**), (**F**), (**Q**), (**i**), (**s**), (**T**), (**d**), (**P**), (**c**) and (**q**) are known, then its Sales Past must be:

$'**= [**$[1-**v**]-**F**-**T**-({**Q**+**P**/[**c**-**q**]}/**/**-**U**)/[1-**d**]]/{**i**[1+**s**]}

Rule-15765:

If both (**/**), (**U**), (**$**), (**v**), (**F**), (**$'**), (**Q**), (**s**), (**T**), (**d**), (**P**), (**c**) and (**q**) are known, then its Interest Portion Planned is:

i= [**$**[1-**v**]-**F**-**T**-({**Q**+**P**/[**c**-**q**]}/**/**-**U**)/[1-**d**]]/{**$'**[1+**s**]}

Rule-15766:

If both (**/**), (**U**), (**$**), (**v**), (**F**), (**$'**), (**i**), (**Q**), (**T**), (**d**), (**P**), (**c**) and (**q**) are known, then its Sales Growth Planned is:

s= **$**/**$'**-1

or it can also be found as

s= [**$**[1-**v**]-**F**-**T**-({**Q**+**P**/[**c**-**q**]}/**/**-**U**)/[1-**d**]]/[**$'i**]-1

Steve Asikin ISBN 14: 978-1514685136, ISBN 10: 1514685132

Rule-15767:

If both (**/**), (**U**), (**$**), (**v**), (**F**), (**$'**), (**i**), (**s**), (**Q**), (**d**), (**P**), (**c**) and (**q**) are known, then its Tax Planned is:

T= **$**[1-**v**]-**F**-**$'i**[1+**s**]-({**Q**+**P**/[**c**-**q**]}/**/U**)/[1-**d**]

Rule-15768:

If both (**/**), (**U**), (**$**), (**v**), (**F**), (**$'**), (**i**), (**s**), (**T**), (**Q**), (**P**), (**c**) and (**q**) are known, then its Dividend Payout Planned is:

d= 1-({**Q**+**P**/[**c**-**q**]}/**/U**)/{**$**[1-**v**]-**F**-**$'i**[1+**s**]-**T**}

Rule-15769:

If both (**/**), (**U**), (**$**), (**v**), (**F**), (**$'**), (**i**), (**s**), (**T**), (**d**), (**Q**), (**c**) and (**q**) are known, then its Procured Inventory Planned is:

P= [**c**-**q**][**/**(**U**+{**$**[1-**v**]-**F**-**$'i**[1+**s**]-**T**}[1-**d**])-**Q**]

Rule-15770:

If both (**/**), (**U**), (**$**), (**v**), (**F**), (**$'**), (**i**), (**s**), (**T**), (**d**), (**P**), (**Q**) and (**q**) are known, then its Current Ratio Planned is:

c= **q**+**P**/[**/**(**U**+{**$**[1-**v**]-**F**-**$'i**[1+**s**]-**T**}[1-**d**])-**Q**]

Rule-15771:

If both (**/**), (**U**), (**$**), (**v**), (**F**), (**$'**), (**i**), (**s**), (**T**), (**d**), (**P**), (**c**) and (**Q**) are known, then its Quick or Acid Test Ratio Planned is:

q= **c**-**P**/[**/**(**U**+{**$**[1-**v**]-**F**-**$'i**[1+**s**]-**T**}[1-**d**])-**Q**]

Steve Asikin ISBN 14: 978-1514685136, ISBN 10: 1514685132

Rule-15772:

If both (**/**), (**U**), (**$**), (**v**), (**F**), (**$'**), (**i**), (**s**), (**T**), (**d**), (**V**), (**p**), (**c**) and (**q**) are known, then its Quoted Longterm Debt Planned is:

Q= **/**(**U**+{**$**[1-**v**]-**F**-**$'i**[1+**s**]-**T**}[1-**d**])
-**Vp**/{360[**c**-**q**]}

Rule-15773:

If both (**Q**), (**U**), (**$**), (**v**), (**F**), (**$'**), (**i**), (**s**), (**T**), (**d**), (**V**), (**p**), (**c**) and (**q**) are known, then its Leverage or Gearing Ratio Planned is:

/= (**Q**+**Vp**/{360[**c**-**q**]})
/(**U**+{**$**[1-**v**]-**F**-**$'i**[1+**s**]-**T**}[1-**d**])

Rule-15774:

If both (**/**), (**Q**), (**$**), (**v**), (**F**), (**$'**), (**i**), (**s**), (**T**), (**d**), (**V**), (**p**), (**c**) and (**q**) are known, then its Utilized or Starting Capital must be:

U= (**Q**+**Vp**/{360[**c**-**q**]})/**/**
-{**$**[1-**v**]-**F**-**$'i**[1+**s**]-**T**}[1-**d**]

Rule-15775:

If both (**/**), (**U**), (**Q**), (**v**), (**F**), (**$'**), (**i**), (**s**), (**T**), (**d**), (**V**), (**p**), (**c**) and (**q**) are known, then its Sales or Revenue Planned is:

$**= {**F**+**$'i[1+**s**]+**T**
+[(**Q**+**Vp**/{360[**c**-**q**]})/**/**-**U**]/[1-**d**]}/[1-**v**]

Steve Asikin ISBN 14: 978-1514685136, ISBN 10: 1514685132

Rule-15776:

If both (**/**), (**U**), (**$**), (**Q**), (**F**), (**$'**), (**i**), (**s**), (**T**), (**d**), (**V**), (**p**), (**c**) and (**q**) are known, then its Variable Portion Planned is:

v= V/$

or it can also be found as

v= 1-{F+$'i[1+s]+T
+[(Q+Vp/{360[c-q]})//U]/[1-d]}/$

Rule-15777:

If both (**/**), (**U**), (**$**), (**v**), (**Q**), (**$'**), (**i**), (**s**), (**T**), (**d**), (**V**), (**p**), (**c**) and (**q**) are known, then its Fixed Cost Planned is:

F= $[1-v]-$'i[1+s]-T-[(Q+Vp/{360[c-q]})//U]
/[1-d]

Rule-15778:

If both (**/**), (**U**), (**$**), (**v**), (**F**), (**Q**), (**i**), (**s**), (**T**), (**d**), (**V**), (**p**), (**c**) and (**q**) are known, then its Sales Past must be:

$'= {$[1-v]-F-T-[(Q+Vp/{360[c-q]})//U] /[1-d]}
/{i[1+s]}

Rule-15779:

If both (**/**), (**U**), (**$**), (**v**), (**F**), (**$'**), (**Q**), (**s**), (**T**), (**d**), (**V**), (**p**), (**c**) and (**q**) are known, then its Interest Portion Planned is:

i= {$[1-v]-F-T-[(Q+Vp/{360[c-q]})//U] /[1-d]}
/{$'[1+s]}

Steve Asikin ISBN 14: 978-1514685136, ISBN 10: 1514685132

Rule-15780:

If both (**/**), (**U**), (**$**), (**v**), (**F**), (**$'**), (**i**), (**Q**), (**T**), (**d**), (**V**), (**p**), (**c**) and (**q**) are known, then its Sales Growth Planned is:

s= {$[1-v]-F-T-[(Q+Vp/{360[c-q]})/**/**U] /[1-d]}
/[$'i]-1

Rule-15781:

If both (**/**), (**U**), (**$**), (**v**), (**F**), (**$'**), (**i**), (**s**), (**Q**), (**d**), (**V**), (**p**), (**c**) and (**q**) are known, then its Tax Planned is:

T= $[1-v]-F-$'i[1+s]-[(Q+Vp/{360[c-q]})/**/**U]
/[1-d]

Rule-15782:

If both (**/**), (**U**), (**$**), (**v**), (**F**), (**$'**), (**i**), (**s**), (**T**), (**Q**), (**V**), (**p**), (**c**) and (**q**) are known, then its Dividend Payout Planned is:

d= 1-[(Q+Vp/{360[c-q]})/**/**U]
/{$[1-v]-F-$'i[1+s]-T}

Rule-15783:

If both (**/**), (**U**), (**$**), (**v**), (**F**), (**$'**), (**i**), (**s**), (**T**), (**d**), (**Q**), (**p**), (**c**) and (**q**) are known, then its Variable Cost Planned is:

V= 360[c-q][**/**(U+{$[1-v]-F-$'i[1+s]-T}[1-d])-Q]/p

Rule-15784:

If both (**/**), (**U**), (**$**), (**v**), (**F**), (**$'**), (**i**), (**s**), (**T**), (**d**), (**V**), (**Q**), (**c**) and (**q**) are known, then its Procured Inventory Days Planned:

p= 360[c-q][**/**(U+{$[1-v]-F-$'i[1+s]-T}[1-d])-Q]/V

Steve Asikin ISBN 14: 978-1514685136, ISBN 10: 1514685132

Rule-15785:

If both (/), (U), ($), (v), (F), ($'), (i), (s), (T), (d), (V), (p), (Q) and (q) are known, then its Current Ratio Planned is:

c= q+Vp
/{360[/(U+{$[1-v]-F-$'i[1+s]-T}[1-d])-Q]}

Rule-15786:

If both (/), (U), ($), (v), (F), ($'), (i), (s), (T), (d), (V), (p), (c) and (Q) are known, then its Quick or Acid Test Ratio Planned is:

q= c-Vp
/{360[/(U+{$[1-v]-F-$'i[1+s]-T}[1-d])-Q]}

Rule-15787:

If both (/), (U), ($), (v), (F), ($'), (i), (s), (T), (d), (p), (c) and (q) are known, then its Quoted Longterm Debt Planned is:

Q= /(U+{$[1-v]-F-$'i[1+s]-T}[1-d])
-$vp/{360[c-q]}

Rule-15788:

If both (Q), (U), ($), (v), (F), ($'), (i), (s), (T), (d), (p), (c) and (q) are known, then its Leverage or Gearing Ratio Planned is:

/= (Q+$vp/{360[c-q]})
/(U+{$[1-v]-F-$'i[1+s]-T}[1-d])

Steve Asikin ISBN 14: 978-1514685136, ISBN 10: 1514685132

Rule-15789:

If both (/), (Q), ($), (v), (F), ($'), (i), (s), (T), (d), (p), (c) and (q) are known, then its Utilized or Starting Capital must be:

U= (Q+$vp/{360[c-q]})//
-{$[1-v]-F-$'i[1+s]-T}[1-d]

Rule-15790:

If both (/), (U), (Q), (v), (F), ($'), (i), (s), (T), (d), (p), (c) and (q) are known, then its Sales or Revenue Planned is:

$= [/(U-{F+$'i[1+s]-T}[1-d])-Q]
/(vp/{360[c-q]}-/[1-v] [1-d])

Rule-15791:

If both (/), (U), ($), (Q), (F), ($'), (i), (s), (T), (d), (p), (c) and (q) are known, then its Variable Portion Planned is:

v= [/(U+{$-F-$'i[1+s]-T}[1-d])-Q]
/[$(p/{360[c-q]}+/[1-d])]

Rule-15792:

If both (/), (U), ($), (v), (Q), ($'), (i), (s), (T), (d), (p), (c) and (q) are known, then its Fix Cost Planned is:

F= $[1-v]-$'i[1+s]-T-[(Q+$vp/{360[c-q]})//-U]
/[1-d]

Steve Asikin ISBN 14: 978-1514685136, ISBN 10: 1514685132

Rule-15793:

If both (**/**), (**U**), (**$**), (**v**), (**F**), (**Q**), (**i**), (**s**), (**T**), (**d**), (**p**), (**c**) and (**q**) are known, then its Sales Past must be:

$'**= {**$[1-**v**]-**F**-**T**-[(**Q**+**$vp**/{360[**c**-**q**]})/**/U**] /[1-**d**]} /{**i**[1+**s**]}

Rule-15794:

If both (**/**), (**U**), (**$**), (**v**), (**F**), (**$'**), (**Q**), (**s**), (**T**), (**d**), (**p**), (**c**) and (**q**) are known, then its Interest Portion Planned is:

i= {**$**[1-**v**]-**F**-**T**-[(**Q**+**$vp**/{360[**c**-**q**]})/**/U**]/[1-**d**]} /{**$'**[1+**s**]}

Rule-15795:

If both (**/**), (**U**), (**$**), (**v**), (**F**), (**$'**), (**i**), (**Q**), (**T**), (**d**), (**p**), (**c**) and (**q**) are known, then its Sales Growth Planned is:

s= **$**/**$'**-1

or it can also be found as

s= {**$**[1-**v**]-**F**-**T**-[(**Q**+**$vp**/{360[**c**-**q**]})/**/U**]/[1-**d**]} /[**$'i**]-1

Rule-15796:

If both (**/**), (**U**), (**$**), (**v**), (**F**), (**$'**), (**i**), (**s**), (**Q**), (**d**), (**p**), (**c**) and (**q**) are known, then its Tax Planned is:

T= **$**[1-**v**]-**F**-**$'i**[1+**s**]-[(**Q**+**$vp**/{360[**c**-**q**]})/**/U**] /[1-**d**]

Steve Asikin ISBN 14: 978-1514685136, ISBN 10: 1514685132

Rule-15797:

If both (**/**), (**U**), (**$**), (**v**), (**F**), (**$'**), (**i**), (**s**), (**T**), (**Q**), (**p**), (**c**) and (**q**) are known, then its Dividend Payout Planned is:

d= 1-[(**Q**+**$vp**/{360[**c**-**q**]})/**/**-**U**]
/{**$**[1-**v**]-**F**-**$'i**[1+**s**]-**T**}

Rule-15798:

If both (**/**), (**U**), (**$**), (**v**), (**F**), (**$'**), (**i**), (**s**), (**T**), (**d**), (**Q**), (**c**) and (**q**) are known, then its Procured Inventory Days Planned:

p= 360[**c**-**q**][**/**(**U**+{**$**[1-**v**]-**F**-**$'i**[1+**s**]-**T**}[1-**d**])-**Q**]
/[**$v**]

Rule-15799:

If both (**/**), (**U**), (**$**), (**v**), (**F**), (**$'**), (**i**), (**s**), (**T**), (**d**), (**p**), (**Q**) and (**q**) are known, then its Current Ratio Planned is:

c= **q**+**$vp**
/{360[**/**(**U**+{**$**[1-**v**]-**F**-**$'i**[1+**s**]-**T**}[1-**d**])-**Q**]}

Rule-15800:

If both (**/**), (**U**), (**$**), (**v**), (**F**), (**$'**), (**i**), (**s**), (**T**), (**d**), (**p**), (**c**) and (**Q**) are known, then its Quick or Acid Test Ratio Planned is:

q= **c**-**$vp**
/{360[**/**(**U**+{**$**[1-**v**]-**F**-**$'i**[1+**s**]-**T**}[1-**d**])-**Q**]}

Steve Asikin ISBN 14: 978-1514685136, ISBN 10: 1514685132

Rule-15801:

If both (**/**), (**U**), (**$**), (**v**), (**F**), (**$'**), (**i**), (**s**), (**T**), (**d**), (**$'**), (**p**), (**s**), (**c**) and (**q**) are known, then its Quoted Longterm Debt Planned is:

Q= **/**(**U**+{**$**[1-**v**]-**F**-**$'i**[1+**s**]-**T**}[1-**d**])
-**$'vp**[1+**s**]/{360[**c**-**q**]}

Rule-15802:

If both (**Q**), (**U**), (**$**), (**v**), (**F**), (**$'**), (**i**), (**s**), (**T**), (**d**), (**$'**), (**p**), (**s**), (**c**) and (**q**) are known, then its Leverage or Gearing Ratio Planned is:

/= (**Q**+**$'vp**[1+**s**]/{360[**c**-**q**]})
/(**U**+{**$**[1-**v**]-**F**-**$'i**[1+**s**]-**T**}[1-**d**])

Rule-15803:

If both (**/**), (**Q**), (**$**), (**v**), (**F**), (**$'**), (**i**), (**s**), (**T**), (**d**), (**$'**), (**p**), (**s**), (**c**) and (**q**) are known, then its Utilized or Starting Capital must be:

U= (**Q**+**$'vp**[1+**s**]/{360[**c**-**q**]})/**/**
-{**$**[1-**v**]-**F**-**$'i**[1+**s**]-**T**}[1-**d**]

Rule-15804:

If both (**/**), (**U**), (**Q**), (**v**), (**F**), (**$'**), (**i**), (**s**), (**T**), (**d**), (**$'**), (**p**), (**s**), (**c**) and (**q**) are known, then its Sales or Revenue Planned is:

$**= [**Q**-**/**(**U**-{**F**+**$'i[1+**s**]+**T**}[1-**d**])
+**$'vp**[1+**s**]/{360[**c**-**q**]}]/{**/**[1-**v**][1-**d**]}

Steve Asikin ISBN 14: 978-1514685136, ISBN 10: 1514685132

Rule-15805:

If both (**/**), (**U**), (**$**), (**Q**), (**F**), (**$'**), (**i**), (**s**), (**T**), (**d**), (**$'**), (**p**), (**s**), (**c**) and (**q**) are known, then its Variable Portion Planned is:

v= [**/**(**U**+{**$**-**F**-**$'i**[1+**s**]-**T**}[1-**d**])-**Q**]
/(**$'p**[1+**s**]/{360[**c**-**q**]} +**$/**[1-**d**])

Rule-15806:

If both (**/**), (**U**), (**$**), (**v**), (**Q**), (**$'**), (**i**), (**s**), (**T**), (**d**), (**$'**), (**p**), (**s**), (**c**) and (**q**) are known, then its Fixed Cost Planned is:

F= **$**[1-**v**]-**$'i**[1+**s**]-**T**
-[(**Q**+**$'vp**[1+**s**]/{360[**c**-**q**]})/**/-U**]/[1-**d**]

Rule-15807:

If both (**/**), (**U**), (**$**), (**v**), (**F**), (**Q**), (**i**), (**s**), (**T**), (**d**), (**$'**), (**p**), (**s**), (**c**) and (**q**) are known, then its Sales Past must be:

$'**= [**/**(**U**+{**$[1-**v**]-**F**-**$'i**[1+**s**]-**T**}[1-**d**])-**Q**]
/([1+**s**]{**vp** /{360[**c**-**q**]+**i/**[1-**d**]})

Rule-15808:

If both (**/**), (**U**), (**$**), (**v**), (**F**), (**$'**), (**Q**), (**s**), (**T**), (**d**), (**$'**), (**p**), (**s**), (**c**) and (**q**) are known, then its Interest Portion Planned is:

i= {**$**[1-**v**]-**F**-**T**-[(**Q**+**$'vp**[1+**s**]/{360[**c**-**q**]})/**/-U**]
/[1-**d**]}/{**$'**[1+**s**]}

Steve Asikin ISBN 14: 978-1514685136, ISBN 10: 1514685132

Rule-15809:

If both (**/**), (**U**), (**$**), (**v**), (**F**), (**$'**), (**i**), (**Q**), (**T**), (**d**), (**$'**), (**p**), (**s**), (**c**) and (**q**) are known, then its Sales Growth Planned is:

s= **$**/**$'**-1

or it can also be found as

s= [**/**U+{**$**[1-**v**]-**F**-**$'i**[1+**s**]-**T**}[1-**d**])-**Q**]
/(**$'**{**vp** /{360[**c**-**q**]+**i/**[1-**d**]})-1

Rule-15810:

If both (**/**), (**U**), (**$**), (**v**), (**F**), (**$'**), (**i**), (**s**), (**Q**), (**d**), (**$'**), (**p**), (**s**), (**c**) and (**q**) are known, then its Tax Planned is:

T= **$**[1-**v**]-**F**-**$'i**[1+**s**]
-[(**Q**+**$'vp**[1+**s**]/{360[**c**-**q**]})/**/U**]/[1-**d**]

Rule-15811:

If both (**/**), (**U**), (**$**), (**v**), (**F**), (**$'**), (**i**), (**s**), (**T**), (**Q**), (**$'**), (**p**), (**s**), (**c**) and (**q**) are known, then its Dividend Payout Planned is:

d= 1-[(**Q**+**$'vp**[1+**s**]/{360[**c**-**q**]})/**/U**]
/{**$**[1-**v**]-**F**-**$'i**[1+**s**]-**T**}

Rule-15812:

If both (**/**), (**U**), (**$**), (**v**), (**F**), (**$'**), (**i**), (**s**), (**T**), (**d**), (**$'**), (**Q**), (**s**), (**c**) and (**q**) are known, then its Procured Inventory Days Planned:

p= 360[**c**-**q**][**/**U+{**$**[1-**v**]-**F**-**$'i**[1+**s**]-**T**}[1-**d**])-**Q**]
/{**$'v**[1+**s**]}

Steve Asikin ISBN 14: 978-1514685136, ISBN 10: 1514685132

Rule-15813:

If both (**/**), (**U**), (**$**), (**v**), (**F**), (**$'**), (**i**), (**s**), (**T**), (**d**), (**$'**), (**p**), (**s**), (**Q**) and (**q**) are known, then its Current Ratio Planned is:

c= **q**+**$'vp**[1+**s**]
/{360[**/**(**U**+{**$**[1-**v**]-**F**-**$'i**[1+**s**]-**T**}[1-**d**])-**Q**]}

Rule-15814:

If both (**/**), (**U**), (**$**), (**v**), (**F**), (**$'**), (**i**), (**s**), (**T**), (**d**), (**$'**), (**p**), (**s**), (**c**) and (**Q**) are known, then its Quick or Acid Test Ratio Planned is:

q= **c**-**$'vp**[1+**s**]
/{360[**/**(**U**+{**$**[1-**v**]-**F**-**$'i**[1+**s**]-**T**}[1-**d**])-**Q**]}

Rule-15815:

If both (**/**), (**U**), (**$**), (**v**), (**F**), (**$'**), (**i**), (**s**), (**T**), (**A**), (**d**) and (**X**) are known, then its Quoted Longterm Debt Planned is:

Q= **/**{**U**+**$**[1-**v**]-**F**-**$'i**[1+**s**]-**T**-**Ad**}-**X**

Rule-15816:

If both (**Q**), (**U**), (**$**), (**v**), (**F**), (**$'**), (**i**), (**s**), (**T**), (**A**), (**d**) and (**X**) are known, then its Leverage or Gearing Ratio Planned is:

/= [**Q**+**X**]/{**U**+**$**[1-**v**]-**F**-**$'i**[1+**s**]-**T**-**Ad**}

Rule-15817:

If both (**/**), (**Q**), (**$**), (**v**), (**F**), (**$'**), (**i**), (**s**), (**T**), (**A**), (**d**) and (**X**) are known, then its Utilized or Starting Capital must be:

U= **F**+**$'i**[1+**s**]+**T**+**Ad**+[**Q**+**X**]/**/**-**$**[1-**v**]

Steve Asikin ISBN 14: 978-1514685136, ISBN 10: 1514685132

Rule-15818:

If both (**/**), (**U**), (**Q**), (**v**), (**F**), (**$'**), (**i**), (**s**), (**T**), (**A**), (**d**) and (**X**) are known, then its Sales or Revenue Planned is:

$**= {**F**+**$'i[1+**s**]+**T**+**Ad**+[**Q**+**X**]/**/U**}/[1-**v**]

Rule-15819:

If both (**/**), (**U**), (**$**), (**Q**), (**F**), (**$'**), (**i**), (**s**), (**T**), (**A**), (**d**) and (**X**) are known, then its Variable Portion Planned is:

v= 1-{**F**+**$'i**[1+**s**]+**T**+**Ad**+[**Q**+**X**]/**/U**}/**$**

Rule-15820:

If both (**/**), (**U**), (**$**), (**v**), (**Q**), (**$'**), (**i**), (**s**), (**T**), (**A**), (**d**) and (**X**) are known, then its Fixed Cost Planned is:

F= **$**[1-**v**]-**$'i**[1+**s**]-**T**-**Ad**}-[**Q**+**X**]/**/U**

Rule-15821:

If both (**/**), (**U**), (**$**), (**v**), (**F**), (**Q**), (**i**), (**s**), (**T**), (**A**), (**d**) and (**X**) are known, then its Sales Past must be:

$'**= {**$[1-**v**]-**F**-**T**-**Ad**-[**Q**+**X**]/**/U**}/{**i**[1+**s**]}

Rule-15822:

If both (**/**), (**U**), (**$**), (**v**), (**F**), (**$'**), (**Q**), (**s**), (**T**), (**A**), (**d**) and (**X**) are known, then its Interest Portion Planned is:

i= {**$**[1-**v**]-**F**-**T**-**Ad**-[**Q**+**X**]/**/U**}/{**$'**[1+**s**]}

Steve Asikin ISBN 14: 978-1514685136, ISBN 10: 1514685132

Rule-15823:

If both (/), (U), ($), (v), (F), ($'), (i), (Q), (T), (A), (d) and (X) are known, then its Sales Growth Planned is:

s= $/$'-1

or it can also be found as

s= {$[1-v]-F-T-Ad-[Q+X]/-U}/[$'i]-1

Rule-15824:

If both (/), (U), ($), (v), (F), ($'), (i), (s), (Q), (A), (d) and (X) are known, then its Tax Planned is:

T= $[1-v]-F-$'i[1+s]-Ad-[Q+X]/-U

Rule-15825:

If both (/), (U), ($), (v), (F), ($'), (i), (s), (T), (Q), (d) and (X) are known, then its After Tax Income Planned is:

A= {$[1-v]-F-$'i[1+s]-T-[Q+X]/-U}/d

Rule-15826:

If both (/), (U), ($), (v), (F), ($'), (i), (s), (T), (A), (Q) and (X) are known, then its Dividend Portion Planned is:

d= {$[1-v]-F-$'i[1+s]-T-[Q+X]/-U}/A

Rule-15827:

If both (/), (U), ($), (v), (F), ($'), (i), (s), (T), (A), (d) and (Q) are known, then its Xpress or Current Debt Planned is:

X= /{U+$[1-v]-F-$'i[1+s]-T-Ad}-Q

Steve Asikin ISBN 14: 978-1514685136, ISBN 10: 1514685132

Rule-15828:

If both (/), (U), (S), (v), (F), (S'), (i), (s), (T), (A), (d), (P), (c) and (q) are known, then its Quoted Longterm Debt Planned is:

Q= /{U+S[1-v]-F-S'i[1+s]-T-Ad}-P/[c-q]

Rule-15829:

If both (Q), (U), (S), (v), (F), (S'), (i), (s), (T), (A), (d), (P), (c) and (q) are known, then its Leverage or Gearing Ratio Planned is:

/= {Q+P/[c-q]}/{U+S[1-v]-F-S'i[1+s]-T-Ad}

Rule-15830:

If both (/), (Q), (S), (v), (F), (S'), (i), (s), (T), (A), (d), (P), (c) and (q) are known, then its Utilized or Starting Capital must be:

U= F+S'i[1+s]+T+Ad+{Q+P/[c-q]}/ /-S[1-v]

Rule-15831:

If both (/), (U), (Q), (v), (F), (S'), (i), (s), (T), (A), (d), (P), (c) and (q) are known, then its Sales or Revenue Planned is:

S= (F+S'i[1+s]+T+Ad+{Q+P/[c-q]}/ /-U)/[1-v]

Rule-15832:

If both (/), (U), (S), (Q), (F), (S'), (i), (s), (T), (A), (d), (P), (c) and (q) are known, then its Variable Portion Planned is:

v= 1-(F+S'i[1+s]+T+Ad+{Q+P/[c-q]}/ /-U)/S

Steve Asikin ISBN 14: 978-1514685136, ISBN 10: 1514685132

Rule-15833:

If both (**/**), (**U**), (**$**), (**v**), (**Q**), (**$'**), (**i**), (**s**), (**T**), (**A**), (**d**), (**P**), (**c**) and (**q**) are known, then its Fixed Cost Planned is:

F= **$**[1-**v**]-**$'i**[1+**s**]-**T**-**Ad**}-{**Q**+**P**/[**c**-**q**]}/**/**-**U**

Rule-15834:

If both (**/**), (**U**), (**$**), (**v**), (**F**), (**Q**), (**i**), (**s**), (**T**), (**A**), (**d**), (**P**), (**c**) and (**q**) are known, then its Sales Past must be:

$'**= (**$[1-**v**]-**F**-**T**-**Ad**-{**Q**+**P**/[**c**-**q**]}/**/**-**U**)/{**i**[1+**s**]}

Rule-15835:

If both (**/**), (**U**), (**$**), (**v**), (**F**), (**$'**), (**Q**), (**s**), (**T**), (**A**), (**d**), (**P**), (**c**) and (**q**) are known, then its Interest Portion Planned is:

i= (**$**[1-**v**]-**F**-**T**-**Ad**-{**Q**+**P**/[**c**-**q**]}/**/**-**U**)/{**$'**[1+**s**]}

Rule-15836:

If both (**/**), (**U**), (**$**), (**v**), (**F**), (**$'**), (**i**), (**Q**), (**T**), (**A**), (**d**), (**P**), (**c**) and (**q**) are known, then its Sales Growth Planned is:

s= **$**/**$'**-1

or it can also be found as

s= (**$**[1-**v**]-**F**-**T**-**Ad**-{**Q**+**P**/[**c**-**q**]}/**/**-**U**)/[**$'i**]-1

Rule-15837:

If both (**/**), (**U**), (**$**), (**v**), (**F**), (**$'**), (**i**), (**s**), (**Q**), (**A**), (**d**), (**P**), (**c**) and (**q**) are known, then its Tax Planned is:

T= **$**[1-**v**]-**F**-**$'i**[1+**s**]-**Ad**-{**Q**+**P**/[**c**-**q**]}/**/**-**U**

Steve Asikin ISBN 14: 978-1514685136, ISBN 10: 1514685132

Rule-15838:

If both (**/**), (**U**), (**$**), (**v**), (**F**), (**$'**), (**i**), (**s**), (**T**), (**Q**), (**d**), (**P**), (**c**) and (**q**) are known, then its After Tax Income Planned is:

A= (**$**[1-**v**]-**F**-**$'i**[1+**s**]-**T**-{**Q**+**P**/[**c**-**q**]}/**/**-**U**)/**d**

Rule-15839:

If both (**/**), (**U**), (**$**), (**v**), (**F**), (**$'**), (**i**), (**s**), (**T**), (**A**), (**Q**), (**P**), (**c**) and (**q**) are known, then its Dividend Payout Planned is:

d= (**$**[1-**v**]-**F**-**$'i**[1+**s**]-**T**-{**Q**+**P**/[**c**-**q**]}/**/**-**U**)/**A**

Rule-15840:

If both (**/**), (**U**), (**$**), (**v**), (**F**), (**$'**), (**i**), (**s**), (**T**), (**A**), (**d**), (**Q**), (**c**) and (**q**) are known, then its Procured Inventory Planned is:

P= [**c**-**q**](**/**{**U**+**$**[1-**v**]-**F**-**$'i**[1+**s**]-**T**-**Ad**}-**Q**)

Rule-15841:

If both (**/**), (**U**), (**$**), (**v**), (**F**), (**$'**), (**i**), (**s**), (**T**), (**A**), (**d**), (**P**), (**Q**) and (**q**) are known, then its Current Ratio Planned is:

c= **q**+**P**/(**/**{**U**+**$**[1-**v**]-**F**-**$'i**[1+**s**]-**T**-**Ad**}-**Q**)

Rule-15842:

If both (**/**), (**U**), (**$**), (**v**), (**F**), (**$'**), (**i**), (**s**), (**T**), (**A**), (**d**), (**P**), (**c**) and (**Q**) are known, then its Quick or Acid Test Ratio Planned is:

q= **c**-**P**/(**/**{**U**+**$**[1-**v**]-**F**-**$'i**[1+**s**]-**T**-**Ad**}-**Q**)

Steve Asikin ISBN 14: 978-1514685136, ISBN 10: 1514685132

Rule-15843:

If both (**/**), (**U**), (**$**), (**v**), (**F**), (**$'**), (**i**), (**s**), (**T**), (**A**), (**d**), (**V**), (**p**), (**c**) and (**q**) are known, then its Quoted Longterm Debt Planned is:

Q= **/**{**U**+**$**[1-**v**]-**F**-**$'i**[1+**s**]-**T**-**Ad**}-**Vp**/{360[**c**-**q**]}

Rule-15844:

If both (**Q**), (**U**), (**$**), (**v**), (**F**), (**$'**), (**i**), (**s**), (**T**), (**A**), (**d**), (**V**), (**p**), (**c**) and (**q**) are known, then its Leverage or Gearing Ratio Planned is:

/= (**Q**+**Vp**/{360[**c**-**q**]})
/{**U**+**$**[1-**v**]-**F**-**$'i**[1+**s**]-**T**-**Ad**}

Rule-15845:

If both (**/**), (**Q**), (**$**), (**v**), (**F**), (**$'**), (**i**), (**s**), (**T**), (**A**), (**d**), (**V**), (**p**), (**c**) and (**q**) are known, then its Utilized or Starting Capital must be:

U= **F**+**$'i**[1+**s**]+**T**+**Ad**+(**Q**+**Vp**/{360[**c**-**q**]})/**/**-**$**[1-**v**]

Rule-15846:

If both (**/**), (**U**), (**Q**), (**v**), (**F**), (**$'**), (**i**), (**s**), (**T**), (**A**), (**d**), (**V**), (**p**), (**c**) and (**q**) are known, then its Sales or Revenue Planned is:

$**= **[F**+**$'i[1+**s**]+**T**+**Ad**+(**Q**+**Vp**/{360[**c**-**q**]})/**/**-**U]**
/[1-**v**]

Steve Asikin ISBN 14: 978-1514685136, ISBN 10: 1514685132

Rule-15847:

If both (/), (**U**), (**$**), (**Q**), (**F**), (**$'**), (**i**), (**s**), (**T**), (**A**), (**d**), (**V**), (**p**), (**c**) and (**q**) are known, then its Variable Portion Planned is:

v= **V/$**

or it can also be found as

v= 1-[**F**+**$'i**[1+**s**]+**T**+**Ad**
+(**Q**+**Vp**/{360[**c**-**q**]})/**/**-**U**]/**$**

Rule-15848:

If both (/), (**U**), (**$**), (**v**), (**Q**), (**$'**), (**i**), (**s**), (**T**), (**A**), (**d**), (**V**), (**p**), (**c**) and (**q**) are known, then its Fixed Cost Planned is:

F= **$**[1-**v**]-**$'i**[1+**s**]-**T**-**Ad**}-(**Q**+**Vp**/{360[**c**-**q**]})/**/**-**U**

Rule-15849:

If both (/), (**U**), (**$**), (**v**), (**F**), (**Q**), (**i**), (**s**), (**T**), (**A**), (**d**), (**V**), (**p**), (**c**) and (**q**) are known, then its Sales Past must be:

$'**= [**$[1-**v**]-**F**-**T**-**Ad**-(**Q**+**Vp**/{360[**c**-**q**]})/**/**-**U**]
/{**i**[1+**s**]}

Rule-15850:

If both (/), (**U**), (**$**), (**v**), (**F**), (**$'**), (**Q**), (**s**), (**T**), (**A**), (**d**), (**V**), (**p**), (**c**) and (**q**) are known, then its Interest Portion Planned is:

i= [**$**[1-**v**]-**F**-**T**-**Ad**-(**Q**+**Vp**/{360[**c**-**q**]})/**/**-**U**]
/{**$'**[1+**s**]}

Steve Asikin ISBN 14: 978-1514685136, ISBN 10: 1514685132

Rule-15851:

If both (/), (U), (S), (v), (F), (S'), (i), (Q), (T), (A), (d), (V), (p), (c) and (q) are known, then its Sales Growth Planned is:

s= [S[1-v]-F-T-Ad-(Q+Vp/{360[c-q]})//U] / [S'i]-1

Rule-15852:

If both (/), (U), (S), (v), (F), (S'), (i), (s), (Q), (A), (d), (V), (p), (c) and (q) are known, then its Tax Planned is:

T= S[1-v]-F-S'i[1+s]-Ad-(Q+Vp/{360[c-q]})/ /U

Rule-15853:

If both (/), (U), (S), (v), (F), (S'), (i), (s), (T), (Q), (d), (V), (p), (c) and (q) are known, then its After Tax Income Planned is:

A= [S[1-v]-F-S'i[1+s]-T -(Q+Vp/{360[c-q]})/ /U]/d

Rule-15854:

If both (/), (U), (S), (v), (F), (S'), (i), (s), (T), (A), (Q), (V), (p), (c) and (q) are known, then its Dividend Payout Planned is:

d= [S[1-v]-F-S'i[1+s]-T -(Q+Vp/{360[c-q]})/ /U]/A

Steve Asikin ISBN 14: 978-1514685136, ISBN 10: 1514685132

Rule-15855:

If both (**/**), (**U**), (**S**), (**v**), (**F**), (**S'**), (**i**), (**s**), (**T**), (**A**), (**d**), (**Q**), (**p**), (**c**) and (**q**) are known, then its Variable Cost Planned is:

V= 360[**c**-**q**](**/**{**U**+**S**[1-**v**]-**F**-**S'i**[1+**s**]-**T**-**Ad**}-**Q**)/**p**

Rule-15856:

If both (**/**), (**U**), (**S**), (**v**), (**F**), (**S'**), (**i**), (**s**), (**T**), (**A**), (**d**), (**V**), (**Q**), (**c**) and (**q**) are known, then its Procured Inventory Days Planned:

p= 360[**c**-**q**](**/**{**U**+**S**[1-**v**]-**F**-**S'i**[1+**s**]-**T**-**Ad**}-**Q**)/**V**

Rule-15857:

If both (**/**), (**U**), (**S**), (**v**), (**F**), (**S'**), (**i**), (**s**), (**T**), (**A**), (**d**), (**V**), (**p**), (**Q**) and (**q**) are known, then its Current Ratio Planned is:

c= **q**+**Vp**/[360(**/**{**U**+**S**[1-**v**]-**F**-**S'i**[1+**s**]-**T**-**Ad**}-**Q**)]

Rule-15858:

If both (**/**), (**U**), (**S**), (**v**), (**F**), (**S'**), (**i**), (**s**), (**T**), (**A**), (**d**), (**V**), (**p**), (**c**) and (**Q**) are known, then its Quick or Acid Test Ratio Planned is:

q= **c**-**Vp**/[360(**/**{**U**+**S**[1-**v**]-**F**-**S'i**[1+**s**]-**T**-**Ad**}-**Q**)]

Rule-15859:

If both (**/**), (**U**), (**S**), (**v**), (**F**), (**S'**), (**i**), (**s**), (**T**), (**A**), (**d**), (**p**), (**c**) and (**q**) are known, then its Quoted Longterm Debt Planned is:

Q= **/**{**U**+**S**[1-**v**]-**F**-**S'i**[1+**s**]-**T**-**Ad**}-**Svp**/{360[**c**-**q**]}

Steve Asikin ISBN 14: 978-1514685136, ISBN 10: 1514685132

Rule-15860:

If both (**Q**), (**U**), (**S**), (**v**), (**F**), (**S'**), (**i**), (**s**), (**T**), (**A**), (**d**), (**p**), (**c**) and (**q**) are known, then its Leverage or Gearing Ratio Planned is:

/= (Q+Svp/{360[c-q]})
/{U+S[1-v]-F-S'i[1+s]-T-Ad}

Rule-15861:

If both (/), (**Q**), (**S**), (**v**), (**F**), (**S'**), (**i**), (**s**), (**T**), (**A**), (**d**), (**p**), (**c**) and (**q**) are known, then its Utilized or Starting Capital must be:

U= F+S'i[1+s]+T+Ad+(Q+Svp/{360[c-q]})//
-S[1-v]

Rule-15862:

If both (/), (**U**), (**Q**), (**v**), (**F**), (**S'**), (**i**), (**s**), (**T**), (**A**), (**d**), (**p**), (**c**) and (**q**) are known, then its Sales or Revenue Planned is:

S= (/{U-F-S'i[1+s]-T-Ad}-Q)
/(vp/{360[c-q]}-S/[1-v])

Rule-15863:

If both (/), (**U**), (**S**), (**Q**), (**F**), (**S'**), (**i**), (**s**), (**T**), (**A**), (**d**), (**p**), (**c**) and (**q**) are known, then its Variable Portion Planned is:

v= (/{U+S-F-S'i[1+s]-T-Ad}-Q)
/[S(p/{360[c-q]}+/)]

Steve Asikin ISBN 14: 978-1514685136, ISBN 10: 1514685132

Rule-15864:

If both (**/**), (**U**), (**$**), (**v**), (**Q**), (**$'**), (**i**), (**s**), (**T**), (**A**), (**d**), (**p**), (**c**) and (**q**) are known, then its Fixed Cost Planned is:

F= **$**[1-**v**]-**$'i**[1+**s**]-**T**-**Ad**}-(**Q**+**$vp**/{360[**c**-**q**]})/**/**-**U**

Rule-15865:

If both (**/**), (**U**), (**$**), (**v**), (**F**), (**Q**), (**i**), (**s**), (**T**), (**A**), (**d**), (**p**), (**c**) and (**q**) are known, then its Sales Past must be:

$'**= [**$[1-**v**]-**F**-**T**-**Ad**-(**Q**+**$vp**/{360[**c**-**q**]})/**/**-**U**]
/{**i**[1+**s**]}

Rule-15866:

If both (**/**), (**U**), (**$**), (**v**), (**F**), (**$'**), (**Q**), (**s**), (**T**), (**A**), (**d**), (**p**), (**c**) and (**q**) are known, then its Interest Portion Planned is:

i= [**$**[1-**v**]-**F**-**T**-**Ad**-(**Q**+**$vp**/{360[**c**-**q**]})/**/**-**U**]
/{**$'**[1+**s**]}

Rule-15867:

If both (**/**), (**U**), (**$**), (**v**), (**F**), (**$'**), (**i**), (**Q**), (**T**), (**A**), (**d**), (**p**), (**c**) and (**q**) are known, then its Sales Growth Planned is:

s= **$**/**$'**-1

or it camn also be found as

s= [**$**[1-**v**]-**F**-**T**-**Ad**-(**Q**+**$vp**/{360[**c**-**q**]})/**/**-**U**]
/ [**$'i**]-1

Steve Asikin ISBN 14: 978-1514685136, ISBN 10: 1514685132

Rule-15868:

If both (**/**), (**U**), (**$**), (**v**), (**F**), (**$'**), (**i**), (**s**), (**Q**), (**A**), (**d**), (**p**), (**c**) and (**q**) are known, then its Tax Planned is:

T= **$**[1-**v**]-**F**-**$'i**[1+**s**]-**Ad**-(**Q**+**$vp**/{360[**c-q**]})/**/U**

Rule-15869:

If both (**/**), (**U**), (**$**), (**v**), (**F**), (**$'**), (**i**), (**s**), (**T**), (**Q**), (**d**), (**p**), (**c**) and (**q**) are known, then its After Tax Income Planned is:

A= **[$**[1-**v**]-**F**-**$'i**[1+**s**]-**T**
-(**Q**+**$vp**/{360[**c-q**]})/**/U]/d**

Rule-15870:

If both (**/**), (**U**), (**$**), (**v**), (**F**), (**$'**), (**i**), (**s**), (**T**), (**A**), (**Q**), (**p**), (**c**) and (**q**) are known, then its Dividend Payout Planned is:

d= **[$**[1-**v**]-**F**-**$'i**[1+**s**]-**T**
-(**Q**+**$vp**/{360[**c-q**]})/**/U]/A**

Rule-15871:

If both (**/**), (**U**), (**$**), (**v**), (**F**), (**$'**), (**i**), (**s**), (**T**), (**A**), (**d**), (**Q**), (**c**) and (**q**) are known, then its Procured Inventory Days Planned:

p= 360[**c-q**](**/**{**U**+**$**[1-**v**]-**F**-**$'i**[1+**s**]-**T**-**Ad**}-**Q**)/[**$v**]

Rule-15872:

If both (**/**), (**U**), (**$**), (**v**), (**F**), (**$'**), (**i**), (**s**), (**T**), (**A**), (**d**), (**p**), (**Q**) and (**q**) are known, then its Current Ratio Planned is:

c= **q**+**$vp**/**[**360(**/**{**U**+**$**[1-**v**]-**F**-**$'i**[1+**s**]-**T**-**Ad**}-**Q**)**]**

Steve Asikin ISBN 14: 978-1514685136, ISBN 10: 1514685132

Rule-15873:

If both (/), (U), (S), (v), (F), (S'), (i), (s), (T), (A), (d), (p), (c) and (Q) are known, then its Quick or Acid Test Ratio Planned is:

q= c-Svp/[360(/{U+S[1-v]-F-S'i[1+s]-T-Ad}-Q)]

Rule-15874:

If both (/), (U), (S), (v), (F), (S'), (i), (s), (T), (A), (d), (p), (c) and (q) are known, then its Quoted Longterm Debt Planned is:

Q= /{U+S[1-v]-F-S'i[1+s]-T-Ad}
-S'vp[1+s]/{360[c-q]}

Rule-15875:

If both (Q), (U), (S), (v), (F), (S'), (i), (s), (T), (A), (d), (p), (c) and (q) are known, then its Leverage or Gearing Ratio Planned is:

/= (Q+S'vp[1+s]/{360[c-q]})
/{U+S[1-v]-F-S'i[1+s]-T-Ad}

Rule-15876:

If both (/), (Q), (S), (v), (F), (S'), (i), (s), (T), (A), (d), (p), (c) and (q) are known, then its Utilized or Starting Capital must be:

U= F+S'i[1+s]+T+Ad
+(Q+S'vp[1+s]/{360[c-q]})//S[1-v]

Steve Asikin ISBN 14: 978-1514685136, ISBN 10: 1514685132

Rule-15877:

If both (/), (U), (Q), (v), (F), (S'), (i), (s), (T), (A), (d), (p), (c) and (q) are known, then its Sales or Revenue Planned is:

$$S= (Q-/\{U-F-S'i[1+s]-T-Ad\} +S'vp[1+s]/\{360[c-q]\})/\{/[1-v]\}$$

Rule-15878:

If both (/), (U), (S), (Q), (F), (S'), (i), (s), (T), (A), (d), (p), (c) and (q) are known, then its Variable Portion Planned is:

$$v= (/\{U+S-F-S'i[1+s]-T-Ad\}-Q) /(S'p[1+s]/\{360[c-q]\}+S/)$$

Rule-15879:

If both (/), (U), (S), (v), (Q), (S'), (i), (s), (T), (A), (d), (p), (c) and (q) are known, then its Fixed Cost Planned is:

$$F= S[1-v]-S'i[1+s]-T-Ad\} -(Q+S'vp[1+s]/\{360[c-q]\})//-U$$

Rule-15880:

If both (/), (U), (S), (v), (F), (Q), (i), (s), (T), (A), (d), (p), (c) and (q) are known, then its Sales Past must be:

$$S'= (/\{U+S[1-v]-F-T-Ad\}-Q) /[[1+s](vp/\{360[c-q]\}+i/)]$$

Steve Asikin ISBN 14: 978-1514685136, ISBN 10: 1514685132

Rule-15881:

If both (**/**), (**U**), (**$**), (**v**), (**F**), (**$'**), (**Q**), (**s**), (**T**), (**A**), (**d**), (**p**), (**c**) and (**q**) are known, then its Interest Portion Planned is:

i= [$[1-v]-F-T-Ad-(Q+$'vp[1+s]/{360[c-q]})/ / U]
/{$'[1+s]}

Rule-15882:

If both (**/**), (**U**), (**$**), (**v**), (**F**), (**$'**), (**i**), (**Q**), (**T**), (**A**), (**d**), (**p**), (**c**) and (**q**) are known, then its Sales Growth Planned is:

s= $/$'-1

or it can also be found as

s= (/{U+$[1-v]-F-T-Ad}-Q)
/[$'(vp/{360[c-q]}+i/)]-1

Rule-15883:

If both (**/**), (**U**), (**$**), (**v**), (**F**), (**$'**), (**i**), (**s**), (**Q**), (**A**), (**d**), (**p**), (**c**) and (**q**) are known, then its Tax Planned is:

T= $[1-v]-F-$'i[1+s]-Ad
-(Q+$'vp[1+s]/{360[c-q]})/ / U

Rule-15884:

If both (**/**), (**U**), (**$**), (**v**), (**F**), (**$'**), (**i**), (**s**), (**T**), (**Q**), (**d**), (**p**), (**c**) and (**q**) are known, then its After Tax Income Planned is:

A= [$[1-v]-F-$'i[1+s]-T
-(Q+$'vp[1+s]/{360[c-q]})/ / U]/d

Steve Asikin ISBN 14: 978-1514685136, ISBN 10: 1514685132

Rule-15885:

If both (/), (U), ($), (v), (F), ($'), (i), (s), (T), (A), (Q), (p), (c) and (q) are known, then its Dividend Payout Planned is:

d= [$[1-v]-F-$'i[1+s]-T
-(Q+$'vp[1+s]/{360[c-q]})/ /U]/A

Rule-15886:

If both (/), (U), ($), (v), (F), ($'), (i), (s), (T), (A), (d), (Q), (c) and (q) are known, then its Procured Inventory Days Planned:

p= 360[c-q](/{U+$[1-v]-F-$'i[1+s]-T-Ad}-Q)
/{$'v[1+s]}

Rule-15887:

If both (/), (U), ($), (v), (F), ($'), (i), (s), (T), (A), (d), (p), (Q) and (q) are known, then its Current Ratio Planned is:

c= q+$'vp[1+s]
/[360(/{U+$[1-v]-F-$'i[1+s]-T-Ad}-Q)]

Rule-15888:

If both (/), (U), ($), (v), (F), ($'), (i), (s), (T), (A), (d), (p), (c) and (Q) are known, then its Quick or Acid Test Ratio Planned is:

q= c-$'vp[1+s]
/[360(/{U+$[1-v]-F-$'i[1+s]-T-Ad}-Q)]

Steve Asikin ISBN 14: 978-1514685136, ISBN 10: 1514685132

Rule-15889:

If both (**/**), (**U**), (**$**), (**v**), (**F**), (**$'**), (**i**), (**s**), (**t**), (**D**) and (**X**) are known, then its Quoted Longterm Debt Planned is:

Q= **/**(**U**+{**$**[1-**v**]-**F**-**$'i**[1+**s**]}[1-**t**]-**D**)-**X**

Rule-15890:

If both (**Q**), (**U**), (**$**), (**v**), (**F**), (**$'**), (**i**), (**s**), (**t**), (**D**) and (**X**) are known, then its Leverage or Gearing Ratio Planned is:

/= [**Q**+**X**]/(**U**+{**$**[1-**v**]-**F**-**$'i**[1+**s**]}[1-**t**]-**D**)

Rule-15891:

If both (**/**), (**Q**), (**$**), (**v**), (**F**), (**$'**), (**i**), (**s**), (**t**), (**D**) and (**X**) are known, then its Utilized or Starting Capital must be:

U= **D**+[**Q**+**X**]/**/**-{**$**[1-**v**]-**F**-**$'i**[1+**s**]}[1-**t**]

Rule-15892:

If both (**/**), (**U**), (**Q**), (**v**), (**F**), (**$'**), (**i**), (**s**), (**t**), (**D**) and (**X**) are known, then its Sales or Revenue Planned is:

$**= (**F**+**$'i[1+**s**]}+{**D**+[**Q**+**X**]/**/**-**U**}/[1-**t**])/[1-**v**]

Rule-15893:

If both (**/**), (**U**), (**$**), (**Q**), (**F**), (**$'**), (**i**), (**s**), (**t**), (**D**) and (**X**) are known, then its Variable Portion Planned is:

v= 1-(**F**+**$'i**[1+**s**]}+{**D**+[**Q**+**X**]/**/**-**U**}/[1-**t**])/**$**

Steve Asikin ISBN 14: 978-1514685136, ISBN 10: 1514685132

Rule-15894:

If both (**/**), (**U**), (**$**), (**v**), (**Q**), (**$'**), (**i**), (**s**), (**t**), (**D**) and (**X**) are known, then its Fixed Cost Planned is:

F= **$**[1-**v**]-**$'i**[1+**s**]}-{**D**+[**Q**+**X**]/**/-U**}/[1-**t**]

Rule-15895:

If both (**/**), (**U**), (**$**), (**v**), (**F**), (**Q**), (**i**), (**s**), (**t**), (**D**) and (**X**) are known, then its Sales Past must be:

$'**= (**$[1-**v**]-**F**-{**D**+[**Q**+**X**]/**/-U**}/[1-**t**])/{**i**[1+**s**]}

Rule-15896:

If both (**/**), (**U**), (**$**), (**v**), (**F**), (**$'**), (**Q**), (**s**), (**t**), (**D**) and (**X**) are known, then its Interest Portion Planned is:

i= (**$**[1-**v**]-**F**-{**D**+[**Q**+**X**]/**/-U**}/[1-**t**])/{**$'**[1+**s**]}

Rule-15897:

If both (**/**), (**U**), (**$**), (**v**), (**F**), (**$'**), (**i**), (**Q**), (**t**), (**D**) and (**X**) are known, then its Sales Growth Planned is:

s= **$**/**$'**-1

or it can also be found as

s= (**$**[1-**v**]-**F**-{**D**+[**Q**+**X**]/**/-U**}/[1-**t**])/[**$'i**]-1

Rule-15898:

If both (**/**), (**U**), (**$**), (**v**), (**F**), (**$'**), (**i**), (**s**), (**Q**), (**D**) and (**X**) are known, then its Tax Rate Planned is:

t= 1-{**D**+[**Q**+**X**]/**/-U**}/{**$**[1-**v**]-**F**-**$'i**[1+**s**]}

Steve Asikin ISBN 14: 978-1514685136, ISBN 10: 1514685132

Rule-15899:

If both (/), (U), (S), (v), (F), (S'), (i), (s), (t), (Q) and (X) are known, then its Dividend Planned is:

D= U+{S[1-v]-F-S'i[1+s]}[1-t]-[Q+X]//

Rule-15900:

If both (/), (U), (S), (v), (F), (S'), (i), (s), (t), (D) and (Q) are known, then its Xpress or Current Debt Planned is:

X= /(U+{S[1-v]-F-S'i[1+s]}[1-t]-D)-Q

Rule-15901:

If both (/), (U), (S), (v), (F), (S'), (i), (s), (t), (D), (P), (c) and (q) are known, then its Quoted Longterm Debt Planned is:

Q= /{U+S[1-v]-F-S'i[1+s]}[1-t]-D)-P/[c-q]

Rule-15902:

If both (Q), (U), (S), (v), (F), (S'), (i), (s), (t), (D), (P), (c) and (q) are known, then its Leverage or Gearing Ratio Planned is:

/= {Q+P/[c-q]}/(U+{S[1-v]-F-S'i[1+s]}[1-t]-D)

Rule-15903:

If both (/), (Q), (S), (v), (F), (S'), (i), (s), (t), (D), (P), (c) and (q) are known, then its Utilized or Starting Capital must be:

U= D+{Q+P/[c-q]}//-{S[1-v]-F-S'i[1+s]}[1-t]

Steve Asikin ISBN 14: 978-1514685136, ISBN 10: 1514685132

Rule-15904:

If both (**/**), (**U**), (**Q**), (**v**), (**F**), (**$'**), (**i**), (**s**), (**t**), (**D**), (**P**), (**c**) and (**q**) are known, then its Sales or Revenue Planned is:

$**= [**F**+**$'i[1+**s**]}+(**D**+{**Q**+**P**/[**c**-**q**]}/**/-U**)/[1-**t**]]/[1-**v**]

Rule-15905:

If both (**/**), (**U**), (**$**), (**Q**), (**F**), (**$'**), (**i**), (**s**), (**t**), (**D**), (**P**), (**c**) and (**q**) are known, then its Variable Portion Planned is:

v= 1-[**F**+**$'i**[1+**s**]}+(**D**+{**Q**+**P**/[**c**-**q**]}/**/-U**)/[1-**t**]]/**$**

Rule-15906:

If both (**/**), (**U**), (**$**), (**v**), (**Q**), (**$'**), (**i**), (**s**), (**t**), (**D**), (**P**), (**c**) and (**q**) are known, then its Fixed Cost Planned is:

F= **$**[1-**v**]-**$'i**[1+**s**]}-(**D**+{**Q**+**P**/[**c**-**q**]}/**/-U**)/[1-**t**]

Rule-15907:

If both (**/**), (**U**), (**$**), (**v**), (**F**), (**Q**), (**i**), (**s**), (**t**), (**D**), (**P**), (**c**) and (**q**) are known, then its Sales Past must be:

$'**= [**$[1-**v**]-**F**-(**D**+{**Q**+**P**/[**c**-**q**]}/**/-U**)/[1-**t**]]/{**i**[1+**s**]}

Rule-15908:

If both (**/**), (**U**), (**$**), (**v**), (**F**), (**$'**), (**Q**), (**s**), (**t**), (**D**), (**P**), (**c**) and (**q**) are known, then its Interest Portion Planned is:

i= [**$**[1-**v**]-**F**-(**D**+{**Q**+**P**/[**c**-**q**]}/**/-U**)/[1-**t**]]/{**$'**[1+**s**]}

Steve Asikin ISBN 14: 978-1514685136, ISBN 10: 1514685132

Rule-15909:

If both (**/**), (**U**), (**$**), (**v**), (**F**), (**$'**), (**i**), (**Q**), (**t**), (**D**), (**P**), (**c**) and (**q**) are known, then its Sales Growth Planned is:

s= $/$'-1

or it can also be found as

s= [$[1-v]-F-(D+{Q+P/[c-q]}//-U)/[1-t]]/[$'i]-1

Rule-15910:

If both (**/**), (**U**), (**$**), (**v**), (**F**), (**$'**), (**i**), (**s**), (**Q**), (**D**), (**P**), (**c**) and (**q**) are known, then its Tax Rate Planned is:

t= 1-(D+{Q+P/[c-q]}//-U)/{$[1-v]-F-$'i[1+s]}

Rule-15911:

If both (**/**), (**U**), (**$**), (**v**), (**F**), (**$'**), (**i**), (**s**), (**t**), (**Q**), (**P**), (**c**) and (**q**) are known, then its Dividend Planned is:

D= U+{$[1-v]-F-$'i[1+s]}[1-t]-{Q+P/[c-q]}//

Rule-15912:

If both (**/**), (**U**), (**$**), (**v**), (**F**), (**$'**), (**i**), (**s**), (**t**), (**D**), (**Q**), (**c**) and (**q**) are known, then its Procured Inventory Planned is:

P= [c-q][/(U+{$[1-v]-F-$'i[1+s]}[1-t]-D)-Q]

Rule-15813:

If both (**/**), (**U**), (**$**), (**v**), (**F**), (**$'**), (**i**), (**s**), (**t**), (**D**), (**P**), (**Q**) and (**q**) are known, then its Current Ratio Planned is:

c= q+P/[/(U+{$[1-v]-F-$'i[1+s]}[1-t]-D)-Q]

Steve Asikin ISBN 14: 978-1514685136, ISBN 10: 1514685132

Rule-15914:

If both (/), (U), ($), (v), (F), ($'), (i), (s), (t), (D), (P), (c) and (Q) are known, then its Quick or Acid Test Ratio Planned is:

q= c-P/[/(U+{$[1-v]-F-$'i[1+s]}[1-t]-D)-Q]

Rule-15915:

If both (/), (U), ($), (v), (F), ($'), (i), (s), (t), (D), (V), (p), (c) and (q) are known, then its Quoted Longterm Debt Planned is:

Q= /{U+$[1-v]-F-$'i[1+s]}[1-t]-D)-Vp/{360[c-q]}

Rule-15916:

If both (Q), (U), ($), (v), (F), ($'), (i), (s), (t), (D), (V), (p), (c) and (q) are known, then its Leverage or Gearing Ratio Planned is:

/= (Q+Vp/{360[c-q]})
/(U+{$[1-v]-F-$'i[1+s]}[1-t]-D)

Rule-15917:

If both (/), (Q), ($), (v), (F), ($'), (i), (s), (t), (D), (V), (p), (c) and (q) are known, then its Utilized or Starting Capital must be:

U= D+(Q+Vp/{360[c-q]})//
-{$[1-v]-F-$'i[1+s]}[1-t]

Steve Asikin ISBN 14: 978-1514685136, ISBN 10: 1514685132

Rule-15918:

If both (**/**), (**U**), (**Q**), (**v**), (**F**), (**S'**), (**i**), (**s**), (**t**), (**D**), (**V**), (**p**), (**c**) and (**q**) are known, then its Sales or Revenue Planned is:

$$S= \{F+S'i[1+s]\}+[D+(Q+Vp/\{360[c-q]\})/ /U] /[1-t]\}/[1-v]$$

Rule-15919:

If both (**/**), (**U**), (**S**), (**Q**), (**F**), (**S'**), (**i**), (**s**), (**t**), (**D**), (**V**), (**p**), (**c**) and (**q**) are known, then its Variable Portion Planned is:

$$v= V/S$$

or it can also be found as

$$v= 1-\{F+S'i[1+s]\}+[D+(Q+Vp/\{360[c-q]\})/ /U] /[1-t]\}/S$$

Rule-15920:

If both (**/**), (**U**), (**S**), (**v**), (**Q**), (**S'**), (**i**), (**s**), (**t**), (**D**), (**V**), (**p**), (**c**) and (**q**) are known, then its Fixed Cost Planned is:

$$F= S[1-v]-S'i[1+s]\}-[D+(Q+Vp/\{360[c-q]\})/ /U] /[1-t]$$

Rule-15921:

If both (**/**), (**U**), (**S**), (**v**), (**F**), (**Q**), (**i**), (**s**), (**t**), (**D**), (**V**), (**p**), (**c**) and (**q**) are known, then its Sales Past must be:

$$S'= \{S[1-v]-F-[D+(Q+Vp/\{360[c-q]\})/ /U]/[1-t]\} /\{i[1+s]\}$$

Steve Asikin ISBN 14: 978-1514685136, ISBN 10: 1514685132

Rule-15922:

If both (/), (U), ($), (v), (F), ($'), (Q), (s), (t), (D), (V), (p), (c) and (q) are known, then its Interest Portion Planned is:

i= {$[1-v]-F-[D+(Q+Vp/{360[c-q]})/ /-U]/[1-t]}
/{$'[1+s]}

Rule-15923:

If both (/), (U), ($), (v), (F), ($'), (i), (Q), (t), (D), (V), (p), (c) and (q) are known, then its Sales Growth Planned is:

s= $/$'-1

or it can also be found as

s= {$[1-v]-F-[D+(Q+Vp/{360[c-q]})/ /-U]/[1-t]}
/[$'i]-1

Rule-15924:

If both (/), (U), ($), (v), (F), ($'), (i), (s), (Q), (D), (V), (p), (c) and (q) are known, then its Tax Rate Planned is:

t= 1-[D+(Q+Vp/{360[c-q]})/ /-U]
/{$[1-v]-F-$'i[1+s]}

Rule-15925:

If both (/), (U), ($), (v), (F), ($'), (i), (s), (t), (Q), (V), (p), (c) and (q) are known, then its Dividend Planned is:

D= U+{$[1-v]-F-$'i[1+s]}[1-t]
-(Q+Vp/{360[c-q]})/ /

Steve Asikin ISBN 14: 978-1514685136, ISBN 10: 1514685132

Rule-15926:

If both (**/**), (**U**), (**S**), (**v**), (**F**), (**S'**), (**i**), (**s**), (**t**), (**D**), (**Q**), (**p**), (**c**) and (**q**) are known, then its Variable Cost Planned is:

V= 360[**c**-**q**][**/**(**U**+{**S**[1-**v**]-**F**-**S'i**[1+**s**]}[1-**t**]-**D**)-**Q**]/**p**

Rule-15927:

If both (**/**), (**U**), (**S**), (**v**), (**F**), (**S'**), (**i**), (**s**), (**t**), (**D**), (**V**), (**Q**), (**c**) and (**q**) are known, then its Procured Inventory Days Planned is:

p= 360[**c**-**q**][**/**(**U**+{**S**[1-**v**]-**F**-**S'i**[1+**s**]}[1-**t**]-**D**)-**Q**]/**V**

Rule-15928:

If both (**/**), (**U**), (**S**), (**v**), (**F**), (**S'**), (**i**), (**s**), (**t**), (**D**), (**V**), (**p**), (**Q**) and (**q**) are known, then its Current Ratio Planned is:

c= **q**+**Vp**
/{360[**/**(**U**+{**S**[1-**v**]-**F**-**S'i**[1+**s**]}[1-**t**]-**D**)-**Q**]}

Rule-15929:

If both (**/**), (**U**), (**S**), (**v**), (**F**), (**S'**), (**i**), (**s**), (**t**), (**D**), (**V**), (**p**), (**c**) and (**Q**) are known, then its Quick or Acid Test Ratio Planned is:

q= **c**-**Vp**
/{360[**/**(**U**+{**S**[1-**v**]-**F**-**S'i**[1+**s**]}[1-**t**]-**D**)-**Q**]}

Steve Asikin ISBN 14: 978-1514685136, ISBN 10: 1514685132

Rule-15930:

If both (**/**), (**U**), (**$**), (**v**), (**F**), (**$'**), (**i**), (**s**), (**t**), (**D**), (**p**), (**c**) and (**q**) are known, then its Quoted Longterm Debt Planned is:

Q= **/**{**U**+**$**[1-**v**]-**F**-**$'i**[1+**s**]}[1-**t**]-**D**)
-**$vp**/{360[**c**-**q**]}

Rule-15931:

If both (**Q**), (**U**), (**$**), (**v**), (**F**), (**$'**), (**i**), (**s**), (**t**), (**D**), (**V**), (**p**), (**c**) and (**q**) are known, then its Leverage or Gearing Ratio Planned is:

/= (**Q**+**$vp**/{360[**c**-**q**]})
/(**U**+{**$**[1-**v**]-**F**-**$'i**[1+**s**]}[1-**t**]-**D**)

Rule-15932:

If both (**/**), (**Q**), (**$**), (**v**), (**F**), (**$'**), (**i**), (**s**), (**t**), (**D**), (**p**), (**c**) and (**q**) are known, then its Utilized or Starting Capital must be:

U= **D**+(**Q**+**$vp**/{360[**c**-**q**]})/**/**
-{**$**[1-**v**]-**F**-**$'i**[1+**s**]}[1-**t**]

Rule-15933:

If both (**/**), (**U**), (**Q**), (**v**), (**F**), (**$'**), (**i**), (**s**), (**t**), (**D**), (**p**), (**c**) and (**q**) are known, then its Sales or Revenue Planned is:

$**= (**Q**-**/**{**U**-**F**-**$'i[1+**s**]}[1-**t**]-**D**)
+**$vp**/{360[**c**-**q**]}/{**/**[1-**v**]}

Steve Asikin ISBN 14: 978-1514685136, ISBN 10: 1514685132

Rule-15934:

If both (**/**), (**U**), (**S**), (**Q**), (**F**), (**S'**), (**i**), (**s**), (**t**), (**D**), (**p**), (**c**) and (**q**) are known, then its Variable Portion Planned is:

v= [**/**{**U**+**S**-**F**-**S'i**[1+**s**]}[1-**t**]-**D**)-**Q**]
/(**p**/{360[**c**-**q**]}+**/**[1-**t**])

Rule-15935:

If both (**/**), (**U**), (**S**), (**v**), (**Q**), (**S'**), (**i**), (**s**), (**t**), (**D**), (**p**), (**c**) and (**q**) are known, then its Fixed Cost Planned is:

F= **S**[1-**v**]-**S'i**[1+**s**]}-[**D**+(**Q**+**Svp**/{360[**c**-**q**]})/**/**-**U**]
/[1-**t**]

Rule-15936:

If both (**/**), (**U**), (**S**), (**v**), (**F**), (**Q**), (**i**), (**s**), (**t**), (**D**), (**p**), (**c**) and (**q**) are known, then its Sales Past must be:

S'= [**/**{**U**+**S**[1-**v**]-**F**}[1-**t**]-**D**)-**Svp**/{360[**c**-**q**]}-**Q**]
/{**i/**[1+**s**] [1-**t**]}

Rule-15937:

If both (**/**), (**U**), (**S**), (**v**), (**F**), (**S'**), (**Q**), (**s**), (**t**), (**D**), (**p**), (**c**) and (**q**) are known, then its Interest Portion Planned is:

i= [**/**{**U**+**S**[1-**v**]-**F**}[1-**t**]-**D**)-**Svp**/{360[**c**-**q**]}-**Q**]
/{**S'/**[1+**s**] [1-**t**]}

Steve Asikin ISBN 14: 978-1514685136, ISBN 10: 1514685132

Rule-15938:

If both (**/**), (**U**), (**$**), (**v**), (**F**), (**$'**), (**i**), (**Q**), (**t**), (**D**), (**p**), (**c**) and (**q**) are known, then its Sales Growth Planned is:

s= $/$'-1

or it can also be found as

s= {$[1-v]-F-[D+(Q+$vp/{360[c-q]})//U]/[1-t]}/[$'i]-1

Rule-15939:

If both (**/**), (**U**), (**$**), (**v**), (**F**), (**$'**), (**i**), (**s**), (**Q**), (**D**), (**p**), (**c**) and (**q**) are known, then its Tax Rate Planned is:

t= 1-[D+(Q+$vp/{360[c-q]})//U]/{$[1-v]-F-$'i[1+s]}

Rule-15940:

If both (**/**), (**U**), (**$**), (**v**), (**F**), (**$'**), (**i**), (**s**), (**t**), (**Q**), (**p**), (**c**) and (**q**) are known, then its Dividend Planned is:

D= U+{$[1-v]-F-$'i[1+s]}[1-t]-(Q+$vp/{360[c-q]})//

Rule-15941:

If both (**/**), (**U**), (**$**), (**v**), (**F**), (**$'**), (**i**), (**s**), (**t**), (**D**), (**Q**), (**c**) and (**q**) are known, then its Procured Inventory Days Planned is:

p= 360[c-q][/(U+{$[1-v]-F-$'i[1+s]}[1-t]-D)-Q]/[$v]

Steve Asikin ISBN 14: 978-1514685136, ISBN 10: 1514685132

Rule-15942:

If both (**/**), (**U**), (**S**), (**v**), (**F**), (**S'**), (**i**), (**s**), (**t**), (**D**), (**p**), (**Q**) and (**q**) are known, then its Current Ratio Planned is:

c= **q**+**Svp**
/{360[**/**(**U**+{**S**[1-**v**]-**F**-**S'i**[1+**s**]}[1-**t**]-**D**)-**Q**]}

Rule-15943:

If both (**/**), (**U**), (**S**), (**v**), (**F**), (**S'**), (**i**), (**s**), (**t**), (**D**), (**p**), (**c**) and (**q**) are known, then its Quoted Longterm Debt Planned is:

q= **c**-**Svp**
/{360[**/**(**U**+{**S**[1-**v**]-**F**-**S'i**[1+**s**]}[1-**t**]-**D**)-**Q**]}

Rule-15944:

If both (**/**), (**U**), (**S**), (**v**), (**F**), (**S'**), (**i**), (**s**), (**t**), (**D**), (**p**), (**c**) and (**q**) are known, then its Quoted Longterm Debt Planned is:

Q= **/**{**U**+**S**[1-**v**]-**F**-**S'i**[1+**s**]}[1-**t**]-**D**)
-**S'vp**[1+**s**]/{360[**c**-**q**]}

Rule-15945:

If both (**Q**), (**U**), (**S**), (**v**), (**F**), (**S'**), (**i**), (**s**), (**t**), (**D**), (**p**), (**c**) and (**q**) are known, then its Leverage or Gearing Ratio Planned is:

/= (**Q**+**S'vp**[1+**s**]/{360[**c**-**q**]})
/(**U**+{**S**[1-**v**]-**F**-**S'i**[1+**s**]}[1-**t**]-**D**)

Steve Asikin ISBN 14: 978-1514685136, ISBN 10: 1514685132

Rule-15946:

If both (/), (Q), ($), (v), (F), ($'), (i), (s), (t), (D), (p), (c) and (q) are known, then its Utilized or Starting Capital must be:

U= D+(Q+$'vp[1+s]/{360[c-q]})//
-{$[1-v]-F-$'i[1+s]}[1-t]

Rule-15947:

If both (/), (U), (Q), (v), (F), ($'), (i), (s), (t), (D), (p), (c) and (q) are known, then its Sales or Revenue Planned is:

$= (Q-/{U-F-$'i[1+s]}[1-t]-D)
+$'vp[1+s]/{360[c-q]}/{/[1-v]}

Rule-15948:

If both (/), (U), ($), (Q), (F), ($'), (i), (s), (t), (D), (p), (c) and (q) are known, then its Variable Portion Planned is:

v=[/{U+$-F-$'i[1+s]}[1-t]-D)-Q]
/($'p[1+s]/{360[c-q]}+$/[1-t])

Rule-15949:

If both (/), (U), ($), (v), (Q), ($'), (i), (s), (t), (D), (p), (c) and (q) are known, then its Fixed Cost Planned is:

F= $[1-v]-$'i[1+s]}
-[D+(Q+$'vp[1+s]/{360[c-q]})//-U]/[1-t]

Steve Asikin ISBN 14: 978-1514685136, ISBN 10: 1514685132

Rule-15950:

If both (/), (U), ($), (v), (F), (Q), (i), (s), (t), (D), (p), (c) and (q) are known, then its Sales Past must be:

$'= (/{U+$[1-v]-F}[1-t]-D)-Q)
/[[1+s](vp/{360[c-q]}+i)

Rule-15951:

If both (/), (U), ($), (v), (F), ($'), (Q), (s), (t), (D), (p), (c) and (q) are known, then its Interest Portion Planned is:

i= [/{U+$[1-v]-F}[1-t]-D)
-$'vp[1+s]/{360[c-q]}-Q]/{$'/[1+s] [1-t]}

Rule-15952:

If both (/), (U), ($), (v), (F), ($'), (i), (Q), (t), (D), (p), (c) and (q) are known, then its Sales Growth Planned is:

s= $/$'-1

or it can also be found as

s= (/{U+$[1-v]-F}[1-t]-D)-Q)
/[$'(vp/{360[c-q]}+i)-1

Rule-15953:

If both (/), (U), ($), (v), (F), ($'), (i), (s), (Q), (D), (p), (c) and (q) are known, then its Tax Rate Planned is:

t= 1-[D+(Q+$'vp[1+s]/{360[c-q]})//U]
/{$[1-v]-F-$'i[1+s]}

Steve Asikin ISBN 14: 978-1514685136, ISBN 10: 1514685132

Rule-15954:

If both (**/**), (**U**), (**$**), (**v**), (**F**), (**$'**), (**i**), (**s**), (**t**), (**Q**), (**p**), (**c**) and (**q**) are known, then its Dividend Planned is:

D= **U**+{**$**[1-**v**]-**F**-**$'i**[1+**s**]}[1-**t**]
-(**Q**+**$'vp**[1+**s**]/{360[**c**-**q**]})/**/**

Rule-15955:

If both (**/**), (**U**), (**$**), (**v**), (**F**), (**$'**), (**i**), (**s**), (**t**), (**D**), (**Q**), (**c**) and (**q**) are known, then its Procured Inventory Days Planned:

p= 360[**c**-**q**][**/**(**U**+{**$**[1-**v**]-**F**-**$'i**[1+**s**]}[1-**t**]-**D**)-**Q**]
/{**$'v**[1+**s**]}

Rule-15956:

If both (**/**), (**U**), (**$**), (**v**), (**F**), (**$'**), (**i**), (**s**), (**t**), (**D**), (**p**), (**Q**) and (**q**) are known, then its Current Ratio Planned is:

c= **q**+**$'vp**[1+**s**]
/{360[**/**(**U**+{**$**[1-**v**]-**F**-**$'i**[1+**s**]}[1-**t**]-**D**)-**Q**]}

Rule-15957:

If both (**/**), (**U**), (**$**), (**v**), (**F**), (**$'**), (**i**), (**s**), (**t**), (**D**), (**p**), (**c**) and (**Q**) are known, then its Quick or Acid Test Ratio Planned is:

q= **c**-**$'vp**[1+**s**]
/{360[**/**(**U**+{**$**[1-**v**]-**F**-**$'i**[1+**s**]}[1-**t**]-**D**)-**Q**]}

Steve Asikin ISBN 14: 978-1514685136, ISBN 10: 1514685132

Rule-15958:

If both (**/**), (**U**), (**S**), (**v**), (**F**), (**S'**), (**i**), (**s**), (**t**), (**A**), (**d**) and (**X**) are known, then its Quoted Longterm Debt Planned is:

$Q = /(U+\{S[1-v]-F-S'i[1+s]\}[1-t]-Ad)-X$

Rule-15959:

If both (**Q**), (**U**), (**S**), (**v**), (**F**), (**S'**), (**i**), (**s**), (**t**), (**A**), (**d**) and (**X**) are known, then its Leverage or Gearing Ratio Planned is:

$/ = [Q+X]/(U+\{S[1-v]-F-S'i[1+s]\}[1-t]-Ad)$

Rule-15960:

If both (**/**), (**Q**), (**S**), (**v**), (**F**), (**S'**), (**i**), (**s**), (**t**), (**A**), (**d**) and (**X**) are known, then its Utilized or Starting Capital must be:

$U = Ad+[Q+X]/ / -\{S[1-v]-F-S'i[1+s]\}[1-t]$

Rule-15961:

If both (**/**), (**U**), (**Q**), (**v**), (**F**), (**S'**), (**i**), (**s**), (**t**), (**A**), (**d**) and (**X**) are known, then its Sales or Revenue Planned is:

$S = (F+S'i[1+s]+\{Ad+[Q+X]/ / -U\}/[1-t])/[1-v]$

Rule-15962:

If both (**/**), (**U**), (**S**), (**Q**), (**F**), (**S'**), (**i**), (**s**), (**t**), (**A**), (**d**) and (**X**) are known, then its Variable Portion Planned is:

$v = 1-(F+S'i[1+s]+\{Ad+[Q+X]/ / -U\}/[1-t])/S$

Steve Asikin ISBN 14: 978-1514685136, ISBN 10: 1514685132

Rule-15963:

If both (/), (U), ($), (v), (Q), ($'), (i), (s), (t), (A), (d) and (X) are known, then its Fixed Cost Planned is:

F= $[1-v]-$'i[1+s]-{Ad+[Q+X]/-U}/[1-t]

Rule-15964:

If both (/), (U), ($), (v), (F), (Q), (i), (s), (t), (A), (d) and (X) are known, then its Sales Past must be:

$'= ($[1-v]-F-{Ad+[Q+X]/-U}/[1-t])/{i[1+s]}

Rule-15965:

If both (/), (U), ($), (v), (F), ($'), (Q), (s), (t), (A), (d) and (X) are known, then its Interest Portion Planned is:

i= ($[1-v]-F-{Ad+[Q+X]/-U}/[1-t])/{$'[1+s]}

Rule-15966:

If both (/), (U), ($), (v), (F), ($'), (i), (Q), (t), (A), (d) and (X) are known, then its Sales Growth Planned is:

s= $/$'-1

or it can also be found as

s= ($[1-v]-F-{Ad+[Q+X]/-U}/[1-t])/[$'i]-1

Rule-15967:

If both (/), (U), ($), (v), (F), ($'), (i), (s), (Q), (A), (d) and (X) are known, then its Tax Rate Planned is:

t= 1-{Ad+[Q+X]/-U}/{$[1-v]-F-$'i[1+s]}

Steve Asikin ISBN 14: 978-1514685136, ISBN 10: 1514685132

Rule-15968:

If both (/), (U), (S), (v), (F), (S'), (i), (s), (t), (Q), (d) and (X) are known, then its After Tax Income Planned is:

A= (U+{S[1-v]-F-S'i[1+s]}[1-t]-[Q+X]/ /)/d

Rule-15969:

If both (/), (U), (S), (v), (F), (S'), (i), (s), (t), (A), (Q) and (X) are known, then its Dividend Payout Planned is:

d= (U+{S[1-v]-F-S'i[1+s]}[1-t]-[Q+X]/ /)/A

Rule-15970:

If both (/), (U), (S), (v), (F), (S'), (i), (s), (t), (A), (d) and (Q) are known, then its Xpress or Current Debt Planned is:

X= /(U+{S[1-v]-F-S'i[1+s]}[1-t]-Ad)-Q

Rule-15971:

If both (/), (U), (S), (v), (F), (S'), (i), (s), (t), (A), (d), (P), (c) and (q) are known, then its Quoted Longterm Debt Planned is:

Q= /(U+{S[1-v]-F-S'i[1+s]}[1-t]-Ad)-P/[c-q]

Rule-15972:

If both (Q), (U), (S), (v), (F), (S'), (i), (s), (t), (A), (d), (P), (c) and (q) are known, then its Leverage or Gearing Ratio Planned is:

/= {Q+P/[c-q]}/(U+{S[1-v]-F-S'i[1+s]}[1-t]-Ad)

Steve Asikin ISBN 14: 978-1514685136, ISBN 10: 1514685132

Rule-15973:

If both (**/**), (**Q**), (**$**), (**v**), (**F**), (**$'**), (**i**), (**s**), (**t**), (**A**), (**d**), (**P**), (**c**) and (**q**) are known, then its Utilized or Starting Capital must be:

U= **Ad**+{**Q**+**P**/[**c**-**q**]}/**/**-{**$**[1-**v**]-**F**-**$'i**[1+**s**]}[1-**t**]

Rule-15974:

If both (**/**), (**U**), (**Q**), (**v**), (**F**), (**$'**), (**i**), (**s**), (**t**), (**A**), (**d**), (**P**), (**c**) and (**q**) are known, then its Sales or Revenue Planned is:

$**= [**F**+**$'i[1+**s**]+(**Ad**+{**Q**+**P**/[**c**-**q**]}/**/**-**U**)/[1-**t**]]/[1-**v**]

Rule-15975:

If both (**/**), (**U**), (**$**), (**Q**), (**F**), (**$'**), (**i**), (**s**), (**t**), (**A**), (**d**), (**P**), (**c**) and (**q**) are known, then its Variable Portion Planned is:

v= 1-[**F**+**$'i**[1+**s**]+(**Ad**+{**Q**+**P**/[**c**-**q**]}/**/**-**U**)/[1-**t**]]/**$**

Rule-15976:

If both (**/**), (**U**), (**$**), (**v**), (**Q**), (**$'**), (**i**), (**s**), (**t**), (**A**), (**d**), (**P**), (**c**) and (**q**) are known, then its Qfixed Cost Planned is:

F= **$**[1-**v**]-**$'i**[1+**s**]-(**Ad**+{**Q**+**P**/[**c**-**q**]}/**/**-**U**)/[1-**t**]

Rule-15977:

If both (**/**), (**U**), (**$**), (**v**), (**F**), (**Q**), (**i**), (**s**), (**t**), (**A**), (**d**), (**P**), (**c**) and (**q**) are known, then its Sales Past must be:

$'**= [**$[1-**v**]-**F**-(**Ad**+{**Q**+**P**/[**c**-**q**]}/**/**-**U**)/[1-**t**]]
/{**i**[1+**s**]}

Steve Asikin ISBN 14: 978-1514685136, ISBN 10: 1514685132

Rule-15978:

If both (**/**), (**U**), (**$**), (**v**), (**F**), (**$'**), (**Q**), (**s**), (**t**), (**A**), (**d**), (**P**), (**c**) and (**q**) are known, then its Interest Portion Planned is:

i= [**$**[1-**v**]-**F**-(**Ad**+{**Q**+**P**/[**c**-**q**]}/**/**-**U**)/[1-**t**]]
/{**$'**[1+**s**]}

Rule-15979:

If both (**/**), (**U**), (**$**), (**v**), (**F**), (**$'**), (**i**), (**Q**), (**t**), (**A**), (**d**), (**P**), (**c**) and (**q**) are known, then its Sales Growth Planned is:

s= **$**/**$'**-1

or it can also be found as

s= [**$**[1-**v**]-**F**-(**Ad**+{**Q**+**P**/[**c**-**q**]}/**/**-**U**)/[1-**t**]]/[**$'i**]-1

Rule-15980:

If both (**/**), (**U**), (**$**), (**v**), (**F**), (**$'**), (**i**), (**s**), (**Q**), (**A**), (**d**), (**P**), (**c**) and (**q**) are known, then its Tax Rate Planned is:

t= 1-(**Ad**+{**Q**+**P**/[**c**-**q**]}/**/**-**U**)/{**$**[1-**v**]-**F**-**$'i**[1+**s**]}

Rule-15981:

If both (**/**), (**U**), (**$**), (**v**), (**F**), (**$'**), (**i**), (**s**), (**t**), (**Q**), (**d**), (**P**), (**c**) and (**q**) are known, then its After Tax Income Planned is:

A= (**U**+{**$**[1-**v**]-**F**-**$'i**[1+**s**]}[1-**t**]-{**Q**+**P**/[**c**-**q**]}/**/**)/**d**

Steve Asikin ISBN 14: 978-1514685136, ISBN 10: 1514685132

Rule-15982:

If both (**/**), (**U**), (**$**), (**v**), (**F**), (**$'**), (**i**), (**s**), (**t**), (**A**), (**Q**), (**P**), (**c**) and (**q**) are known, then its Dividend Payout Planned is:

d= (**U**+{**$**[1-**v**]-**F**-**$'i**[1+**s**]}[1-**t**]-{**Q**+**P**/[**c**-**q**]}/**/**)/**A**

Rule-15983:

If both (**/**), (**U**), (**$**), (**v**), (**F**), (**$'**), (**i**), (**s**), (**t**), (**A**), (**d**), (**Q**), (**c**) and (**q**) are known, then its Procured Inventory Planned is:

P= [**c**-**q**][**/**(**U**+{**$**[1-**v**]-**F**-**$'i**[1+**s**]}[1-**t**]-**Ad**)-**Q**]

Rule-15984:

If both (**/**), (**U**), (**$**), (**v**), (**F**), (**$'**), (**i**), (**s**), (**t**), (**A**), (**d**), (**P**), (**Q**) and (**q**) are known, then its Current Ratio Planned is:

c= **q**+**P**/[**/**(**U**+{**$**[1-**v**]-**F**-**$'i**[1+**s**]}[1-**t**]-**Ad**)-**Q**]

Rule-15985:

If both (**/**), (**U**), (**$**), (**v**), (**F**), (**$'**), (**i**), (**s**), (**t**), (**A**), (**d**), (**P**), (**c**) and (**Q**) are known, then its Quick or Acid Test Ratio Planned is:

q= **c**-**P**/[**/**(**U**+{**$**[1-**v**]-**F**-**$'i**[1+**s**]}[1-**t**]-**Ad**)-**Q**]

Rule-15986:

If both (**/**), (**U**), (**$**), (**v**), (**F**), (**$'**), (**i**), (**s**), (**t**), (**A**), (**d**), (**V**), (**p**), (**c**) and (**q**) are known, then its Quoted Longterm Debt Planned is:

Q= **/**(**U**+{**$**[1-**v**]-**F**-**$'i**[1+**s**]}[1-**t**]-**Ad**)
-**Vp**/{360[**c**-**q**]}

Steve Asikin ISBN 14: 978-1514685136, ISBN 10: 1514685132

Rule-15987:

If both (**Q**), (**U**), (**S**), (**v**), (**F**), (**S'**), (**i**), (**s**), (**t**), (**A**), (**d**), (**V**), (**p**), (**c**) and (**q**) are known, then its Leverage or Gearing Ratio Planned is:

/= (**Q**+**Vp**/{360[**c**-**q**]})
/(**U**+{**S**[1-**v**]-**F**-**S'i**[1+**s**]}[1-**t**]-**Ad**)

Rule-15988:

If both (/), (**Q**), (**S**), (**v**), (**F**), (**S'**), (**i**), (**s**), (**t**), (**A**), (**d**), (**V**), (**p**), (**c**) and (**q**) are known, then its Utilized or Starting Capital must be:

U= **Ad**+(**Q**+**Vp**/{360[**c**-**q**]})//
-{**S**[1-**v**]-**F**-**S'i**[1+**s**]}[1-**t**]

Rule-15989:

If both (/), (**U**), (**Q**), (**v**), (**F**), (**S'**), (**i**), (**s**), (**t**), (**A**), (**d**), (**V**), (**p**), (**c**) and (**q**) are known, then its Quoted Longterm Debt Planned is:

S= **{F**+**S'i**[1+**s**]
+**[Ad**+(**Q**+**Vp**/{360[**c**-**q**]})//-**U]**/[1-**t**]**}**/[1-**v**]

Rule-15990:

If both (/), (**U**), (**S**), (**Q**), (**F**), (**S'**), (**i**), (**s**), (**t**), (**A**), (**d**), (**V**), (**p**), (**c**) and (**q**) are known, then its Variable Portion Planned is:

v= 1-**{F**+**S'i**[1+**s**]
+**[Ad**+(**Q**+**Vp**/{360[**c**-**q**]})//-**U]**/[1-**t**]**}**/**S**

Steve Asikin ISBN 14: 978-1514685136, ISBN 10: 1514685132

Rule-15991:

If both (**/**), (**U**), (**$**), (**v**), (**Q**), (**$'**), (**i**), (**s**), (**t**), (**A**), (**d**), (**V**), (**p**), (**c**) and (**q**) are known, then its Fixed Cost Planned is:

$$\mathbf{F} = \$[1-\mathbf{v}] - \$'\mathbf{i}[1+\mathbf{s}]$$
$$-[\mathbf{Ad} + (\mathbf{Q} + \mathbf{Vp}/\{360[\mathbf{c}-\mathbf{q}]\})/\mathbf{/U}]/[1-\mathbf{t}]$$

Rule-15992:

If both (**/**), (**U**), (**$**), (**v**), (**F**), (**Q**), (**i**), (**s**), (**t**), (**A**), (**d**), (**V**), (**p**), (**c**) and (**q**) are known, then its Sales Past must be:

$$\$' = \{\$[1-\mathbf{v}] - \mathbf{F} - [\mathbf{Ad} + (\mathbf{Q} + \mathbf{Vp}/\{360[\mathbf{c}-\mathbf{q}]\})/\mathbf{/U}]/[1-\mathbf{t}]\}$$
$$/\{\mathbf{i}[1+\mathbf{s}]\}$$

Rule-15993:

If both (**/**), (**U**), (**$**), (**v**), (**F**), (**$'**), (**Q**), (**s**), (**t**), (**A**), (**d**), (**V**), (**p**), (**c**) and (**q**) are known, then its Interest Portion Planned is:

$$\mathbf{i} = \{\$[1-\mathbf{v}] - \mathbf{F} - [\mathbf{Ad} + (\mathbf{Q} + \mathbf{Vp}/\{360[\mathbf{c}-\mathbf{q}]\})/\mathbf{/U}]/[1-\mathbf{t}]\}$$
$$/\{\$'[1+\mathbf{s}]\}$$

Rule-15994:

If both (**/**), (**U**), (**$**), (**v**), (**F**), (**$'**), (**i**), (**Q**), (**t**), (**A**), (**d**), (**V**), (**p**), (**c**) and (**q**) are known, then its Sales Growth Planned is:

$$\mathbf{s} = \$/\$' - 1$$

or it can also be found as

$$\mathbf{s} = \{\$[1-\mathbf{v}] - \mathbf{F} - [\mathbf{Ad} + (\mathbf{Q} + \mathbf{Vp}/\{360[\mathbf{c}-\mathbf{q}]\})/\mathbf{/U}]/[1-\mathbf{t}]\}$$
$$/[\$'\mathbf{i}] - 1$$

Steve Asikin ISBN 14: 978-1514685136, ISBN 10: 1514685132

Rule-15995:

If both (/), (U), (S), (v), (F), (S'), (i), (s), (Q), (A), (d), (V), (p), (c) and (q) are known, then its Tax Rate Planned is:

t= 1-[**Ad**+(**Q**+**Vp**/{360[**c**-**q**]})//-**U**]
/{**S**[1-**v**]-**F**-**S'i**[1+**s**]}

Rule-15996:

If both (/), (U), (S), (v), (F), (S'), (i), (s), (t), (Q), (d), (V), (p), (c) and (q) are known, then its After Tax Income Planned is:

A= [**U**+{**S**[1-**v**]-**F**-**S'i**[1+**s**]}[1-**t**]
-(**Q**+**Vp**/{360[**c**-**q**]})//]/**d**

Rule-15997:

If both (/), (U), (S), (v), (F), (S'), (i), (s), (t), (A), (Q), (V), (p), (c) and (q) are known, then its Dividend Payout Planned is:

d= [**U**+{**S**[1-**v**]-**F**-**S'i**[1+**s**]}[1-**t**]
-(**Q**+**Vp**/{360[**c**-**q**]})//]/**A**

Rule-15998:

If both (/), (U), (S), (v), (F), (S'), (i), (s), (t), (A), (d), (Q), (p), (c) and (q) are known, then its Variable Cost Planned is:

V= 360[**c**-**q**]
[/(**U**+{**S**[1-**v**]-**F**-**S'i**[1+**s**]}[1-**t**]-**Ad**)-**Q**]/**p**

Steve Asikin ISBN 14: 978-1514685136, ISBN 10: 1514685132

Rule-15999:

If both (**ʎ**), (**U**), ($\$$), (**v**), (**F**), ($\$'$), (**i**), (**s**), (**t**), (**A**), (**d**), (**V**), (**Q**), (**c**) and (**q**) are known, then its Procured Inventory Days Planned:

$$p = 360[c-q][ʎ(U+\{\$[1-v]-F-\$'i[1+s]\}[1-t]-Ad)-Q]/V$$

Rule-16000:

If both (**ʎ**), (**U**), ($\$$), (**v**), (**F**), ($\$'$), (**i**), (**s**), (**t**), (**A**), (**d**), (**V**), (**p**), (**Q**) and (**q**) are known, then its Current Ratio Planned is:

$$c = q+Vp/\{360[ʎ(U+\{\$[1-v]-F-\$'i[1+s]\}[1-t]-Ad)-Q]\}$$

Rule-16001:

If both (**ʎ**), (**U**), ($\$$), (**v**), (**F**), ($\$'$), (**i**), (**s**), (**t**), (**A**), (**d**), (**V**), (**p**), (**c**) and (**Q**) are known, then its Quick or Acid Test Ratio Planned is:

$$q = c-Vp/\{360[ʎ(U+\{\$[1-v]-F-\$'i[1+s]\}[1-t]-Ad)-Q]\}$$

Rule-16002:

If both (**ʎ**), (**U**), ($\$$), (**v**), (**F**), ($\$'$), (**i**), (**s**), (**t**), (**A**), (**d**), (**p**), (**c**) and (**q**) are known, then its Quoted Longterm Debt Planned is:

$$Q = ʎ(U+\{\$[1-v]-F-\$'i[1+s]\}[1-t]-Ad)-\$vp/\{360[c-q]\}$$

Steve Asikin ISBN 14: 978-1514685136, ISBN 10: 1514685132

Rule-16003:

If both (**Q**), (**U**), (**$**), (**v**), (**F**), (**$'**), (**i**), (**s**), (**t**), (**A**), (**d**), (**p**), (**c**) and (**q**) are known, then its Leverage or Gearing Ratio Planned is:

/= (Q+$vp/{360[c-q]})
/(U+{$[1-v]-F-$'i[1+s]}[1-t]-Ad)

Rule-16004:

If both (**/**), (**Q**), (**$**), (**v**), (**F**), (**$'**), (**i**), (**s**), (**t**), (**A**), (**d**), (**p**), (**c**) and (**q**) are known, then its Utilized or Starting Capital must be:

U= Ad+(Q+$vp/{360[c-q]})//
-{$[1-v]-F-$'i[1+s]}[1-t]

Rule-16005:

If both (**/**), (**U**), (**Q**), (**v**), (**F**), (**$'**), (**i**), (**s**), (**t**), (**A**), (**d**), (**p**), (**c**) and (**q**) are known, then its Sales or Revenue Planned is:

$= [Q-/(U-{F+$'i[1+s]}[1-t]-Ad)+$vp/{360[c-q]}]
/{/[1-v][1-t]}

Rule-16006:

If both (**/**), (**U**), (**$**), (**Q**), (**F**), (**$'**), (**i**), (**s**), (**t**), (**A**), (**d**), (**p**), (**c**) and (**q**) are known, then itsVariable Portion Planned is:

v= [/(U+{$-F-$'i[1+s]}[1-t]-Ad)-Q]
/[$(p/{360[c-q]}+/[1-t])]

Steve Asikin ISBN 14: 978-1514685136, ISBN 10: 1514685132

Rule-16007:

If both (**/**), (**U**), (**$**), (**v**), (**Q**), (**$'**), (**i**), (**s**), (**t**), (**A**), (**d**), (**p**), (**c**) and (**q**) are known, then its Fixed Cost Planned is:

F= **$**[1-**v**]-**$'i**[1+**s**]-**[Ad**+(**Q**+**$vp**/{360[**c**-**q**]})/**/-U]** /[1-**t**]

Rule-16008:

If both (**/**), (**U**), (**$**), (**v**), (**F**), (**Q**), (**i**), (**s**), (**t**), (**A**), (**d**), (**p**), (**c**) and (**q**) are known, then its Sales Past must be:

$'**= {**$[1-**v**]-**F**-**[Ad**+(**Q**+**$vp**/{360[**c**-**q**]})/**/-U]** /[1-**t**]}/{**i**[1+**s**]}

Rule-16009:

If both (**/**), (**U**), (**$**), (**v**), (**F**), (**$'**), (**Q**), (**s**), (**t**), (**A**), (**d**), (**p**), (**c**) and (**q**) are known, then its Interest Portion Planned is:

i= {**$**[1-**v**]-**F**-**[Ad**+(**Q**+**$vp**/{360[**c**-**q**]})/**/-U]**/[1-**t**]} /{**$'**[1+**s**]}

Rule-16010:

If both (**/**), (**U**), (**$**), (**v**), (**F**), (**$'**), (**i**), (**Q**), (**t**), (**A**), (**d**), (**p**), (**c**) and (**q**) are known, then its Sales Growth Planned is:

s= **$**/**$'**-1

or it can also be found as

s= {**$**[1-**v**]-**F**-**[Ad**+(**Q**+**$vp**/{360[**c**-**q**]})/**/-U]**/[1-**t**]} /[**$'i**]-1

Steve Asikin ISBN 14: 978-1514685136, ISBN 10: 1514685132

Rule-16011:

If both (**/**), (**U**), (**S**), (**v**), (**F**), (**S'**), (**i**), (**s**), (**Q**), (**A**), (**d**), (**p**), (**c**) and (**q**) are known, then its Tax Rate Planned is:

t= 1-[**Ad**+(**Q**+**Svp**/{360[**c**-**q**]})/**/**-**U**]
/{**S**[1-**v**]-**F**-**S'i**[1+**s**]}

Rule-16012:

If both (**/**), (**U**), (**S**), (**v**), (**F**), (**S'**), (**i**), (**s**), (**t**), (**Q**), (**d**), (**p**), (**c**) and (**q**) are known, then its After Tax Income Planned is:

A= [**U**+{**S**[1-**v**]-**F**-**S'i**[1+**s**]}[1-**t**]-(**Q**+**Svp**/{360[**c**-**q**]})/**/**]/**d**

Rule-16013:

If both (**/**), (**U**), (**S**), (**v**), (**F**), (**S'**), (**i**), (**s**), (**t**), (**A**), (**Q**), (**p**), (**c**) and (**q**) are known, then its Dividend Payout Planned is:

d= [**U**+{**S**[1-**v**]-**F**-**S'i**[1+**s**]}[1-**t**]
-(**Q**+**Svp**/{360[**c**-**q**]})/**/**]/**A**

Rule-16014:

If both (**/**), (**U**), (**S**), (**v**), (**F**), (**S'**), (**i**), (**s**), (**t**), (**A**), (**d**), (**Q**), (**c**) and (**q**) are known, then its Procured Inventory Days Planned:

p= 360[**c**-**q**][**/**(**U**+{**S**[1-**v**]-**F**-**S'i**[1+**s**]}[1-**t**]-**Ad**)-**Q**]
/[**Sv**]

Steve Asikin ISBN 14: 978-1514685136, ISBN 10: 1514685132

Rule-16015:

If both (**/**), (**U**), (**S**), (**v**), (**F**), (**S'**), (**i**), (**s**), (**t**), (**A**), (**d**), (**p**), (**Q**) and (**q**) are known, then its Current Ratio Planned is:

c= **q**+**Svp**
/{360[**/**(**U**+{**S**[1-**v**]-**F**-**S'i**[1+**s**]}[1-**t**]-**Ad**)-**Q**]}

Rule-16016:

If both (**/**), (**U**), (**S**), (**v**), (**F**), (**S'**), (**i**), (**s**), (**t**), (**A**), (**d**), (**p**), (**c**) and (**Q**) are known, then its Quick or Acid Test Ratio Planned is:

q= **c**-**Svp**
/{360[**/**(**U**+{**S**[1-**v**]-**F**-**S'i**[1+**s**]}[1-**t**]-**Ad**)-**Q**]}

Rule-16017:

If both (**/**), (**U**), (**S**), (**v**), (**F**), (**S'**), (**i**), (**s**), (**t**), (**A**), (**d**), (**p**), (**c**) and (**q**) are known, then its Quoted Longterm Debt Planned is:

Q= **/**(**U**+{**S**[1-**v**]-**F**-**S'i**[1+**s**]}[1-**t**]-**Ad**)
-**S'vp**[1+**s**]/{360[**c**-**q**]}

Rule-16018:

If both (**Q**), (**U**), (**S**), (**v**), (**F**), (**S'**), (**i**), (**s**), (**t**), (**A**), (**d**), (**p**), (**c**) and (**q**) are known, then its Leverage or Gearing Ratio Planned is:

/= (**Q**+**S'vp**[1+**s**]/{360[**c**-**q**]})
/(**U**+{**S**[1-**v**]-**F**-**S'i**[1+**s**]}[1-**t**]-**Ad**)

Steve Asikin ISBN 14: 978-1514685136, ISBN 10: 1514685132

Rule-16019:

If both (**/**), (**Q**), (**$**), (**v**), (**F**), (**$'**), (**i**), (**s**), (**t**), (**A**), (**d**), (**p**), (**c**) and (**q**) are known, then its Utilized or Starting Capital must be:

U= **Ad**+(**Q**+**$'vp**[1+**s**]/{360[**c-q**]})/**/**
-{**$**[1-**v**]-**F-$'i**[1+**s**]}[1-**t**]

Rule-16020:

If both (**/**), (**U**), (**Q**), (**v**), (**F**), (**$'**), (**i**), (**s**), (**t**), (**A**), (**d**), (**p**), (**c**) and (**q**) are known, then its Sales or Revenue Planned is:

$**= **[Q-/(U-**{**F**+**$'i[1+**s**]}[1-**t**]+**Ad**)
+**$vp**/{360[**c-q**]}**]**/{**/**[1-**v**][1-**t**]}

Rule-16021:

If both (**/**), (**U**), (**$**), (**Q**), (**F**), (**$'**), (**i**), (**s**), (**t**), (**A**), (**d**), (**p**), (**c**) and (**q**) are known, then its Variable Portion Planned is:

v= **[Q-/(U-**{**F**+**$'i**[1+**s**]}[1-**t**]-**Ad**)
+**$'vp**[1+**s**]/{360[**c-q**]}**]**/{**/**[1-**v**][1-**t**]}

Rule-16022:

If both (**/**), (**U**), (**$**), (**v**), (**Q**), (**$'**), (**i**), (**s**), (**t**), (**A**), (**d**), (**p**), (**c**) and (**q**) are known, then its Fixed Cost Planned is:

F= **$**[1-**v**]-**$'i**[1+**s**]
-**[Ad**+(**Q**+**$'vp**[1+**s**]/{360[**c-q**]})/**/-U]**/[1-**t**]

Steve Asikin ISBN 14: 978-1514685136, ISBN 10: 1514685132

Rule-16023:

If both (**/**), (**U**), (**$**), (**v**), (**F**), (**Q**), (**i**), (**s**), (**t**), (**A**), (**d**), (**p**), (**c**) and (**q**) are known, then its Sales Past must be:

$'**= **[/(U**+{**$[1-**v**]-**F**}[1-**t**]-**Ad**)-**Q]**
/**[**[1+**s**](**vp**/{360[**c-q**]}+**i**[1-**t**])**]**

Rule-16024:

If both (**/**), (**U**), (**$**), (**v**), (**F**), (**$'**), (**Q**), (**s**), (**t**), (**A**), (**d**), (**p**), (**c**) and (**q**) are known, then its Interest Portion Planned is:

i= **{$**[1-**v**]-**F-[Ad**+(**Q**+**$'vp**[1+**s**]/{360[**c-q**]})/**/-U]**
/[1-**t**]**}**/{**$'**[1+**s**]}

Rule-16025:

If both (**/**), (**U**), (**$**), (**v**), (**F**), (**$'**), (**i**), (**Q**), (**t**), (**A**), (**d**), (**p**), (**c**) and (**q**) are known, then its Sales Growth Planned is:

s= **$**/**$'**-1

or it can also be found as

s= **[/(U**+{**$**[1-**v**]-**F**}[1-**t**]-**Ad**)-**Q]**
/**[$'**(**vp**/{360[**c-q**]}+**i**[1-**t**])**]**-1

Rule-16026:

If both (**/**), (**U**), (**$**), (**v**), (**F**), (**$'**), (**i**), (**s**), (**Q**), (**A**), (**d**), (**p**), (**c**) and (**q**) are known, then its Tax Rate Planned is:

t= 1-**[Ad**+(**Q**+**$'vp**[1+**s**]/{360[**c-q**]})/**/-U]**
/{**$**[1-**v**]-**F-$'i**[1+**s**]}

Steve Asikin ISBN 14: 978-1514685136, ISBN 10: 1514685132

Rule-16027:

If both (**/**), (**U**), (**$**), (**v**), (**F**), (**$'**), (**i**), (**s**), (**t**), (**Q**), (**d**), (**p**), (**c**) and (**q**) are known, then its After Tax Income Planned is:

A= [**U**+{**$**[1-**v**]-**F**-**$'i**[1+**s**]}[1-**t**]
-(**Q**+**$'vp**[1+**s**]/{360[**c**-**q**]})/**/**]/**d**

Rule-16028:

If both (**/**), (**U**), (**$**), (**v**), (**F**), (**$'**), (**i**), (**s**), (**t**), (**A**), (**Q**), (**p**), (**c**) and (**q**) are known, then its Dividend Planned is:

d= [**U**+{**$**[1-**v**]-**F**-**$'i**[1+**s**]}[1-**t**]
-(**Q**+**$'vp**[1+**s**]/{360[**c**-**q**]})/**/**]/**A**

Rule-16029:

If both (**/**), (**U**), (**$**), (**v**), (**F**), (**$'**), (**i**), (**s**), (**t**), (**A**), (**d**), (**Q**), (**c**) and (**q**) are known, then its Procured Inventory Days Planned is:

p= 360[**c**-**q**][**/**(**U**+{**$**[1-**v**]-**F**-**$'i**[1+**s**]}[1-**t**]-**Ad**)-**Q**]
/{**$'v**[1+**s**]}

Rule-16030:

If both (**/**), (**U**), (**$**), (**v**), (**F**), (**$'**), (**i**), (**s**), (**t**), (**A**), (**d**), (**p**), (**Q**) and (**q**) are known, then its Current Ratio Planned is:

c= **q**+**$'vp**[1+**s**]
/{360[**/**(**U**+{**$**[1-**v**]-**F**-**$'i**[1+**s**]}[1-**t**]-**Ad**)-**Q**]}

Steve Asikin ISBN 14: 978-1514685136, ISBN 10: 1514685132

Rule-16031:

If both (**/**), (**U**), (**$**), (**v**), (**F**), (**$'**), (**i**), (**s**), (**t**), (**A**), (**d**), (**p**), (**c**) and (**Q**) are known, then its Quick or Acid Test Ratio Planned is:

q= **c**-**$'vp**[1+**s**]
/{360[**/**(**U**+{**$**[1-**v**]-**F**-**$'i**[1+**s**]}[1-**t**]-**Ad**)-**Q**]}

Rule-16032:

If both (**/**), (**U**), (**$**), (**v**), (**F**), (**$'**), (**i**), (**s**), (**t**), (**d**) and (**X**) are known, then its Quoted Longterm Debt Planned is:

Q= **/**(**U**+{**$**[1-**v**]-**F**-**$'i**[1+**s**]}[1-**t**][1-**d**])-**X**

Rule-16033:

If both (**Q**), (**U**), (**$**), (**v**), (**F**), (**$'**), (**i**), (**s**), (**t**), (**d**) and (**X**) are known, then its Leverage or Gearing Ratio Planned is:

/= [**Q**+**X**]/(**U**+{**$**[1-**v**]-**F**-**$'i**[1+**s**]}[1-**t**][1-**d**])

Rule-16034:

If both (**/**), (**Q**), (**$**), (**v**), (**F**), (**$'**), (**i**), (**s**), (**t**), (**d**) and (**X**) are known, then its Utilized or Starting Capital must be:

U= [**Q**+**X**]/**/**-{**$**[1-**v**]-**F**-**$'i**[1+**s**]}[1-**t**][1-**d**]

Rule-16035:

If both (**/**), (**U**), (**Q**), (**v**), (**F**), (**$'**), (**i**), (**s**), (**t**), (**d**) and (**X**) are known, then its Sales or Revenue Planned is:

$**= (**F**+**$'i[1+**s**]+{[**Q**+**X**]/**/**–**U**}/{[1-**t**][1-**d**]})/[1-**v**]

Steve Asikin ISBN 14: 978-1514685136, ISBN 10: 1514685132

Rule-16036:

If both (I), (**U**), (**$**), (**Q**), (**F**), (**$'**), (**i**), (**s**), (**t**), (**d**) and (**X**) are known, then its Variable Portion Planned is:

$$v = 1 - (F + \$'i[1+s] + \{[Q+X]/I - U\}/\{[1-t][1-d]\})/\$$$

Rule-16037:

If both (I), (**U**), (**$**), (**v**), (**Q**), (**$'**), (**i**), (**s**), (**t**), (**d**) and (**X**) are known, then its Fixed Cost Planned is:

$$F = \{\$[1-v] - \$'i[1+s]\} - \{[Q+X]/I - U\}/\{[1-t][1-d]\}$$

Rule-16038:

If both (I), (**U**), (**$**), (**v**), (**F**), (**Q**), (**i**), (**s**), (**t**), (**d**) and (**X**) are known, then its Sales Past must be:

$$\$' = (\$[1-v] - F - \{[Q+X]/I - U\}/\{[1-t][1-d]\})/\{i[1+s]\}$$

Rule-16039:

If both (I), (**U**), (**$**), (**v**), (**F**), (**$'**), (**Q**), (**s**), (**t**), (**d**) and (**X**) are known, then its Interest Portion Planned is:

$$i = (\$[1-v] - F - \{[Q+X]/I - U\}/\{[1-t][1-d]\})/\{\$'[1+s]\}$$

Rule-16040:

If both (I), (**U**), (**$**), (**v**), (**F**), (**$'**), (**i**), (**Q**), (**t**), (**d**) and (**X**) are known, then its Sales Growth Planned is:

$$s = \$/\$' - 1$$

or it can also be found as

$$s = (\$[1-v] - F - \{[Q+X]/I - U\}/\{[1-t][1-d]\})/[\$'i] - 1$$

Steve Asikin ISBN 14: 978-1514685136, ISBN 10: 1514685132

Rule-16041:

If both (/), (**U**), (**S**), (**v**), (**F**), (**S'**), (**i**), (**s**), (**Q**), (**d**) and (**X**) are known, then its Tax Rate Planned is:

t= 1-{[**Q**+**X**]/**/**–**U**}/({**S**[1-**v**]-**F**-**S'i**[1+**s**]}[1-**d**])

Rule-16042:

If both (/), (**U**), (**S**), (**v**), (**F**), (**S'**), (**i**), (**s**), (**t**), (**Q**) and (**X**) are known, then its Dividend Payout Planned is:

d= 1-{[**Q**+**X**]/**/**–**U**}/({**S**[1-**v**]-**F**-**S'i**[1+**s**]}[1-**t**])

Rule-16043:

If both (/), (**U**), (**S**), (**v**), (**F**), (**S'**), (**i**), (**s**), (**t**), (**d**) and (**Q**) are known, then its Xpress or Current Debt Planned is:

X= **/**(**U**+{**S**[1-**v**]-**F**-**S'i**[1+**s**]}[1-**t**][1-**d**])-**Q**

Rule-16044:

If both (/), (**U**), (**S**), (**v**), (**F**), (**S'**), (**i**), (**s**), (**t**), (**d**), (**P**), (**c**) and (**q**) are known, then its Quoted Longterm Debt Planned is:

Q= **/**(**U**+{**S**[1-**v**]-**F**-**S'i**[1+**s**]}[1-**t**][1-**d**])-**P**/[**c**-**q**]

Rule-16045:

If both (**Q**), (**U**), (**S**), (**v**), (**F**), (**S'**), (**i**), (**s**), (**t**), (**d**), (**P**), (**c**) and (**q**) are known, then its Leverage or Gearing Ratio Planned is:

/= {**Q**+**P**/[**c**-**q**]}/(**U**+{**S**[1-**v**]-**F**-**S'i**[1+**s**]}[1-**t**][1-**d**])

Steve Asikin ISBN 14: 978-1514685136, ISBN 10: 1514685132

Rule-16046:

If both (I), $(\mathbf{Q})$, $(\mathbf{S})$, $(\mathbf{v})$, $(\mathbf{F})$, $(\mathbf{S'})$, $(\mathbf{i})$, $(\mathbf{s})$, $(\mathbf{t})$, $(\mathbf{d})$, $(\mathbf{P})$, $(\mathbf{c})$ and $(\mathbf{q})$ are known, then its Utilized or Starting capital must be:

$$U= \{Q+P/[c-q]\}/I-\{S[1-v]-F-S'i[1+s]\}[1-t][1-d]$$

Rule-16047:

If both (I), $(\mathbf{U})$, $(\mathbf{Q})$, $(\mathbf{v})$, $(\mathbf{F})$, $(\mathbf{S'})$, $(\mathbf{i})$, $(\mathbf{s})$, $(\mathbf{t})$, $(\mathbf{d})$, $(\mathbf{P})$, $(\mathbf{c})$ and $(\mathbf{q})$ are known, then its Sales or Revenue Planned is:

$$S= [F+S'i[1+s]+(\{Q+P/[c-q]\}/I-U)/\{[1-t][1-d]\}]$$
$$/[1-v]$$

Rule-16048:

If both (I), $(\mathbf{U})$, $(\mathbf{S})$, $(\mathbf{Q})$, $(\mathbf{F})$, $(\mathbf{S'})$, $(\mathbf{i})$, $(\mathbf{s})$, $(\mathbf{t})$, $(\mathbf{d})$, $(\mathbf{P})$, $(\mathbf{c})$ and $(\mathbf{q})$ are known, then its Variable Portion Planned is:

$$v= 1-[F+S'i[1+s]+(\{Q+P/[c-q]\}/I-U)$$
$$/\{[1-t][1-d]\}]/S$$

Rule-16049:

If both (I), $(\mathbf{U})$, $(\mathbf{S})$, $(\mathbf{v})$, $(\mathbf{Q})$, $(\mathbf{S'})$, $(\mathbf{i})$, $(\mathbf{s})$, $(\mathbf{t})$, $(\mathbf{d})$, $(\mathbf{P})$, $(\mathbf{c})$ and $(\mathbf{q})$ are known, then its Fixed Cost Planned is:

$$F= \{S[1-v]-S'i[1+s]\}-(\{Q+P/[c-q]\}/I-U)$$
$$/\{[1-t][1-d]\}$$

Rule-16050:

If both (I), $(\mathbf{U})$, $(\mathbf{S})$, $(\mathbf{v})$, $(\mathbf{F})$, $(\mathbf{Q})$, $(\mathbf{i})$, $(\mathbf{s})$, $(\mathbf{t})$, $(\mathbf{d})$, $(\mathbf{P})$, $(\mathbf{c})$ and $(\mathbf{q})$ are known, then its Sales Past must be:

$$S'= [S[1-v]-F-(\{Q+P/[c-q]\}/I-U)/\{[1-t][1-d]\}]$$
$$/\{i[1+s]\}$$

Steve Asikin ISBN 14: 978-1514685136, ISBN 10: 1514685132

Rule-16051:

If both (**/**), (**U**), (**$**), (**v**), (**F**), (**$'**), (**Q**), (**s**), (**t**), (**d**), (**P**), (**c**) and (**q**) are known, then its Interest Portion Planned is:

i= [**$**[1-**v**]-**F**-({**Q**+**P**/[**c**-**q**]}/**/U**)/{[1-**t**][1-**d**]}]
/{**$'**[1+**s**]}

Rule-16052:

If both (**/**), (**U**), (**$**), (**v**), (**F**), (**$'**), (**i**), (**Q**), (**t**), (**d**), (**P**), (**c**) and (**q**) are known, then its Sales Growth Planned is:

s= **$**/**$'**-1

or it can also be found as

s= [**$**[1-**v**]-**F**-({**Q**+**P**/[**c**-**q**]}/**/U**)/{[1-**t**][1-**d**]}]
/[**$'i**]-1

Rule-16053:

If both (**/**), (**U**), (**$**), (**v**), (**F**), (**$'**), (**i**), (**s**), (**Q**), (**d**), (**P**), (**c**) and (**q**) are known, then its Tax Rate Planned is:

t= 1-({**Q**+**P**/[**c**-**q**]}/**/U**)/({**$**[1-**v**]-**F**-**$'i**[1+**s**]}[1-**d**])

Rule-16054:

If both (**/**), (**U**), (**$**), (**v**), (**F**), (**$'**), (**i**), (**s**), (**t**), (**Q**), (**P**), (**c**) and (**q**) are known, then its Dividend Payout Planned is:

d= 1-({**Q**+**P**/[**c**-**q**]}/**/U**)/({**$**[1-**v**]-**F**-**$'i**[1+**s**]}[1-**t**])

Steve Asikin ISBN 14: 978-1514685136, ISBN 10: 1514685132

Rule-16055:

If both (**ł**), (**U**), (**S**), (**v**), (**F**), (**S'**), (**i**), (**s**), (**t**), (**d**), (**Q**), (**c**) and (**q**) are known, then its Procured Inventory Planned is:

P= [**c**-**q**][**ł**(**U**+{**S**[1-**v**]-**F**-**S'i**[1+**s**]}[1-**t**][1-**d**])-**Q**]

Rule-16056:

If both (**ł**), (**U**), (**S**), (**v**), (**F**), (**S'**), (**i**), (**s**), (**t**), (**d**), (**P**), (**Q**) and (**q**) are known, then its Current Ratio Planned is:

c= **q**+**P**/[**ł**(**U**+{**S**[1-**v**]-**F**-**S'i**[1+**s**]}[1-**t**][1-**d**])-**Q**]

Rule-16057:

If both (**ł**), (**U**), (**S**), (**v**), (**F**), (**S'**), (**i**), (**s**), (**t**), (**d**), (**P**), (**c**) and (**Q**) are known, then its Quick or Acid Test Ratio Planned is:

q= **c**-**P**/[**ł**(**U**+{**S**[1-**v**]-**F**-**S'i**[1+**s**]}[1-**t**][1-**d**])-**Q**]

Rule-16058:

If both (**ł**), (**U**), (**S**), (**v**), (**F**), (**S'**), (**i**), (**s**), (**t**), (**d**), (**V**), (**p**), (**c**) and (**q**) are known, then its Quoted Longterm Debt Planned is:

Q= **ł**(**U**+{**S**[1-**v**]-**F**-**S'i**[1+**s**]}[1-**t**][1-**d**])
-**Vp**/{360[**c**-**q**]}

Rule-16059:

If both (**Q**), (**U**), (**S**), (**v**), (**F**), (**S'**), (**i**), (**s**), (**t**), (**d**), (**V**), (**p**), (**c**) and (**q**) are known, then its Leverage or Gearing Ratio Planned is:

ł= {**Q**+**Vp**/{360[**c**-**q**]})
/(**U**+{**S**[1-**v**]-**F**-**S'i**[1+**s**]}[1-**t**][1-**d**])

Steve Asikin ISBN 14: 978-1514685136, ISBN 10: 1514685132

Rule-16060:

If both (**/**), (**Q**), (**$**), (**v**), (**F**), (**$'**), (**i**), (**s**), (**t**), (**d**), (**V**), (**p**), (**c**) and (**q**) are known, then its Utilized or Starting Capital must be:

U= {Q+Vp/{360[c-q]})//
-{$[1-v]-F-$'i[1+s]}[1-t][1-d]

Rule-16061:

If both (**/**), (**U**), (**Q**), (**v**), (**F**), (**$'**), (**i**), (**s**), (**t**), (**d**), (**V**), (**p**), (**c**) and (**q**) are known, then its Sales or Revenue Planned is:

$= {F+$'i[1+s]+[(Q+Vp/{360[c-q]})/-U]
/{[1-t][1-d]}}/[1-v]

Rule-16062:

If both (**/**), (**U**), (**$**), (**Q**), (**F**), (**$'**), (**i**), (**s**), (**t**), (**d**), (**V**), (**p**), (**c**) and (**q**) are known, then its Variable Portion Planned is:

v= V/$

or it can also be found as

v= 1-{F+$'i[1+s]+[(Q+Vp/{360[c-q]})/-U]
/{[1-t][1-d]}}/$

Rule-16063:

If both (**/**), (**U**), (**$**), (**v**), (**Q**), (**$'**), (**i**), (**s**), (**t**), (**d**), (**V**), (**p**), (**c**) and (**q**) are known, then its Fixed Cost Planned is:

F= {$[1-v]-$'i[1+s]}-[{Q+Vp/{360[c-q]})/-U]
/{[1-t][1-d]}

Steve Asikin ISBN 14: 978-1514685136, ISBN 10: 1514685132

Rule-16064:

If both (**/**), (**U**), (**$**), (**v**), (**F**), (**Q**), (**i**), (**s**), (**t**), (**d**), (**V**), (**p**), (**c**) and (**q**) are known, then its Sales Past must be:

$'= {$[1-v]-F-[{Q+Vp/{360[c-q]}}/ /U]
/{[1-t][1-d]}}/{i[1+s]}

Rule-16065:

If both (**/**), (**U**), (**$**), (**v**), (**F**), (**$'**), (**Q**), (**s**), (**t**), (**d**), (**V**), (**p**), (**c**) and (**q**) are known, then its Interest Portion Planned is:

i= {$[1-v]-F-[{Q+Vp/{360[c-q]}}/ /U]
/{[1-t][1-d]}}/{$'[1+s]}

Rule-16066:

If both (**/**), (**U**), (**$**), (**v**), (**F**), (**$'**), (**i**), (**Q**), (**t**), (**d**), (**V**), (**p**), (**c**) and (**q**) are known, then its Sales Growth Planned is:

s= $/$'-1

or it can also be found as

s= {$[1-v]-F-[{Q+Vp/{360[c-q]}}/ /U]
/{[1-t][1-d]}}/[$'i]-1

Rule-16067:

If both (**/**), (**U**), (**$**), (**v**), (**F**), (**$'**), (**i**), (**s**), (**Q**), (**d**), (**V**), (**p**), (**c**) and (**q**) are known, then its Tax Rate Planned is:

t= 1-[{Q+Vp/{360[c-q]}}/ /U]
/({$[1-v]-F-$'i[1+s]}[1-d])

Steve Asikin ISBN 14: 978-1514685136, ISBN 10: 1514685132

Rule-16068:

If both (/), (U), (S), (v), (F), (S'), (i), (s), (t), (Q), (V), (p), (c) and (q) are known, then its Dividend Payout Planned is:

d= 1-[{Q+Vp/{360[c-q]}}/ /U]
/({S[1-v]-F-S'i[1+s]}[1-t])

Rule-16069:

If both (/), (U), (S), (v), (F), (S'), (i), (s), (t), (d), (Q), (p), (c) and (q) are known, then its Variable Cost Planned is:

V= 360[c-q]
[/(U+{S[1-v]-F-S'i[1+s]}[1-t][1-d])-Q]/p

Rule-16070:

If both (/), (U), (S), (v), (F), (S'), (i), (s), (t), (d), (V), (Q), (c) and (q) are known, then its Procured Inventory Days Planned:

p= 360[c-q]
[/(U+{S[1-v]-F-S'i[1+s]}[1-t][1-d])-Q]/V

Rule-16071:

If both (/), (U), (S), (v), (F), (S'), (i), (s), (t), (d), (V), (p), (Q) and (q) are known, then its Current Ratio Planned is:

c= q+Vp
/{360[/(U+{S[1-v]-F-S'i[1+s]}[1-t][1-d])-
Q]}

Steve Asikin ISBN 14: 978-1514685136, ISBN 10: 1514685132

Rule-16072:

If both (**/**), (**U**), (**$**), (**v**), (**F**), (**$'**), (**i**), (**s**), (**t**), (**d**), (**V**), (**p**), (**c**) and (**Q**) are known, then its Quick or Acid Test Ratio Planned is:

q= c-Vp/{360
[/(U+{$[1-v]-F-$'i[1+s]}[1-t][1-d])-Q]}

Rule-16073:

If both (**/**), (**U**), (**$**), (**v**), (**F**), (**$'**), (**i**), (**s**), (**t**), (**d**), (**p**), (**c**) and (**q**) are known, then its Quoted Longterm Debt Planned is:

Q= /(U+{$[1-v]-F-$'i[1+s]}[1-t][1-d])
-$vp/{360[c-q]}

Rule-16074:

If both (**Q**), (**U**), (**$**), (**v**), (**F**), (**$'**), (**i**), (**s**), (**t**), (**d**), (**p**), (**c**) and (**q**) are known, then its Leverage or Gearing Ratio Planned is:

/= {Q+$vp/{360[c-q]})
/(U+{$[1-v]-F-$'i[1+s]}[1-t][1-d])

Rule-16075:

If both (**/**), (**Q**), (**$**), (**v**), (**F**), (**$'**), (**i**), (**s**), (**t**), (**d**), (**p**), (**c**) and (**q**) are known, then its Utilized or Starting Capital must be:

U= {Q+$vp/{360[c-q]})//
-{$[1-v]-F-$'i[1+s]}[1-t][1-d]

Steve Asikin ISBN 14: 978-1514685136, ISBN 10: 1514685132

Rule-16076:

If both (**/**), (**U**), (**Q**), (**v**), (**F**), (**S'**), (**i**), (**s**), (**t**), (**d**), (**p**), (**c**) and (**q**) are known, then its Sales or Revenue Planned is:

S= [**/U**-{**F**+**S'i**[1+**s**]}[1-**t**][1-**d**])-**Q**]
/(**vp**/{360[**c**-**q**]}- **/**[1-**v**][1-**t**][1-**d**]})

Rule-16077:

If both (**/**), (**U**), (**S**), (**Q**), (**F**), (**S'**), (**i**), (**s**), (**t**), (**d**), (**p**), (**c**) and (**q**) are known, then its Variable Portion Planned is:

v= [**/U**+{**S**-**F**-**S'i**[1+**s**]}[1-**t**][1-**d**])-**Q**]
/[**S**(**p**/{360[**c**-**q**]}-**/**[1-**t**][1-**d**])]

Rule-16078:

If both (**/**), (**U**), (**S**), (**v**), (**Q**), (**S'**), (**i**), (**s**), (**t**), (**d**), (**p**), (**c**) and (**q**) are known, then its Fixed Cost Planned is:

F= {**S**[1-**v**]-**S'i**[1+**s**]}-[{**Q**+**Svp**/{360[**c**-**q**]})/**/U**]
/{[1-**t**][1-**d**]}

Rule-16079:

If both (**/**), (**U**), (**S**), (**v**), (**F**), (**Q**), (**i**), (**s**), (**t**), (**d**), (**p**), (**c**) and (**q**) are known, then its Sales Past must be:

S'= {**S**[1-**v**]-**F**-[{**Q**+**Svp**/{360[**c**-**q**]})/**/U**]
/{[1-**t**][1-**d**]}}/{**i**[1+**s**]}

Steve Asikin ISBN 14: 978-1514685136, ISBN 10: 1514685132

Rule-16080:

If both (/), (U), (S), (v), (F), (S'), (Q), (s), (t), (d), (p), (c) and (q) are known, then its Interest Portion Planned is:

i= {S[1-v]-F-[{Q+Svp/{360[c-q]}})/ /-U]
/{[1-t][1-d]}}/{S'[1+s]}

Rule-16081:

If both (/), (U), (S), (v), (F), (S'), (i), (Q), (t), (d), (p), (c) and (q) are known, then its Sales Growth Planned is:

s= S/S'-1

or it can also be found as

s= {S[1-v]-F-[{Q+Svp/{360[c-q]}})/ /-U]
/{[1-t][1-d]}}/[S'i]-1

Rule-16082:

If both (/), (U), (S), (v), (F), (S'), (i), (s), (Q), (d), (p), (c) and (q) are known, then its Tax Rate Planned is:

t= 1-[{Q+Svp/{360[c-q]}})/ /-U]
/({S[1-v]-F-S'i[1+s]}[1-d])

Rule-16083:

If both (/), (U), (S), (v), (F), (S'), (i), (s), (t), (Q), (p), (c) and (q) are known, then its Dividend Payout Planned is:

d= 1-[{Q+Svp/{360[c-q]}})/ /-U]
/({S[1-v]-F-S'i[1+s]}[1-t])

Steve Asikin ISBN 14: 978-1514685136, ISBN 10: 1514685132

Rule-16084:

If both (/), (U), ($), (v), (F), ($'), (i), (s), (t), (d), (Q), (c) and (q) are known, then its Procured Inventory Days Planned:

p= 360[c-q]
[/(U+{$[1-v]-F-$'i[1+s]}[1-t][1-d])-Q]/[$v]

Rule-16085:

If both (/), (U), ($), (v), (F), ($'), (i), (s), (t), (d), (p), (Q) and (q) are known, then its Current Ratio Planned is:

c= q+$vp/{360
[/(U+{$[1-v]-F-$'i[1+s]}[1-t][1-d])-Q]}

Rule-16086:

If both (/), (U), ($), (v), (F), ($'), (i), (s), (t), (d), (p), (c) and (Q) are known, then its Quick or Acid test Ratio Planned is:

q= c-$vp/{360
[/(U+{$[1-v]-F-$'i[1+s]}[1-t][1-d])-Q]}

Rule-16073:

If both (/), (U), ($), (v), (F), ($'), (i), (s), (t), (d), (p), (c) and (q) are known, then its Quoted Longterm Debt Planned is:

Q= /(U+{$[1-v]-F-$'i[1+s]}[1-t][1-d])
-$'vp[1+s]/{360[c-q]}

Steve Asikin ISBN 14: 978-1514685136, ISBN 10: 1514685132

Rule-16074:

If both (**Q**), (**U**), (**$**), (**v**), (**F**), (**$'**), (**i**), (**s**), (**t**), (**d**), (**p**), (**c**) and (**q**) are known, then its Leverage or Gearing Ratio Planned is:

/= {**Q**+**$'vp**[1+**s**]/{360[**c**-**q**]}) /(**U**+{**$**[1-**v**]-**F**-**$'i**[1+**s**]}[1-**t**][1-**d**])

Rule-16075:

If both (**/**), (**Q**), (**$**), (**v**), (**F**), (**$'**), (**i**), (**s**), (**t**), (**d**), (**p**), (**c**) and (**q**) are known, then its Utilized or Starting Capital must be:

U= {**Q**+**$'vp**[1+**s**]/{360[**c**-**q**]})/**/** -{**$**[1-**v**]-**F**-**$'i**[1+**s**]}[1-**t**][1-**d**]

Rule-16076:

If both (**/**), (**U**), (**Q**), (**v**), (**F**), (**$'**), (**i**), (**s**), (**t**), (**d**), (**p**), (**c**) and (**q**) are known, then its Sales or Revenue Planned is:

$**= [**Q**-**/**(**U**-{**F**+**$'i[1+**s**]}[1-**t**][1-**d**]) +**$'vp**[1+**s**]/{360[**c**-**q**]}]/{**/**[1-**v**][1-**t**][1-**d**]}

Rule-16077:

If both (**/**), (**U**), (**$**), (**Q**), (**F**), (**$'**), (**i**), (**s**), (**t**), (**d**), (**p**), (**c**) and (**q**) are known, then its Variable Portion Planned is:

v= [**/**(**U**+**$**-**F**-**$'i**[1+**s**]}[1-**t**][1-**d**])-**Q**] /(**$'p**[1+**s**]/{360[**c**-**q**]}+**$/**[1-**t**][1-**d**])

Steve Asikin ISBN 14: 978-1514685136, ISBN 10: 1514685132

Rule-16078:

If both (/), (U), (S), (v), (Q), (S'), (i), (s), (t), (d), (p), (c) and (q) are known, then its Fixed Cost Planned is:

F= {S[1-v]-S'i[1+s]}
-[{Q+S'vp[1+s]/{360[c-q]}}/AU]
/{[1-t][1-d]}

Rule-16079:

If both (/), (U), (S), (v), (F), (Q), (i), (s), (t), (d), (p), (c) and (q) are known, then its Sales Past must be:

S'= [AU+{S[1-v]-F}[1-t][1-d])-Q]
/[[1+s](vp/{360[c-q]}+ i/[1-t][1-d])]

Rule-16080:

If both (/), (U), (S), (v), (F), (S'), (Q), (s), (t), (d), (p), (c) and (q) are known, then its Interest Portion Planned is:

i= {S[1-v]-F-[{Q+S'vp[1+s]/{360[c-q]}}/AU]
/{[1-t][1-d]}}/{S'[1+s]}

Rule-16081:

If both (/), (U), (S), (v), (F), (S'), (i), (Q), (t), (d), (p), (c) and (q) are known, then its Sales Growth Planned is:

s= S/S'-1

or it can also be found as

s= [AU+{S[1-v]-F}[1-t][1-d])-Q]
/[S'(vp/{360[c-q]}+ i/[1-t][1-d])]-1

Steve Asikin ISBN 14: 978-1514685136, ISBN 10: 1514685132

Rule-16082:

If both (**/**), (**U**), (**S**), (**v**), (**F**), (**S'**), (**i**), (**s**), (**Q**), (**d**), (**p**), (**c**) and (**q**) are known, then its Tax Rate Planned is:

t= 1-[{**Q**+**S'vp**[1+**s**]/{360[**c**-**q**]}}/**/U**]
/({**S**[1-**v**]-**F**-**S'i**[1+**s**]}[1-**d**])

Rule-16083:

If both (**/**), (**U**), (**S**), (**v**), (**F**), (**S'**), (**i**), (**s**), (**t**), (**Q**), (**p**), (**c**) and (**q**) are known, then its Dividend Payout Planned is:

d= 1-[{**Q**+**S'vp**[1+**s**]/{360[**c**-**q**]}}/**/U**]
/({**S**[1-**v**]-**F**-**S'i**[1+**s**]}[1-**t**])

Rule-16084:

If both (**/**), (**U**), (**S**), (**v**), (**F**), (**S'**), (**i**), (**s**), (**t**), (**d**), (**Q**), (**c**) and (**q**) are known, then its Procured Inventory Days Planned:

p= 360[**c**-**q**]
[**/U**+{**S**[1-**v**]-**F**-**S'i**[1+**s**]}[1-**t**][1-**d**])-**Q**]
/{**S'v**[1+**s**]}

Rule-16085:

If both (**/**), (**U**), (**S**), (**v**), (**F**), (**S'**), (**i**), (**s**), (**t**), (**d**), (**p**), (**Q**) and (**q**) are known, then its Current Ratio Planned is:

c= **q**+**S'vp**[1+**s**]/{360[**/U**+{**S**[1-**v**]-**F**-**S'i**[1+**s**]}
[1-**t**][1-**d**])-**Q**]}

Steve Asikin ISBN 14: 978-1514685136, ISBN 10: 1514685132

Rule-16086:

If both (**/**), (**U**), (**$**), (**v**), (**F**), (**$'**), (**i**), (**s**), (**t**), (**d**), (**p**), (**c**) and (**Q**) are known, then its Quick or Acid test Ratio Planned is:

q= **c**-**$'vp**[1+**s**]/{360[**/**(**U**+{**$**[1-**v**]-**F**-**$'i**[1+**s**]}[1-**t**][1-**d**])-**Q**]}

Rule-16087:

If both (**/**), (**U**), (**$**), (**v**), (**f**), (**I**), (**T**), (**D**) and (**X**) are known, then its Quoted Longterm Debt Planned is:

Q= **/**{**U**+**$**[1-**v**-**f**]-**I**-**T**-**D**}-**X**

Rule-16088:

If both (**Q**), (**U**), (**$**), (**v**), (**f**), (**I**), (**T**), (**D**) and (**X**) are known, then its Leverage or Gearing Ratio Planned is:

/= [**Q**+**X**]/{**U**+**$**[1-**v**-**f**]-**I**-**T**-**D**}

Rule-16089:

If both (**/**), (**Q**), (**$**), (**v**), (**f**), (**I**), (**T**), (**D**) and (**X**) are known, then its Utilized or Starting Capital must be:

U= **I**+**T**+**D**+[**Q**+**X**]/**/**-**$**[1-**v**-**f**]

Rule-16090:

If both (**/**), (**U**), (**Q**), (**v**), (**f**), (**I**), (**T**), (**D**) and (**X**) are known, then its Sales or Revenue Planned is:

$= {**I**+**T**+**D**+[**Q**+**X**]/**/**-**U**}/[1-**v**-**f**]

Rule-16091:

If both (**/**), (**U**), (**$**), (**Q**), (**f**), (**I**), (**T**), (**D**) and (**X**) are known, then its Variable Portion Planned is:

v= 1-**f**-{**I**+**T**+**D**+[**Q**+**X**]/**/**-**U**}/**$**

Steve Asikin ISBN 14: 978-1514685136, ISBN 10: 1514685132

Rule-16092:

If both (**/**), (**U**), (**S**), (**v**), (**Q**), (**I**), (**T**), (**D**) and (**X**) are known, then its Fixed Portion Planned is:

f= 1-**v**-{**I**+**T**+**D**+[**Q**+**X**]/**/-U**}/**S**

Rule-16093:

If both (**/**), (**U**), (**S**), (**v**), (**f**), (**Q**), (**T**), (**D**) and (**X**) are known, then its Interest Expense Planned is:

I= **S**[1-**v**-**f**]-**T**-**D**-[**Q**+**X**]/**/-U**

Rule-16094:

If both (**/**), (**U**), (**S**), (**v**), (**f**), (**I**), (**Q**), (**D**) and (**X**) are known, then its Tax Planned is:

T= **S**[1-**v**-**f**]-**I**-**D**-[**Q**+**X**]/**/-U**

Rule-16095:

If both (**/**), (**U**), (**S**), (**v**), (**f**), (**I**), (**T**), (**Q**) and (**X**) are known, then its Dividend Planned is:

D= **S**[1-**v**-**f**]-**I**-**T**-[**Q**+**X**]/**/-U**

Rule-16096:

If both (**/**), (**U**), (**S**), (**v**), (**f**), (**I**), (**T**), (**D**) and (**Q**) are known, then its Xpress or Current Debt Planned is:

X= **/**{**U**+**S**[1-**v**-**f**]-**I**-**T**-**D**}-**Q**

Rule-16097:

If both (**/**), (**U**), (**S**), (**v**), (**f**), (**I**), (**T**), (**D**), (**P**), (**c**) and (**q**) are known, then its Quoted Longterm Debt Planned is:

Q= **/**{**U**+**S**[1-**v**-**f**]-**I**-**T**-**D**}-**P**/[**c**-**q**]

Steve Asikin ISBN 14: 978-1514685136, ISBN 10: 1514685132

Rule-16098:

If both (**Q**), (**U**), (**$**), (**v**), (**f**), (**I**), (**T**), (**D**), (**P**), (**c**) and (**q**) are known, then its Leverage or Gearing Ratio Planned is:

/= {**Q**+**P**/[**c**-**q**]}/{**U**+**$**[1-**v**-**f**]-**I**-**T**-**D**}

Rule-16099:

If both (**/**), (**Q**), (**$**), (**v**), (**f**), (**I**), (**T**), (**D**), (**P**), (**c**) and (**q**) are known, then its Utilized or Starting capital must be:

U= **I**+**T**+**D**+{**Q**+**P**/[**c**-**q**]}/**/**-**$**[1-**v**-**f**]

Rule-16100:

If both (**/**), (**U**), (**Q**), (**v**), (**f**), (**I**), (**T**), (**D**), (**P**), (**c**) and (**q**) are known, then its Sales or Revenue Planned is:

$= (**I**+**T**+**D**+{**Q**+**P**/[**c**-**q**]}/**/**-**U**)/[1-**v**-**f**]

Rule-16101:

If both (**/**), (**U**), (**$**), (**Q**), (**f**), (**I**), (**T**), (**D**), (**P**), (**c**) and (**q**) are known, then its Variable Portion Planned is:

v= 1-**f**-(**I**+**T**+**D**+{**Q**+**P**/[**c**-**q**]}/**/**-**U**)/**$**

Rule-16102:

If both (**/**), (**U**), (**$**), (**v**), (**Q**), (**I**), (**T**), (**D**), (**P**), (**c**) and (**q**) are known, then its Fixed Portion Planned is:

f= 1-**v**-(**I**+**T**+**D**+{**Q**+**P**/[**c**-**q**]}/**/**-**U**)/**$**

Steve Asikin ISBN 14: 978-1514685136, ISBN 10: 1514685132

Rule-16103:

If both (**/**), (**U**), (**$**), (**v**), (**f**), (**Q**), (**T**), (**D**), (**P**), (**c**) and (**q**) are known, then its Interest Expense Planned is:

I= **$**[1-**v**-**f**]-**T**-**D**-{**Q**+**P**/[**c**-**q**]}/**/**-**U**

Rule-16104:

If both (**/**), (**U**), (**$**), (**v**), (**f**), (**I**), (**Q**), (**D**), (**P**), (**c**) and (**q**) are known, then its Tax Planned is:

T= **$**[1-**v**-**f**]-**I**-**D**-{**Q**+**P**/[**c**-**q**]}/**/**-**U**

Rule-16105:

If both (**/**), (**U**), (**$**), (**v**), (**f**), (**I**), (**T**), (**Q**), (**P**), (**c**) and (**q**) are known, then its Dividend Planned is:

D= **$**[1-**v**-**f**]-**I**-**T**-{**Q**+**P**/[**c**-**q**]}/**/**-**U**

Rule-16106:

If both (**/**), (**U**), (**$**), (**v**), (**f**), (**I**), (**T**), (**D**), (**Q**), (**c**) and (**q**) are known, then its Procured Inventory Planned is:

P= [**c**-**q**](**/**{**U**+**$**[1-**v**-**f**]-**I**-**T**-**D**}-**Q**)

Rule-16107:

If both (**/**), (**U**), (**$**), (**v**), (**f**), (**I**), (**T**), (**D**), (**P**), (**Q**) and (**q**) are known, then its Current Ratio Planned is:

c= **q**+**P**/(**/**{**U**+**$**[1-**v**-**f**]-**I**-**T**-**D**}-**Q**)

Rule-16108:

If both (**/**), (**U**), (**$**), (**v**), (**f**), (**I**), (**T**), (**D**), (**P**), (**c**) and (**Q**) are known, then its Quick or Acid test Ratio Planned is:

q= **c**-**P**/(**/**{**U**+**$**[1-**v**-**f**]-**I**-**T**-**D**}-**Q**)

Steve Asikin ISBN 14: 978-1514685136, ISBN 10: 1514685132

Rule-16109:

If both (**/**), (**U**), (**$**), (**v**), (**f**), (**I**), (**T**), (**D**), (**V**), (**p**), (**c**) and (**q**) are known, then its Quoted Longterm Debt Planned is:

Q= **/**{**U**+**$**[1-**v**-**f**]-**I**-**T**-**D**}-**Vp**/{360[**c**-**q**]}

Rule-16110:

If both (**Q**), (**U**), (**$**), (**v**), (**f**), (**I**), (**T**), (**D**), (**V**), (**p**), (**c**) and (**q**) are known, then its Leverage or Gearing Ratio Planned is:

/= (**Q**+**Vp**/{360[**c**-**q**]})/{**U**+**$**[1-**v**-**f**]-**I**-**T**-**D**}

Rule-16111:

If both (**/**), (**Q**), (**$**), (**v**), (**f**), (**I**), (**T**), (**D**), (**V**), (**p**), (**c**) and (**q**) are known, then its Utilized or Starting Capital must be:

U= **I**+**T**+**D**+(**Q**+**Vp**/{360[**c**-**q**]})/**/**-**$**[1-**v**-**f**]

Rule-16112:

If both (**/**), (**U**), (**Q**), (**v**), (**f**), (**I**), (**T**), (**D**), (**V**), (**p**), (**c**) and (**q**) are known, then its Sales or Revenue Planned is:

$= **[I**+**T**+**D**+(**Q**+**Vp**/{360[**c**-**q**]})/**/**-**U]**/[1-**v**-**f**]

Rule-16113:

If both (**/**), (**U**), (**$**), (**Q**), (**f**), (**I**), (**T**), (**D**), (**V**), (**p**), (**c**) and (**q**) are known, then its Variable Portion Planned is:

v= **V**/**$**

or it can also be found as

v= 1-**f**-**[I**+**T**+**D**+(**Q**+**Vp**/{360[**c**-**q**]})/**/**-**U]**/**$**

Steve Asikin ISBN 14: 978-1514685136, ISBN 10: 1514685132

Rule-16114:

If both (**/**), (**U**), (**$**), (**v**), (**Q**), (**I**), (**T**), (**D**), (**V**), (**p**), (**c**) and (**q**) are known, then its Fixed Cost Planned is:

f= 1-**v**-[**I**+**T**+**D**+(**Q**+**Vp**/{360[**c**-**q**]})/**/**-**U**]/**$**

Rule-16115:

If both (**/**), (**U**), (**$**), (**v**), (**f**), (**Q**), (**T**), (**D**), (**V**), (**p**), (**c**) and (**q**) are known, then its Interest Expense Planned is:

I= **$**[1-**v**-**f**]-**T**-**D**-(**Q**+**Vp**/{360[**c**-**q**]})/**/**-**U**

Rule-16116:

If both (**/**), (**U**), (**$**), (**v**), (**f**), (**I**), (**Q**), (**D**), (**V**), (**p**), (**c**) and (**q**) are known, then its Tax Planned is:

T= **$**[1-**v**-**f**]-**I**-**D**-(**Q**+**Vp**/{360[**c**-**q**]})/**/**-**U**

Rule-16117:

If both (**/**), (**U**), (**$**), (**v**), (**f**), (**I**), (**T**), (**Q**), (**V**), (**p**), (**c**) and (**q**) are known, then its Dividend Planned is:

D= **$**[1-**v**-**f**]-**I**-**T**-(**Q**+**Vp**/{360[**c**-**q**]})/**/**-**U**

Rule-16118:

If both (**/**), (**U**), (**$**), (**v**), (**f**), (**I**), (**T**), (**D**), (**Q**), (**p**), (**c**) and (**q**) are known, then its Variable Cost Planned is:

V= 360[**c**-**q**](**/**{**U**+**$**[1-**v**-**f**]-**I**-**T**-**D**}-**Q**)/**p**

Rule-16119:

If both (**/**), (**U**), (**$**), (**v**), (**f**), (**I**), (**T**), (**D**), (**V**), (**Q**), (**c**) and (**q**) are known, then its Procured Inventory Days Planned is:

p= 360[**c**-**q**](**/**{**U**+**$**[1-**v**-**f**]-**I**-**T**-**D**}-**Q**)/**V**

Steve Asikin ISBN 14: 978-1514685136, ISBN 10: 1514685132

Rule-16120:

If both (**/**), (**U**), (**S**), (**v**), (**f**), (**I**), (**T**), (**D**), (**V**), (**p**), (**Q**) and (**q**) are known, then its Current Ratio Planned is:

c= **q**+**Vp**/[360(**/**{**U**+**S**[1-**v**-**f**]-**I**-**T**-**D**}-**Q**)]

Rule-16121:

If both (**/**), (**U**), (**S**), (**v**), (**f**), (**I**), (**T**), (**D**), (**V**), (**p**), (**c**) and (**Q**) are known, then its Quick or Acid Test Ratio Planned is:

q= **c**-**Vp**/[360(**/**{**U**+**S**[1-**v**-**f**]-**I**-**T**-**D**}-**Q**)]

Rule-16122:

If both (**/**), (**U**), (**S**), (**v**), (**f**), (**I**), (**T**), (**D**), (**p**), (**c**) and (**q**) are known, then its Quoted Longterm Debt Planned is:

Q= **/**{**U**+**S**[1-**v**-**f**]-**I**-**T**-**D**}-**Svp**/{360[**c**-**q**]}

Rule-16123:

If both (**Q**), (**U**), (**S**), (**v**), (**f**), (**I**), (**T**), (**D**), (**p**), (**c**) and (**q**) are known, then its Leverage or Gearing Ratio Planned is:

/= (**Q**+**Svp**/{360[**c**-**q**]})/{**U**+**S**[1-**v**-**f**]-**I**-**T**-**D**}

Rule-16124:

If both (**/**), (**Q**), (**S**), (**v**), (**f**), (**I**), (**T**), (**D**), (**p**), (**c**) and (**q**) are known, then its Utilized or Starting Capital Planned is:

U= **I**+**T**+**D**+(**Q**+**Svp**/{360[**c**-**q**]})/**/**-**S**[1-**v**-**f**]

Steve Asikin ISBN 14: 978-1514685136, ISBN 10: 1514685132

Rule-16125:

If both (**/**), (**U**), (**Q**), (**v**), (**f**), (**I**), (**T**), (**D**), (**p**), (**c**) and (**q**) are known, then its Sales or Revenue Planned is:

$**= {**/**[**U**-**I**-**T**-**D**]-**Q**}/(**vp**/{360[**c**-**q**]}-**$/[1-**v**-**f**])

Rule-16126:

If both (**/**), (**U**), (**$**), (**Q**), (**f**), (**I**), (**T**), (**D**), (**p**), (**c**) and (**q**) are known, then its Variable Portion Planned is:

v= (**/**{**U**+**$**[1-**f**]-**I**-**T**-**D**}-**Q**)/[**$**(**p**/{360[**c**-**q**]}+**/**)]

Rule-16127:

If both (**/**), (**U**), (**$**), (**v**), (**f**), (**I**), (**T**), (**D**), (**p**), (**c**) and (**q**) are known, then its Fixed Portion Planned is:

f= 1-**v**-[**I**+**T**+**D**+(**Q**+**$vp**/{360[**c**-**q**]})/**/**-**U**]/**$**

Rule-16128:

If both (**/**), (**U**), (**$**), (**v**), (**f**), (**Q**), (**T**), (**D**), (**p**), (**c**) and (**q**) are known, then its Interest Expense Planned is:

I= **$**[1-**v**-**f**]-**T**-**D**-(**Q**+**$vp**/{360[**c**-**q**]})/**/**-**U**

Rule-16129:

If both (**/**), (**U**), (**$**), (**v**), (**f**), (**I**), (**Q**), (**D**), (**p**), (**c**) and (**q**) are known, then its Tax Planned is:

T= **$**[1-**v**-**f**]-**I**-**D**-(**Q**+**$vp**/{360[**c**-**q**]})/**/**-**U**

Rule-16130:

If both (**/**), (**U**), (**$**), (**v**), (**f**), (**I**), (**T**), (**Q**), (**p**), (**c**) and (**q**) are known, then its Dividend Planned is:

D= **$**[1-**v**-**f**]-**I**-**T**-(**Q**+**$vp**/{360[**c**-**q**]})/**/**-**U**

Steve Asikin ISBN 14: 978-1514685136, ISBN 10: 1514685132

Rule-16131:

If both (**/**), (**U**), (**$**), (**v**), (**f**), (**I**), (**T**), (**D**), (**Q**), (**c**) and (**q**) are known, then its Procured Inventory Days Planned is:

p= 360[c-q](/{U+$[1-v-f]-I-T-D}-Q)/[$v]

Rule-16132:

If both (**/**), (**U**), (**$**), (**v**), (**f**), (**I**), (**T**), (**D**), (**p**), (**Q**) and (**q**) are known, then its Current Ratio Planned is:

c= q+$vp/[360(/{U+$[1-v-f]-I-T-D}-Q)]

Rule-16133:

If both (**/**), (**U**), (**$**), (**v**), (**f**), (**I**), (**T**), (**D**), (**p**), (**c**) and (**Q**) are known, then its Quick or Acid test Ratio Planned is:

q= c-$vp/[360(/{U+$[1-v-f]-I-T-D}-Q)]

Rule-16134:

If both (**/**), (**U**), (**$**), (**v**), (**f**), (**I**), (**T**), (**D**), (**$'**), (**p**), (**s**), (**c**) and (**q**) are known, then its Quoted Longterm Debt Planned is:

Q= /{U+$[1-v-f]-I-T-D}-$'vp[1+s]/{360[c-q]}

Rule-16135:

If both (**Q**), (**U**), (**$**), (**v**), (**f**), (**I**), (**T**), (**D**), (**$'**), (**p**), (**s**), (**c**) and (**q**) are known, then its Leverage or Gearing Ratio Planned is:

/= (Q+$'vp[1+s]/{360[c-q]})/{U+$[1-v-f]-I-T-D}

Steve Asikin ISBN 14: 978-1514685136, ISBN 10: 1514685132

Rule-16136:

If both (/), (Q), (S), (v), (f), (I), (T), (D), (S'), (p), (s), (c) and (q) are known, then its Utilized or Starting Capital must be:

U= I+T+D+(Q+S'vp[1+s]/{360[c-q]})/-S[1-v-f]

Rule-16137:

If both (/), (U), (Q), (v), (f), (I), (T), (D), (S'), (p), (s), (c) and (q) are known, then its Sales or Revenue Planned is:

S= (Q-/{U-I-T-D}+S'vp[1+s]/{360[c-q]})
/{/[1-v-f]}

Rule-16138:

If both (/), (U), (S), (Q), (f), (I), (T), (D), (S'), (p), (s), (c) and (q) are known, then its Variable Cost Planned is:

v= (/{U+S[1-f]-I-T-D}-Q)
/(S'p[1+s]/{360[c-q]}+S/)]

Rule-16139:

If both (/), (U), (S), (v), (Q), (I), (T), (D), (S'), (p), (s), (c) and (q) are known, then its Fixed Cost Planned is:

f= 1-v-[I+T+D+(Q+S'vp[1+s]/{360[c-q]})/-U]/S

Rule-16140:

If both (/), (U), (S), (v), (f), Q), (T), (D), (S'), (p), (s), (c) and (q) are known, then its Interest Expense Planned is:

I= S[1-v-f]-T-D-(Q+S'vp[1+s]/{360[c-q]})/-U

Steve Asikin ISBN 14: 978-1514685136, ISBN 10: 1514685132

Rule-16141:

If both (**/**), (**U**), (**S**), (**v**), (**f**), (**I**), (**Q**), (**D**), (**S'**), (**p**), (**s**), (**c**) and (**q**) are known, then its Tax Planned is:

T= **S**[1-**v**-**f**]-**I**-**D**-(**Q**+**S'vp**[1+**s**]/{360[**c**-**q**]})//-**U**

Rule-16142:

If both (**/**), (**U**), (**S**), (**v**), (**f**), (**I**), (**T**), (**Q**), (**S'**), (**p**), (**s**), (**c**) and (**q**) are known, then its Dividend Planned is:

D= **S**[1-**v**-**f**]-**I**-**T**-(**Q**+**S'vp**[1+**s**]/{360[**c**-**q**]})//-**U**

Rule-16143:

If both (**/**), (**U**), (**S**), (**v**), (**f**), (**I**), (**T**), (**D**), (**Q**), (**p**), (**s**), (**c**) and (**q**) are known, then its Sales Past must be:

S'= (**/**{**U**+**S**[1-**v**-**f**]-**I**-**T**-**D**}-**Q**)/(**vp**[1+**s**]/{360[**c**-**q**]})

Rule-16144:

If both (**/**), (**U**), (**S**), (**v**), (**f**), (**I**), (**T**), (**D**), (**S'**), (**Q**), (**s**), (**c**) and (**q**) are known, then its Procured Inventory Days Planned is:

p= 360[**c**-**q**](**/**{**U**+**S**[1-**v**-**f**]-**I**-**T**-**D**}-**Q**)/{**S'v**[1+**s**]}

Rule-16145:

If both (**/**), (**U**), (**S**), (**v**), (**f**), (**I**), (**T**), (**D**), (**S'**), (**p**), (**Q**), (**c**) and (**q**) are known, then its Sales Growth Planned is:

s= **S**/**S'**-1

or it can also be found as

s= (**/**{**U**+**S**[1-**v**-**f**]-**I**-**T**-**D**}-**Q**)/(**S'vp**/{360[**c**-**q**]})-1

Steve Asikin ISBN 14: 978-1514685136, ISBN 10: 1514685132

Rule-16146:

If both (**/**), (**U**), (**$**), (**v**), (**f**), (**I**), (**T**), (**D**), (**$'**), (**p**), (**s**), (**Q**) and (**q**) are known, then its Current Ratio Planned is:

c= **q**+ **$'vp**[1+**s**]/[360(**/**{**U**+**$**[1-**v**-**f**]-**I**-**T**-**D**}-**Q**)]

Rule-16147:

If both (**/**), (**U**), (**$**), (**v**), (**f**), (**I**), (**T**), (**D**), (**$'**), (**p**), (**s**), (**c**) and (**Q**) are known, then its Quick or Acid Test Ratio Planned is:

q= **c**-**$'vp**[1+**s**]/[360(**/**{**U**+**$**[1-**v**-**f**]-**I**-**T**-**D**}-**Q**)]

Rule-16148:

If both (**/**), (**U**), (**$**), (**v**), (**f**), (**I**), (**T**), (**d**) and (**X**) are known, then its Quoted Longterm Debt Planned is:

Q= **/**(**U**+{**$**[1-**v**-**f**]-**I**-**T**}[1-**d**])-**X**

Rule-16149:

If both (**Q**), (**U**), (**$**), (**v**), (**f**), (**I**), (**T**), (**d**) and (**X**) are known, then its Leverage or Gearing Ratio Planned is:

/= [**Q**+**X**]/(**U**+{**$**[1-**v**-**f**]-**I**-**T**}[1-**d**]

Rule-16150:

If both (**/**), (**Q**), (**$**), (**v**), (**f**), (**I**), (**T**), (**d**) and (**X**) are known, then its Utilized or Starting capital must be:

U= [**Q**+**X**]/**/**-{**$**[1-**v**-**f**]-**I**-**T**}[1-**d**]

Rule-16151:

If both (**/**), (**U**), (**Q**), (**v**), (**f**), (**I**), (**T**), (**d**) and (**X**) are known, then its Sales or Revenue Planned is:

$= (**I**+**T**+{[**Q**+**X**]/**/**-**U**}/[1-**d**])/[1-**v**-**f**]

Steve Asikin ISBN 14: 978-1514685136, ISBN 10: 1514685132

Rule-16152:

If both (**/**), (**U**), (**$**), (**Q**), (**f**), (**I**), (**T**), (**d**) and (**X**) are known, then its Vartiable Portion Planned is:

v= 1-**f**-(**I**+**T**+{[**Q**+**X**]/**/-U**}/[1-**d**])/**$**

Rule-16153:

If both (**/**), (**U**), (**$**), (**v**), (**Q**), (**I**), (**T**), (**d**) and (**X**) are known, then its Fixed Portion Planned is:

f= 1-**v**-(**I**+**T**+{[**Q**+**X**]/**/-U**}/[1-**d**])/**$**

Rule-16154:

If both (**/**), (**U**), (**$**), (**v**), (**f**), (**Q**), (**T**), (**d**) and (**X**) are known, then its Interest Portion Planned is:

I= **$**[1-**v**-**f**]-**T**-{[**Q**+**X**]/**/-U**}/[1-**d**]

Rule-16155:

If both (**/**), (**U**), (**$**), (**v**), (**f**), (**I**), (**Q**), (**d**) and (**X**) are known, then its Tax Planned is:

T= **$**[1-**v**-**f**]-**I**-{[**Q**+**X**]/**/-U**}/[1-**d**]

Rule-16156:

If both (**/**), (**U**), (**$**), (**v**), (**f**), (**I**), (**T**), (**Q**) and (**X**) are known, then its Dividend Payout Planned is:

d= 1-{[**Q**+**X**]/**/-U**}/{**$**[1-**v**-**f**]-**I**-**T**}

Rule-16157:

If both (**/**), (**U**), (**$**), (**v**), (**f**), (**I**), (**T**), (**d**) and (**Q**) are known, then its Xpress or Gearing Ratio Planned is:

X= **/**(**U**+{**$**[1-**v**-**f**]-**I**-**T**}[1-**d**])-**Q**

Steve Asikin ISBN 14: 978-1514685136, ISBN 10: 1514685132

Rule-16158:

If both (**/**), (**U**), (**S**), (**v**), (**f**), (**I**), (**T**), (**d**), (**P**), (**c**) and (**q**) are known, then its Quoted Longterm Debt Planned is:

Q= **/**(**U**+{**S**[1-**v**-**f**]-**I**-**T**}[1-**d**])-**P**/[**c**-**q**]

Rule-16159:

If both (**Q**), (**U**), (**S**), (**v**), (**f**), (**I**), (**T**), (**d**), (**P**), (**c**) and (**q**) are known, then its Leverage or Gearing Ratio Planned is:

/= {**Q**+**P**/[**c**-**q**]}/(**U**+{**S**[1-**v**-**f**]-**I**-**T**}[1-**d**]

Rule-16160:

If both (**/**), (**Q**), (**S**), (**v**), (**f**), (**I**), (**T**), (**d**), (**P**), (**c**) and (**q**) are known, then its Utilized or Starting Capital must be:

U= {**Q**+**P**/[**c**-**q**]}/**/**-{**S**[1-**v**-**f**]-**I**-**T**}[1-**d**]

Rule-16161:

If both (**/**), (**U**), (**Q**), (**v**), (**f**), (**I**), (**T**), (**d**), (**P**), (**c**) and (**q**) are known, then its Sales or Revenue Planned is:

S= **[I**+**T**+({**Q**+**P**/[**c**-**q**]}/**/**-**U**)/[1-**d**]**]**/[1-**v**-**f**]

Rule-16162:

If both (**/**), (**U**), (**S**), (**Q**), (**f**), (**I**), (**T**), (**d**), (**P**), (**c**) and (**q**) are known, then its Variable Portion Planned is:

v= 1-**f**-**[I**+**T**+({**Q**+**P**/[**c**-**q**]}/**/**-**U**)/[1-**d**]**]**/**S**

Steve Asikin ISBN 14: 978-1514685136, ISBN 10: 1514685132

Rule-16163:

If both (**/**), (**U**), (**$**), (**v**), (**Q**), (**I**), (**T**), (**d**), (**P**), (**c**) and (**q**) are known, then its Fixed Portion Planned is:

f= 1-**v**-[**I**+**T**+({**Q**+**P**/[**c**-**q**]}/**/**-**U**)/[1-**d**]]/**$**

Rule-16164:

If both (**/**), (**U**), (**$**), (**v**), (**f**), (**Q**), (**T**), (**d**), (**P**), (**c**) and (**q**) are known, then its Interest Expense Planned is:

I= **$**[1-**v**-**f**]-**T**-({**Q**+**P**/[**c**-**q**]}/**/**-**U**)/[1-**d**]

Rule-16165:

If both (**/**), (**U**), (**$**), (**v**), (**f**), (**I**), (**Q**), (**d**), (**P**), (**c**) and (**q**) are known, then its Tax Planned is:

T= **$**[1-**v**-**f**]-**I**-({**Q**+**P**/[**c**-**q**]}/**/**-**U**)/[1-**d**]

Rule-16166:

If both (**/**), (**U**), (**$**), (**v**), (**f**), (**I**), (**T**), (**Q**), (**P**), (**c**) and (**q**) are known, then its Dividend Payout Planned is:

d= 1-({**Q**+**P**/[**c**-**q**]}/**/**-**U**)/{**$**[1-**v**-**f**]-**I**-**T**}

Rule-16167:

If both (**/**), (**U**), (**$**), (**v**), (**f**), (**I**), (**T**), (**d**), (**Q**), (**c**) and (**q**) are known, then its Procured Inventory Planned is:

P= [**c**-**q**][**/**(**U**+{**$**[1-**v**-**f**]-**I**-**T**}[1-**d**])-**Q**]

Rule-16168:

If both (**/**), (**U**), (**$**), (**v**), (**f**), (**I**), (**T**), (**d**), (**P**), (**Q**) and (**q**) are known, then its Current Ratio Planned is:

c= **q**+**P**/[**/**(**U**+{**$**[1-**v**-**f**]-**I**-**T**}[1-**d**])-**Q**]

Steve Asikin ISBN 14: 978-1514685136, ISBN 10: 1514685132

Rule-16169:

If both (**/**), (**U**), (**S**), (**v**), (**f**), (**I**), (**T**), (**d**), (**P**), (**c**) and (**Q**) are known, then its Quick or Acid test Ratio Planned is:

q= **c**-**P**/[**/**(**U**+{**S**[1-**v**-**f**]-**I**-**T**}[1-**d**])-**Q**]

Rule-16170:

If both (**/**), (**U**), (**S**), (**v**), (**f**), (**I**), (**T**), (**d**), (**V**), (**p**), (**c**) and (**q**) are known, then its Quoted Longterm Debt Planned is:

Q= **/**(**U**+{**S**[1-**v**-**f**]-**I**-**T**}[1-**d**])-**Vp**/{360[**c**-**q**]}

Rule-16171:

If both (**Q**), (**U**), (**S**), (**v**), (**f**), (**I**), (**T**), (**d**), (**V**), (**p**), (**c**) and (**q**) are known, then its Leverage or Gearing Ratio Planned is:

/= (**Q**+**Vp**/{360[**c**-**q**]})/ (**U**+{**S**[1-**v**-**f**]-**I**-**T**}[1-**d**]

Rule-16172:

If both (**/**), (**Q**), (**S**), (**v**), (**f**), (**I**), (**T**), (**d**), (**V**), (**p**), (**c**) and (**q**) are known, then its Utilized or Starting Capital must be:

U= (**Q**+**Vp**/{360[**c**-**q**]})/**/**-{**S**[1-**v**-**f**]-**I**-**T**}[1-**d**]

Rule-16173:

If both (**/**), (**U**), (**Q**), (**v**), (**f**), (**I**), (**T**), (**d**), (**V**), (**p**), (**c**) and (**q**) are known, then its Sales or Revenue Planned is:

S={**I**+**T**+[(**Q**+**Vp**/{360[**c**-**q**]})/**/**-**U**]/[1-**d**]}/[1-**v**-**f**]

Steve Asikin ISBN 14: 978-1514685136, ISBN 10: 1514685132

Rule-16174:

If both (Ⱡ), (**U**), (**$**), (**Q**), (**f**), (**I**), (**T**), (**d**), (**V**), (**p**), (**c**) and (**q**) are known, then its Variable Portion Planned is:

v= **V**/**$**

or it can also be found as

v= 1-**f**-{**I**+**T**+[(**Q**+**Vp**/{360[**c**-**q**]})/Ⱡ**U**]/[1-**d**]}/**$**

Rule-16175:

If both (Ⱡ), (**U**), (**$**), (**v**), (**Q**), (**I**), (**T**), (**d**), (**V**), (**p**), (**c**) and (**q**) are known, then its Fixed Portion Planned is:

f= 1-**v**-{**I**+**T**+[(**Q**+**Vp**/{360[**c**-**q**]})/Ⱡ**U**]/[1-**d**]}/**$**

Rule-16176:

If both (Ⱡ), (**U**), (**$**), (**v**), (**f**), (**Q**), (**T**), (**d**), (**V**), (**p**), (**c**) and (**q**) are known, then its Interest Expense Planned is:

I= **$**[1-**v**-**f**]-**T**-[(**Q**+**Vp**/{360[**c**-**q**]})/Ⱡ**U**]/[1-**d**]

Rule-16177:

If both (Ⱡ), (**U**), (**$**), (**v**), (**f**), (**I**), (**Q**), (**d**), (**V**), (**p**), (**c**) and (**q**) are known, then its Tax Planned is:

T= **$**[1-**v**-**f**]-**I**-[(**Q**+**Vp**/{360[**c**-**q**]})/Ⱡ**U**]/[1-**d**]

Rule-16178:

If both (Ⱡ), (**U**), (**$**), (**v**), (**f**), (**I**), (**T**), (**Q**), (**V**), (**p**), (**c**) and (**q**) are known, then its Dividend Payout Planned is:

d= 1-[(**Q**+**Vp**/{360[**c**-**q**]})/Ⱡ**U**]/{**$**[1-**v**-**f**]-**I**-**T**}

Steve Asikin ISBN 14: 978-1514685136, ISBN 10: 1514685132

Rule-16179:

If both (**/**), (**U**), (**$**), (**v**), (**f**), (**I**), (**T**), (**d**), (**Q**), (**p**), (**c**) and (**q**) are known, then its Variable Cost Planned is:

$$V= 360[c-q][/(U+\{S[1-v-f]-I-T\}[1-d])-Q]/p$$

Rule-16180:

If both (**/**), (**U**), (**$**), (**v**), (**f**), (**I**), (**T**), (**d**), (**V**), (**Q**), (**c**) and (**q**) are known, then its Procured Inventory Days Planned is:

$$p= 360[c-q][/(U+\{S[1-v-f]-I-T\}[1-d])-Q]/V$$

Rule-16181:

If both (**/**), (**U**), (**$**), (**v**), (**f**), (**I**), (**T**), (**d**), (**V**), (**p**), (**Q**) and (**q**) are known, then its Current Ratio Planned is:

$$c= q+Vp/\{360[/(U+\{S[1-v-f]-I-T\}[1-d])-Q]\}$$

Rule-16182:

If both (**/**), (**U**), (**$**), (**v**), (**f**), (**I**), (**T**), (**d**), (**V**), (**p**), (**c**) and (**Q**) are known, then its Quick or Acid Test Ratio Planned is:

$$q= c-Vp/\{360[/(U+\{S[1-v-f]-I-T\}[1-d])-Q]\}$$

Rule-16183:

If both (**/**), (**U**), (**$**), (**v**), (**f**), (**I**), (**T**), (**d**), (**p**), (**c**) and (**q**) are known, then its Quoted Longterm Debt Planned is:

$$Q= /(U+\{S[1-v-f]-I-T\}[1-d])-Svp/\{360[c-q]\}$$

Steve Asikin ISBN 14: 978-1514685136, ISBN 10: 1514685132

Rule-16184:

If both (**Q**), (**U**), (**$**), (**v**), (**f**), (**I**), (**T**), (**d**), (**p**), (**c**) and (**q**) are known, then its Leverage or Gearing Ratio Planned is:

/= (Q+$vp/{360[c-q]})/ (U+{$[1-v-f]-I-T}[1-d]

Rule-16185:

If both (/), (**Q**), (**$**), (**v**), (**f**), (**I**), (**T**), (**d**), (**p**), (**c**) and (**q**) are known, then its Utilized or Starting Capital must be:

U= (Q+$vp/{360[c-q]})//-{$[1-v-f]-I-T}[1-d]

Rule-16186:

If both (/), (**U**), (**Q**), (**v**), (**f**), (**I**), (**T**), (**d**), (**p**), (**c**) and (**q**) are known, then its Sales or Revenue Planned is:

$= (/{U-[I+T][1-d]}-Q)/(vp/{360[c-q]}-/[1-v-f])

Rule-16187:

If both (/), (**U**), (**$**), (**Q**), (**f**), (**I**), (**T**), (**d**), (**p**), (**c**) and (**q**) are known, then its Variable Portion Planned is:

v= [/(U+{$[1-f]-I-T}[1-d])-Q]
/[$(p/{360[c-q]}+/[1-d])]

Rule-16188:

If both (/), (**U**), (**$**), (**v**), (**Q**), (**I**), (**T**), (**d**), (**p**), (**c**) and (**q**) are known, then its Fixed Portion Planned is:

f= 1-v-{I+T+[(Q+$vp/{360[c-q]})/-U]/[1-d]}/$

Steve Asikin ISBN 14: 978-1514685136, ISBN 10: 1514685132

Rule-16189:

If both (/), (U), ($), (v), (f), (Q), (T), (d), (p), (c) and (q) are known, then its Interest Expense Planned is:

I= $[1-v-f]-T-[(Q+$vp/{360[c-q]})/ /U]/[1-d]

Rule-16190:

If both (/), (U), ($), (v), (f), (I), (Q), (d), (p), (c) and (q) are known, then its Tax Planned is:

T= $[1-v-f]-I-[(Q+$vp/{360[c-q]})/ /U]/[1-d]

Rule-16191:

If both (/), (U), ($), (v), (f), (I), (T), (Q), (p), (c) and (q) are known, then its Dividend Payout Planned is:

d= 1-[(Q+$vp/{360[c-q]})/ /U]/{$[1-v-f]-I-T}

Rule-16192:

If both (/), (U), ($), (v), (f), (I), (T), (d), (Q), (c) and (q) are known, then its Procured Inventory Days Planned is:

p= 360[c-q][/(U+{$[1-v-f]-I-T}[1-d])-Q]/[$vp]

Rule-16193:

If both (/), (U), ($), (v), (f), (I), (T), (d), (p), (Q) and (q) are known, then its Current Ratio Planned is:

c= q+$vp/{360[/(U+{$[1-v-f]-I-T}[1-d])-Q]}

Rule-16194:

If both (/), (U), ($), (v), (f), (I), (T), (d), (p), (c) and (Q) are known, then its Quick or Acid test Ratio Planned is:

q= c-$vp/{360[/(U+{$[1-v-f]-I-T}[1-d])-Q]}

Steve Asikin ISBN 14: 978-1514685136, ISBN 10: 1514685132

Rule-16195:

If both (**/**), (**U**), (**S**), (**v**), (**f**), (**I**), (**T**), (**d**), (**S'**), (**p**), (**s**), (**c**) and (**q**) are known, then its Quoted Longterm Debt Planned is:

Q= **/**(**U**+{**S**[1-**v**-**f**]-**I**-**T**}[1-**d**])-**S'vp**[1+**s**]/{360[**c**-**q**]}

Rule-16196:

If both (**Q**), (**U**), (**S**), (**v**), (**f**), (**I**), (**T**), (**d**), (**S'**), (**p**), (**s**), (**c**) and (**q**) are known, then its Leverage or Gearing Ratio Planned is:

/= (**Q**+**S'vp**[1+**s**]/{360[**c**-**q**]})
/(**U**+{**S**[1-**v**-**f**]-**I**-**T**}[1-**d**]

Rule-16197:

If both (**/**), (**Q**), (**S**), (**v**), (**f**), (**I**), (**T**), (**d**), (**S'**), (**p**), (**s**), (**c**) and (**q**) are known, then its Utilized or Starting Capital must be:

U= (**Q**+**S'vp**[1+**s**]/{360[**c**-**q**]})/**/**
-{**S**[1-**v**-**f**]-**I**-**T**}[1-**d**]

Rule-16198:

If both (**/**), (**U**), (**Q**), (**v**), (**f**), (**I**), (**T**), (**d**), (**S'**), (**p**), (**s**), (**c**) and (**q**) are known, then its Sales or Revenue Planned is:

S= (**Q**-**/**{**U**-[**I**+**T**][1-**d**]}+**S'vp**[1+**s**]/{360[**c**-**q**]})
/{**/**[1-**v**-**f**] [1-**d**]}

Steve Asikin ISBN 14: 978-1514685136, ISBN 10: 1514685132

Rule-16199:

If both (**/**), (**U**), (**$**), (**Q**), (**f**), (**I**), (**T**), (**d**), (**$'**), (**p**), (**s**), (**c**) and (**q**) are known, then its Variable Portion Planned is:

v= [**/**(**U**+{**$**[1-**f**]-**I**-**T**}[1-**d**])-**Q**]
/(**$'p**/{360[**c**-**q**]}+**$/**[1-**d**])

Rule-16200:

If both (**/**), (**U**), (**$**), (**v**), (**Q**), (**I**), (**T**), (**d**), (**$'**), (**p**), (**s**), (**c**) and (**q**) are known, then its Fixed Portion Planned is:

f= 1-**v**-{**I**+**T**+[(**Q**+**$'vp**[1+**s**]/{360[**c**-**q**]})/**/**-**U**]
/[1-**d**]}/**$**

Rule-16201:

If both (**/**), (**U**), (**$**), (**v**), (**f**), (**Q**), (**T**), (**d**), (**$'**), (**p**), (**s**), (**c**) and (**q**) are known, then its Interest Expense Planned is:

I= **$**[1-**v**-**f**]-**T**-[(**Q**+**$'vp**[1+**s**]/{360[**c**-**q**]})/**/**-**U**]
/[1-**d**]

Rule-16202:

If both (**/**), (**U**), (**$**), (**v**), (**f**), (**I**), (**Q**), (**d**), (**$'**), (**p**), (**s**), (**c**) and (**q**) are known, then its Tax Planned is:

T= **$**[1-**v**-**f**]-**I**-[(**Q**+**$'vp**[1+**s**]/{360[**c**-**q**]})/**/**-**U**]
/[1-**d**]

Steve Asikin ISBN 14: 978-1514685136, ISBN 10: 1514685132

Rule-16203:

If both (/), (U), (S), (v), (f), (I), (T), (d), (S'), (p), (s), (c) and (q) are known, then its Quoted Longterm Debt Planned is:

$$d= 1-[(Q+S'vp[1+s]/\{360[c-q]\})/\text{/}U]/\{S[1-v-f]-I-T\}$$

Rule-16204:

If both (/), (U), (S), (v), (f), (I), (T), (d), (Q), (p), (s), (c) and (q) are known, then its Sales Past must be:

$$S'= [\text{/}U+\{S[1-v-f]-I-T\}[1-d])-Q]/(vp[1+s]/\{360[c-q]\})$$

Rule-16205:

If both (/), (U), (S), (v), (f), (I), (T), (d), (S'), (Q), (s), (c) and (q) are known, then its Procured Inventory Days Planned is:

$$p= 360[c-q][\text{/}U+\{S[1-v-f]-I-T\}[1-d])-Q]/\{S'v[1+s]\}$$

Rule-16206:

If both (/), (U), (S), (v), (f), (I), (T), (d), (S'), (p), (Q), (c) and (q) are known, then its Sales Growth Planned is:

$$s= S/S'-1$$

or it can also be found as

$$s= [\text{/}U+\{S[1-v-f]-I-T\}[1-d])-Q]/(S'vp/\{360[c-q]\})-1$$

Steve Asikin ISBN 14: 978-1514685136, ISBN 10: 1514685132

Rule-16207:

If both (**/**), (**U**), (**S**), (**v**), (**f**), (**I**), (**T**), (**d**), (**S'**), (**p**), (**s**), (**Q**) and (**q**) are known, then its Current Ratio Planned is:

c= **q**+**S'vp**[1+**s**]/{360[**/**(**U**+{**S**[1-**v**-**f**]-**I**-**T**}[1-**d**])-**Q**]}

Rule-16208:

If both (**/**), (**U**), (**S**), (**v**), (**f**), (**I**), (**T**), (**d**), (**S'**), (**p**), (**s**), (**c**) and (**Q**) are known, then its Quick or Acid Test Ratio Planned is:

q= **c**-**S'vp**[1+**s**]/{360[**/**(**U**+{**S**[1-**v**-**f**]-**I**-**T**}[1-**d**])-**Q**]}

Rule-16209:

If both (**/**), (**U**), (**S**), (**v**), (**f**), (**I**), (**T**), (**A**), (**d**) and (**X**) are known, then its Quoted Longterm Debt Planned is:

Q= **/**{**U**+**S**[1-**v**-**f**]-**I**-**T**-**Ad**}-**X**

Rule-16210:

If both (**Q**), (**U**), (**S**), (**v**), (**f**), (**I**), (**T**), (**A**), (**d**) and (**X**) are known, then its Leverage or Gearing Ratio Planned is:

/= [**Q**+**X**]/{**U**+**S**[1-**v**-**f**]-**I**-**T**-**Ad**}

Rule-16211:

If both (**/**), (**Q**), (**S**), (**v**), (**f**), (**I**), (**T**), (**A**), (**d**) and (**X**) are known, then its Utilized or Starting Capital must be:

U= **I**+**T**+**Ad**+[**Q**+**X**]/**/**-**S**[1-**v**-**f**]

Steve Asikin ISBN 14: 978-1514685136, ISBN 10: 1514685132

Rule-16212:

If both (**/**), (**U**), (**Q**), (**v**), (**f**), (**I**), (**T**), (**A**), (**d**) and (**X**) are known, then its Sales or Revenue Planned is:

$= {**I**+**T**+**Ad**+[**Q**+**X**]/**/**-**U**}/[1-**v**-**f**]

Rule-16213:

If both (**/**), (**U**), (**$**), (**Q**), (**f**), (**I**), (**T**), (**A**), (**d**) and (**X**) are known, then its Variable Portion Planned is:

v= 1-**f**-{**I**+**T**+**Ad**+[**Q**+**X**]/**/**-**U**}/**$**

Rule-16214:

If both (**/**), (**U**), (**$**), (**v**), (**Q**), (**I**), (**T**), (**A**), (**d**) and (**X**) are known, then its Fixed Portion Planned is:

f= 1-**v**-{**I**+**T**+**Ad**+[**Q**+**X**]/**/**-**U**}/**$**

Rule-16215:

If both (**/**), (**U**), (**$**), (**v**), (**f**), (**Q**), (**T**), (**A**), (**d**) and (**X**) are known, then its Interest Portion Planned is:

I= **$**[1-**v**-**f**]-**T**-**Ad**-[**Q**+**X**]/**/**-**U**

Rule-16216:

If both (**/**), (**U**), (**$**), (**v**), (**f**), (**I**), (**Q**), (**A**), (**d**) and (**X**) are known, then its Tax Planned is:

T= **$**[1-**v**-**f**]-**I**-**Ad**-[**Q**+**X**]/**/**-**U**

Rule-16217:

If both (**/**), (**U**), (**$**), (**v**), (**f**), (**I**), (**T**), (**Q**), (**d**) and (**X**) are known, then its After Tax Income Planned is:

A= {**$**[1-**v**-**f**]-**I**-**T**-[**Q**+**X**]/**/**+**U**}/**d**

Steve Asikin ISBN 14: 978-1514685136, ISBN 10: 1514685132

Rule-16218:

If both (**/**), (**U**), (**S**), (**v**), (**f**), (**I**), (**T**), (**A**), (**Q**) and (**X**) are known, then its Dividend Payout Planned is:

d= {**S**[1-**v**-**f**]-**I**-**T**-[**Q**+**X**]/**/**+**U**}/**A**

Rule-16219:

If both (**/**), (**U**), (**S**), (**v**), (**f**), (**I**), (**T**), (**A**), (**d**) and (**Q**) are known, then its Xpress or Current Debt Planned is:

X= **/**{**U**+**S**[1-**v**-**f**]-**I**-**T**-**Ad**}-**Q**

Rule-16220:

If both (**/**), (**U**), (**S**), (**v**), (**f**), (**I**), (**T**), (**A**), (**d**), (**P**), (**c**) and (**q**) are known, then its Quoted Longterm Debt Planned is:

Q= **/**{**U**+**S**[1-**v**-**f**]-**I**-**T**-**Ad**}-**P**/[**c**-**q**]

Rule-16221:

If both (**Q**), (**U**), (**S**), (**v**), (**f**), (**I**), (**T**), (**A**), (**d**), (**P**), (**c**) and (**q**) are known, then its Leverage or Gearing Ratio Planned is:

/= {**Q**+**P**/[**c**-**q**]}/{**U**+**S**[1-**v**-**f**]-**I**-**T**-**Ad**}

Rule-16222:

If both (**/**), (**Q**), (**S**), (**v**), (**f**), (**I**), (**T**), (**A**), (**d**), (**P**), (**c**) and (**q**) are known, then its Utilized or Starting Capital must be:

U= **I**+**T**+**Ad**+{**Q**+**P**/[**c**-**q**]}/**/**-**S**[1-**v**-**f**]

Steve Asikin ISBN 14: 978-1514685136, ISBN 10: 1514685132

Rule-16223:

If both (/), (**U**), (**Q**), (**v**), (**f**), (**I**), (**T**), (**A**), (**d**), (**P**), (**c**) and (**q**) are known, then its Sales or Revenue Planned is:

$\$= (I+T+Ad+\{Q+P/[c-q]\}/\text{/}\text{-}U)/[1-v-f]$

Rule-16224:

If both (/), (**U**), (**$**), (**Q**), (**f**), (**I**), (**T**), (**A**), (**d**), (**P**), (**c**) and (**q**) are known, then its Variable Portion Planned is:

$v= 1-f-(I+T+Ad+\{Q+P/[c-q]\}/\text{/}\text{-}U)/\$$

Rule-16225:

If both (/), (**U**), (**$**), (**v**), (**Q**), (**I**), (**T**), (**A**), (**d**), (**P**), (**c**) and (**q**) are known, then its Fixed Portion Planned is:

$f= 1-v-(I+T+Ad+\{Q+P/[c-q]\}/\text{/}\text{-}U)/\$$

Rule-16226:

If both (/), (**U**), (**$**), (**v**), (**f**), (**Q**), (**T**), (**A**), (**d**), (**P**), (**c**) and (**q**) are known, then its Interest Expense Planned is:

$I= \$[1-v-f]-T-Ad-\{Q+P/[c-q]\}/\text{/}+U$

Rule-16227:

If both (/), (**U**), (**$**), (**v**), (**f**), (**I**), (**Q**), (**A**), (**d**), (**P**), (**c**) and (**q**) are known, then its Tax Planned is:

$T= \$[1-v-f]-I-Ad-\{Q+P/[c-q]\}/\text{/}+U$

Steve Asikin ISBN 14: 978-1514685136, ISBN 10: 1514685132

Rule-16228:

If both (/), (U), ($), (v), (f), (I), (T), (Q), (d), (P), (c) and (q) are known, then its After Tax Income Planned is:

A= ($[1-v-f]-I-T-{Q+P/[c-q]}//+U)/d

Rule-16229:

If both (/), (U), ($), (v), (f), (I), (T), (A), (Q), (P), (c) and (q) are known, then its Dividend Payout Planned is:

d= ($[1-v-f]-I-T-{Q+P/[c-q]}//+U)/A

Rule-16230:

If both (/), (U), ($), (v), (f), (I), (T), (A), (d), (Q), (c) and (q) are known, then its Procured Inventory Planned is:

P= [c-q](/{U+$[1-v-f]-I-T-Ad}-Q)

Rule-16231:

If both (/), (U), ($), (v), (f), (I), (T), (A), (d), (P), (Q) and (q) are known, then its Current Ratio Planned is:

c= q+P/(/{U+$[1-v-f]-I-T-Ad}-Q)

Rule-16232:

If both (/), (U), ($), (v), (f), (I), (T), (A), (d), (P), (c) and (Q) are known, then its Quick or Acid test Ratio Planned is:

q= c-P/(/{U+$[1-v-f]-I-T-Ad}-Q)

Steve Asikin ISBN 14: 978-1514685136, ISBN 10: 1514685132

Rule-16233:

If both (**/**), (**U**), (**$**), (**v**), (**f**), (**I**), (**T**), (**A**), (**d**), (**V**), (**p**), (**c**) and (**q**) are known, then its Quoted Longterm Debt Planned is:

Q= **/**{**U**+**$**[1-**v**-**f**]-**I**-**T**-**Ad**}-**Vp**/{360[**c**-**q**]}

Rule-16234:

If both (**Q**), (**U**), (**$**), (**v**), (**f**), (**I**), (**T**), (**A**), (**d**), (**V**), (**p**), (**c**) and (**q**) are known, then its Leverage or Gearing Ratio Planned is:

/= (**Q**+**Vp**/{360[**c**-**q**]})/{**U**+**$**[1-**v**-**f**]-**I**-**T**-**Ad**}

Rule-16235:

If both (**/**), (**Q**), (**$**), (**v**), (**f**), (**I**), (**T**), (**A**), (**d**), (**V**), (**p**), (**c**) and (**q**) are known, then its Utilized or Starting Capital must be:

U= **I**+**T**+**Ad**+(**Q**+**Vp**/{360[**c**-**q**]})/**/**-**$**[1-**v**-**f**]

Rule-16236:

If both (**/**), (**U**), (**Q**), (**v**), (**f**), (**I**), (**T**), (**A**), (**d**), (**V**), (**p**), (**c**) and (**q**) are known, then its Sales or Revenue Planned is:

$= **[I**+**T**+**Ad**+(**Q**+**Vp**/{360[**c**-**q**]})/**/**-**U]**/ [1-**v**-**f**]

Steve Asikin ISBN 14: 978-1514685136, ISBN 10: 1514685132

Rule-16237:

If both (**/**), (**U**), (**$**), (**Q**), (**f**), (**I**), (**T**), (**A**), (**d**), (**V**), (**p**), (**c**) and (**q**) are known, then its Variable Portion Planned is:

v= **V**/**$**

or it can also be found as

v= 1-**f**-[**I**+**T**+**Ad**+(**Q**+**Vp**/{360[**c**-**q**]})/**/**-**U**]/**$**

Rule-16238:

If both (**/**), (**U**), (**$**), (**v**), (**Q**), (**I**), (**T**), (**A**), (**d**), (**V**), (**p**), (**c**) and (**q**) are known, then its Fixed Cost Planned is:

f= 1-**v**-[**I**+**T**+**Ad**+(**Q**+**Vp**/{360[**c**-**q**]})/**/**-**U**]/**$**

Rule-16239:

If both (**/**), (**U**), (**$**), (**v**), (**f**), (**Q**), (**T**), (**A**), (**d**), (**V**), (**p**), (**c**) and (**q**) are known, then its Interest Expense Planned is:

I= **$**[1-**v**-**f**]-**T**-**Ad**-(**Q**+**Vp**/{360[**c**-**q**]})/**/**+**U**

Rule-16240:

If both (**/**), (**U**), (**$**), (**v**), (**f**), (**I**), (**Q**), (**A**), (**d**), (**V**), (**p**), (**c**) and (**q**) are known, then its Quoted Longterm Debt Planned is:

T= **$**[1-**v**-**f**]-**I**-**Ad**-(**Q**+**Vp**/{360[**c**-**q**]})/**/**+**U**

Rule-16241:

If both (**/**), (**U**), (**$**), (**v**), (**f**), (**I**), (**T**), (**Q**), (**d**), (**V**), (**p**), (**c**) and (**q**) are known, then its After Tax Income Planned is:

A= [**$**[1-**v**-**f**]-**I**-**T**-(**Q**+**Vp**/{360[**c**-**q**]})/**/**+**U**]/**d**

Steve Asikin ISBN 14: 978-1514685136, ISBN 10: 1514685132

Rule-16242:

If both (**/**), (**U**), (**$**), (**v**), (**f**), (**I**), (**T**), (**A**), (**Q**), (**V**), (**p**), (**c**) and (**q**) are known, then its Dividend Payout Planned is:

$$d= [\$[1-v-f]-I-T-(Q+Vp/\{360[c-q]\})//+U]/A$$

Rule-16243:

If both (**/**), (**U**), (**$**), (**v**), (**f**), (**I**), (**T**), (**A**), (**d**), (**Q**), (**p**), (**c**) and (**q**) are known, then its Variable Cost Planned is:

$$V= 360[c-q](/\{U+\$[1-v-f]-I-T-Ad\}-Q)/p$$

Rule-16244:

If both (**/**), (**U**), (**$**), (**v**), (**f**), (**I**), (**T**), (**A**), (**d**), (**V**), (**Q**), (**c**) and (**q**) are known, then its Procured Inventory Days Planned:

$$p= 360[c-q](/\{U+\$[1-v-f]-I-T-Ad\}-Q)/V$$

Rule-16245:

If both (**/**), (**U**), (**$**), (**v**), (**f**), (**I**), (**T**), (**A**), (**d**), (**V**), (**p**), (**Q**) and (**q**) are known, then its Current Ratio Planned is:

$$c= q+Vp/[360(/\{U+\$[1-v-f]-I-T-Ad\}-Q)]$$

Rule-16246:

If both (**/**), (**U**), (**$**), (**v**), (**f**), (**I**), (**T**), (**A**), (**d**), (**V**), (**p**), (**c**) and (**Q**) are known, then its Quick or Acid Test Ratio Planned is:

$$q= c-Vp/[360(/\{U+\$[1-v-f]-I-T-Ad\}-Q)]$$

Steve Asikin ISBN 14: 978-1514685136, ISBN 10: 1514685132

Rule-16247:

If both (**/**), (**U**), (**$**), (**v**), (**f**), (**I**), (**T**), (**A**), (**d**), (**p**), (**c**) and (**q**) are known, then its Quoted Longterm Debt Planned is:

$$Q= /\{U+\$[1-v-f]-I-T-Ad\}-\$vp/\{360[c-q]\}$$

Rule-16248:

If both (**Q**), (**U**), (**$**), (**v**), (**f**), (**I**), (**T**), (**A**), (**d**), (**p**), (**c**) and (**q**) are known, then its Leverage or Gearing Ratio Planned is:

$$/= (Q+\$vp/\{360[c-q]\})/\{U+\$[1-v-f]-I-T-Ad\}$$

Rule-16249:

If both (**/**), (**Q**), (**$**), (**v**), (**f**), (**I**), (**T**), (**A**), (**d**), (**p**), (**c**) and (**q**) are known, then its Utilized or Starting Capital must be:

$$U= I+T+Ad+(Q+\$vp/\{360[c-q]\})/ /-\$[1-v-f]$$

Rule-16250:

If both (**/**), (**U**), (**Q**), (**v**), (**f**), (**I**), (**T**), (**A**), (**d**), (**p**), (**c**) and (**q**) are known, then its Sales or Revenue Planned is:

$$\$= (/\{U-I-T-Ad\}-Q)/(vp/\{360[c-q]\}-/[1-v-f])$$

Rule-16251:

If both (**/**), (**U**), (**$**), (**Q**), (**f**), (**I**), (**T**), (**A**), (**d**), (**p**), (**c**) and (**q**) are known, then its Variable Portion Planned is:

$$v= (/\{U+\$[1-f]-I-T-Ad\}-Q)/[\$(p/\{360[c-q]\}+/)]$$

Steve Asikin ISBN 14: 978-1514685136, ISBN 10: 1514685132

Rule-16252:

If both (**/**), (**U**), (**$**), (**v**), (**Q**), (**I**), (**T**), (**A**), (**d**), (**p**), (**c**) and (**q**) are known, then its Fixed Portion Planned is:

f= 1-**v**-[**I**+**T**+**Ad**+(**Q**+**$vp**/{360[**c**-**q**]})/**/**-**U**]/**$**

Rule-16253:

If both (**/**), (**U**), (**$**), (**v**), (**f**), (**Q**), (**T**), (**A**), (**d**), (**p**), (**c**) and (**q**) are known, then its Interest Expense Planned is:

I= **$**[1-**v**-**f**]-**T**-**Ad**-(**Q**+**$vp**/{360[**c**-**q**]})/**/**+**U**

Rule-16254:

If both (**/**), (**U**), (**$**), (**v**), (**f**), (**I**), (**Q**), (**A**), (**d**), (**p**), (**c**) and (**q**) are known, then its Tax Planned is:

T= **$**[1-**v**-**f**]-**I**-**Ad**-(**Q**+**$vp**/{360[**c**-**q**]})/**/**+**U**

Rule-16255:

If both (**/**), (**U**), (**$**), (**v**), (**f**), (**I**), (**T**), (**Q**), (**d**), (**p**), (**c**) and (**q**) are known, then its After Tax Income Planned is:

A= [**$**[1-**v**-**f**]-**I**-**T**-(**Q**+**$vp**/{360[**c**-**q**]})/**/**+**U**]/**d**

Rule-16256:

If both (**/**), (**U**), (**$**), (**v**), (**f**), (**I**), (**T**), (**A**), (**Q**), (**p**), (**c**) and (**q**) are known, then its Dividend Payout Planned is:

d= [**$**[1-**v**-**f**]-**I**-**T**-(**Q**+**$vp**/{360[**c**-**q**]})/**/**+**U**]/**A**

Steve Asikin ISBN 14: 978-1514685136, ISBN 10: 1514685132

Rule-16257:

If both (/), (U), ($), (v), (f), (I), (T), (A), (d), (Q), (c) and (q) are known, then its Procured Inventory Days Planned:

p= 360[c-q](/{U+$[1-v-f]-I-T-Ad}-Q)/[$v]

Rule-16258:

If both (/), (U), ($), (v), (f), (I), (T), (A), (d), (p), (Q) and (q) are known, then its Current Ratio Planned is:

c= q+$vp/[360(/{U+$[1-v-f]-I-T-Ad}-Q)]

Rule-16259:

If both (/), (U), ($), (v), (f), (I), (T), (A), (d), (p), (c) and (Q) are known, then its Quick or Acid Test Ratio Planned is:

q= c-$vp/[360(/{U+$[1-v-f]-I-T-Ad}-Q)]

Rule-16260:

If both (/), (U), ($), (v), (f), (I), (T), (A), (d), ($'), (p), (s), (c) and (q) are known, then its Quoted Longterm Debt Planned is:

Q= /{U+$[1-v-f]-I-T-Ad}-$'vp[1+s]/{360[c-q]}

Rule-16261:

If both (Q), (U), ($), (v), (f), (I), (T), (A), (d), ($'), (p), (s), (c) and (q) are known, then its Leverage or Gearing Ratio Planned is:

/= (Q+$vp/{360[c-q]})/{U+$[1-v-f]-I-T-Ad}

Steve Asikin ISBN 14: 978-1514685136, ISBN 10: 1514685132

Rule-16262:

If both (/), (Q), (S), (v), (f), (I), (T), (A), (d), (S'), (p), (s), (c) and (q) are known, then its Utilized or Starting Capital must be:

U= I+T+Ad+(Q+Svp/{360[c-q]})//-S[1-v-f]

Rule-16263:

If both (/), (U), (Q), (v), (f), (I), (T), (A), (d), (S'), (p), (s), (c) and (q) are known, then its Sales or Revenue Planned is:

S= (/{U-I-T-Ad}-Q)/(vp/{360[c-q]}-/[1-v-f])

Rule-16264:

If both (/), (U), (S), (Q), (f), (I), (T), (A), (d), (S'), (p), (s), (c) and (q) are known, then its Variable Portion Planned is:

v= (/{U+S[1-f]-I-T-Ad}-Q)/[S(p/{360[c-q]}+/)]

Rule-16265:

If both (/), (U), (S), (v), (Q), (I), (T), (A), (d), (S'), (p), (s), (c) and (q) are known, then its Fixed Portion Planned is:

f= 1-v-[I+T+Ad+(Q+Svp/{360[c-q]})//-U]/S

Rule-16266:

If both (/), (U), (S), (v), (f), (Q), (T), (A), (d), (S'), (p), (s), (c) and (q) are known, then its Interest Expense Planned is:

I= S[1-v-f]-T-Ad-(Q+Svp/{360[c-q]})//+U

Steve Asikin ISBN 14: 978-1514685136, ISBN 10: 1514685132

Rule-16267:

If both (**/**), (**U**), (**$**), (**v**), (**f**), (**I**), (**Q**), (**A**), (**d**), (**$'**), (**p**), (**s**), (**c**) and (**q**) are known, then its Tax Planned is:

T= **$**[1-**v**-**f**]-**I**-**Ad**-(**Q**+**$vp**/{360[**c**-**q**]})/**/**+**U**

Rule-16268:

If both (**/**), (**U**), (**$**), (**v**), (**f**), (**I**), (**T**), (**Q**), (**d**), (**$'**), (**p**), (**s**), (**c**) and (**q**) are known, then its After Tax Income Planned is:

A= **[$**[1-**v**-**f**]-**I**-**T**-(**Q**+**$vp**/{360[**c**-**q**]})/**/**+**U]**/**d**

Rule-16269:

If both (**/**), (**U**), (**$**), (**v**), (**f**), (**I**), (**T**), (**A**), (**Q**), (**$'**), (**p**), (**s**), (**c**) and (**q**) are known, then its Dividend Payout Planned is:

d= **[$**[1-**v**-**f**]-**I**-**T**-(**Q**+**$vp**/{360[**c**-**q**]})/**/**+**U]**/**A**

Rule-16270:

If both (**/**), (**U**), (**$**), (**v**), (**f**), (**I**), (**T**), (**A**), (**d**), (**$'**), (**Q**), (**s**), (**c**) and (**q**) are known, then its Procured Inventory Days Planned is:

p= 360[**c**-**q**](**/**{**U**+**$**[1-**v**-**f**]-**I**-**T**-**Ad**}-**Q**)/[**$v**]

Rule-16271:

If both (**/**), (**U**), (**$**), (**v**), (**f**), (**I**), (**T**), (**A**), (**d**), (**$'**), (**p**), (**s**), (**Q**) and (**q**) are known, then its Current Ratio Planned is:

c= **q**+**$vp**/**[**360(**/**{**U**+**$**[1-**v**-**f**]-**I**-**T**-**Ad**}-**Q**)**]**

Steve Asikin ISBN 14: 978-1514685136, ISBN 10: 1514685132

Rule-16272:

If both (**/**), (**U**), (**S**), (**v**), (**f**), (**I**), (**T**), (**A**), (**d**), (**S'**), (**p**), (**s**), (**c**) and (**Q**) are known, then its Quick or Acid Test Ratio Planned is:

q= **c**-**Svp**/[360(**/**{**U**+**S**[1-**v**-**f**]-**I**-**T**-**Ad**}-**Q**)]

Rule-16273:

If both (**/**), (**U**), (**S**), (**v**), (**f**), (**I**), (**t**), (**D**) and (**X**) are known, then its Quoted Longterm Debt Planned is:

Q= **/**(**U**+{**S**[1-**v**-**f**]-**I**}[1-**t**]-**D**)-**X**

Rule-16274:

If both (**Q**), (**U**), (**S**), (**v**), (**f**), (**I**), (**t**), (**D**) and (**X**) are known, then its Leverage or Gearing Ratio Planned is:

/= [**Q**+**X**]/(**U**+{**S**[1-**v**-**f**]-**I**}[1-**t**]-**D**)

Rule-16275:

If both (**/**), (**Q**), (**S**), (**v**), (**f**), (**I**), (**t**), (**D**) and (**X**) are known, then its Utilized or Starting Capital:

U= **D**+[**Q**+**X**]/**/**-**S**[1-**v**-**f**]-**I**}[1-**t**]

Rule-16276:

If both (**/**), (**U**), (**Q**), (**v**), (**f**), (**I**), (**t**), (**D**) and (**X**) are known, then its Sales or Revenue Planned is:

S= (**I**+{**D**+[**Q**+**X**]/**/**-**U**}/[1-**t**])/[1-**v**-**f**]

Rule-16277:

If both (**/**), (**U**), (**S**), (**Q**), (**f**), (**I**), (**t**), (**D**) and (**X**) are known, then its Variable Portion Planned is:

v= 1-**f**-(**I**+{**D**+[**Q**+**X**]/**/**-**U**}/[1-**t**])/**S**

Steve Asikin ISBN 14: 978-1514685136, ISBN 10: 1514685132

Rule-16278:

If both (**/**), (**U**), (**$**), (**v**), (**Q**), (**I**), (**t**), (**D**) and (**X**) are known, then its Fixed Portion Planned is:

f= 1-v-(I+{D+[Q+X]//U}/[1-t])/$

Rule-16279:

If both (**/**), (**U**), (**$**), (**v**), (**f**), (**Q**), (**t**), (**D**) and (**X**) are known, then its Interest Portion Planned is:

I= $[1-v-f]-{D+[Q+X]//U}/[1-t]

Rule-16280:

If both (**/**), (**U**), (**$**), (**v**), (**f**), (**I**), (**Q**), (**D**) and (**X**) are known, then its Tax Rate Planned is:

t= 1-{D+[Q+X]//U}/{$[1-v-f]-I}

Rule-16281:

If both (**/**), (**U**), (**$**), (**v**), (**f**), (**I**), (**t**), (**Q**) and (**X**) are known, then its Dividend Planned is:

D= U+{$[1-v-f]-I}[1-t]-[Q+X]//

Rule-16282:

If both (**/**), (**U**), (**$**), (**v**), (**f**), (**I**), (**t**), (**D**) and (**Q**) are known, then its Xpress or Current Debt Planned is:

X= /(U+{$[1-v-f]-I}[1-t]-D)-Q

Rule-16283:

If both (**/**), (**U**), (**$**), (**v**), (**f**), (**I**), (**t**), (**D**), (**P**), (**c**) and (**q**) are known, then its Quoted Longterm Debt Planned is:

Q= /(U+{$[1-v-f]-I}[1-t]-D)-P/[c-q]

Steve Asikin ISBN 14: 978-1514685136, ISBN 10: 1514685132

Rule-16284:

If both (**Q**), (**U**), (**$**), (**v**), (**f**), (**I**), (**t**), (**D**), (**P**), (**c**) and (**q**) are known, then its Leverage or Gearing Ratio Planned is:

/= {Q+P/[c-q]}/(U+{$[1-v-f]-I}[1-t]-D)

Rule-16285:

If both (**/**), (**Q**), (**$**), (**v**), (**f**), (**I**), (**t**), (**D**), (**P**), (**c**) and (**q**) are known, then its Utility or Starting Capital must be:

U= D+{Q+P/[c-q]}//-$[1-v-f]-I}[1-t]

Rule-16286:

If both (**/**), (**U**), (**Q**), (**v**), (**f**), (**I**), (**t**), (**D**), (**P**), (**c**) and (**q**) are known, then its Sales or Revenue Planned is:

$= [I+(D+{Q+P/[c-q]}//-U)/[1-t]]/[1-v-f]

Rule-16287:

If both (**/**), (**U**), (**$**), (**Q**), (**f**), (**I**), (**t**), (**D**), (**P**), (**c**) and (**q**) are known, then its Variable Portion Planned is:

v= 1-f-[I+(D+{Q+P/[c-q]}//-U)/[1-t]]/$

Rule-16288:

If both (**/**), (**U**), (**$**), (**v**), (**Q**), (**I**), (**t**), (**D**), (**P**), (**c**) and (**q**) are known, then its Fixed Portion Planned is:

f= 1-v-[I+(D+{Q+P/[c-q]}//-U)/[1-t]]/$

Steve Asikin ISBN 14: 978-1514685136, ISBN 10: 1514685132

Rule-16289:

If both (/), (U), ($), (v), (f), (Q), (t), (D), (P), (c) and (q) are known, then its Interest Portion Planned is:

I= $[1-v-f]-(D+{Q+P/[c-q]}//U)/[1-t]

Rule-16290:

If both (/), (U), ($), (v), (f), (I), (Q), (D), (P), (c) and (q) are known, then its Tax Rate Planned is:

t= 1-(D+{Q+P/[c-q]}//U)/{$[1-v-f]-I}

Rule-16291:

If both (/), (U), ($), (v), (f), (I), (t), (Q), (P), (c) and (q) are known, then its Dividend Planned is:

D= U+{$[1-v-f]-I}[1-t]-{Q+P/[c-q]}//

Rule-16292:

If both (/), (U), ($), (v), (f), (I), (t), (D), (Q), (c) and (q) are known, then its Procured Inventory Planned is:

P= [c-q][/(U+{$[1-v-f]-I}[1-t]-D)-Q]

Rule-16293:

If both (/), (U), ($), (v), (f), (I), (t), (D), (P), (Q) and (q) are known, then its Current Ratio Planned is:

c= q+P/[/(U+{$[1-v-f]-I}[1-t]-D)-Q]

Rule-16294:

If both (/), (U), ($), (v), (f), (I), (t), (D), (P), (c) and (Q) are known, then its Quick or Acid Test Ratio Planned is:

q= q-P/[/(U+{$[1-v-f]-I}[1-t]-D)-Q]

Steve Asikin ISBN 14: 978-1514685136, ISBN 10: 1514685132

Rule-16295:

If both (**/**), (**U**), (**$**), (**v**), (**f**), (**I**), (**t**), (**D**), (**V**), (**p**), (**c**) and (**q**) are known, then its Quoted Longterm Debt Planned is:

$$Q= /(U+\{S[1-v-f]-I\}[1-t]-D)-Vp/\{360[c-q]\}$$

Rule-16296:

If both (**Q**), (**U**), (**$**), (**v**), (**f**), (**I**), (**t**), (**D**), (**V**), (**p**), (**c**) and (**q**) are known, then its Leverage or Gearing Ratio Planned is:

$$/= (Q+Vp/\{360[c-q]\})/(U+\{S[1-v-f]-I\}[1-t]-D)$$

Rule-16297:

If both (**/**), (**Q**), (**$**), (**v**), (**f**), (**I**), (**t**), (**D**), (**V**), (**p**), (**c**) and (**q**) are known, then its Utilized or Starting Capital Planned is:

$$U= D+(Q+Vp/\{360[c-q]\})/ /-S[1-v-f]-I\}[1-t]$$

Rule-16298:

If both (**/**), (**U**), (**Q**), (**v**), (**f**), (**I**), (**t**), (**D**), (**V**), (**p**), (**c**) and (**q**) are known, then its Sales or Revenue Planned is:

$$S= \{I+[D+(Q+Vp/\{360[c-q]\})/ /-U]/[1-t]\}/[1-v-f]$$

Rule-16299:

If both (**/**), (**U**), (**$**), (**Q**), (**f**), (**I**), (**t**), (**D**), (**V**), (**p**), (**c**) and (**q**) are known, then its Variable Portion Planned is:

$$v= V/S$$

or it can also be found as

$$v= 1-f-\{I+[D+(Q+Vp/\{360[c-q]\})/ /-U]/[1-t]\}/S$$

Steve Asikin ISBN 14: 978-1514685136, ISBN 10: 1514685132

Rule-16300:

If both (**/**), (**U**), (**$**), (**v**), (**Q**), (**I**), (**t**), (**D**), (**V**), (**p**), (**c**) and (**q**) are known, then its Fixed Portion Planned is:

f= 1-**v**-{**I**+[**D**+(**Q**+**Vp**/{360[**c**-**q**]})/**/**-**U**]/[1-**t**]}/**$**

Rule-16301:

If both (**/**), (**U**), (**$**), (**v**), (**f**), (**Q**), (**t**), (**D**), (**V**), (**p**), (**c**) and (**q**) are known, then its Interest Expense Planned is:

I= **$**[1-**v**-**f**]-[**D**+(**Q**+**Vp**/{360[**c**-**q**]})/**/**-**U**]/[1-**t**]

Rule-16302:

If both (**/**), (**U**), (**$**), (**v**), (**f**), (**I**), (**Q**), (**D**), (**V**), (**p**), (**c**) and (**q**) are known, then its Tax Rate Planned is:

t= 1-[**D**+(**Q**+**Vp**/{360[**c**-**q**]})/**/**-**U**]/{**$**[1-**v**-**f**]-**I**}

Rule-16303:

If both (**/**), (**U**), (**$**), (**v**), (**f**), (**I**), (**t**), (**Q**), (**V**), (**p**), (**c**) and (**q**) are known, then its Dividend Planned is:

D= **U**+{**$**[1-**v**-**f**]-**I**}[1-**t**]-(**Q**+**Vp**/{360[**c**-**q**]})/**/**

Rule-16304:

If both (**/**), (**U**), (**$**), (**v**), (**f**), (**I**), (**t**), (**D**), (**Q**), (**p**), (**c**) and (**q**) are known, then its Variable Portion Planned is:

V= 360[**c**-**q**][**/**(**U**+{**$**[1-**v**-**f**]-**I**}[1-**t**]-**D**)-**Q**]/**p**

Steve Asikin ISBN 14: 978-1514685136, ISBN 10: 1514685132

Rule-16305:

If both (/), (**U**), (**$**), (**v**), (**f**), (**I**), (**t**), (**D**), (**V**), (**Q**), (**c**) and (**q**) are known, then its Procured Inventory Planned is:

$$p= 360[c-q][/(U+\{\$[1-v-f]-I\}[1-t]-D)-Q]/V$$

Rule-16306:

If both (/), (**U**), (**$**), (**v**), (**f**), (**I**), (**t**), (**D**), (**V**), (**p**), (**Q**) and (**q**) are known, then its Current Ratio Planned is:

$$c= q+Vp/\{360[/(U+\{\$[1-v-f]-I\}[1-t]-D)-Q]\}$$

Rule-16307:

If both (/), (**U**), (**$**), (**v**), (**f**), (**I**), (**t**), (**D**), (**V**), (**p**), (**c**) and (**Q**) are known, then its Quick or Acid Test Ratio Planned is:

$$q= q-Vp/\{360[/(U+\{\$[1-v-f]-I\}[1-t]-D)-Q]\}$$

Rule-16308:

If both (/), (**U**), (**$**), (**v**), (**f**), (**I**), (**t**), (**D**), (**p**), (**c**) and (**q**) are known, then its Quoted Longterm Debt Planned is:

$$Q= /(U+\{\$[1-v-f]-I\}[1-t]-D)-\$vp/\{360[c-q]\}$$

Rule-16309:

If both (**Q**), (**U**), (**$**), (**v**), (**f**), (**I**), (**t**), (**D**), (**p**), (**c**) and (**q**) are known, then its Leverage or Gearing Ratio Planned is:

$$/= (Q+\$vp/\{360[c-q]\})/(U+\{\$[1-v-f]-I\}[1-t]-D)$$

Steve Asikin ISBN 14: 978-1514685136, ISBN 10: 1514685132

Rule-16310:

If both (**/**), (**Q**), (**$**), (**v**), (**f**), (**I**), (**t**), (**D**), (**p**), (**c**) and (**q**) are known, then its Utilized or Starting Capital must be:

U= **D**+(**Q**+**$vp**/{360[**c**-**q**]})/**/**-**$**[1-**v**-**f**]-**I**}[1-**t**]

Rule-16311:

If both (**/**), (**U**), (**Q**), (**v**), (**f**), (**I**), (**t**), (**D**), (**p**), (**c**) and (**q**) are known, then its Sales or Revenue Planned is:

$= (**/**{**U**-**I**[1-**t**]-**D**}-**Q**)/(**vp**/{360[**c**-**q**]}-**/**[1-**v**-**f**][1-**t**])

Rule-16312:

If both (**/**), (**U**), (**$**), (**Q**), (**f**), (**I**), (**t**), (**D**), (**p**), (**c**) and (**q**) are known, then its Variable Portion Planned is:

v= **[/(U**+{**$**[1-**f**]-**I**}[1-**t**]-**D**)-**Q]**
/**[$**(**p**/{360[**c**-**q**]}-**/**[1-**t**])**]**

Rule-16313:

If both (**/**), (**U**), (**$**), (**v**), (**Q**), (**I**), (**t**), (**D**), (**p**), (**c**) and (**q**) are known, then its Fixed Portion Planned is:

f= 1-**v**-**{I**+**[D**+(**Q**+**$vp**/{360[**c**-**q**]})/**/**-**U]**/[1-**t**]**}**/**$**

Rule-16314:

If both (**/**), (**U**), (**$**), (**v**), (**f**), (**Q**), (**t**), (**D**), (**p**), (**c**) and (**q**) are known, then its Interest Portion Planned is:

I= **$**[1-**v**-**f**]-**[D**+(**Q**+**$vp**/{360[**c**-**q**]})/**/**-**U]**/[1-**t**]

Steve Asikin ISBN 14: 978-1514685136, ISBN 10: 1514685132

Rule-16315:

If both (/), (U), ($), (v), (f), (I), (Q), (D), (p), (c) and (q) are known, then its Tax Rate Planned is:

t= 1-[D+(Q+$vp/{360[c-q]})//-U]/{$[1-v-f]-I}

Rule-16316:

If both (/), (U), ($), (v), (f), (I), (t), (Q), (p), (c) and (q) are known, then its Dividend Planned is:

D= U+{$[1-v-f]-I}[1-t]-(Q+$vp/{360[c-q]})//

Rule-16317:

If both (/), (U), ($), (v), (f), (I), (t), (D), (Q), (c) and (q) are known, then its Procured Inventory Days Planned:

p= 360[c-q][/(U+{$[1-v-f]-I}[1-t]-D)-Q]/[$v]

Rule-16318:

If both (/), (U), ($), (v), (f), (I), (t), (D), (p), (Q) and (q) are known, then its Current Ratio Planned is:

c= q+$vp/{360[/(U+{$[1-v-f]-I}[1-t]-D)-Q]}

Rule-16319:

If both (/), (U), ($), (v), (f), (I), (t), (D), (p), (c) and (Q) are known, then its Quick or Acid Test Ratio Planned is:

q= q-$vp/{360[/(U+{$[1-v-f]-I}[1-t]-D)-Q]}

Steve Asikin ISBN 14: 978-1514685136, ISBN 10: 1514685132

Rule-16320:

If both (/), (U), ($), (v), (f), (I), (t), (D), ($'), (p), (s), (c) and (q) are known, then its Quoted Longterm Debt Planned is:

Q= /(U+{$[1-v-f]-I}[1-t]-D)-$'vp[1+s]/{360[c-q]}

Rule-16321:

If both (Q), (U), ($), (v), (f), (I), (t), (D), ($'), (p), (s), (c) and (q) are known, then its Leverage or Gearing Ratio Planned is:

/= (Q+$'vp[1+s]/{360[c-q]})
/(U+{$[1-v-f]-I}[1-t]-D)

Rule-16322:

If both (/), (Q), ($), (v), (f), (I), (t), (D), ($'), (p), (s), (c) and (q) are known, then its Utilized or Starting Capital must be:

U= D+(Q+$'vp[1+s]/{360[c-q]})//-$[1-v-f]-I}[1-t]

Rule-16323:

If both (/), (U), (Q), (v), (f), (I), (t), (D), ($'), (p), (s), (c) and (q) are known, then its Sales or Revenue Planned is:

$= (Q-/{U-I[1-t]-D}+$'vp[1+s]/{360[c-q]})
/{[1-v-f][1-t])

Steve Asikin ISBN 14: 978-1514685136, ISBN 10: 1514685132

Rule-16324:

If both (**/**), (**U**), (**$**), (**Q**), (**f**), (**I**), (**t**), (**D**), (**$'**), (**p**), (**s**), (**c**) and (**q**) are known, then its Variable Cost Planned is:

v= [**/**(**U**+{**$**[1-**f**]-**I**}[1-**t**]-**D**)-**Q**]
/(**$'p**[1+**s**]/{360[**c**-**q**]}+**$/**[1-**t**])

Rule-16325:

If both (**/**), (**U**), (**$**), (**v**), (**Q**), (**I**), (**t**), (**D**), (**$'**), (**p**), (**s**), (**c**) and (**q**) are known, then its Fixed Portion Planned is:

f= 1-**v**-{**I**+[**D**+(**Q**+**$'vp**[1+**s**]/{360[**c**-**q**]})/**/**-**U**]
/[1-**t**]}/**$**

Rule-16326:

If both (**/**), (**U**), (**$**), (**v**), (**f**), (**Q**), (**t**), (**D**), (**$'**), (**p**), (**s**), (**c**) and (**q**) are known, then its Interest Portion Planned is:

I= **$**[1-**v**-**f**]-[**D**+(**Q**+**$'vp**[1+**s**]/{360[**c**-**q**]})/**/**-**U**]
/[1-**t**]

Rule-16327:

If both (**/**), (**U**), (**$**), (**v**), (**f**), (**I**), (**Q**), (**D**), (**$'**), (**p**), (**s**), (**c**) and (**q**) are known, then its Tax Rate Planned is:

t= 1-[**D**+(**Q**+**$'vp**[1+**s**]/{360[**c**-**q**]})/**/**-**U**]
/{**$**[1-**v**-**f**]-**I**}

Steve Asikin ISBN 14: 978-1514685136, ISBN 10: 1514685132

Rule-16328:

If both (/), (U), ($), (v), (f), (I), (t), (Q), ($'), (p), (s), (c) and (q) are known, then its Dividend Planned is:

D= U+{$[1-v-f]-I}[1-t]
-(Q+$'vp[1+s]/{360[c-q]})//

Rule-16329:

If both (/), (U), ($), (v), (f), (I), (t), (D), (Q), (p), (s), (c) and (q) are known, then its Sales Past must be:

$'= [/(U+{$[1-v-f]-I}[1-t]-D)-Q]
/([1+s]vp/{360[c-q]})

Rule-16330:

If both (/), (U), ($), (v), (f), (I), (t), (D), ($'), (Q), (s), (c) and (q) are known, then its Procured Inventory Days Planned is:

p= 360[c-q][/(U+{$[1-v-f]-I}[1-t]-D)-Q]
/{$'v[1+s]}

Rule-16331:

If both (/), (U), ($), (v), (f), (I), (t), (D), ($'), (p), (Q), (c) and (q) are known, then its Sales Growth Planned is:

s= $/$'-1

or it can also be found as

s= [/(U+{$[1-v-f]-I}[1-t]-D)-Q]
/($'vp/{360[c-q]})-1

Steve Asikin ISBN 14: 978-1514685136, ISBN 10: 1514685132

Rule-16332:

If both (/), (U), (S), (v), (f), (I), (t), (D), (S'), (p), (s), (Q) and (q) are known, then its Current Ratio Planned is:

$$c= q+S'vp[1+s]/\{360[/(U+\{S[1-v-f]-I\}[1-t]-D)-Q]\}$$

Rule-16333:

If both (/), (U), (S), (v), (f), (I), (t), (D), (S'), (p), (s), (c) and (Q) are known, then its Quick or Acid test Ratio Planned is:

$$q= q-S'vp[1+s]/\{360[/(U+\{S[1-v-f]-I\}[1-t]-D)-Q]\}$$

Steve Asikin ISBN 14: 978-1514685136, ISBN 10: 1514685132

Will be continued in next Math Fin Law-5 and after

Chapter-19: **Q**= QUOTED LONGTERM, Debt Optimization,
Chapter-20: **G**= GOODS or Trade Payable Optimization,
Chapter-21: **Z**= ZERO TRADE or Current Debt Optimization,
Chapter-22: **J**= JOB or Trade Account Receivable Optimization,
Chapter-23: **C**= CURRENT ASSETS, Optimization,
Chapter-24: **K**=KIND of CASH, Optimization,
Chapter-25: **N**=NON CURRENT ASSETS, Optimization,

Steve Asikin ISBN 14: 978-1514685136, ISBN 10: 1514685132

REFERENCES:

A. Books:

1)**Amin, A.R. & Asikin, S.**(2011a- 110 pages), *CEO Time Matrix: Productivity Management Skill that Increase CEO Effectiveness by 12400%,* Seattle (US): Amazon Createspace. http:/www.amazon.com/CEO-Time-Matrix-Productivity-Effectiveness/dp/1461066069/ref=sr_1_1?s=books&ie=UTF8&qid=1387091880&sr=1-1&keywords=ceo+time+matrix

2)**Amin, A.R. & Asikin, S.** (2011b- 292 pages), *The Celestial Management: Spiritual Wisdom for All Human Beings,* Seattle (US): Amazon Createspace. http:/www.amazon.com/Celestial-Management-Islamic-Wisdom-Beings/dp/1456450700/ref=sr_1_1?s=books&ie=UTF8&qid=1387092036&sr=1-1&keywords=celestial+management

3)**Amin, A.R. & Asikin, S.** (2014- 76 pages), *No Urgency CEO: A High-Tech Book in Modern Management,* Seattle (US): Amazon Createspace. http:/www.amazon.com/No-Urgency-CEO-Hi-Tech-Management/dp/1489509356/ref=sr_1_1?s=books&ie=UTF8&qid=1396440688&sr=1-1&keywords=no+urgency

4)**Asikin, S.** (2009- 548 pages), *Arigato, Obrigado... (Thanks): Portuguese-Japan 1550-1647 Historic Novel,* Seattle (US): Amazon Createspace. http:/www.amazon.com/Arigato-Obrigado-Thanks-Portuguese-Japan-1550-1647/dp/1453786368/ref=sr_1_1?s=books&ie=UTF8&qid=1396442162&sr=1-1&keywords=arigato

5)**Asikin, S.** (2010a- 290 pages), *Investment ALGEBRA: Strategic Profitable Comprehensive Business Finance Engineering Technology,* Seattle (US): Amazon Createspace. http:/www.amazon.co.uk/Investment-ALGEBRA-Profitable-Comprehensive-Engineering/dp/1453856811/ref=tmm_pap_title_0?ie=UTF8&qid=1385371309&sr=1-1,

6)**Asikin, S.** (2010b- 224 pages), *FINANCE Architecture: VISUAL art, for Corporate Finance's Beauty, Safety and Simplicity,* **Seattle** (US): Amazon Createspace. http:/www.amazon.co.uk/FINANCE-Architecture-Corporate-Finances-Simplicity/dp/1453829547/ref=sr_1_14?s=books&ie=UTF8&qid=1385371842&sr=1-14&keywords=finance+architecture

7)**Asikin, S.** (2010c- 150 pages), *Geometric FINANCE: Useful concepts that work better than Economic Noble Laureates'* Seattle (US): Amazon Createspace. http:/www.amazon.co.uk/Magic-Geometric-FINANCE-concepts-Laureates/dp/1453790284/ref=tmm_pap_title_0,

Steve Asikin ISBN 14: 978-1514685136, ISBN 10: 1514685132

8)**Asikin, S.** (2010d- 294 pages), *TREASURY Planology: Strategic Profitable Comprehensive Business Finance Engineering Technology,* Seattle (US): Amazon Createspace. http:/www.amazon.co.uk/TREASURY-PLANOLOGY-Steve-Asikin-ebook/dp/B00G88MTFW/ref=sr_1_1?s=books&ie=UTF8&qid=1385403690&sr=1-1&keywords=Treasury+Planology.
9)**Asikin, S.** (2011a- 352 pages), *State Economics: Comprehensive Macro-Micro Economics' Simple Fiscal-Monetary Export-Import Accouting, Integrated Supply-Demand Managerial ... Mathematical Engineering Visual .* Seattle (US): Amazon Createspace. http:/www.amazon.co.uk/STATE-ECONOMICS-Steve-Asikin-ebook/dp/B00G89Q5SS, http:/www.amazon.co.uk State-Economics-Steve-Asikin-ebook/dp/B00B8PLQDI,
10)**Asikin, S.** (2011b- 540 pages), *Terima Kasih: Novel Internasional Sejarah Jepang-Portugis 1550-1647,* Seattle (US): Amazon Createspace. http:/www.amazon.com/Terima-Kasih-Internasional-Jepang-Portugis-1550-1647/dp/146351106X/ref=sr_1_1?s=books&ie=UTF8&qid=1396442361&sr=1-1&keywords=terima,
11)**Asikin, S.** (2013a- 824 pages), *Econometric Rex: 888 Marketing Promo 6666 Six Sigma Parametric Theories (of 10 Data),* Seattle (US): Amazon Createspace. http:/www.amazon.co.uk/Econometric-Rex-Marketing-Parametric-Theories/dp/1482745259,
12)**Asikin, S.** (2013b- 260 pages), *Technoeconomistat: Economic &Statistical Technology for Financial Planning,* Seattle (US): Amazon Createspace. http:/www.amazon.co.uk/Technoeconomistat-Economic-Statistical-Technology-Financial/dp/1482304244,
13)**Asikin, S.** (2014a- 190 pages), *Calculus Accountancy: Leibnitz Newton Pacioli's Polynomial Quantitative Finance Risk Modelling Optimization,* Seattle (US): Amazon Createspace. http:/www.amazon.com/Calculus-Accountancy-Polynomial-Quantitative-Optimization/dp/1495463028/ref=sr_1_1?s=books&ie=UTF8&qid=1396441338&sr=1-1&keywords=calculus+accountancy,
14)**Asikin, S.**(2014b- 318 pages), *No Parametric: No Parametric: Hyperbolic Octahedron Central 3-Dimensional Orthogonal Risk Distribution Topology*, Seattle (US): Amazon Createspace. http:/www.amazon.com/Parametric-Hyperbolic-Octahedron-3-Dimensional-Distribution/dp/1496089731/ref=sr_1_1?s=books&ie=UTF8&qid=1396441543&sr=1-1&keywords=No+Parametric,
15)**Asikin, S.** (2014c- 138 pages), *Quick Help: Religion vs Real-Legion Philosophy,* Seattle (US): Amazon Createspace. http:/www.amazon.com/Quick-Help-Religion-Real-Legion-Philosophy/dp/1497466520/ref=sr_1_2?s=books&ie=UTF8&qid=1396441740&sr=1-2&keywords=quick+help,

Steve Asikin ISBN 14: 978-1514685136, ISBN 10: 1514685132

16)**Asikin, S.** (2014d- 824 pages), *Anne of Denmark: King James I the Bible, Historic Plays Drama*, Seattle (US): Amazon Createspace. http:/www.amazon.com/Anne-Denmark-James-Bible-Drama/dp/149299734X/ref=sr_1_1?s=books&ie=UTF8&qid=1396441857&sr=1-1&keywords=anne+asikin,

17)**Asikin, S.** (2014e- 184 pages), *Constructive Spells: Positive Spiritual Mentality Personal Development*: Seattle (US): Amazon Createspace. http:/www.amazon.com/Constructive-Spells-Spiritual-Mentality-Development/dp/1497498899/ref=sr_1_1?s=books&ie=UTF8&qid=1397664512&sr=1-1&keywords=constructive+spells;

18) **Asikin, S.** (2014f- 466 pages), *Elizabeth of Russia: Another Virgin Queen Gloariana: A Romanov Plays Drama,* Seattle (US): Amazon Createspace. http:/www.amazon.com/Elizabeth-RUSSIA-Another-Gloriana-Romanov/dp/1505542480/ref=sr_1_4?s=books&ie=UTF8&qid=1423888482&sr=1-4&keywords=elizabeth+russia;

19) **Asikin, S.** (2014f- 708 pages), *Shinmen Munisai: Miyamoto Musashi's Father, A Sword Samurai Plays Drama,* Seattle (US): Amazon Createspace. http:/www.amazon.com/Shinmen-Munisai-Miyamoto-Musashis-Samurai/dp/1507857799/ref=sr_1_1?s=books&ie=UTF8&qid=1428094280&sr=1-1&keywords=munisai;

20) **Asikin, S.** (2015-767 pages), *Math Fin Law-1: Mathematical Financial Law for Public Listed Firms Rule 1-4441,* Seattle (US): Amazon Createspace. http://www.amazon.com/Math-Fin-Law-Mathematical-Financial-ebook/dp/B00WLNX3Y4/ref=asap_bc?ie=UTF8;

21) **Asikin, S.** (2015-813 pages), *Math Fin Law-2: Mathematical Financial Law for Public Listed Firms Rule 4442-8712,* Seattle (US): Amazon Createspace. http://www.amazon.com/Math-Fin-Law-Mathematical-No-4442-8712/dp/1512285080/ref=sr_1_1?s=books&ie=UTF8&qid=1432732199&sr=1-1&keywords=math+fin+law+2;

22) **Asikin, S.** (2015-813 pages), *Math Fin Law-3: Mathematical Financial Law for Public Listed Firms Rule 8713-12575,* Seattle (US): Amazon Createspace. http://www.amazon.com/Math-Fin-Law-Mathematical-8712-12575/dp/1514276763/ref=sr_1_1?s=books&ie=UTF8&qid=1433982149&sr=1-1&keywords=math+fin+law+3

23)**Asikin, T.& Asikin, S.** (2013- 828 pages), *448 Poems: KARAOKE Song Book: 24 Hours Story of 30 Languages*, Seattle (US): Amazon Createspace. http:/www.amazon.com/448-Poems-KARAOKE-Song-Book/dp/1480034363/ref=sr_1_1?s=books&ie=UTF8&qid=1396442034&sr=1-1&keywords=448+poems,

Steve Asikin ISBN 14: 978-1514685136, ISBN 10: 1514685132

24)**Brigham, EF &Gapenski, LC,** . (1988- Textbook), *Financial Management: Theory and Practice (Study Guide),* NY (US): Dryden Press. http:/www.amazon.com/Financial-Management-Theory-Practice-Study/dp/0030125391/ref=sr_1_1?s=books&ie=UTF8&qid=1429380835&sr=1-1 ;
25)**Nikisa, Evets.** (2014g- 828 pages), *Wonderful Live Spells: 144-Steps Recreative Nice Children Story, Succesful Development:* Seattle (US): Amazon Createspace. http:/www.amazon.com/WONDERFUL-Live-Spells-Recreative-Development/dp/1499248423/ref=sr_1_1?s=books&ie=UTF8&qid=1418698379&sr=1-1&keywords=wonderful+asikin
26)**Niswonger CR, Fess PE & Warren CE.** (1993-Textbook), *Accounting Principles:* Seattle (US): South Western Publishing Co. http:/www.amazon.com/Accounting-Principles-Chapters-C-Rollin-Niswonger/dp/0538818603
27)**Paccioli, L. (1989-Textbook),** *Summa de Arithmetica Geometria Proportioni et Proportionalita,* Venice 1494. http:/www.amazon.com/Arithmetica-Geometria-Proportioni-Proportionalita-Venice/dp/B00T4HXT1W/ref=sr_1_fkmr0_2?s=books&ie=UTF8&qid=1429597966&sr=1-2-fkmr0&keywords=summa+mhematica+arithmetica
28)**Weston JF, Copeland TE & Shastri, K.** (2004-Textbook), *Financial Theory and Corporate Policy:* NY (US): Addison Wesley. http:/www.amazon.com/Financial-Corporate-Copeland-Kuldeep-Paperback/dp/B008YSUE4M/ref=sr_1_1?s=books&ie=UTF8&qid=1429381700&sr=1-1&keywords=weston+copeland

B. Intellectual Properties:

1)**Indonesian Banker Certification 1993**, *Sertifikasi Bankir Indonesia 1993*;
2)**MATHFINPLAN** (Mathematical Financial Planning); *Perencanaan Keuangan Matematis*
3)**BANK MATHPLAN** (Bank's Mathematical Planning); *Perencanaan Matematis Bank*
4)**TECHNO-ECONOMISTAT** (Technology for Economical Statistics); *Teknologi Statistik untuk Ekonomi*
5)**FINANCIAL GEOMETRY** (IT Program); *Ilmu Geometri Keuangan*
6)**BAK DOBEL** (Car Gasoline Tank: Left and Right Side Fuel Fill-in). *Tangki Mobil yang bisa diisi dari Kiri atau Kanan*
7)**Mister GO-CHENG** (Cheap Food Franchising). *Waralaba Pangan Murah*

Steve Asikin ISBN 14: 978-1514685136, ISBN 10: 1514685132

C. SCIENTIFIC PAPERS:

1)**Eichhornia Crassipes**, as Anti Polutant and Fertile Sediment, *Pemanfaatan Eceng Gondok untuk Menetralkan Polusi dan Pengurukan Subur pada Genangan Air,* LKIR, Lomba Karya Ilmiah Remaja, LIPI (1979)

2)**Archimedes Catesian Buoyancy Law**, *Penyimpangan Hukum Archimedes pada Model Pelampung Cartesian,* Lomba Karya Ilmiah Remaja, Depdikbud-RI (1980)

3)**SURAMADU**, Surabaya-Madura Inter Island Bridge, Design Supervisor, *Sumarsono Sumali Engineering Contractor*, Tambak-V, Tandes, Surabaya (1982),

4)**GAZOLINE Taxi-Pool Depot**, Jemur Handayani, Surabaya, *Sumarsono Sumali Engineering Contractor*, Tambak-V, Tandes, Surabaya (1982),

5)**SOEKARNO-HATTA**, Jakarta International Airport, Garuda Maintenance Facilities, *Junior Architect*, As Bulit Drawings, Coinstruction Management, Aerospatiale-Encona Engineering (1983-1985),

6)**ADISUCIPTO**, Yogya International Airport, *Junior Architect*, Preliminary Technical Drawings, Budiono Surasno-Encona Engineering (1986),

7)**BLOK-M PLAZA**, "New Garden Hall", Kebayoran Baru Supermall, *Preliminary Design Architect*, Danisworo-James Ferry- Jasa Ferry-Encona (1986),

8)**NORTH JAKARTA PUBLIC LIBRARY**, Semper, Tanjung Priok, *Chief Architect*, Agung Darma Sarana-Asep Hermawan (1987),

9)**The Banker's Trainer**: A Manual for The Training Head in Banking. *Profesi Bankir Pelatih, Tuntunan bagi Kepala Diklat Perbankan,* (1200 Pages, 1993),

10)**Indonesian Bankers Certification**, a Strategic Move for the National Banking Improvement. *Sertifikasi Bankir Indonesia, Langkah Strategis Pembenahan Perbankan Nasional,* (KOMPAS, 20 Juli 1993).

11)**Indonesian Rupiah's Exchange Rate**, on a Quiet but Steady Movement. *Nilai Tukar Rupiah, Diam-diam Meningkat Pesat,* (MANAJEMEN, Jan-Feb 1994).

12)**The Bank's Credit**, for the Best Practices. *Kredit Praktis* (200 pages, 1994),

13)**Down to Earth Credit Meeting Credo**, *Kiat Rapat Pemberian Kredit yang Realistik.* (MANAJEMEN, Sep-Oct 1994).

14)**SCHIPHOL**, Amsterdam International Airport, 1st Anthony Fokker Weg, Sloten, *Airport Engineer*, Samsung- Fokker-Beurlin, BV (1996),

15)**Ultrapeneurship: The Secret of Starting Business Without Capital**. Ultrapreneur: *Kiat Usaha Tanpa Modal,* (MANAJEMEN, Nov-Dec 1996),

16)**Recipe from the Corporate Raider's Kitchen**, *Menguak Dapur Corporate Raiders,* (MANAJEMEN, Jan-Feb 1997).

17)**The 10-Tips and Tricks of Engineered Conglomeration**. *Sepuluh Rekayasa Keuangan Konglomerat,* (MANAJEMEN, Mar-Apr 1997),

18)**The Management of Succession in Five Parts of the World**. *Manajemen Suksesi di Lima bagian Dunia,* (MANAJEMEN, Mei-Jun 1997),

19)**Auditing**: The Global Language for the 2003's Human Resources. *Audit : Bahasa Global untuk SDM 2003,* (MANAJEMEN, Jul-Aug 1997),

Steve Asikin ISBN 14: 978-1514685136, ISBN 10: 1514685132

20)**The Elected Parliament 1997-2002**, and the Hope of Professionals. *DPR 1997–2002 dan Harapan Profesional,* (MANAJEMEN, Sep-Okt 1997),
21)**Disappearance of World's Soft Currencies**, and Proposition for Global Economic Thinking. *Lenyapnya Mata Uang Lunak Dunia dan Usulan Pemikiran Ekonomi Global,* (MANAJEMEN, Nov-Dec 1997),
22)**The Human Resources and Its Accountability**, in Critical Situation. *Akuntabilitas Sumberdaya Manusia di Masa Krisis,* (MANAJEMEN, Jan-Feb 1998), '
23)**The Investment Economic Model**, and the Failure of Growth Economic Theories. *Model Ekonomi Investasi dan Kegagalan Teori Ekonomi Pertumbuhan,* (MANAJEMEN, Nov 1998),
24)**The Exact Equilibrium of Capital Budgeting**, *Persamaan-persamaan akurat dalam Penganggaran Modal Perusahaan.* (MANAGEMENT EXPOSE, November 1998),
25)**The Corruptive Economic Model**, and Its Solution for Long Term Debt's Problem. *Model Ekonomi Korupsi dan Penyelesaian Utang Jangka Panjang,* (MANAJEMEN, Dec 1998),
26)**The Self Equity Budgeting**, and Operations Research in Investments. *Penganggaran Modal Sendiri dan Riset Operasi dalam Investasi,* (MANAJEMEN, Jan 1999),
27)**Actual Impact of Inflation**, and the Need of Geometric Approach in Statistic. *Pengaruh Riil Inflasi dan Perlunya Statistika Geometri,* (MANAJEMEN, Feb 1999),
28)**Multiplication of Money**, to Overcome the Crisis. *Melipat Gandakan Uang untuk Mengusir Krisis,* (MANAJEMEN, Apr 1999),
29)**The Interactive Mathematical Model**, for Multi Constraints Financial Budgeting, *Model Matematika Ekonomi Interaktif bagi Penganggaran Keuangan Multi Kendala.* (SCRIPTA ECONOMICA, Vol. 2, April 1999)
30)**Application of Ethics**, for the Management Accounting Practices. *Aplikasi Etika dalam Praktek Akuntansi Manajemen,* (MANAJEMEN, May 1999),
31)**The Curriculum of Magistrate of Management in Finance**: Its Needs and Practices. *Kurikulum Magister Manajemen Keuangan 1988-1998: Antara Kebutuhan dan Praktek,* (MANAJEMEN, Jun 1999),
32)**The Attractiveness**, Personal Power and Management. *Pesona Pribadi: "Personal Power" dan Manajemen,* (MANAJEMEN, Jul 1999),
33)**The Market Research**, for Magistrate Degree in Management. *Soal Jawab Riset Pasar untuk Pascasarjana Manajemen* (126 Pages, 1999),
34)**The Thesis Manual**, Question and Answer for Magistrate Degree in Management. *Soal Jawab Tugas Akhir untuk Pascasarjana Manajemen* (126 Pages, 1999),
35)**The Risk Management**, Question and Answers for Magistrate Degree in Management. *Soal Jawab Manajemen Resiko untuk Pascasarjana Manajemen* (126 Pages, 1999),

Steve Asikin ISBN 14: 978-1514685136, ISBN 10: 1514685132

36)**The Credit Manual**, Question and Answers for Magistrate Degree in Management. *Soal Jawab Manajemen Kredit untuk Pascasarjana Manajemen* (126 Pages, 1999),

37)**Enhancing the Morality Structure**. *Solidkan Struktur Moralitas,* (MANAJEMEN, Oct 1999)

38)**Performance of PT. Astra Otoparts**, Plc, in its Press Releases and Public Financial Reports. *Kinerja PT. Astra Otoparts dalam Siaran Pers dan Laporan Keuangan Publik,* (MANAJEMEN, Oct 1999)

39)**The Management of Conflicts**, for the New Indonesian's "Rainbow Colored" Cabinet. *Manajemen Konflik dan Kabinet Pelangi,* (MANAJEMEN, Dec 1999),

40)**Technoeconometric**: An Engineering Elaboration to Matrix Multi Variate Econometrics, *Tekno Ekonometrik sebagai sumbangan teknologi untuk Ekonometrika Matrik Multi Variat.* (MANAGEMENT EXPOSE, November 1999),

41)**Technoeconometric** : A Mathematical Shortcut for Non Linear Bivariate Econometrics, *Tekno Ekonometrik sebagai Usulan Jalan Pintas Matematik untuk Ekonometrika non Linear Bivariat.* (SCRIPTA ECONOMICA, Vol. 2, No. 3, Desember 1999),

42)**Developing "New Image"**, for the New Indonesia. *Membangun Citra Indonesia Baru,* (MANAJEMEN, Jan 2000),

43)**Segregations of Duties**, the Needs of Both Indonesian Governments and Private Sectors. *Pemerintah dan Swasta Perlu Berbagi Tugas,* MANAJEMEN, Mar 2000),

44)**The 812.500 Financial Models**, for the Autonomous Country Governments. *Tentang 812.500 Model Keuangan Otonomi Daerah,* (MANAJEMEN, Mar 2000),

45)**The Forecast of 10 Most Attractive Share Values**, in the Jakarta Stock Exchange, April 2000. *Prakiraan Harga Sepuluh Saham Unggulan di Bursa Efek Jakarta, April 2000,* (MANAJEMEN, Mar 2000),

46)**Let Us Merchandise the Intelectual Properties!**. *Jual "Intelectual Property" Saja!* (MANAJEMEN, Apr 2000),

47)**The Forecast of ten Most Attractive Share Values**, in the Jakarta Stock Exchange May 2000. *Prakiraan Harga Sepuluh Saham Unggulan di Bursa Efek Jakarta Mei 2000,* (MANAJEMEN, Apr 2000),

48)**HEATHROW**, London International Airport, Second Runway, *Q-Basic Algorithmic Engineer*, Beasley-Krishnamoorthy, Imperial College, Cambridge, Post Doctoral Project (2000),

49)**Why the Banking Sector Should be Healthy**, Before Others?. *Mengapa Perbankan Harus Sehat?* (MANAJEMEN, Mei 2000),

50)**Recycling of Wastes as a Negative Costing**. *Memanfaatkan Sampah atau Negative Costing,* (MANAJEMEN, Agt 2000),

51)**The Guidance for Indonesia Corporate Restructuring**. *Kiat Restrukturisasi Perusahaan Indonesia,* (MANAJEMEN, Agt 2000),

52)**Redefinition of the Ethical Comprehension**. *Pemaknaan Kembali atas Pemahaman Etika,* (MANAJEMEN, Sep 2000),

Steve Asikin ISBN 14: 978-1514685136, ISBN 10: 1514685132

53)**The Critical View the 1999's Indonesians Central Bank Financial Statement**. *Analisa Laporan Keuangan Bank Indonesia 1999,* (MANAJEMEN, Oct 2000),
54)**Becoming an "Achiever"**, not Just Common "Worker". *Bukan Sekadar "Kerja", Tapi Menghasilkan Karya!* (MANAJEMEN, Feb 2001),
55)**Ten Steps to Real Wealth!** *Sepuluh Langkah Menjadi Kaya,* (MANAJEMEN, Jun 2001),
56)**Creative Financial Breakthrough**, *Seminar Sehari tentang Terobosan Keuangan yang Kreatif,* (Gramedia-PPM dan MANAJEMEN, Jun 2001),
57)**The Marketing Practices of Financial Softwares**, for Publics Listed Corporation. *Praktek Pemasaran Program Perangkat Lunak untuk Keuangan Perusahaan Publik* (Univ Satyagama, Jakarta Juni 24, 2001).
58)**The Mathematical Financial Planning**, *Perancangan Keuangan Secara Eksak dengan Matematik.* (MANAJEMEN, Jul 2001),
59)**The Real Research and Development, not Rework and Destruction**. *Litbang yang Bukan "Sulit Berkembang",* (MANAJEMEN, Jul 2001),
60)**The Treatise on the Advantages and Disadvantages**, of Revolutionary Learning Methods. *Baik Buruknya: Revolusi Cara Belajar,* (MANAJEMEN, Jul 2001),
61)**Better Make an Indonesian "Corruption Prevention" Body**, not Just a "Corruption Watch". *Buat Indonesian Corruption Prevention (ICP), Bukan Indonesian Corruption Watch, ICW!* (MANAJEMEN, Okt 2001),
62)**Preparation of the Worlds Business Leaders**, for our Century and the Century After. *Mempersiapkan Manusia-manusia yang akan menjadi Khafilah Bisnis Dunia Abad ini dan Abad Mendatang.* (Univ Satyagama – Gregorio Araneta Univ, Jakarta, Oktober 27, 2001).
63)**The Good Corporate Governance**, is not just a Morale Condition. *Kepemerintahan Perusahaan yang Baik, Bukan Sekedar Moralitas,* (MANAJEMEN, Nov 2001),
64)**Corporate Healthiness and Cleanliness as a Challenge**. *Bersih, Sehat Perusahaan, sebuah Tantangan* (MANAJEMEN, Jan 2002),
65)**Developing the Cultural Base for Future Competitions**. *Budaya untuk Bersaing di Masa Depan* (MANAJEMEN, Jan 2002),
66)**"Catapult", "Bullet" or "Ballistic Missiles" Approach**, for Corporate Healthiness and Cleanliness. *Pola "Ketapel", "Peluru Biasa" atau "Rudal" untuk Keuangan Perusahaan yang Bersih Sehat.* (MANAJEMEN, Feb 2002),
67)**The Mathematical Invesment Solution Model**, for Indonesian Foreign Debt and Its Corruption, *Solusi Matematika Investasi bagi Penyelesaian Masalah Utang Luar Negeri dan Korupsi di Indonesia.* (MANAGEMENT EXPOSE, Mar 2002),
68)**The Attitude, Skills and Knowledge (ASK)**, in Educational System Approach. *Sikap, Keterampilan dan Pengetahuan Menurut Pendekatan Sistemik Pendidikan,* (MANAJEMEN, Apr 2002),

Steve Asikin ISBN 14: 978-1514685136, ISBN 10: 1514685132

69)**Strategic Treatise on Regional Autonomy**, in Futurization, Digitalization, Securitization and Globalization Era in Accordance to the Awakening of the Mercantile Nations. *Kajian Strategik Otonomi Daerah : Era Futurisasi, Digitalisasi, Strukturisasi, dan Globalisasi dalam Kebangkitan Negara-Negara Dagang.* (Training for Provincial Government Regional Management, Indonesian Association of Provincial Governments, *Pelatihan Pengembangan Manajemen Pemerintah Daerah bagi Perangkat Pemerintah Propinsi, Angkatan II,* (APPSI), June 27, 2002.

70)**The Suggestion for the Doctoral Course**, for Islamic Financial Management System. *Usulan-usulan Perbaikan dalam Kurikulum Doktor Manajemen Keuangan Islam.* (December, 17-19, 2002, Canadian International Development Agency - CIDA, ADSGM, Universiti Utara Malaysia, Kedah, Da'arul Sintok).

71)**The Defensive Strategy for the Market Leader**, a Nokia Cellular Telephone's Case. *Strategi Bertahan Sang Pemimpin Pasar dengan Menggunakan Multi Merk, Kasus Telepon Seluler Nokia.* (March, 12, 2004, Univ Tarumanegara, Jakarta).

Steve Asikin ISBN 14: 978-1514685136, ISBN 10: 1514685132

Steve Asikin ISBN 14: 978-1514685136, ISBN 10: 1514685132

Steve Asikin ISBN 14: 978-1514685136, ISBN 10: 1514685132

Steve Asikin ISBN 14: 978-1514685136, ISBN 10: 1514685132

Steve Asikin ISBN 14: 978-1514685136, ISBN 10: 1514685132

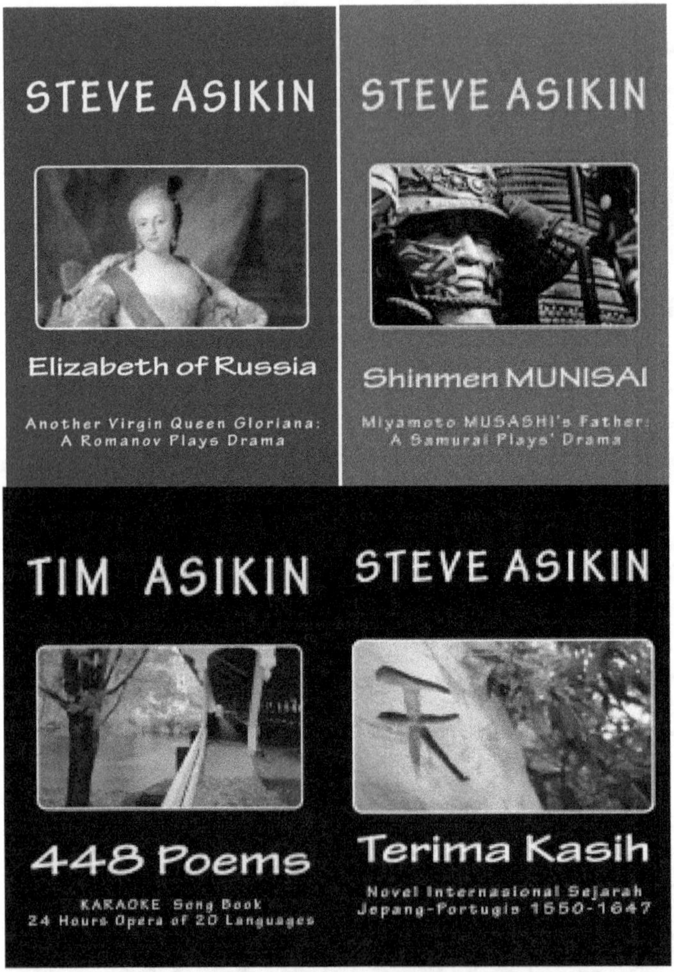

Steve Asikin ISBN 14: 978-1514685136, ISBN 10: 1514685132

WONDERFUL
LIVE SPELLS
Evets Nikisa

CELESTIAL
MANAGEMENT
ISLAMIC MANAGEMENT
ZIKR, PIKR, MIKR
DR. A. RIAWAN
AMIN

NO URGENCY
CEO
A HI-TECH BOOK IN
MODERN TIME MANAGEMENT
DR A RIAWAN
AMIN

CEO TIME
MATRIX
PRODUCTIVITY MANAGEMENT
SKILL THAT INCREASE CEO
EFFECTIVENESS BY 12400%
DR. A . RIAWAN
AMIN

www.ingramcontent.com/pod-product-compliance
Lightning Source LLC
Chambersburg PA
CBHW051846170526
45168CB00001B/1
* 9 7 8 1 5 1 4 6 8 5 1 3 6 *